U0197655

占『新』为民
兴『材』报國

王占国院士文集

中国科学院半导体研究所 编

科学出版社

北京

内 容 简 介

本书梳理和总结了中国科学院院士、半导体材料及材料物理学家王占国院士近60年从事半导体材料领域科研活动的历程。主要包括王占国院士生活和工作的珍贵照片、有代表性的研究论文、科研和工作事迹、回忆文章、获授奖项以及育人情况等内容。王占国院士是我国著名的半导体材料及材料物理学家，对推动我国半导体材料科学领域的学术繁荣、学科发展、技术创新、产业振兴以及人才培养做出了重要贡献。

本书可供从事半导体材料等相关专业的科研人员和管理工作者参考。

图书在版编目(CIP)数据

占"新"为民　兴"材"报国：王占国院士文集／中国科学院半导体研究所编．—北京：科学出版社，2018. 11

ISBN 978-7-03-059106-7

Ⅰ. ①占…　Ⅱ. ①中…　Ⅲ. ①半导体材料-物理性能-文集
Ⅳ. ①TN304. 01-53

中国版本图书馆 CIP 数据核字（2018）第 238064 号

责任编辑：李　敏／责任校对：彭　涛
责任印制：肖　兴／封面设计：王　浩

科学出版社 出版
北京东黄城根北街16号
邮政编码：100717
http://www.sciencep.com

中国科学院印刷厂 印刷

科学出版社发行　各地新华书店经销

*

2018年11月第　一　版　开本：787×1092　1/16
2018年11月第一次印刷　印张：40 1/4　插页：24
字数：1 380 000

定价：480. 00 元

（如有印装质量问题，我社负责调换）

王占国院士

1959年，南开大学物理系57级篮球队合影（前排右二）

1960年，于南开大学

1961年，南开大学物理系57级部分同学在主楼前合影
（后排左二）

1968年夏，于武汉

1965年，在半导体研究所民兵登香山鬼见愁活动中获得第一名（前排右一）

1973年，与母亲张秉芝在天安门前

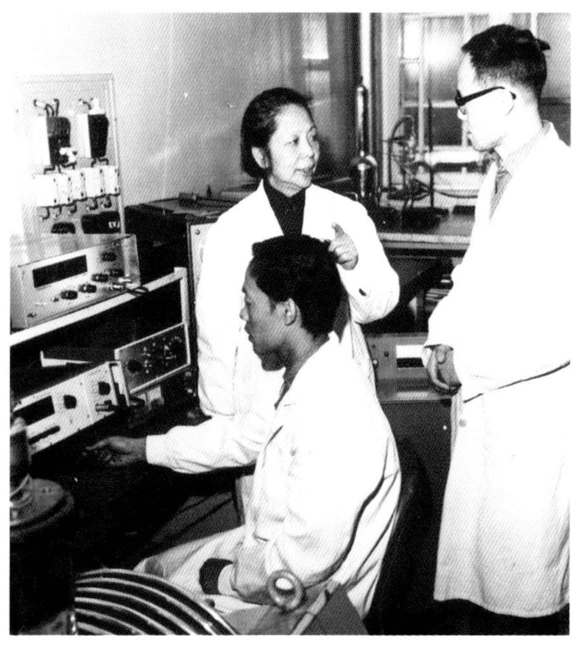

1978年，与林兰英（后）和徐寿定（右）在实验室

1980年10月，出国进修前于北京全家合影

1982年9月，在法国参加国际半导体物理会议时与Bo Monemar（右二）和H. P. Gislason（中）等合影

1981年，在瑞典隆德大学固体物理系实验室工作

1982年9月，在意大利Cagliari参加三元-四元化合物半导体材料国际会议

1983年11月，结束进修回国前与Grimmeiss教授（前排左二）等合影

1984年1月，于丹麦哥本哈根

1984年，与父亲王克佑在半导体研究所宿舍

1986年10月，在广州黄埔军校前与黄昆教授（前排左四）等合影

1988年7月，与褚一鸣（左一）、林兰英（右二）和范提文（右一）在半导体研究所合影

1988年10月，与林兰英（左后）和Pankov（右）在美国科罗拉多州

1989年9月，在日本东京DRIP-3国际会议上作大会邀请报告

1990年，与保加利亚籍博士研究生库意沃在半导体研究所实验室

1991年5月，参加在匈牙利布达佩斯举行的欧洲第三届晶体生长会议

1991年11月，与郑有炓（左一）、小川智哉（右三）、麦振洪（右一）等在美国夏威夷

1992年2月，于印度新德里

1992年10月，南开大学57级毕业30周年于母校行政楼前合影（第四排左一）

1992年10月，于俄罗斯莫斯科红场

1993年，与刘巽琅（左一）、林兰英（右二）和郁元桓（右一）在慕田峪长城

1994年2月，同蒋民华（左一）访问美国佛罗里达大学与材料系教授以及半导体研究所访问学者周伯骏（右一）等合影

1994年6月11日，于冰岛死火山口边

1994年9月，与俄罗斯籍博士后V. L. Dostov在半导体研究所

1994年10月，与熊家炯（左）和蒋民华（中）在俄罗斯圣彼得堡夏宫

1995年7月17日，与Sasaki教授（中）和蒋民华（左）在日本大阪

1995年8月，与梁基本在英国伦敦唐人街

1996年，与师昌绪先生在珠江渡船上

1996年5月，于西班牙马德里

1996年12月，过生日时在家招待研究生

1997年12月，与陈浩明（左二）、刘治国（右三）和郑敏政（右四）等在澳大利亚悉尼大学与悉尼大学教授合影

1998年3月，与D. C. Look及夫人于北京

1998年4月，在国际会议上与瑞典皇家科学院院士Lars Samuelson 交谈

1998年7月，于美国夏威夷

1998年10月3日，与马燕合（左一）、石力开（右二）等在井冈山

1998年10月，在白春礼（中）等视察中国科学院半导体研究所时为其介绍研究工作

1998年11月3日，与夫人刘建文于拉萨

1998年12月29日，60寿辰时与夫人刘建文、学生及同事在一起

1999年4月，与韩国科学技术研究院金长弘（左一）、李精一（右一）等在韩国光州

1999年10月1日，国庆50周年阅兵现场王占国、王圩、王守武、王守觉、王启明5位王姓院士合照

1999年10月，与夏建白（左一）、Bo Monemar（左二）、陈涌海（右一）在瑞典

2000年，与陈涌海在实验室

2001年7月，与孙中哲（左）和周伟（右）在新加坡

2001年9月，于意大利里米尼参加DRIP-4国际会议

2001年，国家“863”新材料领域专家委员会委员合影

2002年2月，在国家科学技术奖励颁奖大会上合影

2002年4月，与夫人刘建文游长江三峡

2002年5月，与来访的诺贝尔物理学奖获得者Zhores I. Alferov在实验室

2002年6月，在西安主持IUMRS-ICEM国际会议

2002年7月，与杜武青夫妇于斯洛伐克古城堡

2002年9月，在红外傅里叶光谱实验室

2002年9月，与夫人刘建文在九寨沟

2002年9月，于九寨沟

2002年12月，于巴西福塔莱萨

2003年10月，与叶小玲（左一）、何军（左二）、龚谦（右一）在荷兰Einheven

2003年11月，于西昌卫星发射中心

2003年12月，《中国材料工程大典》编委会成员合影

2004年4月13日，在美国加利福尼亚大学洛杉矶分校与王康隆教授在校园合影

2004年4月，与学生李瑞钢博士于美国洛杉矶

2004年9月，与方兆强（右一）等在SIMC-13国际会议上

2004年9月23日，与江德生（左）、方兆强（右）于北京友谊宾馆

2004年10月29日，半导体材料科学重点实验室第三届三次学术委员会会议委员合影

2005年4月5日，与秦国刚（左一）、樊志军（左二）、梁骏吾（右二）和陈良惠（右一）
在北京友谊宾馆召开的中俄会议上

2005年6月22日，与博士研究生史桂霞（左一）、郭瑜（左二）、韩修训（右二）、
张春玲（右一）合影

2005年9月，在DRIP-11国际会议
开幕式上作为大会主席致辞

2005年10月9日，与郑有炓院士在敦煌鸣沙山

2005年10月12日，与蒋民华院士在酒泉东风航天城“神舟六号”发射现场

2006年6月21日，与研究组全体师生合影

2006年9月21日，与刘国治司令员在21基地

2006年9月22日，与黄琳（左一）、蒋民华（左二）、都有为（右一）参加院士西部行合影

2007年1月13日，与夫人刘建文在海南省万宁市兴隆热带植物园

2006年11月10日，与夫人刘建文在越南下龙湾

2007年6月2日，与白春礼（中）和徐洪起（右）于瑞典合影

2007年6月6日，与诺贝尔物理学奖获得者Albert Fert教授于法国

2007年6月，与已毕业博士研究生陆沅于法国巴黎

2007年6月，与Skonic教授（中）和刘会赟（右）在英国谢菲尔德

2007年6月，与博士研究生刘宁（前排左一）、周慧英（前排左二）、车晓玲（后排左）、魏鸿源（后排中）、任芸芸（前排右二）、刘俊岐（后排右）及邵烨（前排右一）合影

2007年7月，于英国伦敦

2007年7月，国家自然科学基金评审会成员于哈尔滨合影

2007年9月19日，于青海金银滩原子城

2008年6月21日，于北京石林峡景区

2008年6月18日，与博士研究生梁凌燕（左一）、石礼伟（左二）、范海波（右一）
及王志成（右二）合影

2008年8月，于母校南阳二中与师生合影

2008年9月，在美国西北大学参加中美自然科学基金会论坛

2008年9月，随国家自然科学基金委员会代表团访问美国亚特兰大佐治亚理工学院

2008年10月2日，与杨宝华于美国洛杉矶

2008年11月16日，在海南休假

2009年4月，接受英国天地广播公司记者采访

2009年9月，参加国家自然科学基金评审会与张荣教授于哈尔滨合影

2009年6月7日，与夫人刘建文在新疆喀纳斯湖

2010年9月18日，与夫人刘建文于贵州马岭河峡谷

2009年12月24日，国家“973”计划材料领域专家咨询组成员合影

2011年5月13日，在江西永新县参加科技咨询座谈会

2011年5月，于江西永新县

2011年7月27日，在办公室

2011年11月29日，于加拿大渥太华国会大厦前

2011年12月，在北京西郊宾馆参加中国科学院半导体材料科学重点实验室学术委员会年会

2012年8月8日，与夫人刘建文于黑龙江扎龙国家自然保护区

2013年12月29日，于北京全家福

2014年6月13日，于北京京西宾馆
参加院士大会

2015年11月11日，参加院士增选投票会

2016年9月22–23日，参加570次香山科学会议合影

2016年10月13日，与同事们在分子束处延实验室

2016年10月13日，在办公室

2017年12月29日，79岁生日刘峰奇代表实验室献花

2017年12月29日，79岁生日与学生和同事合影

序

2018 年 12 月 29 日，值此中国科学院院士王占国先生 80 华诞之际，中国科学院半导体研究所编撰出版《**占“新”为民　兴“材”报国**　王占国院士文集》一书，我有幸作序，谨在此向王占国院士表达崇高的敬意和深切的仰慕之情。

王占国院士是我国著名的半导体材料及材料物理学家。1938 年 12 月 29 日生于河南省南阳市镇平县，1962 年毕业于南开大学物理系，同年到中国科学院半导体研究所工作至今。近 60 年来，王占国院士奋发拼搏，孜孜以求，为我国半导体学科建设、技术创新、产业振兴以及人才培养做出了杰出贡献。

王占国院士在半导体材料及材料物理学领域辛勤耕耘、造诣颇深，并取得了一系列重要科研成果。20 世纪 60 年代末至 70 年代初，他主要从事半导体材料、器件辐照效应研究。其中，硅太阳电池电子、质子辐照效应研究成果为我国人造卫星用硅太阳电池定型（由 *PN* 改为 *NP*）投产起到关键作用；受国防科委第 14 研究院的委托，他负责制定了我国电子材料、元件、器件和组件辐照效应研究方案和实施计划，电子材料、元件、器件和组件的电子、质子、中子和 γ 射线的静态、动态和核爆瞬态辐照实验结果为中国航天事业、核加固、核突围和电子对抗等国防工程做出了贡献。1971－1976 年，他主要从事体 GaAs 材料电学和光学性质研究，指出材料纯度、均匀性和热稳定性是研制 GaAs 体效应器件的关键制约因素，为研究所把高纯度 GaAs 研制工作重点由体材料转移到外延材料起到了重要的作用。1976－1980 年，他主要开展了外延 GaAs 和其他Ⅲ－Ⅴ族材料电学和光学性质研究，在国内首先建成了 4. 2－400K GaAs 变温霍尔系数测量系统，并首先建成了国内最先进的光致发光实验室，提出了材料质量（当时）仍由“杂质控制”的基本观点，为半导体研究所高纯 GaAs 外延材料 20 世纪 80 年代初达到国际先进水平贡献了力量。自 1980 年起，他主要从事半导体深能级物理和光谱物理研究，提出了识别两个深能级共存系统两者是否同一缺陷不同能态的新方法，解决了国际上对 GaAs 中 A、B 能级和硅中金受主及金施主能级本质的长期争论；提出了混晶半导体中深能级展宽和光谱谱线分裂的物理模型，解释了它们的物理起因；提出了 GaAs 电学补偿五能级模型和电学补偿新判据。协助林兰英先生，首次在太空从熔体中生长出 GaAs 单晶，并对其光、电性质做了系统研究。从 1993 年开始，他主要致力于半导体低维结构和量子器件研究，与 MBE 组的同事一起，成功地生长了国内领先、国际先进水平的电子迁移率（4. 8K）高达百万的 2DEG 材料和高质量、器件级 HEMT 和 *P*-HEMT 结构材料。近年来，他领导的实验组又在应变自组装 In(Ga)As/GaAs, In(Ga)As/InAlAs/InP 等量子点（线）与量子点（线）超晶格材料生长、大功率量子点激光器、量子点超辐射发光管、量子级联激光器和探测器研制方面获得突破，初步在纳米尺度上实现了对量子点（线）尺寸、形状和密度的可控生长；首次发现 InP 基 InAs 量子线空间斜对准的新现象，被国外评述文章大段引用；成功地制备了从可见光到近红外的量子点（线）材料，并研制成功室温连续工作输出光功率达 4W（双面之和）的大功率量子点激光器，为目前国际上报道的最好结果之一；红光量子点激光器的研究水平也处在国际的前列；在国际上首次提出量子点超辐射发光管的设想，并研制出高性能量子点超辐射发光管和宽光谱可调谐量子点激光器；生长出高质量量子级联材料，并研制出中红外至太赫兹波段多种量子级联激光器和中远红外量子级联探测器，相关材料和器件处于国际先进、国内领先水平。提出了柔性衬底的概念，开拓了大失配材料体系研制的新方向。在我国率先提出开展超宽禁带半导体材料和器件研究的建议，受到国家有关部门和相关科研单位的高度重视。

王占国院士胸怀大志、脚踏实地、勇攀科学高峰，先后获得国家自然科学奖二等奖和国家科学技术进步奖三等奖，中国科学院自然科学奖一等奖和中国科学院科学技术进步奖一、二和三等奖，何梁何利科学与技术进步奖以及国家重点科技攻关奖（个人）等；先后在国外著名学术刊物发表论文 700 余篇。1995 年当选为中国科学院院士。

王占国院士教书育人、甘为人梯、提携后进，为国家培养和造就了一大批优秀科技人才，对我国半导体事业的发展和半导体领域人才的培养做出了重要贡献。他坚持真理、创新不止的科学精神，无私奉献、报效祖国的爱国情操和虚怀若谷、淡泊名利的高尚品德，更是广大科技工作者和师生学习的榜样。

本书的主书名为“占‘新’为民　兴‘材’报国”，这是王占国院士追求真理、勇于创新的写照，也是他对我国半导体材料科学领域做出突出贡献的体现，更是对他无私奉献、报效祖国道德风范的展现。

本文集收录了王占国院士工作和生活的珍贵照片、代表性论著、科研和工作事迹、获授奖项以及育人情况等，这是对王占国院士科研成果和学术思想的总结，也是他教书育人、奉献祖国半导体事业的生动记录。文集在王占国院士 80 华诞之际出版，是对他从事半导体科研和教学工作付出辛勤劳动的回报和取得丰硕成果的肯定；同时，也为从事半导体事业的中青年科技工作者提供了一份学习和参考的珍贵资料。

最后，敬祝王占国院士健康长寿，科技之树常青！

中国科学院副院长

中国科学院大学党委书记、校长

李树深

2018 年 10 月

目　录

第一篇　自　传

第二篇　论著选编

第三篇 学术贡献

第四篇 回 忆

第五篇　附　录

第一篇

自　　传

幼 年 时 光

明末最后一次大移民发生在永乐十五年（1417）。据村里现存的族谱推算，大约在300年前，我们的祖先是从山西洪洞县大槐树迁到河南的，家谱中记录的王氏家族第一代的名字叫王均斗，王均斗育有2个儿子：王子行和王子文，子文也有王国聚和王国兴2个儿子，到第四代的王国兴只有一子，名王朝举。王朝举的前妻生兴邦、安邦和平邦3个儿子、第二个妻子生定邦和正邦2个儿子，共5个儿子，是一个很富的大家庭。5个儿子长大成人后，在村里盖起了5处大宅子，随后就分了家。王兴邦是第5代老大，也是我爷爷的祖父，王兴邦夫妇生3个儿子：大儿子王修奇是我的曾祖父（老二和老三分别是修已和修学），即我爷爷王书典和大爷王修光的父亲，大爷王修光生有一子叫王丙堂（我叫他大伯）。爷爷育有丙道（克佑）和丙远（三大克森）两子，已是第8代了。大爷王修光在外做生意，不幸客死在他乡。我记忆中的大奶奶两腿已经瘫痪，只能坐在垫子上艰难移动，因其子王丙堂尚小，没能分家。大伯成亲后就同爷爷分了家，这时家境已经没落，解放时，我家靠租地主的地勉强度日，已沦为贫农。

1938年12月29日（农历11月初8日），我（已是移居河南第9代了）出生在河南省镇平县贾宋区马庄乡东黄楝扒村的一个家境已经没落的贫苦农民家庭。爷爷王书典、奶奶彭氏（距我们村庄六七里[①]的邓县彭家人，家境比较富裕，奶奶能做一手好菜，村里的婚丧大事都常请她掌勺），父亲王克佑、母亲张氏、三大王克森（我父亲的弟弟，在最亲近的家族中排行老三，我叫他三大）和姐姐一家7口人靠种田为生。父亲王克佑生于1905年，9岁时到彭家外婆家上学，由于学不进去，辍学在家，直到11岁时，才进村里的庙上读了一年多的《三字经》和《论语》，后辍学在家务农。民国初年，土匪猖獗，常常绑票索要钱财，村里修起了高达20余米的小土寨，寨内每家都有一间临时住房，房顶上修有走道，备有土制手雷，每到晚上全村人都进寨防匪。十七八岁的父亲跟着村里的年轻人晚上看寨、打更防土匪来扰。21岁经彭家大舅介绍，到湖北老河口的京子关镇，在三舅开的铺子里做伙计。父亲24岁（1929年）成亲，母亲是邓县前张村人，隔着一条严陵河，离我们庄上只有三四里路。母亲虽不识字，但很精明，听说大舅被土匪杀害后，母亲不顾危险为大舅收尸下葬。1930年姐姐出生，3年后随母亲去了京子关。但京子关的生意也不好做，后来把铺子和家具变卖后仍不能还完借债（直到中华人民共和国成立后还有人来家

---

① 1里=0.5公里

讨债)，全家又回到老家。父亲在农闲时靠染布补贴家里，母亲纺花织布到深夜，勉强维持生计（中华人民共和国成立后家里分了地，生活有所改善。父亲为人忠厚，当选过农业互助组组长、生产大队队长和村人民公社社长等，工作任劳任怨，受到村民们的信任）。当时还是单身汉的三大仅上一年多学，就辍学在家，他怕苦不愿干农活，在家闲着。后被抓了壮丁，当了兵，19岁（1930年）时分派到镇平县彭禹廷的民团当兵，当时土匪猖獗，民团的任务主要是剿匪；彭禹廷被刺后民团解散，20出头的三大回到家中，闲了几年。后来经人介绍，当了个贾宋区区长的勤务兵，平时就住在区政府里，到也自在；只是时间不长，日本人来了，区长逃往陕西，他从此回到了家里，再也没有出去过。

父亲王克佑（1905-1998）

母亲张秉芝（1906-1985）

（于1964年秋拍摄）

听父亲说我在一岁多时得过一场"大病"，多处寻医医治无效，一个活蹦乱跳的小胖子，不到两个月已是骨瘦如柴，眼看着不行了，正在一家人一筹莫展时，经人介绍请到了赵集的一名老中医。老中医骑着毛驴来了，经诊治说是得了百日咳，并说"不要紧，死不了！吃三服药就好了"。第一服药吃完后，要去拿第二服。尽管爷爷起得很早，可是赶到赵集医生家时，不料他已到外村看病去了。爷爷又急忙跑到那里，取来了第二服药；拿第三服药时，爷爷干脆半夜就到赵集，睡在医生家住宅的后面，顺利拿回来了最后一服药。三服药吃完后，果然病好了。衷心感谢这位医生，救了我的小命。在我快到两岁时，弟弟出生了。母亲看我太瘦，就让我和弟弟一起吃奶，一边一个吃着母亲的奶汁！过年时，我又能活蹦乱跳了，家里人都很高兴。

家乡处于南阳盆地的镇平县西南和邓县的交界地带，距湖北襄樊大约200公里。

虽然家乡道路崎岖，交通闭塞，外出不便，但土地肥沃，尤其宁静美丽的田园风光，令人神往。严陵河水绕村东约百米自北向南流去，时急时缓，在村北、村南形成两个深潭，潭水清澈透底，成群结队的鱼虾自由自在地在水中戏耍，甚是可爱。秋冬之间，常有划着小船带着鱼鹰的捕鱼人出入这里，我最喜欢看鱼鹰捉鱼，为它们不时地捉到大鱼叫好，也为它们的脖子常被绳子扎住而感到不平。两大深潭之间的河道很窄，水流湍急，哗哗的水响声震耳欲聋，被乡亲们称作“响水口”，传说这响声是来自藏在水下的“金鸭子”的叫声。沿河的两岸，生长着茂密的森林和茅竹，走进树林几步，就会消失得无影无踪，儿时常在这里玩捉迷藏的游戏，乡亲们也曾利用这一天然屏障躲避过日寇的铁蹄。那是1944年初秋的一个傍晚，突然传来日本鬼子出城下乡扫荡的消息，当村里的老百姓听到了鬼子的装甲车隆隆的响声时受到惊吓，扶老携幼都向河两岸的丛林中跑去；那时我尚不满6岁，惊恐地跟着父母家人一起躲在河岸边茂密的丛林里不敢出声，随着鬼子装甲车隆隆的响声越来越近，大伙紧张的心都提到了嗓子眼，所幸的是鬼子那晚未敢闯入这块抗日游击队十分活跃的地区。

严陵河鱼类资源丰富，下河捉鱼也是我童年的一大趣事。春夏之交，鱼类非常活跃，晚上我跟着大人，举着火把，拿着锋利的大砍刀下河捉鱼，鱼在夜晚见到火光就会径直朝火把游来，我们则趁机砍杀，一个晚上可捉到鲤鱼、鲫鱼和鲶鱼等数十斤。冬天常学着大人与村里的小伙伴一起拿着鱼叉，站在河岸上叉鱼，由于常常不能得手，干脆就脱去鞋袜卷起裤子下到河里，将用竹子编制的小簸箕放在河套里，堵着鱼虾逃跑的路，另一个小伙伴在河套上方不远处用竹竿驱赶鱼虾，见机提起簸箕捕捉，尽管捉不到大鱼，但半天下来也能捉到一大碗小虾，足够小伙伴们美餐一顿了。

我家在的村子不大，只有30余户人家，除两个从他乡流落到这里的暂住户外，都是不出5代的王姓，听说我们的祖先来自山西洪洞县的大槐树（见前述）。村子的四周被枝叶茂盛、粗而高大的榆树和楸树包围，村内和每个农家的小院内外生长着枣树、柿树、花椒和香椿树等，牛羊成群、鸡犬相闻，莺歌燕舞，其乐融融！我很喜爱小动物如猫、兔和小狗等，更是养小鸟的“能手”。沿河两岸茂密的森林和村里参天的大树是鸟类的天堂，斑鸠、喜鹊、乌鸦、鹞子、老鹰、鹌鹑、黄莺、燕子和麻雀等多种鸟类一直在这里繁衍生息。我对喂养斑鸠，特别是小麻雀很有兴趣。每当春末夏初鸟类繁殖的高峰季节，我常常爬上高树，从树洞的鸟窝里掏出刚刚孵出的雏鸟。这些眼睛尚未睁开、身上无毛的幼鸟怕冷，需要放在续有棉花的棉布套内喂养，晚上还要放在被窝里保暖，尽管我小心翼翼，但小鸟有时还会被压死，使人心痛；后来想出了一个法子，晚上先把小鸟放进小土罐，然后再将土罐放进被窝，这个办法果然有效，绝大多数幼鸟都能成活长大。从未睁眼的幼鸟喂养到成熟的麻雀，它们不再像野雀那样胆小怕人，易与人亲近。我时常将它们放飞到院子里的树上，只要你一声口哨，它们就会飞回到你的手上或肩上，当然这时你要奖赏给它们一些好吃的小米或昆虫。麻雀另一个习性是好斗，只要把村里小伙伴喂养的麻雀放在一块，就会打起架来，不过它们不像公鸡之间的争斗，嘴和爪子一起上阵，非打得遍体鳞伤、血流不止才算分出高低。麻雀之间争斗很有意思，它们扇着翅膀，翘起分开的尾巴，吱吱地叫着冲上前去，用尖嘴叮啄对方，一个回合之后，双方都向后退，准备第二回合的厮杀，几个回合后，弱者一方则逃之夭夭，随之战斗结束。除了我养的麻雀之外，家里还有只大花猫，两者就像水火一样，难以相容，所以每当我出去割草、拾柴时，总要把鸟笼挂在很高的树枝上，即使这样还担心出现意外，生怕心爱的小鸟被猫捉住吃掉。

村旁埋葬祖先的墓地也是最受村民尊敬的地方，它不仅是后辈怀念故人的场地，也是村人联络感情的纽带，每当清明时节，全村偕老扶幼到墓地祭祖。每家墓地大小不等，但都被高耸入云的粗壮松柏大树笼罩，不同取向的坟头一字排开，石刻碑文清楚地记载着每个家族世代的兴衰史。

插　　曲

1964 年的三大王克森（1911-2011）和
娘王花庭（1920-2001）

大约是 1941 年，我父亲的弟弟王克森［2011 年 11 月去世，享年 101 岁（虚岁）］成了亲。农村有个习惯，家中兄弟们在都成家后，就要分开过。我的父亲和三大也不例外，就在这年底分了家，当时我还不到三周岁。记得在分家后的一个晚上，奶奶带着我在灶火里玩，我拿着三大家的烧火夹子到处乱画，正玩的欢时，三大从外边走了进来，当他看见我用火夹子到处乱画时，便从我手中把火夹子夺了下来摔在地下，并大声吼叫："滚出去!"我吓得哇的一声哭了起来。父亲闻声跑来把我抱了出去，还同三大吵了一架，弄得奶奶张口结舌，毫无办法。这件事虽在我幼小的心灵里留下了难忘的记忆、并对他产生了不好印象，然而这种不好印象很快就被另外一件事所冲淡了。不知什么原因，吵架几天后，三大把我接到他在贾宋区政府的临时家中，改嫁过来不久的娘（即婶子，我们那里叫娘），对我不错。记得一天早上，她给我端来一碗白花花的泡馍，碗里面还漂着香气扑鼻的芝麻油珠儿，一片受宠若惊的感激之情油然而生。说实在话，虽然我妈（母亲）同娘之间相处的关系不太好，可我一直对娘怀有好感。娘的名字叫王花庭，虽不识字，但却很能干；解放初期，她曾当过乡妇女联合会主任，在土地改革、抗美援朝和"三反""五反"等运动中，都做出了贡献。不幸的是，晚年由于病痛折磨，于 2001 年服毒自杀身亡，时年 80 有余。

初入小学

1964 年秋的外婆（1885–1972）
和她的孙女张心平

我五岁时，已经懂事了，看到别家的孩子背着书包去上学，很是羡慕；回家就闹着也要去上学。当时，一是因为穷，二是年纪小，家里没有同意；为此，还大哭了一场呢！后来，由外地迁来暂住的邻居申老七的儿子申天恩看我求学心切，便用草纸抄了一个课本送给了我。我如获至宝，就拿着这个手抄书本，跟着别的孩子一起去上学，所幸的是，老师马振国收留了我。仅半年左右时间，我不但学会了课本上所有的字，而且算术也学得呱呱叫。次年，父亲与外祖父商量后，我便正式到距我家不到两公里的外婆家的前张村红庙小学上学。红庙小学是我家乡附近有名的正规学校，不但老师水平高，而且对学生的衣食住行都要求很严。入学后的第一件难事是校服，做一身校服要花很多钱，再加上童子军军杠和小绿背包等，就更出不起了，怎么办呢？家里东拼西凑才做了个制服短裤，好在舅舅张秉中也在这个学校念书，只得捡他的旧衣裳充数了。舅舅是我外祖父和外祖母最小的儿子，比我母亲要小 20 多岁，比我也只大几岁。舅舅的哥哥（我的大舅）的外号叫"猫娃"，因偶然听得当时土匪头子的一些活动计划，而被杀害，死时刚刚 20 出头，母亲冒着危险将弟弟收敛埋葬。民国初年，镇平县和邓县等境内土匪猖獗，几乎所有村子都修筑了高寨，晚上一家老小都躲进寨子，寨子顶层有人行道和枪眼，便于青年壮丁在上巡夜，其上备有土制辫子手雷和火枪，以防土匪来犯。

我的启蒙老师是个老先生，名字叫张吉堂，他很有学问，教学有方，对学生管教极严，我们都很怕他；同学们也常常因早自习时书背得不熟而挨板子。有一天下午的上课前，学校的操场上来了几个玩猴子的艺人，锣鼓一响，大家都蜂拥而上，围成了一个圈子。这只猴子聪明灵巧，一会儿骑狗，一会儿牵羊，还不时地爬竿耍鬼脸，逗得大家哈哈大笑。正看得入迷，上课铃响了，可是没有一个同学去上课。忽然，教导主任带着各班的上课老师，直奔玩猴场。只听嘟嘟一声哨响，顿时大家被吓得不知所措。老师让我们按班级排好队之后，示意一个接一个走过大门，在大门的两旁，站着手里拿着大板子的老师，进一个打一个，无一幸免。这是我第一次尝到挨板子的滋味。每个人的小手都被打得又红又肿，手一攥就像针扎一样，疼痛难忍。事后，张吉堂先生到外婆家喝酒时，还告了我的状。外婆家当时开了个黄酒小店，能赚点零花钱，日子过得比我们家要好些。记得外祖父包角子（是一种较大的饺子）时，除要放猪肉丁外，还要放上不少芝麻香油，吃起来满嘴是油，非常香。他的另一个拿手好菜是别有风味的芹菜蒸肉。不少来喝酒的人常点这两样饭菜——煮“大角子”和“芹菜蒸肉”，有时我也能沾点客人的光，享点口福。

远足与讲演

入学后第二年的春天，学校照例要进行远足活动，远足对学生来说，是一件非常开心的事。学生都要穿上校服，脖子上系着绿领巾，背上背着盛着干粮的绿色硬背包，肩上扛着童子军军杠，十分威武。行进时，校旗引路，校乐队的鼓号齐鸣，很是壮观。途经的大路两旁围观的老百姓很多，热闹非凡。远足的过程中还要开展多种有趣的游戏，先遣队员除了画出前进的路标之外，为了不让同学们在短暂的休息中感到单调，还事先在休息地附近的草地里、树枝上、石头缝里和花丛中藏些小礼品，如铅笔、粉笔、糖疙瘩、几颗大枣、花生、核桃或一个柿饼等。大队人马到后，休息号一吹过，三五成群的小伙伴一起立刻在附近可能藏匿小礼品的地方找呀找呀，忘记了疲倦；就这样走一段，玩一会，不知不觉就到了目的地。在休息大约十几分钟后，大家重新集合起来，开始一年一度的讲演比赛。参加比赛的同学，事先都把讲演的内容背得滚瓜烂熟；老师告诫的窍门是，讲演的语调不仅要把握好抑扬顿挫，还要表达得生动有趣，因而表情和手势的配合也很重要。

我讲演的内容是从模范作文上抄下来的，主题是要珍惜时间。讲演开始后，同学们一个接一个地上台，评分的老师十分认真地听着、观察着，手里还拿着小本记录着什么。最后一个讲演的是我，我上台后，直挺挺地站在大石台子的中间，台下鸦雀无声，很是紧张。听到老师“开始”的口令后，我便两眼朝天，机械地背诵着“光阴如箭，日月如梭，不知不觉，已过半月……”这引起了台下同学们的哈哈大笑，我趁势向台下扫了一眼，站在台前的我的一个要好的同学，挤眉弄眼地小声说“打手势，打手势!”我恍然大悟，我开始一边有节奏地轮起双臂打着手势，一边大声地朗诵“一寸光阴一寸金，寸金难买寸光阴……”最后，以“少壮不努力，老大徒伤悲”的诗句结束，出乎意料地赢得了老师同学们的热烈的掌声！我跳下石台，走进队里。这时，风吹一冷，才发现汗已湿透了衬衫。讲演评比结果，我得了第一名。

夜“逃”红军

1947年的春天，由于学校每天都要上早自习和晚自习，我和同村的王占德只得住在学校里。宿舍是一个大庙，里面还有一个很大的石菩萨。大庙白天是教室，晚上当宿舍。夏天跳蚤很多，觉得痒痒时，用手一摸便是一个；冬天四面漏风，寒冷刺骨，加上虱子不时地发起攻击，实在难熬。

一个漆黑的夜晚，半夜里，王占德的父亲王书礼，慌慌张张地跑进我们住的大庙里，用着颤抖的声音把我们叫醒，不由分说把我俩拉出庙外，翻过厕所不高的土围墙，一溜烟地跑回了家，并且神秘地对我们说“红军要打来了！”中华人民共和国成立前，由于国民党的宣传，老百姓眼里的红军个个都是“杀人不眨眼”的妖魔鬼怪，心里都很害怕。王占德是个独生子，他的父母怕他被红军抓走，所以连夜把他叫回家，顺带也把我带回家里。第二天，我们才知道，其实红军只是从这里经过，并没在我们这里扎下。当然，学校仍是照常上课，而我们俩为此还挨了板子。从此之后，三天两头“逃”红军，学也上不成了。

辍学回家后不久的一个深夜，震耳欲聋的一声炮声把我从睡梦中惊醒，只觉得房子仍在颤动，人们议论纷纷。原来是红军一炮轰开了马庄寨门，攻占了马庄乡政府所在地，驻守马庄寨的国民党地方武装的王营长被活捉。这位营长是我们村里的人，家境中下，兄弟四人，排行老二，平时不摆架子，没有仗势欺人的劣迹。第二天一大早，在他亲属的苦苦哀求下，附近几个村子里十几个有头面的老年人一起到处寻找红军为他求情。其实，当天夜里红军就已撤走，而这位营长则就地被秘密处决了，其原因是他迫害过红军的地下工作人员。红军走后，国民党的队伍又卷土重来，一方面，他们为王营长大办了丧事，听说还发放20担麦子的“抚恤金”；另一方面，他们到处寻找为红军带路或帮助过红军的人，网罗罪名，伺机报复。不久，一个名叫张传举的乡村医生不明不白地被暗杀在菜地里，听说他为路过的红军伤病员看过病。令人发指的是，就连一个平时做事有点疯癫、外号叫张“疯子”的小货郎也不放过，他以莫须有的通匪罪名被枪决。张“疯子”在即将被行刑前，向张庄乡长恶霸地主张法庭哭叫冤枉，张则说，冤枉也就冤枉你这一次了！话音刚落，可怜的张“疯子”就被打死在村边的土路旁，暴尸多天，无人敢为这个辛苦一辈子无儿无女、无依无靠的老人下葬。这种白色恐怖的日子一直持续到1948年春解放为止，这个罪恶累累的恶霸地主张法庭没有逃脱人民的法网，1950年土地改革时被人民政府镇压，也为张“疯子”报了仇。

崭新的小学生活

2007 年 4 月 21 日，姐姐（1930–2009）
在北京天安门前留影

随着解放大军由河南向湖北、湖南推进，驻扎在河南南阳地区的国民党军队王凌云部已溃不成军，在襄樊会战中被全部歼灭，1948 年 5 月南阳地区获得全部解放。1949 年秋天，我考入了马庄完小 5 年级，开始了崭新的学生生活。马庄离我的家东黄楝扒村 5 里路，但隔着一条十多米宽、半米多深的严陵河，特别在夏季，雨后常常暴发洪水，恶浪汹涌，有时几天都过不了河。为了不耽误学习，我只能在学校住宿；碰巧这一年，我姐姐也出嫁到了马庄，这样吃饭问题也就得到了解决。两年的小学生活经历了许许多多难忘的事情，如诉苦运动，抗美援朝，镇反和土地改革等。

入学后学校开展的第一个政治运动是发动学生控诉旧社会地主恶霸和国民党统治者欺压贫苦大众的罪行。不少农民被盘剥得一无所有，流落村头，以乞讨为生，哪里还敢有上学的奢望！时年与我差不多年龄的王大汗兄弟俩，父亲眼瞎了，就被地主赶出家门，母亲只得带着他们三人四处讨饭为生，住在我村的一个四面漏风的破庙里。中华人民共和国成立后，他家分了房子和土地，也能像其他孩子一样上学了，心里有说不出的高兴。在大会上，我也痛哭流涕地控诉了因我爷爷没钱交苛捐杂税，只能眼睁睁地看着三大被国民党抓去当壮丁的旧社会罪行，引起了同学们的共鸣。1950 年朝鲜战争爆发，学校里开展了轰轰烈烈的抗美援朝运动。河南省豫剧名演员常香玉个人捐献一架飞机的爱国盛举极大地激励着我们这些十二三岁的青年学生，我自告奋勇地捐出了我最喜欢的那只老公鸡，受到学校老师的表扬。学校的捐献运动很快就扩展到社会上，那种有钱捐钱、无钱捐物的感人场面至今令人难忘！不少群众赶着猪、羊，牵着牛，提着鸡、鸭和成篮子的鸡、鸭蛋涌向政府接待站……翻身农民热爱新中国、为保家卫国不惜一切代价的大无畏精神，预示着站起来的中国人民必将会赢得这场战争。随着土地改革的深入和朝鲜战争的爆发，国内被打倒的、暗藏的敌特分子也在蠢蠢欲动，伺机配合国民党反攻大陆，甚嚣尘上。为了打击国内敌对势力的嚣张气焰，从 1950 到 1953 年，开展了轰轰烈烈的“三反”“五反”和镇压反革命分子的群众运动。一天上午，我们五六年级的学生在老师带领下，紧急集合出发到校外打麦场上参加听都未听过的公审大会，会场上人山人海，我们被安排在会场主席台的左侧前面坐下。不一会儿，几个全副武装的民兵荷枪实弹地押着两个被五花大绑的罪犯入场，会场一片轰动，人们都伸长了脖子向前挤，想看个清楚。待群众安静下来后，公审大会开始，乡农会主席愤怒地宣读了这两个地主所犯的罪行。一个是泰山庙村人，他趁天热乡亲们在大水坑里洗澡之机，将分了他家土地的一个青年民兵按入水里活活憋死；另一个是名叫王子阳的马庄财主，暗地里向他堂弟反攻倒算，威胁其堂弟把分得到他家的财产原封不动地还给他等。这时会场群情激昂，“打倒地主恶霸，坚决镇压反革命分子”的吼声震天动地，吓得两个地主浑身发抖。农会主席接着大声宣判：“经县人民政府批准，现将这两个罪大恶极的犯人押赴刑场，立即执行枪决。”会场顿时鸦雀无声，气氛显得十分紧张，不少坐在后面的群众开始站起来向前拥来；再看这两个平日作威作福、鱼肉百姓的地头蛇，已被吓得面无人色，瘫软在了地上。几个民兵只得将他们拖到主席台的左边不远的“刑场”上，令两人并排跪下。“刑场”就在我们这些小学生的正前面，当时的情景看得清清楚楚，至今记忆犹新。只见两个民兵同时举起了枪，枪口几乎顶住了罪犯的后脑勺，随着扳机扣动的一刹那，泰山庙村的那个地主脑袋开了花，栽倒在地。大家奇怪的是瘫在地上的财主王子阳，却丝毫无损，原来他是被拉出来赔罪的！就是这个当时被吓得屎尿拉满裤裆的不法财主，仍不思改过、不重新做人，而是继续与人民为敌；次年，在土改复查运动中罪加一等，被判死刑。

公审大会后的当天晚上，我们在校住宿的几个小伙伴都很害怕，一闭眼那种恐怖的场面就浮现在眼前，为了壮胆，大家都挤在一个由几条被子叠成的大被窝里睡觉，就连半夜上厕所也要同行；尽管如此，我们还是经常从噩梦中惊醒。让青少年学生参加处决犯人的公审大会一直持续到20世纪的50年代中期。人们在问，难道非要通过这种手段才能教育这些尚不太懂事的青少年学生吗？从我的亲身感受来看，在青少年幼稚的心灵里产生的负面影响是不容忽视的。

在两年的小学生活中，除了唱国歌、升国旗，学打霸王鞭，排演“四洋湾”和“小二黑结婚”等话剧倍感新鲜外，还有两件与学习有关的事令人难忘。一是每天早自习的背书课，教我们班的语文老师有绝招，凡是背不下或背不全指定课文的学生都要受到惩罚。他的惩罚办法很独特，让几个同班学生围着你，在用手恶狠狠地指点着你的同时，口中还喋喋不休地唱着“八只脚，一对鹗，走路横着走，口里吐白沫……”来喻指你就像螃蟹一样横行霸道，不努力学习。虽然我的学习成绩在全年级一直名列前茅，但也有偷懒、偶尔失误的时候，享受这种“待遇”的经历，怎敢忘记！二是在地理课的小测试中，有一道问及“贯穿我国东西大动脉的铁路名称”的试题。由于马虎，未加思考地就写了“浙赣铁路”，但当看到老师在自己的卷子上打的大红叉子时，突然想到了“陇海铁路”，这时心中的后悔劲儿真是难以言表。

侯集镇的 3 年初中生活

1951 年的夏天，我以优异的成绩，读完了小学，家境虽然贫寒，但父母仍坚持要我报考初中，因为他们知道这是孩子走出农村的唯一办法。当时的镇平县只有两所初中和一所初级职业学校，两所初中距家乡都很远，位于贾宋镇的初级职业学校离家较近，约 15 里。由于交通闭塞，在难以看到这三所学校的招生广告的情况下，只能跑到贾宋镇去碰运气，可巧赶在初级职业学校入学考试报名截止日期之前报上了名，高兴的心情，难于言表。但考试时间已迫在眉睫，回到家后的第二天，就又冒着大雨赶回贾宋镇的考场，当晚就和衣睡在学校里的课桌上，熬过了一夜。考试的成绩自我感觉还不错，但还没等到发榜，又看到了侯集中学第二批招生的广告，侯集中学是县里的两所名校之一，是我梦中向往的地方。在近半月的时间里，我夜以继日地抓紧时间复习功课备考，最终我在几百人中以第 9 名的成绩考取了该校，从这年的秋季开始了三年的初中生活。当时学校还没有通电，晚自习用的是棉花捻子的油灯，每人一盏，虽然条件艰苦，但大家的学习劲头仍然是很高的。

夜　惊

为纪念在中华人民共和国成立前夜英勇牺牲的共产党员侯庠生烈士，侯集中学在解放初曾更名为“庠生中学”，不久又改为“镇平二中”。学校离家约25华里，只得在校住宿，几十个人睡在由约1米走道隔开的南北两个大通土炕上，冬天挤在一起暖暖和和，其乐融融；可是到了夏天，通风不好的宿舍内闷热似蒸笼，加上跳蚤的叮咬，久久难以入眠。记得一个炎热夏天的夜晚，我们4个同学贪图凉快，偷偷睡在操场东边的一棵大树下，由于当时社会上流传有老虎在学校附近出没伤人，弄得大家胆战心惊，睡不踏实。半夜里，睡得迷迷糊糊时，突然听到令人毛骨悚然的惊叫声：“老虎来了，老虎来了！”处在半睡半醒中的我，只见其他几个同学不自主地纵身跳了起来，光着膀子无意识地上下抖着被单子，嘴中还嗷嗷叫个不停，加之来自附近乱成一片的男女宿舍里震耳欲聋的尖叫声，甚是吓人！就在大家处于惊恐不安之时，学校教导主任提着灯笼一边大声喊着“同学们安静，安静！”一边急急地向我们走来，看见老师，我壮了胆，主动迎向前去，与老师一起来安定大家的情绪，大约经过10分钟才逐渐平静了下来。不少人在惊恐的奔跑中擦破了皮，碰伤了头，当时学校卫生室的纱布、胶带和红药水都告了急。后来知道这种现象叫“夜惊”，常发生在神经处于高度紧张的人群之中，特别在军营中出现“夜惊”是很危险的。为了防止“夜惊”的再次发生，造成不测，学校在一段时间里，还特地安排了巡夜员，来回于学生宿舍之间，以安定人心。

雪夜宿房营

1951年9月，我考上侯集中学的消息很快就传遍了全村，乡亲们都来祝贺，家人自然也都高兴，然而一想到尚无着落的学费和生活费，父母还是发愁。为了尽可能减轻家里的经济负担，我不得不自己起火做饭。我和几个同学在学校附近大街上的一家商铺后院的走廊上支起了锅灶，黄泥做的小灶台是买来的，我的饭锅是一个大铜瓢。每周从家里带来花卷（即用白面和红薯面交替卷在一起蒸成的馒头），熬稀饭的玉米面、红薯，当菜用的辣酱或芝麻盐，烧饭用的柴火也是利用课余时间从树林里拾来的。每天的早餐是一个烤花卷、一碗玉米红薯粥，午餐和晚饭吃的都是两个热好的馒头就酱菜，高兴时打个鸡蛋白菜汤。虽然每天花去不少时间在做饭上，但能为家里节约点钱而感到欣慰。

学校离家20多里路，每周六下午，我们几个要好的同乡同学都要结伴回家，第二天下午再返回学校。回校时大多都要背上或挑上二三十斤[①]的东西。在校公共食堂入伙的学生，主要带的是要上交的柴和面，像我这样自己搭伙做饭的学生，则要背上一周的干粮和生活用品，总共也要有20多斤。在平常的风暖日和的天气里，带上这点东西，走上20多里路，对于我们这些生长在农村的青年学生来说是不成问题的，可是遇到炎热的夏季或滴水成冰的寒冬就难说了。记得是1952年年初的一个冬天，我背着20多斤的东西独自一人返校时，半路上突然遇到了暴风雪，呼啸的寒风迎面吹来，简直使人寸步难行，雪花打在脸上就像刀割一样难受。我在大风雪里苦苦挣扎了两个多小时，才走了四五里路，这时钻入靴子里的雪化了，靴袜都已湿透，眼看着天慢慢地黑了下来，心中十分焦虑不安。无奈间，我高一步、低一步，摇摇晃晃、艰难地向邻近的房营村奔去。短短几百米的路就走了半个多小时，还摔倒在雪地里多次，最后总算走到了村子里的一个小学校旁边，可是这时天已黑，我心里一点主意也没有，只是站在雪地里张望。非常幸运，一位住校老师出门倒水，看见了我。在他知道我是侯集中学的学生后，立即把我带到他住的屋里，屋里生了一个火炉子，很暖和。他给我倒了一茶缸热水，让我先喝点暖暖身子，还关心地问我饿不饿，吃饭了没有？我说"我带有干粮，不饿"。其实，我是不愿意过于麻烦这位素不相识的好人，才忍住了饥饿。夜里就在他住的屋里为我打了一个地铺，当我脱了靴袜准备睡觉时，这位细心的老师发现了我的靴袜都湿透了，于是亲自动手为我烘烤。这时，我心里暖乎乎的，充满感激之情，热泪夺眶而出，这是我有生以来的第一次！这一夜我睡得很香、很甜。

---

① 1斤=0.5千克

第二天一大早我就起了床，千谢万谢地告别了这位善良的老师，急急忙忙地向学校奔去。大雪虽已停了，但刺骨的北风仍然吹个不停，我在半尺深的雪地里行走起来仍然非常吃力，等我赶到学校的时候，第一节上课的铃声刚刚响过。我顾不得了什么，竟背着东西跨进了教室，同学们都以好奇的眼光看着我，原来我的头发和眉毛上都挂满了白花花的冰碴，活像个“圣诞老人”。上课的语文老师趁机说，我们每一个人都应当有那种不畏艰险、克服困难和勇往直前的精神，只有这样才能做出一番事业来。

“雪夜宿房营”之事虽已经过去了60多个年头，回想起当时的情景仍历历在目，感激之情油然而生；但由于当时年少无知，竟未问及这位老师的名字，后经多方打听而无果，遗憾终生，我也只能向这位善良的好人表示深深的歉意了。

入　伙

1952 年秋天，家乡鼓励种植经济作物，父亲开始种烟。烟苗要从烟农那里买，如果烟苗栽好后雨水不足，还需人工挑水浇灌，以确保烟苗成活。烟苗生长期容易生虫，那时不用农药，每天早晨由人下地逐棵捕捉虫子，十分费时；在暑假或每个星期天的上午，我都要同家人一起下地捉虫。烟苗长得很好，经过烤制获得上等烟叶几十斤，前后共卖了近 40 元，除去烟苗、烤制等的花费外，还有 30 元左右剩余，有了入伙的基础。

学校的学生食堂分为两类：饭量大的（高年级、大个子学生）每月 5 元，饭量小的（女生与小个子学生）每月 4 元，没钱可用粮食和柴火等顶替。初中二年级第一学期我在饭量小的食堂入伙，每月的烧柴从家里挑来，故每月大致需交 3 元伙食费，卖烟叶剩的钱基本上都用在我的伙食上了。第二年又多种植了半亩烤烟，收入也有所增加，初三上学的花销基本上也得到了解决。

每月 4 元的伙食费吃得还不错，早饭是稀饭加红薯面窝头，午饭吃花卷和一小盆蔬菜，晚饭红薯稀饭、花卷和咸菜。每周两三次吃肉包子或大肉熬面来改善生活；吃面条有规定，10 人一组，每人盛好面条后要放在地下，不能先吃，要等大家都盛好、管伙食的老师哨响后才能进食。只听老师的一声哨响，鸦雀无声的饭场里顿时响起了阵阵吃面条的呼呼如风的声音，甚是有趣。聪明的同学曾传授过吃面经验，第一碗不要盛得太满，争取先吃完，这样就有机会再盛第二碗时捞得更多的肉和面条。我虽也想多吃几块肉，但对这些在吃饭上下功夫同学的做法却不敢恭维，并不在意的说了句“会吃的人不一定能学得好、考得好”的话，没有想到这句话竟刺痛了一些原来与我关系不错的同学。要好同学的远离使我在精神上备受打击，总疑神疑鬼地觉得同学们在议论自己，第一次感到了被孤立的难受滋味；但也从这个教训中学会了“严于律己、宽以待人”的做人准则。

紧张有趣的课外活动

学校建在一个土寨子里，寨子四周的围墙壕沟里的水已经干枯，长满了杂草，学校朝北的大门是唯一可以出入的地方。两排整齐的教室建在北大门的两侧，距寨墙10多米，中间生长着枝叶茂盛的槐树，春末夏初盛开的槐花，引来了成群的蜜蜂，校长吴雨田就养了两笼蜂，他不仅教学生如何因地制宜地养蜂增加家庭收入，而且也很重视培养学生的劳动观点。寨子的西侧是学生食堂、宿舍和学校职工与家属住处，南边则是苗圃和花房。学校中路东边是一大片约25亩①的实验田，田间有灌溉水井；按年级、班组把地分到每个学生，每人不到半分地，27平方米，用于种植蔬菜瓜果。多数学生都爱栽培番茄，因为它可生吃；但也有种南瓜、扁豆什么的，秧苗大都从学校的苗圃里移植过来。生长管理技术，包括施肥、打掐枝芽、人工授粉等，由植物课老师教授，吴校长也常下菜地进行实际指导。番茄又叫西红柿，新鲜的番茄有一种难闻的气味，刚吃起来感到不太可口，但习惯后还真是离不开呢！尝到了甜头的我还试着把番茄移植到家乡里栽培，可惜的是没成功，主要是虫子咬断幼苗问题没能解决好。这些菜蔬的生长期都较长，大家常为暑假离校无人照管挂心；每当过完暑假返校的第一件事，总是急急忙忙地跑到我的那一小块地里，看看番茄生长得如何，才能放下心来。

学校的大操场位于学校大中路西侧和学生食堂之间，场内设有两个篮球场，还有跳高、跳远的沙坑和单、双杠等，运动器械虽简陋，但还算齐全。篮球比赛最热闹，特别是双方的啦啦队都很卖劲，喊声不断；初中时我又小又瘦，常被分配在男女生混编的队里参加比赛，心中总感不快。大操场除用作升旗、早操和课间操外，也是一个露天大礼堂，学校的一些重大活动，如开学与毕业典礼、看无声电影、文艺演出、演讲比赛等也在这里举行。学习英语对大多数学生来说，都是一件枯燥无味、觉得头疼的事，为了激发大家的学习兴趣，英语老师在这里组织召开了一个经验交流大会，一个高年级同学自告奋勇上台介绍记英语单词的经验。他从在英语单词旁标注中文到通过谐音联想读记等介绍得十分“有趣”，例如英语“good-bye”，他的注音是“狗蛋白”，因为白色的“狗蛋”容易记；又如英语“China”，他当场示范说：要先从“天那，天那！”开始，逐渐过渡到“欠拿！”如此等。虽然当时有不少老师和同学不以为然，但还是在一定程度上引起了学习英语的兴趣。

大操场还是平时同学们一起吃饭的地方，特别是改善生活的时候，饭后丢一地用树枝做成的“筷子”，这些“筷子”曾帮我解决了做饭少柴烧的难题，当然，我也心甘情愿地做了一年的义务“清场夫”。

---

① 1亩≈666.67平方米

南阳第二高中

1954年7月我顺利地考取了南阳第二高中（简称南阳二中），开始了紧张而又难忘的3年高中生活。南阳二中始建于1953年，次年秋季开始招收首届新生约200人，按照甲、乙、丙、丁、戊、己次序分为6个班，我被分在甲班。当时实行男女分校，二中没有女生。学校建在距南阳市中心约3.5公里外的卧龙岗上，紧邻苍松翠柏环绕着的诸葛亮庵，白河水静静地从岗下流过，风景十分优美；然而，在初建的校园里却光秃秃的没有一棵树，也没有一条像样的路，一下雨校内泥泞不堪，难以行走。为了改善学校环境和校园里的道路，同学们在老师的带领下，几乎是每周六的下午，都要步行5里多路到位于学校西边的十二里河挖沙子，或在学校周围的荒山岗上捡石头，用于铺路。经过半年多的辛勤劳动，校内通往教室、宿舍、食堂和操场的大路、小道都得到修整，从此再也不愁下雨走路难了。在修路的同时，大家还动手在学校周围、路旁、教室和宿舍的前后，种上了法国梧桐树。时隔45年后的1999年年初，当我重返学校旧址时，它们已长成了参天大树，而我当时也已满了60周岁。

初建学校的教学条件自然没有老校好，然而崭新的教室、宿舍、课桌和教学、生活设施与用具，对于从未见过大世面、来自偏远农村的我，也感到很满足了。首任校长崔溪萍是个老革命，他的夫人贺毓秀，听说与贺龙将军是同乡，大家都很尊重和敬仰他（她）们。学校的老师多是从其他兄弟学校抽调来的，教学经验都很丰富，特别是他们各具特色的授课风格，至今仍深深地印在我的脑海里，难以忘怀。教物理的张照普老师，同我是仅有一河（严陵河）之隔的同乡，他常常结合实验，深入浅出，或采用形象比喻的方法，把不易理解的物理概念和难记的物理定律讲得头头是道，引起了我对物理学的浓厚兴趣；张老师待人也很和气，大家都很喜欢他，遗憾的是后来他被调到了别的学校。高文鹤老师教授的历史课，最受同学欢迎。他对历史故事绘声绘色的描述，常在不知不觉中就将你置身于历史的事件中，就好像亲临其境一样，一幕幕活生生的历史画卷展现在你的眼前，使你入神；这时，同学们最怕听到下课的铃声，那样你就不得不等待高老师的下回分解了。尽管我很喜欢高老师的历史课，但我心中一直对他给我历史课的年度总评为4分，感到不解和不满，因为那一学年平时考核和期终考试我得的都是5分呀！

在南阳二中学习期间对我影响最大的是白鸿鹤老师，他主讲的数学课最令我难忘。上课时，他首先把当堂课要讲解的主要内容，包括公式与图例，写（画）在黑板上；然后，就背靠黑板旁的墙壁，闭起双眼，开始用他那特有的幽默语言和抑扬顿挫的音调讲述定理或证明例题。在讲到重点地方，他会从不同角度或从正反两个方面反复论证，时而，还加上点乍看起来有些“滑稽”的表演动作，使本来抽象难懂的概念，或枯燥的演算，常在轻松的笑声中留在了我们的脑海中。白老师不仅有高超的讲课艺术，而且他还对中学教学研究付出了很多心血，他常将教学的心得，数学难题求解或证明技巧等写成短文投稿。我则经常利用课余或星期天，主动帮助白老师整理和抄写他的手稿，在他的身教和言传下，使我对数学产生了浓厚的兴趣。在南阳二中学习期间，我数学课的成绩一直很好，几乎全是5分。1957年高考时，起初我填写的第一志愿就是北京大学数学系；后来，为了确保学校首届毕业生的升学率，我接受了时任学校教导主任马恒昌老师的劝导，改为报考南开大学物理系，并被录取。教化学、语文、政治、俄语和体育等课程的老师也很优秀，同学们都非常满意，大家的学习劲头很足，学习成绩一点也不逊色于老牌兄弟学校，从南阳二中1957年近70%的高考升学率可以得到证实。

远离市区学校的文娱活动不多，但也经常在周末或节假日时，以班级为单位，自编、自排和自演一些小节目活跃生活。除此之外，学校每隔两三周还组织全校师生员工，兴高采烈、浩浩荡荡地步行几公里到市内电影院看上一场电影，或到戏院看上一场河南梆子，算是最大的享受了！

当时学校内还没有通电，冬天的早自习和晚自习用的是汽灯照明，遇到汽油质量不好时，汽灯的喷（油）孔，经常被汽油中的渣子堵塞而使灯熄灭，这时就要拿上手电筒，小心地爬上高凳子，用一根很细的钨丝针，将被堵的喷（油）孔通开，非常不易。有时，一个晚上汽灯会熄灭多次，每当这种情况出现时，同学们都静静地耐心等待着，从未出现过急躁不安情绪。冬天用炭火盆取暖，教室的后面放置两张床，有两位同学住在那里守夜。每天按时出早操和课间操，从不间断。同学们还要轮流值日，负责打扫教室、宿舍，担水供大家早晚洗漱，生活过得井井有条。

1956年秋季，不知道是什么原因，南阳二中和南阳师范学校地址对调，搬到了位于南阳市内（京武门外）的南阳师范学校校址。绿荫成林的校园，宽敞的大马路、明亮的电灯和漂亮的两层教学楼以及设施比较完善的运动场等，使同学们感到新鲜。学校的南边紧邻原南阳女中，那时已改为男女合校。当时我是班级的体育“部长”，常组织两校篮球比赛，我心中明白，原女中的男生不多，比赛时我们获胜的机会就比较大，为了“鼓舞”士气和为校“争光”，我常常要这个“小把戏”，心里还得意得很呢！学校的东边紧邻南阳市广场，每年“五一”国际劳动节和“十一”国庆节等大型节日庆典活动就在这里举行，倒也热闹；但它也用来作为处决犯人的刑场。同学们常被要求集体参加处决犯人的公审大会，当五花大绑的犯人当场被执行枪决的那一片刻，会场里笼罩着令人窒息的“恐怖”气氛！诚然，罪犯被处决是罪有应得，但留给我们这些青少年学生的印象是什么？恐怕没有人认真思考过。这当然不是南阳市的“发明”，我在小学六年级读书时，就曾被组织参加过这样的公审会。

20世纪50年代，南阳地区的交通十分闭塞，既不通铁路，也没有像样的公路，市内没有公共汽车，就是自行车也极少见。尽管当时有几条从南阳通往各县城的长途汽车，但对我们这些来自偏僻农村的学生，不要说没有长途车可乘，就是有车也没有钱坐。我和同乡同学（王正寅、王世哲、王保志和侯永奇）的家离学校约90华里[①]，通常每月回家一次，为了不耽搁功课，都是在周六下午3点左右出发，为了壮胆，我们几个常结伴同行，通常大约夜里2点左右才能回到家里。在夜行中，常要通过一些“鬼火”时隐时现的坟地，走进村庄时，还要提防被狗咬，这时大家紧紧跟随，走得很快，谁都不肯落后。在精神高度集中的情况下，长途跋涉近10个小时，倒也不觉得很累。回家的途中，不仅要渡过水流湍急的辽河，而我和王保志还要赤脚蹚水渡过严陵河，特别是在夏季涨水时，过河是很危险的。记得有一次是在麦子快要收割的夏季回家，大约是夜里1点多，我们已走到了不得不分开的交叉路口，虽然离各自的家只有几里路，急切回家的心情也自不用说，但谁都犯愁独自走向自己的庄村。俗话说“远怕水，近怕鬼”。我回家的路就必须经过为村里死去的人送葬的那个路口，故人生前的衣、被、褥子和席子等，也都要在这里焚烧，说是送程。虽然我知道人死如灯灭，没有什么鬼、神，但在这种特殊的环境下，心里总是不踏实，只得硬着头皮、神经高度紧张地向村北边的那个路口走去；为了壮胆，嘴里还不停地唱着小曲，手中拿着一条棍棒乱舞。说也奇怪，当我刚走近那个路口时，两旁一人高的麦地便突然地哗哗作响，好像有个“怪物”向我猛扑过来，瞬时吓得我魂飞魄散，不由地惊叫了一声，出了一身冷汗，头发都全竖了起来！幸亏这时，我看到了来自村里的灯光和听到了狗的叫声，狂跳的心才平静了下来。定睛细看，原是碰巧刮来的一阵微风掀起的麦浪。回到家里，倒头便睡，待一觉美梦醒来时，已是星期天上午8点多钟，又开始忙着整理返校的行装了。来回约90公里的路程要在20个小时左右走完，返校时还要负重，实在是一件苦差事，没有良好的身体和克服困难的毅力是难以坚持下去的。我在大学里和工作后参加的长跑、爬山和游泳比赛中，常能取得良好的成绩，大概就是得益于青少年时代的这种磨炼。今非昔比，但同学们千万不要忽视锻炼身体，因为没有一个健康的身体，是难以为振兴中华做出自己的贡献的。

① 1华里=500米

只身北上南开求学

1957年8月初的一天，我被南开大学物理系录取的喜讯在家乡传开了，我和家人自然高兴，乡亲和亲友们也来祝贺；因为这毕竟是有史以来这个偏僻的农村走出的第一个大学生。乡亲们对我寄托了深切的厚望，希望我能够成就一番事业，为家乡争光。爷爷记起了很久以前一个风水先生的话，老王家的祖坟是埋在贯穿豫西南两支龙脉的一支上，这支龙脉的龙头就向着你们这个小村庄。爷爷信心十足地说："是该出人物的时候了。"

父亲东拼西借勉强凑足了去天津的17元路费，发愁的是以后学校的生活费如何办？记得是9月10日这一天的早上，家人和我都起得很早，送我起程去天津，这是我第一次单独出远门，家里人多少还是有些不放心，特别是年过80的爷爷和奶奶站在村头，望着我远去的背影，久久不肯离去；当时的情景，至今仍历历在目，难以忘怀。经过一天的长途跋涉，晚上到了母校南阳二中，受到了老师们的热情欢迎。按照当地政府对有关被外地大学录取的贫困生酌情给予路费补助的规定，我得到了7元的补贴金。有了这点钱，心里踏实多了，至少使我摆脱了在入学时身无分文的尴尬处境。

9月12日一大早，由南阳汽车站搭乘长途客车去许昌，"客车"实际上就是一个大卡车，车上没有座位，上车后有的坐在自己的行李上，有的则席地而坐。车在高低不平的公路上行驶，大家就像煤球一样被摇得东倒西歪，迎面飞驰而来的卡车掀起的阵阵尘土，弄得大家满身是土，苦不堪言。当天傍晚，客车按时到达许昌火车站。我顾不得劳累，背起行李赶快跑到火车站售票处排队，幸好还是买上了当天晚上去北京的车票。有了票，紧张的心情便平静了下来，好容易才在候车室找到了一个地方坐了下来，啃着事先准备好的干粮充饥，喝着候车室为大家准备的凉白开水解渴。9月中旬的候车室仍然很热，挤满了来往的旅客，空气污浊。尽管如此，但想到即将就要坐上这个从未见过的"大怪物"去首都北京，心情还是很激动的，一天的劳累瞬时便消失得无影无踪！

晚上9点钟左右，从车站的广播喇叭里传来了由武汉去北京的火车即将到达许昌站的消息，人群顿时乱了起来，个个都拼命地冲向前。我赶紧背起行装，尽力挤入进站口的长队里，紧紧地靠着前面的旅客，唯恐不小心被挤出队外，误了上车时间。许昌是京汉铁路线上的一个中间站，出售的车票都没有固定的座位，能不能找到空位，就看你是否能先上车。进了站台的那些身强力壮的乘客都做好了第一个冲上车去的准备，我背着一个大包袱，人又瘦弱，自不敢拼命，只求能及时上了车就心满意足了。火车一声长鸣，呼哧呼哧地喷着蒸汽，慢慢地停了下来，我只顾欣赏我梦中的"怪物"

长龙，竟没看到争先恐后的上车情景！我急忙地上了车，不要说没有空位，就连车厢的过道上也挤满了人，我只好在两节车厢的连接处的一边找个地方，把包袱放下，坐在上面，深深地松了一口气。天已经慢慢地暗了下来，窗外一片漆黑，什么也看不见，只听见车轮压过铁轨接缝时发出的有节奏的咣当、咣当声，不知不觉便进入了梦乡。也不知过了多久，一阵喧杂声把我从梦中惊醒，原来车已经到了郑州。郑州是个大站，上下车的人很多，我趁机找到了一个位子，坐下来依附在桌上又睡了起来。

火车经过近 20 个小时的运行，次日下午到达了首都前门车站。前门热闹非凡，商铺、小吃店到处都是，可我就是找不到售票处；虽然已很疲惫，但仍不敢怠慢，只好叫了辆三轮车，三轮车师傅拉着我饶过前门城楼就看到了售票处，心里很后悔，因为短短的 100 米路，就花了两毛五分钱，实在不值！买好当天下午去天津的车票后，这位好心的师傅看出我是初到京城，怕我迷路，故仍要送我回候车室时，我婉言谢绝了他。在去天津的候车室里，很巧遇到了一位同被南开大学物理系录取的新同学杨玉芬（我们后来不仅被分到同一个班级，而且毕业后又一起到中国科学院半导体研究所工作），我喜出望外，几句话，我们变成了好像相识已久的老朋友。他是北京人，自然是个北京通，我问他，天安门在哪里？他告诉我穿过前门城楼，一直向北走，10 分钟左右就可到达。天安门是我做梦都想看的，如此良机，岂能错过！我即刻沿着通往天安门的巷子一边跑、一边张望，原来天安门就在位于“丁字型街”北边的长安街上。雄伟的天安门和门前的玉带河，白色大理石砌成的金水桥和高耸入云的华表，使我流连忘返。东看看，西摸摸，不知不觉半个多小时就过去了，差一点忘了要赶去天津的火车。我一路小跑地向车站奔去，远远就看见了新同学正在朝这个方向焦急地张望，当我们赶到检票口时，距开车时间只剩下不到一刻钟了。进站后，我们一溜烟地跑向去天津的站台，在开车铃声拉响的前一分钟，我们跨上了车门，事后想起来，真有点后怕呢！但让我欣慰的是，我这位新结识的同学并没有因此难为我。

大约两个半小时后，火车到达天津东站，在我们出站口的广场上，一眼便看见了一条红色横幅上写着南开大学新生接待站的大标语，心中热乎乎地走向接待站，在接待站同学们的热情安排下，我们上了车。站在我旁边的是一位穿着漂亮、戴着一副黑色边框眼镜的女同学（后来知道她是来自西安的崔玉凤同学）；再看看自己，身穿粗布衣裤，脚下是一双由姐姐赶做的土布鞋，加上两天一夜的奔波，满身是尘土，来自偏远农村、从未见过大世面的我，心中还真有点不是滋味呢！就在这个时候，我就暗暗地下了决心，在学习上一定不能落在他（她）们的后头。正在浮想联翩的时候，校车沿“8 路车”路线已经到了校园，下车后报了到，我被分在物理系 57 级一班。提起“8 路车”使我又想起了录取通知书中说的：从天津东站可直接搭乘 8 路车来校报到的“8 路车”，这里为什么同 8 路军的“8 路”联系在一起呢？我百思不得其解；直到后来看到市里还有 1 路、2 路等公共汽车路线时，才恍然大悟。由此可见解放初期身处闭塞农村学生的社会知识是多么的贫乏！

入学后，我按照要求填写了甲等助学金申请书，甲等助学金每月可得到国家 19 元的补助，除了每月 15 元的伙食费外，还有 4 元的零用钱，冬天还可申请一套棉衣。我的申请很快得到了批准，解除了我和家人之忧，感激党和政府之情难以言表。可以毫不夸张地说，没有新中国，就不会有今天的我。南开的伙食办得很好，早餐花样很多，鸡蛋、馒头、花卷、油条、豆浆和米粥等，午餐和晚餐则更为丰盛，每桌四菜一汤，鸡、鸭、鱼、肉齐全，凑够 8 个人就吃，吃完便走，无须自带筷子和饭碗，进饭厅吃饭也无须出示证件。南开这种模式的食堂一直办到 1959 年的下半年，国家经济发展遇到严重困难时为止，至今还令我怀念。

胆战心惊的高等数学课

1957 年，国家公布的大学招生名额仅为 10.7 万人，然而实招的新生只有 8 万多，在这种情况下，能考入全国重点大学的学生的质量应当是不错的。然而填鸭式的中学教学毕竟与以自学为主的大学不同，尽管我中学的数学成绩一直很好，也很喜欢数学，但在大学一年级的高等数学学习中，我与多数同学一样，遇到了预想不到的困难。一方面，大家刚接触极限和函数这些抽象的概念，理解不深，存在畏难情绪；另一方面，主讲高等数学的老师的突袭小考的方式，弄得人人胆战心惊。数学课一般都是两节连着上，上一节讲的内容还未来得及消化吸收，15 分钟的课间休息后的第二节一开始，这位老师常常便出题小考，考题就来自上节课他讲的内容，而且要在 10 分钟内交卷，多数同学答不好，成绩当然是不及格。这位老师还有一个“理论”，那就是说，成绩不及格就是不及格，要想及格，短时间里不可能；考 3 分的也难变成 4 分，更不要想 5 分了。他的这种逻辑更加重了大家的心理负担，这年物理系一年级学生的上学期数学期终成绩被评为不及格者多达 40% 以上，我虽侥幸及格，但心里也对高等数学产生了恐惧感，尽管我从小对数学就怀有特殊的兴趣。

1958 年，南开大学为帮助新建的山西大学，抽调一批青年教师到山西大学任教，我们这位教数学的老师也被选中，在被破格晋升为副教授后，调往该校去了。新任的数学老师是著名的数学家严志达教授（后当选为中国科学院数理学部的院士），他常常采用深入浅出的简练语言，把抽象的数学概念讲得生动有趣，有时还穿插一些数学家生活方面的小故事，逐渐消除了大家的畏难情绪，增强了学好数学的信心，当然从此也取消了令大家心惊胆战的突击小测验了。第一学年数学终评成绩不及格者寥寥无几，总评为 4 分和 5 分的同学占了绝大多数，又激起了我对数学的兴趣。数学成绩的提高，为大学四大力学（理论力学、热力学、电动力学和量子力学）的学习打下了良好的基础。

共产主义暑假

1958年，为了赶美、超英，南开大学领导号召同学们不要回家，留在学校过一个共产主义的暑假。一开始，我被分配去推轱辘马，把装满土的轱辘马从学校西边的“西伯利亚”顺着小铁轨推到正在建设的“校办工厂”工地；虽然劳动时间长，强度大，但不觉累，大家的干劲十足，因为谁也不甘落后，都想尽快跑步进入共产主义呀！当时有一个著名的口号：“甩开膀子干活，敞开肚皮吃饭。”的确如此，在整个50多天的暑假里，学校食堂每天24小时都是开放的，你可随时去用餐。

在工地劳动不到一周，我又被分配去搞人工降雨研究，这是一个与天津大学化工系同学合作的研究项目，实验室就设在天津大学主楼里。我们建了简单的云雾室，模拟云雾凝结下雨的过程和条件，除常用的试剂如氯化银、镁粉等外，还探讨过超声增雨的可能性。这种初生牛犊不怕虎的精神固然可嘉，但我们这些未曾受过相关专业教育的学生，如何能解决这一世界性的复杂的问题呢？折腾了近1个月后，也就不了了之了。

这之后，我又被分配去天津市的炼钢厂参加大炼钢铁的群众运动，我提出在向炼钢炉送风、送氧的通道上安装超声发生器，来提高炼钢效率的建议，得到认可；实验虽有效果，但不显著。共产主义暑假之后，学校根据教育部要求进行了教学改革，其中一个重要的措施是要增加在校学生参加劳动的时间（也叫作勤工俭学），物理系决定由56级学生做试点。从1958年秋季开始到1959年暑假整整一年，56级同学不再上课，而是在校从事勤工俭学劳动，这一改革随着1959年下半年而来的3年困难时期而停止。56级同学不得不推迟毕业半年，而我们其他年级则幸免。

毛主席视察南开大学

1958 年 8 月初的一个早上，在学校领导丝毫不知的情况下，毛主席在天津市和河北省领导的陪同下，出其不意地出现在南开大学的校园里，下车后，他径直地走向化学楼东侧的校办工厂，与同学们亲切交谈，详细地了解同学们暑期勤工俭学、科研和教学改革的情况。随后他又兴致勃勃地视察了化学系，杨石先校长领导的南开大学化学系当时在国内高校中名列前茅，特别在农药、离子交换树脂等研究方面成绩突出，这或许是毛主席特别光顾化学系的原因吧！毛主席视察南开的消息迅速传遍了南开园，渴望一睹领袖风采的学生像洪水一样从四面八方涌向化学系前的小花园，就连正在天津大学合作做人工降雨研究的我，也被叫回。这时毛主席正巧从化学教学楼出来，热情高涨的学生把毛主席围得水泄不通，难以走动。此时，有一位随从（大概是保卫人员）大声宣布："同学们，请到芝琴楼前广场集合，毛主席要给大家讲话……"话音刚落，人们就转而争先恐后地奔向那里，希望能占个好位置。毛主席一行趁机立即上了小轿车，绕过芝琴楼门前的广场，穿过南开大中路，沿着马蹄湖和新开湖之间的小路，直向北边的天津大学驶去。同学们看着逐渐远去的车队，知道中了脱身之计，大家觉得这也许是为了领袖的安全不得已而为之，并无太多怨言。许多未曾亲眼看到毛主席的同学又向天津大学跑去，希望在那里能看见他。

在天津大学主楼前的广场上，已经挤满了两校的师生员工，都焦急地等待着毛主席的出现。当毛主席高大的身影出现在主楼的小凉台上时，顿时响起了长久的掌声和“毛主席万岁”的口号声，毛主席不时地向欢呼的群众挥手致意，并慢慢地绕着凉台走了一周，之后便走进了大楼。沉浸在幸福之中的人群仍站在广场上，久久都不肯离去……

毛主席在酷暑季节视察南开大学和天津大学的消息立即传遍了全国，极大地鼓舞了全国人民赶美、超英的雄心壮志。也激励着我们要好好学习，将来报效祖国。

3 年困难时期的大学生活

1959 年的下半年，我们 57 级学生进入了大三，微分方程、数理方程和四大力学等基础课都将陆续开设，能不能学好这些课程，是关系到个人前途的大事，大家都鼓足了劲，希望取得优异成绩。然而令人料想不到的事情发生了，首先是开始实行粮食定量，我们这些吃惯了“四菜一汤”的毫无经验的年轻学生，男生自报的定量一般在 32 斤或 32 斤以下，女生多在 30 斤以下。学生食堂也由包伙制改为食堂制，每人每月的主食不能超出自己的定量标准，开始时由于菜的花样和油水尚多，自然感觉不到粮食的不足。过了国庆节，食堂菜的花样越来越少，最后只剩下了一个菜，而且也难找到一片肉。出生在农村的我起初还能忍受，可生长在大城市的同学早已叫苦连天了。1960 年的新年，学校为了改善同学们的生活，将校内新开湖的水抽干，捉到几千斤鱼，大家总算过了个愉快的新年。

随着中苏关系的恶化和苏联专家的撤走，严重自然灾害造成的粮食大幅度减产，使国家经济建设和人民的生活遇到了极大的困难。这时的直辖市天津，不要说食油，就连酱油、醋等也难以买到，副食商店的货架上空空如也，由此可以想到学校食堂的困境了。到街上小饭铺吃饭只收天津市地方粮票，而且全国通用粮票每人每月在天津只能使用 2 斤，且须兑换成天津市粮票；此外，每人每月半斤平价糕点票，两张可到饭店享受油水较多的高价菜票。后来，在一些商店里有高价糖出售，然而价格之高是我们这些穷学生想都不敢想的。学校食堂采取了一些措施，如按定量确定每人每天的主食饭卡，以保证计划不周者月底能有饭吃。领取主食时，值班员在你的饭卡当天相应的空格（每个空格为二两主食）里打×，以示用过；当天未用的空格，可在它日再用。在使用的过程中发现少数学生饭卡的空格较多，而且多出自化学系的学生，原来他们是用化学试剂将×涂掉了，不注意很难发现。为了杜绝这种多吃、多占事情的再次发生，食堂采用了在饭卡空格上打孔的办法，这个办法很有效。

在这一困难期间，一方面，学校领导也想了不少办法，如租用船只去黄海、渤海湾打捞毛蚶，到内蒙古草原捕捉黄羊，购买杨柳青的大白菜等，对改善同学们的生活起到了积极作用。尽管如此，仍约有 40% 的学生由于营养不良而出现了浮肿，这些学生每人每月可得到两斤黄豆的营养补助。另一方面，学校要求同学们注意“劳逸结合”，并在授课、体育锻炼等安排上也做了调整，一些课程的期末考试取消了，这在一定程度上减轻了同学们的精神压力。

就在这年（1960 年）的秋末，家里传来奶奶病重、爷爷去世的噩耗，爷爷的离去，深深地刺痛了我的心。1959 年暑假爷孙俩一起睡在大柿树下的情景历历在目，那是我离家回校的最后一个晚上，次日一大早我刚要起来，爷爷就忧伤地看着我说：“不知道明年我们爷孙还能不能再见面！”我看着流淌在爷爷饱经风霜脸上的伤心泪水，心如刀绞。我强忍泪水说：“明年我一定再回来看您！”想不到这竟是我们爷孙的最后诀别。这时我的心早已飞回了家乡，急切地盼望寒假的到来。我向学校申请回家看望病重的奶奶，但当时学校有重灾区的学生不准回家的规定，经我再三要求，才得到了学校的批准。我用省下来的生活费买了火车票后，就着手准备上路的干粮。当时由于全国粮食紧张，火车上只有开水，不再有饭菜供应。我将3天3斤的15个窝窝头切成薄片，在火炉上烤干后和半斤点心一起装进一个小布带，以备车上享用。从天津到许昌没有直达车，须在北京转，好在北京是起始站，总算能有个座位。春节期间火车非常紧张，就连车厢的走道上都挤满了旅客，行走困难，要点水则更难。

第二天下午火车到达许昌站，由于错过了去南阳的长途汽车，不得不在汽车站旁边的旅店住下。旅店设备简陋，东西两排头对头的地铺一字排开，铺位一个连一个，一个大房间可住三四十人。每人一床被头发黑的被子和一个沾满了头油膏子的枕头，人到这时也顾不得许多，和衣倒下便睡。其实每个人睡得并不踏实，因为一是怕丢了随身带的东西，二是怕误了第二天的汽车。

汽车在第三天的下午经南阳到达镇平县晁陂车站，离家还有20余华里，随身所带的干粮早已吃光，剩下只有半斤点心了。是我准备孝敬奶奶的，自然不能动。然而劳累和饥饿无情地折磨着我，使我难以忍受，在万般无奈情况下，我从包装点心的牛皮纸的一侧打开了一个小口，取出一小块点心，边走边吃了起来。当晚上赶到家时，这一包点心只剩下一块了，我带着内疚和自责走到病危的奶奶睡的地铺旁，将一小块点心放在她的嘴里，奶奶慢慢地咽了下去，嘴动了动好像要说什么似的，看到这里，真如万箭穿心，难过的眼泪夺眶而出，大声哭了起来。

奶奶在我到家的几天后便与世长辞，骨瘦如柴的奶奶的遗体装在由四块薄木板拼凑而成的棺材里，出殡那天竟难找到几个抬棺木的壮年人来。我们村不大，共有30多户100余人。一年多来，年过60的老人多因饥饿而死，其中我的伯父、伯母、爷爷和奶奶以及外祖父都是在这年去世的。村里几乎所有的青壮年都患上了浮肿病，哪里还有劲抬棺材呢！这时的农村仍然是吃人民公社的大锅饭，每人每天定量说的是半斤粮食，但实际上每家每顿能吃到的除了一瓦罐稀饭，别的什么也没有，如何能填饱肚子呢。这年春节，全村只分到2斤猪肉，生产队长为了照顾我这个大学生，就给了我家，当时我的心里既感激又不安，因为这毕竟是全村百十口父老乡亲过年的肉啊！

进入大四，我被分在固体物理专业，半导体专门化，心中自然高兴，因为1957年10月苏联发射了人造卫星，开辟了航天的新时代，我认识到半导体在发展航空、航天中的重要作用，激起了我为中国的航天事业奋斗的兴趣。半导体专门化主要设置的课程有：固体物理、半导体物理、固体电子学电路、半导体材料和第二外语（英文）等。劳逸结合为同学们提供了一个宽松的学习环境，每天上完课做完不多的作业后，还剩下不少空闲的时间，于是同学们围坐在一起，毫无顾忌地从稀奇古怪的见闻到科学上的“奇思妙想”无所不谈，受益匪浅。在讨论中，我提出了能带扰动模型、空气窒息弹和意识波动论 3 个设想，引起了大家的兴趣和热烈的争论。我毕业论文的题目是“半导体的辐照效应”，这是一篇综述性的评论文章。在查阅了大量文献后我发现，在中子、γ射线和电子的辐照下，随着辐照剂量的增加，空穴导电的 *P* 型硅（*P*–Si）的电导率逐渐变小，最终会转变为电子导电的 *N* 型硅。为了解释这一实验结果，我提出了“能带扰动模型”。高能粒子照射半导体材料，会在晶体中产生俘获载流子的缺陷中心，如果辐照在 *P*–Si 中产生施主中心，那么这个中心将从价带俘获一个空穴，使 *P*–Si 的导电性下降；随着辐照产生的施主中心密度增加，*P*–Si 将经历本征导电（高阻）后而转变为 *N* 型导电，圆满地解释了上述实验结果。“能带扰动模型”是基于 *P*–Si 在高能粒子辐照下，首先在材料的局部产生 *N*–Si 区，进而在材料体内形成由无数个 *PN* 结组成的 *P*–*N*–*P*–*N*–*P*–*N*……结构而提出来的。当 *N*–Si 区体积和密度小时，*P*–Si 导电占优势（下图左），当两者的体积和密度相当时，导电最差（本征，下图中）；相反当 *N*–Si 占优势时，则表现为电子导电（下图右）。论文中还对不同势垒厚度和高度的电导做了计算，当时光明日报记者就此访问了学校指导老师，并作为敢想、敢于提出新见解的典型做了报道。在 1964 年的南开大学建校 45 周年的校庆上，老师替我宣读了此文（当时我不在校，1962 年毕业后被分配到中国科学院半导体研究所工作），受到欢迎。后来想把全文在国内学术刊物发表，但由于“文化大革命”而未如愿。

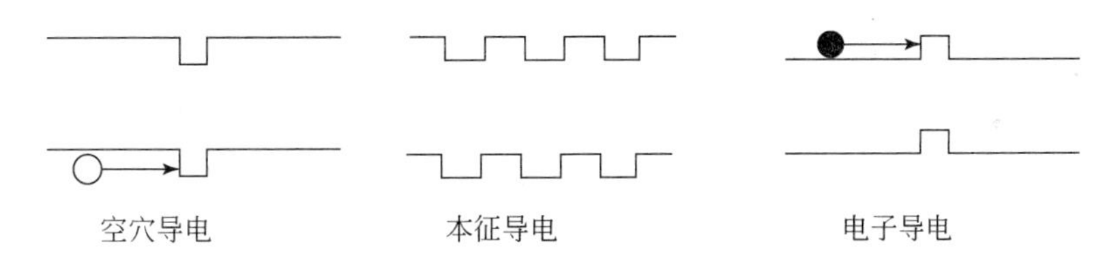

“能带扰动模型”的示意图。这个模型与 1970 年江琦和朱兆祥提出的调制掺杂超晶格结构非常相似

我提出的第二个设想是“空气窒息弹”。这个想法是基于炸弹在空中爆炸产生的冲击波可将具有强还原性的可燃物质（颗粒）抛向空中，借助爆炸产生的高温而导致的剧烈燃烧，在某个特定的时间间隔里将耗尽一定范围内空气中的氧气，形成真空区，使该区中所有的战车无法开动，战斗人员（当然包括人和动物等）窒息，从而失去战斗能力。这是一种干净的大规模杀伤武器，它的作用半径可根据炸弹的爆炸力确定，500–1000 米应当不成问题。近年来从新闻报道里得知：英国已研制成功“云爆弹”；最近俄罗斯又试爆成功“真空弹”，然而这些武器系统的工作原理与我 50 多年前提出的“真空窒息弹”有什么不同呢!？看来如何创造一个能最大限度地发挥青年人聪明才智的环境，已成为建设创新型国家首先要解决的重大问题。

“意识波动论”是我提出的第三个观点，也是当时同学们争论最多的一个，有同学认为我是陷入了唯心主义，要好好改造世界观等。“意识波动论”中的“意识”指的是人的思维。人的思维过程是一个极其复杂的物质运动过程，思维过程要消耗能量是人所共知的；既然要消耗能量，那么这些能量是以什么形式向外发射（耗散）呢？我以为是波。基于这个假设，思维过程所发出的“意识波”必然携带着人们想要表达的信息，这些信息也许是公开的，也许是不愿为人所知的。如果人在思维过程中发出的信息能被探知和还原（尽管目前还不能，但我相信总有一天能实现），那么思维的秘密，特别是人类社会的种种丑恶现象将被及时发现和化解，人和人、人和动物、人与自然的和谐共处将成为可能，人类社会将进入一个更高的层次。

早期科研工作概述

1962 年 9 月初，在未征求 57 级毕业生任何意见的情况下，学校宣布了分配方案，我、周旋、史一京和杨玉芬 4 位同学被分配到中国科学院半导体研究所工作。当时，无条件服从组织的分配是天经地义的事，大家都很平静地接受了党和组织的安排。9 月中旬我们 4 人乘天津去北京的火车到半导体研究所报了到，第一次领得了 46 元的工资，心情是很激动的。入所后，我被分配到半导体材料室（一室），史一京去了测试中心（七室），周旋去了电子学室（四室），杨玉芬去了微波器件室（五室）。当时对进入中国科学院的大学生有不少要求，如在 3 年之内要求通过两门外语（第一外语需 85 分以上，第二外语需达 80 分以上），否则要调出中国科学院，等等，这给新入所的我们这些刚出校门的大学生不少精神上的压力。我们刚入所不久，就要求参加了半导体物理的摸底考试，我们几个还好都过了这一关。1964 年的外语考试，我的第一外语英语 89 分过了关，第二外语俄语 79 分差一分未过关。后来由于“文化大革命”，上述的规定就被搁置了起来，“文化大革命”后也无执行。

入所后，我和同时到所的佘觉觉（北京大学 62 届毕业生）被分到一室（半导体材料室）电学测量组，周洁组长分配我们两人负责研制光电导寿命测试仪。我对电子学，特别是电路，只有一些书本上的知识，没有实践经验，能否完成任务，心中没底。所幸的是曾任北京大学广播站负责人的佘觉觉对电子电路很熟悉，经过调研，他很快就拿出了一个研制方案，我在他的帮助下，负责激发光源和样品架的研制，他则负责难度较大的低噪声放大器的设计和研制。两人合作比较愉快，经过一年多的紧张工作，于 1963 年年底建成了光电导寿命测试仪，很好地完成了预期任务。

集成电路的发展，对半导体硅材料的纯度提出了要求，我们电学测量组接受了硅材料纯度测量的研究任务。硅材料纯度可通过变温霍尔系数测量得出，由于高纯的基磷、基硼硅（高纯硅）中的浅施主和浅受主的激活能都比较浅，测量温度需要低达 20K。在低温下，由于硅中磷施主和硼受主中心上的电子和空穴消电离，样品的电阻率可高达 100MΩ · cm 以上，这就需要具有高输入阻抗的测量仪器，如英国制造的振动簧片静电计，但价格十分昂贵，无力购买。我就是在此种情况下于 1964 年年初授命研制高阻静电计的。研制高阻静电计的关键是静电电子管和对其良好的电磁屏蔽，国产的静电管质量不很稳定，恰好林兰英先生访苏，她利用出国省下的零花钱购买了一只苏制的真空静电管，为高阻静电计的研制创造了条件。静电管和相关元件安放在电磁屏蔽盒里，电磁屏蔽盒由铜和铁材料加工而成，具有良好的电磁屏蔽效果。此外，从样品架到测量端的所有连线彼此之间和对地的漏电阻都要远大于低温下样品的电阻。常规霍尔测量用的连线和选择开关以及分压电阻等都需改换成由高绝缘的导线和开关替代。经过近 1 年的辛勤劳动，于 1964 年年底研制成功高阻静电计及霍尔测试系统，解决了低温高阻测量问题，并在国内首先实现了 20–400K 硅的变温霍尔系数测试，并对高纯硅的低温电学性质进行了研究，研究成果发表在《物理学报》[1966，22（4）：404] 和《科学通报》外文版 [1966，17（5）：206]。

1965 年后，我加入时任党支部委员的尹永龙的半导体辐照效应研究小组，这个小组除尹永龙和我之外，还有何良，共 3 个人。主要从事硅材料中子、γ 射线的辐照效应研究。1965 年在硅的中子辐照效应研究中，基于对辐照缺陷引入率的分析，发现 *P* 型和 *N* 型 CZ–Si 具有明显不同的抗辐照能力；这为后来空间用硅光伏电池的选型打下了基础。热中子和 γ 射线对 *P* 型 Si 的辐照效应研究发现随着辐照剂量的增加，*P* 型 Si 的电阻逐渐增加，在达到本征导电后，电阻开始变小，并反转成 *N* 型导电，验证了文献报道的结果（见前面我提出的“能带扰动模型”的描述）。

1965 年年底李永常和彭万华同志加入到尹永龙领导的小组，经过认真的调研，大家感到单一的电学测量研究难以弄清半导体材料的基本性质，自力更生建立光学测量装置是需要的，也是有条件的。于是大家分工协作开始了低温红外光谱仪的研制，我负责液氦（4. 2K）到室温的温度控制仪和测温热电偶、低温恒温器和样品架的研制。后因“文化大革命”，造反派把我们的这个计划作为只专不红、走白专道路的典型，进行了批判，计划不得不中断。不过我研制的低温恒温器和金铁热电偶对后来的变温电学和光学测量起到了重要作用。金铁热电偶是在 99. 99% 以上纯度的纯金中掺入少量铁制成；为此，我曾经 3 次到云南仪器仪表厂检验、核查和验收，终获满意的结果。但令人遗憾的是，锁在实验室抽屉里的重约 500 克、价值数十万元的金铁电偶丝，加上我从隆德大学带回来的宝贵样品等，在 20 世纪末不翼而飞。此外，为了弥补金铁电偶在 30K 以下灵敏度较低的缺点，还委托昆明贵金属研究所研制铑铁电阻温度计，也得到了试样。

1967 年我负责承担 651 任务人造卫星用硅太阳电池电子和质子辐照效应研究（参加这一研究的还有李涛和陈书南）。通过对硅太阳电池的电子辐照效应的系统研究，发现 *NP* 硅电池抗辐能力比 *PN* 硅太阳电池好数十倍以上。在 651 设计院［即中国科学院卫星设计院，是 1965 年中国科学院专门为中国第一颗人造卫星任务成立的，1968 年由国防部国防科学技术委员会（简称国防科委）接管］召开的太阳电池定型会上，这一结果改变了原定生产 *PN* 电池的计划，并为我国定型 *NP* 电池生产起了关键作用。1971 年 3 月 3 日我国发射的第一颗科学实验卫星“实践一号”的太阳能 *NP* 结光伏电池板就是我所提供的，这不仅在经济上，而更重要的是在政治上具有重要意义，但我也付出了右手受到严重辐射损伤的沉重代价。出院时，医生给了 1–2 年不要再接触射线的忠告。

从 1967 年 9 月到 1968 年 6 月，这段时间正是“文化大革命”两派激烈对立和争斗的时期，我因“文化大革命”开始时受到一些压力，而不愿参加任何一派，成了名符其实的“逍遥派”。如何打发时间呢？萌生了装半导体收音机的念头。说干就干，那时，几乎每天上午都在自行车上，四处奔跑寻找无线电商店，去买便宜的、或是处理的用于装收音机的半导体元器件和收音机外壳。大约1个月的时间，一只4管收音机便在宿舍里响了起来，但它只能收听中央的几个电台；为了收听其他地区和短波电台，我开始筹划组装6管超外差收音机。这种收音机市场上有卖的，但价钱昂贵（近百元人民币），对我们这些参加工作不久的人来说是买不起的。这种收音机电路比较复杂，没有现成的电路板可买，于是我开始钻图书馆，查阅有关这类收音机的电路，经过挑选对比，选取了美制6管（锗晶体管）外差机型，因为当时国内市场上可买到的都是锗晶体管，国内尚无硅晶体管出售。首先根据市场上收音机元器件的大小尺寸，如：机壳、可变电容器、中周、变压器、喇叭等，设计印刷电路板。对设计好的电路板的刻蚀是个难题，只能土法上马，先用胶将连线保护起来，然后放到废弃的酸溶液中，腐蚀去掉无用的铜板，然后用清水冲洗干净。在线路板上打孔、开窗要用到电钻或什锦锉，只能假公济私了，好在“文化大革命”中头头们都在“闹革命”，顾不得管这些琐事了。3个多月过去了，6管超外差收音机装成了，为使收音机在整个波段上灵敏度达到最佳，需要在长波、中波和短波3个波长选择3个电台来跟踪调试，长波电台选择苏联的“和平与进步”电台最合适。这个电台经常广播一些讽刺“文化大革命”的内容，引起了一些人的注意，并汇报给造反派头头；造反派头头借以收听敌台的“罪名”，要我做检查。我只得按照他们的要求做了“深刻”检查，总算过了关。检讨是检讨，收音机还是要调的，我只好晚上到没有人的地方，或者干脆到实验室用信号发生器来调试。经过调试的收音机灵敏度很好，可同市场上的产品媲美，花钱又少；于是有不少同事要我帮装这种收音机，我也的确帮了不少同事的忙。但使用的锗晶体管漏电大，工作温度特性差，不理想。一次偶然的机会，得知109厂正在处理高频硅晶体管，几毛钱一只，比较便宜，我就买了几只试试看，用一般的测量方法来测管子的放大倍数，很令人失望，放大倍数不超过20！进一步的测量发现，如果基极注入电流增大，放大倍数随之增大，而且多数管子都高于50以上。这个发现使我产生了采用硅晶体管装外差收音机的想法，虽说当时国内还没有厂家和个人装过这种收音机，我就根据硅晶体管的特点，先设计了一款6管超外差机的线路图，试装效果不错。为了节约开支，中周用买来磁芯等零件自己绕制，采取晶体管电路替代输出变压器设计思路，进一步降低了费用，就这样一个高灵敏度的三波段8管超外差收音机试制成功了。记得1967年春节回家探亲，这个收音机曾在我的家乡轰动一时，不少人跑到我家来看这个新鲜玩意，之后只好为老家也装了一台。后来，那台8管机为我学习英语，特别是练习听力和口语，起了很大作用。

1968年6月到1970年6月，我受国防科委第14研究院委托，负责了调研和制定14研究院电子材料、元件、器件和组件辐照效应研究方案和实施计划，并任院辐照实验组业务组长。对14研究院研制的电子材料、元件、器件及组件进行了系统的电子、质子、中子和γ射线辐照效应研究。实验结果由我等3位同志汇编成我国第一本《电子材料、元件、器件和组件抗辐照性能手册》，为我国航天事业、电子对抗以及核突围等国防工程做出了积极贡献。

1970年，我作为国防科委第14研究院辐照实验组的业务组长，在21基地，负责组织、实施了氢弹空爆瞬态核辐射对电子材料、元件、器件和组件性能影响的研究工作。在国内首次自动记录了核爆炸光、电磁辐射对太阳电池、MOS器件的瞬态效应。后被电子工业部13所吴××诬陷。参加不参加第二期试验对我并不重要，遗憾的是那些由我设计和实施的项目，因无法对实验结果进行合理的分析而流失令人痛心。

什刹海黑夜救同事

大约是20世纪60年代末70年代初，为响应毛主席畅游长江和到大风大浪中去锻炼的号召，半导体研究所工会经常组织全所职工集体下河、下湖学游泳，多数人都学会了游泳，但不熟练。在一个闷热的夏天的晚上，9点左右，我和同事龙泽民到什刹海游泳乘凉，老龙刚刚学会“狗爬式”的游泳，我怕他出问题，问他：“你现在能游多远？”他说：“200来米吧。”由于从岸边到什刹海中心岛的距离小于200米，而且从我们开始下水到中心岛之间还有一个由木桩围成的四方浮标，可供游泳者休息，他游上小岛应该不成问题。我告诉他，咱们的目标是湖心的小岛，我在前面游，您紧跟在我后，不要乱游；他一边答应着，一边开始游向岛心；然而当我在前面游出10多米左右时，听到身后有噼里啪啦的打水声，我急忙扭头看，龙泽民双手在原地胡乱打水，头一下冲出水面，一下又沉入水下，充满了恐惧和紧张。说时迟，那时快，我竭尽全力向他游过去，当我靠近他时，他猛扑过来，一下就死死地抱住我，我们两人都向水下沉去，在这非常危险的时候，我挣扎着大喊一声“救人啦!”晚上9点左右，游泳的人很少，在我们的附近根本就没有一个人。我感觉到继续喊救人会增加龙泽民的恐惧，弄不好两人会同归于尽；于是我对他说：“有我在你就不会有问题。”我一边说一边试着用力把他的头露出水面，要他放松点，不要死抓着我……这一招很管用，他恐惧紧张的心逐渐平静下来，听从我的指令并配合我的施救动作，就这样我把他的头托出水面，慢慢地向岸边游去，到了岸边，他长出了一口气，说：“他妈的！肚子喝满了水。”我笑着说，这都是你吹牛的结果。

天津小站劳动锻炼

1971 年 5 月，我同所内数十人被派到天津小站国防科委第 14 研究院所属的一个农场参加劳动锻炼，当时来到这个农场的还有天津 703 厂和电子部 46 所的科研人员，总共近百人。我被“任命”为司务长，除了为大家烧菜做饭外，还要抽空参加一些其他如建房和泥、脱坯等劳动。炊事班的大多数人都没有做过饭菜，只有我所后勤管房产的马士贤同志有过做饭和炒菜的经验，但老马却要摆架子，需要去求他方才下手，大家心里很不是滋味。有一次，我们在水沟和稻田里捉了不少鱼，想改善一下生活。老马是做红烧鱼的好手，大家很快就把鱼收拾干净了，就等马师傅来下锅，可是到处也找不到他，他躲了起来，想难为我们。我们只得自己动手干，其实并不难，先将鱼下锅炸了，然后再将炸好的鱼放入配好酱油、醋和红糖以及花椒、大料等浸泡后，下入热锅中，盖上盖子约半小时，待浇汁将干时出锅便可，大家都说味道不错；当然，老马倒觉得有点不好意思了，从此之后，他再也没有难为我们了。

炊事班的另一个比较劳累的工作是挑水，水井离厨房约 200 米，两桶水重约 50 公斤，挑起来走在窄而高低不平的乡间稻田的小道上，实在吃力，特别遇上路滑的下雨天，真是苦不堪言。我作为司务长，别人不愿干的只有自己去干，没处说理。

按规定参加劳动锻炼的伙食标准不能超过当地老百姓的生活水平，每月的饭菜金不得超过 11 元，由于劳动强度大，劳动时间长，11 元的标准很难维持大家的需求。于是我们采取多吃粗粮，如玉米面窝窝头和高粱米饭等来减少开支；另外，有时间下河抓些鱼虾，改善大家的生活，或买便宜的大白菜和豆腐做菜等，采购员胡二（胡金兆）花费了不少脑筋；但月底一核算，还是要超支。记得有一个月超支了 5 元，703 厂的一些同志不愿补交超支的费用，一直拖了很长时间才交齐菜金。

1971 年 9 月 13 日，发生在蒙古人民共和国温都尔汗的一次空难，林彪、叶群和他们的儿子林立果死了，在中国的政坛上掀起了大地震。不知什么原因，在小站农场劳动的我们，很快就接到回所的通知，就这样劳动锻炼就仓促地结束了。

1971 年 10 月回所后，我被分配到 104 组（GaAs 组），和向贤碧、徐寿定一起做测量工作，刚被“解放”了的林兰英先生也经常到我们这个小组来讨论工作。这之前，我未做过 GaAs 性质的研究工作，从查阅 104 组原有测试记录着手，看能否从中找出一些改进材料质量的线索，经过分析和实验对比，意外地发现了在石英舟生长的某些体 GaAs 样品中，确实存在正温度系数（$d\rho/dT>0$）样品，这之前大家认为这种体 GaAs 样品是不可能存在正温度系数的；这个结果为研制 $d\rho/dT>0$ 的高纯度体 GaAs 材料提供了实验依据。

1972–1974 年，针对 GaAs 体效应器件（Gunn 和 LSA）对材料的要求，对 GaAs 材料的强场性质做了系统实验研究（包括建立强场实验方法），指出体 GaAs 由于纯度、均匀性等严重问题而导致的击穿是研制 LSA 器件的致命弱点之一，它不能满足器件研制的要求。我执笔撰写的《GaAs 材料质量的初步探讨》一文作为 1972 年在上海召开的 GaAs 学术会议特邀报告，由林兰英先生宣读。

1975–1976 年，继 GaAs 强场特性研究后，进行了 GaAs 材料热学稳定性的研究，这是借助于测量 GaAs 材料在 n^+工艺（体效应器件所必需的）热循环处理前后电学性质变化来研究材料质量的一个课题，也是国内最早注意到 GaAs 热稳定性对器件有直接影响的一个工作。这个工作分析了体 GaAs 热学不稳定的原因，提出了相应的物理模型，进一步指出了体 GaAs 热学不稳定性是造成体 GaAs 器件（LSA）成品率极低的基本原因。由于这两方面的工作，为我室高纯度 GaAs 研制工作重点由体材料转移到外延起了重要的作用。其中《杂质和缺陷在 GaAs 中的行为》一文（此文由我执笔成文）在 1977 年 GaAs 及其他Ⅲ–Ⅴ族化合物半导体会议上（广西柳州）作为特邀报告由我代林兰英先生宣读。

1978 年中国物理学会年会趣事

1978 年 8 月 1–15 日，在江西庐山召开了中国物理学会年会，与会代表 602 人，这是“文化大革命”后首次物理学界的盛会，也是众多物理学家及其领导人物的亮相大会（周培源致开幕词，钱三强作工作报告，王竹溪作修改会章报告，施汝为致闭幕词）。半导体研究所的主要业务领导人黄昆所长，王守武、林兰英副所长等以及我们几个青年科技工作者（郑厚植、沈光地和我）也有幸获准参加了会议。我们首先乘由北京去武汉的火车到达武汉后，改乘由武汉去上海的轮船，在九江站下船上岸，在岸上等待我们的是一字排开的汽车长龙，约有 1 公里长，路的两旁站满了好奇的老乡，浩浩荡荡的车队在高低不平的土路上向前慢慢驶去，尘土滚滚，甚是壮观！经过两小时左右，车队到达庐山。我在小组分会上作了“*N*-GaAs 补偿度的计算和散射机制的分析”的报告。

在会议期间，安排了自由登山活动。知名教授像黄昆先生等本来可以先坐车到山下，然后是否爬山，由自己根据身体状况决定。黄先生拒绝坐车，跟着我们（沈光地、郑厚植和我）向五老峰山顶爬去。上山比较顺利，大约两个小时就到了山顶，山上非常凉爽，神工鬼斧所造就的奇特的大自然的风光尽收眼底，让人心旷神怡，爬山的辛苦早已忘得无影无踪。要下山了，是从原路返回还是另辟捷径，我的抄近路下山的建议得到了大家的认可。自然由我带头沿着一条陡直的小道开始下山，起初还算顺利，然而没走多远，路愈来愈窄，需要双手拨开密密麻麻的灌木树枝才能前行，加上茂密丛林中的闷热，个个大汗淋漓，苦不堪言。这时已是下午两三点了，走回头路已不可能，只得硬着头皮走下去。当我们挣扎着走出丛林时，大家又渴又饿，精疲力竭；为了弥补由我的建议带来的麻烦，我让大家坐在路旁休息，我则快步跑回住地，拿来了一瓶水和面包，供大家解渴和充饥。事后，不苟言笑的黄昆先生感慨地说：“这是我有生以来受的最大的一次苦！”直到黄先生晚年，提起当年在庐山登五老峰之事，他还记忆犹新。

变温霍尔系数测量系统建设

1976–1978 年，主要开展外延 GaAs 电学性质的研究。在国内首先建成了 4.2–400K GaAs 变温霍尔系数测量系统；研制了高精度大电流磁场电源，高稳定度小电流恒流源以及高灵敏温控单元。计算了 GaAs 中总电离杂质浓度，分析了影响迁移率提高的原因；提出了材料质量（当时）仍由杂质控制的基本观点。这为我室液相和汽相 GaAs 外延组进一步提高外延材料质量指明了方向，也为我室高纯 GaAs 外延材料 20 世纪 80 年代初达到国际先进水平贡献了力量。其中“*N*-GaAs 补偿度的计算和散射机制的分析”在 1978 年第二届中国物理学年会上作报告。1978–1980 年，进行了 GaAs 及其他Ⅲ–Ⅴ族化合物半导体光学性质研究。为鉴别高纯 GaAs 中的剩余杂质，为深入了解深中心杂质对半导体材料性质的影响，在国内最先提出了建立光致发光（PL）实验室和深能级谱仪设备（后由一室阮圣央和五室邓兆阳等实现）的建议。从 1977 年年底到 1978 年，从无到有建成了国内最先进的 PL 实验室，并对 GaAs、InP 的光学性质进行了研究。这些设备为我所材料物理研究工作打下了基础。此外，还筹备了开展液氦实验的必需条件。1971–1980 年共完成十多个研究课题，其中 8 篇论文在有关会议和学术刊物上发表。

留学瑞典隆德大学固体物理系

1978年年底英语口语班结业后，研究室领导原打算让我以访问学者身份出国半年开展半导体材料物理研究；当时由于深能级谱仪的研制成功，半导体材料中深能级的研究成为热点研究方向，黄昆所长考虑到半年时间太短，学不到什么东西，于是把我的访问学者身份改为两年进修，纳入教育部派出留学人员队列，留学身份的变化比较复杂，原来填的表格都得重新填过，记得仅照片就提交了20余张！出国时间拖后了1年多。在克服重重困难联系好去瑞典隆德大学进修后，又遇到了新问题，教育部要求参加英语统考，尽管黄昆所长给我的英语能力打了"保票"，但教育部坚持必须通过英语考核的原则。这时正巧瑞典驻北京大使馆举办赴瑞留学生英语测验（1980年10月8–9日），教育部同意我参加考试；我顺利通过了两天的英语笔试和口试，过了教育部这一关。1980年10月11日到教育部领取了2美元的零花钱，10月12日凌晨乘国航到达莫斯科后，换乘斯堪的纳维亚（SAS）航空公司的飞机经哥本哈根到达瑞典首都斯德哥尔摩。在外航飞机上我小心地观察送来的饮料和快餐是否收费，因为口袋里只有2美元！到达莫斯科机场后的几个小时，由于中苏关系紧张，留在机场候机室，不敢外出，唯恐出事。到达斯德哥尔摩机场后，急忙去领取行李，等了很久，传输线上已无行李传出，这时我意识到行李可能丢失了。我找到机场相关部门官员申诉，填写了行李尺寸、颜色和里面存放的东西以及邮寄地址等；之后机场海关人员给我一份刷洗用具包，我忐忑不安地走出机场出口，已经是晚上10点多了。来接机的是中国驻瑞典大使馆的张德安同志，他看见我后，就直截了当地问为什么这么长时间才出来？我说明了理由，随后上车到使馆暂住。第二天上午，丢失的行李送到了大使馆，一切完好，一颗着急的心才平静下来。

1980年10月14日，我从斯德哥尔摩乘火车来到了瑞典南部的大学城隆德车站，到车站接我的是系主任助理莱德堡博士。我的住处是在离固体物理系不远（步行大约10分钟）的斯巴达学生公寓，厨房为该层住户公用，但每人有自己的厨灶和厨具以及分格的冰箱。厨房的另一侧是客厅，是大家吃饭、休息和交流的地方，客厅还有共租、公用的电视机。公寓的下面就是一个大超市，吃、穿和生活用品齐全，非常方便。瑞典建筑通风极好，卧室内有办公桌椅，淋浴和卫生间一应齐全，房内干湿适中，就是夏天，也无须空调，我在这里度过了终生难忘的1120余天。

我到隆德后的第2天，就来到了学校固体物理系的实验室，期望能会见自己的导师，听候工作安排。然而，在系办公室里，等到的还是到车站接我的莱德堡博士，他委婉地说："教授要你先看看资料，熟悉熟悉实验室的仪器操作。"我心里明白：我是来这儿进修的第一个中国人，而导师是国际深能级学术方面的权威，我能否适应这里的工作，能否做出成绩，导师对此不能不打个问号。也许，这就是今天对我约而不见的真正原因吧！

隆德大学固体物理系，虽然规模不大，仪器设备却很先进，测试分析手段齐备，处处给人耳目一新之感。先后来此工作、学习的，有芬兰、英国、匈牙利、德国的科研人员。遗憾的是我被莱德堡领进一个条件很差（甚至不如我在所里的）的实验室，并说“你就在这儿工作，有事可以找我”，说完，这位主任助理就告辞了。到系里已经3天了，没有研究人员和读学位的学生跟我攀谈、交流。就连技术员受命来讲深能级谱仪操作规程，也只是演示了一遍便匆匆离去。又过了2天，工程师送来的实验样品是液相外延砷化镓（LPE-GaAs）材料，研究内容是其中的A、B能级的性质；因为没有*P*型GaAs样品，受主中心A、B能级性质研究是一个当时尚未解决的在多子系统里研究少子行为的难题，这儿的一些研究生也曾尝试过，但都是半途而废。一些有名望的学者对此也甚感头疼，因而被搁置了多年。我查阅了一些有关文献，一方面希望能找到解决这个难题的实验方法，另一方面在简陋的实验室里开始实验摸索，但由于缺乏温控和自动记录设备，实验结果非常分散。为了取得重复性较好的实验结果，有时记录一条低温热发射率实验曲线需要长达20多个小时，我就带上干粮吃住在实验室里。经过两周的苦干，并尝试过利用系里的大型计算机中处理多指数函数问题的软件，对实验曲线进行求解，但得到的结果仍不理想。我分析了*PN*结少子深能级电容随时间多指数函数衰减的起因，利用*PN*结耗尽层宽度随偏压而改变的基本原理，总结出一套克服源于*PN*结区边缘的自由载流子尾俘获而导致的慢瞬态过程的实验新方法，解决了在多子系统中研究少子陷阱性质的难题。这一实验方法上的创新，既缩短了测量时间，也提高了实验精度和可靠性。这个难题的解决，引起了系里同行的关注。莱德堡博士不解地问：“这么一个新的实验方法，你是如何想出来的呢？”我说：“还记得哥伦布立鸡蛋的故事吧，用他的话说：‘事情就这么简单。’”我的回答，使在场的人都笑了。

圣诞节前夕，95%以上的人信奉基督教的瑞典王国，一派欢乐热烈的气氛。我的导师——兼任位于斯德哥尔摩的爱立信电子公司副总裁职务的哥尔马斯教授，这时也回到了隆德，准备与家人团聚。我的初步工作想必他已从莱德堡口中略知一二。次日一上班，他就把我叫到他的办公室，我把2个月来的研究工作作了简要汇报，特别是如何解决在多子系统里研究少子行为的方法。他沉思片刻，突然站起身来说：“走，去看看你的实验室。”进了工作室，哥尔马斯显得有些吃惊，说：“怎么能在这样的实验条件下做出一流的工作呢？我给你另建一个比我的实验室还要好的实验室。”我还以为他是随便说说，没有料到的是他立即要莱德堡找工程师同我商量，开列要购买的仪器设备清单，马上着手建立新实验室。他又回头对我说：“新的实验室未建成之前，可先用我的实验室。”瑞典人做事非常认真，不到2个月，花费200万克朗，新实验室就建成了，并成了我专用的实验室。为此，那些瑞典学生还真有点嫉妒呢！之后来自国内的访问学者，就没有这么幸运了，在我回国之前，他们只能在那个设备条件很差的实验室苦拼了。

赴瑞典隆德大学固体物理系（国际深能级研究中心）进修的3年时间，与瑞典同事合作共完成15个研究项目（发表的论文全部是基于我的实验工作），参加过两次国际会议，在深能级和光谱物理研究方面取得的主要成果如下：

通过对多个深中心共存系统的理论分析和计算机模拟，提出了一个用于直接鉴别两个深能级共存系统是否相关的光电容实验方法，解决了在深能级研究中经常遇到而又一直未曾解决的难题。这一方法已成功地应用到识别硅中金施主和受主能级（《应用物理快报》，1983）以及液相GaAs中A、B能级上，证明它们都属于同一缺陷的不同荷电状态，而不是人们认为的两个独立能级（如深能级权威Lang，Grimmeiss和Jaros等在《物理评论》上发表的有分量的文章，对硅中金施主和受主能级起源的看法）。

利用*PN*结耗尽层宽度随偏压而改变的基本原理，提出了一整套克服来自自由载流子尾的俘获而导致的慢瞬态过程的新方法，从而解决了在多子系统中研究少子缺陷性质的问题。这些方法既大大节省了测量时间，提高了实验精度，也简化了对实验样品的制备要求。

基于深中心近邻原子的无序分布导致的围绕着宏观平均组分的涨落，提出了一个关于混晶半导体中深能级展宽的新模型。这个模型不但成功地解释了混晶半导体中所固有的载流子通过深能级的非指数发射和俘获瞬态过程的物理实质，而且通过对实验结果的理论拟合运算还将给出描述混晶半导体中深能级的重要参数，如能级展宽（ΔE）、平均热激活能（$\overline{E_T}$）以及给理论工作者提供缺陷波函数扩展范围等重要信息。

半导体光谱物理研究取得进展：在MOCVD生长Cu掺杂$GaAs_{1-x}P_x$的低温发光研究中，首先观察到由混晶效应导致的与铜相关光谱线的分裂，并用混晶中缺陷近邻组态不同成功地解释了这种分裂的物理实质，引起了有关学者的兴趣和重视（论文发表在1984年《特理评论快报》上）。在掺Cu、Cu-Li、Cu-Zn的GaAs和InP光致发光研究中，澄清了一些长期被错误指派的GaAs中与铜相关的缺陷中心（如错误地把C、F线看成是束缚在0.15eV和0.45eV受主中心的束缚激子），首次识别了C和F中心的受主特征，即每一个激子络合物都是由强束缚的空穴和一个由这个空穴的库仑势场弱束缚的电子所组成，并提出了GaAs中与Cu相关中心的微观结构模型。观察到一些新的发光谱线如Li-Cu发光中心和GaAs：Zn中与Cu相关的电吸引中性络合物发光及其精细结构。发现了Li具有强烈抑制GaAs中络合物形成的作用以及合金起伏势束缚激子发光等。在上述基础上，提出了GaAs中束缚在Cu等相关缺陷中心上的激子局域化理论模型，解释并成功地预示了GaAs中束缚激子出现的规律。在GaAs(InP)中首次观察到的“分子型”等电子中心发光（束缚激子），不但具有重要理论价值，而且通过进一步的深入研究，有可能在实验上实现强的红外发光。其中7篇论文在国际一流刊物（如《物理评论》《应用物理》）上发表。

1984 年回国后的研究工作

1984–1993 年，主要从事半导体材料生长及性质研究，先后负责承担多项国家自然科学基金、国家重点科技攻关和国家“863”高技术新材料研究课题。提出了 SI（半绝缘）–GaAs 电学补偿五能级模型和电学补偿新判据：SI–GaAs 分别由一个（或多个）浅和中等深施主和一个（或多个）浅和中等深受主以及一个位于禁带中心左右的深施主中心组成五能级模型，其中施主中心浓度之和应小于受主中心总浓度（两者均控制在 $10^{15}cm^{-3}$左右），浓度高于 $10^{16}cm^{-3}$深施主中心浓度将补偿未电离的受主中心，使其呈半绝缘特性。为提高 GaAs 质量器件与电路的成品率提供了依据。与人合作，提出了直拉硅中新施主微观结构新模型，摒弃了新施主微观结构直接与氧相关的传统观点（提出了新施主不仅同 CZ–Si 中氧、碳杂质有关，而且微缺陷的存在对新施主的形成也有十分重要的作用新观点；并且通过足够长时间热处理，即当新施主浓度饱和时，测得了 3 种新施主的电子能级位置，避免了已发表结果的极其分散性），成功地解释了现有的实验事实，预示了它的新行为；在国内率先开展了超长波长锑化物基材料生长和性质研究，并首先在国内研制成功 InGaAsSb、AlGaAsSb 材料及红外探测器和激光器原型器件。协助林兰英先生开拓了我国微重力半导体材料科学研究新领域，首次在太空从熔体中生长出 GaAs 单晶并对其光、电性质作了系统研究，受到国内外同行的高度评价。

从 1993 开始，我的工作重点集中在半导体低维结构和量子器件这一国际前沿研究方面，先后主持和参与负责十多个国家“863”及国家重点科技攻关、国家自然科学基金重大与重点和面上项目以及中国科学院重点与重大等研究项目。我和 MBE 组的同事一起在成功地生长了国内领先、国际先进水平的电子迁移率（4. 8K）高达百万的 2DEG 材料和高质量、器件级 HEMT 和 *P*–HEMT 结构材料的基础上，近年来，又发展了应变自组装 In（Ga）As/GaAs，InAlAs/AlGaAs/GaAs，InAs/InAlAs/InP 和 InAs/InGaAs/InP 等量子点、量子线和量子点（线）超晶格材料生长技术，并初步在纳米尺度上实现了对量子点（线）尺寸、形状和密度的可控生长，并研制成功国内首个量子点激光器；首次发现 InP 基 InAs 量子线空间斜对准的新现象，被国外评述文章大段引用；成功地制备了从可见光到近红外的量子点（线）材料，并研制成功室温连续工作输出光功率达 4W（双面之和）的大功率量子点激光器，为目前国际上报道的最好结果之一；红光量子点激光器的研究水平也处在国际的前列；InP 基和 GaAs 基量子级联激光材料和器件也取得了国际先进水平的工作，并在国际上首次研制成功量子点量子级联激光器；在国际上首先提出并研制成功国际首只量子点超辐射发光管。2001 年我作为国家“973”项目“信息功能材料相关基础研究”的首席科学家，又提出了柔性衬底的概念，为大失配异质结构材料体系研制开辟了一个可能的新方向。2010年率先建议有关单位立项开展超宽禁带半导体材料如半导体金刚石、立方BN、BeZnO和钙钛矿氧化物材料的基础研究，引起了重视。

上述研究成果曾获国家自然科学奖二等奖和国家科学技术进步奖三等奖，中国科学院自然科学奖一等奖和中国科学院科技进步奖一、二和三等奖，何梁何利科学与技术进步奖，国家“863”计划15周年先进个人、国家重点科技攻关奖（个人）以及优秀研究生导师奖等多项。从1983年以来，先后在国外著名学术刊物发表论文数百篇，培养博士、硕士和博士后170余人。

第二篇

论 著 选 编

硅的低温电学性质

周　洁，王占国，刘志刚，王万年，尤兴凯

（中国科学院半导体研究所，北京，100083）

摘要　本文在20–300K研究了室温载流子浓度$2\times10^{12}-1\times10^{20}cm^{-3}$含硼或磷（砷）Si的电学性质。对一些$p$-Si样品用弱场横向磁阻法及杂质激活能法进行了补偿度的测定，并进行了比较。从霍尔系数与温度关系的分析指出，对于较纯样品，硼受主能级的电离能为0.045eV，磷施主能级为0.045eV。在载流子浓度为$10^{18}-10^{19}cm^{-3}$时发现了费米简并。对载流子浓度为$2\times10^{17}-1\times10^{18}cm^{-3}$的$p$-Si及$5\times10^{17}-4\times10^{18}cm^{-3}$的$n$-Si观察到了杂质电导行为。从霍尔系数与电导率计算了非本征的霍尔迁移率。在100–300K，晶格散射迁移率μ满足关系式$AT^{-\alpha}$，其中$A=2.1\times10^9$，$\alpha=2.7$（对空穴）；或$A=1.2\times10^8$，$\alpha=2.0$（对电子）。

另外，根据我们的材料（载流子浓度为$5\times10^{11}-1\times10^{20}cm^{-3}$），分别建立了一条电阻率与载流子浓度及电阻率与迁移率的关系曲线，以提供制备材料时参考之用。

Low temperature electrical properties of silicon material

Zhou Jie, Wang Zhanguo, Liu Zhigang, Wang Wannian, You Xingkai

(Institute of Semiconductors, Chinese Academy of Sciences, Beijing 100083, China)

Abstract　In this paper electrical properties of silicon containing Boron and phosphor (or arsenic) have been measured from 20 to 300K in the range of carrier concentration 2×10^{12} to $1\times10^{20}cm^{-3}$ at room-temperature. The weak-field transverse magneto-resistance and impurity activation energy methods were used to determine the degree of impurity compensation for some p-type silicon samples and these methods have been compared.

Analyses of Hall coefficient and electrical conductivity vs temperature curves indicate the ionization energy of Boron (phosphor) acceptor (donor) levels to be 0.045 eV for low impurity concentration; Fermi degeneracy is found to occur in the range of 10^{18} to $10^{19}cm^{-3}$. Impurity conduction has been observed for carrier concentration 2×10^{17} to $1\times10^{18}cm^{-3}$ for p-type silicon and 5×10^{17} to $4\times10^{18}cm^{-3}$ for n-type silicon respectively. Extrinsic Hall mobility is computed from Hall coefficient and conductivity. The temperature dependence of lattice-scattering mobility is

原载于：物理学报，1966，22（4）：404–411.

found: $\mu_L = 2.1\times10^9 T^{-2.7}$ for holes; $\mu_L = 1.2\times10^8 T^{-2.0}$ for electrons.

Carrier concentration vs resistivity and Hall mobility vs resistivity curves had been plotted for our silicon material in the range of carrier concentration 5×10^{11} to $1\times10^{20}\text{cm}^{-3}$. It is aimed to give as reference for preparing silicon material.

1　引言

关于含硼与磷（或砷）Si 的低温电学性质已进行了较为广泛的研究，如 Morin，Maita[1]，Swartz[2]和 Putley[3]等分别对一定杂质浓度范围的材料进行了杂质能级、散射机构、杂质导电等行为的研究。我们在 20–300K 间，对更广的杂质浓度范围（$2\times10^{12}-1\times10^{20}\text{cm}^{-3}$）进行了电学参数的测量。试图利用霍尔系数与温度的关系曲线来决定杂质元素在 Si 中激活能的数值，以进一步鉴定材料的纯度，与我们已建立的弱场横向磁阻法[4]求得的补偿度进行比较。另外，我们亦试图观察一下不同杂质元素及不同杂质浓度的样品在上述温度范围的电学行为，以进一步了解材料的简并行为、杂质电导及迁移现象。为了提供制备材料时的需要，根据我们大量数据的积累，分别对不同电导类型的材料建立了电阻率与载流子浓度及电阻率与迁移率的关系曲线。

2　实验方法

Si 单晶样品是由直拉法与浮带法所制备的单晶锭上从垂直于拉晶的方向切下的，样品均切成 $16\times4\times2\text{mm}^3$，电极用冷压金丝或铝合金化丝带有弹簧的铜针压触来达到欧姆接触。电阻 $<10^5\Omega$ 的样品，用直流补偿法来进行测量。电阻 $>10^5\Omega$ 的样品则采用 DC-1 静电计电路来进行测量，这样可顺利测量到 $10^{10}\Omega$ 的电阻。前者的实验误差为 ±5%，后者为±10%。实验装置见图 1 所示，具体测量线路见图 2 所示。

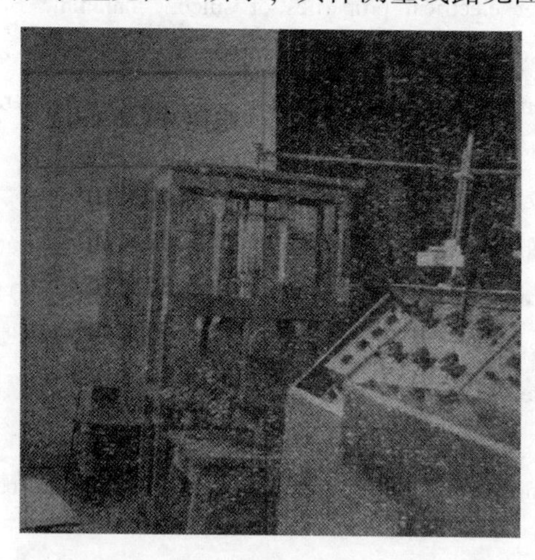

图 1　实验装置

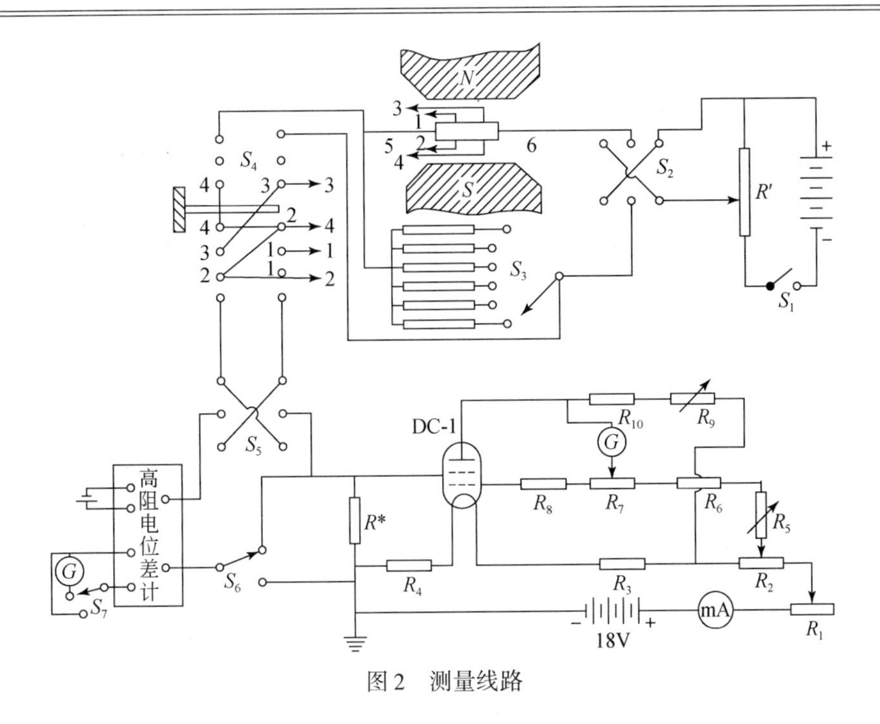

图 2　测量线路

图 3 为低温恒温器，其中样品 1 放在塑料板 2 上，样品及电极引线间的绝缘电阻均可达 $1\times10^{11}\Omega$ 以上。电极引线（见图 3 的 B）通过德银管 3 引出作为测量引线，样品周围套以黄铜管 4，作为加热电源用。黄铜管外再套一铜管 5，此管与德银管 3 可以密封，作为真空室用，便于抽空或充气以调节样品室的温度。温度是用铜–康铜热电偶配以 10^{-8}V 的低阻电位计来进行测量的，足以测出 0. 1μV 的变化。温度计从 20–80K 是利用已定标的铂电阻温度计校正得到的，80–300K 是由国家计量局进行定标。

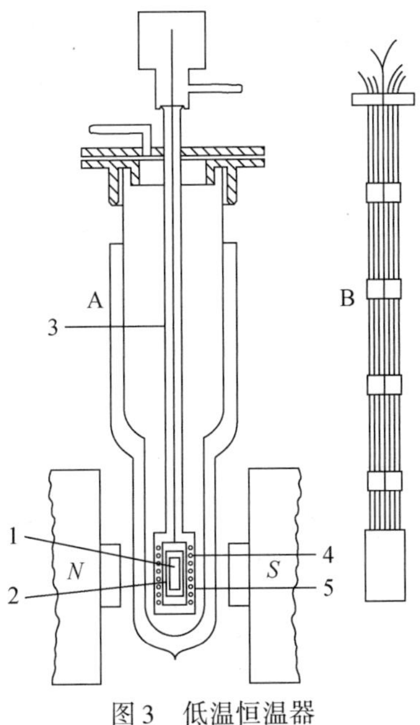

图 3　低温恒温器

3　实验结果与讨论

样品室温时的电学参数及杂质激活能的数据见表 1。

表 1　样品的电学参数及杂质激活能

样品号	电阻率（Ω · cm）	(N_A-N_D) (cm^{-3})	迁移率 $\left(\frac{cm^2}{V \cdot sec}\right)$	电离能（eV）	掺杂元素
B-1	1.00×10^4	1.26×10^{12}	3.81×10^2	4.51×10^{-2}	硼
B-2	5.00×10	3.49×10^{14}	3.77×10^2	4.50×10^{-2}	硼
B-3	2.80×10	6.79×10^{14}	3.80×10^2	4.50×10^{-2}	硼
B-4	4.45×10^0	4.44×10^{15}	3.70×10^2	4.53×10^{-2}	硼
B-5	3.50×10^{-1}	7.57×10^{16}	2.77×10^2	4.45×10^{-2}	硼
B-6	1.51×10^{-1}	2.00×10^{17}	2.43×10^2	3.16×10^{-2}	硼
B-7	5.38×10^{-2}	1.06×10^{18}	1.32×10^2	1.72×10^{-2}	硼
B-8	1.44×10^{-2}	7.27×10^{18}	7.40×10	简并	硼
B-9	2.01×10^{-3}	8.25×10^{19}	3.78×10	简并	硼
P-1	4.48×10^{-1}	1.31×10^{16}	1.26×10^3	4.48×10^{-2}	磷
P-2	2.01×10^{-1}	3.44×10^{16}	9.22×10^2	3.98×10^{-2}	磷
P-3	2.77×10^{-2}	5.08×10^{17}	5.22×10^2	1.72×10^{-2}	磷
As-1	1.09×10^{-2}	4.60×10^{18}	1.47×10^2	1.40×10^{-2}	砷
As-2	1.12×10^{-3}	8.93×10^{19}	6.26×10	简并	砷

含硼或磷(砷)的 Si 的电阻率和霍尔系数与温度倒数的关系曲线分别如图 4，图 5，图 6，图 7 所示。

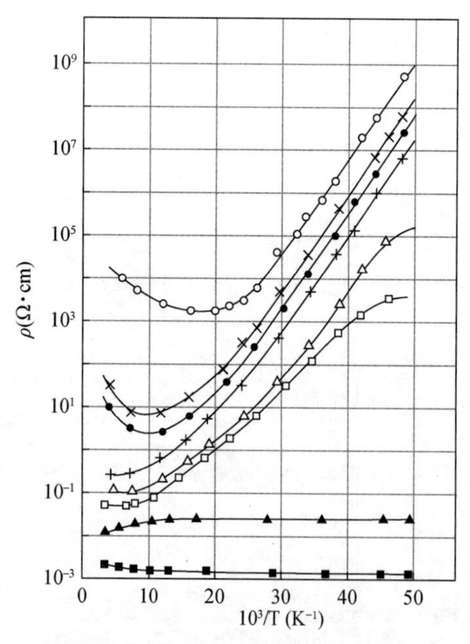

图 4　含硼硅的电阻率与绝对温度倒数的关系曲线

—○— B1　—×— B2　—●— B3　—+— B5
—△— B6　—□— B7　—▲— B8　—■— B9

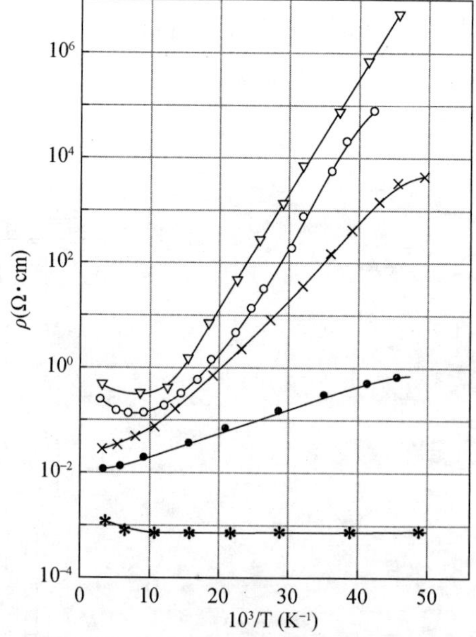

图 5　含磷(砷)硅的电阻率与绝对温度倒数的关系曲线

—▽— P1　—○— P2　—×— P3　—●— As1　—✱— As2

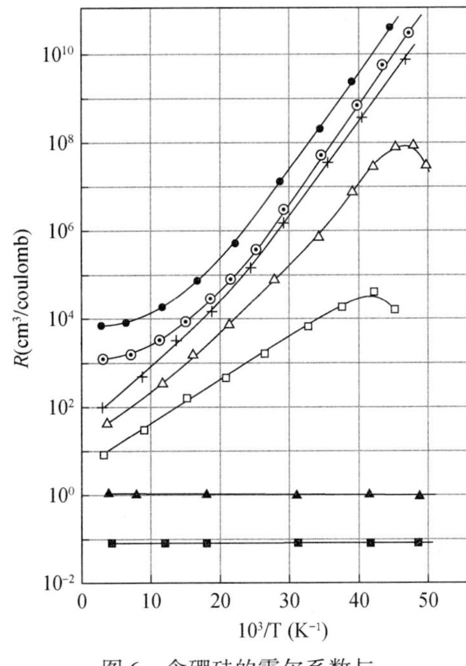

图 6　含硼硅的霍尔系数与绝对温度倒数的关系曲线

—●— B3　—⊙— B4　—+— B5　—△— B6
—□— B7　—▲— B8　—■— B9

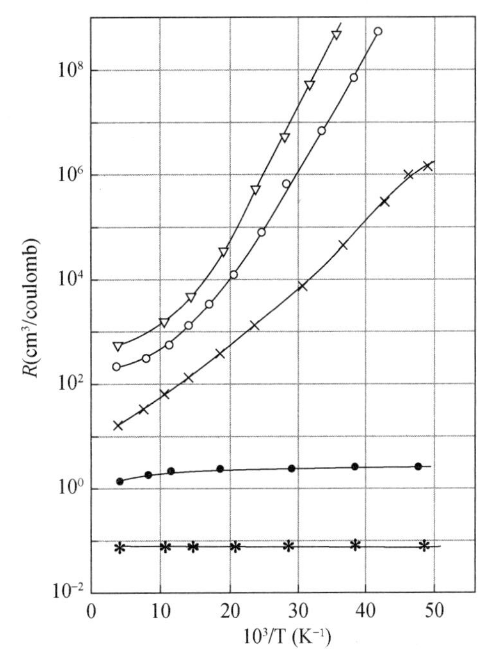

图 7　含磷(砷)硅的霍尔系数与绝对温度倒数的关系曲线

—▽— P1　—○— P2　—×— P3　—●— As1　—✱— As2

3.1　补偿度的测定

(1) 杂质激活能法　根据质量作用定律可得

$$\frac{p(p+N_D)}{(N_A-N_D)-p}=\frac{2f}{g}\left(\frac{2\pi m^* kT}{\hbar^2}\right)^{\frac{3}{2}}\exp\left(\frac{-E_A}{kT}\right); \tag{1}$$

其中，g 是简并度；f 为导带（或价带）极值的数目；m^* 为电子（或空穴）的有效质量；E_A 为杂质电离能。当温度足够低，$p \ll N_D$ 及（N_A-N_D）时，上式可简化为

$$p=\frac{N_A-N_D}{N_D}\left(\frac{2\pi m^* kT}{\hbar^2}\right)^{\frac{3}{2}}\exp\left(\frac{-E_A}{kT}\right). \tag{2}$$

设 $m^*=0.6m_0$（空穴）和 $m^*=m_0$（电子）[5,6]，则可从 $\lg(pT^{-\frac{3}{2}})-1/T$ 图的斜率求出电离能 E_A 的数值。

对 B-1，B-2，B-3，B-4 四块样品求得了硼受主杂质在 Si 中的电离能为 0.045eV（其中 B-1，B-2，两块样品的电离能是从 $\lg\rho-1/T$ 图上得出的，并对 $\mu\propto T^{-2.7}$ 进行了修正）。从 P-1 样品得到磷施主能级为 0.045eV，已知 E_A（或 E_D）后，结合耗尽区的 $p=N_A-N_D$，则可分别得到 N_A 及 N_D，由此法求得的数据见表 2。

(2) 弱场横向磁阻法　利用液氮温度时，弱场横向磁阻的大小与空穴（或电子）为所有电离杂质中心的散射十分有关这一特点，可用来作为测量总杂质浓度的工具。根据这一原理，我们在 1962 年建立了一条 p-Si 的主曲线[4]（见图 8），即液氮时磁阻与电离杂质浓度和的关系曲线。因此，只需测出 77K 时的磁阻，即由图 8 得到 $(N_A+N_D)_{77K}$，并由下述关系式

$$(N_A+N_D)=(N_A+N_D)_{77K}+\left(\frac{1}{(\rho e\mu)_{300K}}-\frac{1}{(\rho e\mu)_{77K}}\right), \tag{3}$$

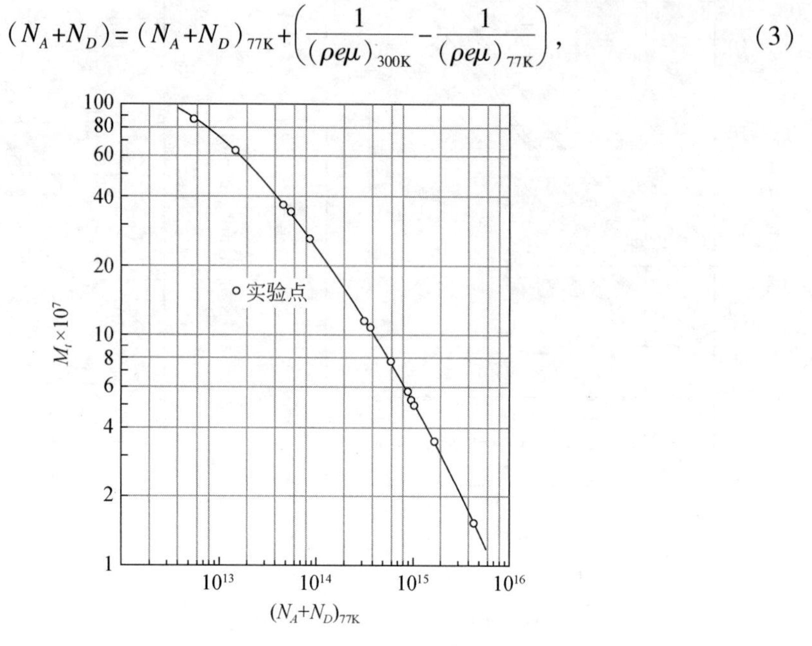

图 8 77K 时的磁阻与总电离杂质浓度和的关系曲线

结合 300K 及 77K 时的霍尔系数与电阻率的测量，即可测出补偿度。

用此法对 B-1，B-2，B-3，B-4 四块样品进行了测量，结果亦列于表 2。

表 2 不同方法求得的补偿度

样品号	磁阻法			杂质激活能法		
	$N_A(cm^{-3})$	$N_D(cm^{-3})$	N_D/N_A (%)	$N_A(cm^{-3})$	$N_D(cm^{-3})$	N_D/N_A (%)
B-1	4.93×10^{12}	3.67×10^{12}	74.5	4.00×10^{12}	2.74×10^{12}	68.5
B-2	4.50×10^{14}	1.01×10^{14}	22.4	4.42×10^{14}	0.93×10^{14}	21.0
B-3	7.77×10^{14}	9.80×10^{13}	12.7	7.35×10^{14}	5.57×10^{13}	7.6
B-4	4.48×10^{15}	4.00×10^{13}	0.89	4.45×10^{15}	1.88×10^{13}	0.42

用上述两种方法求得的补偿度，由表 2 可见，得到了较为一致的结果。从方法的简便及精度来说，以磁阻法为最好（如果已建立好主曲线的话）。因弱场横向磁阻与迁移率的平方成正比，而迁移率在 77K 时随杂质浓度的增加而迅速下降，所以利用磁阻来测量杂质补偿，是较为灵敏的。同时，磁阻正比于 $\Delta\rho/\rho$，与样品的几何尺寸无关，所以能较精确地测出。但此法仅适用于杂质浓度 $<10^{16}cm^{-3}$ 的材料的测量。杂质激活能法要测量到比液氮更低的温度才能进行计算，尤其对于高阻样品，要在 40K 以下，载流子浓度随温度变化才具有明显的斜率，同时又必须进行连续的温度测量，比较繁杂。

3.2 杂质电导

如同洪朝生等[7]在Ge中液氦温度附近观察到的电学行为一样，Swartz[2]，Carlson[8]等在Si中也观察到了类似的现象。我们在载流子浓度为$2\times10^{17}-1\times10^{18}cm^{-3}$的$p$-Si中及$5\times10^{17}-4\times10^{18}cm^{-3}$的$n$-Si中，在液氢温度附近也观察到了类似的行为（图4～图7），即此时霍尔系数随着温度的降低出现一个极值，而电阻率逐渐趋于饱和。电阻率开始缓慢增加的温度与霍尔系数出现极值的温度刚好对应。随着杂质浓度的增加，杂质电导出现的温度也相应提高。这一行为首先为洪朝生等进行了解释，他们认为，由于杂质浓度足够高时相邻原子的波函数有明显的重叠，以致形成一个带，构成杂质带导带。在一般温度范围，由于此带很窄，带中电子的迁移率较小，所以呈现不了杂质带导电的作用；但当导带中电子浓度变得十分小时，杂质带中导电就不可忽略。图6中R极值的出现，就是两种导电机构过渡的结果，而此时ρ也开始以另一斜率缓慢增加（图4）。

3.3 迁移率的行为

20－300K的迁移率行为见图9所示。由图可见，除B-9，As-2两块重掺杂样品外，其他非简并的样品显示相同的规律性。在100K以上为晶格散射起主要作用，在40－50K以下电离杂质散射才明显的起作用。随着杂质浓度的增加，电离杂质散射起作用的范围逐渐向高温移动。在100－300K，晶格散射迁移率$\mu_L=2.1\times10^9T^{-2.7}$（对空穴）和$\mu_L=1.2\times10^8T^{-2.0}$（对电子），这与前人[3]所得的结果是较为一致的。对于两块简并样品的迁移行为，其μ与T的关系呈现了平坦的特点。

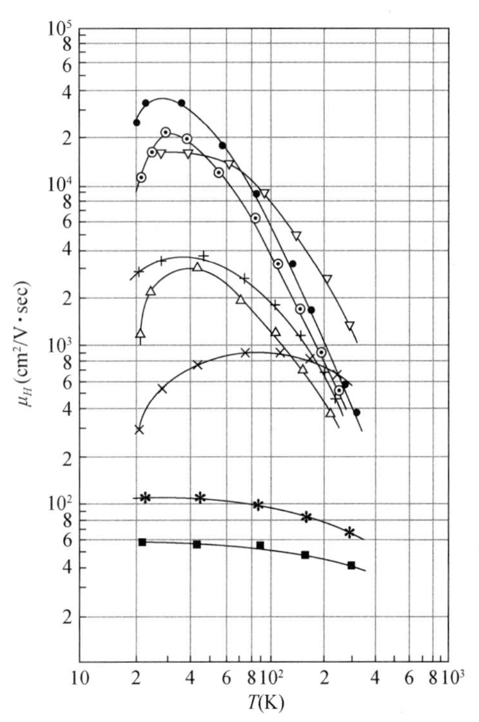

图9 霍尔迁移率与绝对温度的关系曲线

3.4 电阻率与载流子浓度及迁移率关系曲线的建立

为了便于在测出电阻率之后，就能立即知道材料的载流子浓度与迁移率的数值，故结合我们的材料（直拉法及浮带法制备的）分别建立了P型与N型硅的电阻率（$1\times10^{-3}-5\times10^{4}\Omega\cdot cm$）与载流子浓度及迁移率的关系曲线，分别如图10，图11所示。

在由霍尔系数计算载流子浓度时，对于$\rho\geqslant1\times10^{-2}\Omega\cdot cm$的样品，我们取$\gamma=3/8\pi$；对于$\rho\leqslant1\times10^{-2}\Omega\cdot cm$的样品，取$\gamma=1$；这曲线是我们依靠大量样品建立来的，以提供制备材料时参考之用。

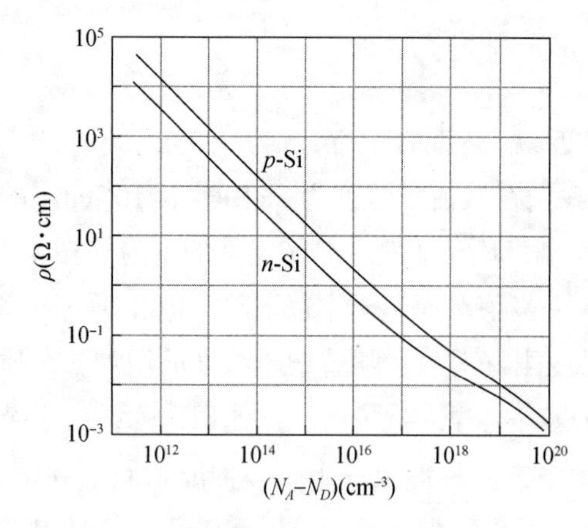

图 10　电阻率与载流子浓度的关系曲线

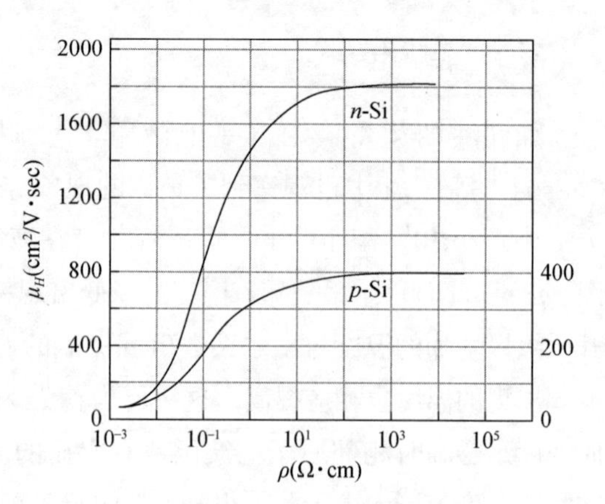

图 11　电阻率与霍尔迁移率的关系曲线

4　结论

（1）采用了弱场横向磁阻法、杂质激活能法求出了杂质补偿度，结果较为一致。

（2）在载流子浓度为 $2\times10^{17}-1\times10^{18}cm^{-3}$ 的 p-Si 及 $5\times10^{17}-4\times10^{18}cm^{-3}$ 的 n-Si 中，观察到了杂质电导行为。

（3）在晶格散射范围（100-300K）得到

$$\mu_L=2.1\times10^9T^{-2.7}(\text{空穴}),$$
$$\mu_L=1.2\times10^8T^{-2.0}(\text{电子}),$$

这与前人的结果颇为一致。

（4）在较纯硅的材料中，得到硼的受主能级的电离能为 0. 045eV，磷施主能级为 0. 045eV。

在工作中得到林兰英先生的全面指导，和洪朝生先生关于低温方面的指导，在此一并致谢。

参 考 文 献*

[1] Morin, F. J., Maita, J. P., Phys. Rev., 96(1954), 28.
[2] Swartz, G. A., J. Phys. Chem. Solids, 12(1960), 245.
[3] Putley, E. H, Mitchell. W. H. Proc. Phys. Soc. London, 72(1958), 193.
[4] 周洁, 刘志刚, 王万年. 第一次全国半导体会议上的报告.
[5] Benjamin Lax, Mavroides, J. G., Phys. Rev., 100(1955),1650.
[6] Dexter, R. N., Zeiger, H. J., Benjamin Lax, Phys. Rev. 104(1956), 637.
[7] 洪朝生, Gliessman, J. R., Phys. Rev., 96(1954), 1226.
[8] Carlson, R. D., Phys. Rev., 100(1956), 1075.

* 本书论著部分均为已发表过的文章，尊重原期刊著录形式，参考文献格式不变，特此说明。下同，不再标注。

Evidence that the gold donor and acceptor in silicon are two levels of the same defect

L-Å. Ledebo[1], Zhan-Guo Wang[2]

([1]Innovance AB, Magle Stora Kyrkogata 8, 5-223 50 Lund, Sweden; [2]Department of Solid State Physics, University of Lund, S-220 07 Lund, Sweden; Present address: Institute of Semiconductors, Chinese Academy of Sciences, Beijing 100083, China)

Abstract A photocapacitance method was used to monitor the time dependences of the occupation numbers for the gold-related donor and acceptor in silicon during optical excitation. The experimental data give strong evidence that the donor level corresponds to the +/0 transition and the acceptor to the 0/− transition for one single defect.

Many deep level dopants in semiconductors create two or more energy levels within the band gap. Since the early studies of impurities, a recurrent question has been whether or not two simultaneously introduced levels are caused by the same defect, however, no general method for direct measurement has so far been found. The two levels related to Au in Si are often without any particular evidence assumed to belong to the same defect. Even though their concentrations are difficult to determine accurately, it can at least be safely stated that they are not very different. From an argument essentially based on differences in the measured concentrations, it was, however, recently argued that the gold donor and acceptor levels are not related to the same gold center.[1] The present study was undertaken to resolve the fundamental question of coupling for the technologically important gold donor(E_v+0.35 eV) and acceptor(E_v+0.62 eV) in silicon. The procedure that has been used is general, and can be applied to many other systems.

We start with the equations for the time dependence of the occupation numbers at optical excitation. First, assume that two energy levels in the lower half of the band gap are not coupled; i.e., they each represent two charge states of two separate defects [see insert in Fig. 1(a)]. Assume that we use a Schottky barrier on p-type material. The levels are first filled with holes by a zero bias pulse, after which the reverse bias is restored. The temperature must be sufficiently low to prevent thermal emission. The initial conditions

原载于: Appl. Phys. Lett., 1983, 42(8): 680–682.

then are

$$n_1^- = n_2^- = 0; \quad n_1^0 = N_1; \; n_2^0 = N_2, \tag{1}$$

where n_i is the electron occupation of level i with total concentration N_i and 0, and $-$ the charge state of the defect.

If the sample is illuminated with photons of an energy sufficiently large to excite holes from both levels to the valence band, then levels 1 and 2 independent of each other will change occupation numbers (so far we assume that no electron transitions to the conduction band take place):

$$\begin{aligned} &n_1^0(t) + n_1^-(t) = N_1; \quad n_2^0(t) + n_2^-(t) = N_2, \\ &\mathrm{d}n_1^0(t)/\mathrm{d}t = -e_1 n_1^0(t); \quad \mathrm{d}n_2^0(t) = -e_2 n_2^0(t), \end{aligned} \tag{2}$$

where e_i is the optical emission rate for holes for level i. The simple solution is

$$n_1^-(t) = N_1[1 - \exp(-e_1 t)]; \quad n_2^-(t) = N_2[1 - \exp(-e_2 t)]. \tag{3}$$

The capacitance variation will include both transitions:

$$\Delta C(t) = C_0[n_1^-(t) + n_2^-(t)]/2N_S \quad (N_T \ll N_S), \tag{4}$$

where C_0 is the total capacitance and N_S the shallow level concentration.

If the two levels belong to the same defect we instead have [see insert in Fig. 1(b)]:

$$n^+(t) + n^0(t) + n^-(t) = N. \tag{5}$$

By filling the levels with holes in the initial state we obtain

$$n^+ = N; \quad n^0 = n^- = 0. \tag{6}$$

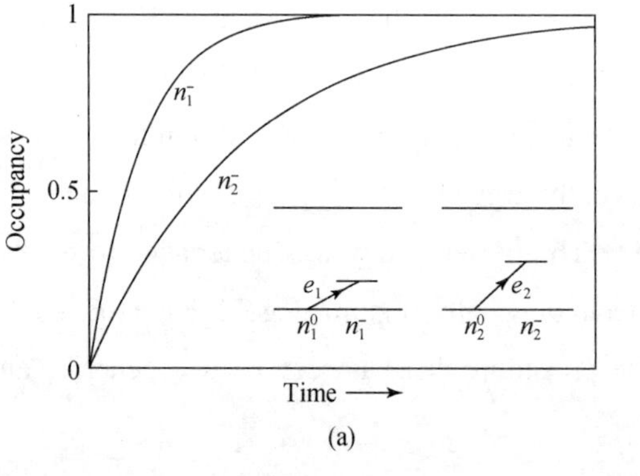

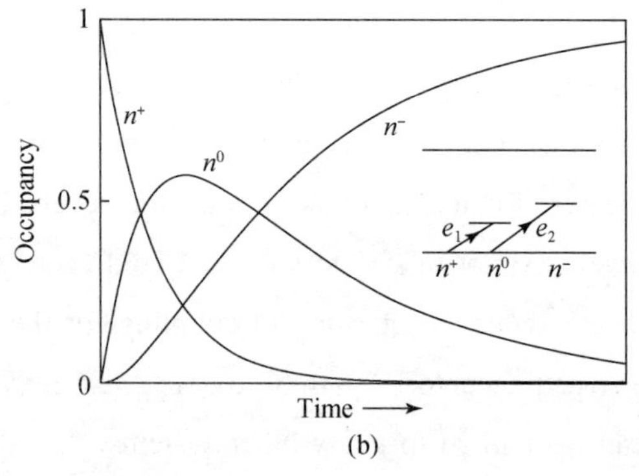

Fig. 1 Time dependence of the electron occupancies for (a) two independent deep levels and (b) two coupled levels under optical excitation. Equations (3), (5), (8), and (9) have been used for the calculation with $e_1/e_2 = 3$. At $t = 0$ all levels are filled with holes

The differential equations to solve are

$$\begin{aligned} &\frac{\mathrm{d}n^0(t)}{\mathrm{d}t} = e_1 n^+(t) - e_2 n^0(t), \\ &\frac{\mathrm{d}n^-(t)}{\mathrm{d}t} = e_2 n^0(t), \end{aligned} \tag{7}$$

with the solutions

$$n^0(t)=Ne_1[\exp(-e_1t)-\exp(-e_2t)]/(e_2-e_1), \tag{8}$$

$$n^-(t)=N\{1+[e_1\exp(-e_2t)-e_2\exp(-e_1t)]/(e_2-e_1)\}. \tag{9}$$

The capacitance change associated with the optical excitation is

$$\Delta C(t)=C_0[n^0(t)+2n^-(t)]/2N_S, \tag{10}$$

Eq. (3) is plotted in Fig. 1(a), and Eqs. (8) and (9) in Fig. 1(b). The occupation number time dependences are markedly different in the coupled and the uncoupled cases.

After having established that the time dependence of the occupation numbers is a suitable entity for study in order to demonstrate coupling, we have to find a practical method to monitor these numbers. This was done with a photocapacitance method for the two gold-related levels in silicon. The sample was a Schottky diode on p-type material. The shallow level doping was $1.6\times10^{16}\text{cm}^{-3}$, and the gold concentration was $3.8\times10^{14}\text{cm}^{-3}$ as determined by deep level transient spectroscopy. The temperature used was 96K, which resulted in a sufficiently slow thermal emission rate of holes from the donor so as not to compete with the optical emission.

An initial hole occupancy was set by a zero bias pulse, after which a reverse bias of 3V was restored. The levels in the larger part of the depletion region were thus filled with holes. In this particular experiment the levels in the rest of the depletion region can be neglected. The sample was then illuminated a time t with light of photon energy 0.72eV. This energy is large enough to excite holes from both levels to the valance band. (There is also a small cross section for the excitation of electrons from the acceptor to the conduction band. For the sake of clarity we temporarily neglect this cross section, and instead make a correction when presenting the final results.) The excitation is accompanied by a capacitance variation ΔC_{A+D} (Fig. 2). Then 0.50eV is used to completely fill the donor with electrons. A capacitance step $\Delta C_D(t)=C_0\Delta n_D/2N_S$ is observed, giving information about the concentration of donors not filled with electrons during the initial 0.72eV illumination. Finally, 0.72eV is used to fill also the acceptor with electrons. The capacitance step is now $\Delta C_A(t)=C_0\Delta n_A/2N_S$, which contains information about the concentration of acceptors that were not filled with electrons during the initial 0.72eV excitation. The experiment was repeated for different times t. In particular, we have $\Delta C_D(0)=C_0N_D/2N_S$ and $\Delta C_A(0)=C_0N_A/2N_S$ with $N_A=N_D$ if the levels are coupled. From Eqs. (4) and (10) and the relations in the preceding paragraph we get

$$n_A(t)/N_A=1-\Delta C_A(t)/\Delta C_A(0), \tag{11}$$

in the coupled as well as the uncoupled case, and

$$n_D(t)/N_D=\{\Delta C_{A+D}(t)-2[\Delta C_A(0)-\Delta C_A(t)]\}/\Delta C_D(0), \tag{12}$$

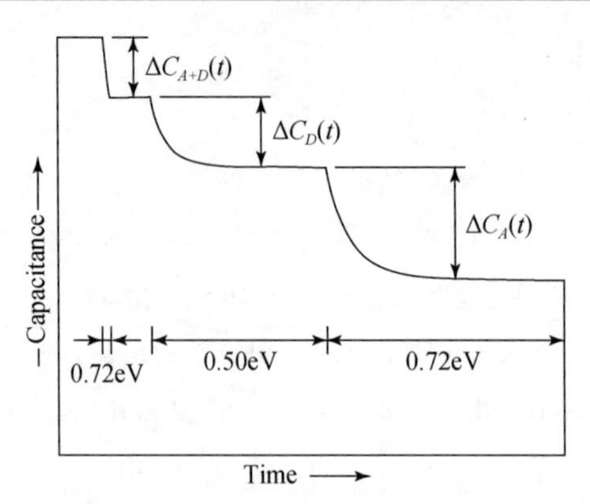

Fig. 2 Actual output from a transient recorder of the capacitance variation during the illumination cycle. Both the donor and the acceptor were initially filled with holes

for coupled levels. For uncoupled levels we instead have

$$n_D(t)/N_D=\{\Delta C_{A+D}(t)-[\Delta C_A(0)-\Delta C_A(t)]\}/\Delta C_D(0). \tag{13}$$

As Eqs. (12) and (13) are alternatively applicable the experimental data from Eq. (11) should be compared with Eqs. (3) and (9) for a pertinent test of coupling. The difference between Eqs. (3) and (9) appears predominantly for short times. The comparison is made in Fig. 3. Data for longer times are shown in Fig. 4. Even though the concentration of the levels is the same, the capacitance step from the acceptor is smaller than that from the donor because of a nonzero optical cross section for the excitation of electrons to the conduction band. It is, however, not sufficient to correct for this by making use of the experimentally determined optical cross sections, since in addition the difference in effective excitation width for the donor and the acceptor modifies the capacitance amplitudes with some tens of percent.[2] Instead of making an intricate calculation of this correction, we chose to use an empirical normalization by multiplying $C_A(t)$ by 1.15. This factor was obtained from the requirement that before putting the experimental data into our equations, $C_A(0)$ must be equal to $C_D(0)$. It is important to note that the correction is unimportant for the general shape of the curve, and in no way influences the conclusion.

Mainly based on the existence of a lag in the initial part of the time dependence for n_A, we thus conclude that the two gold-related levels are coupled and correspond to three different charge states of the same defect as in the insert of Fig. 1(b). The charge designation is the one traditionally used, originating from compensation studies, but independently supported by the magnitudes of the various cross sections for capture of free carriers.[3] The present result is in conflict with conclusions by Lang et al.,[1] which were based on differences in the experimentally obtained concentrations for the donor and the acceptor. The accurate measurement of concentration of the gold levels in silicon is, however, a very delicate task, and even though saturation was observed for the donor level during injection,[1] the actual occupation number at saturation could not be accurately known. Even though we feel that our demonstration of coupling is an important step towards

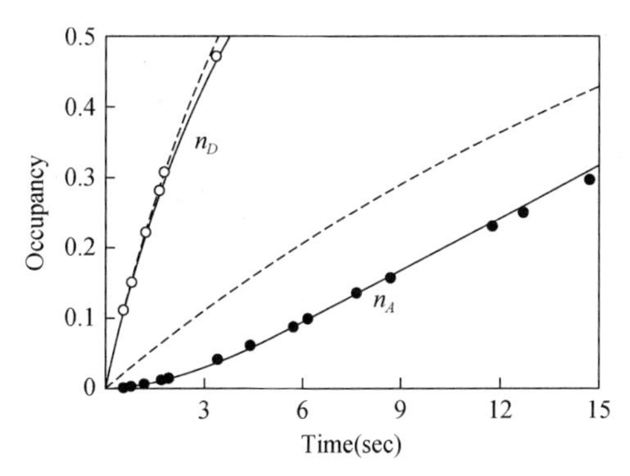

Fig. 3 Electron occupancy time dependences for the acceptor A and the donor D with optical excitation of 0.72eV. The points are the experimental data as determined by experiments described in Fig. 2 and in text. The solid lines result from a calculation using Eqs. (8) and (9), and the dashed lines are the expected time dependence for uncoupled levels using Eq. (3)

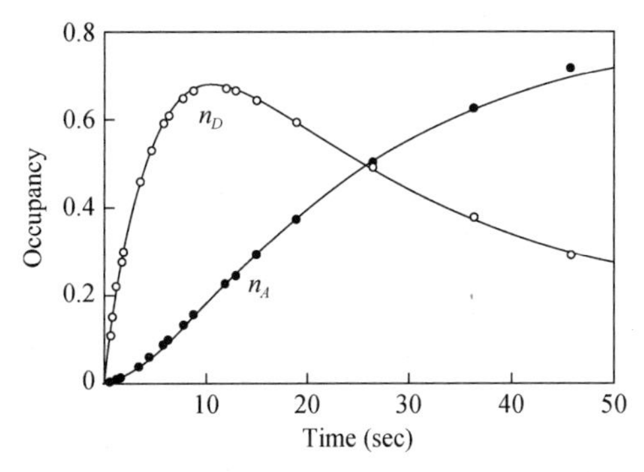

Fig. 4 Electron occupancy time dependences for longer times. The solid lines are the result from the same calculation as in Fig. 3

an identification of the defect causing the gold-related levels, we are not yet in a position to suggest a detailed identification for this center.

The method of measurement described above has also been applied to the so-called A and B levels at E_v+0.40 and E_v+0.70eV, often present in liquid phase epitaxial GaAs. The levels were found to be coupled, and from this and other electronic properties it has been suggested that the identity of the defect is the Ga_{As} antisite.[4]

We are grateful to K. Nideborn for help with the sample preparation. Wang thanks Professor H. G. Grimmeiss for his kind hospitality and continuous support. Ledebo thanks the Swedish Board for Technical Development for financial support during the initial part of this study.

References

[1] D. V. Lang, H. G. Grimmeiss, E. Meijer, and M. Jaros, Phys. Rev. B 22, 3917(1980).
[2] H. G. Grimmeiss, L-Å. Ledebo, and E. Meijer, in E. Meijer: thesis, Lund 1982, paper Ⅳ.
[3] R. H. Wu and A. R. Peaker, Solid-State Electron. 25, 643(1982).
[4] Zhan-Guo Wang, L-Å. Ledebo, and H. G. Grimmeiss(unpublished).

Optical properties of iron doped $Al_xGa_{1-x}As$ alloys

Zhan-Guo Wang[1], L-Å. Ledebo[2], H. G. Grimmeiss[1]

([1]Department of Solid State Physics, University of Lund, S-220 07 Lund, Sweden;
Present address: Institute of Semiconductors, Chinese Academy of Sciences, Beijing 100083, China)

([2]Present address: Innovance AB, Magle Stora. Kyrkogata 8, 5-223 50 Lund, Sweden)

Abstract Photocapacitance measurements were used to study the photoionization cross-section spectra of holes in iron doped $Al_xGa_{1-x}As$. The spectra are well described by the crystal field theory giving a nearly constant crystal field splitting of about 0.36eV for all compositions studied. As a result of an extended range of composition($0 \leqslant x \leqslant 0.76$), it was possible to show that the binding energy of the dominant iron center increases linearly with composition and that the relative position of the center is not related to any of the conduction band minima. The variation of the energy for $\Gamma_1(^5E) \to \Gamma_5(^5T_2)$ transition in $Fe^{2+}(3d^6)$ is discussed for different host crystals.

1 Introduction

Transition metal impurities with partly filled $3d$ shells are known to produce deep centers in the forbidden energy gap of Ⅲ–Ⅴ compounds. Several papers have been published on the electronic properties of iron in GaAs,[1-5] GaP,[6-8] and InP.[9] Since the electronic properties of such centers are closely related to the band structure, their characteristic parameters have been studied in different matrices. The number of compounds available for such studies are, however, rather limited. Furthermore, some of them have direct band gaps, whereas the others have indirect band gaps which makes it still more difficult to understand the general pattern of the data obtained. This is one of the reasons for the growing interest in the study of alloys doped with particular impurities.

The purpose of this paper is to present optical measurements which have been performed in iron-doped $Al_xGa_{1-x}As$ using photocapacitance techniques. Iron has been studied earlier in both GaAs[5] and $Al_xGa_{1-x}As$.[10] However, the data for $Al_xGa_{1-x}As$ have been obtained using thermal measurements in a small range of compositions. It will be shown that the results of thermal measurements in alloys are more difficult to analyze than the results of

原载于: J. Appl. Phys., 1984, 56(10): 2762-2767.

optical measurements due to nonexponential transients, and that both previous and our own thermal measurements are not in agreement with the optical data presented in this paper. As a result of an extended range of composition it is now possible to give evidence that the binding energy of the dominant iron center increases linearly with composition and that the relative position of the center is not related to any of the conduction band minima. Assuming that Fe occupies cation sites in the compounds, the optical spectra obtained are well understood within the framework of crystal field theory implying that the ^5D free iron level of $Fe(3d^6)$ is split by the cubic crystal field into the levels $^5T_2$ and ^5E. The ground state ^5E is further split into five equidistant levels by a second order spin-orbit and first order spin-spin interaction[2, 11] but these splittings are too small to be seen in our experiments.

2 Experimental details

The samples used in this study were layers of n-type Al_xGa_{1-x} As alloys with different composition x, grown on n^+-GaAs substrates by liquid phase epitaxy(LPE) (table 1). The layers were iron diffused in the temperature range between 725 and 855℃. The diffusion time was chosen such that homogeneous iron concentrations in the range between 2×10^{14} and 2×10^{15} cm^{-3} were obtained in the layers. The free electron concentration of the epitaxial layers was kept within the range $(2-6)\times10^{16}$ charge carriers per cm^3 by doping the layers with Sn during LPE. p^+n junctions were obtained by zinc diffusion at temperatures below 710℃ after the samples had been doped with iron. The composition x of the alloy layers was determined using electron excited X-ray analysis in a scanning electron microscope.

Different techniques of junction space charge spectroscopy were used for our investigations. All experiments were performed in a temperature controlled cryostat in which the sample was cooled by direct contact with an exchange gas and which had a thermal stability of better than 0.1K at all temperatures. Owing to the weak signals and the larger difference between intrinsic and extrinsic absorption coefficients all kinds of stray light had to be avoided. An incandescent lamp with a Zeiss MM-12 prism double monochromator was therefore used as a light source. In the cases where a second light source of high intensity was needed a Bausch & Lomb high intensity double grating monochromator was employed. Silicon and germanium filters were used to suppress stray light and higher orders of diffracted light.

In order to measure optical cross sections of holes, the iron center had to be at least partly filled with holes. This was achieved by illuminating the samples with photon energies close to the band gap such that electron-hole pairs were generated in the n-type layer. After removing the light source the dark capacitance of the junction usually changed slowly with time probably due to the capture of electrons from the free-carrier tail of the neutral n-type region.[12] This change in the capacitance considerably reduced the sensitivity of the measurements. To avoid the long relaxation times for establishing the initial conditions for the experiment, the reverse bias was increased to −5V after the centers had been partly emptied of electrons using optical inter-band excitation at −4V.

3 Experimental results

All optical cross sections were measured using the photocapacitance technique. In the cases where thermal emission processes did not interfere with the experiment the initial slope technique[13] was employed. In order to check whether or not the transients are single exponentials, full capacitance transients were recorded at several representative photon energies for every spectrum measured. Some of the results are presented in Fig. 1, showing that good exponentiality is observed for different compositions.

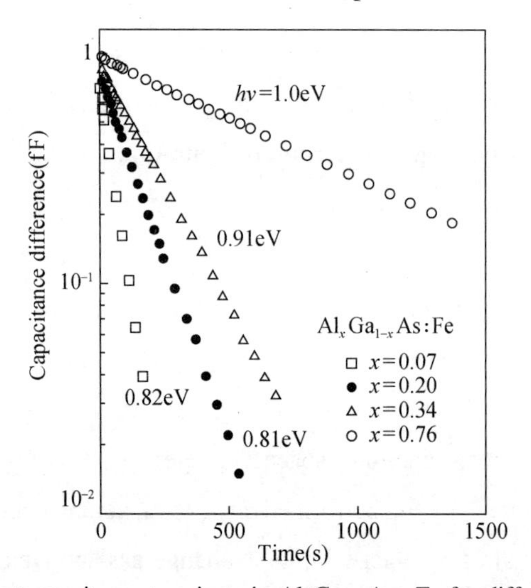

Fig. 1 Typical photocapacitance transients in $Al_xGa_{1-x}As$: Fe for different compositions x

When measuring the temperature dependence of the spectra the constant photocapacitance technique[14] was employed in all the experiments where single shot transient measurements could not be used owing to the interference by thermal excitation processes at elevated temperatures. Typical normalized spectra of photoionization cross sections for holes measured below the freeze-out temperature of the iron center are presented in Fig. 2 for two compositions $x = 0.07$ and 0.76, respectively. In order to show the similarity of the two spectra the $x = 0.76$ curve has been shifted 0.348eV towards lower energies. It is readily seen that (1) the shape of the spectra are similar for samples of different compositions, (2) two thresholds are observed, (3) both threshold energies increase with x, and (4) the phonon broadening in the low energy part of the spectra is very moderate. The spectral distributions of the optical cross section for all compositions studied are shown in Fig. 3 in absolute values.

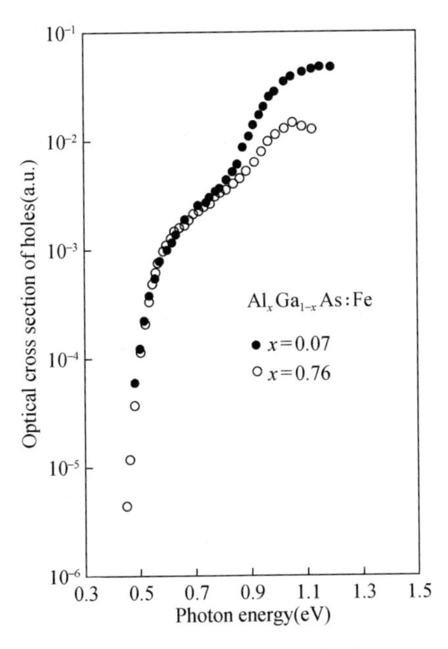

Fig. 2 Normalized spectra of photoionization cross sections of holes for two compositions $x=0.07$ (full circles) and $x=0.76$ (open circles). The curve for $x=0.76$ has been shifted 0.348eV towards lower energies

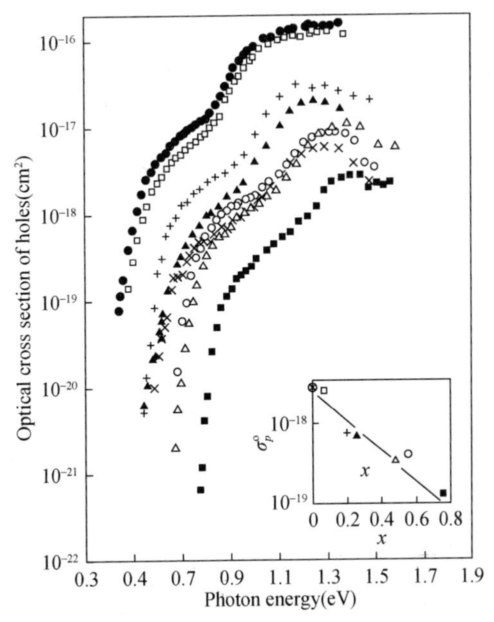

Fig. 3 Spectral distributions of optical cross sections for holes in $Al_xGa_{1-x}As$: Fe with different compositions x. ●: $x=0$, ▫: $x=0.07$, +: $x=0.2$, ▲: $x=0.26$, ×: $x=0.34$, △: $x=0.49$, ○: $x=0.56$, and ■: $x=0.76$. The insert shows the absolute values of the photoionization cross section taken 0.1eV above the low energy thresholds

The spectra have been measured at the lowest temperature possible in order to make the analysis of the data easier. A closer inspection of the results suggests that the energy difference between the two thresholds seems to be independent of composition within the accuracy of the analysis. An exact determination of the two threshold energies for different compositions is difficult since to the best of our knowledge no theoretical relation describing the spectral distributions of such spectra is presently available. Different empirical evaluation methods have therefore been used. In order to determine the relative change of the two threshold energies with composition x, the spectra have been shifted towards lower energy until they coincided with the GaAs:Fe spectrum. The energies obtained for the relative positions are summarized in table 1. The absolute values have been obtained by fitting the low energy part of the GaAs:Fe spectrum (Fig. 4, full circles) with a curve given by the relation[15]

$$\sigma_p^0 = B(h\nu - E_0)^{5/2}(h\nu)^{-3} \tag{1}$$

and adjusting the constant B and the optical threshold E_0 such that the best fit was obtained (Fig. 4). For the transitions from the valence band into the ^5E levels a value of 0.43eV for E_0 in GaAs:Fe has been deduced. This value is an empirical estimate of the threshold energy and probably differs from the binding energy of the ground state. The fitted curve has then been subtracted from the rest of the data, giving the spectrum for the transitions from the valence band into the $^5T_2$ states (Fig. 4, open circles). Fitting the low energy part of this spectrum with Eq. (1) (Fig. 4, dotted line) the energy of the second threshold (0.79eV) has been obtained giving an energy difference of about 0.36eV between the two threshold energies of the spectrum. Once the values for the two threshold energies are obtained in GaAs:Fe the corresponding values of all other compositions studied are readily calculated from the relative positions of the thresholds (table 1).

Table 1　Summary of some parameters given in this paper

Composition (x)	T measured (K)	Relative position ^5E(eV)	Relative position $^5T_2$(eV)	Threshold energies, E_0 ^5E(eV)	Threshold energies, E_0 $^5T_2$(eV)	Difference of threshold energies (eV)	σ_{pA}* ($\times 10^{-18}$cm^2)
0	128	0	0	0.43	0.79	0.36	2.78
0.07	96	0.03	0.03	0.46	0.82	0.36	2.50
0.20	88	0.09	0.11	0.52	0.90	0.38	0.75
0.26	130	0.12	0.12	0.55	0.91	0.36	0.69
0.34	172	0.14	0.14	0.57	0.93	0.36	0.26
0.49	139	0.23	0.24	0.66	1.03	0.37	0.34
0.56	140	0.24	0.24	0.67	1.03	0.36	0.42
0.76	134	0.35	0.35	0.78	1.14	0.36	0.13

* Data taken 0.1eV above the low energy threshold

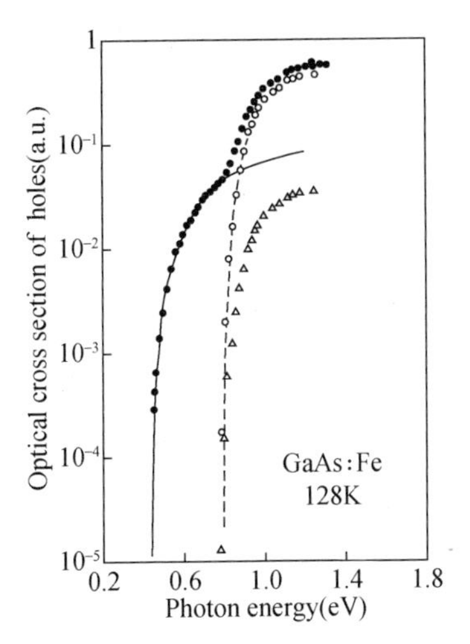

Fig. 4 Optical cross section for holes in GaAs:Fe at 128K. The full circles are experimental data, the solid curve is the fit obtained using Eq. (1), and the open circles show the second branch determined by subtracting the extrapolation of the low energy part of the spectrum from the distribution at higher energies. The broken curve is also a fit obtained using Eq. (1). Shifting the new curve(open circles) downward until the peak value of this curve(triangles) is comparable with the peak value of the low energy part of the spectrum, the splitting between ^5E and $^5T_2$(0.36eV) was obtained

Fig. 4 shows that agreement between the fitted curve and the spectrum of transitions into the $^5T_2$ states could not be obtained for the highest photon energies measured. Since the low energy part of the two spectra has the same shape for both branches the disagreement between the fitted curve and the spectrum at higher energies should have no influence on the evaluation of the second threshold energy. This is readily shown by extrapolating the low energy part of the spectrum presented in Fig. 4 to higher energies and subtracting this part of the spectrum from the distribution at higher energies. Shifting the new curve (Fig. 4, open circles) downward until the peak value of this curve (Fig. 4, triangles) coincides roughly with the peak value of the low energy part of the spectrum, it is found that the two branches form parallel curves close to their threshold energies with constant energy difference of about 0.36eV, in agreement with the first evaluation method. Since the latter one does not depend on any fitting of the spectrum (at higher photon energies) and gives similar results to the first method, it is believed that the relative values given in table 1 for the second threshold energies are not unreasonable.

Since the best spectra have been obtained at low temperatures, the results presented in Fig. 3 have been measured at different temperatures because of the different freeze-out temperatures of the shallow donors in different alloys. In order to be able to compare the spectra measured for different compositions at different temperatures some of the samples have been investigated at several temperatures. Within the temperature range studied, it has been found that the spectra only showed a very weak temperature dependence in the low energy range of the spectral distribution, whereas any change for other energies was within the experimental error. This behavior can be visualized from the data obtained in $Al_{0.49}Ga_{0.51}As$ which are presented in Fig. 5 as an example. Though the phonon broadening in the low energy part of the spectrum is weak it may nevertheless explain why our threshold energies are somewhat smaller than those previously published for lower temperatures and in a more limited range of photoionization cross sections. [5] As mentioned earlier, our values given for the threshold energies ought to be considered only as relative values since most of our samples could not be studied at low temperatures and there is obviously no unambiguous way to deduce binding energies from spectra presented in this paper.

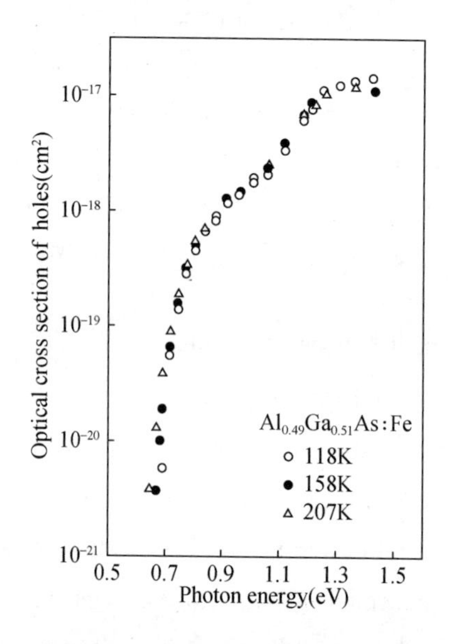

Fig. 5　Absolute values of optical cross section for holes versus photon energy measured at different temperatures

It can be seen from Fig. 3 that the absolute values of the photoionization cross section decrease with increasing x as illustrated in the insert in Fig. 3 where the optical cross sections for a photon energy 0.1eV above the low energy thresholds are plotted as a function of the composition x. These data can only indicate trends since the samples did not have identical geometries and it is therefore possible that the excitation density may vary in different samples. For similar reasons we did not analyze data in Fig. 3 taken at photon energies larger than 1.35eV because at these energies disturbances caused by the GaAs substrates may occur.

No detectable change in the photocapacitance caused by the excitation of electrons from the iron center to the conduction band could be observed (Fig. 6) and, hence, no information on the photoionization cross section of electrons can be given.

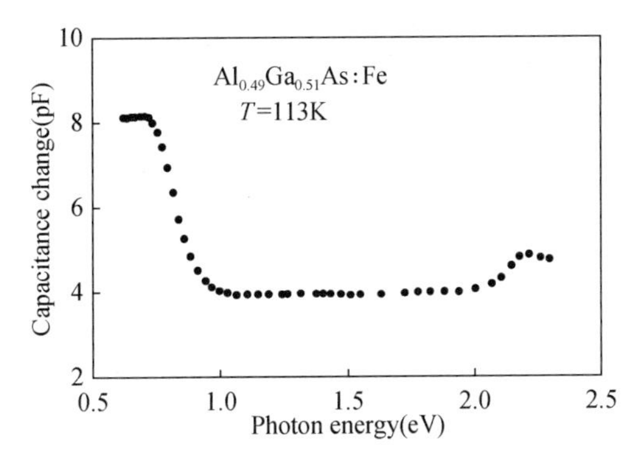

Fig. 6 The steady-state photocapacitance change as a function of photon energy at 113K

4 Discussion

Transition metals in GaAs have been studied previously using optical absorption,[2,4] electron spin resonance,[1] or junction space charge techniques.[5] The spectral distribution of the photoionization cross section for holes observed in iron-doped GaAs is well described by the crystal field theory giving a crystal field splitting of 0.374eV. Optical transitions are likely to occur from the valence band to levels of the ground state ^5E and to the energy states of $^5T_2$ The similarity of our spectra observed for alloys with those obtained for GaAs:Fe suggests that the spectral distribution of the optical cross section in alloys is generated by the same transitions between the valence band and the ground state ^5E and the level $^5T_2$, respectively, i. e. , by different energy levels of the same center. This assignment is supported by the data presented in Fig. 6, and the fact that only one time constant has been obtained in the whole spectral region of the optical cross section. It can also be seen from Fig. 3 that though the threshold energies increase with increasing values of x the shape of the spectra are similar for all samples independent of the composition. Plotting the threshold energies obtained for different values of x (see also table 1) as a function of alloy composition one finds that although both energies increase linearly with x none of them is coupled directly to one of the conduction band minima (Fig. 7). Using deep level transient spectroscopy (DLTS) Lang et al.[10] studied several deep centers in $Al_xGa_{1-x}As$ in the range $0<x<0.38$ including an iron-related level with similar threshold energies as our iron-related center. The authors concluded from their data (see Fig. 7) that the iron center has nearly the same fractional shift with x as the direct band gap of $Al_xGa_{1-x}As$. It has been pointed out previously that the analysis of thermal capacitance transients may not be straightforward since thermal emission and capture processes are often strongly nonexponential in alloys.[16, 17] Non-exponential thermal emission and capture processes have also been observed in our alloy samples. We therefore preferred not to use dark measurements for our studies but employed instead photocapacitance techniques which revealed exponential transients for all compositions. Dark measurements have, however, been performed for comparison. Typical dark capacitance transients are shown in Fig. 8(a) for $x=0.2$ and 0.76. Arrhenius plots of the thermal emission rate for holes obtained from DLTS measurements are seen in Fig. 8(b). Nearly straight lines are obtained. The T^2-corrected apparent thermal activation energies estimated from these plots are 0.59 and 0.72eV which are different from the optical threshold energies of 0.52 and 0.78eV, respectively, obtained in the same samples. It should be noted, however, that the optical thresholds are only relative values and that the thermal data are not very accurate since they have been obtained without taking the temperature dependence of the capture cross section for holes into account. The apparent thermal activation energies obtained for different compositions are shown in Fig. 7 for comparison. These values seem to increase linearly with x and obviously show a different composition dependence than the optical data.

In spite of detailed studies no optical excitation of electrons from the iron center into the conduction band could be detected in any of our samples. Since all measurements were performed in n-type material it has to be assumed that the optical cross section for electrons is very small. These conclusions are in agreement with results previously obtained in GaAs by Mitonneau et al. [18] and Kleverman et al. [5]

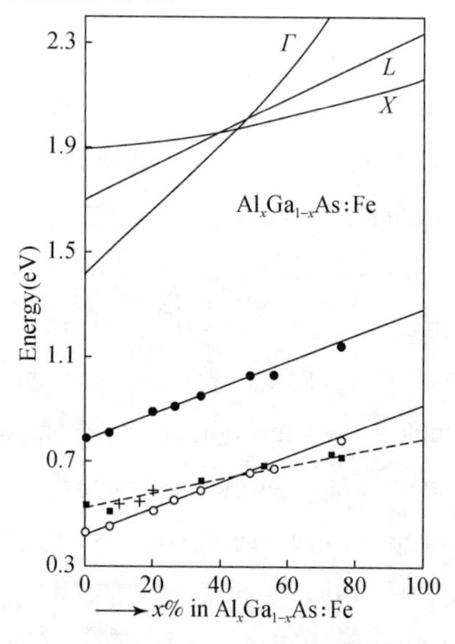

Fig. 7　Optical threshold energies for holes excitation from the gound state ^5E(open circles) and the excited state $^5T_2$(full circles) of the iron center in $Al_x Ga_{1-x}$ As to the valence band as a function of composition x. The conduction band minima Γ, X, and L are also plotted. [19] For comparison + are data taken from Ref. [10] and are results obtained from dark capacitance transients

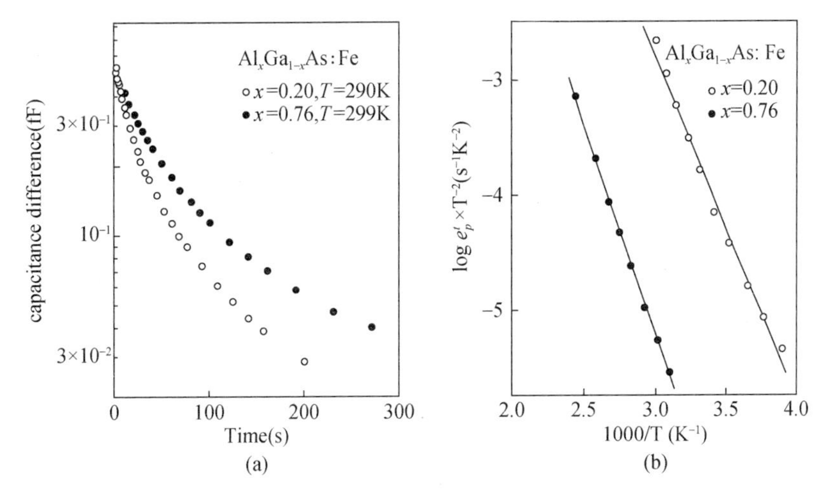

Fig. 8 (a) Dark capacitance transients for $Al_x Ga_{1-x} As$: Fe.

(b) Arrhenius plots of e_p^t/T^2 for similar compositions as in (a)

It is interesting to note that within the accuracy of our measurements the difference in the excitation energies to the ^5E and $^5T_2$ states seems to be similar for all compositions. These observations are in agreement with the crystal field splitting parameters Δ expected in GaAs and AlAs. It has been shown previously by Baranowski et al. [2] that for Fe(d^6) in Ⅱ-Ⅵ compounds Δ decreases as the crystal spacing increases. Since it is not unreasonable to assume a similar behavior for $Fe^{2+}(3d^6)$ in Ⅲ-Ⅴ compounds the energy for the $\Gamma_1(^5E)\rightarrow\Gamma_5(^5T_2)$ transition of $Fe^{2+}(3d^6)$ in GaP, [7] GaAs, [4] and InP[9] has been plotted versus the nearest-neighbor spacing of the host crystal (Fig. 9). A similar dependence to that observed for Ⅱ-Ⅵ compounds is found suggesting that the corresponding transients in GaAs and AlAs should not differ by more than about 5 meV which is within our experimental accuracy.

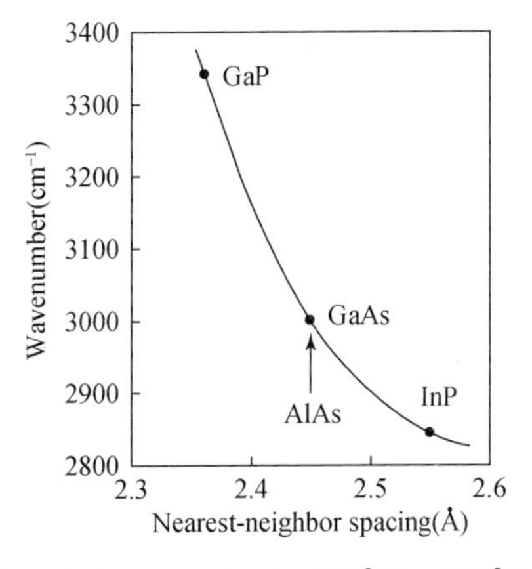

Fig. 9 Variation of the energy for the $\Gamma_1(^5E)\rightarrow\Gamma_5(^5T_2)$ transition in $Fe^{2+}(3d^6)$ with atom spacing of different host crystals

Acknowledgments

The authors would like to thank the Swedish Board for Technical Development for their financial support. One of us (Z-G. Wang) would like to thank the staff and students of the Department of Solid State Physics of the Lund Institute of Technology for their support and hospitality during a visit to Lund. We are also grateful to M. Ahlström for sample preparation and epitaxial growth.

References

[1] M. DeWit and T. L. Estle, Phys. Rev. 132, 195(1963).

[2] J. M. Baranowski, J. W. Allen, and G. L. Pearson, Phys. Rev. 160, 627(1967).

[3] W. H. Koschel, S. G. Bishop, and B. D. McCombe, Proceedings of the 13th International Conference on the Physics of Semiconductors, edited by F. G. Fumi (F. G. Fumi, Roma, 1976), p. 1065.

[4] G. K. Ippolitova and E. M. Omel'yanowskii, Sov. Phys. Semicond. 9, 156(1975).

[5] M. Kleverman, P. Omling, L-Å. Ledebo, and H. G. Grimmeiss, J. Appl. Phys. 54, 814(1983).

[6] V. Kaufmann and J. Schneider, Solid State Commun. 21, 1073(1977).

[7] R. F. Brunwin, B. Hamilton, J. Hodkinson, A. R. Peaker, and P. J. Dean, Solid State Electron. 24, 249(1981).

[8] X. Z. Yang, H. G. Grimmeiss, and L. Samuelson, Solid State Commun. 48, 427(1983).

[9] W. H. Koschel, V. Kaufmann, and S. G. Bishop, Solid State Commun. 21, 1069(1977).

[10] D. V. Lang, R. A. Logan, and L. C. Kimerling, in Proceedings of the 13th International Conference on the Physics of Semiconductors, edited by F. G. Fumi (F. G. Fumi, Roma, 1976), 615.

[11] G. A. Slack, F. S. Ham, and R. M. Chrenko, Phys. Rev. 152, 376(1966).

[12] H. G. Grimmeiss, L-Å. Ledebo, and E. Meijer, Appl. Phys. Lett. 36, 307(1980).

[13] H. G. Grimmeiss and C. Ovrén, J. Phys. E 14, 1032(1981).

[14] H. G. Grimmeiss and N. Kullendorff, J. Appl. Phys. 50, 5852(1980).

[15] H. G. Grimmeiss, L-Å. Ledebo, and C. Ovrén, in Proceedings of the 12th International Conference on the Physics of Semiconductors, edited by M. H. Pilkuhn (B. G. Teubner, Stuttgart, 1974), p. 386.

[16] L. Jansson, Z-G. Wang, L-Å. Ledebo, and H. G. Grimmeiss, Nuovo Cimento 2D, 1718(1983).

[17] P. Omling, L. Samuelson, and H. G. Grimmeiss, J. Appl. Phys. 54, 5117(1983).

[18] A. Mitonneau, G. M. Martin, and A. Mircea, Electron. Lett. 13, 666(1977).

[19] R. Dingle, R. A. Logan, and J. K. Arthur, Jr., Inst. Phys. Conf. Ser. No. 33a, edited by C. Hilsum (The Institute of Physics, London, 1977), 210.

Electronic properties of native deep-level defects in liquid-phase epitaxial GaAs

Zhan-Guo Wang[1], L-Å. Ledebo[2] and H. G. Grimmeiss[1]

([1]Department of Solid State Physics. University of Lund. S-220 07 Lund. Sweden;
Present address: Institute of Semiconductors, Chinese Academy of Sciences, Beijing 100083, China)
([2]Innovance AB, Magle Stora Kyrkogata 8, 5-223 50 Lund. Sweden)

Abstract GaAs grown under Ga-rich conditions often contains two dominant deep levels. sometimes referred to as the A and B levels. We have investigated the electronic properties of these levels in *n*-type and *p*-type material. The apparent activation energies for thermal hole emission were measured over eight orders of magnitude of thermal emission rate. and found to be 0. 40eV for level A and 0. 70eV for level B after the T^2 correction. All four optical cross sections have been determined. The thresholds for electron excitation are 1. 15 and 0. 90eV for the A and B levels. respectively, at 80K. A photocapacitance method is used to demonstrate that the two levels are coupled and thus due to one single defect. The electronic properties, in combination with the conditions under which the defect is formed, lead to the suggestion that the defect may be the Ga_{As} antisite.

1 Introduction

It is often found that GaAs grown by liquid-phase epitaxy contains two impurity levels with about the same concentration, often as high as in the $10^{14}cm^{-3}$ range(Lang 1974, Lang and Logan 1975, 1976 Hasegawa and Majerfeld 1975, Uji and Nishida 1976, Vasudev et al. 1977, Mitonneau et al. 1977, Okumura and Ikoma 1978). The levels were originally referred to as the A and B levels(Lang 1974). Isolated papers have appeared where these levels have also been claimed to be present in vapour-phase epitaxial material(Bois and Boulou 1974, Humbert et al. 1976). The capture coefficients for holes for both levels are high(Lang and Logan 1975), and their presence may be detrimental to minority carrier devices.

The defect identity of the levels has not yet been established. Humbert et al. (1976) suggested that they are related to vacancies. Most of the published measurements have

原载于: J. Phys. C: Solid State Phys. , 1984, 17(2): 259-272.

concentrated on determining the thermal emission rates in order to establish apparent activation energies or signatures of the defects. These investigations thus merely serve the purpose of establishing that levels obtained in different laboratories are the same, but do not contribute towards an understanding of them. To gain further information, we have made measurements for the first time to determine accurately the optical properties of the levels. This is particularly important for an understanding of the capture of free carriers (Pässler 1981, Burt 1982). Previous optical work has been based on simple photocapacitance experiments which are only qualitative, and furthermore cannot distinguish between signals from level A and from level B (Vasudev et al. 1977, Humbert et al. 1976, Bois and Boulou 1974). The interpretation of such investigations is additionally complicated since level A has an activation energy close to that of Cu in GaAs. Furthermore, level B is close in energy to both Fe and *EL*2, the so-called oxygen level. Some of the results previously claimed to be due to A and B may therefore be caused by Cu, Fe or *EL*2, all of which are common deep-level defects in GaAs.

2 Experimental details

The samples used for this investigation were p^+n junctions and n^+p junctions grown by liquid-phase epitaxy. The p^+n junctions were made on an n^+ substrate with a Sn-doped n-layer with a shallow level concentration of about 2×10^{16} cm^{-3}. The top layer was doped p^+ to about 2×10^{18} cm^{-3} using Ge. In the liquid-phase process normally used at our laboratory, the total concentration of deep-level impurities including A and B is below the limit of detection ($<10^{11}$ cm^{-3}). To obtain the A and B levels, the process had to be manipulated either by rapid cooling after the epitaxial growth, or by the addition of about one atomic percent of Ni, Co or Cr to the gallium melts. Upon cooling from about 800 to 100℃ in less than one minute, a concentration of about 1×10^{15} cm^{-3} of A and B was obtained in n-type material, also without the addition of the transition metals. This procedure was however not successful for p-type material, and in no case did we manage to obtain the A and B levels in p-type samples grown at our laboratory. The n^+p junctions were therefore obtained from the Institute of Semiconductors, Chinese Academy of Sciences. Beijing, where a p-type layer without deliberate doping was grown on an n^+ substrate. The shallow-level doping was about 2×10^{15} cm^{-3} and the concentration of A and B was 2×10^{14} cm^{-3}. The n^+ substrate had a silicon doping of about 1×10^{18} cm^{-3}. Evaporation and alloying of ohmic contacts, mounting and bonding were carried out in a standard fashion for both types of diodes.

The experimental techniques used were various forms of well known junction space-charge spectroscopy (Grimmeiss and Ovrén 1981). Some of the methods employed in this work are modifications of previously known techniques. The photocapacitance method used to test whether or not the two levels are coupled has previously been used for the gold donor and acceptor in silicon (Ledebo and Wang 1983). For the optical and thermal transient measurements, we used a cryostat in which the sample was cooled by direct contact with an exchange gas which had a thermal stability of better than 0.1K. The light sources were a Zeiss M3 prism monochromator with high spectral purity, and a Bausch and Lomb grating monochromator for high-intensity light. All measurements were performed on both p^+n and n^+p junctions. No significant differences were found in the properties of the levels in n-type and p-type material with the exception of a slight shift in the absolute values of the thermal emission rates.

3 The measurements

Both deep-level transient spectroscopy (DLTS) and single-shot dark capacitance transients were used to measure the thermal emission rates for holes. When p^+n junctions are used for thermal or optical hole emission studies, care must be taken to minimise electron capture from the free-electron tail subsequent to hole injection (Meijer et al. 1983). We therefore used a special procedure to avoid the spurious capacitance transients caused by refilling from the tail. Taking thermal emission from level B as an example, at each temperature level A was kept filled with electrons by a strong light source of 0.55eV. Level B could then be studied separately. A second light source with photon energy 1.0eV was used to set a certain hole population in level B. After a time sufficient to obtain a steady state, this light source was removed, and the thermal capacitance transient due to hole emission from level B was recorded at different temperatures. The thermal emission rates are proportional to the inverse time constant for this transient, and are shown as a function of inverse temperature with T^2 correction in Fig. 1. The apparent activation energies found are 0.40 and 0.70eV. Data from the literature have also been included in Fig. 1 to demonstrate how important it is to measure the thermal emission rates over many orders of magnitude if an accurate activation energy is to be deduced.

When the experimentally determined concentrations for two levels in a semiconductor are similar, the two levels may be due to the same defect. However, no general method to demonstrate the coupling has been found until recently (Ledebo and Wang 1983), when a photocapacitance method was used to demonstrate that the gold donor and acceptor in silicon are two levels of the same defect. It is difficult to determine concentrations of deep-level impurities accurately, but from our DLTS data it is clear that the concentration of level A in all samples studied is very similar to that of level B. Fig. 2 is a representative DLTS spectrum for a p-type sample with almost identical peak heights. The same method as used for the two

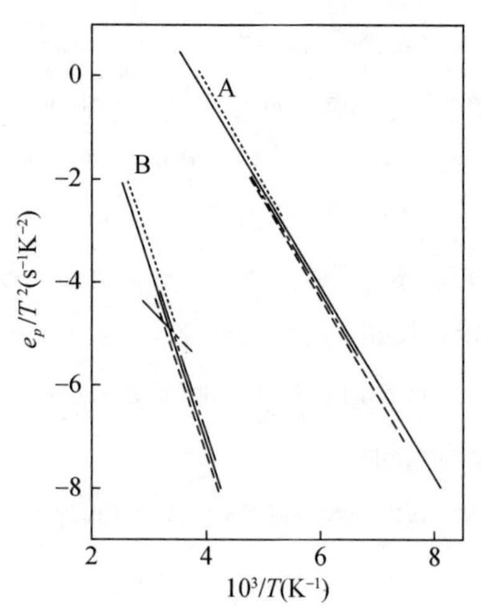

Fig. 1　The thermal emission rate as a function of reciprocal temperature. A, e_{pA}; B, e_{pB}. After T^2 correction the activation energies obtained are $E_A - E_V = 0.40$eV and $E_B - E_V = 0.70$eV. The full curve shows data for n-type material and the broken curve for p-type material. For comparison, previously published data have been included: —·—·, Vasudev et al. (1977); ······, Lang(1974); —··—··, Hasegawa and Majerfeld(1975)

gold levels in silicon was therefore applied to the A and B levels in an n^+p junction to test whether they are coupled or not. The time dependence of the occupancy for the A and B levels under optical excitation with 0.72eV light is seen in Fig. 3. Light with photon energy 0.59eV was used to fill level A selectively, which is necessary for a determination of the occupancies (Ledebo and Wang 1983). The important part of this plot is shown magnified in Fig. 4, where the full curves are the theoretical dependences for coupled levels, and the broken curve is the dependence for uncoupled levels. We consider the data to be understood only if it is assumed that the A and B levels are coupled, and that the two levels thus correspond to three different charge states of the same defect. It is essential that this is known for a proper evaluation of the optical cross sections from photocapacitance transients.

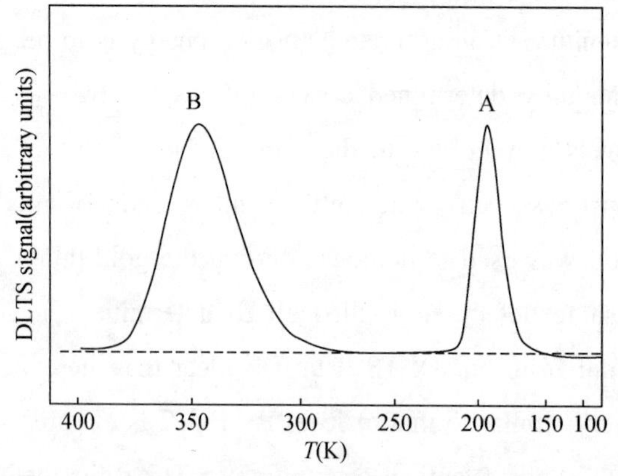

Fig. 2　The DLTS spectrum for the A and B levels in an n^+p sample The rate window was $120s^{-1}$

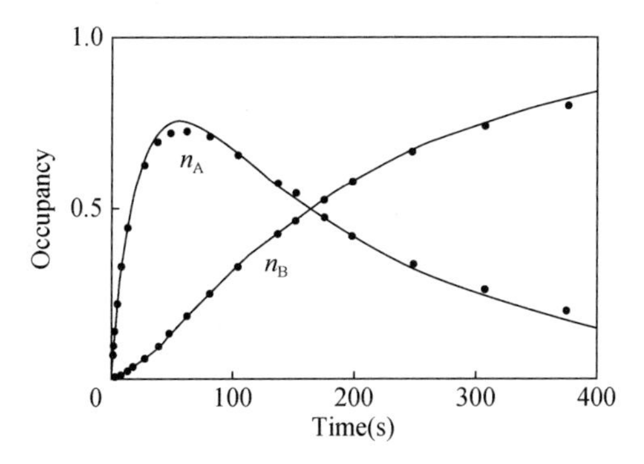

Fig. 3 The time dependence of the occupancies under illumination with 0. 72eV. At time zero both levels are filled with holes. The dots are the experimental data and the full curves are the calculated dependences with the assumption of coupled levels

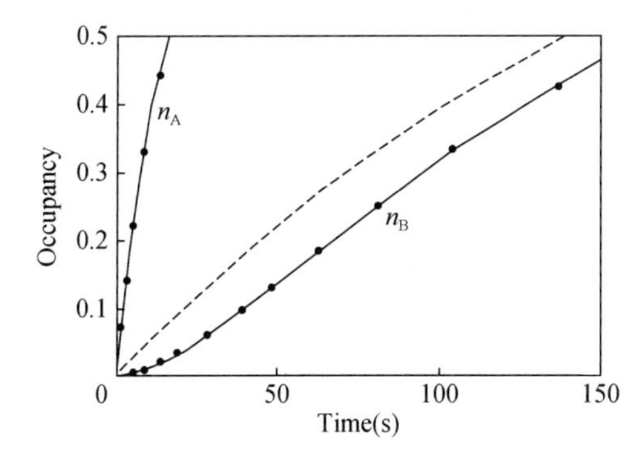

Fig. 4 Part of Fig. 3 magnified. The initial lag in the electron occupancy of level B clearly demonstrates that level A must be occupied before level B can start to be filled with electrons. The full curves are calculated with the assumption of coupled levels. whereas the broken curves are calculated assuming uncoupled levels

The spectral distributions of all four photoionisation cross sections for the two levels are shown in Fig. 5. The optical emission rate of holes e_{pA}^0 was directly evaluated using the time constant of the full transient for photon energies small then $E_B - E_V$. For larger photon energies, holes can also be excited from level B. The full transients were then fitted by a computer program for analysis of multi-exponential transients. With similar results, the transients were also evaluated using the initial slope technique. Since the levels are coupled, the optical emission rate e_{pA}^0 is proportional to the initial slope of the capacitance transient for all photon energies up to the band-gap energy:

$$\left.\frac{dC(t)}{dt}\right|_{t=0} = \frac{C(0)}{2N_S}\left.\frac{d(n_A+n_B)}{dt}\right|_{t=0} = -\frac{C(0)}{2N_S}n_A e_{pA}^0, \tag{1}$$

since $n_B(0)=0$, and where n_A and n_B are the concentrations of levels A and B, respectively, filled with electrons. N_S is the concentration of the shallow acceptor.

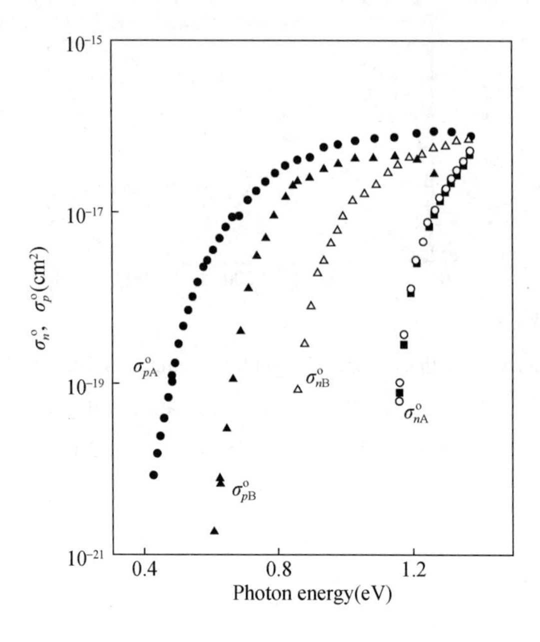

Fig. 5　The photoionisation cross sections at 80K. σ_{nA}^0 has been measured with two different methods as described in the text. which are indicated by the different symbols

For photon energies such that $E_C - E_B < h\nu < E_C - E_A$, both e_{pB}^0 and e_{nB}^0 were calculated by determining the ratio e_{pB}^0 / e_{nB}^0 from the magnitude of the capacitance transient and by measuring the time constant $\tau = 1/(e_{pB}^0 + e_{nB}^0)$. The optical emission rate e_{nB}^0 was also obtained using a dual-light steady-state photocapacitance method (Braun and Grimmeiss 1974). The results from the two different methods were in good agreement.

The measurement of e_{nA}^0 is rather difficult, since this emission rate has the highest threshold energy, and all four optical emissions take place simultaneously. A large reverse bias was applied to minimise capture from free-carrier tails. It was verified that the cross section was not significantly dependent on the electric field in the junction. For each photon energy, a pumping light of 0.55eV was used simultaneously with the probe light of variable photon energy ($E_C - E_A < h\nu < E_g$). When the steady state was obtained the probe light was turned off, and a new steady state was obtained with level A filled with electrons. This procedure does not change the electron occupancy of level B. The measurement is then performed by recording the transient resulting when the 0.55eV light is turned off, and the probe light is turned on. As the final occupancy of level B is already set, the change of the capacitance does not contain any disturbances from level B, and the transient amplitude is proportional to $e_{nA}^0 / (e_{nA}^0 + e_{pA}^0)$. To simplify the evaluation, we measured the initial slope which is proportional to e_{nA}^0. Another way of making the experiment is not to turn off the strong 0.55eV light in the latter step of the measuring sequence. Then the capacitance transient amplitude is simply proportional to e_{nA}^{0} (provided the intensity of the 0.55eV light is high enough). For the results presented in Fig. 5, both methods of measurement were used.

The magnitudes of the capture cross sections are an important clue in determining the charge states of a deep-level defect, since attractive centres are expected to have larger cross sections than repulsive ones. The cross sections for the A and B levels have previously been measured by Lang and Logan (1976) and by Lang (1980), but since their evaluation did not take the coupling into consideration, we found it worthwhile to remeasure the capture cross section.

The capture cross sections for holes are high for both levels, and therefore difficult to measure accurately with zero-bias pulse techniques, since the bias pulses must be as short as a few nanoseconds for the shallow-level doping of our samples. At low temperatures we used instead a photocurrent technique (Partin et al. 1979) in an $n^{+}p$ junction. Both levels were initially filled with holes by a zero-bias pulse. For the measurement of σ_{pA}, a strong light of 0.55eV was used to fill completely level A with electrons. At steady-state conditions the 0.55eV light was removed and the diode was illuminated from the p side with interband light (1.5eV). Since the p layer is thin, by using weakly absorbed light the photocurrent mainly consists of holes. Capture of holes by level A results in an exponential capacitance transient with a time constant τ. As level B is already filled with holes it does not interfere in the measurement. The hole capture cross section was calculated using Eq. (2) (Partin et al. 1979):

$$\sigma_{p}=\frac{qv_{\mathrm{d}}A}{v_{\mathrm{th}}I\tau}, \tag{2}$$

where q is the electron charge, A is the area of the diode, I is the photocurrent caused by the interband light illumination, and τ is the experimentally determined time constant for capture. v_{d} is the hole drift velocity under a high electrical field, and v_{th} the thermal velocity of holes, defined as

$$v_{\mathrm{d}}=\left(\frac{8}{3\pi}\frac{E_{\mathrm{ph}}}{m_{p}^{*}}\right)^{1/2} \tag{3}$$

and

$$v_{\mathrm{th}}=(3kT/m_{p}^{*})^{1/2}, \tag{4}$$

where E_{ph} is the optical phonon energy of 0.036eV and m_{p}^{*} is the hole density-of-states effective mass. Since in equation (2) we have neglected the photocurrent caused by electrons, this equation gives a lower bound for the capture cross section. From the experimental conditions we estimate the corrected value to be less than a factor of two larger.

The hole capture cross section for level B, σ_{pB}, was measured in a way similar to that for σ_{pA}. Level A was kept filled with electrons by continuously illuminating the sample with the strong light source of 0.55eV. Level B was initially filled with electrons by light of 0.8eV, and then allowed to capture holes injected by the interband illumination. Eq. (2) was again used. The electron capture cross sections for both levels were determined using a pulse-train technique (Henry et al. 1973). The results for all four cross sections are shown in Fig. 6 together with previously published data. The two sets of data are in reasonably good agreement. In our measurements we have extended the previous data to lower temperatures which may be important particularly for the interpretation of the hole capture mechanisms.

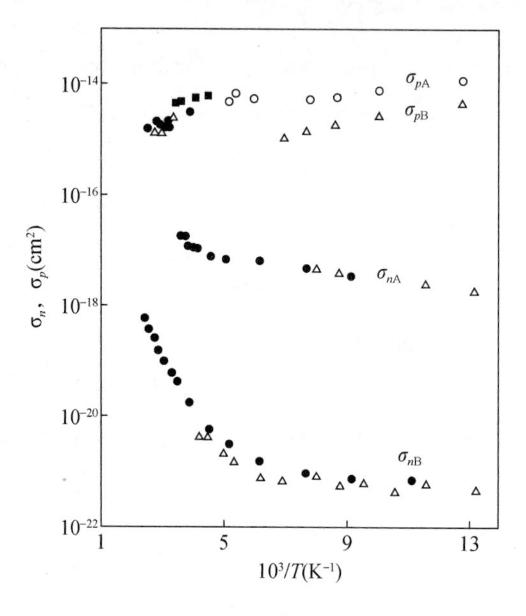

Fig. 6 The capture cross sections for electrons and holes. The open symbols show data from the present work whereas the full symbols show data from Lang and Logan (1976) and from Lang (1980)

4 What is the possible origin of levels A and B?

The A and B levels are present only in liquid-phase epitaxial material, and with pure Ga melts only if the cooling after epitaxial growth is sufficiently rapid. It thus seems that the centre is introduced into the material during epitaxial growth, but anneals out during slow cooling. From studies of deliberate transition-metal doping by liquid-phase epitaxy (Ledebo, unpublished), we can exclude contaminants such as Cr, Fe, Ni and Cu. It is unlikely that any other deep-level impurity is ubiquitous in epitaxial systems all over the world. The defect is therefore likely to be of native origin. The addition of about one atomic percent of Ni, Co or Cr to the Ga melts caused the A and B levels to remain even if the cooling was comparatively slow (about 1h from 800 to 400℃). The transition metals are believed to occupy Ga sites, and their presence may drastically change the equilibria between the various defects possible, in this case stimulating the formation of defects containing extra Ga atoms and preventing the formation of defects involving Ga vacancies. It is important to note that deep levels related to Ni have not been observed in liquid-phase epitaxial GaAs (Kumar and Ledebo 1981). A similar observation is valid for Co in the liquid-phase epitaxial cycle used in this investigation. This, of course, does not exclude the possibility that those impurities are present in the lattice, and signals from optical absorption and emission are readily seen from an internal transition in isolated Ni atoms, presumably substitutional on a Ga site (Ennen et al. 1981).

We have seen single samples, where the Fermi level is controlled by the shallower level A, which indicates that the levels are acceptors. If a Schottky diode made of such material is used in a DLTS measurement, the paradoxical thing happens that the signal from level A almost disappears and only level B is seen, even if the concentrations are equal. The reason is that at the temperature where the peak from level A is expected, the junction does not function since the occupancy of the free-carrier controlling level A cannot follow the capacitance-measuring signal which by necessity is faster than the DLTS rate window. In the few cases where we have seen a much larger signal from level B than from level A, an admittance measurement of the sample confirmed the above interpretation, and gave a level A thermal emission energy which was identical to that found by DLTS in other samples with lower deep-level doping.

A further indication that the levels are acceptors is given by the order of magnitude of the capture cross sections. The capture cross sections for holes are high for both levels whereas the electron capture cross sections are smaller (Fig. 6). This ordering is qualitatively in agreement with the charge designation scheme in Fig. 7, since an attractive potential increases the capture cross section, and a neutral potential gives a larger cross section than a repulsive one. Equivalent simple trends in the capture cross sections are observed for the chalcogens in silicon (Janzén, unpublished), which are considered to be double donors (Grimmeiss and Skarstam 1981). The order-of-magnitude trends also hold for the gold donor and acceptor in silicon (Ledebo and Wang 1983, Wu and Peaker 1982).

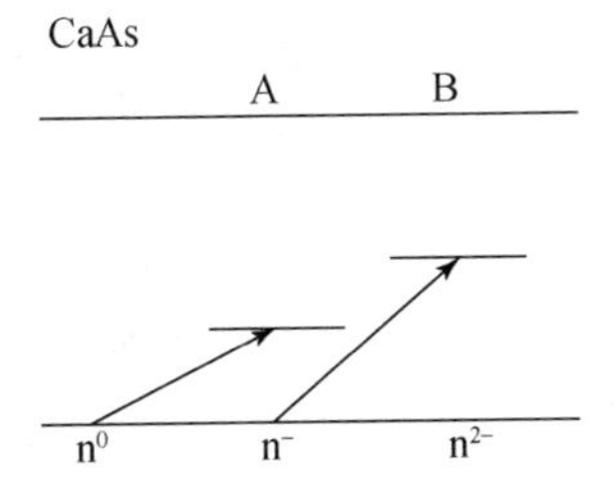

Fig. 7 The charge designation and coupling scheme for the A and B levels as suggested in this paper

The results so far discussed suggest that the defect causing levels A and B is a double acceptor and of native origin. The simplest candidate expected to have these properties is the Ga_{As} antisite defect. The A and B levels are only present in material grown under Ga-rich conditions, and anneal out during slow cooling from 800℃ to 400℃. No thorough investigation of annealing of the Ga_{As} antisite is known to us, but the As_{Ga} antisite has been identified and studied by ESR and found to anneal at about 500℃ (Wörner et al. 1982). Since As and Ga atoms are similar, the energies of formation and the annealing mechanisms are expected to be similar for the As_{Ga} and Ga_{As} antisites (van Vechten 1975). It is therefore likely that Ga_{As} also anneals at temperatures close to 500℃, which is within the experimental range of temperatures for which the A and B levels are removed. All our data are thus well understood by assuming that the defect identity of the A and B levels, is the Ga_{As} antisite defect. By necessity our suggestion for identity is of a speculative nature. In order to explain our data it may well be possible to conceive of other but more complicated defects including the Ga_{As} antisite complexing with various impurities.

Yu et al. (1982) have suggested that two levels at 0.077 and about 0.23eV in meltgrown material are energy states of the Ga_{As} antisite. With a scaling argument based on the effective-mass approximation, the energy values were found to be in agreement with the energy levels of the double acceptor Zn in Ge, which is isochoric with Ga_{As} in GaAs. Since the binding energies are considerably larger than predicted by effective-mass theory, it is however doubtful whether this theory can be applied to a scaling. Furthermore, it was not verified that the two levels are coupled-on the contrary the suggestion of a common origin for the two levels is contradicted by a DLTS spectrum (Yu et al. 1982), in which only one level is seen.

5 The photoionisation cross sections and their temperature dependences

The photoionisation cross sections at 80K are collected in Fig. 6. The threshold energies for electron excitation are approximately 1.15 and 0.90eV for levels A and B respectively. The threshold energies for hole excitation are difficult to estimate because of the comparatively slow variation close to the threshold, but are close to 0.40 and 0.60 eV for levels A and B respectively. Using these values, the sum of the thresholds for hole and electron excitation is 1.50eV for level B and 1.55eV for level A at 80K. This should be compared with the band gap of 1.50eV, demonstrating that the zero-phonon optical matrix element gives a considerable contribution to the photoionisation cross section.

According to the charge designation scheme of Fig. 7, Rydberg-like excited hole states are expected both for levels A and B. When the optical excitation of holes results in a negatively charged centre, such excited states might introduce large temperature dependences in the optical cross sections because of a photothermal process via the excited state (Rosier and Sah 1971). The spectral distributions of σ^0_{pA} and σ^0_{pB} are shown at 40 and 90K in Fig. 8 The cross sections were determined from the time constants of the capacitance transients since the initial-slope technique is sensitive to a possible temperature-dependent filling. It is readily seen that the temperature dependence of level A is much larger than that of level B. In order to characterise this large temperature dependence more carefully, a more detailed measurement was made by varying the temperature for some constant photon energies and determining the time constant of the optical transient. An energy shift of the threshold of 2meVK^{-1} was found, which is much larger than the temperature variation of the band gap, which is 0.25meVK^{-1} in the same temperature region. Such an anomalous temperature dependence may indeed be explained by a photothermal excitation via excited states merging with the continuum part of the cross section. To test this hypothesis, the cross section at a constant photon energy of 0.517eV is plotted as a function of inverse temperature in Fig. 9. The data can be characterised with an activation energy of 25meV, moving towards lower values at the lower temperatures. This is consistent with a deeper excited state at E_V+25meV, and shallower excited states which become rate-limiting at the lower temperatures. The evaluation of the energy of the excited state is inexact because the photothermal contribution is not clearly separable from the continuum part of the cross section, and depending on the photon energy chosen the energy value varies between about 25 and 30meV. In any case the energy value is too large to be explained by effective-mass theory predicting 11meV for the deepest excited state (Baldereschi and Lipari 1973). In our interpretation, a certain amount of central-cell correction would therefore be needed to explain the energy value.

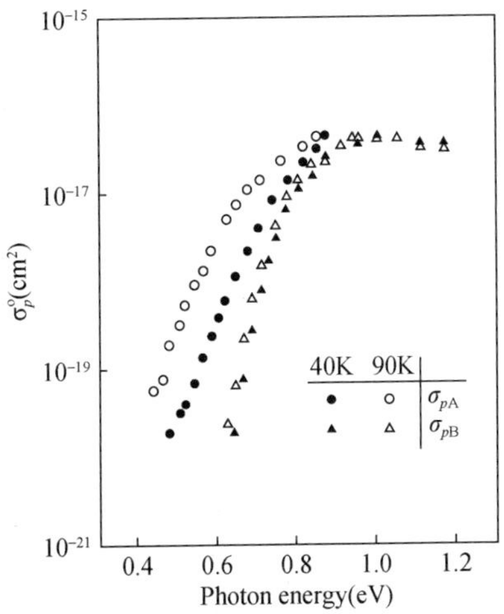

Fig. 8 The hole photoionisation cross sections at 40 and 90K. The temperature dependence for level A is much larger than that for level B

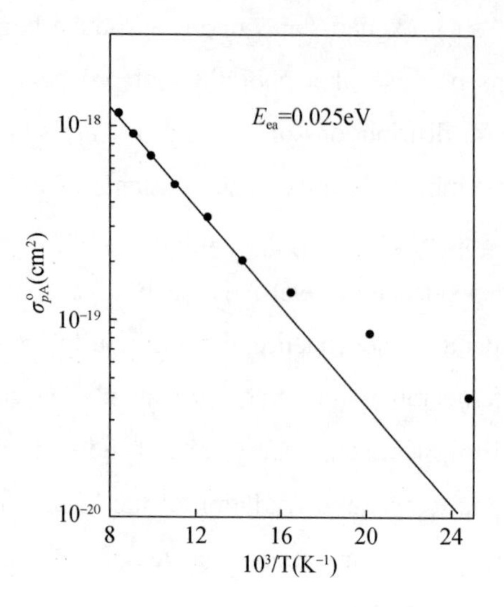

Fig. 9　The optical cross section for hole excitation to level A at 0.517eV as a function of inverse temperature. The straight line corresponds to an apparent activation energy E_{ea} of 25meV

Peaks due to excited states in the cross section for optical hole excitation to level B were not observed. However, since level B is doubly charged, the rate-limiting state may become too deep to be thermally ionised sufficiently rapidly to compete with the transition back to the ground state.

For an understanding of the mechanisms for the capture of electrons, it is of interest to know the temperature dependence of the photoionisation cross sections for electrons since detailed balance should be applicable. The temperature dependence of σ^{0}_{nB} is presented in Fig. 10. A dual-light steady-state photocapacitance method made it possible to perform the

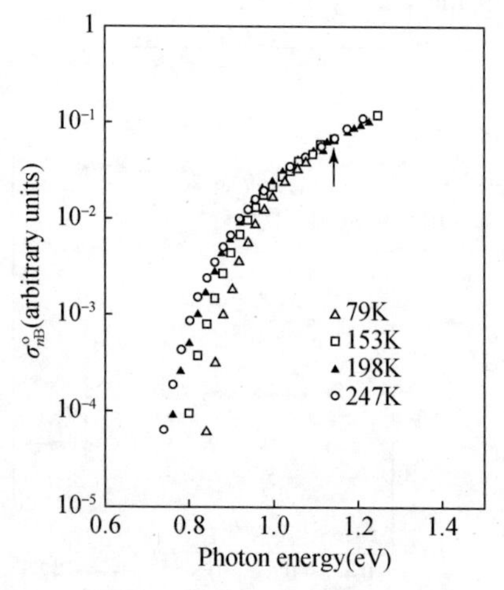

Fig. 10　The photoionisation cross section σ^{0}_{nB} as a function of temperature. The spectra have been normalised at the arrow, and are measured in arbitrary units

measurements at the relatively high temperatures for which thermal hole excitation is active. The cross section can, however, only be determined in arbitrary units in the temperature range of interest, and the spectral distributions have been normalised at 1.15eV. It was not possible to measure σ_{nA}^{0} at higher temperatures, and we can therefore not present any temperature dependence for this cross section.

6 The capture cross sections

The ordering of the capture cross sections as presented in Fig. 6 is in good agreement with the charge designation scheme of Fig. 7 in a simple electrostatic argument. According to this model the hole capture is large and takes place at an attractively charged centre. Thus the mechanism of cascade capture through excited states is probable (Lax 1960, Abakumov et al. 1978, Gibb et al. 1977). The assumption of cascade capture is further supported by the temperature dependence of the capture cross section. Since the probability for re-excitation from an excited state is larger at high temperatures, the cross section decreases with increasing temperature. The quality of the measurement of capture cross sections with the photocurrent technique using Eq. (2) may be questioned, since the capture takes place in an electric field, and the ratio of electron to hole current is not known. The photocurrent capture data however fit fairly well with the data taken by zero-bias pulse techniques at higher temperatures where DLTS can be used, and it is probable that they can be trusted, at least on a qualitative scale. The ratio σ_{pA}/σ_{pB} is consistently close to three for all temperatures. With level B doubly attractive it may be expected that σ_{pB} would be larger than σ_{pA} in contrast with the experimental data. The magnitude of the capture cross section, however, depends on the position and density of the excited states, and the conditions close to the valence band edge are complicated because of the perturbation of the Rydberg-like states by the band structure.

The electron capture by level B has been the subject for extensive theoretical work since the temperature dependence of this cross section readily lends itself to calculations using theory for multiphonon emission. An important parameter in the calculations is the so-called Huang-Rhys factor S, which is a measure of the strength of the coupling to the lattice. Even though we have no detailed model to interpret the spectral distributions of the optical cross sections, it is difficult to understand the general behaviour unless it is assumed that the zero-phonon transitions have large amplitudes. Previous fits to the temperature dependence for capture of electrons to level B have used $S=0.5-1$ (Ridley 1978), $S=3$ (depending on phonon frequency) (Burt 1979), $S=5.9$ (Lang 1980) and $S=7.7$ (Pässler 1978). The higher of these values indicates a strong coupling to the lattice, which is difficult to reconcile with the present optical data.

A difficulty when making fits with multiphonon emission theories is that the radiative part of the recombination must be subtracted from the total capture cross section. Previously no optical data were available, and the radiative part was only estimated on the basis of the magnitude that optical cross sections often have (Lang 1980). By fitting our optical data for σ_{nB}^{0} at 80K with a simple model (Grimmeiss and Ledebo 1975), we obtain a value of the capture cross section of $\sigma_{nB} = 1.44 \times 10^{-20} (T/100)^{1/2} cm^2$ from detailed-balance arguments (Blakemore 1967) (T in degrees Kelvin). In the absence of anything better we assumed the degeneracy factor to be one. The experimental capture cross section is the sum of radiative and non-radiative capture. The radiative capture cross section from detailed balance is however more than one order of magnitude higher than the experimentally determined capture cross section. One may question the applicability of the principle of detailed balance, since the capture is made in neutral material and the optical emission takes place in a depletion region. Before abolishing the principle of detailed balance in this context it is however desirable to carry out more detailed theoretical work on the photoionisation cross section.

7 The thermal activation energies

We find the thermal activation energies for holes to be 0.40 and 0.70eV for levels A and B respectively after the T^2 correction. The standard deviation in the least-squares fit to the data was less than 0.01eV for both levels. The result for level B is remarkable since we not only find that the sum of the hole and electron optical excitation thresholds equals the band gap, but also that the thermal activation energy exceeds the optical threshold energy, all in contrast with the previously assumed strong coupling to the lattice for level B (Lang 1980). A possible explanation for this kind of deviation between thermal activation energy and optical threshold has been suggested for InP:Cu and GaAs:Cu by Kullendorff et al. (1983) in a model with negligible lattice relaxation in connection with the electronic transition. A similar observation has been made for GaAs:Fe (Kleverman et al. 1983). Taken together, the observations in the various systems indicate that strong coupling to the lattice is the exception rather than the rule for defects in Ⅲ-Ⅴ semiconductors.

8 Conclusions

The A and B levels often present in liquid-phase epitaxial GaAs have been found to be two levels caused by the same defect. After having established the coupling, capacitance transient measurement techniques could be modified to measure all four optical cross sections. Based on the compensation behaviour and the order of magnitude of the capture cross sections it is concluded that the defect is a double acceptor, which is only introduced when excess gallium is present during the growth. The simplest defect expected to have these properties is the Ga_{As} antisite, which we suggest is likely actually to be the identity of the defect. The annealing behaviour is in agreement with that expected from independent studies by electron spin resonance of the As_{Ga} antisite (Wörner et al. 1982).

Even though we have not made a detailed fit, it is difficult to understand the general behaviour of the photoionisation cross sections if strong coupling to the lattice is assumed. Furthermore, calculations of the radiative part of the capture cross section σ_{nB} using detailed balance suggest that a substantial fraction of the low-temperature capture cross section is radiative. Those two observations may necessitate a re-evaluation of previous theoretical calculations using multiphonon emission to explain the capture cross section σ_{nB}.

The temperature dependence of the optical cross section σ^0_{pA} suggests the presence of a fairly deep excited state with a binding energy of about 25meV. No indications of excited states are seen in the cross section σ^0_{pB}. The capture cross section for holes to both levels are in agreement with cascade theory lending further support to the assumption of a 25meV excited state associated with level A.

Acknowledgments

Discussions with D V Lang and E R Weber were important for the suggestion of the identity of the defect. We are grateful to M Ahlström for sample preparation and epitaxial growth, and to K Nideborn for assistance with instrumentation. We also gratefully acknowledge the supply of p-type samples by Professor L-Y Lin and the Liquid-Phase Epitaxy Group in Beijing. Comments on the manuscript by B K Ridley, M G Burt and L Samuelson are highly appreciated. This investigation was partly supported by the Swedish Board for Technical Development.

References

[1] Abakumov V N, Perel V I and Yassievich I N 1978 Sov. Phys.-Semicond. 12 1.

[2] Baldereschi A and Lipari N O 1973 Phys. Rev. B 8 2697.

[3] Blakemore J S 1967 Phys. Rev. 163 809.

[4] Bois D and Boulou M 1974 Phys. Status Solidi a 22 671.

[5] Braun S and Grimmeiss H G 1974 J. Appl. Phys. 45 2658.

[6] Burt M G 1979 J. Phys. C: Solid State Phys. 12 4827.

[7] ——1982 J. Phys. C: Solid State Phys. 15 L965.

[8] Ennen H, Kaufmann U and Schneider J 1981 Appl. Phys. Lett. 38 355.

[9] Gibb R M, Rees G J, Thomas B W, Wilson B L H, Hamilton B, Wight D R and Mott N F 1977 Phil. Mag. 36 1021.

[10] Grimmeiss H G and Ledebo L-Å 1975 J. Phys. C: Solid State Phys. 8 2615.

[11] Grimmeiss H G and Ovrén C 1981 J. Phys. E: Sci. Instrum. 14 1032.

[12] Grimmeiss H G and Skarstam B 1981 Phys. Rev. B 23 1947.

[13] Hasegawa F and Majerfeld A 1975 Electron. Lett. 11 286.

[14] Henry C H, Kukimoto H and Merritt G L 1973 Phys. Rev. B 7 2499.

[15] Humbert A, Holland L and Bois D 1976 J. Appl. Phys. 47 4137.

[16] Kleverman M, Omling P. Ledebo L-Å and Grimmeiss H G 1983 J. Appl. Phys. 54 814.

[17] Kullendorff N, Jansson L and Ledebo L-Å 1983 J. Appl. Phys. 54 3203.

[18] Kumar V and Ledebo L-Å 1981 J. Appl. Phys. 52 4866.

[19] Lang D V 1974 J. Appl. Phys. 45 3023.

[20] ——1980 J. Phys. Soc. Japan Suppl. A 49 215.

[21] Lang D V and Logan R A 1975 J. Electron. Mater. 4 1053.

[22] ——1976 J. Appl. Phys. 47 1533.

[23] Lax M 1960 Phys. Rev. 119 1502.

[24] Ledebo L-Å and Wang Zhan-Guo 1983 Appl. Phys. Lett. 42 680.

[25] Meijer E, Grimmeiss H G and Ledebo L-Å 1983 J. Appl. Phys. to be published.

[26] Mitonneau A, Martin G M and Mircea A 1977 Gallium Arsenide and Related Compounds(Edinburgh) 1976(Inst. Phys. Conf. Ser. 33a) 73.

[27] Okumura T and Ikoma T 1978 J. Cryst. Growth 45 459.

[28] Partin D L, Chen J W, Milnes A G and Vassamillet L F 1979 J. Appl. Phys. 50 6845.

[29] Pässler R 1978 Phys. Status Solidi b 86 K39.

[30] ——1981 Phys. Status Solidi b 103 673.

[31] Ridley B K 1978 J. Phys. C: Solid State Phys. 11 2323.

[32] Rosier L L and Sah C T 1971 J. Appl. Phys. 42 4000.

[33] Uji T and Nishida K 1976 Japan. J. Appl. Phys. 15 2247.

[34] Vasudev P K, Mattes B L and Bube R H. 1977 Gallium Arsenide and Related Compounds(St Louis) 1976(Inst. Phys. Conf. Ser. 33b) 154.

[35] van Vechten J A 1975 J. Electrochem. Soc. 122 423.

[36] Wörner R, Kaufmann U and Schneider J 1982 Appl. Phys. Lett. 40 141.

[37] Wu R H and Peaker A R 1982 Solid-State Electron. 25 643.

[38] Yu P W, Mitchel W C, Mier M G, Li S S and Wang W L 1982 Appl. Phys. Lett. 41 532.

Direct evidence for random-alloy splitting of Cu levels in $GaAs_{1-x}P_x$

L. Samuelson, S. Nilsson, Z. G. Wang, and H. G. Grimmeiss

(Department of Solid State Physics, University of Lund, S-220 07 Lund, Sweden;
Present address: Institute of Semiconductors, Chinese Academy of Sciences, Beijing 100083, China)

Abstract A splitting of the Cu-related 1.36eV luminescence in GaAs into a set of lines in the $GaAs_{1-x}P_x$ alloy system is interpreted in a model with only the nearest-neighbor group-V shell influencing the defect levels. Experiental peak intensities are in good agreement with simple calculations and the binding energies of the defects are found to follow calculated chemical trends for defect pairs. Results for the ternary alloys are extrapolated to and compared with the case of Cu in GaP.

In spite of considerable theoretical and experimental efforts,[1,2] the understanding of deep impurities in semiconductors is still far from satisfactory. One approach to investigate such defects systematically has been to use a well-defined host semiconductor, like Si, and study the electronic properties of different impurity centers. Unfortunately, even closely related elements in the periodic table often behave too differently to allow interpretation of trends in their electronic properties. A different approach is to use the same impurity but gradually change the host material. We use this method in order to study the influence of variations in the composition of the host crystal on the properties of a specific deep-impurity center and report, as far as we know, for the first time on resolved randomalloy splitting of a defect due to different local surroundings of the defect core.

One important property of a deep-impurity level which is difficult to investigate experimentally is the extent of the wave function in space. In the one extreme, 4 *f*-shell wave functions of rare-earth impurities are strongly localized. In the other extreme, wave functions of shallow hydrogenlike impurities sample a very large crystal volume, averaging out local fluctuations in, for example, the composition of an alloy. In this paper we are able to estimate the extent of the luminescent Cu-related state in GaAs (the 1.36eV emission line)[3] from a detailed study of photoluminescence lines arising when an increasing number of As

原载于: Phys. Rev. Lett., 1984, 53(15): 1501-1503.

atoms are replaced by P atoms in the $GaAs_{1-x}P_x$ alloy.

Cu in GaAs is thought to occupy a Ga site which results in a defect level responsible for the well known 1.36eV luminescence band.[3] The center seems to be distorted although the exact symmetry of the defect is still under debate.[4, 5] A substitutional Cu_{Ga} center in GaAs is surrounded by four more or less symmetrically oriented nearest-neighbor As atoms, twelve next-nearest-neighbor Ga atoms, etc. In an alloyed ternary material like $GaAs_{1-x}P_x$, each Ga-site defect can be characterized by a great number of neighbor configurations. The number of different surroundings is, however, reduced, if only the first or the first few shells of surrounding atoms are sampled by the defect wave function. In such a case a set of discrete levels, characteristic for each differently alloyed surrounding, may be expected.

If we assume random mixing of the two alloying atoms, in our case As and P, the statistical occurrence of the two types of atoms among N sites is simply given by the binomial distribution,

$$P(m, N-m)=\binom{N}{m}x^{m}(1-x)^{N-m}. \tag{1}$$

This is the probability that, in an alloy with composition x, m of N sites will be occupied by that atom species which appears with probability x, e. g., P in $GaAs_{1-x}P_x$. The expression in Eq. (1) becomes exceptionally simple if the defect is only influenced by the first four nearest neighbors ($N=4$), since then only five configurations can occur, i. e., with $m=0, 1, \cdots, 4$ P atoms adjoining the Cu atom.

The samples used in this study are $GaAs_{1-x}P_x$ alloys grown by metal-organic vapor-phase epitaxy.[6] The composition, x, of each sample studied was determined with microprobe analysis. Into this material, with background shallow doping levels of $1\times10^{15}-1\times10^{16}cm^{-3}$, Cu was introduced by diffusion from an evaporated film of $\approx$ 1000Å Cu on the sample surface. Diffusion was carried out in an evacuated quartz ampoule at temperatures in the range 500–700℃ for 30–100min. Prior to optical studies the samples were mechanically and chemically polished to remove the remaining Cu on the surfaces. Photoluminescence measurements were carried out at 2K with 10–100mW of 5145Å excitation light from an Ar^+-ion laser, with a detection system including a 0.75cm double-grating monochromator and an S1 photomultiplier.

Photoluminescence spectra of Cu-doped GaAs, $GaAs_{0.95}P_{0.05}$, $GaAs_{0.89}P_{0.11}$, and $GaAs_{0.80}P_{0.20}$ are shown in Fig. 1. The $x=0$ spectrum shows band-edge emission lines around 1.5eV consisting of excitonic as well as free-to-bound and donor-acceptor luminescence lines. The Cu-related luminescence band appears with a no-phonon line at 1.36eV and LO-phonon replicas on the low-energy side. With increasing P content the band-edge emission shifts to higher energies following the Γ band gap.[6] Similarly, the Cu-related peak occurring at 1.36eV in GaAs shifts with increasing x to higher energies. However, the energy position relative to the band-edge luminescence peaks increases with x which means that the $E_A \approx 0.15$eV Cu level in GaAs becomes deeper in the alloy. The marked feature of the ternary-alloy system is clearly found to be the appearance of new luminescence lines on the high-energy side of the 1.36eV Cu line. Two peaks can be clearly resolved, one shifted by $\approx$ 35meV and the other by $\approx$ 70meV towards higher energy. The $\approx$ 35meV satellite is found to appear first as x is increased from zero. For $x \geqslant 0.2$, individual luminescence lines can no longer be resolved. It is, anyway, obvious that more peaks appear on the high-energy side when x is increased.

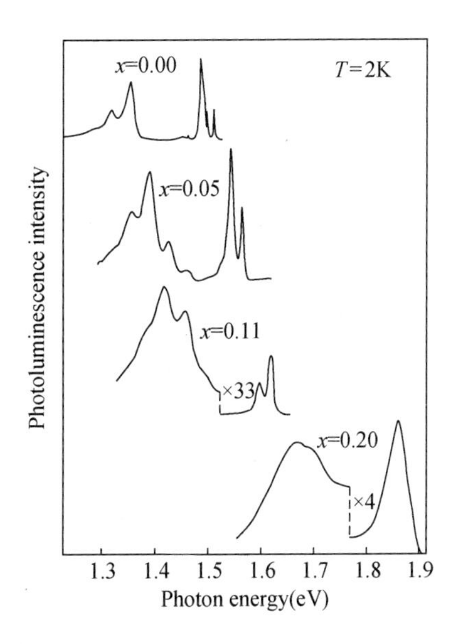

Fig. 1 A set of low-temperature photoluminescence spectra of Cu-doped $GaAs_{1-x}P_x$ obtained for four different values of x

If we tentatively assign values of $m = 0, 1, \cdots$ to the series of peaks starting with $m = 0$ for the 1.36eV peak in GaAs, we can compare calculated occurrences of $CuAs_{4-m}P_m$ molecules according to Eq. (1) with the measured spectra. The results are shown in Fig. 2 where each of the spectra for $x>0$ has been shifted in energy to make the shallow acceptor peaks coincide. The agreement between the measured and calculated peak intensities is quite satisfactory and may indicate that the different defect levels have similar radiative probabilities. We thus conclude that the observed structure of Cu-related emission bands in $GaAs_{1-x}P_x$ can be explained by a simple model where each composition of the four-atom nearest-neighbor shell is assumed to give rise to one type of energy level. Since the energy levels observed appear to be governed in zeroth approximation by the composition of the nearest-neighbor shell, we conclude that the impurity wave function is effectively limited to the volume inside the next-nearest group-V shell. The relatively strong broadening of the Cu luminescence peaks in the alloy may indicate that the hole wave function does have a nonnegligible amplitude even at more distant neighbors. The remaining differences between calculated and observed peak intensities are, at least partly, due to overlap between the different phonon-broadened contributions.

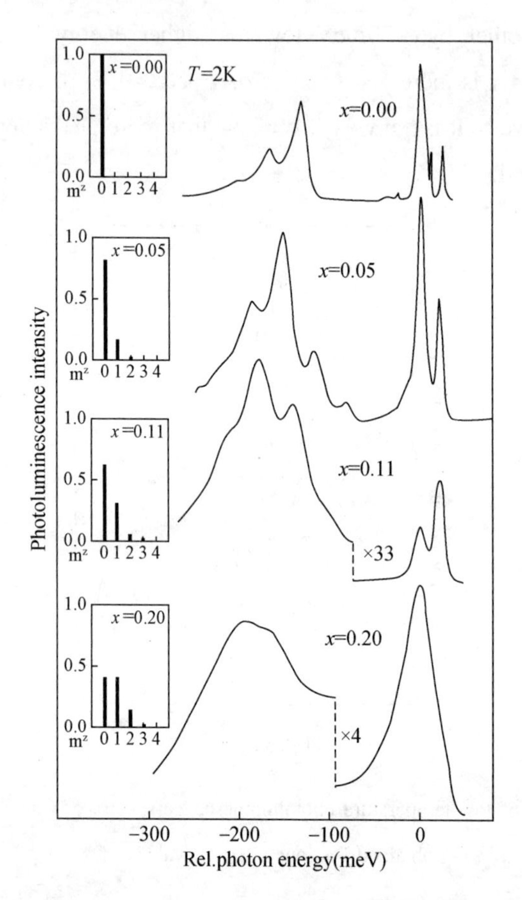

Fig. 2　Comparison of the calculated statistical occurrence (the inset) of different $CuAs_{4-m}P_m$ molecules, Eq. (1) with $N=4$, and the measured photoluminescence spectra for various alloy compositions x in $GaAs_{1-x}P_x$. The spectra have been shifted in energy to make the shallow acceptor peaks coincide. The identification of the dominant band-edge emission for $x=0.20$ as being free to bound is made from a comparison with undoped reference samples where free to bound and excitonic peaks are clearly resolved

Within the framework of this model the observed energy position of the Cu levels for each composition is plotted in Fig. 3. Obviously, for GaAs only the $E_A^0(x=0)\approx 0.15$eV level can occur. The binding energy of this level relative to the valence-band edge is found to vary approximately linearly with x, as

$$E_A^0(x)\approx E_A^0(x=0)+0.43x. \tag{2}$$

Each of the states occurring for $x>0$ and corresponding to $m=1$, 2, 3, and 4, $E_A^m(x)$, is displaced to lower energies by $m\Delta E$ (where $\Delta E \approx 35-40$meV), with a slope approximately equal to that for the $m = 0$ level. For a comparison with GaP data we can estimate the expected position of the $m=4$ level for $x=1$, which should occur at

$$E_A^4(x=1) \approx E_A^0(x=0) - 4\Delta E + 0.43 \approx 0.44\text{eV}$$

from the valence-band edge. The close agreement with the accepted binding energy of the radiative Cu level in GaP, $E_A \approx 0.5$eV,[7] is an additional support of our defect identification. It has so far not been possible to determine the identity of this Cu-related level in GaP, although it has been suggested that it might be due to substitutional Cu on a Ga site. This assumption is thus supported by the present investigation where the luminescence due to substitutional Cu_{Ga} in GaAs has been traced through the $GaAs_{1-x}P_x$ alloy to GaP.

The data and interpretation presented above can be compared with theoretical estimates of chemical trends in the binding energies of substitutional defect pairs.[8] Theory suggests that Cu_{Ga} produces a dangling-bond-like T_2 acceptor level above the valence-band edge.[9] Chemical trends in T_2 energies of substitutional defect pairs calculated in Ref. [8] can be used to estimate the expected change in binding energy when a simple Cu_{Ga} defect is transferred into a(Cu_{Ga}, P_{As}) defect pair in GaAs. Because of the difference in p potentials of As and P, the acceptor level of the defect pair(Cu_{Ga}, P_{As}), $E_A^1(x=0)$, is expected (Fig. 2 of Ref. [8]) to be positioned closer to the valence-band edge than that of isolated Cu_{Ga}, $E_A^0(x=0)$. The qualitative agreement with the binding energies of different defect configurations in $GaAs_{1-x}P_x$ extrapolated to GaAs as deduced in Fig. 3 lends further support to our interpretation of the data.

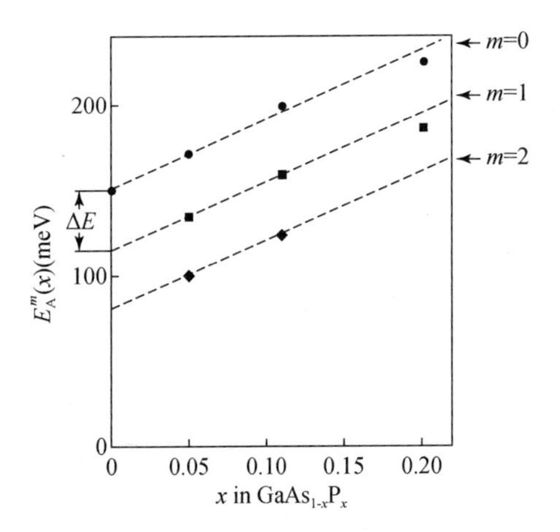

Fig. 3 Estimated energetic positions relative to the top of the valence band, $E_A^m(x)$, of the different Cu defects vs the alloy composition x in $GaAs_{1-x}P_x$. $E_A^m(x)$ is the binding energy for Cu_{Ga} surrounded by mP atoms and $4-m$ As atoms

We want to thank P. Vogl and A. Baldereschi for useful discussions. This work was supported by the Swedish Natural Science Research Council and the Swedish Board for Technical Development.

References

[1] S. T. Pantelides, Rev. Mod. Phys. 50, 797(1978).

[2] H. G. Grimmeiss, Annu. Rev. Mater. Sci. 7, 341(1977).

[3] H. J. Queisser and C. S. Fuller, J. Appl. Phys. 37, 4895(1966).

[4] F. Willmann, M. Blätte, and H. J. Queisser, Solid State Commun. 9, 2281(1971).

[5] N. S. Averkiev, T. K. Ashirov, and A. A. Gutkin, Fiz. Tekh. Poluprovodn. 17, 97(1983)[Sov. Phys. Semicond. 17, 61(1983)].

[6] L. Samuelson, P. Omling, H. Titze, and H. G. Grimmeiss, J. Phys. (Paris), Colloq. 12, C5-323 (1982).

[7] H. G. Grimmeiss, B. Monemar, and L. Samuelson, Solid State Electron. 21, 505(1973).

[8] O. F. Sankey and J. D. Dow, Appl. Phys. Lett. 38, 685(1981).

[9] L. A. Hemstreet, Phys. Rev. B 22, 4590(1980).

Acceptor associates and bound excitons in GaAs:Cu

Z. G. Wang

(Department of Solid State Physics, University of Lund, S-220 07 Lund, Sweden;

Present address: Institute of Semiconductors, Chinese Academy of Sciences, Beijing 100083, China)

H. P. Gislason and B. Monemar

(Linköping University, Department of Physics and Measurement Technology, S-581 83 Linköping, Sweden)

Abstract The effect of several parameters in the diffusion conditions for copper into GaAs on the 1.36eV photoluminescence (PL) band and the C_0 (1.5026eV) and F_0 (1.4832eV) bound excitons (BE) has been investigated using photoluminescence techniques. Also, the different behavior of the corresponding centers under heat treatment and cool down after diffusion is reported. The results are discussed in relation to the vast literature on the GaAs: Cu, which is confusing on the relation between the 1.36eV PL band, C_0 and F_0 bound excitons and the well-known 0.156 and 0.45eV acceptors. Here it is concluded that there exists no simple relation between the C_0 BE and the 1.36eV PL band nor is the F_0 BE related to any other Cu-related PL band in our samples. Models for the identities of these centers are suggested on the basis of the new experimental data presented in this paper.

1 Introduction

Being one of the most frequent contaminants in GaAs, Cu has been the subject of numerous investigations for at least two decades. Despite extensive effort, no satisfactory models for the identities of even the most common Cu-related centers have been presented yet. Instead the literature is extremely incoherent on this subject.

The introduction of Cu in GaAs gives rise to a number of complex Cu associates, i. e., complexes of Cu and other impurities or lattic defects.[1] The two best known Cu centers are the acceptor levels at 0.156 and 0.45eV which have been studied by electrical measurements.[2-4] In early work these levels were attributed to the two ionization levels of the double Cu_{Ga} acceptor.[2, 5] This interpretation was suggested on the basis of the observation of double compensation of n-type material upon Cu doping.[2]

Cu_{Ga} is indeed expected to be a double acceptor from valence bond considerations. No

原载于: J. Appl. Phys., 1985, 58(1): 230-239.

clear evidence is available, on the other hand, for the assignment of the 0.156 and 0.45eV acceptor levels to the two ionization levels. On the contrary, studies of corresponding Cu levels in $Al_xGa_{1-x}As$ and InP suggest that the concentrations of the centers causing these levels are different.[4] The double acceptor interpretation seems to be widely accepted in the literature, however. A number of other possibilities for the identities of the two Cu acceptors have been suggested. They include, for the 0.156 and 0.45eV levels, respectively: Cu_{Ga} and $Cu_{Ga}V_{As}$[3, 6-8] as well as $V_{As}Cu_{Ga}V_{As}$ and $Cu_{Ga}V_{As}$.[1]

Furthermore, there is a persistent source of confusion in the literature regarding the relation of the bound exciton lines C_0 at 1.5026eV and F_0 at 1.4832eV to the two mentioned acceptor levels. These BE lines have been studied in some detail ever since the original experimental discover.[9] Piezospectroscopic studies showed the anisotropy of the Cu centers binding the excitons and led to the assignment of a trigonal symmetry to the C center (binding the C_0 exciton) and an orthorhombic symmetry to the F center (binding the F_0 exciton).[10] Models for the identities include $Cu_{Ga}V_{As}$ and $V_{Ag}Cu_{Ga}V_{As}$[11] for the C center and $Cu_{Ga}V_{As}$ and $Cu_{Ga}V_{As}Cu_{Ga}$[12] for the F center.

Unfortunately, by tradition these excitons have been tentatively attributed to the two main acceptor levels introduced by Cu. The C_0 line has been assumed to be bound to the 0.156eV acceptor and the F_0 line to the 0.45eV acceptor.[1, 13, 14] In retrospect, however, it turns out that it has never even been properly questioned whether these excitons are really bound to acceptors or to some neutral isoelectronic associates.

A contributing factor to the association of the C_0 line to the 0.156eV acceptor is the interpretation of an absorption spectrum of GaAs: Cu.[13] Fine structure in the photoionization cross section of holes, corresponding to the 0.156eV level was interpreted as having trigonal symmetry.

Thus both the C center and the 0.156eV acceptor were assumed to have the same symmetry and hence the centers were considered identical.[13] Recent piezospectroscopic investigations[15, 16] on the center giving rise to the 1.36eV PL band in Cu-doped GaAs suggest a lower symmetry, however.

It is generally accepted that the 1.36eV photoluminescence band represents radiative transitions between the conduction band or shallow donors and the 0.156eV Cu acceptor. We shall support this assignment below. Similarly, a PL band peaking at 1.02eV is sometimes ascribed to the 0.45eV acceptor (Ref. [1] and references therein). We shall give evidence against this view from simple binding energy considerations.

We shall also give evidence that there exists no simple relationship between the 0.156eV acceptor (via the 1.36eV PL band) and the C center via its bound exciton. Nor is the F

center related to a PL band at about 1eV. This is deduced from our investigation which is based on doping procedure, relative intensities, and annealing behavior employing widely different bulk and epitaxial GaAs starting materials. Results from annealing behavior of Czochralski-type GaAs material in an independent study similarly indicate the absence of relationship between the C_0 and 1.36eV emissions, in agreement with our results.[17]

In Sec. Ⅱ we describe the doping procedure which provides the main evidence for the hypotheses arising from this work. We also describe briefly the experimental techniques employed. In Sec. Ⅲ we give the experimental results which are discussed in the discussion section, Sec. Ⅳ. Here we also relate our findings to the various existing models for the identity of the Cu centers. The models we put forward are not to be regarded as definite with respect to microscopic details, however.

2 Experiment

2.1 Doping procedure

In order to avoid systematic errors in the doping procedure a wide range of different starting materials were used in this study. These materials are listed in table 1 which also summarizes the different Cu-diffusion conditions and the most important results.

The starting materials include bulk material of the liquid-encapsulated Czochralski (LEC) and the horizontal Bridgman (HB) types. The former was undoped semi-insulating (SI) material, while the latter was *n*-type through Te doping. Also *p*-type Zn-doped HB material was used for the Cu doping. The high-purity epitaxial layers were grown on either Cr-doped semi-insulating or Te-and Si-doped substrates, giving similar results. Both *n*-and *p*-type liquid-phase-epitaxial (LPE) and vapor-phase-epitaxial (VPE) wafers were used.

The Cu-diffusion procedure was as follows. First a thin film (about 600 Å) of Cu was evaporated onto the polished, etched sample surface. This is the Cu source in the diffusion step. The evaporated samples were immediately sealed in evacuated high-purity spectrosil quartz tubes. All quartz tubes employed in this work were cleaned for at least 2h in a HNO_3 : HF : H_2O mixture prior to use. Annealing of a GaAs reference sample in an ampoule from such a tube at 800℃ for 1h produced no PL bands related to Cu. Therefore, Cu contamination from the quartzware can be excluded in the present study.

Sometimes a Ga-GaAs melt was included in the ampoule (see table 1). Typical As vapor pressures at the diffusion temperature for vacuum annealing and Ga-GaAs melt annealing, respectively, can be estimated from the volumes of ampoule, sample, and melt.[18, 19] With the

Table 1　Electrical and optical properties of some representative GaAs samples before and after Cu diffusion

Sample notation	Material and dominating impurity	Epilayer thickness (μm)	Before Cu diffusion			After Cu diffusion	
			n_{300} (cm^{-3})	μ_{300} ($cm^2/V \cdot s$)	Cu diffusion conditions	Conductivity type	Photoluminescence bands
HB 1	Horizontal Bridgman Te doped		1×10^{16}	2690	800℃, 30min, vacuum	p	Strong 1.36eV PL Weak C_0 and F_0 BE See Fig. 1(a)
LEC 1	Czochralski semi-insulating undoped		1.8×10^{12}	4200	850℃, 60min, melt	p	Strong 1.36eV PL and C_0 BE. No F_0 BE. See Fig. 1(b).
LEC 2(1)					750℃, 62min, vacuum	p	Strong F_0 BE. Weak 1.36eV PL. No C_0 BE
LEC 2(2)					Etched 20μm	p	Strong C_0 and F_0 BE. Rather strong 1.36eV PL
LEC 3					800℃, 30min, vacuum	p	Strong F_0 BE and 1.36eV PL. Weak C_0 BE
LEC 4					700℃, 20min, vacuum	p	Strong F_0 and C_0 BE. Rather strong 1.36eV PL
LEC 5					520℃, 40h, vacuum	p	Strong 1.36eV PL and shallow DAP. Weak C_0 BE No F_0 BE
LPE 81623	LPE wafer undoped	28.5	1.05×10^{15}	8300	700℃, 25min, vacuum	p	Strong C_0 BE. Weak F_0 BE. Very weak 1.36eV PL. Rather strong DAP. See Fig. 1(d)
LPE 81516		27	$P_{300}=4.65\times10^{14}$	401	730℃, 36min, melt	p	Strong C_0 BE and 1.36eV PL. No F_0 BE
LPE 80756-1		24	2.3×10^{14}	8690	750℃, 36min, melt	p	C_0 BE and 1.36eV PL. No F_0 BE
LPE 80756-2		24	2.3×10^{14}	8690	625℃, 2h, melt	p	C_0 BE and 1.36eV PL. No F_0 BE
LPE 80763		33	2.6×10^{13}	7775	750℃, 36min, melt	p	C_0 BE and 1.36eV PL. No F_0 BE
VPE 8062	VPE wafer	40	not measured		625℃, 1h, melt	p	C_0 BE and 1.36eV PL. No F_0 BE
VPE 120	VPE wafer undoped	18.2	3×10^{14}	not measured	750℃, 20min, vacuum	p	Strong C_0 and F_0 BE. Weak 1.36eV PL. See Fig. 1(c)
VPE 037 VPE 028		130 86	3×10^{14}	not measured	850℃, 18min, melt	p	C_0 BE and 1.36eV PL. No F_0 BE
						p_{300}(cm^{-3})	μ_{300}($cm^2/V \cdot s$)
HB 2	Horizontal Bridgman		1×10^{16}	2690	850℃, 12h	9.9×10^{16}	242
HB 3			1×10^{16}	2690	850℃, 1h	2.3×10^{16}	272
LEC 6	Czochralski semi-insulating undoped		1.8×10^{12}	4200	750℃, 20min	4.5×10^{16}	365

melt present the equilibrium pressure P_{As_2} in the ampoule is estimated to be about 10^{-6} atm, which should be sufficient to suppress the As vacancy formation at the surface of the GaAs sample. The equilibrium arsenic pressure eventually reached in vacuum anneal is about 10^{-5} atm[18] which would be sufficient to create an average As vacancy concentration $V_{As}>10^{16}cm^{-3}$, probably larger than the original V_{As} concentration in the GaAs samples.[20] Significant differences between vacuum annealing and melt annealing will be reported below. The Cu diffusions were carried out over various periods of time, ranging from a few minutes to 40h. The temperature range was between 500 and 850℃ as summarized in table 1.

A careful investigation on the effect of different cooldown processes on the Cu-related spectra under study were also made. Therefore the samples were sometimes left in the furnace to cool down slowly to room temperature. The other extreme was to quench the ampoules rapidly to liquid nitrogen temperature. Most often we quenched the samples to room temperature in water, however.

After Cu diffusion the usual cleaning procedure includes removal of residual Cu from the surface by polishing with a diamond polishing compound (1μm grains). Afterwards the samples were etched in a solution of a few drops of bromine in $10cm^3$ methyl alcohol for a few seconds to 1min. This procedure was made as shortly before the photoluminescence measurements as possible. Between measurements the Cu-diffused samples were kept in the dark and immersed in liquid nitrogen to avoid degradation of the Cu spectra which otherwise occurred, as reported below.

2.2 Experimental techniques

The samples were mounted in an immersion cryostat for the photoluminescence measurements and excited with the 5145Å line of an Ar^+ laser. The measurements were made at He temperatures below the lambda point, typically 2K. The photoluminescence was dispersed in a 0.75m Jarrell Ash double grating monochromator and detected with an *S* 1 photomultiplier cooled with dry ice (RCA 7102) or a North Coast Optics Ge detector. For data acquisition a Nicolet 1174 Signal Averager was used.

Conductivity and Hall measurements at 0.4 T in the van der Pauw configuration were performed either at room temperature or in the temperature interval 75-340K. The atomic contacts were alloyed to the samples using In and In-Sn alloy for *n*-and *p*-type GaAs, respectively.

Some typical samples were measured before and after Cu diffusion. The carrier concentration and mobility for these samples are listed in table 1.

3 Experimental Results

3.1 Photoluminescence measurements

Fig. 1 illustrates the different types of starting material employed by the PL spectra of four samples from table 1. These samples are a representative selection from the numerous samples investigated. In Fig. 1(a) the HB bulk material shows a strong 1.36eV peak after Cu diffusion at 800℃/30min, without Ga melt in the ampoule. Both the C_0 and F_0 BE lines at 1.5026 and 1.4832eV, respectively (at 2K), are weak in this sample. A prominent broad

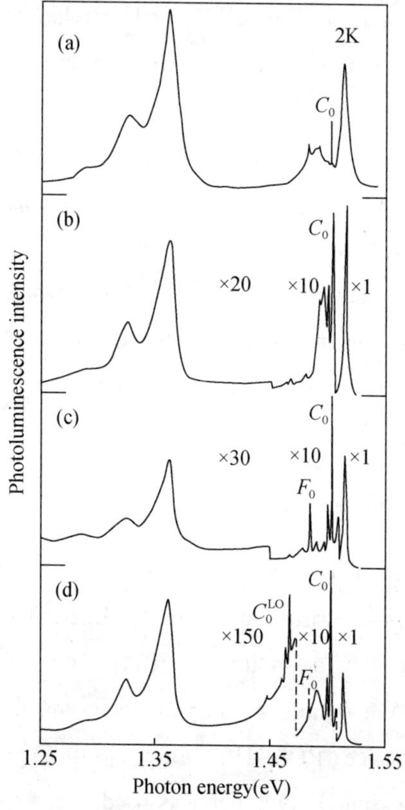

Fig. 1　Photoluminescence spectra of Cu-doped GaAs at 2K obtained for different types of starting material. (a) Horizontal Bridgman bulk material Cu diffused at 800℃ for 30min in vacuum. (b) Semi-insulating Czochralski-type bulk material, Cu diffused at 850℃ for 60min with Ga-GaAs melt in the ampoule. (c) Vapor-phase-epitaxial wafer, Cu diffused at 750℃ for 20 min in vacuum. (d) Liquid-phase-epitaxial wafer, Cu diffused at 700℃ for 25min in vacuum. In all cases (a)–(d) the samples were rapidly quenched in water after the diffusion. Note the presence of the 1.36eV PL band in all samples and the strong C_0 BE line with local mode phonon replicas in (b), (c), and (d). The F_0 BE is absent in (b), strong in (c), and rather weak in (d)

band peaking around 1.20eV in the reference sample (possibly due to a V_{Ga}-D complex)[21] disappears after the Cu-diffusion step.

The LEC semi-insulating sample of Fig. 1(b) (Cu diffusion 850℃/60min, with Ga melt) is the only one of the samples in Fig. 1 diffused with Ga melt in the ampoule. Here both the C_0 line and the 1.36eV band are strong. The F_0 line is absent as a consequence of the presence of the Ga melt as discussed below. No deeper PL bands are consistently present down to 1.0eV, which applies to all samples shown in Figs. 1(a)-1(d). A faint band peaking at 1.04eV has occasionally been observed, however.

The VPE wafer of Fig. 1(c) (Cu diffusion 750℃/20min, without Ga melt) shows strongly the C_0 and F_0 lines and a rather strong 1.36eV peak. Broad PL bands around 1.22eV in the reference samples are not affected by the Cu diffusion.

Finally, the LPE wafer of Fig. 1(d) (Cu diffusion 700℃/25min, no Ga melt) shows a strong C_0 line, a weak F_0 line, and a relatively weak 1.36eV band. This sample, which in fact is a p^+n layer, was measured down to 0.89eV without any traces of other PL bands. An n^+p diode, on the other hand, showed rather strong PL peaks around 0.96 and 1.2eV, which were unaffected by the Cu doping.

The spectrum of Fig. 2 was produced with Cu diffusion at 850℃ for 20min with Ga melt in the ampoule. A prominent C_0 line is observed after such a diffusion. The 3.6meV phonon replica C_1 is also strong. The LO replicas of the C_0 line and its C_1 replica are clearly seen in the spectrum, displaced 36meV towards lower phonon energies. The C_0 line can easily be produced strongly in all types of starting material (most weakly in the HB bulk material) at temperatures between 650 and 850℃.

It seems to be impossible to obtain the F_0 line under the circumstances described above. If the Ga melt is instead omitted, the F_0 line is easily produced in the otherwise similar doping procedure which produces the C_0 line. An example of this is illustrated in Fig. 3, where the only significant difference from Fig. 2 is the absence of Ga melt in the ampoule. Here a rather strong F_0 line appears after Cu diffusion at 730℃/32min. The previously reported 5.1meV replica F_1, and the LO replica of the F_0 line are also clearly observed in the spectrum. The C_0 line is strong in this case as well as in Fig. 2, (it is always stronger relative to the band-edge emission after vacuum diffusion). Three local mode replicas of C_0 labelled C_1, C_2, and C_3 are observable in the well-resolved spectrum. The third replica C_3 has to our knowledge not been reported previously. The same is true for the LO replicas C_0^{LO}, C_1^{LO}, F_0^{LO} and F_1^{LO}.

Increasing the diffusion temperature (and in general the diffusion time) the F_0 line can be made to dominate the luminescence close to the band edge. The spectrum in Fig. 4 was

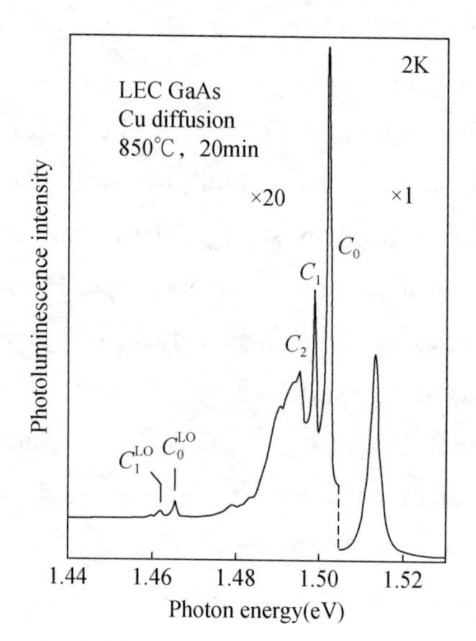

Fig. 2　Low-temperature photoluminscence spectrum for a semi-insulating Czochralski-type GaAs, Cu diffused at 850℃ for 20min with melt in the ampoule. The C_0 BE line at 1.5026eV and its phonon replica C_1 at 1.4990eV are strong in this spectrum. The second replica C_2 at 1.4954eV is also seen as well as LO replicas of C_0 and C_1, C_0^{LO} and C_1^{LO}, respectively

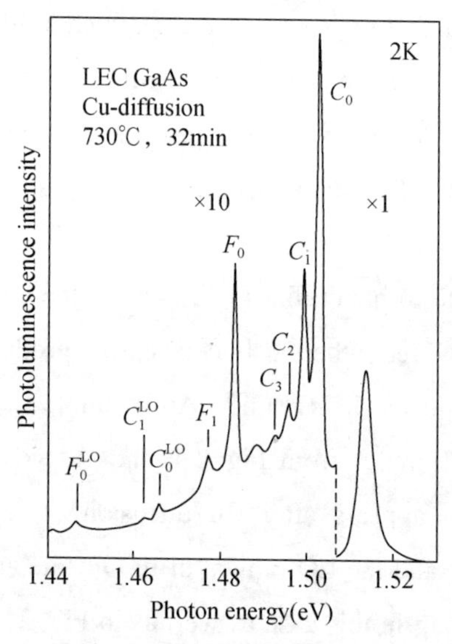

Fig. 3　Low-temperature photoluminscence spectrum of semi-insulating Czochralski-type GaAs. Cu diffused at 730℃ for 32min in vacuum. The C_0 line is strong with three clear phonon replicas. Also, the F_0 bound exciton at 1.4832eV and its replica F_1 at 1.4771eV are strong in this spectrum. The F spectrum is absent if Ga-GaAs melt is included in the ampoule

observed after a Cu diffusion at 800℃/30min in the same type of starting material as the samples of Figs. 2 and 3 (semi-insulating LEC crystals). In extreme cases we have obtained a

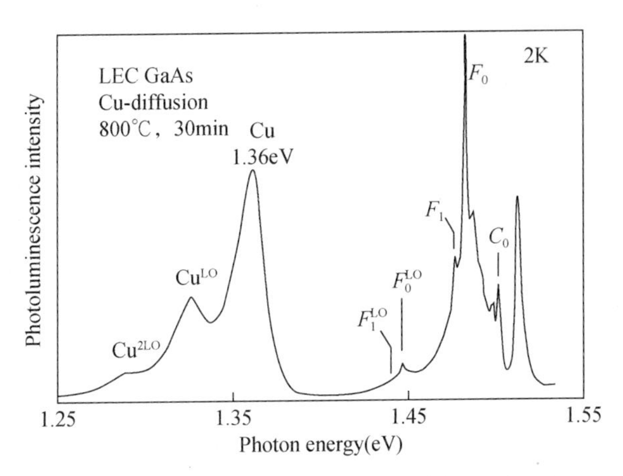

Fig. 4 Low-temperature photoluminescence spectrum of semi-insulating Czochralski-type GaAs, Cu diffused at 800℃ for 30min in vacuum. In this case the F_0 line dominates the spectrum close to the surface of the sample. This line can be made to dominate the spectrum in the surface region with hardly and 1.36eV PL band or C spectrum present

very strong F_0 line but hardly any C_0 line of 1.36eV peak. In such cases the linewidth of the F_0 line is large, exceeding the spacing between the no-phonon F_0 line and the first replica F_1, which merges into the broad line. A slight etching normally reestablishes the usual spectrum with a rather strong C_0 line and a 1.36eV peak, however. The strength of the remaining F_0 line depends on the etching depths as already reported.[11] This confirms that the F lines are strongest near the sample surface.

In order to test the hypothesis that the F_0 line is related to a deeper PL band, peaking about 1.02eV,[1] we measured the photoluminescence down to 0.885eV in samples showing strong F lines. As already noted in connection with Fig. 1, no such PL bands appear at the same time as the F lines. Consequently, there is no evidence for the F lines being related to the 0.45eV Cu acceptor from PL data. Actually, the assignment of the 1.02eV PL band to the 0.45eV acceptor level (Ref. [1] and references therein) is by no means certain (see below in the Discussion section).

Several observations in the course of this work give reason to doubt the association between the C_0 BE line and the 1.36eV PL band which is frequently encountered in the literature (Ref. [1] and references therein). Firstly, as obvious from Fig. 1, there is no correlation in intensity between the 1.36eV PL band and the C_0 BE line. This is not unambiguous evidence since different doping levels can easily affect the linewidths and thus apparent intensities of BE lines. In fact, we have never observed the C_0 line without the persistent 1.36eV band being present in the spectra. An important observation is, however, that the same samples diffused with and without Ga-GaAs melt show different relative intensity of the C_0 line and the

1.36eV PL band(normalized to the band-edge emission). The 1.36eV PL band is relatively much weaker in the presence of melt than the C_0 line. Additional evidence from the doping procedure such as the observed quenching behavior also indicates that there is no relation between these spectral features.

In Fig. 5, the result of two different Cu diffusions is shown, which normally create strong C and F spectra in LEC material. In Fig. 5(a), the sample was immersed into LN_2 immediately after diffusion, and in Fig. 5(b) the sample was slowly cooled down to room temperature in the furnace, which was turned off. The integrated intensity of the C_0 line is much larger relative to the band edge emission when quenching in LN_2. The quenching rate of the samples was the only significant parameter which differed in Fig. 5. The typical behavior shown in the figure is highly reproducible. We note that the band edge emission is not an unambiguous reference, since different excitation intensities, focussing, and spatial variation may affect the ratio between the band-edge emission (mainly excitons bound to neutral donors) and the Cu-related spectra as much as a factor of 2. We are monitoring qualitative behavior, however, with larger amplitudes than this uncertainty. Using the same reference the rather weak F spectrum of Fig. 5(a) may be enhanced in intensity when quenching more slowly. The 1.36eV peak is largely unaffected by the quenching rate, as it seems from our study. This is strong evidence for the C spectrum being independent of the 0.156eV acceptor.

Similar, but not identical results are observed with regard to the annealing behavior. We observe that annealing in vacuum of previously Cu-diffused samples at any temperature

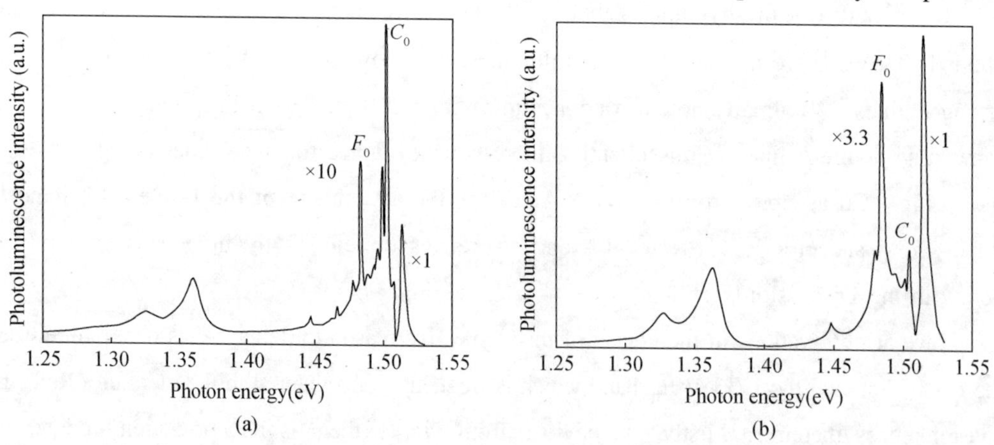

Fig. 5 Photoluminescence spectra of semi-insulating Czochralski-type GaAs at 2K, illustrating the effect of the quenching rate to room temperature after Cu diffusion under the following conditions: (a) 730℃, 32min, quenching in liquid nitrogen. (b) 730℃, 32min, slow quenching. In general, the C_0 line is relatively stronger the faster the quenching rate. The F_0 line shows a tendency to the opposite, i. e., increase in intensity with slower quenching rate. The 1.36eV PL band is almost unaffected

between room temperature and 500℃ makes the C spectrum nearly completely vanish. At room-temperature storage the C_0 line turns out to be quite unstable, almost disappearing in a couple of weeks. In order to conserve the original C spectrum, we kept the samples in the dark immersed in LN_2. In Fig. 6 a sample which was Cu doped at 700℃ for 30 min was kept at room temperature for 26 days. Comparison between the spectra in Figs. 6(a) and 6(b) shows that the C_0 line has almost disappeared, while the F_0 line has grown considerably stronger(with reference to the band edge emission). The 1.36eV PL band(not shown in the figure) is found to decrease in intensity, almost as much as the C_0 line under exended room-temperature storage. This is in sharp contrast with the effect of heat treatment at 300℃ for 36min on the relative intensity of the different spectra. As shown in Fig. 6(c) such a heat treatment severely weakens the C spectrum, while the F lines are significantly enhanced. In this case, the effect on the 1.36eV PL band and the F_0 line is similar, the former grows con-

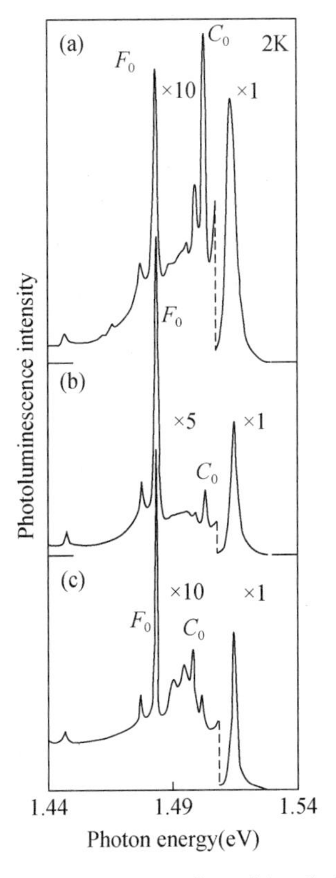

Fig. 6 Low-temperature photoluminescence spectra of semi-insulating Czochralski GaAs, Cu diffused at 700℃ for 30min. In(a) the sample is measured immediately after the Cu-diffusion procedure. In(b) the same sample was measured after room-temperature storage for 26 days. In(c) a piece of the same sample as in(a) was measured after heat treatment at 300℃ for 36min in vacuum. In both(b) and(c) the relative intensity of the C_0 line is very much reduced compared with the band-edge emission. The intensity of the F_0 line is significantly enhanced in both cases

siderably stronger upon heat treatment. Relative to the band-edge emission the F_0 line increases more, however. The centers binding the C_0 exciton are evidently more unstable than the centers binding the F_0 exciton. The same is true for the 0.156eV acceptor, despite the somewhat surprising correlation to the decreasing intensity of the C_0 line after room-temperature storage. Dissociation during both heat treatment and cooldown is able to break up the C centers, whereas the threshold for dissociation of the other Cu-related centers is clearly higher.

3.2 Electrical measurements

The room-temperature carrier concentration and mobility for some typical samples before and after Cu diffusion are listed in table 1. After the Cu diffusion process all samples have p-type conductivity. In order to obtain a homogeneous distribution of the Cu centers throughout the bulk material, an n-type Te-doped HB sample was Cu diffused at 850℃ for 12h. Fig. 7 shows the hole concentration (with $T^{-3/2}$ correction) of this sample as measured by Hall measurements in the van der Pauw configuration plotted against $1/T$. It is important to notice that the curve suggests only one Cu-related level in the sample. Therefore, in a two-level compensation model (a shallow donor and a deep Cu acceptor) the temperature dependence of the carrier concentration at low temperatures can be approximated by the well-known expression

$$p \approx N_v \exp\left(-\frac{E_a}{KT}\right) \approx T^{3/2}\exp\left(-\frac{E_a}{KT}\right),$$

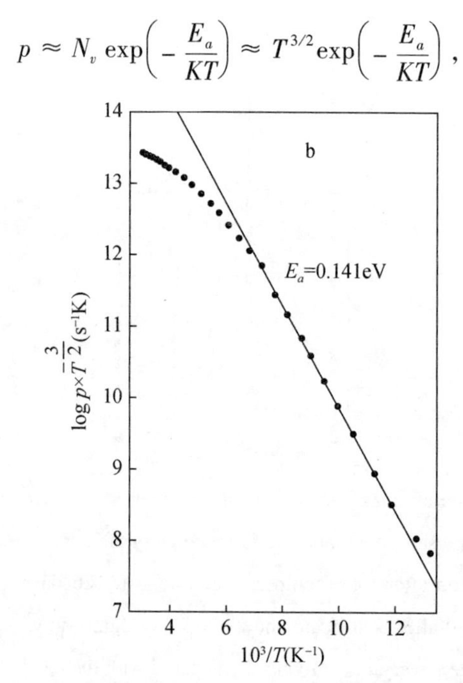

Fig. 7　Arrhenius plot of the hole concentration with a $T^{3/2}$ correction for a horizontal Bridgman bulk GaAs sample, Cu diffused at 850℃ for 12h to ensure homogeneous Cu concentration in the bulk

where N_v is the effective density of states in the valence band and E_a the activation energy of the acceptor center. A semi-log plot of $p \times T^{-3/2}$ vs $1/T$ as in Fig. 7 gives a straight line with a slope corresponding to an activation energy $E_a = 141\text{meV}$. This value is in fair agreement with the reported thermal value of the 0.156eV Cu acceptor.[4] The strong 1.36eV PL band of this sample supports the generally accepted relation between the 0.156eV acceptor and this PL band.

4 Discussion

4.1 General

Ever since the bound exciton lines C_0 and F_0 in GaAs were first discovered by Gross and Safarov[9] they have been attributed to Cu defects. Besides the evidence from doping procedure which always correlates these lines with Cu doping, the distribution of Cu in a GaAs sample as determined by a radioactive tracer method agrees with the depth profile of the total intensities of the C and F spectra together.[11] The literature is much more divergent on the admittedly difficult question of the defect identities, however.

For a long time it has been assumed that the 0.156eV acceptor has trigonal symmetry.[13, 14, 22] This was concluded from the symmetry of the C_0 bound exciton center (arbitrarily assigned to the 0.156eV acceptor level[14, 22]) as well as from infrared excitation measurements.[13] This is, however, in conflict with recent piezospectroscopic investigations, which suggest a C_2 symmetry axis parallel to the ⟨100⟩ lattice cube edge for this Cu center.[15, 16] If these piezospectroscopic results are correctly interpreted and the measurements themselves are not affecting the symmetry, they strongly argue against any connection between the C center and the 0.156eV acceptor. It is therefore important to investigate the relation between the Cu centers independently, in order to eventually reveal their identities.

We have shown in this study that there exists no correlation in the presence of the 1.36eV PL band and the C spectrum. Ultimate proof would be to produce a sample with the C spectrum alone, but the persistent appearance of the 1.36eV PL band even after inadvertent doping makes this unfeasible. Also, it is possible that the 0.156eV acceptors are necessary prerequisites for the formation of the C centers. In that case, an explanation for the instability of the C centers may be that they tend to dissociate to form these acceptors again, even at low temperatures.

We find no correlation between the F_0 line and the frequently observed (although not in our samples) 1.02eV PL band. We simply do not observe this band, regardless of the strength of the F spectrum. Regarding the attribution of the 1.02eV PL band to the 0.45e V acceptor level, a discrepancy in the energy positions allows doubt about any correlation. A radiative recombination to an acceptor level at 0.45eV should have a high-energy edge around 1.07eV. The PL band peaking at 1.02eV has a half-width of 120meV[1] and a high-energy edge above 1.1eV at low temperatures.[23] Consequently, it may be related to an unidentified shallower level with a binding energy of about 0.40eV.

The quenching procedure room-temperature storage and heat treatment as performed in this study all seem to have the opposite effect on the C and F centers. While the C centers tend to disappear during slow cooldown and heat treatment (including room temperature), the F_0 line grows stronger. The center binding the F_0 exciton consequently seems to be in an energetically stable configuration, whereas the C centers are unstable.

Although we here claim that the F spectrum and the 1.02eV PL band are not related, we cannot state anything about the possible relation between the F_0 line and the 0.45eV acceptor from PL measurements alone. Since we do not believe that the 1.02eV band originates from that acceptor level, it is necessary to make additional electrical measurements in order to clarify any such relation. We have made Hall coefficient measurements for a few typical samples, but never found signs of the 0.45eV acceptor level. The dominating impurity level from these measurements is the 0.156eV level. The 0.45eV acceptor must occur at much lower concentrations, which is in agreement with recent photocapacitance data.[4]

4.2 The acceptor nature of the C and F centers

As already mentioned, the appealing simplicity of assigning the C_0 and F_0 lines to the 0.156 and 0.45eV Cu-related acceptor levels in GaAs may have led to a rather uncautious acceptance of this assignment. In fact, not even the acceptor nature of the centers binding these excitons seems to have been properly examined yet. We shall therefore discuss this point in some detail, in relation to novel data to be reported separately.[24]

The binding energy of the C_0 bound exciton is 17.4 meV. Energetically, this puts the C_0 exciton into the range of excitons bound to neutral acceptors in GaAs. The corresponding value for the F_0 BE is 36.8eV, or much larger than usually found for excitons bound to the substitutional neutral acceptors in GaAs. In GaP, for example, it is found that localized excitons with good PL efficiency are usually bound to neutral centers of the isoelectric molecular type.[25] Several systems of excitons bound to Cu-related such centers,[26, 27] as well as neutral complexes of Cu and Li,[28] have recently been investigated in GaP.

Willmann et al.[14] have reported the Zeeman splitting of the C_0 and F_0 photoluminescence lines in GaAs. They interpret the splitting as a BE recombination at a neutral acceptor. Negligible thermalization is observed in these PL spectra. This is not commented on in Ref. [14], nor is a large diamagnetic shift (1.5meV for 5.73T) noticed in the Zeeman spectra. Also, as shown below, both the Zeeman splitting and the diamagnetic shift are consistent with the electronic configuration of excitons bound to isoelectronic associates as well as complex acceptors.

In Ref. [14] the interpretation of the C_0 and the F_0 lines as excitons recombining at neutral acceptors suggests a transition between two Kramers' doublets, the $|J, M_J\rangle = |1/2, \pm 1/2\rangle$ state of the bound exciton and the $|3/2, \pm 1/2\rangle$ state of the acceptor in question. This requires that the $|3/2, \pm 3/2\rangle$ acceptor state is split off into the valence band, which in turn requires a strong compressive local strain field.[24]

Under such circumstances a hole-attractive isoelectronic defect would bind a hole of pure spin character with g value close to $g=2$[27] as well as an electron in the Coulomb potential of the primary hole. The resulting BE states would give a very similar Zeeman pattern as the one observed in Ref. [14], as has recently been demonstrated for an isoelectronic Cu complex in InP.[29] Thus the previously reported Zeeman data[14] are not conclusive on this point.

In the course of the present study, magneto-optical experiments were carried out on the C and F centers. Transmission measurements were made in magnetic field of 0–10T at liquid He temperatures to study the thermalization behavior between the magnetic subcomponents of the BE lines. A strong thermalization effect is observed in transmission, indicating a magnetic splitting in the ground state of both the C and the F centers. Such a splitting is incompatible with excitons bound to isoelectronic associates, which can show magnetic splitting only in the excited state. Consequently, the C and F centers are in fact acceptors. These new magneto-optical data as well as the electronic configuration of the C_0 and F_0 bound excitons will be discussed in more detail in a separate publication.[21]

4.3 Possible identities of the Cu-related centers

To summarize the above discussion, we are left with four independent Cu centers of interest here:

(i) the ⟨100⟩ oriented 0.156eV acceptor center giving rise to the 1.36eV PL band and dominating the electrical characteristic of most samples;

(ii) the 0.45eV acceptor level with unknown symmetry, which is apparently not related to the 1.02eV PL band reported in the literature;

(iii) the trigonal ⟨111⟩-oriented C acceptor center binding the C_0 exciton; and

(iv) the orthorhombic ⟨100⟩-oriented F acceptor center binding the F exciton.

In this section we shall discuss, for each center in turn, the most plausible identification of these defects in the light of available experimental data.

(1) The 0.156eV acceptor and the 1.36eV PL band

As illustrated in Fig. 1, our experiments show that the 1.36eV PL band is easily introduced into all types of GaAs, especially the Te-doped HB and the LEC bulk material. In fact, it is difficult to avoid. However, none of our reference samples show this PL band prior to doping. At the same time as the 1.36eV PL band dominates the homogeneously Cu-doped bulk samples, we have observed that the 0.156eV acceptor level controls the electrical properties as anticipated. Conductivity and Hall measurements prove that the n-type samples become p type after Cu doping through the introduction of this acceptor level.

It may be argued that a center which is so easily formed in a wide temperature range (down to 500℃) and even at very low Cu concentration levels must be simple in its nature, perhaps even a simple substitutional impurity. Indeed, Cu_{Ga} has frequently been suggested to be responsible for the 1.36eV emission.[3,6-8,15,16] In view of the evidence showing that the 0.156 and 0.45eV Cu-related acceptor levels in GaAs are caused by centers of independent concentrations,[4] these two levels cannot be assigned to the two ionization levels of the double Cu_{Ga} acceptor as frequently observed in the literature.[5,30,31] On the other hand, we do not reject a double acceptor model as such for the Cu_{Ga} acceptor.

The erroneous assignment of the C_0 bound exciton to the 0.156eV acceptor has contributed to the confusion in the literature about the identity of this acceptor. Thus it is often encountered in the literature that models originally suggested for the C center are thought to be valid for the 0.156eV acceptor as well. For example, $Cu_{Ga}V_{As}$ is interpreted as a model for the latter,[4] although originally proposed solely for the C center.[11] But sometimes the suggested defects such as $V_{As}Cu_{Ga}V_{As}$[1] do not even have the trigonal symmetry, adapted from the C center. A model of a Cu_{Ga} acceptor in a ⟨100⟩-oriented Jahn-Teller distorted environment was proposed on the basis of the recent piezospectroscopic investigations of the 1.36eV PL band.[15,16]

From calculations Cu_{Ga} is expected to behave primarily as a conventional d^{10} acceptor, in contrast to most other transition metals which give rise to acceptor states of d-like character.[32] For Cu the position of the Cu d^9 hole state was calculated to be about 5eV below the band gap.[32] Accordingly, in this discussion we assume that Cu has a filled d shell ($3d^{10}$).

If the hypothesis of a substitutional Cu_{Ga} being responsible for the 1.36eV PL band is correct,[15,16] it is not obvious which charge state of this double acceptor would correspond to the 0.156eV acceptor level. Either this level corresponds to transitions between the neutral and the singly ionized charge state of the Cu_{Ga} acceptor, in which case transitions to the doubly ionized charge state are of much higher energy, or above the band gap. Alternatively, transitions between the singly and doubly ionized charge states are observed as the 0.156eV level. Hence, transitions to the neutral charge state would be very shallow and neither seen in electrical measurements nor PL measurements (merging into the near edge emission caused by other shallow impurities).

It must be noted that the spacing between the effective masslike excited states of an acceptor should be in principle distinguish between the two charge states of a double acceptor. Published data on the 0.156eV level[13] are not unambiguous on this point in our opinion. The fine structure observed has been interpreted as a Rydberg series of a single acceptor, split by a trigonal crystal field. However, transitions which are normally observed in infrared excitation spectra for shallower acceptors in GaAs are reported missing.[13] Considering the recent data on the orthorhombic symmetry of the center causing the 1.36eV PL band,[16] a reinterpretation of the infrared excitation data seems justified. In absence of such data in a wider energy range, no firm conclusion can be drawn from such studies as to which charge states of the Cu acceptor cause the 0.156eV level.

An interesting piece of information is provided when the Cu-doped GaAs samples are co-doped with Li. A gradual decrease of the 1.36eV PL band is observed with increasing Li doping. At the same time a new peak appears around 1.41eV. In a separate publication we present the details of the Cu-Li co-doping and propose that the 1.41eV PL band originates from recombinations to a Cu_{Ga}-Li_i single acceptor associate.[33] This interpretation implies that the Cu_{Ga} acceptor is doubly ionized at the low Li-diffusion temperatures employed (300–500℃). In view of this we tend to favor a model of the 0.156eV acceptor level as caused by transitions between the singly and the doubly ionized charge states of Cu_{Ga} with the first ionization level being much shallower. If the second ionization level were very deep instead (and the 0.156eV level corresponded to the first one) the formation of a Cu_{Ga}-Li_i acceptor associate would seem to require an activation energy on the order of the band gap of GaAs (the binding energy of the hole to the singly ionized Cu_{Ga} acceptor).

(2) The *C* center

The C_0 exciton is easily produced with excellent intensity, particularly in high-purity material such as the VPE and LPE expitaxial wafers as well as in undoped semi-insulating LEC material. Even though the C_0-line intensity was weaker in the Te-doped HB bulk material, it was comparable with most published spectra for bulk material.[10-12]

We do not agree with models suggesting As vacancies to be involved in the *C* center. One piece of evidence against this is a marked difference in the migration of the *C* and *F* centers, during heat treatment and cooldown after diffusion, suggesting the possible involvment of V_{As} in the latter but not in the C centers. Thus, the depth profile of the C centers is homogeneous throughout the sample[11] in agreement with our observations. The concentration of V_{As} has a very different profile with highest concentration at the surface.[33] Furthermore, the C centers seem to dissociate homogeneously throughout the bulk during heat treatment. We see no signs of accumulation at the surface, as expected for vacancy-related centers. On the contrary, we have occasionally observed the absence of the C spectrum in the surface layer, where the vacancy concentration should be high. In summary, we disregard models for the C center which include V_{As} such as $Cu_{Ga}V_{As}$[11, 12] and $V_{As}Cu_{Ga}V_{As}$.[1]

Evidently the C acceptor center is a complex consisting of donorlike and acceptorlike parts, which are ionized at the high diffusion temperatures. The complex acceptor is formed due to Coulomb attraction of the differently charged species, the simplest case being realized with a double Cu_{Ga} acceptor together with a single donor. The complex defect is then caused to freeze in by the fast quenching, evidently in a rather unstable configuration. The uniform concentration of the C center in the sample and its instability, suggests in our opinion the participation of the fast-diffuser Cu_i in the complex. This species diffuses faster than both V_{As} and V_{Ga}.[12]

A possible identity of the C center would be a⟨111⟩oriented $Cu_{Ga}Cu_i$ acceptor associate. A model involving an interstitial Cu atom seems to be in a much better agreement with the compressive sign of the local crystal field deduced for the center[24] than models involving vacancies.[1, 11, 12] (The latter configurations would in fact be expected to create a tensional strain field locally at the defect sites). Our model may also explain the sensitive formation and instability of the center. The details in the instable behavior of the center are not clear, however. Furthermore, it is important in future work to positively identify the corresponding acceptor level.

(3) The F center

The striking difference in the doping conditions for producing the F and the C spectra is the necessary absence of Ga-GaAs melt in the ampoule for the F_0 BE to appear. In all types of starting material the presence of Ga melt effectively prevented the appearance of the F_0 BE. Assuming that the Ga melt prevents the in-diffusion of As vacancies into the sample, it seems natural to assume the participation of V_{As} in either the in-diffusion of the F centers or even the associates themselves. It has in fact often been assumed that V_{As} is a part of the F center. The most common hypotheses are $V_{As}Cu_{Ga}V_{As}$,[11] $Cu_{Ga}V_{As}Cu_{Ga}$,[12] and $Cu_{Ga}V_{As}$.[1]

The F center has an orthorhombic symmetry with one of the axes⟨100⟩oriented.[10] It has also been shown to be an acceptor in this work.[24] The suggested identity must therefore agree with both these criteria. The 0.156eV acceptor might perhaps seem a proper candidate for binding the F_0 exciton, being a ⟨100⟩-oriented acceptor. However, we have never observed an unequivocal correlation between the appearance of the F_0 line and the 1.36eV PL band. On the contrary, either one can be present in absence of the other. Usually the 1.36eV PL band appears without the F spectrum being observed, but as mentioned above the latter can be very strong in the surface region without any signs of the 1.36eV peak. In addition, there is an obvious difference between the F spectrum and 1.36eV PL band regarding the presence of Ga-GaAs melt under Cu diffusion. Hence, nothing is known about the acceptor level corresponding to the F center, since it has neither been observed via radiative transitions nor in electrical measurements, probably due to a low average concentration.

Were it not for the evidence from the material prepration and doping for the presence of V_{As} in the F center other possibilities might be considered, such as a ⟨100⟩-oriented $Cu_{Ga}Cu_i$ pair (as was discussed for the 0.156eV acceptor) or other configurations involving impurities or native defects besides Cu, such as Ga_i, for example.

It appears impossible to design a model of the F-center associate which both involves V_{As} and has the correct symmetry. Neither $V_{As}Cu_{Ga}V_{As}$ [11, 12] nor $V_{As}Cu_{Ga}$ [1] possess the ⟨100⟩ symmetry axis postulated by Gross et al. [10] Our data, on the other hand, together with the literature seem to suggest involvement of V_{As} in some stage of the defect formation. The logical conclusion of this fact is that V_{As} need not necessarily participate in the molecular associate, rather it could be a necessary prestage in its formation.

Instead, we tentatively suggest that the F center consists of a substitutional Cu_{Ga} associated with an interstitial native defect, such as Ga_i.

It has been concluded that the electronic configuration of the F_0 bound exciton corresponds to a compressive sign of the crystal field created locally by the defect binding the exciton. [24] This suggests interstitial atoms rather than vacancies being involved in the F center as was mentioned in Sec. 4.3(2) (for the C center).

5 Conclusions

A study of several doping parameters in the diffusion of Cu into GaAs and the relating photoluminescence spectra has been performed. A wide range of GaAs materials was used, horizontal Bridgman and semi-insulating liquid-encapsulated Czochralski bulk material, as well as liquid-phase and vapor epitaxial wafers. Samples were made from the different GaAs material give a uniform picture of the role of Ga-GaAs melt in the ampoule under diffusion. While both the C_0 BE and the 1.36eV PL band appear with reduced intensities under these conditions, the F_0 BE does not appear. Heat treatment in the temperature range up to 500℃

shows a considerable difference in the stability of these centers. This investigation has led to the first positive identification of the *C* and *F* centers as acceptors. On the basis of this fact and from the accumulated data we have discussed tentative models for the microscopic identities of the 0.156eV acceptor, the *C* center, and the *F* center. We arrive at the conclusion that none of these centers is likely to involve vacancies, which hitherto has been a common factor in the heterogeneous hypotheses for their identities. We suggest that the 1.36eV PL band is caused by a radiative recombination to the doubly ionized Cu_{Ga} acceptor, which consequently corresponds to the 0.156eV acceptor level. The *C* and the *F* centers are found likely to involve interstitial atoms rather than vacancies, in view of the electronic configurations of the excitons bound to these centers as well as for symmetry reasons. Possible candidates include Cu_i and the native Ga_i.

Acknowledgements

We gratefully acknowledge the supply of GaAs material from the LPE and VPE GaAs Group, Institute of Semiconductors, Chinese Academy of Sciences, Beijing and from L-Å. Ledebo and L. Samuelson. We also wish to thank L-Å. Ledebo for critically reading the manuscript and K. Nideborn for technical assistance. Wang is very grateful to Professor H. G. Grimmeiss for his hospitality during his stay in Lund. This investigation was supported by the National Swedish Board for Technical Development, STU, and the Swedish Natural Science Council, NFR.

References

[1] H. J. Guislain, L. De Wolf, and P. Clauws, J. Electron. Mater. 7, 83(1978).

[2] R. N. Hall and J. H. Racette, J. Appl. Phys. 35, 379(1964).

[3] H. J. Queisser and S. S. Fuller, J. Appl. Phys. 37, 4895(1966).

[4] N. Kullendorff, L. Jansson, and L-Å. Ledebo, J. Appl. Phys. 54, 3203(1983).

[5] J. S. Blakemore and S. Rabimi, Semiconductors and Semimetals, Vol. 20(Academic, New York, 1984), 233.

[6] E. Fabre, Phys. Status Solidi A 9, 259(1972).

[7] F. M. Vorobkalo, K. D. Glichuk, A. V. Prokhorovich, and G. John, Phys. Status Solidi A 15, 287 (1973).

[8] K. D. Glichuk, K. Kukat, and V. I. Vovenenko, Phys. Status Solidi A 69, 521(1982).

[9] E. F. Gross and V. I. Safarov, Sov. Phys. Semicond. 1, 241(1967).

[10] E. F. Gross, V. I. Safarov, V. E. Sedov, and V. A. Maruschak, Sov. Phys. Solid State 11, 277 (1969).

[11] V. I. Safarov, V. E. Sedov, and T. G. Yugova, Sov. Phys. Semicond. 4, 119(1970).

[12] M. G. Mil'vidskii, V. B. Osvenskii, V. I. Safarov, and T. G. Yugova, Sov. Phys. Solid State 13, 1144(1971).

[13] F. Willmann, M. Biätte, H. J. Queisser, and J. Treusch, Solid State Commun. 9, 2281(1971).

[14] F. Willmann, D. Bimberg, and B. Blätte, Phys. Rev. B 7, 2473(1973).

[15] N. S. Averkiev, T. K. Ashirov, and A. A. Gutkin, Sov. Phys. Semicond. 15, 1145(1983); 17, 61(1983).

[16] N. S. Averkiev, T. K. Ashirov, and A. A. Gutkin, Sov. Phys. Solid State 24, 1168(1982).

[17] T. K. Ashirov and A. A. Gutkin, Sov. Phys. Semicond. 16, 99(1982).

[18] R. M. Logan and D. T. J. Hurle, J. Phys. Chem. Solids 32, 1739(1971).

[19] J. R. Arthur, J. Phys. Chem. Solids 28, 2257(1967).

[20] G. M. Blom, J. Crystal Growth 36, 125(1976).

[21] E. W. Williams, Phys. Rev. 168, 922(1968).

[22] K. D. Glichuk and A. Prokborovich, Phys. Status Solidi A 29, 339(1975).

[23] T. Risbaev, I. M. Fishman, and Yu. G. Shreter, SOY. Phys. Semicond. 6, 1709(1973).

[24] H. P. Gislason, B. Monemar, Z. G. Wang, Ch. Uihlein, and P. L. Liu(unpublished).

[25] P. J. Dean and D. C. Herbert, in Topics in Current Physics. VoL 14. Excitons, edited by K. Cho (Springer, Berlin, 1979).

[26] B. Monemar, H. P. Gislason, P. J. Dean, and D. C. Herbert, Phys. Rev. B 25, 7719(1982).

[27] H. P. Gislason, B. Monemar, P. J. Dean, D. C. Herbert, S. Depinna. B. C. Cavenett, and N. Killoran, Phys. Rev. B 26, 827(1982).

[28] H. P. Gislason, B. Monemar, M. E. Pistol, P. J. Dean, D. C. Herbert, A. Kana'ah, and B. C. Cavenett, Phys. Rev. B 31, 3774(1985).

[29] M. S. Skolnick, P. J. Dean, A. D. Pitt, Ch. Uihlein, H. Krath, B. Deveaud, and E. J. Foulkes, J. Phys. C 16, 1967(1983).

[30] J. M. Whelan and C. S. Fuller, J. Appl. Phys. 31, 1507(1969).

[31] R. Hall and J. H. Racette, Bull. Amer. Phys. Soc. 7, 234(1962).

[32] L. A. Hemstreet, Phys. Rev. B 22, 4590(1980).

[33] H. P. Gislason, Z. G. Wang, and B. Monemar, J. Appl. Phys. 58, 249(1985).

[34] S. Y. Chiang and G. L. Pearsson, J. Appl. Phys. 46, 2986(1975).

Localization of excitons to Cu-related defects in GaAs

B. Monemar and H. P. Gislason
(Department of Physics and Measurement Technology, Linköping Institute of Technology, S-58183 Linköping, Sweden)
Z. G. Wang
(Institute of Semiconductors, Chinese Academy of Sciences, Beijing 100083, China)

Abstract The possibility of binding excitons to complex-type defects in GaAs is discussed, with reference to recent photoluminescence data on Cu-related centers. The apparent absence of bound excitons(BE's) associated with certain complex-type acceptors may be explained as a consequence of a local compressional strain field at the defect. Such a field will decrease the binding energy of electron states derived from the Γ_1 conduction-band minimum in GaAs, so that they ultimately become resonant with the band in the limit of a strong field. A similar effect on the BE electron state is expected for neutral-complex defects, particularly if they involve two interstitial species. Only one example of an exciton binding to a neutral complex was found in GaAs so far. It involves a tensional local strain field, in which case the BE electron localization becomes stabilized.

1 Introduction

Cu introduces several deep acceptor states when incorporated in GaAs;[1-4] additional states of this type are introduced by co-doping with Li.[5] In addition, neutral complexes of isoelectronic character involving Cu are expected to form, as observed in other Ⅲ-Ⅴ compounds, such as GaP(Refs. [6]-[8]) and InP(Ref. [9]), and recently also in GaAs (Ref. [10]). The latter class of defects is detectable in photoluminescence(PL) via bound-exciton(BE) spectra at low temperatures. In GaAs very few examples of Cu-related BE spectra are found, suggesting that there must be fundamental reasons why these excitations cannot be observed. In this paper we discuss the possibilities of binding excitons to deep-level defects in GaAs, with particular emphasis on Cu-related defects, which are, in most cases, of a complex nature. The optical spectra of these Cu-related defects are discussed in considerable detail separately.[4, 5, 10] Therefore the experimental data here will be shown only for easy reference, when necessary for the discussion. A simple model is provided for the physical explanation of the lack of binding of BE states observed in most of the cases discussed.

原载于：Phys. Rev. B, 1985, 31(12): 7919-7924.

2 Acceptor states and the binding of excitons

Several Cu-related acceptor levels are reported for GaAs, the most prominent being those at 0.15 and 0.45eV.[3,4] The 0.15eV acceptor level is usually dominant, and rather convincing arguments have been put forward for its identification as substitutional Cu_{Ga}, distorted in the⟨100⟩direction.[3,4,11] No bound excitons have been found to be associated with this level.[4] The 0.45eV acceptor level was previously thought to be the second ionization level(2−/−) of the double acceptor Cu_{Ga} [the 0.15eV level was then suggested to be the first(−/0)].[1] Since these levels were recently reported to occur at different unrelated concentrations,[3] this assignment seems to be in error. The 0.45eV acceptor level is therefore referred to here as an unidentified Cu-related acceptor complex. Two prominent BE lines are seen in Cu-doped GaAs (Fig. 1): the so-called *C* and *F* lines,[4,12-14] which have recently been shown to be associated with acceptors.[15] The *C*-line acceptor has a⟨111⟩-oriented symmetry axis,[12] and has been shown to be unrelated to the 0.15eV acceptor.[4] The *F* line has a lower symmetry (orthorhombic),[12] and is also associated with a Cu-related acceptor complex (i. e., different from the 0.15eV one[4]). The association of the *F* line with the 0.45eV acceptor level cannot be definitely ruled out, but is unlikely from studies of diffusion profiles. The Cu-Li co-doping creates a 0.11eV acceptor-complex level as evidenced via the 1.41eV PL band in Fig. 2. No bound-exciton state is found to be associated with this acceptor (Fig. 2). We therefore have the somewhat disturbing experimental situation that of the three Cu-related acceptor levels (0.11, 0.15, and 0.45eV) discussed here none seems to bind excitons. On the other hand, bound excitons for deep acceptor states do indeed exist in GaAs, as manifested by the observation of the strong *C* and *F* lines[15] (Fig. 1).

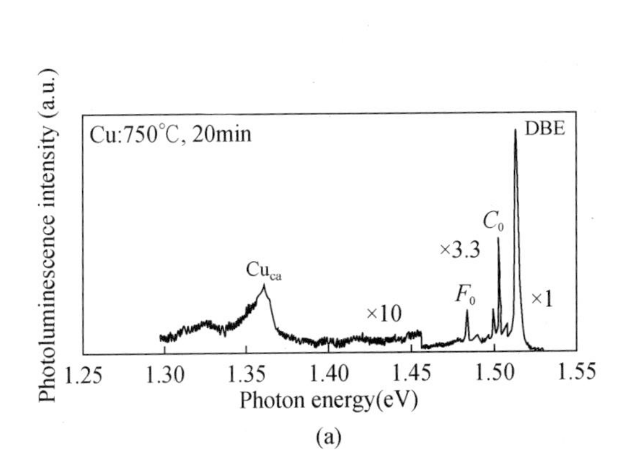

(a)

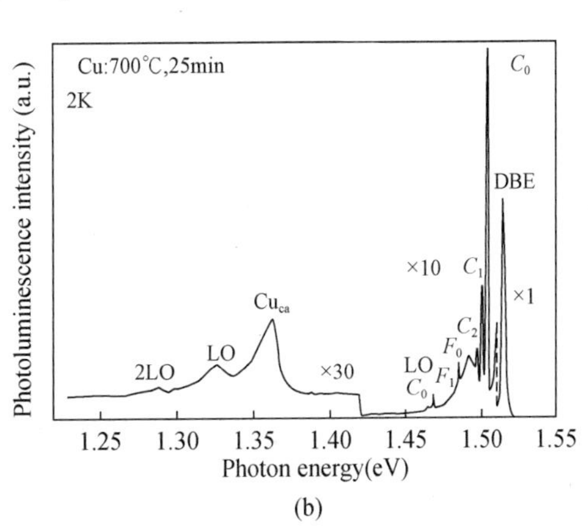

(b)

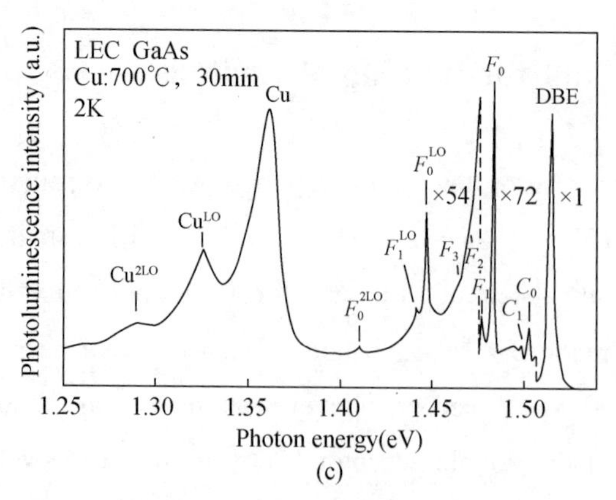

(c)

Fig. 1 (a) Near-band-gap photoluminescence (PL) spectrum (2K) from an originally n-type vapor-phase-epitaxy (VPE) GaAs sample, with an 18μm thick epilayer of an original uncompensated doping level $n_{300K} = 3\times10^{14}\,cm^{-3}$, which has been Cu-diffused at 750℃ for 20min in vacuum and rapidly quenched in water. Apart from the slightly broadened DX band at 1.5135eV due to residual shallow donors, the spectrum is dominated by the rather strong C_0 and F_0 lines and their phonon sidebands. These lines are both connected with Cu-related deep acceptors (Ref. [4]). The presence of the Cu_{Ga} double acceptor is manifested by the broad band at 1.36eV. No additional Cu-related BE lines apart from the C and F series are observed. (b) Similar near-band-gap PL spectrum (2K) as for an originally n-type liquid-phase-epitaxy (LPE) GaAs sample, with a 28μm thick epilayer of an original uncompensated doping level $n_{300K} = 1\times10^{15}\,cm^{-3}$. The sample was Cu-diffused in vacuum at 700℃ for 20min and rapidly quenched in water. The spectrum is very similar to the one in (a), apart from a higher radiative efficiency in this LPE sample. The C and F BE spectra are seen, together with the ≈1.5135eV broadened donor BE band and the 1.36eV Cu_{Ga} band. No additional bound excitons are seen that could be related to the Cu_{Ga} acceptor. (c) Similar near-band-gap PL spectrum (2K) as in (a) and (b) for an originally semi-insulating undoped liquid-encapsulated-Czochralski (LEC) bulk GaAs sample, which was diffused with Cu in vacuum at 700℃ for 30min and rather slowly quenched in air. Similar spectra as in (a) and (b) are produced in this case also, and no bound exciton is observed associated with the Cu_{Ga} acceptor

The binding of excitons to neutral single acceptors in GaAs should be possible even in the case of low-symmetry defects, since two holes should still (via incomplete screening) be attracted to a hole-attractive (Cu-induced) central-cell potential for such an acceptor.[16] The electron in the BE state should be bound by a secondary Coulomb attraction to the center, once it is charged by the presence of two bound holes.[16] This "pseudodonor" model has been experimentally verified for excitons bound to acceptors (acceptor-bound excitons, or ABE's) in several cases for GaAs.[15, 17, 18] Therefore, the presence of the C and F lines is not surprising, if the corresponding center are single acceptors. Rather, the lack of observation of ABE lines associated with other deep acceptors will have to be understood in terms of the electronic properties of hole and electron states bound to complex defects in GaAs.

2.1 Double-acceptor case

We shall first discuss the special case of the Cu_{Ga} acceptor, which is a double acceptor. There is substantial evidence that the 0.15eV level is associated with the(−/2−)transition of the defect,[4] i. e., the capture of the first hole into the doubly ionized acceptor. Furthermore, there is a static Jahn-Teller(JT)relaxation taking place upon this hole capture, so that the Cu atom is distorted in a [100] direction.[19] This will affect the binding energy of the 0.15eV acceptor state, being partly derived from this static relaxation.

Binding of a second hole by the singly ionized Cu acceptor is, in general, expected to be possible. However, an electron would not easily bind to the resulting neutral Cu acceptor state, due to the repulsive effect of the strongly hole-attractive core potential of the Cu_{Ga}. In general, binding of excitons to charged acceptors (which is the case discussed here) is not expected to be possible for GaAs, due to the large m_h^*/m_e^* effective-mass ratio,[16] and has not been reported in the literature. With this assignment of the 0.15eV level to the Cu_{Ga}(−/2−) transition, the absence of a BE state associated with this state is therefore simply what is expected.

A related problem is whether Cu_{Ga} actually binds a second hole. Only one level has been observed in our experimental data. The(0/−)level is expected to be much shallower than the 0.15eV level, of the same order as the shallow acceptor levels in GaAs(25−35meV).[20] This is inferred from recent studies of other double acceptors in GaAs, such as Ga_{As}(Ref. [21])and Li_{Ga}(Ref. [5]). We see no such additional acceptor level revealed as, e. g., donor-level—acceptor-level (*D-A*) or conduction-band—acceptor-level transitions (free to bound) in the photoluminescence spectra. Therefore the binding of the second hole by the defect is not observed so far, but cannot be excluded for lack of more extensive experimental data from sufficiently pure starting material. A speculative solution to the possibility that the second hole is not bound would be that the neutral Cu^0 state is actually resonant with the valence band. The first charge state Cu^- is stabilized in the gap by a distortion off the tetrahedral lattice site to produce a level in the band gap, as observed in the(2−/−)0.15eV level. The piezospectroscopic data on the 1.36eV emission only indicate a small (JT) relaxation, however.[19]

The case of binding an exciton to a neutral double Cu_{Ga}, acceptor in GaAs is of particular interest, if there is an axial distortion connected with this state as well. In tetrahedral symmetry a neutral double acceptor is expected to be able to accommodate a third bound hole (actually, four holes are possible in a closed-shell multiple-bound-exciton model[16, 22]). This has been verified in other materials, such as Ge (Ref. [23]) and Si (Ref. [24]). Once an axial stress field is applied, this hole shell is split up, so that the lowest-energy branch can hold only two holes, as has been demonstrated by piezospectroscopic data in Ge.[25] A local axial strain field at the defect usually has a large effect on hole states, as observed, e. g., in GaP (Refs. [7] and [8]) and ZnTe (Ref. [26]). Consequently, the strain field of an axial defect could easily drive the split-off hole state (accommodating two holes) to higher hole energies, i. e., into the valence band. Therefore the fact that no bound exciton has so far been observed connected with a neutral Cu_{Ga} acceptor in GaAs might not give any indication as to whether the corresponding acceptor level [the (−/0) transition] is in the band gap. In addition, the possibility of a strong Auger effect in the recombination of excitons bound to a neutral double acceptor cannot be ruled out as an alternative explanation of the absence of an ABE associated with neutral Cu_{Ga} in PL spectra. The Auger effects will be further discussed below.

2.2 Single-acceptor case

The absence of binding of excitons for neutral single acceptors seems to call for a different explanation. We have interpreted the Cu-Li 0.11eV acceptor level as due to the (−/0) transition of an axial Cu_{Ga}-Li_i single acceptor.[5] The case of the deep 0.45eV acceptor level mentioned above is more uncertain and will therefore not be discussed in detail here. [We just note that it is most likely to be a (−/0) transition of a single acceptor complex, since the attribution of the 0.15eV level to the Cu_{Ga} acceptor exhausts the choice of simple Cu-related double acceptors.] No bound exciton is seen associated with the 0.11eV Cu-Li acceptor (Fig. 2). We cannot completely rule out that an ABE state could be resonant with other shallow ABE states in this case, but the overall weakness of such shallow ABE spectra in Cu-Li—doped GaAs does not support this point. We shall therefore search for a model to explain the absence of a bound ABE state in this case. We believe that the second hole would be bound comfortably in this case, as normally expected for a BE bound to a neutral acceptor. The binding of the electron to the positive $(Cu\text{-}Li)^+$ state could be more difficult for an axial defect, since the electron binding energy for donor states is so small in GaAs (<6meV).[27] Furthermore, the electron states are very sensitive to a strain field.

The Cu-Li acceptor complex is expected to cause an overall compressive local strain field as a result of the presence of an interstitial Li in the complex. A field of this sign is expected to raise the shallow bound-electron states in energy, in the same way as the Γ_1 con-

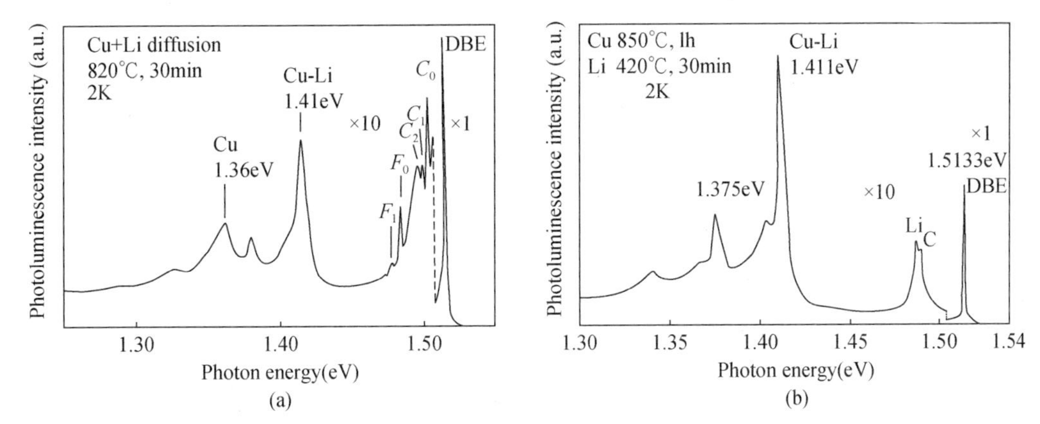

Fig. 2 (a) Photoluminescence spectrum at 2K from a Horizontal-Bridgman (HB) GaAs sample diffused with Cu and Li simultaneously at 820℃ for 30min in an evacuated ampoule. Note the presence of the Cu_{Ga} acceptor from the 1.36eV emission, and likewise, the Cu-Li acceptor from the 1.41eV emission. BE spectra from the C and F centers dominate, together with the ≈1.5135eV peak related to shallow donors. (b) PL spectrum from another HB sample doped with Cu at 850℃ for 1h and subsequently with Li at 420℃ for 30min. The dominant deep acceptor is the Cu-Li acceptor associated with the 1.41eV peak. Other shallow acceptors give rise to the peak at ≈1.49eV, but the only BE related peak is the near-band-gap peak at ≈1.5135eV due to residual donors

duction-band minimum behaves with stress in GaAs.[28] Local strain fields around defects in semiconductors are found to be quite large.[29] Furthermore, it has been demonstrated in GaP that such local fields may have substantial effects on the binding energy of both electrons and holes to complex defects.[6-8] We therefore suggest that a compressional local strain field could render a shallow Γ_1 electron state in GaAs resonant with the conduction band, in which case no BE state can be seen in optical spectra for such defects.

All substitutional acceptor dopants from groups Ⅱ and Ⅳ in GaAs are found to produce bound excitons. Even the deep Sn_{As} acceptor with a binding energy of 167meV produces a strong bound-exciton line at 1.507eV.[17] The case of Mn_{Ga} is interesting in connection with this work, however, since it was previously reported not to produce a bound exciton.[30] We have repeated the experiment on Mn doping, and indeed no ABE state related to the Mn was found in our case either; see Fig. 3. The Mn acceptor has a binding energy very close to the Cu-Li acceptor discussed above, just ≈3meV deeper.[5] Mn has been shown to be a single substitutional acceptor on a Ga site,[31] and thus represents an anomalous case. For Mn the influence of the d-like character of localized hole states in the band gap may well cause an increase in hole-hole repulsion energy in cases where more than one hole is to be bound to the acceptor (as in an ABE), eventually resulting in an instable ABE state.[32] This problem should not be important for Cu_{Ga}, where the d-like hole states are predicted to be resonant very deep down in the valence band.[33]

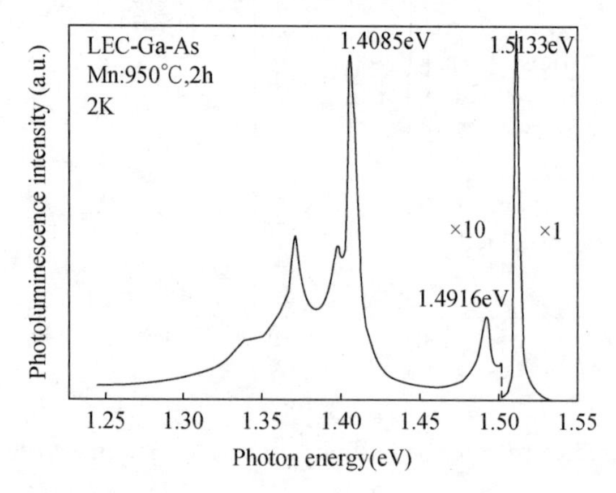

Fig. 3 Near-band-gap photoluminescence spectrum (2K) from an originally nominally undoped semi-insulating sample of LEC GaAs, Mn-diffused at 950℃ for 2h. The peak at ≈1.4085eV is due to the Mn acceptor, occurring at about 3meV lower energy than the corresponding Cu-Li peak in Fig. 2. [It has been carefully checked that no Mn contamination was present in the (Cu-Li)-doping experiments (Ref. [5])] No bound-exciton line is seen associated with the Mn acceptor, only the broad line at ≈1.5135eV related to shallow donors

At this point it seems appropriate to return to the *C* and *F* complex acceptors in GaAs, where indeed deeply bound ABE's are observed.[15] In addition, the acceptor hole state is in this case found to be a spinlike hole, which is only possible in a rather strong compressive local field of symmetry lower than tetrahedral.[15] The electron of the ABE is found to be loosely bound, like a shallow donor electron, from the observed diamagnetic shift in Zeeman data for both these ABE's.[15] This means that the binding of the electron is delicate, since a rather strong compressive local stress field does not necessarily render the electron unbound. In addition, the local complex-defect potential, in this case thought to be dominated by the hole-attractive Cu_{Ga} site, may, in these cases, contain substantial electron-attractive contributions from a localized donorlike part (such as a rather deep interstitial donor level),[4] which could compensate for the action of the local compressive strain field on the electron state of the ABE, discussed above.

3 Neutral (Isoelectronic) complexes

In GaP a large number of neutral-complex defects occur with Cu doping[6-8] and an additional number of such centers are created upon Li co-doping, as manifested from BE spectra taken at low temperature.[8] There is no reason to believe that the defect chemistry upon Cu and Li diffusion is drastically different in GaAs and GaP, and we consequently believe that neutral Cu complexes and Cu-Li complexes are formed also in GaAs upon proper diffusion treatments, as discussed separately.[4, 5] No bound-exciton spectra associated with such defects in GaAs are observed, however. This should thus be taken as evidence that BE's are, in general, not bound to such defects in GaAs, while the corresponding neutral-complex defects indeed are expected to exist in the material. In fact, no BE's related to simple substitutional isovalent atoms have been observed in GaAs at normal pressures.

Since these neutral-complex Cu-related defects should have an overall hole-attractive central-cell potential being dominated by Cu_{Ga}, a hole may be bound to such a defect at low temperature, making it positively charged. An electron might subsequently be bound as a result of the Coulomb attraction from the bound hole. The usual effective-mass-like binding energy of $\leqslant$6meV of an electron in GaAs will certainly in such a case be reduced by the electron-repulsive (hole-attractive) Cu_{Ga} central cell, but this effect might be insufficient to render the electron state unbound. The above-mentioned local strain field might be much more potent in this respect, however. As explained above, it should take just a moderate local strain field of a compressive sign to raise a shallow bound-electron state a few meV in energy to make it resonant with the Γ_1 conduction band in GaAs.[28] It is believed that most Cu-related (or Cu-Li-related) neutral-complex defects in GaAs actually give rise to a strong compressive strain field, as they are deduced to do in GaP.[6-8] This is easy to imagine for the model case of a Cu_{Ga} (double acceptor) and two additional interstitial atoms accommodated around the same site. Therefore our conclusion is that such "compressive" neutral complexes exist in GaAs, although they are unable to bind excitons in the cases studied in our work.

In contrast, we believe that there is a case in Cu-diffused Zn-doped GaAs where a neutral complex actually binds an exciton deeply. As described in detail separately, an emission with the lowest electronic line at 1.429eV is ascribed to an exciton bound to a neutral complex involving both Cu and Zn (Ref. [10]) (Fig. 4). The electronic structure revealed by Zeeman data shows that the defect gives rise to a local strain field of tensional sign.[10] In such a case the binding energy of the electron is actually increased by the strain field. This is consistent with the strongly reduced diamagnetic shift observed in Zeeman data for this BE state,[10] which is only expected when the electron state is strongly localized.

Therefore a consistent picture emerges from these data for the possibility of binding excitons to neutral-complex defects with a hole-attractive central-cell potential in GaAs. If the local strain field at the defect is tensional, BE states are expected to be bound. On the contrary, defects causing a compressive local strain field may, in general, not bind excitons.

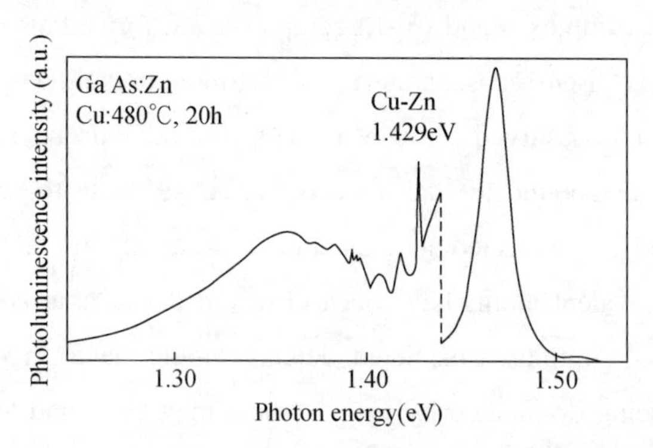

Fig. 4 Near-band-gap photoluminescence spectrum (2K) from a bulk GaAs sample originally doped with Zn to a concentration ≈ 1.5×10^{16} cm^{-3}, and subsequently Cu-diffused at 480℃ for 20h. This procedure produces a rather strong line at 1.429eV, a bound-exciton line related to a neutral defect containing Cu and Zn (Ref. [10]). A rich spectrum of phonon replicas accompany this line. The broad band at ≈ 1.47eV is related to Zn_{Ga} acceptors

4 Absence of strong auger effects in bound-exciton spectra

Auger effects in BE recombination are known to be quite important in indirect-band-gap materials, both for donors and acceptors. [16] The effect is particularly strong for deeply bound excitons where the bound particles are more localized. For direct-band-gap materials the radiative lifetime is usually much shorter (≈1ns, or even less), meaning that Auger processes could be of less importance. Detailed investigations on the Auger effects in DBE or ABE spectra for GaAs are not known to the authors, but have recently been reported for ZnTe. [34] In the ZnTe case Auger effects were concluded to strongly influence the observed BE lifetime for ABE's with rather high binding energy. [29] Returning to GaAs, our PL spectra show that the rather deep *C* and *F* ABE's are indeed quite strong in intensity (Fig. 1), indicating that Auger effects are not important for excitons bound to a single-acceptor ABE's in GaAs. This is contrary to the predictions made in recent literature. [16] As discussed above, the possible role of Auger effects for the double acceptor Cu_{Ga} is less obvious. Naturally, no Auger effects can occur with BE recombination for neutral defects, and therefore we would expect to see BE spectra from neutral defects in GaAs in PL spectra once they are able to bind excitons.

5 Summary and conclusions

Bound excitons in GaAs have previously been studied in detail mainly for shallow substitutional donors and acceptors. From a study of Cu-related defects in GaAs, several conclusions can be drawn on the ability of deep-level defects as well as complex-type defects to bind excitons in GaAs. For double acceptors such as Cu_{Ga} no ABE's are observed in PL spectra. For the case of the 0.15eV acceptor level, interpreted as being connected with the second ionization level of Cu_{Ga} the absence of a corresponding BE state is consistent with the general rule that charged acceptors do not bind excitons in GaAs. No evidence of a BE associated with the neutral state of Cu_{Ga} is seen either, which can be explained if this state also experiences a Jahn-Teller distortion such as that experienced by the singly ionized charge state. In any case, the third hole necessary for creation of a bound-exciton state for neutral Cu_{Ga} might easily be split off to be resonant with the valence band by the action of an axial strain field. With other Cu-related acceptors, BE's are observed in two cases, the so-called *C* and *F* lines, due to two different Cu-related complex-type acceptors. These are believed to be single acceptors, and the excitons are bound to their neutral charge states. Other Cu-related complex-type acceptors observed in our study probably do not bind excitons, however. This is believed to be caused by the delicate shallow binding of the electron in the BE state. This electron state is believed to be easily driven up in the conduction band by the action of a compressional local strain field at the defect.

For the case of neutral (so-called isoelectronic) defects, it is well known that such single substitutional atoms do not bind excitons in GaAs, i. e., the central-cell potentials for binding the primary particles are not strong enough to localize a state in the band gap. For complex-type defects consisting of more than one impurity atom, we have found one case where a neutral-(Cu-Zn)-related defect binds an exciton deeply (1.429eV) in GaAs, in a tensional local strain field. This sign of the crystal field seems to be required in GaAs, in order to stabilize the electron state in the band gap for neutral complexes with hole-attractive central-cell potentials. In the cases where neutral complex defects are believed to give rise to a compressive local strain field, BE states are not seen in our study.

The experimental data for the cases where bound excitons associated with neutral acceptors are actually observed in this work are interesting since a high intensity of ABE lines is consistently seen. This means that Auger effects for ABE's associated with deep acceptors are not very important in GaAs, contrary to previous belief.

Acknowledgments

We would like to acknowledge the assistance of M. Ahlström in the preparation of some of the GaAs samples involved in this study. In addition, we thank the Swedish Natural

Science Research Council as well as the Swedish Board for Technical Development for financial support.

References

[1] R. N. Hall and J. H. Racette, J. Appl. Phys. 35, 379(1964).

[2] H. J. Queisser and C. S. Fuller, J. Appl. Phys. 37, 4895(1966).

[3] N. Kullendorf, L. Jansson, and L. Å. Ledebo, J. Appl. Phys. 54, 3202(1983).

[4] Z. G. Wang, H. P. Gislason, and B. Monemar, J. Appl. Phys. (to be published).

[5] H. P. Gislason, Z. G. Wang, and B. Monemar, J. Appl. Phys. (to be published).

[6] B. Monemar, H. P. Gislason, P. J. Dean, and D. C. Herbert, Phys. Rev. B 25, 7719(1982).

[7] H. P. Gislason, B. Monemar, P. J. Dean, D. C. Herbert, S. Depinna, B. C. Cavanett, and N. Killoran, Phys. Rev. B 26, 827(1982).

[8] H. P. Gislason, B. Monemar, M. E. Pistol, P. J. Dean, D. C. Herbert, S. Depinna, A. Kanaah, and B. C. Cavenett, Phys. Rev. B 31, 3774(1985).

[9] M. S. Skolnick, J. Dean, A. D. Pitt, Ch. Uihlein, H. Krath, B. Deveaud, and E. J. Foulkes, J. Phys. C 16, 1967(1983).

[10] H. P. Gislason, B. Monemar, and Z. G. Wang(unpublished).

[11] L. Samuelson, S. Nilsson, Z. G. Wang, and H. G. Grimmeiss, Phys. Rev. Lett. 53, 1501 (1984).

[12] E. F. Gross, V. I. Safarov, V. E. Sedov, and V. A. Maruschak, Fiz. Tverd. Tela (Leningrad) 11, 348(1969)[Sov. Phys. Solid State 11, 277(1969)].

[13] M. G. Milvidskii, V. B. Osvenskii, V. I. Safarov, and T. G. Yugova, Fiz. Tverd. Tela (Leningrad)13, 1367(1971)[Sov. Phys. Solid State 13, 1144(1971)].

[14] F. Willman, D. Bimberg, and B. Blätte, Phys. Rev. B 7, 2473(1973).

[15] H. P. Gislason, B. Monemar, Z. G. Wang, Ch. Uihlein, and P. L. Liu(unpublished).

[16] For a review of basic properties of bound excitons in semiconductors, see P. J. Dean and D. C. Herbert, Excitons, Vol. 14 of Topics in Current Physics, edited by K. Cho (Springer, Berlin, 1979).

[17] W. Schairer, D. Bimberg, W. Kottler, K. Cho, and M. Schmidt, Phys. Rev. B 13, 3452(1976).

[18] A. M. White, P. J. Dean, K. H. Fairhurst, W. Bardsley, and B. Day, J. Phys. C 7, L35 (1974).

[19] V. Gutkin, Fiz. Tverd. Tekh. Poluprovodn. 15, 659 (1981); 17, 97 (1983) [Sov. Phys. Sernicond. 15, 1145(1981); 17, 61(1983)].

[20] R. F. Kirkman, R. A. Stradling, and P. J. Lin-Chung, J. Phys. C11, 419(1978).

[21] K. R. Elliot, Appl. Phys. Lett. 42, 474(1983).

[22] G. Kirczenow, Solid State Commun. 21, 713(1977).

[23] H. Nakata and E. Otsuka, Phys. Rev. B29, 2347(1984).

[24] R. Sauer, Phys. Rev. Lett. 31, 376(1973).

[25] E. E. Hailer, in Proceedings of the XⅧth International Conference on the Physics of Semiconductors, San Francisco, 1984(unpublished).

[26] B. Monemar, H. P. Gislason, and P. O. Holtz(to be published).

[27] G. E. Stillman, D. M. Larsen, C. M. Wolfe, and R. C. Brandt, Solid State Commun. 9, 2245 (1971).

[28] D. J. Wolford, J. A. Bradley, K. Fry, and J. Thompson, in Proceedings of the XⅧth International Conference on the Physics of Semiconductors, San Francisco, 1984, Ref. 25.

[29] U. Lindefelt(unpubished).

[30] W. Schairer and M. Schmidt, Phys. Rev. B 10, 2501(1974).

[31] M. Ilegems, R. Dingle, and L. W. Rupp, Jr., J. Appl. Phys. 46, 3059(1975).

[32] J. W. Allen, P. J. Dean, and A. M. White, J. Phys. C 9, L113(1976).

[33] R. L. Hemstreet, Phys. Rev. B 22, 4590(1980).

[34] W. Schmid and P. J. Dean, Phys. Status Solidi B 110, 591(1982).

Direct evidence for the acceptorlike character of the Cu-related *C* and *F* bound-exciton centers in GaAs

H. P. Gislason, B. Monemar, and Z. G. Wang

(Department of Physics and Measurement Technology, Linköping University, S-581 83 Linköping, Sweden; Present address: Institute of Semiconductors, Chinese Academy of Science, Beijing 100083, China)

Ch. Uihlein and P. L. Liu

(Max-Planck Institut für Festkörperforschung, Hochfeld Magnetlabor, Grenoble, 166X 38042 Grenoble, Cedex, France; Present address: Shanghai Institute of Technical Physics, Chinese Academy of Science, Shanghai, China.)

Abstract Zeeman measurements have been performed on Cu-doped GaAs in order to reveal the nature of the centers binding the Cu-related excitons C_0 at 1.5030eV and F_0 at 1.4839eV. Thermalization between the magnetic subcomponents in Zeeman transmission spectra shows that a splitting occurs in the ground state of the bound-exciton system in each case. In photoluminescence Zeeman spectra no thermalization is observed. From the diamagnetic shift observed for both centers it is concluded that each excitonic complex contains a weakly bound electron, localized by the Coulomb field of more tightly bound holes. Consequently, we definitely identify the *C* and *F* centers as neutral acceptor centers, which has hitherto been arbitrarily, albeit correctly assumed. The magnetic measurements lead to effective *g* values for the C_0 and F_0 bound excitons, $g_{\text{eff}}=2.44$ and $g_{\text{eff}}=2.30$, respectively. In both cases the splitting is almost isotropic. A model is proposed for the electronic configurations of these excitons in which the isotopic *g* value is explained by a quenching of the orbital angular momentum of the bound hole states through the action of a strong compressive strain field locally at the defect site. In this model the bound exciton consists of two spinlike holes with their spins coupled off, and an electron. The bound-exciton transition occurs between the $S=1/2$ state of the exciton to a spin-only $j_z=\pm1/2$ hole state at the neutral acceptor. This state has a hole *g* value close to the electron value $g=2$ for both acceptor centers. In addition we report a new unrelated line in the transmission spectra close to the C_0 line. We assign this one to an unidentified neutral isoelectronic donor complex from the thermalization behavior and the diamagnetic shift measured for this center.

1 Introduction

Cu is known to give rise to several defect centers in GaAs, the interrelation of which has

原载于: Phys. Rev. B, 1985, 32(6): 3723-3729.

been the cause of a long-standing confusion in the literature.[1] The most persistent of these defect centers are the 0.156eV and the 0.45eV acceptors, which have been studied using junction techniques,[2] Hall measurements,[3] and in the case of the former, infrared-absorption measurements. These acceptors are generally accepted to be related to Cu, although their microscopic identities have been subject to various hypotheses.[1,5] The 0.156eV acceptor gives rise to a photoluminescence (PL) band peaking at 1.36eV upon recombination of an electron from the conduction-band or shallow donor states. Through piezospectroscopic measurements of this PL band the 0.156eV acceptor level has been assigned to a ⟨100⟩-oriented Jahn-Teller-distorted Cu_{Ga}.[6] On the other hand, the identity of the 0.45eV acceptor center is more uncertain.

In this paper we focus on the two bound-exciton (BE) lines C_0 and F_0 at 1.5030eV and 1.4839eV, respectively.[7] (Other values have been reported, differing less than 0.1meV from ours, which were checked with calibration lines) In the literature the centers binding these excitons have often arbitrarily been assumed to be the two Cu-related acceptors discussed above. However, it has recently been shown that there exists no correlation between the occurrence of the C_0 line and the 1.36eV PL band in the photoluminescence spectra of Cu-doped GaAs.[1,8] Furthermore, the symmetry found for the C_0 line (trigonal ⟨111⟩-oriented[7]) does not agree with the recently reported symmetry of the 1.36eV center[6]. Similarly, it is by no means certain or even likely that the 0.45eV acceptor center is responsible for the F_0 bound exciton. In fact, it has never been proved that the centers binding these excitons are just acceptors.

In this paper we report new experimental results from magneto-optical spectroscopy on the C_0 and F_0 lines. Our Zeeman data for magnetic fields up to 10T definitely prove that both BE systems involve acceptor associates. Further, we discuss the electronic structure of the BE states for these acceptors. In particular, the occurrence of a spinlike hole in the ground state of both BE systems is explained in a simple perturbation scheme. It is proposed that a local strain field is the primary perturbation, dominating over the spin-orbit splitting at the neutral Cu-acceptor associates.

2 Sample preparation and experimental procedure

The GaAs samples used in this part of a more extended investigation of Cu-doped GaAs (Ref. [1]) were made from liquid-encapsulated Czochralski (LEC)-type bulk material. The starting material was semi-insulating GaAs, nominally undoped with electron concentration $n_{300}=1.8\times10^{12}cm^{-3}$ and mobility $\mu_{300}=4200cm^2/Vs$[1]. To create strong Cu-related BE lines the Cu diffusion procedure was as follows: A thin film of Cu was evaporated on the polished and etched sample surface, typically 600Å, and the samples were immediately sealed in evacuated quartz tubes. The C_0 and F_0 lines are strongest after a Cu diffusion at about 700℃ for typically 30min. A crucial condition for a strong C_0 line is a fast quenching to room temperature either in water or liquid nitrogen, while the F_0 line tends to become stronger if the quenching rate is slower. A necessary condition for the appearance of the F_0 line is the absence of Ga-GaAs melt in the ampoule during the Cu diffusion. The intensity of the C_0 line is also affected by the presence of melt, although less. The samples used in this work show simultaneously strong C_0 and F_0 lines. This is accomplished through an optimization of the above conditions, that is, Cu diffusion at about 700℃ for 30min in vacuum followed by a rapid quenching to room temperature. The most favorable doping conditions for the different Cu centers in GaAs are discussed in more detail in a separate publication.[1]

The magneto-optical measurements were performed at the Max Planck Hochfeld Magnetlabor, Grenoble, employing a 10T superconducting magnet in the Voigt configuration. Measurements were usually made at 2K, but also in the temperature range up to 10K in order to study the thermalization behavior between the magnetic subcomponents. The signal was dispersed and recorded through a 1.5m Jobin-Yvon monochromator with high resolution. The photoluminescence was excited with the 5145Å Ar^+ laser line, and for zero-field PL measurements the signal was recorded through a 0.75m Jarrell-Ash double-grating monochromator.

3 Experimental results and discussion

Fig. 1 illustrates the PL spectrum of a sample showing strong C and F lines, that is, the electronic lines C_0 and F_0 with phonon replicas. A phonon replica C_1 represents a low-energy phonon mode of 3.6meV coupling to the C_0 line, with second and third replicas clearly resolved in the spectrum, C_2 and C_3, respectively. The LO replicas C_0^{LO} and C_1^{LO} are also observed in the spectrum. In the F spectrum a low-energy mode of 6.2meV, F_1, and the LO replica, F_0^{LO}, are present.

In order to tell whether an exciton is bound to an isoelectronic center or a neutral donor or acceptor associate, the thermalization behavior in both the excited state of the system (the center with the bound exciton) and the ground state of the system (the center without the bound exciton) has to be investigated. In addition, for the latter type of BE system it must be possible to distinguish between a loosely bound and a tightly bound particle to decide whether a center is an acceptor or a donor.

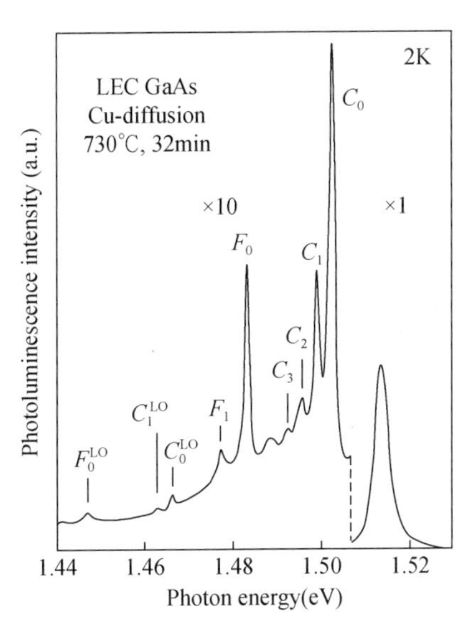

Fig. 1 Typical photoluminescence spectrum of a semi-insulating GaAs sample of Czochralski type, Cu-diffused at 730℃ for 32min in vacuum. Both the C_0 and F_0 BE lines are strong in this spectrum. Three local-mode replicas of the C_0 line are resolved, labeled C_1, C_2, and C_3, as well as the LO replicas C_0^{LO} and C_1^{LO}. In the F spectrum a local-mode replica F_1 is observed together with the LO replica of F_0

The C center binding the C_0 exciton has previously been shown to possess a trigonal ⟨111⟩-oriented symmetry axis, while the F center binding the F_0 exciton is orthorhombic with one axis oriented in the ⟨100⟩ direction. [7] A magnetic field in the ⟨100⟩ direction should give the simplest splitting pattern for the trigonal C center since all four possible lattice orientations of the defect are equivalent in that case. For the F center this direction of the magnetic field is parallel with one of the defect orientations, e. g., [100], and perpendicular with two, [010] and [001].

The transmission spectrum of the same crystal as in Fig. 1 is shown in Fig. 2. The zero-field spectrum is shown as well as the Zeeman spectrum for magnetic field H ∥ ⟨100⟩ and k ⊥ H for fields up to 10T. In the zero-field transmission spectrum the C_0 line and the F_0 line are clearly resolved at the same energy positions as in the PL spectrum of Fig. 1. In addition, a new line of unknown identity appears in the spectrum at 1.5022eV, or about 1meV below the C_0 line position. This line does not appear in the PL spectrum of Fig. 1, nor has it been observed in PL. spectra of higher resolution. Consequently, it cannot be an electronic line belonging to the C spectrum, since it is not observed in emission despite its energy being lower than that of the C_0 line. It is not a phonon replica either because it appears in the Stokes wing on the low-energy side of the C_0 line in absorption. Further, as discussed below, its thermalization behavior rules out any connections with the C_0 line.

As is obvious from Fig. 2 and, in greater detail, Fig. 4 below, the splitting and

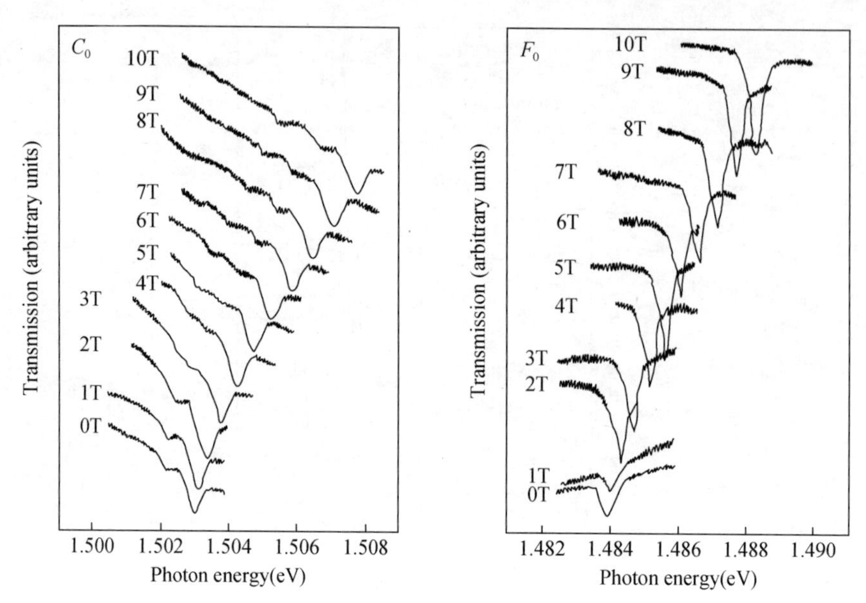

Fig. 2　Transmission spectra of the sample in Fig. 1 in magnetic fields from 0 to 10T, with H ∥ 〈100〉 and k ⊥ H. Three different lines are observed at zero field, the Cu-related BE lines C_0 at 1.5030eV and F_0 at 1.4839eV, together with a new unidentified line at 1.5022eV. Strong thermalization is observed for the Cu-related lines, where only one magnetic subcomponent is observed at all fields (the one at highest energy). The 1.5022eV line shows negligible thermalization

thermalization of this new line in magnetic field differ from those of the F_0 line, which makes it unlikely that it is a high-energy component of the F spectrum. The 1.5022eV line is not detectable in the absorption spectrum reported by Gross et al. [7]

At first sight perhaps one does not realize the strong thermalization between the magnetic subcomponents of both the C_0 and F_0 lines in Fig. 2. Only a single highest-energy component is observable in each case, whereas two components of the 1.5022eV line can be traced up to 10T. The thermalization behavior becomes obvious upon comparison with the PL spectrum measured at 10T for the same magnetic field configuration as presented in Fig. 3. Here two strong components of approximately equal strength are present. The energy position of the high-energy component in PL coincides with the transmission line at 10T for both centers.

An immediate consequence of this thermalization behavior is a firm exclusion of a hypothetical isoelectronic center binding either of the C_0 or F_0 excitons. Since the ground state of such a system contains no electronic particles, the magnetic splitting must occur in the excited state and thermalization will therefore be observable only in luminescence. The assignment of both centers to acceptors but not donors is not possible without taking into account the diamagnetic shift of the Zeeman components, however. This shift is obvious already in Fig. 2, and will be discussed in connection with Fig. 4 below.

The arbitrary assignment of the C and F lines to the 0.156eV and the 0.45eV acceptor centers[9] in previous work possibly made comments on the negligible thermalization in the Zeeman split PL spectrum unnecessary. The published spectra of Willmann et al.[9] are, however, not decisive as to whether the centers are, in fact, acceptors or isoelectronic associates. They observe four lines at the relatively small magnetic field of 5.7T for H ∥ ⟨100⟩ in the case of the C center and H ∥ ⟨110⟩ for the F center. In our measurements we only resolve two rather broad components. This is due to the lower resolution of our experiment, which was designed for transmission measurements but not for PL measurements. Consequently, bulk material was used in this work instead of epitaxial wafers, which give smaller PL linewidths.

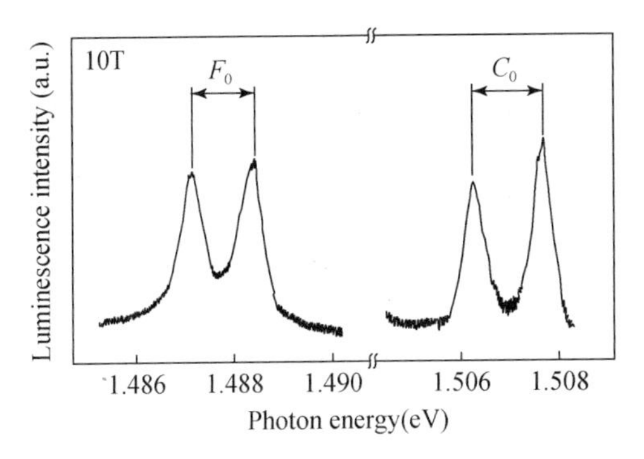

Fig. 3 Zeeman splitting of the C_0 and the F_0 lines at 10T in photoluminescence(H ∥ ⟨100⟩, k ⊥ H). Obviously no thermalization between the two resolved magnetic subcomponents is observed in contrast to the transmission spectra. This splitting, consequently, occurs in the ground state of the BE system

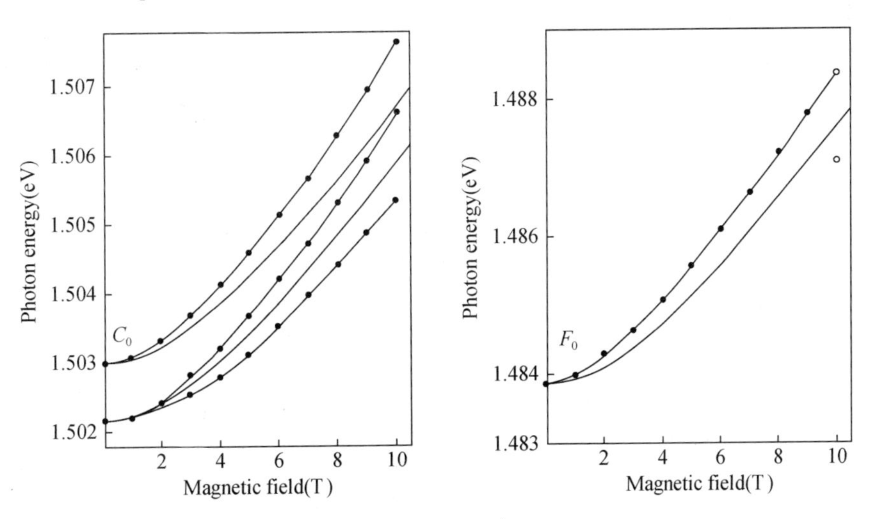

Fig. 4 Zeeman splitting of the transmission spectrum of Fig. 2 for magnetic fields from 0 to 10T, with H ∥ ⟨100⟩ and k ⊥ H. Solid circles are experimental points from the transmission measurements, whereas the open circles inserted for 10T are the corresponding experimental points from PL measurements. The solid lines are the diamagnetic shift for an effective-mass donor electron calculated from Ref. [12]

The two broad components of the C_0 line in PL at 10T for H ∥ ⟨100⟩ in Fig. 3 have half-widths of 0.48meV and consist of two unresolved subcomponents each. These are separated by approximately 0.2meV at 10T, the inner components being weaker according to Willmann et al. [9] The separation between the two outer Zeeman components of the C_0 line in Fig. 3 is 1.4meV. This can be interpreted as the splitting of the hole states of the neutral acceptor in a magnetic field of 10T at an angle of 54.7° to the defect axis. The smaller unresolved splitting within each broad component in Fig. 3 would then be the corresponding splitting of the electron, as will be postulated below in view of the collected evidence.

This interpretation of the PL data is by no means the only possible one, since a similar pattern of Zeeman components is expected for an exciton bound to an isoelectronic center in case both the electrons and holes are spin-1/2 particles. The former has a negative g value, as is usual for shallow donors in GaAs, and the latter a g value close to the spin value $g=2$. In the Paschen-Back limit, where the Zeeman splitting exceeds any electron-hole exchange splitting, the transition probabilities are governed by the selection rule $\Delta m_s=0$, and a four-line pattern with the outer pair of subcomponents being the strongest is expected. Such a splitting has been reported for an isoelectronic Cu center in InP (where the g values of both particles are positive, however). [10] In the case of an isoelectronic center, thermalization would be expected in PL measurements given that the sample heating from the laser excitation is negligible and that the spin relaxation time is shorter than the recombination rate. This possibility has to be disregarded for the C and the F centra in view of the transmission measurements reported here, however.

The Zeeman pattern for the F_0 line is similar to that of the C_0 line. Here the direction of the magnetic field is not equivalent for all defect orientations and different from that in Ref. [9]. This is unimportant, however, because of the isotropic behavior of the splitting observed. We return to the question of g values below.

In Fig. 4 the Zeeman splitting of the three BE lines in the transmission spectrum is plotted for magnetic fields from 0 to 10T, H ∥ ⟨100⟩ and k ⊥ H, as in Fig. 1. The experimental points at 10T from the PL measurements are included in the figure. A striking feature of the magnetic field splitting pattern is the large diamagnetic shift observed for all three lines, about 4meV at 10T. This is close to the shift of electrons bound to effective-mass-like donors in GaAs. [11] In Fig. 4 the diamagnetic shift for such electrons in GaAs as obtained from a calculation of the diamagnetic shift of the 1s hydrogenic ground state of shallow donors[12] has been included. For scaling this calculation to fit shallow donors in GaAs, the effective Rydberg $R^* = m^* e^4/2\varepsilon^2\hbar^2 = 5.52\text{meV}$, with the conduction band effective mass $m^*=0.067m_0$, was used as scaling parameters.

The shift of the center of gravity of the Zeeman patterns for the C_0 and the F_0 lines (measured from the PL data at 10T) agree well with the calculated diamagnetic shift of a shallow donor electron.

For the F_0 BE the diamagnetic shift is about 0.16meV larger than for the effective-mass donor electron as calculated above. This corresponds to a slightly smaller effective mass, $m^*=0.064m_0$, if m^* is taken as an adjustable parameter. For the C_0 BE the observed shift is 0.25meV larger than the donor-electron value, corresponding to $m^*=0.063m_0$. This close agreement strongly suggests that a pseudodonor model is valid for both centers. In this model an electron is loosely bound in the Coulomb field of more tightly bound holes at the center, which, consequently, is an acceptor in view of the thermalization behavior. Hence, the combined information from the thermalization and the diamagnetic shift allows us to reach the conclusion that both the C and the F centers are, in fact, acceptors. Both the Zeeman splitting in luminescence and the diamagnetic shift could, however, be in agreement with isoelectronic donors, that is, isoelectronic centers with hole-attractive central cells. The transmission measurements thus provide the necessary piece of evidence in form of the thermalization behavior (which in itself is insufficient, since it is consistent with a model of excitons bound to either donors or acceptors).

The diamagnetic shift of the C_0 and the F_0 lines has been reported previously[13] from Zeeman PL measurements giving a similar value to the one we find. In that work it was, however, presupposed that both excitons were bound to acceptors. The interpretation was also different, since the diamagnetic shift was ascribed to a free exciton which has been pointed out to be incorrect if the exciton-localization energy exceeds the free-exciton binding energy.[11]

As far as the 1.5022eV line is concerned, no thermalization is observed between its two magnetic subcomponents in the transmission measurements. No corresponding BE line was resolved in PL measurements in our samples. In the absence of thermalization in the ground state of the BE system, it is then natural to assume that the center binding this exciton is of isoelectronic nature. In luminescence from an excited state of a BE system, spin-lattice relaxation has to be faster than the recombination rate for thermalization to occur. Thermalization in the ground state of an acceptor-related BE system should always be observable, on the other hand, since the ground state has a practically infinite lifetime.

In view of the diamagnetic shift of the Zeeman components of the 1.5022eV line it can be concluded that a pseudodonor model is also applicable in this case. The agreement with the expected value for an effective-mass donor is even closer in this case than in the previous ones, a perfect fit to the center of gravity corresponds to $m^* = 0.065m_0$. The center consequently is an isoelectronic donor, consisting of a hole-attractive central cell binding the hole in a short-range potential and the electron by the Coulomb potential of the hole. The effective g value for this exciton is $g=2.2$.

In Fig. 5 a schematic level diagram for the case of excitons bound to neutral acceptors is shown. The small g value for weakly bound electrons in CaAs, $g_e=-0.46$, causes only a small splitting, about 0.27meV at 10T or below our resolution limit. Thus the major part of the splitting is caused by the hole bound at the neutral acceptor. This is illustrated in Fig. 5. The outer components are shown stronger in the diagram, in agreement with the spin electron rule $\Delta m_s=0$, which is valid in the electron picture (the hole corresponds to a missing electron with $m_e=-m_h$). The level diagram is drawn in this picture.

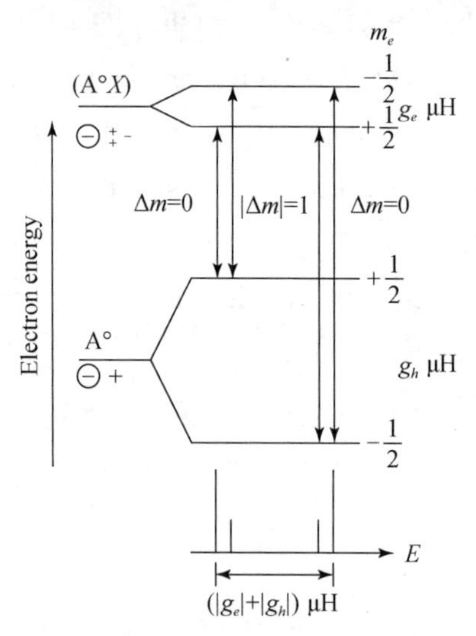

Fig. 5　Schematic level diagram illustrating the Zeeman splitting of the C_0 and the F_0 BE lines. The small (unresolved) splitting in our data is ascribed to the electrons with g value $g_e=-0.46$, while the holes have g value close to $g=2$. The figure illustrates the spin selection rule $\Delta m_s=0$ in the electron picture ($m_e=-m_h$)

The Zeeman splitting of the C_0 and the F_0 lines has been reported to be isotropic in PL data with a small anisotropy present in the case of the C_0 line. [9] Since both centers have noncubic symmetry, [7] it is natural to assume that the angular orbital momentum of the hole states is quenched as a result of a compressive low-symmetry strain field. [10, 14] This would explain the absence of anisotropy in the Zeernan splitting. For the C center the angular dependence of the hole states as a function of the angle θ between the direction of the magnetic field and the trigonal axis can be written $\Delta E=g(\theta)\mu H$, where the g value for the holes, $g(\theta)$, involves the angular dependence. A similar expression is valid for the orthorhombic symmetry of the F center. In the case of a compressive trigonal field, the

Kramers doublet $|3/2, \pm 1/2\rangle$ in $|J, M_J\rangle$ notation would be the ground state of the neutral acceptor (see below) with the g value $g(\theta) = (g_{\parallel}^2 \cos^2\theta + g_{\perp}^2 \sin^2\theta)^{1/2}$. For a tensional axial crystal field the ground state of the holes would be $|3/2, \pm 3/2\rangle$ and the angular dependence of the g value is then simply $g(\theta) = g_h \cos\theta$.[15]

From the absence of anisotropy in the hole g value in the Zeeman data, it is clear that neither of these two cases applies to the two acceptors of this study. Instead, the simplest expression for isotropic splitting is valid both for electrons and holes: $E = g_{\text{eff}} \mu H$, where g_{eff} is the effective g value. The total splitting is therefore, as is obvious from Fig. 5, $E = (|g_e| + |g_h|)\mu H$, which gives for the C center $g_h = 2.44 - 0.46 = 1.98$ and for the F center $g_h = 2.30 - 0.46 = 1.84$. Both g_h values are relatively close to the spin value $g = 2$, which agrees with the hypothesis of a quenched angular orbital momentum of the holes bound to the defects. A necessary condition for this to occur is the combination of a strong noncubic strain field of a compressive sign and a localized hole wave function, overlapping the strain field created locally at the defect. This is reasonable for the case of Cu-related defects, which are expected to have strongly hole-attractive central cells. Similar effects have been observed for complex Cu-related defects in GaP (Refs. [16] and [17]) and in InP (Ref. [10]). Another condition for a complete quenching of the hole orbital angular momentum is that the spin-orbit interaction of the hole states derived from the valence band can be regarded as a small perturbation of the dominating local strain field. In GaAs the spin-orbit splitting is 340meV, or 4 times larger than that of GaP. This requires a strong local strain field and highly localized hole wave functions. In ZnTe, however, a similar phenomenon has been observed for Cu-related centers, despite the much larger spin-orbit splitting 0.9eV.[18] Also, InP has a much larger spin-orbit splitting, or close to 1eV.[19]

For the case of a compressive sign of the local strain field, but of a smaller magnitude, so that the spin-orbit interaction would be the primary perturbation, the hole states of the neutral acceptor are properly described in terms of the projection M_J of the total angular momentum J along the defect axis. Then the $P_{3/2}$ hole state of cubic symmetry splits into the Kramers' doublets $|3/2, \pm 3/2\rangle$ and $|3/2, \pm 1/2\rangle$ in $|J, M_J\rangle$ notation. The sign of the strain field raises the energy of the former doublet and lowers the energy of the latter one, $|3/2, \pm 1/2\rangle$.[20] Willmann et al.[9] ascribe the $|3/2, \pm 1/2\rangle$ state to the bound hole state of the neutral acceptor for both the C and F centers, while the $|3/2, \pm 3/2\rangle$ state is suggested to merge into the valence band by the action of the strain field. This model does not explain the isotropic behavior of the Zeeman pattern, however, but would instead give the anisotropic g value given above. In view of the spinlike hole states expressed by the isotropic g value for both centers, it must instead be concluded that the strain field is the primary perturbation, with the smaller spin-orbit interaction being superimposed, as illustrated in Fig. 6 for both the C and the F centers. In the level diagram the BE energy is drawn so that transitions to acceptor states of higher binding energy have higher transition energy and, consequently, the BE energy increases vertically downwards as the hole energy. This simplified diagram (which is valid only for the $J=0$ hole state of the BE, where the hole spins are coupled off) accounts for the increase of the magnetic subcomponents of higher transition energy upon thermalization to hole states of lowest hole energy in transmission measurements at low temperatures. In this picture we assume that the compressive strain field increases the hole energy of the $p_{\pm}$ states, which are proposed to merge into the valence band. The ground state of the neutral acceptor is derived from the p_0 state, which is lowered in energy by the strain field. Upon spin-orbit coupling these states are denoted $|0, \pm 1/2\rangle$ in the $|l_z, j_z\rangle$ notation, [14] where I_z and j_z are projections along the defect axis z. The exciton is formed by the coupling of two such spinlike holes, which pair off their spins, in addition to an electron, resulting in a $S=1/2$ BE state. The electronic transitions occur between this BE state and the spinlike $|0, \pm 1/2\rangle$ hole states of the neutral acceptor.

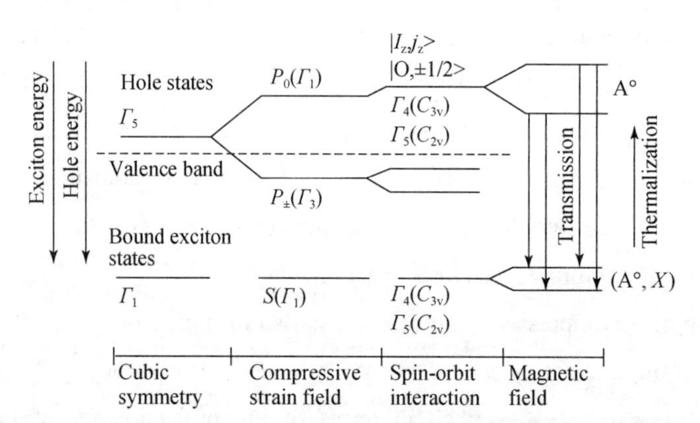

Fig. 6　Schematic level diagram for the C_0 and the F_0 centers as explained in the text. The primary perturbation is ascribed to the crystal field with the spin-orbit interaction of the valence band being superimposed. The BE states are drawn towards higher hole energy, which illustrates the thermalization in the ground state of the BE system favoring higher-energy components at low temperature

This picture supports the model for the C and the F centers put forward in a separate publication, suggesting the involvement of Cu interstitials in both. Such defect associates can naturally be assumed to create compressive strain locally at the defect sites through the presence of the intersitial species. [1]

To summarize, we have shown that both the C_0 and F_0 excitons in Cu-doped GaAs are bound to complex acceptors, creating strong compressive local strain field of noncubic symmetry. We have also shown that previous assignments of these spectral features as excitons bound to acceptors were only coincidentally true, since the formerly available data were not conclusive on this point. The erroneous assignment of these bound excitons to well-known Cu acceptors was possibly misleading in this respect as well.

References

[1] Z. G. Wang, H. P. Gislason, and B. Monemar, J. Appl. Phys. 58, 240(1985).

[2] N. Kullendorff, L. Jansson, and L. Å. Ledebo, J. Appl. Phys. 54, 3203(1983).

[3] R. N. Hall and J. H. Racette, J. Appl. Phys. 35, 379(1964).

[4] F. Willmann, M. Blätte, H. J. Queisser, and J. Treusch, Solid State Commun. 9, 2281(1971).

[5] H. J. Guislain, L. De Wolf, and P. Clauws, J. Electron. Mater. 7, 83(1978).

[6] N. S. Averkiev, T. K. Ashirov, and A. A. Crutkin, Fiz. Tekh. Poluprovodn. 15, 1970(1981); 17, 97(1983)[Sov. Phys. Semicond. 15, 1145(1981); 17, 61(1983)].

[7] E. F. Gross, V. I. Savarov, V. E. Sedov, and V. A. Maruschak, Fiz. Tverd. Tela(Leningrad)2, 348(1969)[Sov. Phys. Solid State 11, 277(1969)].

[8] T. K. Ashirov and A. A. Gutkin, Fiz. Tekh. Poluprovodn. 16, 163(1982)[Sov. Phys. Semicond. 16, 99(1982)].

[9] F. Willmann, D. Bimberg, and M. Blätte, Phys. Rev. B 7, 2473(1973).

[10] M. S. Skolnick, P. J. Dean, A. D. Pitt, Ch. Uihlein, H. Krath, B. Deveaud, and E. J. Foulkes, J. Phys. C 16, 1967(1983).

[11] W. Schairer, D. Bimberg, W. Kottler, K. Cho, and M. Schmidt, Phys. Rev. B 13, 3452(1976).

[12] D. Cabib, E. Fabri, and G. Fiorio, Nuovo Cimento 10B, 185(1972).

[13] F. Willmann, W. Dreybrodt, and M. Bettini, Phys. Rev. B 8, 2891(1973).

[14] H. P. Gislason, B. Monemar, P. J. Dean, and D. C. Herbert, Physica 117&118B, 269(1983).

[15] A. Abragam and B. Bleaney, in Paramagnetic-Resonance in Transition Ions (Clarendon, Oxford, 1970).

[16] B. Monemar, H. P. Gislason, P. J. Dean, and D. C. Herbert, Phys. Rev. B 25, 7719(1982).

[17] H. P. Gislason, B. Monemar, P. J. Dean, D. C. Herbert, S. Depinna, B. C. Cavenett, and N. Killoran, Phys. Rev. B 26, 827(1982).

[18] B. Monemar(unpublished).

[19] K. J. Bachmann, Annu. Rev. Mater. Sci. 11, 441(1981).

[20] J. van W. Morgan and T. N. Morgan, Phys. Rev. B 1, 739(1970).

混晶半导体中深能级的展宽及其有关效应*

王占国

（中国科学院半导体研究所，北京，100083）

摘要　本文以 $GaAs_xP_{1-x}$ 为例，在假定 As-P 原子在Ⅴ族点阵位上随机分布情况下，从基本统计理论出发，计算了找到 k 个 As 原子的概率 F 以及能级展宽的宽度与组分的依赖关系。在进一步假定能级展宽服从高斯分布的情况下，计算了混晶半导体中载流子通过深能级中心的热、光发射电容瞬态过程。理论计算和实验结果的很好一致，不但成功地解释了混晶中来自深中心载流子的非指数热发射和俘获电容瞬态过程的物理实质，而且通过对实验结果的拟合还能给出表征混晶半导体中深能级特征的两个重要参数：深能级的展宽宽度 E_B 和平均热激活能 E_T。

Deep level broadening and related effects in semiconductor alloys

Wang Zhanguo

(Institute of Semiconductors. Chinese Academy of Science, Beijing 100083, China)

Abstract　The probability of finding k arsenic atoms in a region of the As-P sublattice containing n atoms has been calculated by using fundamental statistical theory, and the composition dependence of deep level broadening is also determined.

Further, assuming that the deep level broadening can be described by a Gaussian distribution function centered around a mean value E_T, thermal and optical emission capacitance transient processes of carriers via a deep level in semiconductor alloys have been simulated. A good agreement between the experimental results and calculations shows that the particular nonexponential transient behaviour for thermal emission and capture of carriers in semiconductor alloys can be successfully explained by a model in which the energy levels are broadened. Two characterizing parameters of deep level in semiconductor alloy, E_B (a measure of deep level broadening) and E_T (average thermal activation energy) are also given.

---

* 原载于：半导体学报，1985，6（2）：132-141.
本文实验工作是在瑞典隆德大学固体物理系完成的。

1 引言

近年来，对混晶半导体中所特有的载流子通过深能级中心的非指数俘获和发射过程的物理实质的研究，引起了实验和理论工作者的广泛兴趣。众所周知：高缺陷浓度（N_T/N_S>0.1）[1,2]、自由载流子带尾[3]、电场及微观势垒效应[4,5]等都会导致深中心的非指数俘获和发射。然而，这些理论都不能满意地用来解释在混晶中所观察到的基本实验结果[6,7]。

Ngai[8]进一步发展了Wigner的所谓“红外发散瞬态响应”理论（即在一个具有大量相关态的体系中，电子态的突然改变会引起低能激发和消激发，这种过程导致非指数的发射和俘获瞬态）[9]，并用来解释在电容介质里普遍存在的所谓低频涨落、耗散和弛豫性质。例如：在电容瞬态和深能级瞬态能谱里，单指数瞬态 $\exp(-t/\tau)$ 将被相关态的红外发散响应（幂数为 n）修正为 $\exp(-Kt^{\alpha}/\tau)$，其中 $K=\exp(-nr)/(1-n)E_c^n$，$\alpha=1-n$，E_c 为相关态激发能量的上限，$r=0.5722$，如果这个理论成立，那么深能级暗电容瞬态值与时间必有 t^{α} 的依赖关系。然而对 $Al_xGa_{1-x}As$：Fe 和 $GaAs_xP_{1-x}$：Cu 中与铁和铜相关的深中心暗电容发射瞬态的实验研究表明，不存在上述所预示的 t^{α} 关系[7]。

另外，从统计观点来看，在 $Al_xGa_{1-x}As$、$GaAs_xP_{1-x}$ 等混晶半导体中，Al-Ga、As-P原子分别在Ⅲ族和Ⅴ族子点阵上的分布是随机的，由此而导致的偏离宏观值的局部组分的涨落必将反映到局域电子（或空穴）势能上。实验上已观察到束缚在杂质上的激子线的非均匀展宽[10,11]；同时，不少理论工作者[12-14]对此也做了相应的理论分析。

已经知道，束缚在深能级杂质上电子（或空穴）的能量依赖于组分[15]，围绕着宏观平均组分的涨落则必然引起束缚能的展宽，但这与所谓的“集合”现象不同，因为每个杂质位仍具有自己确定的束缚能。可以预料，能级的展宽对深能级杂质特别重要，因为束缚在深能级上的电子（或空穴）相对地被局域在晶体中非常小的区域里，而区域越小，偏离宏观平均组分的概率就越大。

本文基于As-P或Al-Ga等系统的随机分布，从基本的统计理论出发，计算了能级展宽宽度与组分的依赖关系。在假定能级展宽服从高斯分布情况下，对来自深能级到一个带的载流子热、光发射电容瞬态实验曲线成功地进行了理论拟合，并给出了表征混晶中深能级中心的两个重要参数：E_B 和 E_T。

2 样品制备和实验现象的观察

实验采用的样品是 N^+P 和 P^+N 二极管。不同组分 x 的 P 和 N 型 $Al_xGa_{1-x}As$ 和 $GaAs_xP_{1-x}$ 是用LPE法分别生长在GaAs或GaP低阻衬底上。在575－855℃温度范围内进行铜和铁扩散，选择扩散时间以控制深能级浓度远小于浅杂质浓度（$N_T/N_S \leq 0.1$），并保证其均匀分布。组分 x 用带有电子激发 x 射线分析仪的扫描电镜测得，误差小于±2％。

实验样品安放在可变温恒温器中，并被直接与其接触的交换气体所冷却。在整个温度范围内（10－400K）系统的热稳定性优于±0.1K。光电容和暗电容瞬态讯号用 Boonton 72B 电容计测量，模拟输出直接用函数记录仪或瞬态存储示波器记录。对实验获得的不同温度下电容瞬态过程的理论拟合是在 9825A 计算机和 9872A 绘图仪上进行。

典型的来自 $Al_xGa_{1-x}As$: Fe 深中心上的空穴热发射和电子俘获电容瞬态过程的实验曲线如图 1（a）和（b）所示。很显然，强烈的非指数电容瞬态过程使我们不可能简单地用一个确定的时间常数来表征它，从而也无法精确地计算深能级的激活能。不少作者[16,17]仍沿用 DLTS 技术来研究混晶半导体中的深能级。应当指出，基于单指数瞬态过程的 DLTS 技术，从原则上来说是不适用的。因为（1）深能级态密度（N_T）不能简单地按照 DLTS 讯号峰值高度来计算；（2）即使在较小的温度范围内，热发射率的 Arrhenius 作图，其斜率也不是很好的直线，这给研究混晶半导体中深能级的行为带来了很大困难。但对某些特殊情况，如 $GaAs_xP_{1-x}$ 中的 *EL*2 能级，在假定俘获截面不依赖温度或具有热激活过程情况下，Omling 等[18]基于我们早先提出的能级展宽模型[19]对 DLTS 资料的模拟表明：它仍能给出深能级的平均热激活能。关于这一点，我们还将在下文中进行讨论。

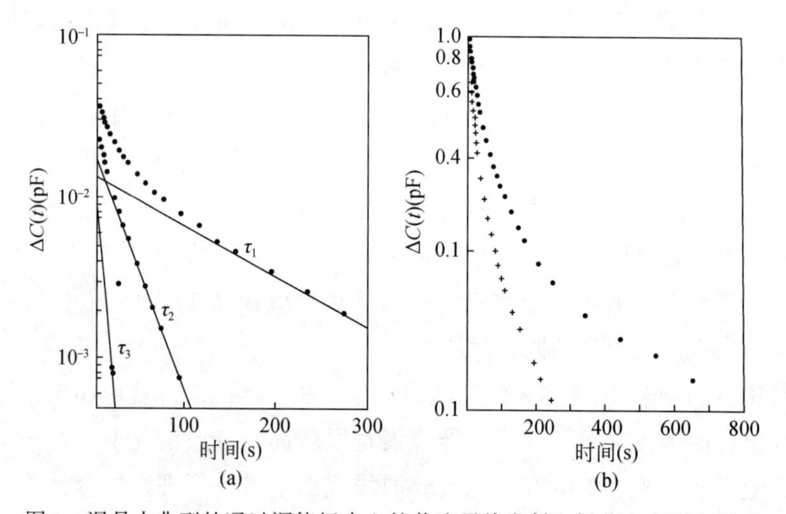

图 1　混晶中典型的通过深能级中心的载流子热发射和俘获电容瞬态过程

（a）$Al_{0.07}Ga_{0.93}As$: Fe 中与 Fe 相关的深能级的空穴在 T＝198.5K 热发射电容瞬态过程，显然，它不能用一个确定的时间常数来描述

（b）$Al_xGa_{1-x}As$: Fe 中与 Fe 相关的深能级的电子俘获电容瞬态过程

● x＝0.49，T＝172K；＋x＝0.20，T＝97K

3 深能级展宽效应及其描述

混晶半导体如 $GaAs_xP_{1-x}$、$Al_xGa_{1-x}As$ 等每立方厘米包含着大约 5×10^{22} 个原子。就 $GaAs_xP_{1-x}$ 而言，其中一半属于 As-P 子点阵。对包含有 n 个原子的 As-P 系统，假定 As、P 原子混合在一起，根据基本的统计理论，找到 k 个 As 原子的概率可由下式表示

$$F = x^k(1-x)^{n-k}\frac{n!}{k!\,(n-k)!}。\tag{1}$$

其中，k 表示在包含有 n 个 As、P 原子的系统中 As 的原子个数。显然，在计算 F 时 k 的数值应从 0 到 n 变化。x 是 As 原子所占的平均分数。对于不同的晶体体积（即不同的 n），由式（1）计算得到的 $x=0.5$ 时的 F 函数如图 2 所示，从图 2 可以看到，即在较大范围内找到偏离其宏观平均组分（$x=0.5$）的概率仍然很高。另外，F 函数的半宽度随混晶的组分 x 而改变，并近似地服从简单的二项式规律。对此，将在下面进行讨论。

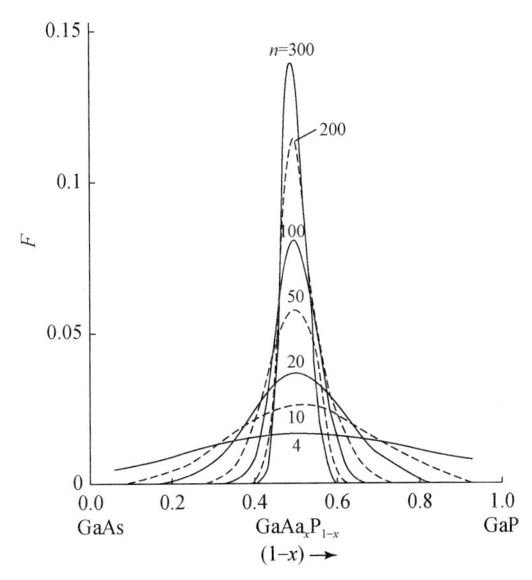

图 2 对包含有 n 个原子的 As-P 子点阵系统，按照式（1）算得的围绕平均组分 $x=0.5$ 找到 k 个 As 原子的概率

为了简单，本文进一步假定：在混晶中每个杂质位相对于一个带边具有唯一确定的束缚能，而这个束缚能是由局部组分决定的。由于混晶中局部组分从一个点阵位到另一个点阵位是涨落变化的，所以不同杂质位的集合所具有的束缚能也应是在一个相应能量范围内的某种分布。从上述简单的统计分析得知，局部组分涨落服从高斯分布，加之束缚能线性的依赖组分的实验事实，我们有理由认为能级的展宽也将服从高斯分布[19]（图 3）。

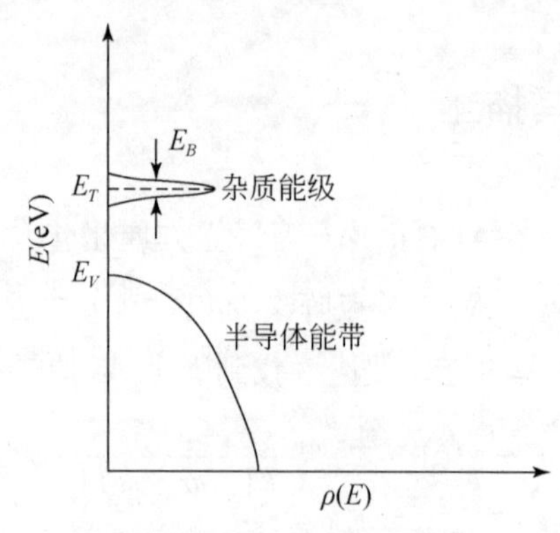

图 3　混晶半导体中深能级展宽模型示意图

$$\rho_T(E) = \rho_0 \exp - \left(\frac{\Delta E}{E_B}\right)^2 。\tag{2}$$

来自深能级载流子的热发射率可由细致平衡原理来描述

$$e^t = \frac{1}{\tau(E,T)} = e_0^t \exp - \left(\frac{E}{KT}\right) \tag{3}$$

$$E = E_T + \Delta E。\tag{4}$$

其中，$\rho_T(E)$为高斯分布函数，这里量纲为 $[\mathrm{ev}]^{-1}$。E_B 为度量能级展宽效应的参数，单位是 [eV]；E_T 为深能级平均能量。$e_0^t = \sigma_{p(n)}^t \langle V_{p(n)} \rangle N_{v(c)}$ 假定不依赖于混晶局部组分；$\sigma_{p(n)}^t$、$\langle V_{p(n)} \rangle$ 和 $N_{v(c)}$ 分别表示空穴（或电子）的俘获截面、热速度和价带（或导带）的有效态密度。

对光发射情况（低温，$e^t \approx 0$），方程（3）应由下式来代替，

$$e^0 = \frac{1}{\tau(E,hv)} = \phi\sigma_0\sigma(hv)。\tag{5}$$

其中，ϕ 为光通量（光子数/cm^2），σ_0 为不依赖于光子能量（hv）的常数。关于 $\sigma(hv)$的函数形式，已有不少文章做了详细讨论。为了方便，本文选用由 Lucovsky 最早建议的表达式[20]

$$\sigma(hv) = \frac{(hv - E)^{3/2}}{(hv)^3}。\tag{6}$$

另外，我们知道，对于简单的单指数瞬态情况，由于深能级上电子（或空穴）发射导致的结电容 $C(t)$ 与时间的依赖关系可由下式来描述

$$\Delta C(t) = C_0 \exp(-et),\tag{7}$$

这样，在混晶半导体中，来自深能级中心上载流子的热、光发射电容瞬态过程便可看成个以高斯分布函数为权重因子的、具有不同时间常数的电容瞬态过程的叠加，并可用下面的积分来表达，

$$\Delta C(t) = C_0 \int_E \rho_T(E) \exp - \left(\frac{t}{\tau(E,T)}\right) \mathrm{d}E,\tag{8}$$

对暗、光电容瞬态，$\tau(E,T)$分别用式（3）、式（4）、式（5）和式（6）代入式（8）

求解。

4 非指数电容瞬态的理论拟合

利用公式（8），对来自深能级的载流子的热、光发射电容瞬态的理论计算曲线如图4（a）和（b）及图5（a）和（b）所示。很明显，它们具有以下特征：①在混晶半导体中，由于能级的展宽（即使 E_B 较小）也将导致非指数的电容瞬态过程，随着 E_B 的增加，热、光电容瞬态偏离单指数过程越明显。②对相同的 E_B，展宽效应对热发射过程的影响远较比光发射过程为强（除靠近光阈值能量 E_{th}^0 外）。③在光电容测量中，初始斜率技术仍可使用［图5（b）］，但不适用于热瞬态；因为不但展宽效应对热瞬态初始部分有明显影响［图4（b）］，而且在较高温度下深能级的初始填充状态也将改变。

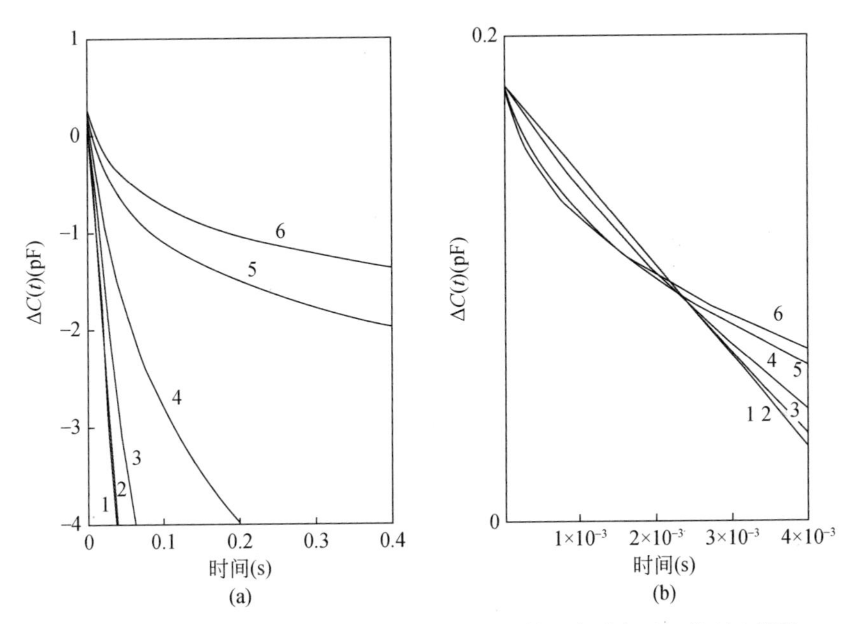

图4 混晶半导体中来自深能级载流子的热发射电容瞬态过程的理论模拟

（T=200K，E_T=0.4eV）

（a）图中编号1-6分别对应于能级展宽宽度 E_B 为3、5、10、20、40和50meV

（b）图（a）初始瞬态过程放大

为了验证上述假定的可靠性，应用公式（8），本文对 $GaAs_xP_{1-x}$：Cu 和 $Al_xGa_{1-x}As$：Fe 样品与铜和铁相关深能级的热、光发射电容瞬态的实验结果分别做了拟合运算，实验结果的理论拟合是非常成功的，如图6和图7所示。应当着重指出，这种拟合运算还能给出表征混晶半导体中深能级的两个重要参数——能级展宽宽度 E_B；平均热（$\bar{e}^t$）、光发射率（$\bar{e}^0$）。再者，从图7还可看出，愈接近光阈值能量的电容瞬态，受 E_B 的影响也就越大。当然，这会给精确计算光发射率带来一定困难，但由于光电容瞬态的初始部分仍基本上保持线性，所以用初始斜率技术所带来的误差不大。

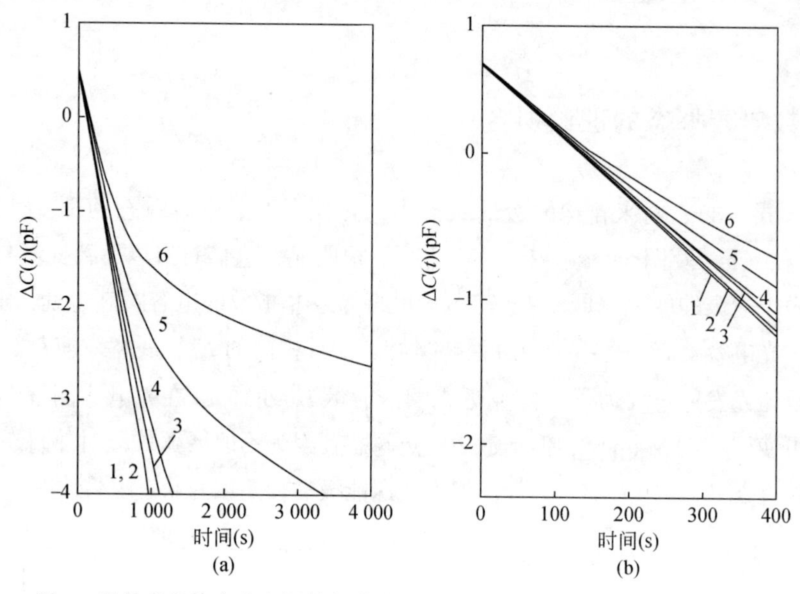

图 5　混晶半导体中来自深能级载流子的光发射电容瞬态过程的理论模拟

（平均光阈值 $E_{th}^{0}=0.63$eV）

（a）图中编号 1–6 分别对应于能级展宽宽度 E_B 为 10、20、40、50、75 和 100meV

（b）图（a）初始瞬态过程放大

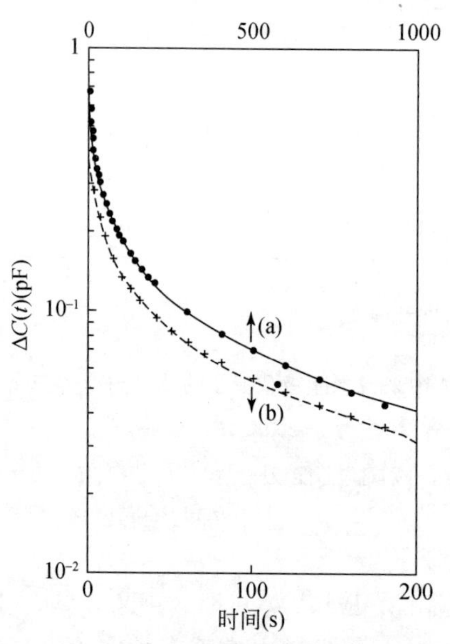

图 6　混晶半导体中深能级热发射电容瞬态实验结果的理论拟合

○，×，实验点；-----理论计算

（a）$GaAs_{0.6}P_{0.4}$：Cu，$T=208$K；拟合参数 $E_B=63$meV，$\bar{e}^t=5.50\times10^{-2}s^{-1}$

（b）$Al_{0.34}Ca_{0.66}As$：Fe，$T=168$K；拟合参数 $E_B=47$meV，$\bar{e}^t=1.51\times10^{-1}s^{-1}$

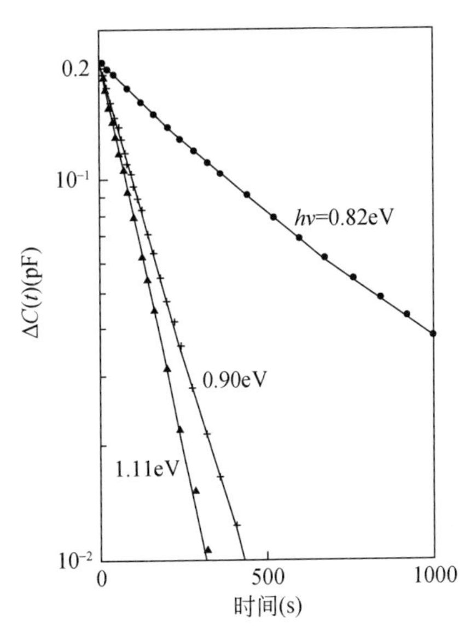

图 7 混晶 $Al_{0.49}Ga_{0.51}As$: Fe 中与铁相关深能级的光发射电容瞬态实验曲线（相应不同光子能量）的理论拟合，拟合参数 $E_B=50\text{meV}$

●，×，▲ 实验点；——计算

5 讨论

5.1 深能级展宽宽度 E_B 同组分 x 的依赖关系

以 $Al_xGa_{1-x}As$ 为例，对于体积大约为 40 个晶体原胞的局域中心（Al-Ga 子点阵系统），利用方程（1）计算了不同组分 x 时找到 k 个 Ga（或 Al）原子的概率 $F(x)$。如果用 F 分布函数的半宽度 E_B 同组分 x 作图（图 8 实线，并在 $x=0.5$ 时用实验值 ~50meV 归一化），发现两者的依赖关系近似地服从简单的二项式规律。

应用方程（8），对 $Al_xGa_{1-x}As$: Fe 样品与铁相关的深能级的空穴热发射电容瞬态实验曲线进行了理论拟合，求出了与不同 x 值相应的能级展宽宽度 $E_B(x)$（图 8 中黑点）。计算值和实验结果的拟合值之间的很好符合，再次表明上述的假定反映了混晶半导体中所观察到的非指数瞬态的物理实质。上述结果与 Wolford[21] 等所观察到的 PL（与浅能相关的发光带）展宽实验现象以及王永良[14] 等提出的晶格弛豫机构所给出的 S 因子对混晶组分的依赖关系基本一致。

应当指出，上述规律也适用于混晶 $GaAs_xP_{1-x}$: Cu 与铜相关的深能级中心。但考虑到 Cu_{Ga} 和 Fe_{Ga} 的近邻组态不同（Cu_{Ga} 的最近邻为 As-P 子点阵系统，而 Fe_{Ga} 的最近邻是 As，次近邻才是 Al-Ga 子点阵系统），所以，对于相同的 x，$(E_B)_{Cu}$ 大于 $(E_B)_{Fe}$ 是预料之内的。这一点已由我们的实验结果所证实（图 6）。

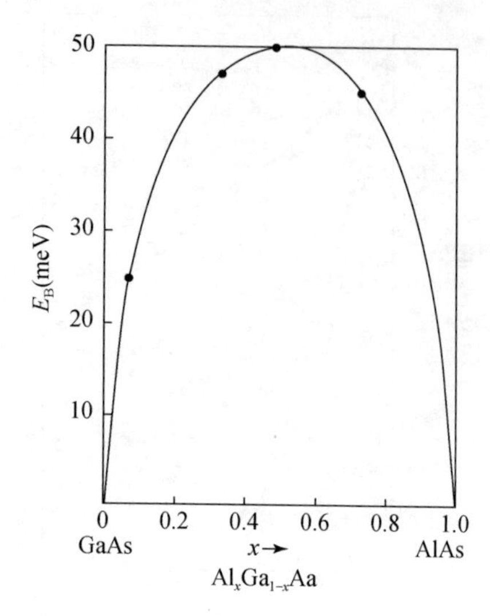

图 8 混晶中深能级展宽宽度 E_B 和混晶组分 x 的依赖关系。

实线和黑点分别表示计算和实验曲线拟合值

5.2 混晶半导体中深能级的平均激活能

前面已经说过，由于混晶半导体中能级展宽导致的非指数热发射和热俘获瞬态，使我们不能简单地应用基于分析单指数过程的 DLTS[22] 或求不同温度下单次瞬态时间常数方法来计算深能级在禁带中的位置。尽管可以通过对光电离截面与光子能量关系实验点的理论拟合来获得深能级的光阈值 E_{th}^0。但这个方法不仅耗时，而且在一定程度上还依赖于实验系统的灵敏度和所采用的拟合公式。再者，E_{th}^0 也不一定等于热激活能 E_T。

为了克服上述困难，本文基于能级展宽模型，利用公式（8）对 p-$Al_{0.73}Ga_{0.23}$ As：Fe 样品与铁相关的深能级在不同温度下，对记录的空穴热发射电容瞬态进行了拟合运算。得到了能级展宽宽度 E_B=45meV 和一组相应不同温度的平均热发射率$\bar{e}_i^t$。平均热发射率$\bar{e}_i^t$可的 Arrhenius 作图（T^2 修正）显示了所期望的直线关系（图 9）。由直线的斜率可求得该组分下铁能级的热激活能 E_T = 0.755 ±0.010eV（即空穴由基态^5E 到价带的跃迁能量[15]）。应当注意，这里我们假定了空穴的热俘获截面不依赖温度。我们已经知道，分析混晶中深能级的俘获资料同样是困难的，但用 1/e 法对一些组分的掺铁 $Al_xGa_{1-x}As$ 样品的俘获资料的估算表明，空穴俘获截面在本文研究的温度范围内，不存在明显的温度依赖关系[23]。另外，如果用光阈值能量同组分 x 的依赖关系[15]来修正x=0.76样品的光阈值 E_{th}^0=0.78eV，相应于 x=0.73 的光阈值能量应为 0.766eV。这个值在实验误差范围内同上述分析得到的平均热活能的 E_T=0.755±0.010eV 符合得很好。光阈值与热激活能的一致，意味着在混晶 $Al_xGa_{1-x}As$ 中与铁相关的深中心不存在大的晶格弛豫过程。

5.3 能级展宽效应的温度依赖关系

上述未考虑能级展宽效应的温度依赖关系。为了考察这个关系：（1）对 x=0.49 的

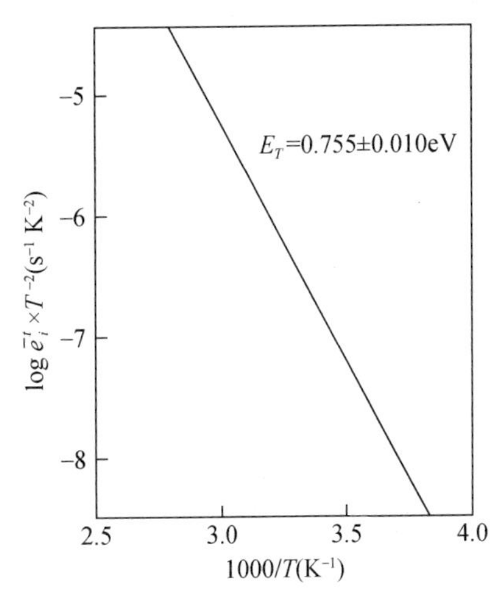

图 9　在 $Al_{0.73}Ga_{0.27}As$：Fe 样品中，与铁相关的空穴陷阱平均热发射率$\bar{e}_i^t$ 的 Arrhenius 作图。能级展宽宽度 $E_B=45$meV

$Al_xGa_{1-x}As$：Fe 样品，在不同温度下测量了绝对空穴光电离截面 $\sigma_p^0(h\nu)$。结果表明，在 120–210K 温度范围里，在实验误差之内，未曾观察到接近阈值附近光电离截面曲线的温度展宽效应。(2) 对 $x=0.73$ 样品，应用相同的能级展宽拟合参数 E_B（=45meV），对不同温度下（270–340K）热发射电容瞬态实验曲线进行了拟合运算，实验结果非常成功的拟合表明，能级展宽效应的温度依赖关系可以忽略不计。

5.4　混晶中深能级的载流子俘获

利用公式（8）对实验资料的拟合，虽不像对 DLTS 曲线拟合那样要求俘获截面不依赖于温度或具有热激活过程，以及拟合参数多等[18]，但能级在禁带中的准确位置（而不是表观激活能），仍需要俘获截面的温度依赖关系资料。众所周知，混晶中深能级的俘获电容瞬态如同热发射瞬态一样，是一个更为复杂的非指数过程。应用上述类似的考虑，对实验记录的俘获电容瞬态过程的理论拟合，可望得到有益的资料。这部分工作尚在尝试中。

6　结论

(1) 本文应用能级展宽效应，成功地解释了混晶半导体中深能级所特有的非指数热、光发射和热俘获过程的物理实质。通过对实验结果的理论拟合，可获得表征混晶半导体中深能级的一些重要参数，如能级展宽宽度 E_B 和平均表观热激活能 E_T 等。

(2) 混晶半导体中深能级展宽宽度 E_B 与组分 x 的依赖关系近似服从简单的二项式规律。

(3) 混晶中能级展宽效应对热瞬态过程的影响远比光瞬态过程为强，这是因为热发射率更强烈地依赖于深能级在禁带中的能量位置之故。

本文的实验工作是在 H. G. Grimmeiss 教授支持下，在瑞典 Lund 大学固体物理系完成的。同 L-Å. Ledebo 博士进行过多次有益的讨论，实验样品是由 M. Alhström 制备的，作者表示诚挚的感谢。

参 考 文 献

[1] R. A. Craven and D. Finn, J. Appl. Phys, 50, 6334(1979).

[2] D. V. Lang, R. A. Logan and M. Jaros, Phys. Rev, 13, 19, 1015(1979).

[3] H. G. Grimmeiss, L-Å Ledebo and E. Meijer, Appl. Phys. Lett, 36, 307(1980).

[4] D Pons and S. Makram-Ebeid, J. de Phys(Paris), 40, 1161(1979).

[5] T. Figielski, Solid, State. Electronics, 21, 1403(1978).

[6] G. Ferenczi, F. Belezmay and L. Dózsa, The 3d"Lund"Conf. on Deep Level Impurities, U. S. A., May, 26-29(1981).

[7] 王占国,(未发表)

[8] K. L. Ngai, Comments on Solid State Physics, 9, 127(1979).

[9] E. P. Wigner, Gatlinberg Conf on Neutron Physics, Oak Ridge National Lab. Report No. ORNL. 2309,59.

[10] D. J. Wolford, B. G. Streetman, Shui Lai and M. V. K'lein, Solid State Commun. 32, 51(1979).

[11] Shui Lai and M. V. K'Lein, Phys. Rev. Lett., 44, 1087(1980).

[12] H. Mariette, J. Chevallier and P. Leroux-Hugon, Phys. Rev. B, 21, 5706(1980).

[13] Charles W. Myles, John D. Dow and Otto F. Sankey, Phys. Rev. B, 24, 1137(1981).

[14] 王永良，顾宗权和黄昆，科学通报，9, 531(1981).

[15] Zhan-Guo Wang(王占国), H G. Grimmeiss and L-Å. Ledebo, To be published in J. Appl. Phys., (1984).

[16] D. V Lang, R. A. Logan and L. C. Kimerling., Proc. XⅢ Inst. Conf. Phys. Semicond, Roma (1976).

[17] E. Calleja, E. Munoz and F. Garcia, Appl. Phys. Lett, 42, 15(1983).

[18] P. Omling, L. Samuelson and H. G. Grimmeiss, J. Appl. Phys., 54, 5117(1983).

[19] J. Jansson, Zhan-Guo Wang(王占国), L-Å. Ledebo and H. G. Grimmeiss, Proc. 5th Inter. Conf. on Ternary and Multinary Compounds, Cagliari, Italy, Sept(1982), IL NUOVO CIMENTO, 2D, 6, 1718(1983).

[20] G. Lucovsky, Solid, State. Commun., 3, 299(1965).

[21] D. L. Wolford, W. Y Hsu, J. D Dow and B. G. Etreetman, J. Luminescence, 18/19, 863 (1979).

[22] D. V. Lang, J. Appl Phys, 45, 3023(1974).

[23] 王占国,(未发表).

Electronic properties of an electron-attractive complex neutral defect in GaAs

B. Monemar, H. P. Gislason, and W. M. Chen
(Linköping University, Department of Physics and Measurement Technology, S-581 83 Linköping, Sweden)
Z. G. Wang
(Institute of Semiconductors, Chinese Academy of Sciences, Beijing 100083, China)

Abstract This Rapid Communication reports the first detailed optical study of a complex neutral defect in GaAs(probably Cu_{Ga}-As_{Ga}), binding an exciton at 1.429eV. Zeeman data at 10T for the bound exciton are analyzed in detail, considering both the electron-hole exchange interaction and the local strain field, The defect has an electron-attractive local potential and a tensional local strain field. The g value for the deeply bound electron $g_e = 0.9$ is strongly modified from the value $g_e = -0.46$ for a shallow donor, while the bound hole remains effective-mass-like.

The properties of neutral "isoelectronic"[1] defects in GaAs have not been discussed previously, since such defects are usually not able to localize excitons in the band gap in this material.[2] This is in sharp contrast to most other Ⅲ-Ⅴ materials, where excitons bound to neutral defects have been observed and studied in detail.[3] The study of residual complex defects in GaAs is of considerable technological importance though. In this Rapid Communication we present the first detailed optical study of the electronic structure of a complex neutral defect in GaAs, which gives rise to a bound exciton(BE) at 1.429eV at 2K. From the magnetic properties of this BE it is concluded that the defect has C_{1h}, symmetry, and a dominantly electron-attractive local potential.

The defect studied in this work is created by long-term diffusion (20h at ~500℃ in evacuated ampoules) of Cu into bulk horizontal Bridgman-grown GaAs, Zn doped during growth to a level of $\approx 1.3\times10^{16}cm^{-3}$. After the diffusion the ampoules were rapidly quenched in water. More details of the preparation procedure are given elsewhere.[4,5] The Cu-doping procedure does not drastically change the electrically active Zn doping, presumably because of a low concentration of the produced complex defects. The 1.429eV defect described

原载于: Phys. Rev. B, 1986, 33(6): 4424-4427.

below has only been observed when Zn-doped p-type GaAs was used as starting material, and only after a rather long-term Cu diffusion. Consequently, both Zn and Cu are active in the creation of the defect.

Optical absorption associated with the 1.429eV BE was too weak to be detected in the transmission spectra of a 3mm thick sample, and therefore the experimental data are restricted to emission spectra [photoluminescence (PL)]. Magneto-optical data for fields up to 10T were obtained in the Voigt configuration with a superconducting magnet at the Max Planck Hochfeld Magnetlabor in Grenoble.

A PL spectrum of the 1.429eV emission at 2K is shown in Fig. 1. As is obvious from the inset in Fig. 1 three electronic zero-field lines are resolved in PL at somewhat elevated temperatures at 1.4285eV ($L3$), 1.4299eV ($L2$), and 1.4308eV ($L1$). This manifold of PL electronic lines is induced by the electron-hole (e-h) exchange interaction and the local strain field at the defect (see below). The temperature dependence of the relative PL intensity for these lines is consistent with thermalization in the excited (BE) state only, as expected for a neutral defect with no electronic particle bound in the ground state.

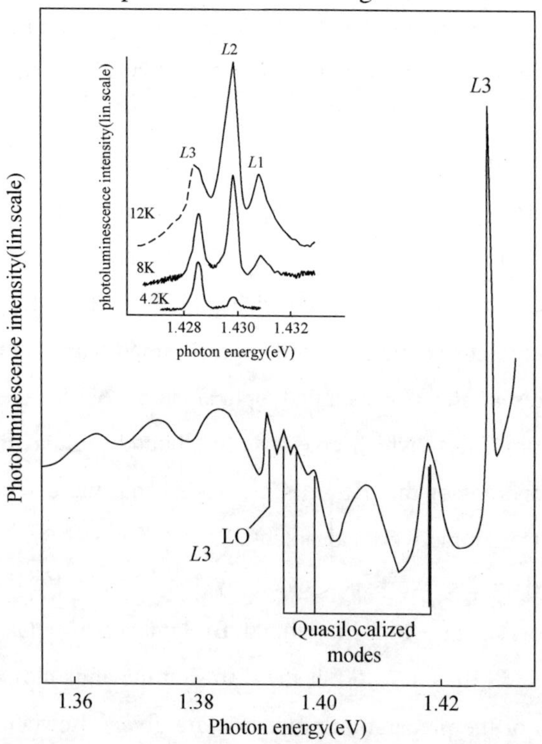

Fig. 1 Photoluminescence (PL) spectrum at 2K for a Zn-doped horizontal Bridgman-grown (HB) bulk GaAs sample ([Zn]≈1.3×10^{16} cm^{-3}), Cu diffused at 500℃ for 19h, and rapidly quenched in water. Only the lowest-energy energy electronic line $L3$ at 1.4285eV is seen at 2K, together with a phonon wing towards lower energies, In the inset PL spectra are also shown for higher temperatures, where three electronic lines can be resolved

The electronic structure of the 1.429eV defect is revealed from a careful analysis of Zeeman data. In Fig. 2(a) we show the development of the electronic lines with magnetic field along the [110] direction up to 10T. Obviously, the *L*3 and *L*2 lines are each composed of two unresolved lines at zero field, where the weak lines only appear as shoulders. Another observation is the negligible quadratic Zeeman shift of the center of gravity of these lines, at least an order of magnitude smaller than for shallow donorlike electrons in GaAs.[6, 7]

The full angular dependence of the Zeeman splitting of the electronic lines was recorded, and is shown in Fig. 2(b) for a rotation in the $(1\bar{1}0)$ plane. A theoretical fit is also included in Fig. 2(b) according to the following perturbation Hamiltonian for a BE consisting of one electron and one hole, both bound in a localized neutral potential of low symmetry,[8, 9]

$$H' = H_{\mathrm{ex}} + H_{\mathrm{LCF}} + H_{\mathrm{LZ}},$$

where

$$H_{ex} = -aS_h \cdot S_e - b(S_{hx}^3 S_{ex} + S_{hy}^3 S_{ey} + S_{hz}^3 S_{ez})$$

denotes the *e*-*h* exchange interaction.

$$H_{\mathrm{LCF}} = -D\left[S_{h\xi}^2 - \frac{1}{3}S_h(S_h + 1)\right]$$

is the axial strain perturbation on the hole states from the local crystal field (LCF) at the defect,[8] while

$$H_{LZ} = \mu_B[g_e S_e \cdot H + KS_h \cdot H + L(S_{hx}^3 H_x + S_{hy}^3 H_y + S_{hz}^3 h_z)]$$

is the linear Zeeman (LZ) term.[9] S_e is the electron spin, S_h the effective hole spin (including angular momentum), and μ_B the Bohr magneton. *x*, *y*, and *z* refer to the usual cubic axes of the zinc-blende lattice, while ξ refers to the defect axis. The possible influence of localization on the symmetry of the bound electron state as well as quadratic Zeeman effects are neglected. The major effect of both the *e*-*h* exchange and the low-symmetry crystal field are included in this perturbation Hamiltonian H'.

In the theoretical simulation of experimental data [Figs. 2(a) and 2(b)] the above three contributions to H' are treated on the same level, and are diagonalized simultaneously. The best fit to the data partly shown in Fig. 2 is obtained with the following parameters: $a=1.75\pm0.05$meV, $b=-0.30\pm0.05$meV, $D=1.15\pm0.05$meV, $g_e=+0.90\pm0.02$, $K=0.80\pm0.02$, and $L=0.04\pm0.02$. Further, the symmetry of the defect is determined as being C_{1h} from Fig. 2(b), i.e., the defect is linear with a [110] axis.

As a consequence of the above theoretical fit [full lines in Figs. 2(a) and 2(b)] the zero-field lines *L*1-*L*3 in Fig. 1 are identified as deriving from the "$J=2$" quintuplet of a BE, as expected for the case of a large *e-h* exchange interaction and as found previously for electron-attractive neutral complexes in GaP, e. g. , in the cases of GaP : Zn, O(Refs. [10] and [11]) and GaP : Li, Li, O.[12] In the present case with low C_{1h} symmetry, the entire degeneracy of the quintuplet is already lifted at zero magnetic field. The lowest line *L3* is identified as the $|\overline{2, \pm 2}\rangle$ pair in the $|\overline{J, m_j}\rangle$ notation,[13] consistent with a tensional local strain field ($D>0$). Here the notation $|\overline{J, m_j}\rangle$ means that we no longer have a pure $|J, m_j\rangle$ state, but a mixed one in this low symmetry.[4] *L3* is split even at zero magnetic field due to the anisotropic exchange interaction [parameter *b* above, see Fig. 2(a)]. A similar splitting occurs for the *L2* line, identified as the $|\overline{2, \pm 1}\rangle$ pair, while the high-energy singlet line *L1* is assigned to the $|\overline{2, 0}\rangle$ state. The $J=1$ triplet is expected to be well above the $J=2$ lines, and not seen in low-temperature PL spectra due to thermalization.

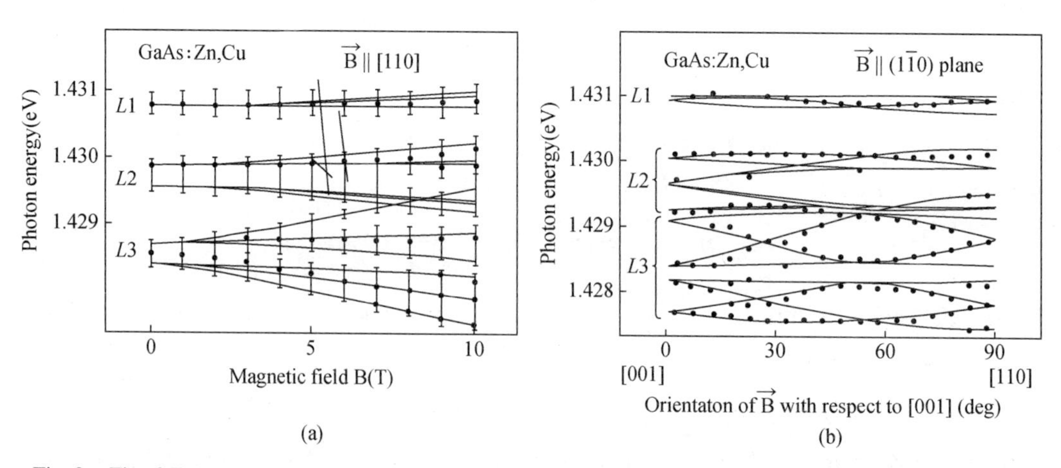

Fig. 2 Fit of Zeeman data for the 1.429eV PL emission, taken in the Voigt configuration at 5K, to the Hamiltonian described in the text. In (a) a fan diagram is shown for fields up to 10T, for B || [110]. No significant quadratic Zeeman shift is seen, and *L2* and *L3* are split even at zero field. In (b) the angular dependence of the lines at 10T is shown, together with the fit to the model Hamiltonian. Lines with no experimental points are seen as weak shoulders experimentally, and they are also expected to be weak from theory (Ref. [4])

The electronic structure of the 1.429eV BE is summarized in Fig. 3. The dominant perturbation is the *e-h* exchange interaction (as evident from the parameters evaluated above), which leaves the $J=2$ quintuplet at lowest energy. The effect of the local [110]-oriented axial tensional strain field is comparable, though, and gives rise to the five zero-field lines. The value $g_e=+0.9$ is drastically different from $g_e=-0.46$ observed for a shallow donorlike electron in GaAs.[6] This proves the strong localization of the bound electron; a limiting value of $g_e=+2$ is expected when the effect of the conduction-band structure is lost because of strong localization, as observed, e. g. , for O_{Te}, in ZnTe. [14] The absence of a quadratic Zeeman shift is further evidence for strong localization of the bound electron.

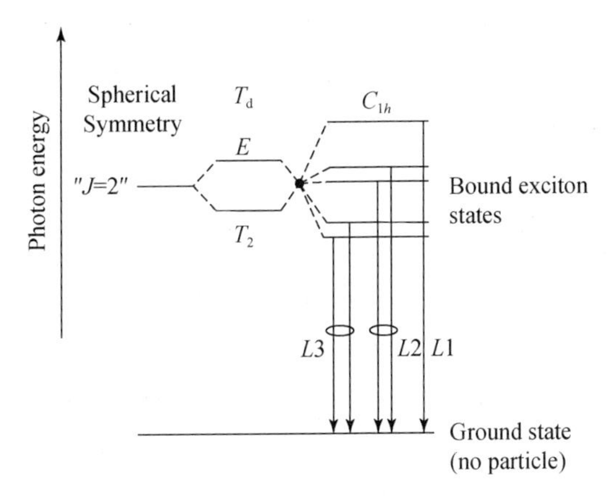

Fig. 3 Perturbation scheme at zero magnetic field for the experimentally observed lowest $J = 2$ quintuplet of the bound exciton, showing the lifting of electronic degeneracy by the C_{1h} crystal field. The experimentally observed PL transitions $L1$-$L3$ are indicated

The K and L values for the bound-hole state are rather close to the values found for previously studied acceptor hole states in GaAs, such as the Sn acceptor. [6] The observed anisotropy of the Zeeman-split components is fully derived from the bound-hole, [8] since the electron g factor is isotropic. This indicates an effective-mass-like bound hole state, but with a moderate amount of localization to maintain the observed value of the e-h exchange splitting (factors a and b above).

On the other hand, a dominantly hole-attractive local potential, would be expected to have strong effects on the g factors of the bound hole, as has recently been observed for hole-attractive Cu-related complex defects in GaP. [15-19] A completely quenched hole angular momentum is observed in this case for low-symmetry defects, with an approximately isotropic hole g value $g_h = +2$. [15-19] The lowest BE state would behave as an isotropic triplet in this case. [15-19]

The proposed identity of the defect responsible for the 1.429eV BE spectrum has to be consistent with the [110] axial direction, a tensional local strain field, and a local strongly electron-attractive neutral potential. An activation energy of about 1.65eV for the creation of this defect, [5] together with the involvement of both Cu and Zn, suggest a slow defect reaction during the long-term Cu diffusion, where Zn_{Ga} is replaced by Cu_{Ga} i. e. , $Zn_{Ga} + Cu_i \rightarrow Cu_{Ga} + Zn_i$. A $Cu_{Ga}Zn_i$ pair would presumably create a compressive local strain field. In addition, a [110] axial symmetry for a $Cu_{Ga}Zn_i$ pair is most unexpected. The Zn_i may migrate away after the Cu-Zn exchange has been completed on a Ge site, however.

The [110] symmetry of the neutral complex is suggestive of a Ga-site pair, i. e. , a pair of Cu_{Ga}-D_{Ga}, where D_{Ga} is a deep Ga-site double donor compensating the Cu_{Ga} double acceptor. The only deep intrinsic Ga-site double donor known so far is the As_{Ga} antisite defect,[20, 21] which has deep states at midgap in GaAs. Such an As_{Ga} donor could be produced in this case via local defect reactions involving both Ga-and As-site vacanices V_{Ga} and V_{As}.[22] It has recently been shown that Zn-doping induces a remarkable enhancement of vacancy diffusion in GaAs,[23] so that transport of V_{Ga} and V_{As} to the sites of formation of the 1. 429eV defect is facilitated (during 20h at ≈ 500℃). Clearly, the strongly electron-attractive As_{Ga} donor potential would dominate the moderately deep Cu_{Ga} acceptor 0. 15eV (Ref. [24]), so that the total Cu_{Ga}-As_{Ga} defect potential would be strongly electron attractive, as observed. The observed g value $g_e = +0.9$ for the electron bound to Cu_{Ga}-As_{Ga} is consistent with a reduction of the electron-attractive As_{Ga} potential by the adjacent Cu_{Ga}, as compared to the isolated As_{Ga} donor, where a g value $g_e = 2.04$ has been reported.[25]

In summary, we have reported the first detailed study of a neutral defect binding an exciton in GaAs. An appropriate perturbation Hamiltonian is employed for the analysis of Zeeman data of this 1. 429eV bound exciton, showing a very good fit to the data. The defect is electron attractive, with an anomalous g value $g_e = +0.9$ for the deeply bound electron, while the hole is effective-mass-like. The established [110] symmetry of the defect suggests a Ga-site pair, probably Cu_{Ga}-As_{Ga}, as the identity of this complex.

We are grateful to Ch. Uihlein and P. L. Liu for collaboration in Zeeman experiments, and to U. Lindefelt for critical comments on the manuscript.

References

[1] The notation neutral defect is used here as a synonym of "isoelectronic" defects. Such defects have no electronic particles in the ground state, and a bound electron-hole pair ("bound exciton") as the lowest-energy electronic excitation at low temperature.

[2] B. Monemar, H. P. Gislason, and Z. G. Wang, Phys. Rev. B 31, 7919(1985).

[3] P. J. Dean and D. C. Herbert, in Excitons, edited by K. Cho, Topics in Current Physics, Vol. 14 (Springer, Berlin, 1979), 55.

[4] H. P. Gislason, B. Monemar, W. M. Chen, Z. G. Wang, Ch. Uihlein, and P. L. Liu(unpublished).

[5] C. J. Hwang, J. Appl. Phys. 39, 4313(1968); Phys. Rev. 180, 827(1969).

[6] W. Schairer, D. Bimberg, W. Kottler, K, Cho, and M. Schmidt, Phys. Rev. B 13, 3452(1976).

[7] H. P. Gislason, B. Monemar, Z. G. Wang, Ch. Uihlein, and P. L. Liu, Phys. Rev. B 32, 3723 (1985).

[8] A. Abragam and B. Bleaney, in Paramagnetic Resonance in Transition Ions (Clarendon, Oxford,

1970), Chap. 9.

[9] J. M. Luttinger, Phys. Rev. 102, 1030(1955).

[10] C. H. Henry, P. J. Dean, and J, D. Cuthbert, Phys. Rev. 166, 754(1968).

[11] T. N. Morgan, B. Welber, and R. N. Bhargava, Phys. Rev. 166, 751(1968).

[12] P. J. Dean, Phys. Rev. B 4, 2596(1971).

[13] J. van W. Morgan and T. N. Morgan, Phys. Rev. B 1, 739(1970).

[14] P. J. Dean, H. Venghaus, J. C. Pfister, B. Schaub, and J. Marine, J. Lumin. 16, 363(1978).

[15] B. Monemar, H. P. Gislason, P. J. Dean, and D. C. Herbert, Phys. Rev. B 25, 7719(1982).

[16] H. P. Gislason, B. Monemar, P. J. Dean, D. C. Herbert, S. Depinna, B. C. Cavenett, and N. Killoran, Phys. Rev. B 26, 837(1982).

[17] H. P, Gislason, B. Monemar, M. E. Pistol, P. J. Dean, D. C. Herbert, A. Kanáah, and B. C. Cavenett, Phys. Rev. B 31, 3734(1985).

[18] H. P. Gislason, B. Monernar, M. E. Pistol, P. J. Dean, D. C. Herbert, S. Depinna, A. Kanáah, and B. C. Cavenett, Phys. Rev. B32, 3958(1985).

[19] H. P. Gislason, B. Monemar, M. E. Pistol, A. Kanáah, and B. C. Cavenett, Phys. Rev. B 33, 1233(1986).

[20] B. K. Meyer, J. M. Spaeth, and M. Scheffler, Phys. Rev. Lett. 52, 851(1984).

[21] M. Kaminska, M. Skowronski, J. Lagowski, J. M. Parsey, and H. C. Gatos, Appl. Phys. Lett. 43, 302(1983).

[22] J. A. van Vechten, in Handbook on Semiconductors, edited by S. P. Keller (North-Holland, Amsterdam, 1980), Vol. 3, Chap. 1.

[23] W. D. Laidig, N, Holonyak, M. D. Camras, K. Hess, J. J. Coleman, P. D. Dapkus, and J. Bardeen, Appl. Phys. Lett. 38, 776(1981).

[24] Z. G. Wang, H. P. Gislason, and B. Monemar, J. Appl. Phys. 58, 230(1985).

[25] R. J. Wagner, J. J. Krebs, G. H. Stauss, and A. M. White, Solid State Commun. 36, 15(1980).

硅中金施主和受主光电性质的系统研究

王占国

（中国科学院半导体研究所，北京，100083）

摘要　本文利用光电容和光电流瞬态技术结合 DLTS 测量对 P 和 N 型掺金硅（不同补偿比 $K=N_{Au}/N_s=0.02\sim0.9$）中金施主和受主的行为进行了全面、系统地实验研究，得到如下结果：（1）进一步证实了硅中金施主和受主是同一个缺陷的两个不同能态（$Au^{+/0}$ 和 $Au^{0/-}$）；（2）金受主和施主对应的深中心不是金同浅杂质的络合物（Au-Ds）；（3）实验结果拟合运算得到的金施主电子和空穴光电离截面阈值能量之和基本等于硅的禁带宽度表明：载流子通过金施主激发和俘获时不存在明显的晶格弛豫效应。金受主空穴光电离截面阈值能量大的温度依赖关系可能与空穴通过金受主激发态的光热激发相关。此外，本文还研究了接近阈值能量附近的金施主和受主能级光发射率的电场依赖关系。

1　引言

在过去的近 30 年里，硅中金施主(D)和受主(A)的行为一直是个最受重视和广泛被研究的课题之一。然而令人遗憾的是，至今它的很多重要性质仍纷说不一。最近 D. V. Lang[1] 等根据：①DLTS 测得的金 D 和 A 浓度不同；②金 A 电子俘获截面依赖于 Au 同浅施主浓度补偿比（$K=N_{Au}/N_s$）；③外延硅和 CZ-硅中金 A 的光电离截面相差 10 多倍等声称，金 D 和 A 能级不是一个缺陷的两个能态，而是两个独立的缺陷，它们多半是金同浅杂质的络合物。他们的观点引起了很多学者的关注。Morante 等[2] 在 300K，应用电流 DLTS 谱（CDLTS）测得了不同 K 值时金受主能级的电子俘获截面证明：金 A 电子俘获截面（σ_{nA}）不依赖于补偿比 K（直到 0.85）。进而 Ledebo 和 Wang[3] 对金 D 和 A 能级上电子占有率（在适当光子能量激发下）的实验测量和分析结果表明：金 D 和 A 是同一个缺陷的两个不同能态。另外，金 A 能级温度依赖关系的研究结果也相互矛盾[4-6]。

为了澄清上述问题，本工作通过控制硅样样品（包括 P^+N 结和 P 型肖特基势垒）中的金浓度获得不同的补偿比 K（0.016-0.93），系统地测量了金 D 和 A 的电子和空穴绝对光电离截面以及金 A 电子俘获截面的温度依赖关系（80-280K）。结果进一步证实硅中金 A 和 D 是同一个缺陷的两个不同能态，其缺陷的化学组成尚待进一步研究。

原载于：半导体学报，1987，8（3）：236-245.

但本文已证明它不是金同浅杂质的络合物。此外，某些掺硼的 P 型硅，在金扩散后除金 D 和 A 外，还经常观察到另一个与金相关的络合物“Au-X”中心。

2 实验细节和结果

2.1 样品制备

本文采用两种不同导电类型的掺金硅样品，P 肖特基和 P^+N 二极管。P^+N 结构如图 1 所示。在重掺锑的 CZ-n^+ 硅衬底上外延约 40μm 的掺磷 N 型硅外延层，浓度为 $4.1\times10^{14}\text{cm}^{-3}$。用标准平面工艺制备 P^+N 结。在 P^+N 结背面 N^+ 衬底上蒸发 600Å 的高纯金，控制扩散温度和时间以保证金在外延层中均匀分布。金的浓度从 $6.6\times10^{12}\text{cm}^{-3}$ 到 $3.7\times10^{14}\text{cm}^{-3}$，相应的 K 为 0.016–0.93。对 P 肖特基二极管，金扩散工艺相同，但温度稍高一些。浅受主（硼）浓度在 $3\times10^{14}\text{cm}^{-3}$ 到 $1.6\times10^{16}\text{cm}^{-3}$ 之间，K 由 0.02 到 0.2。管芯用导电胶粘在 TO-5 管座上，用超声压焊引出电极。

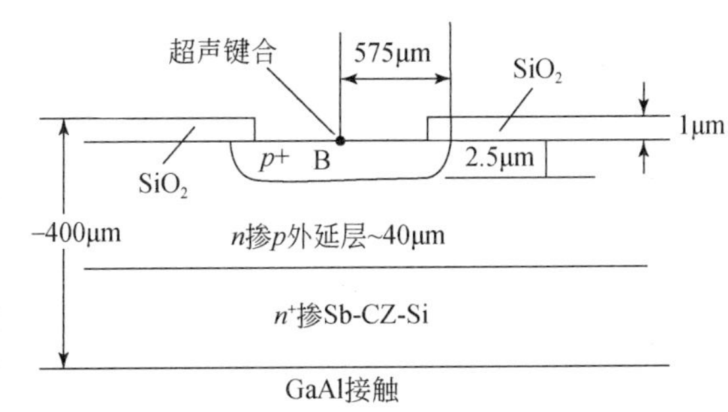

图 1 掺 Au 硅 P^+N 二极管的结构简图

2.2 硅中金 D 和 A 的电子和空穴光电离截面谱

对 P 型硅肖特基和 P^+N 二极管，典型的金 D 和 A 电子和空穴光电离截面谱如图 2 (a)和 (b) 所示。如果金 D 和 A 上的电子的初始态是完全抽空的，又因为两者是同一个缺陷的两个不同能态[3]，则很容易证明，在整个研究的光子能最范围内，金 D 空穴发射率（$e^0_{p\text{D}}$）正比于电容瞬态的初始斜率[7]：

$$\left.\frac{\mathrm{d}\Delta C(t)}{\mathrm{d}t}\right|_{t=0}=\frac{C(0)}{2N_s}\left.\frac{\mathrm{d}(n_\text{D}+n_\text{A})}{\mathrm{d}t}\right|_{t=0}=-\frac{C(0)}{2N_s}N_T e^0_{p\text{D}}。\tag{1}$$

这里，$\left.\frac{\mathrm{d}n_\text{A}}{\mathrm{d}t}\right|_{t=0}\equiv0$，$n_\text{A}(t)$ 和 $n_\text{D}(t)$ 分别为金 A 和 D 上在时刻 t 占有的电子数，N_T 为金的总浓度，N_s 为浅杂质浓度。

对 P 型硅肖特基和 P^+N 二极管（先用带间光激发使金 A 抽空），金 A 空穴光电离截面 $\sigma^0_{p\text{A}}$ 可用上述的初始斜率法直接测量，不过金 D 必须事先用强光源（$E_\text{D}-E_\text{V}<h\nu<E_\text{A}-E_\text{V}$）将其充满电子。当光子能量大于 E_C-E_D 时，电容的改变或许部分来自金 D 上电子被激发到导带（$e^0_{n\text{D}}$）的贡献。幸运的是在实验上发现 $e^0_{n\text{D}}\ll e^0_{p\text{A}}$，故 $e^0_{n\text{D}}$ 的贡献可以不计。

金 A 的电子光发射率（$e^0_{n\text{A}}$），除了可用初始斜率法外，当光子能量小于 $E_\text{A}-E_\text{V}$ 时，还可以用光电容瞬态的时间常数求得。但当光子能量 $h\nu\geqslant E_A-E_V$ 时，由于存在着两步激

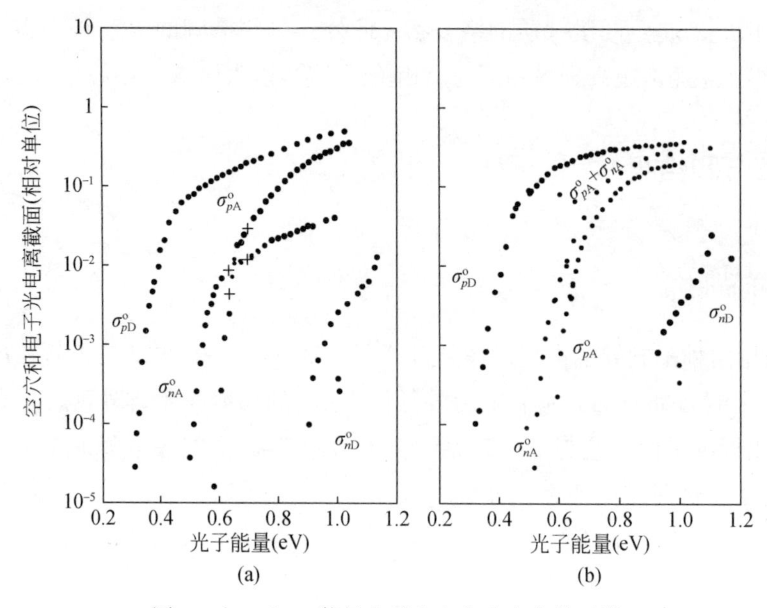

图 2　金 D 和 A 能级电子和空穴光电离截面谱

（a）P^+N，T=77K　Si:Au；（b）P 肖特基　T=99K　Si:Au

发过程，用这种方法求 e^0_{nA} 的测量变得复杂起来。对 P^+N 结，首先用初始斜率法求出 $e^0_{nA}(h\nu)/e^0_{pA}(h\nu)$ 的比值，然后结合该光子能量下光电容瞬态的时间常数（$\tau(h\nu))^{-1}=e^0(h\nu)=e^0_{nA}+e^0_{pA}$，从而求出 e^0_{nA} 和 e^0_{pA} 的值。这里应当指出，为了精确测 e^0_{nA}/e^0_{pA} 比值，金 A 初始空穴占有率必须知道。测量方法如下：首先用足够长的零偏使金 A 和 D 充满电子，接着用带间光照射样品产生电子–空穴对。金 A 和 D 将同时从价带俘获空穴和从导带俘获电子（低温下，热发射率≈0），达到稳态时电容变化记作 ΔC_{A+D}，再用 $h\nu\simeq0.5$eV 的强光使金 D 完全充满电子，此时电容的减最为 ΔC_D，最后用 $E_C-E_A<h\nu<E_A-E_v$ 光将金 A 上电子抽空并引起电容增量 $\Delta C'_A$。因而当用带间光激发时（创造初始条件），金 A 上空穴占有率 f_{Ah}。可由 $\dfrac{\Delta C_{A+D}-\Delta C_D}{\Delta C_{A+D}-\Delta C_D+\Delta C'_A}$ 求出，实测值为 ~0.95。知道了 f_{Ah}；$e^0_{nA}+e^0_{pA}$ 之值便可准确得出。另外，用电流瞬态方法同样可区分 e^0_{nA} 和 e^0_{pA}。

对于 P 型硅肖特基二极管，应用初始斜率和时间常数法我们只能求出金 A 的空穴相对光发射率 e^0_{pA} 和（$e^0_{nA}+e^0_{pA}$）之和［图 2(b)］；因为金 A 被电子充满的初始条件无法实现。

金 D 的电子光发射率 e^0_{nD} 的实验测量比较困难。因为 e^0_{nD} 有着较高的阈值能量，在这种情况下，必须同时考虑来自金 D 和 A 电子和空穴四个激发过程。本文利用 Wang（王占国）等[7]提出的方法，成功地测得了金 D 的 e^0_{nD}（$h\nu$）同光子能量的关系谱图［图 2(a)］。

对同一背景材料，结构、制备工艺完全相同而补偿比 K 不同的 P^+N 样品，实验测得的（几个典型 K 值下）金 D 绝对空穴光电离截面谱和金 A 绝对电子和空穴光电离截面谱如图 3 所示。为了减小可能的实验误差，对不同 K 值的样品，在相同的条件下

(温度、光子能量和光强相同)，用实际测得的光电容瞬态时间常数进行校准。由图3可见，金D和A的电子和空穴光电离截面谱不依赖于补偿比K。

2.3 热发射率

当金中心浓度和浅杂质浓度N_s之比$K \leqslant 0.1$时，本文用恒压电容DLTS和单次瞬态法测量金D和A的表观激活能。为了减小实验误差，本文尽可能地扩大了热发射率的测量范围，实际测量高达8个数量级之多。当$K>0.1$时，由于恒压电容瞬态的非指数特性，难于精确测量它的时间常数，故采用恒容DLTS[8]，和电流DLTS[9]测之。可以证明，对高缺陷浓度的样品，CDLTS与恒压电容DLTS相比，具有明显的优越性[10]，即使当K接近1时，CDLTS仍能正确的给出深中心的表观热激活能。用上述方法测得的金A表观激活能（不同K值）及其他参数见表1。金D及所谓“Au-X”中心的热激活能分别为E_v+0.343eV和E_v+0.226eV；利用单次电流瞬态测得的金A电子和空穴热激活能（考虑俘获截面温度依赖关系修正，D取1.6）如图4所示。E_C-E_A与E_A-E_V之和为1.163eV，这同已发表的结果符合得很好[11]。

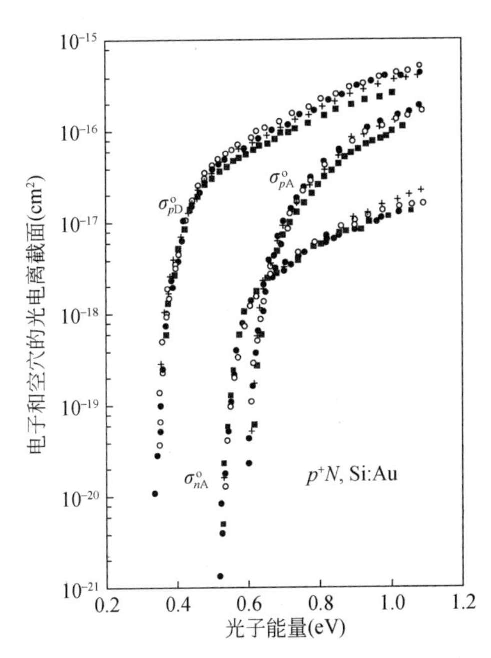

图3 几个典型K值的金D和A电子和空穴绝对光电离截面谱T=80K

■0.11 +0.23 ○0.61 ●0.76

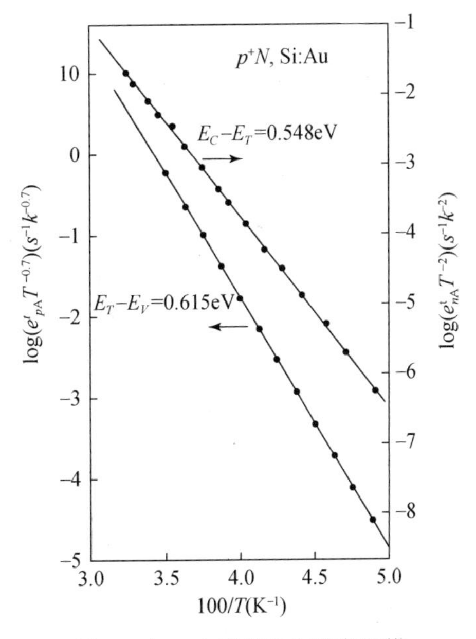

图4 金A电子和空穴热激活能（俘获截面温度依赖关系修正后）

表1 不同K值的金A电子俘获和发射等动力学参数

序号	$K=N_{Au}/N_s$	e^t_n (250K)（1/s）	σ_n(80–280K) (cm^2) ×10^{-14}	ΔH_n(eV)	X_n	$\Delta S_n/k$	ΔG_n(250K) (eV)
0992-2	0.016	12.23	1.21	0.546	23.4	3.15	0.478

续表

序号	$K=N_{Au}/N_s$	e_n^t (250K) (1/s)	σ_n(80–280K) (cm^2) ×10^{-14}	ΔH_n(eV)	X_n	$\Delta S_n/k$	ΔG_n(250K) (eV)
0825-4	0. 046	11. 19	0. 86	0. 551	38. 0	3. 64	0. 473
0829-6	0. 11	12. 81	0. 85	0. 553	48. 3	3. 88	0. 469
0889-2	0. 23	12. 48	1. 34	0. 544	19. 7	2. 98	0. 48
1014-5	0. 42	13. 48	1. 61	0. 546	19. 4	2. 96	0. 482
0998-5	0. 61	11. 61	1. 73	0. 552	20. 6	3. 03	0. 479
1013-3	0. 68	16. 19	1. 64	0. 549	26. 3	3. 27	0. 479
0990-3	0. 76	13. 4	2. 00	0. 553	21. 5	3. 07	0. 487
0991-3	0. 94	20. 8	2. 48 *	0. 551	24. 6	3. 20	0. 482
平均值		13. 8	1. 52	0. 549	2. 69	3. 24	0. 479

* 仅在 288K 测得

2. 4 俘获截面

金 D 和 A 多数载流子俘获截面采用 Henry[12] 和 Lang[13] 提出的方法测量。但当温度降低热发射率趋近零时，上述方法不再适用。这时可用系列脉冲法[14] 记录并计算不同温度下俘获瞬态时间常数，从而在较宽的温度范围内取得俘获截面的信息。当K>0. 1 时，要正确测量金 A 的电子俘获截面，就必须同时知道金和浅杂质的浓度，本节我们将只作扼要的描述，详文则另外发表。

2. 4. 1 补偿比 K 的计算

由于实验样品为突变 P^+N 结（或 N-肖特基），故结在反偏下耗尽区基本位于 N 区（图 5）。E_A 为金 A 能级，浓度为 N_T，在稳态情况下（$T=80K$，$e_{nA}^t \approx 0$），应用耗尽层近似有：

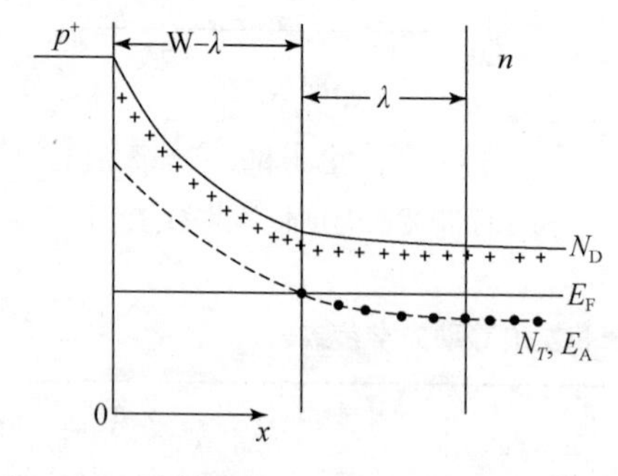

图 5 用于计算补偿比 K 的 P^+N 结耗尽区模型

（1）$X>W-\lambda$，金 A 被电子充满，荷负电；$X<W-\lambda$，金 A 全部被抽空（如用适当光能量激发）呈中性。

（2）$X>0$，金 A 用足够长时间的零偏而被电子充满，荷负电。

相应于上述两种情况的泊松方程的积分分别给出：

$$C_e=\frac{\varepsilon A}{W_e}=\frac{\varepsilon A}{\left[\left(\lambda^2K^2-K\lambda^2+\frac{2\varepsilon(V_D+V)}{eN_D}\right)^{\frac{1}{2}}+\lambda K\right]} \tag{2}$$

$$C_f=\frac{\varepsilon A}{W_f}=\varepsilon A\left[\frac{e(1-K)N_D}{2\varepsilon(V_D+V)}\right]^{\frac{1}{2}}, \tag{3}$$

其中，

$$\lambda=\left[\frac{2\varepsilon(E_F-E_A)}{e^2N_D(1-K)}\right]^{\frac{1}{2}}, \tag{4}$$

$$E_F=-kT\ln\left[\frac{N_D}{N_C}(1-K)\right]。 \tag{5}$$

式中，C_e、C_f 和 W_e、W_f 分别为金 A 空着和充满电子时相应的结电容和耗尽层宽度；A 为结面积；ε 为介电常数；E_F 为费米能级；N_D 为浅施主浓度以及内建电压 V_D 和偏置电压 V；e 为正电子电荷。

补偿比 K 可借助计算机拟合下面方程得到

$$\frac{C_f}{C_e}=\frac{W_e}{W_f}=\left[\frac{e(1-K)N_D}{2\varepsilon(V_D+V)}\right]^{\frac{1}{2}}\left\{\left[\lambda^2K^2-K\lambda^2+\frac{2\varepsilon(V_D+V)}{eN_D}\right]^{\frac{1}{2}}+\lambda K\right\}。 \tag{6}$$

式中，C_f、C_e，可由实验直接测得，自由载流子浓度 n 可由参考样品用 C-V 或 Hall 系数测出，对低补偿的外延样品 $n\simeq N_D$。当然这里假定了浅施主磷浓度在金扩散前后保持不变，即金不同磷形成络合物。这一点将在后面进一步讨论。

如果应用恒容法测量金 A 抽空和充满电子时的结电容值 C_f、C_e，类似上述处理，K 也可用计算机拟合下式获得

$$\left[\lambda^2K^2-K\lambda^2+\frac{2\varepsilon(V_D+V_1)}{eN_D}\right]^{\frac{1}{2}}-\left[\frac{2\varepsilon(V_D+V_2)}{e(1-K)N_D}\right]^{\frac{1}{2}}+\lambda K=0。 \tag{7}$$

式中，V_1、V_2 分别对应于金 A 电子抽空和电子充满时的电压。用上述两种方法求得的 K 如表 2 所示，两者符合得很好。

表 2　用恒压和恒容稳态光电容测量结合图 5 计算模型求得的补偿比 K 值

样品序号	$K=N_{Au}/N_s$ 恒压电容法	$K=N_{Au}/N_s$ 恒容法
0988-2	0.226	0.240
1014-5	0.453	0.510
0998-5	0.607	0.670
1013-3	0.714	0.757

2.4.2　大 K 值时金 A 电子俘获截面测量

我们已经知道，用恒压电容 DLTS 测量 $K>0.1$ 的样品时，DLTS 峰值对应的温度

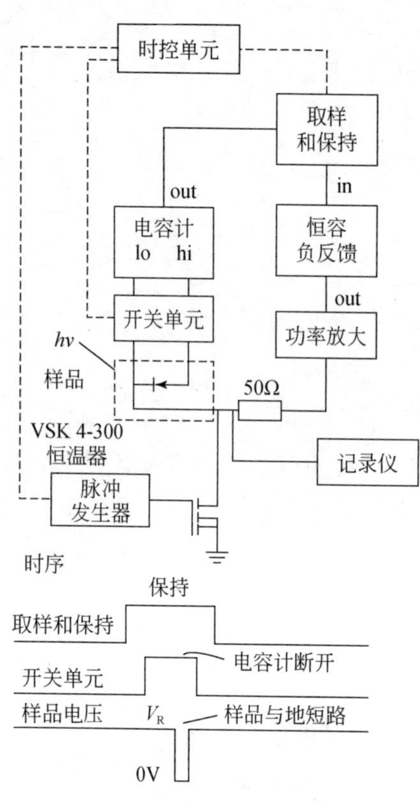

图6　测量大补偿比 K 样品的电子（空穴）俘获截面方法简图

将向低温漂移（对相同的率窗与 $K \ll 1$ 的情况相比）。以金 A 为例，当 $K=0.94$ 时，峰值温度向低温漂移达 25℃之多！不但表观激活能改变，而且还造成俘获截面显著增加的假象。相反，不同 K 值的 CDLTS 讯号计算机模拟表明，即使 K 接近 1 时，它仍能给出正确的结果[10]。然而 CDLTS 仅在较高的温度和大缺陷浓度时才具有高的灵敏度，随着测量温度的降低其 CDLTS 讯号将迅速减小，致使测量十分困难。为了把金 A 电子俘获截面的测量扩展到低温，本文则使用经过改进的恒容单次瞬态法。样品初始条件用带间光将金 A 抽空，后用强光（$h\nu=0.5$eV）将金 D 填满，以保证俘获过程只发生在金 A。测量过程中，在 P^+N 结上施加填充脉冲的前后，样品立即通过一个恒容反馈回路和 Boonton 72B 电容计相连（图 6），这时恒定空间电荷区宽度里电荷的改变直接反映在外加偏压的变化上；通过测量零偏填充脉冲时间为 t，∞ 时相应的反偏电压改变量 $\Delta V(t)$，$\Delta V(\infty)$，金 A 电子俘获时间常数便可由 $\ln\left[\dfrac{\Delta V(\infty)}{\Delta V(\infty)-\Delta V(t)}+\dfrac{K}{1-K}\right]$ vs. t 的斜率求出[5]。用上述方法以及 $K \leqslant 0.1$ 时通常方法在 80–280K 温度范围内测得的金 A 电子俘获截面（不同 K 值）如表 1 所示。显然，金 A 的 σ_{nA} 在研究的温度范围内基本上保持不变且不依赖于补偿比 K。当 K 由 0.016 到 0.94 改变时，σ_{nA} 的平均值为 $(1.6\pm0.8)\times10^{-16}\text{cm}^2$，这同已发表的结果相一致[2]。

3　讨论

3.1　硅中金 D 和 A 是同一个缺陷的两种不同能态

文献[3]中已经证明了硅中金 D 和 A 是同一个缺的两个不同能态，这里再提出一些支持上述结论的实验证据。众所周知，如果上述结论成立，那么金 D 和 A 浓度必须相等。Lang 等[1]在其很有影响的文章里也曾根据金 D 和 A 的 DLTS 峰高明显不同而认为两者不可能是相关的同一中心。在 DLTS 测量中，由于金 A 和 D 能级分别位于禁带中心两侧，对 P^+N 结，必须有少子注入时才能观察到金 D 能级。在实际测量中，由于很难保证金 D 为少子（空穴）完全填满以及无法完全克服注入脉冲后沿的下降时间内金 D 对电子的俘获[15]，所以表观上金 A 峰的 DLTS 讯号总比金 D 少子讯号高。为了证实上述判断，本文在低温下($e^t \simeq 0$)，用稳态光电容方法测量了掺 Au P 型硅肖特基和不同 K 值的 P^+N 样品中金 D 和 A 浓度，毫无例外地发现它们的浓度在实验误差范围内保持相等。最近 Van. Staa 等[16]用 DLTS 测量了掺 Au 硅（P 型，$K=0.64$）的金 D 和 A DLTS 峰高比，并同理论作了比较，同样证明了金 D 和 A 具有相等的浓度，两者同属一个深中心缺陷。另外一些工作如 Brotherton 等[17]，在硅中 Au-Fe 络合物电学性质的研究中发现，伴随着低温退火（<350℃）Au-Fe 对的形成而金 A 和 D 浓度等量同时减少；反之，在高温退火 Au-Fe 对分解时金 D 和 A 又以等量同时增加。另外，硅中金 D 和 A 电子和空穴俘获截面的相对数值测量研究等结果，还从另一个方面支持了金 D 和 A 是同一个缺陷的两个不同能态[18]。

3.2 硅中金 A 是金同浅施主（Au-Ds）的络合物吗?

文献[1]中认为金 A 是（Au-Ds）的络合物的一个重要证据是金 A 的电子俘获截面强烈地依赖于补偿比 K（有 30 倍不同）。应当指出的是，他们所引证的结果多来自不同的背景材料、不同扩散工艺以及不同的测试和计算方法。这些差别将会引起结果的差异。本文充分考虑了这个问题，采用浅施主浓度相同的高质量外延层(~40μm)作背景材料，利用成熟的平面工艺制成 P^+N 结，控制金扩散温度和时间制备了补偿比 K 不同的样品。在 80-280K 温度范围内测量了不同 K 值（0.016-0.94）的金 A 电子俘获截面。实验结果表明，金 A 的 σ_{nA} 在研究的温度范围内基本保持不变，其值为（1.6±0.8)×10^{-16}cm^2，不依赖于补偿比 K。另外，当补偿比 K 保持不变，而浅施主浓度 N_s 由 10^{14}cm^{-3}改变到 5×10^{15}cm^{-3}时，金 A 的 σ_{nA} 仍保持上述的常数，而不受 N_s 浓度的影响。这就是说 Lang[1]等基于金 A 的 σ_{nA} 依赖于 K 而得出金 A 是（Au-Ds）络合物的结论是靠不住的。

Lang[1]等认为金 A 是同浅施主的络合物的第二个依据是不同样品的金 A 光电离截面的差异。我们知道，深中心的电子（或空穴）绝对光电离截面值受研究样品几何形状（器件结构）以及光在样品里由于反射率不同而造成光强变化等影响。为克服上述可能引起的误差，本文除选用结构、制备工艺完全相同的样品外，还细心地保证入射到不同 K 值样品上的光强均匀而且相等；并在温度相同、样品室气氛和压力不变情况下，对不同 K 值的样品，在几个具有代表性的光子能量激发下记录相应的光电容瞬态并计算其光发射时间常数 τ_i^0。电子（或空穴）的绝对光电离截面 σ_{nA}^0(σ_{pA}^0）由($\tau_i^0\phi$)$^{-1}$求出并对整个光电离截面谱进行校准（图 3)，其中 ϕ 为光子流密度，用经过校准的热电堆探测器测量。为了清楚起见，图 3 只给出了 4 个 K 值的金 D 和 A 的空穴和电子绝对光电离截面谱。显然，无论是谱线的形状还是它们的绝对值，在实验误差范围内彼此符合的很好。这就说明 Lang[1]等的第二个依据也是不可靠的。另外，在某些P肖特基样品里，由于样品几何结构、势垒面积和厚度等的差异，在实验中发现其光电离截面可相差10多倍；这说明在引用不同的实验结果时必须十分小心。

另一个证明金不同浅施主杂质 N_s 形成络合物的实验证据是将两个背景完全相同的 P^+N 二极管，一个扩金，另一个只经历和扩金相同温度的热处理（参考样品），在低温下用 C–V 法测量参考样品的 N_s，在同样的温度下（$e^t_{A,D}\simeq 0$），用 C–V 法测量金A电子充满和完全抽空时相应的自由载流子浓度，比较参考样品和金A抽空时计算出的浓度，发现两者很好一致。这说明金A只是起补偿浅施主 N_s 的作用，而金不同浅施主杂质形成络合物；否则当金A抽空时测得的自由载流子浓度应当小于由参考样品测得的 N_s，因为金A这时要补偿两个自由电子。

再者，金补偿样品的Hall系数和电阻率实验研究也否定了金同浅施主形成络合物的看法。假定在扩散过程中金原子同浅施主杂质形成络合物，那么每一个金原子要补偿两个来自浅施主杂质的电子，这应反映在掺金后导带自由电子迅速减小（电阻率很快增加），而当金原子为浅施主杂质一半时，*N*-Si便被补偿成为本征高阻。但实际情况恰巧相反，实验发现[19]当 *N*-Si 被金补偿为高阻时所需的Au原子浓度总比 N_s，高一些（1.5–2.0），其原因可能同金在晶体缺陷处的部分沉淀相关。

综上所述，虽然硅中金A是（Au-Ds）络合物的看法是没有根据的，但由于扩金条件、淬火速度等不同，硅中金原子同其他杂质形成络合物如Au-Fe、Au-B等[17]却是实验实事。本文在部分 *P* 型扩金硅中除金A和D外，也观察到另一个与金相关的络合物，"Au-X"中心，这个中心位于禁带下半部（E_v+0.226eV），具有大的空穴俘获截面（$\sigma_{Px}\geqslant 6\times 10^{-15}\text{cm}^2$），但不依赖于温度（90–120K）。

3.3 金中心光电离截面谱及其温度依赖关系

低温下深中心光电离谱的实验研究不但可用来比较精确地确定深能级的束缚能，提供杂质势的性质，而且还可以获得有关电–声子互作用以及深中心激发态等方面的重要信息。在过去的20年里，尽管不少理论工作者应用不同的杂质势[20,21]从理论上计算了光电离截面并同实验结果比较上也取得了一些进展，但由于这个问题的复杂性，至今没有解决根本问题；即使对于不存在电–声子作用的简单情况，也没有一个统一的理论模型可循。本文利用[22,23]所报道的方法拟合了硅中金D和A光电离截面的实验资料。在80K，金D和A的电子和空穴光电离截面谱的阈值能量之和分别为1.170eV和1.120eV；金A电子和空穴热发射率与温度依赖关系的分析（考虑俘获截面的温度依赖关系）得到的电子和空穴热激活能之和为1.163eV，这同硅在该温度下的禁带宽度 E_g（1.167eV）基本一致。这说明对金中心光电离截面贡献主要是来自于零声子光学矩阵元；也就是说，硅中金杂质中心的晶格弛豫效应是很小的。

值得注意的是金A电子和空穴光电离截面谱阈值能量之和在80K为1.120eV（E^0_{nA}=0.51eV，E^0_{pA}=0.61eV），比该温度下硅禁带宽度要小47meV。为了弄清楚引起上述差异的原因。本文首先测量了金 A 电子、空穴光发射率阈值能量的温度依赖关系（80－135K），发现当温度升高时金 A 空穴和电子光发射阈值能量都减小，且空穴减小要比电子的多。空穴光发射阈值能量随温度增加而减少原因之一是金 A 中心可能存在着激发态。根据金 A 的电荷态设定，当激发金 A 空穴到价带时，金 A 成为负电中心。故金受主类 Rydberg 空穴激发态的形成是可以预见的。通过激发态的空穴光热激发则可导致光电离截面的大的温度依赖关系。本文因受 P^+N 结样品（金 A 很难完全抽空）的限制，未能从实验上直接检验上述看法是否合理。至于金 A 电子激发态问题，一方面由于中性金 A 从原理上来说不会存在由库仑引力而形成的激发态；另一方面从实验上（光导耐谱）也确实未观察到。另一个引起电子和空穴光阈值能量随温度变化的原因可能是电场效应。为此我们在 78K，在不同电场作用下（1. 75kV/cm 到 43. 2kV/cm）测量了接近光阈值时电子和空穴光发射率，实验结果表明，电场效应可以忽略不计［图 2（a）+点］。

比较金 A 电子和空穴的热激活能和光阈值能量就会发现，同具有大的晶格弛豫效应的深中心相反，金 A 的热激活能反而大于光阈值能量。这种反常结果在 GaAs:Fe[24]，GaAs 中 B 能级[7]都曾观察到，除激发态可以定性解释这种反常现象外，还没有一个满意的理论模型。

3. 4　硅中金中心的本质

硅中金中心的化学本质一直是个争议的问题，早期通过金的扩散行为研究[25,26]认为金基本上是以代位形式进入点阵的，这就很自然地把金 A 和 D 能级看成是同一个缺陷（Au_{si}）的不同能态（D，$Au_{Si}^{+/0}$；A，$Au_{Si}^{0/-}$）。这个观点曾被广泛的接受，直到最近还有一些实验证据[16]支持这个模型。本文的实验结果只是直接证明了金中心是同一个缺陷的两个不同能态，否定了 Lang 等的（Au-Ds）络合物论断，但未排除金中心是它种形式的络合物的可能性。事实上，近几年来掺金硅的光致发光研究[27]和单轴应力效应的结果[28]都暗示了金中心不像是简单的代位缺陷。在 Thebanlt 等[27]的硅中金 D 高分辨光致发光实验中发现第一激发态能级包含四个子能级，这意味着金 D 中心元胞势引起 $1s(^2T_2)$ 或 $1s(^2T_2+^3E)$ 态的进一步分裂［这种分裂仅用自旋轨道微扰导致的 $1s(^2T_2)$ 分裂为 T_d 群 T_r+T_s 是无法解释的。因为在这种情况下 ^2E 不分裂］，也就是说金缺陷的对称性不是四面体对称，而是被降低了。李名复等[29]认为，具有 T_d 对称势的缺陷能级单轴应力系数应是各向同性的，且等于相应的流体静压系数的 1/3。从这个论点出发，他们测量了金 A 的流体静压系数并同姚秀琛等[28]获得的硅中金 A 单轴应力系数作了比较，发现金 A 缺陷势远偏离于 T_d 对称性；因此硅中金 A 缺陷不像具有简单的 Au_i 或 Au_{si} 组态。

前面曾详细地讨论了金中心不可能是（Au-Ds）的络合物，因为它既不依赖于浅施主或浅受主掺杂浓度、施主或受主的掺杂类型，也不依赖于金同浅杂质的补偿比 K；

那么究竟硅中金受主和施主的化学本质是什么呢？目前尚不清楚。代位金和晶格缺陷络合物或偏离硅点阵的金代位缺陷或许是一个合适的候选人。当然这仅是个假设，还待进一步的实验结果来证实。

本文是在 H. G. Grimmeiss 教授大力支持下完成的，并进行过多次十分有益的讨论；样品是由 K. Nideborn 制备的。作者表示诚挚的感谢。

参 考 文 献

[1] D. V. Lang, H. G. Grimmeiss, E. Meijer and M. Jaros, Phys. Rev., B22, 3917(1980).
[2] J. R. Morante, J. E. Carceller, A. Hermas, and P. Cartujo, Appl. Phys. Letter, 41, 656(1982).
[3] L-Å. Ledebo and Zhan-guo Wang(王占国), Appl. Phys. Letter., 42, 680(1983).
[4] O. Engstrom and H. G. Grimmeiss, J. Appl. Phys., 46, 831(1975).
[5] S. D. Brotherton and J. Bicknell, J. Appl. Phys., 49, 667(1978).
[6] V. Kalyanaraman and V. Kumar, Phys. Stat. Sol. (a), 70.317(1982).
[7] Zhan-guo Wang(王占国)L-Å. Ledebo and H. G. Grimmeiss, J of Phys. C: Solid State Physics., 17.259(1984).
[8] J. A. Pale, Solid State Electronics., 17, 1139(1974).
[9] D. V. Lang. Topics in Appl. Phys., Chap. 3, Springer Verlay(1979).
[10] 王占国, 大缺陷浓度的电容 DLTS 和电流 DLTS 讯号的计算机拟合,(待发表)
[11] H. G. Grimmeiss, Anu. Rev. Mater. Sci., 7.341(1977).
[12] C. H. Henry. H. Kukimoto and G. L. Merritt, Phys. Rev., B7, 2499(1973).
[13] D. V. Lang, J. Appl. Phys., 45.3023(1974).
[14] H. G. Grimmeiss, E. Jansen and B. Skaratam, J. Appl. Phys., 51 3740(1980).
[15] E. Meijer, L-Å. Ledebo and Zhan-guo Wang(王占国), Solid State Commun., 48.255(1983).
[16] P. Van, Stas and R. Kassing, Solid State Commun., 50, 1051(1984).
[17] S. D. Brotherton, P. Bradley, A. Gill and E. R. Weber, J. Appl. Phys., 4, 952(1984).
[18] R. H. Wu and A. R. Peaker, Solid State Electronics., 25, 643(1982).
[19] W. M. Bullis, Solid State Electronics., 9, 143(1966)and O. Engstrom,(私人通信).
[20] G. Lucovsky, Solid State Commun., 3, 299(1965).
[21] T. H. Ning and C. T. aah, Phys, Rev., B4, 3468(1971).
[22] H. G. Grimmeiss, L-Å. Ledebo, J. Phys., C: S. S. Phys., 8, 2615(1975).
[23] L-Å. Ledebo, J. Phys., C: Solid State Phys., 14, 3279(1981).
[24] M. Kleverman, P. Omling, L-Å. Ledebo and H. G. Grimmeiss, J. Appl. Phys., 54, 814(1983).
[25] W. R. Wilcox et al., J. Electrochem. Soc., 111, 1377(1964).
[26] J. Martin et al., Solid State Electronics, 8, 83(1966).
[27] D. Thebault, J. Barrau, M. Brousseau, D. Xuen Thanh, J. C. Brabant, F. Voillot and Mme. Ribault, Solid State Commun., 45, 645(1983).
[28] 姚秀琛、秦国刚、曾树荣和元民华, 物理学报. 33, 377(1984).
[29] 李名复等, J. Appl. Phys.,(待发表).

A novel model of "new donors" in Czochralski-grown silicon

J. J. Qian

(Institute of Semiconductors, Chinese Academy of Sciences, Beijing 100083, China;
Beijing Laboratory of Electron Microscopy, Chinese Academy of Sciences,
Beijing 100080, China)

Z. G. Wang, S. K. Wan, and L. Y. Lin

(Institute of Semiconductors, Chinese Academy of Sciences, Beijing 100083, China)

Abstract A new model of "new donors" is presented, based on electrical, infrared measurements, transmission electron microscopy, and high-resolution electron microscopy observations on Czochralski-grown silicon single crystals containing "new donors." In this model, the electrical activity of "new donors" originates from the uncoordinated Si dangling bonds on small dislocation loops resulting from oxygen precipitation. In comparison with other models, the present model can better explain the experimental results of the heat treatment Czochralski-grown Si wafers.

1 Introduction

Czochralski-(Cz-) grown silicon crystal is at the present time the most widely used semiconductor material for integrated-circuit fabrication. The oxygen content in Cz-Si may reach concentrations as high as $1\times10^{18}cm^{-3}$, approaching the saturated solubility of oxygen in silicon at the melting point of Si. Oxygen is incorporated into the silicon lattice on electrically inactive interstitial site in the growth process. During subsequent heat treatment, supersaturated oxygen will react with silicon to form $Si-O_x$ clusters or precipitates which apparently act as donor traps. For silicon wafers annealed at different temperatures, two types of oxygen-related donors can be formed. The "thermal donor" (TD) and the "new donors" (ND) are generated in the temperature ranges of 350–500 and 550–850℃, respectively.

原载于：J. Appl. Phys., 1990, 68(3): 954–957.

As for the nature of the ND, Kanamori and Kanamori[1] suggested that the ND are possibly formed by combination of a substitutional oxygen and a vacancy, but it can hardly be proved by other experimental facts. Cazcarra and Zunino[2] considered that the ND probably relate to Si_yO_x clusters containing several hundred oxygen atoms, while Gaworzewski and Schmalz[3] proposed that C–O complexes possibly are the core of the ND. However, all these models cannot well explain the origin of the ND electrical activity. Recently, Hölzlein et al.[4] supposed that the ND originate from states at the surface of SiO_x precipitates, while Henry, Pautrat, and Saminadayar[5] concluded that it is resulted from the inversion layer around the SiO_2 precipitates. Though such models based on the fixed positive charge of the SiO_2/Si interface can better explain the passivating action of the hydrogenation treatment on the ND, yet it cannot interpret why the density of the ND decreases and even approaches zero in spite of the growth of SiO_2 precipitates and therefore the SiO_2/Si interface under the thermal treatment at elevated temperatures(e. g., >850℃). Up to date, there has not yet been a model which can better elucidate the main experimental facts related to the ND. The present work focuses on the microprecipitates of oxygen in silicon and, particularly, the related high-density small defects with a size smaller than 10nm. Based on the observations of transmission electron microscopy(TEM) and high-resolution electron microscopy(HREM) in this paper, a new model of the ND is proposed in which the electrical activity of the ND does not originate from the oxygen atom itself, but from the uncoordinated Si dangling bonds on the dislocation loops which result from the Si self-interstitials produced by oxygen precipitation during the heat treatment process. In comparison with other models, the present model can better explain the experimental results already obtained.

2 Experiments

In this work, *p*-type(100) B-doped Cz-Si wafers are used; the typical parameters are as shown in table 1. The silicon wafers are preannealed at 450℃ for 16h and then annealed at 650℃ for 4–5days under an Ar atmosphere. The resistivity of Si wafers is measured by the four-probe method and is turned into the carrier concentration by means of the Irvin curve. The oxygen concentration ($[O_i]$) and the carbon concentration ($[C_s]$) of the samples are measured by the infrared(IR) method. The samples for TEM and HREM observation are first ground and polished, then thinned by Ar^+ at 4 – 5kV up to an electronic transparency thickness. The electron diffraction images and the HREM images are obtained by EM-420 and JEOL-200CX at 120 and 200kV, respectively.

Table 1 Some parameters of representative Cz-Si samples before and after annealing

Sample No.	Before annealing				After annealing				
	Conductivity type	p 10^{15} (cm^{-3})	$[O_i]^*$ 10^{17} (cm^{-3})	$[C_s]$ 10^{17} (cm^{-3})	Conductivity type	ND 10^{15} (cm^{-3})	$\Delta[O_i]$ 10^{17} (cm^{-3})	Rodlike defects (cm^{-2})	Small loops (cm^{-3})
A	p	1.81	9.3	<0.1	n	2.6	7.2	10^6-10^7	$10^{13}-10^{14}$
B	p	0.05	6.7	<0.1	n	0.5	2.4	10^4-10^5	$10^{13}-10^{14}$
C	p	1.5	6.6	3.8	n	2.2	4.4	none	$10^{13}-10^{14}$
D	p	2.0	4.6	0.4	p	none	0	none	none

* According to ASTM F 121-81

3 Experimental results

The electrical measurements show that three kinds of samples, A, B, and C, are turned into n type after being annealed at 650℃ for a long enough time, whereas the resistivity of annealed sample D has no obvious change. The donor concentrations for samples A, B, and C calculated from their resistivities are as shown in table 1. According to IR measurement, oxygen precipitation is observed for samples of A, B, and C, and the more interstitial oxygen $[O_i]$ decreased($\Delta[O_i]$), the more ND generated. For sample D, there is no oxygen precipitation($\Delta[O_i] \sim 0$); there is no ND produced. For samples B and C, their original $[O_i]$ contents are nearly equal, but their $[C_s]$ are different. For sample C, the $[C_s]$ content is higher; the ND produced are also a bit more. TEM observations indicate that in samples A, B, and C containing the ND, there are several types of defects. Among them, in the samples A and B with lower $[C_s]$ content, there are dislocation dipoles as well as $\langle 110 \rangle$ oriented rodlike defects[6-8] with a length of a few micrometers, and their densities are $10^4-10^7/cm^2$(table 1). But in sample C($[C_s]>10^{17}/cm^3$), there are no such rodlike defects or dislocation dipoles. In addition, the observations by chemical preferential etching and optical microscopy indicate that all three samples A, B, and C contain a kind of microdefect clusters with extremely high density and nonuniform distribution. But in the sample D containing no ND, there is neither rodlike defects and dislocation dipoles nor microdefect clusters and other defects of high density. The TEM observations show that the microdefects of high density and nonuniform distribution in samples A, B, and C will assume to have black-white lobe contrast under certain diffraction conditions. Based on the spatial orientation of the dislocation dipoles in the samples as well as their projection length on the electron micrographs, the thickness of the thin films can be calculated and thus the density of the microdefects is obtained to be about $10^{13}-10^{14}/cm^3$(Fig. 1). The HREM observations show that the microdefects have a size of ~10nm with a stress field expanding outwards and the central portion of which has a platelike precipitate[8, 9] with a thickness of 1–2 atom layers (P in Fig. 2). Nearby the precipitate there is a pair of the extra half {111} silicon planes marked by arrow in Fig. 3, and this obviously is an interstitial 60° dislocation loop. The diameter of the loops is of 2–7nm with a {113} habit plane. The structure schematic of such dislocation loop is shown in Fig. 4, with the dislocation core D having dangling bonds.

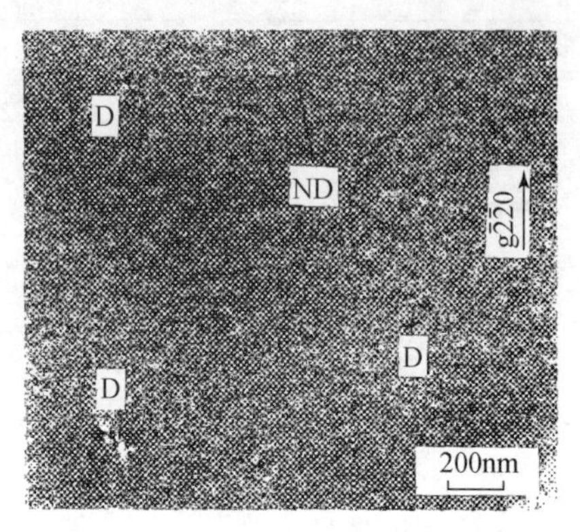

Fig. 1 TEM micrograph of the ND and dislocation dipoles (D) in Cz-Si annealed at 650℃ for ~5days. B ~ [001], g ~ 220, and $s>0$

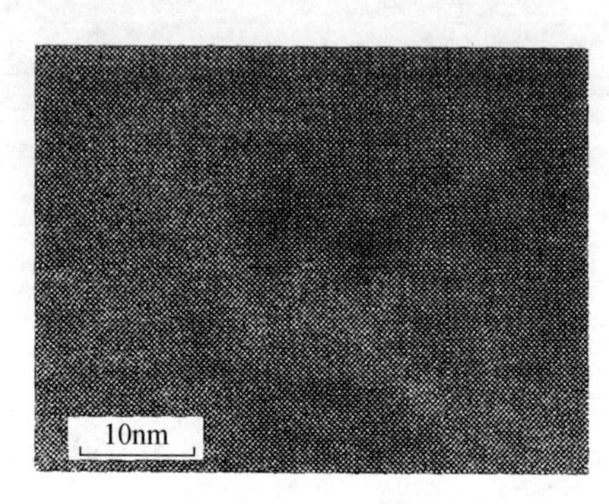

Fig. 2 Lattice image of the platelike precipitate (P)

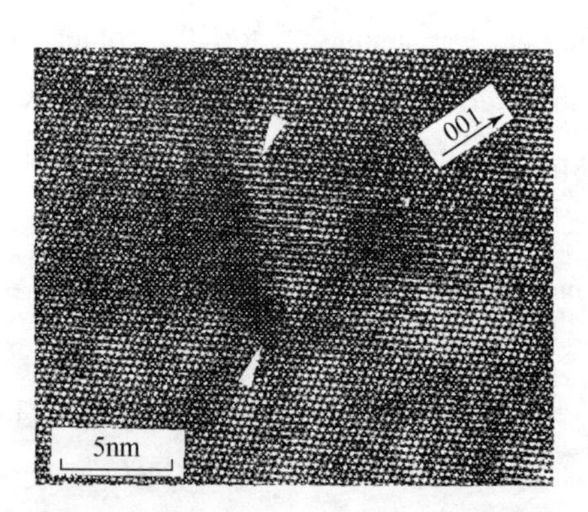

Fig. 3 Lattice image of the small interstitial dislocation loop. A pair of the extra half {111} silicon planes are denoted by arrow. Observations are at 200kV, $[1\bar{1}0]$ zone axis

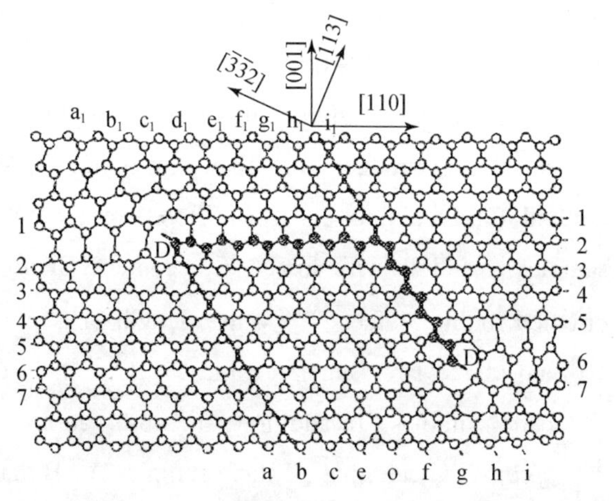

Fig. 4 Schematic structure of small interstitial dislocation loop, heavy lines denote a pair of the extra half {111} Si planes with the dangling bonds (D). The atoms with shadow are excess Si atoms

4 Discussion

The above experimental results demonstrate that after heat treatment at 650℃ for a long

period of time, the silicon wafers containing ND must have small dislocation loops with a certain density; conversely, such dislocation loops have never been observed in the samples of containing no ND. Now, we turn to discuss the formation of the small dislocation loops. During an annealing at 650℃, the supersaturated interstitial oxygen atoms in Cz-Si will agglomerate on certain cores such as carbon atoms or point defect clusters to form SiO_x precipitates, The shape of the precipitates is in general platelike, leading to a minimum deformation energy. As the mole volume ratio between SiO_2 and Si (V_{SiO_2}/V_{Si}) is ~2, the volume of the interstitial oxygens after changing into precipitates increases about 100%. Therefore, when a precipitate grows up to a certain volume, the Si self-interstitials are emitted from growing oxygen precipitates to release the excessive stresses caused by a large volume increase of the precipitates.[10-13] Obviously, the number of Si self-interstitials is proportional to the volume of all oxygen precipitates already formed and, thus, is proportional to the decreasing of the interstitial oxygen concentration ($\Delta[O_i]$) during the annealing process. However, if Si self-interstitial concentration reaches a certain level, they begin to agglomerate, which results in the formation of the interstitial-type dislocation cores.[13-15] For silicon, each dangling bond already amounts to about 1eV of free energy; thus the dislocation cores may be constructed with the least number of dangling bonds only if the dislocation line is in a⟨110⟩direction. For this reason, the linear configuration will be favored.[15] Along with the increase of heat treatment time, newly generated Si self-interstitials will continuously agglomerate to the dislocation lines; then the latter can slip on {113} plane[16, 17] under the action of the strain field to form the interstitial-type 60° dislocation loops with a {113} habit plane and a number of dangling bonds in their cores.

In homopolar covalent bonding Si crystal, the dangling bonds possess donor properties;[18] thus the ND electrical activity produced during annealing at 650℃ is considered to be related to the dangling bonds of the small dislocation loops formed by Si self-interstitials. From the results of TEM observations, it can be seen that the density of small dislocation loops is of $10^{13}-10^{14}/cm^3$ (table 1) and the diameter of them is of 2-7nm, if the average diameter is about 5nm, and each Si atom spacing in⟨110⟩direction is $\sqrt{2}/2a$ ($a=0.543$nm); then the number of the dangling bonds along the dislocation core of each loop will be about 40, and thus all of them are $4\times10^{14}-4\times10^{15}/cm^3$. This result is quite similar to the ND density obtained by electrical measurements (table 1). Similarly, it can readily be seen from Fig. 4 that the atoms with shadow are excess Si atoms in comparison with perfect lattice surrounding the small dislocation loop, which are transformed from Si self-interstitials caused by oxygen precipitates. As the mentioned above, only such atoms occupied on dislocation cores have dangling bonds; the others are all four coordinated with neighboring lattice atoms and leave no dangling bonds. For a loop with 40 dangling bonds (~ 5nm diameter), the excess Si atoms are calculated to be about $(3-5)\times10^2$. Therefore, the 10^{13}–10^{14} cm^{-3} small loops experimentally obtained from this work should be involved in $(0.3-5)\times10^{16}cm^{-3}$ extra Si atoms (N_{ex}). Assuming that the oxygen precipitates are simply crystalline SiO_2 (coesite), and that the fraction (β) of the stress not relieved by Si interstitial emission is 0.2,[9] the Si interstitial concentration (N_{Si_i}) of $(1.6-5.0)\times10^{17}cm^{-3}$ generated by a volume increase resulting from $(2.4-7.2)\times10^{17}$ [O]/cm^3 (table 1) oxygen precipitation is then derived. Obviously, there still exists some difference between the N_{ex} and N_{Si_i}; however, this apparent difference could be understood if one takes into account other microdefects such as clusters, dipoles, and rodlike defects, etc., are also formed by Si interstitials during annealing at 650℃. On the other hand, the sample with a typical ND density of 4×10^{15} cm^3 should contain about 1×10^{14} cm^{-3} dislocation loops of ~ 5nm diameter, and the average distance between the loops is ~ 200nm. The formation of such a dislocation loop can approximately be considered as a consequence of oxygen diffusion and precipitation in a sphere with a radius of ~ 100nm. According to the oxygen diffusion coefficient data at 650℃,[19] the time calculated for oxygen atoms passing through 100nm is about 12h. This value coincides with the intensity dependence of the 0.902eV line on the 650℃ annealing time.[13] In Ref. [13], this intensity dependence is suggested in relation to the behavior of Si self-interstitials emitted during the oxygen precipitation process. Based on this, we believe that the interstitial-type small dislocation loops caused by oxygen precipitation upon annealing at 650℃ for a long time are possibly the ND themselves, and the donor trap electrical activity originates from the uncoordinated Si dangling bonds on the small dislocation loops.

By means of the present model, the main experimental facts related to ND can be very well interpreted.

(1) ND are related to oxygen precipitation.[1,2,20] The interstitial-type small dislocation loops are formed by the agglomeration of Si self-interstitials generated in oxygen precipitation; the more the dislocation loops produced, the larger the concentration of ND. Conversely, for the samples without oxygen precipitation after annealing, there are no small dislocation loops therein; thus no ND is formed.[20]

(2) Effect of carbon content in Si on the formation of ND.[1,3,20,21] As is well known, this impurity acts as a nucleus for oxygen precipitation in Si and promotes the precipitation of supersaturated oxygen; therefore, the existence of carbon is favorable for the formation of the small dislocation loops, and hence it can increase the generation of ND.

(3) The electrically active effect of hydrogen passivated ND.[4,5] In accordance with our model, the hydrogen passivation effect on ND is not caused by the hydrogen passivated interface

states of SiO_2/Si. As a matter of fact, while annealing at 300℃, hydrogen very easily diffuses into Si crystal and combines with the unsaturated dangling bonds on the small dislocation loops to form Si–H bonds;[22] thus the dangling bonds lose their electrical activity of the donor.

(4) The inhomogeneity of ND distribution.[5,21] It is generally accepted that the oxygen distribution in Cz-Si wafer is inhomogeneous and depends on the growing conditions and the thermal history.[1] During heat treatment, supersaturated oxygen interstitials tend to cluster at some sites such as carbon atoms, point defects, etc., and to form oxide precipitates. As our experimental results, the interstitial-type small dislocation loops generated by silica precipitates have not only different sizes, but also an inhomogeneous distribution. This has better explained why the formation of ND is related to the original material and the thermal history of crystal growth. On the other hand, a space-charge region surrounding the dislocation loop must be formed for electrical neutrality condition, It is obvious that such a space-charge scattering region with a large scattering cross section and high density will reduce the mobility of carriers.[23] Therefore, the lower electron mobility of the ND samples can also be readily understood according to our model.

(5) Thermal treatment at elevated temperature (e, g., >900℃) leads to a lower ND concentration,[1,2] This experimental result can also be well explained by the present model. The temperature for the elastic deformation of Si is around 900℃, when annealed above this temperature, the 60° dislocations may dissociate gradually (e. g., dissociated into 30° and 90° partial dislocations with lower energy) and after the reconstructing of the bonds,[24] the number of the dangling bonds will decrease, and thus the density of ND will decrease too.

5 Conclusion

This paper proposes a new model for the microstructure of the new donor traps in Cz-Si based on the electrical, IR measurements, and TEM and HREM observations. According to this model, the electrical activity of the ND is not directly originating from oxygen itself, but from the uncoordinated Si dangling bonds on the interstitial-type dislocation loops, and such interstitial-type dislocation loops are formed by Si self-interstitials caused by oxygen precipitates. The main experimental results related to the ND can be better interpreted by means of this model.

Acknowledgments

The authors extend their thanks herewith to the helpful discussions with Professor Fang-hua Lee and Professor Zhon-qian Zhou, and to the much help of the HREM photography from senior engineer Da-yu Yang. This work was supported by the National Natural Science Foundation of China, Grant No. 85267.

References

[1] K. Kanamori and M. Kanamori, J. Appl Phys. 50, 8095(1979).

[2] V. Cazcarra and P. Zunino, J. Appl. Phys. 51,4206(1980).

[3] P. Gaworzewski and K. Schmalz, Phys. Status Solidi A 77, 571(1983).

[4] K. HÖlzlein, G. Pensl, M. Schulz, and N. M. Johnson, Appl. Phys. Lett. 48, 916(1986).

[5] A. Henry, J. L. Pautrat, and K. Saminadayar, J. Appl. Phys. 60, 3192(1986).

[6] K. Tempelhoff and F. Spiegelberg, in Semiconductor Silicon 1977, edited by H. Huff and E. Sirtl (The Electrochemical Society, Pennington, NJ, 1977), 585.

[7] N. Yamamoto, P. M. Petroff, and J. R. Patel, J. Appl. Phys. 54, 3475(1983).

[8] K. Tempelhoff, R. Gleichmann, F. Spiegelberg, and D. Wruck, Phys, Status Solidi A 56, 213 (1979).

[9] A. Bourrent, J. Thibault-Desseaux, and D. N. Seidman, J. Appl. Phys. 55, 825(1984).

[10] P. E. Freeland, K. A. Jackson, C. W. Lowe, and J. R. Patel, Appl. Phys. Lett. 30, 31(1977).

[11] S. Mahajan, G. A. Rozgonyi, and D. Brasen, Appl. Phys. Lett. 30, 73(1977).

[12] J. R. Patel, in Semiconductor Silicon 1981, edited by H. R. Huff, R. J. Kriegler, and Y. Takeishi (The Electronchemical Society, Pennington, 1981), 189.

[13] M. Tajima, U. Gösele, J. Weber, and R. Sauer, Appl. Phys. Lett. 43, 270(1983).

[14] D. M. Maher, A. Staudinger, and J. R. Patel, J. Appl. Phys. 47, 3813(1976).

[15] T. Y. Tan, Philos. Mag. A 44, 101(1981).

[16] C. A. Ferreira Lima and A. Howie, Philos. Mag. 34, 1057(1976).

[17] I. G. Salisbury, J. Appl. Phys. 52, 1108(1981).

[18] V. Heine, Phys, Rev, 146, 568(1966).

[19] M. Stavola, J. R. Patel, L. C. Kimerling, and P. E. Freeland, Appl Phys. Lett. 42, 73(1983).

[20] L. Y. Lin, Z. G, Wang, J. J. Qian, W. K. Ge, S. K. Wan, and R. G. Lin, Mater. Sci. Prog. 2, 56(1988)(in Chinese).

[21] A. Ohsawa, R. Takizawa, K. Honda, A. Shibatomi, and S. Ohkawa, J. Appl. Phys. 53, 5734 (1982).

[22] S. J. Pearton, A. M. Chantre, L. C. Kimerling, K. D. Cummings, and W. C. Dautremont-Smith, MRS Symp. Proc. 59, 475(1986).

[23] The discussion in detail to be published elsewhere.

[24] P. B. Hirsch, J. Microsc. 118, 3(1980).

Electrical characteristics of GaAs grown from the melt in a reduced-gravity environment

Z. G. Wang, C. J. Li, F. N. Cao, Z. W. Shi, B. J. Zhou, X. R. Zhong, S. K. Wan, S. D. Xu, and L. Y. Lin

(Institute of Semiconductors, Chinese Academy of Sciences, Beijing 100083, China)

Abstract The electrical properties and structural defects of Te-doped GaAs grown in space have been investigated by using various techniques. The experimental results confirm that the microgravity conditions offer some advantages for the melt growth of Ⅲ-Ⅴ compound semiconductor materials; improvements of homogeneity and perfection as well as purity of the space GaAs single crystal are expected.

1 Introduction

Thermal instability caused by gravitational convections is a main factor preventing us from growing GaAs with high quality on Earth. Therefore, in recent years, growth of semiconductor materials in space has been receiving considerable interest, and significant results have been obtained.[1-4] Unfortunately, studies of GaAs crystals grown from the melt in space are not yet seen in publications.

A systematic study of the electrical and optical properties and structural defects of GaAs melt-grown in a Chinese recoverable satellite by using various analytical techniques is presented. The distribution of the tellurium dopant, dislocations throughout the bulk, and the behavior of deep centers as well as the possibility of high-quality GaAs single-crystal growth in space are discussed in detail.

2 Single-crystal growth and sample preparation

Growth of a GaAs single crystal in space makes use of the remelting and recrystallizing method. A Te-doped GaAs single crystal with an average electron concentration $3.5\times10^{18}/cm^3$ grown in the laboratory is processed into an ingot of 9mm diameter and 100mm length; it is

原载于: J. Appl. Phys., 1990, 67(3): 1521-1524.

sealed in a quartz tube of 17mm diameter and 130mm length. The space growth of the GaAs crystal is carried out in a resistance tube furnace, with the power supply turned on for 90min. The temperature is raised to 1250℃ for 20min, maintained at this point for 25min and decreased with a cooling rate of 0.2℃/min for 45min to achieve the crystal growth. During the initial heating stage of the space experiment, a molten zone is formed in the middle of the single-crystal GaAs ingot. Suspended by the unmelted seed crystals at both sides, the GaAs is regrown as a result of the molten zone through normal solidification, The molten zone was found to break up in space, and GaAs crystals grew at both seed ends as two portions, 1 and 2. Part 1 is found to be a single crystal in the shape of a torch with a size of 10mm maximum diameter and 15mm length; part 2 is also a single crystal with a size of 10mm maximum diameter and 7mm length but enclosed by an outer polycrystalline shell.

Three samples (A, C, and B) are investigated in this study. Samples A and C are longitudinal cuts along the growth axis from single crystals 1 and 2 respectively, in which there exists GaAs both grown on Earth and in space; sample B is a segment from the single crystal 2 which depicts only space-grown material.

Electrochemical capacitance-voltage (ECCV) and scanning photoluminescence were employed to study the electrical homogeneity point by point with a spatial resolution of about 500μm for samples A, B, and C; cathodoluminescence imaging (CLI) was adopted to reveal the structural defects in these samples. In addition, electroluminescence spectra of single heterojunction diodes (SHDs) with the space crystal as substrate are also used to characterize the homogeneity of the space materials. In order to understand better the defects of gravity on the behavior of impurities and point defects, junction capacitance techniques, such as deep-level transient spectroscopy (DLTS) were then employed to investigate the properties of deep centers in space materials.

3　Results and discussions

3.1　Experimental results

Table 1 shows the electrical characteristics of two samples (B, C) as measured by the Hall effect and ECCV methods. The spatial distribution of the free-carrier concentration (n) for sample A (taken from space crystal 1) is shown in Fig. 1. The following observations can be made: (a) The free-carrier concentration of GaAs grown in space is nearly 1 order of magnitude smaller than that of GaAs grown on Earth. (b) There is a transition region between the space and the Earth materials that shows n decreasing abruptly (see Fig. 1, upper left inset) and then increasing with distance from the seed-space-crystal interface, because the segregation coefficient of the Te dopant in GaAs, K, is smaller than 1.

Table 1 Electron concentration and electron Hall mobility of space- and Earth-grown GaAs determined on the crystal 2, wherein a and b indicate Earth seed samples, c and d Earth and space samples, and e-h space samples

Sample	n(300K) (10^{18}/cm^3)	μ(300K) (cm^2/Vs)	n(77K) (10^{18}/cm^3)	μ(77K) (cm^2/Vs)	Remarks
a	5.85	1740	5.63	1858	Hall
b	5.07	1790	5.12	1923	Hall
c	0.36-3.8	-	-	-	ECCV
d	0.72-0.96	-	-	-	ECCV
e	0.36	-	-	-	CV
f	0.588	613	-	-	Hall
g	0.397	3005	0.385	3413	Hall
h	0.468	2361	0.470	2529	Hall

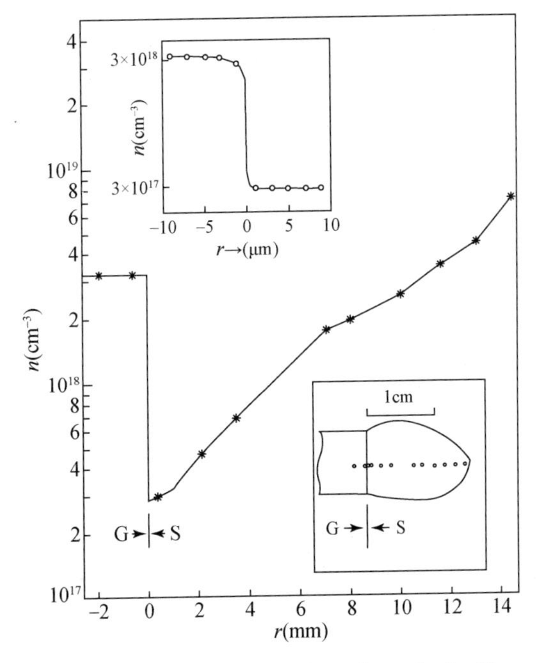

Fig. 1 Spatial distribution of the electron concentration of sample A. The geometry of sample A and the locations of the measurement points are shown in the lower right inset, The upper left inset is the electron concentration distribution near the seed-space GaAs interface obtained by CL. G is the ground(Earth) seed GaAs crystal and S the space GaAs single crystal

The CLI studies of the Te-doped GaAs indicate that the high density of defects(mainly

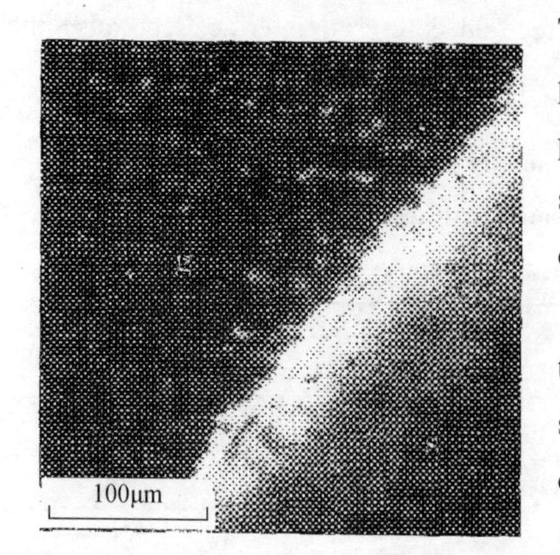

Fig. 2　CL topography in the neighborhood of space-ground crystal interface for space crystal 2. G labels the ground-(Earth-) grown crystal and S the space-grown crystal

Te-dopant precipitates) seen in the seed-crystal portion are initially absent in the space regrown portion(Fig. 2), However, the defect density in the space-grown crystal increases along the regrowth direction with distance from the interface.

In order to show the dislocation distribution of the space crystal and the Earth crystal, three samples were analyzed from three different portions of crystal 1. The orientation of the sample face is (100). Dislocations are revealed by etching in molten KOH, The results (see table 2) clearly demonstrate that the average dislocation density in the space-grown sample is nearly 1 order magnitude higher than that of the seed crystal, which is in contradiction with the earlier experimental data on space-grown InSb and GaAs.[5]

Table 2　Sampling locations for dislocation counts and average dislocation density etch pits density(cm^{-2})(EPD) measurements for the crystal 1

Crystal type	Seed crystal	Seed-space crystal	Space crystal
Location	−12mm	0	5mm
Average EPD	11 000	97000 64000	96000

In an earlier communication,[6] it was shown that no periodical impurity striations are formed in space-grown crystals (see Fig. 3), but that a cellular structure and discontinuous nonperiodical impurity striations are found in the tail region of the space-grown GaAs due to constitutional supercooling. CLI and anodic etching results also prove that close-spaced periodical impurity striations with 10μm periodicity still exist close to the surface of the space GaAs crystal ingot.

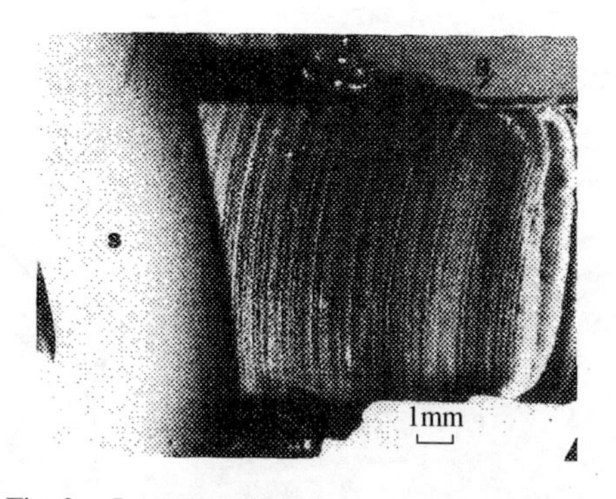

Fig. 3　Impurity striations revealed after 2min anodic etching at a current density 100mA/cm^2 for the ground-seed crystal (G). No impurity striations were found after 2min separate etching at a current density 250mA/cm^2 for the space crystal(S). The sample is taken from the space crystal 2

Deep-level transient spectroscopy (DLTS) for samples B and C are shown in Figs. 4(a) and 4(b), respectively. It seems that the density of deep centers in the space GaAs samples is greatly reduced in comparison with the Earth reference sample. However, for sample C, two broad DLTS peaks are observed, which indicate that the quality of this crystal is worse due to fast quenching when the power supply was suddenly turned off.

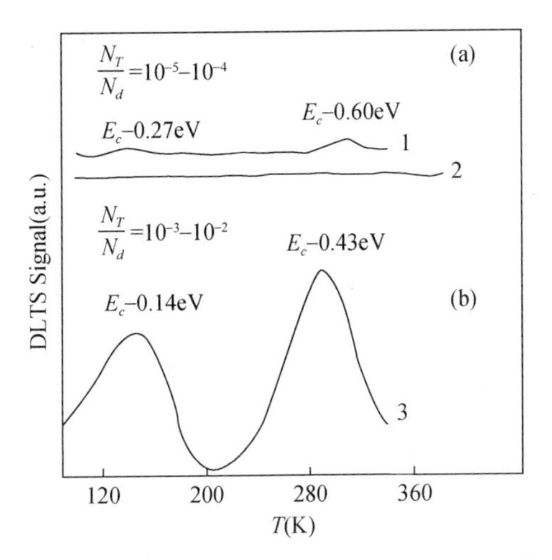

Fig. 4 Typical DLTS spectra of the space GaAs samples. (a) The diode was fabricated on the space sample B, whereas in (b) the diode was fabricated on the location (sample C) very close to the interface of the space single crystal and the space polycrystalline shell. Curves 1 and 3 are the majority-carrier pulse, and curve 2 is the minority-carrier pulse

Single heterojunction emitting diodes with space GaAs as a substrate were fabricated, using liquid-phase epitaxy (LPE) growth. The very similar *I–V* characteristics and spatial distributions of electroluminescence (EL) intensities of these diodes clearly indicate that sample B has rather good homogeneity, whereas sample C has poorer quality, which agrees well with its very broad DLTS peaks. The wavelength of EL peaks from these single heterojunction diodes is around 8300Å, which confirms that the emission is from the space GaAs crystal itself.

3.2 Discussions

3.2.1 The equilibrium segregation coefficient *K* of the Te dopant in the space GaAs crystal

It is worth noting that the carrier concentration (n) of the space Te-doped GaAs is 1 order magnitude smaller than that of the Earth samples; it changes abruptly from the Earth-grown region to that grown in space, then increases with continuing growth (see Fig. 1). The spatial distribution of the free-carrier concentration of the space GaAs can be explained as due to impurity segregation associated with normal solidification. The equilibrium segregation coefficient of the Te dopant in space GaAs, K, can be estimated from $C_s = KC_o$ where C_o is the

Te concentration in the GaAs melt, which is approximately the same as that of the seed GaAs crystal($3.5\times10^{18}/cm^3$), and C_s is the Te concentration of the space GaAs crystal in the "clean" region($2.9\times10^{17}/cm^3$). The value of $K(0.1)$ there obtained is much smaller than 0.3, reported in the literature.[7] In addition to the segregation of the Te dopant in GaAs and its evaporation during GaAs growth in space, the decrease of the segregation coefficient of the Te dopant in GaAs under microgravity may be an important factor that is responsible for the lower free-carrier concentration of the space sample. This finding is in agreement with an earlier report on the segregation coefficient of the Zn dopant in space-grown GaAs crystals.[5]

3.2.2 Impurity striations in the space GaAs single crystal

It is generally accepted that striations in crystals grown on Earth are caused by temperature fluctuations that lead to changes in growth conditions and to a variation of impurity incorporation into the solid. The temperature fluctuations can be attributed to buoyancy-driven convection. This was proved experimentally in the early Skylab experiments on doped lnSb by Witt et al.[8] and also in our recent report on the Te-doped GaAs single crystal grown in space,[6] where no striations were found (see Fig. 3). However, in the neighborhood of the edge and in the tail portion of the space GaAs ingot, cellular structure and discontinuous nonperiodical impurity striations are still found. The cellular structure is probably due to constitutional supercooling that resulted from very fast crystallization of the space GaAs crystal. The discontinuous nonperiodical impurity striations may be related to random acceleration with the amplitude[9] 2×10^{-4} g caused by accidental disturbance during crystal growth. On the other hand, the presence of close-spaced periodical impurity striations that only exist close to the surface of the space GaAs crystal suggests, since gravity-driven convection is suppressed, that they can also be generated by time-dependent thermocapillary flows driven by surface-tension gradients. The absence of striations in other parts of the space crystal may be caused by a decrease of the free melt surface, so that the Marangoni number fell below its critical value for oscillatory thermocapillary flow.

3.2.3 DLTS spectra of the space GaAs crystal

It can be clearly seen from Fig. 4(a) that deep centers and their densities are reduced in the space crystal. Since solutal convection, driven by the gravitational force, is suppressed, the melt stoichiometry of Ⅲ-Ⅴ compound materials can be more precisely controlled. This is the primary reason for the reduction of deep centers and their densities in the space crystal. Furthermore, thermal convection in the melt and that in the vapor phase in space are eliminated, and the crystal does not contact the growth container during crystal growth (the introduction of crystal defects by supercooling is avoided). These two factors are also favorable for improving the space crystal quality. The origins of these deep centers [as shown in Fig. 4(a)] possibly relate to native defects.

Comparing Figs. 4(a) with 4(b), a significant difference of DLTS spectra is observed between samples B and C. First, the latter has a rather high deep-level concentration, and second, its DLTS peaks are very broad, These indicate that the DLTS signal is not a simple exponential transient. This kind of DLTS spectrum can be well understood by the model in which the energy levels are broadened.[10] Furthermore, assuming that deep-level broadening can be described by a Gaussian distribution centered around a mean value E_0, then the DLTS spectra are simulated, and E_b [full width at half maximum (FWHM)] and E_0 are obtained. The broadening DLTS peaks demonstrate deep centers originating from an impurity-defect complex that has the same core but different atomic surroundings. It should be pointed out that the diode studied was fabricated on sample C, very close to the outer shell of the space polycrystalline region. Obviously, the thermal stress must extend to the inner part of the space crystal during the formation of the polycrystalline outer shell and must cause formation of a high-density dislocation network and other lattice defects that result in very broad DLTS peaks. It must be pointed out that this behavior is not inherent to the space crystal.

In summary, we conclude that the microgravity environment provides some advantages for GaAs growth. These can be attributed to the following: (i) Thermal convections in the melt and solutal convection are suppressed under microgravity conditions, so impurity striations are eliminated,[6] leading to a uniform distribution and enhancing homogeneity of the crystals. (ii) The stoichiometry fluctuations caused by convection in the melt and in the vapor phase can be eliminated in space; deep centers and native point defects are reduced. (iii) Since the space crystal does not contact the growth container, an improvement of the purity of the space material is expected.

Acknowledgment

This project was supported by the National Natural Science Foundation of China.

References

[1] K. W. Benz and H. Weiss, Mater. Sci. Space, Proc. Eur. Symp. 2nd, 1976, Eur. Space Agency [Spec. Publ.], 217.

[2] H. C. Gatos, in Materials Processing in Reduced Gravity Environment of Space, edited by G. Rindone (Elsevier, New York, 1982).

[3] H. Rodot, J. C. Guillaume, J. Chevallier, M. Boulou, Y. T. Kriapov, F. R. Kashimov, T. I. Makkova, and I. A. Zoubridski, Physica 116B, 168(1983).

[4] A. Eyer, M. Harsy, T. Gorog, I. Gyvro, I. Pozsgai, F. Koltal, J. Gyulai, T. Lohner, G. Mezey, E. Kotai, F. Paszti, Y. I. Hrjapov, N. A. Kultchisky, and L. L. Regel, J. Cryst. Growth 71,173 (1985).

[5] Manufacturing in Space: Processing Problems and Advances, edited by V. S. Avduyevsky, translated from the Russian by M. Edelev(MIR, Moscow, 1985),88.

[6] B. J. Zhou, X. R. Zhong, F. N. Cao, L. Y. Lin, D. A. Da, K. L. Wu, L. F. Huang, S. H. Zheng, and X. Xie, Chin. J. Semicond. 5, 548(1988).

[7] S. Scalski, Segregation in GaAs, Vol. 1 of Compound Semiconductors, edited by R. K. Willardson and H. L. Goering(Reinhold, New York, 1962),385.

[8] A. F. Witt, H. C. Gatos, M. Lichtensteiger, M. C. Larine, and C. J. Herman, J. Electrochem. Soc. 122, 276(1975).

[9] L. F. Huang and C. T. Luo, in Proceedings of the 1st Chinese Symposium on Microgravity Science and Space Experiments, Beijing, 1987, edited by L. Y. Lin(China Science and Technology, Beijing, 1988),280.

[10] Wang Zhanguo, Chin. J. Semicond. 6, 133(1985).

SI-GaAs 单晶热稳定性及其电学补偿机理研究

王占国 戴元筠 徐寿定 杨锡权 万寿科 孙 虹 林兰英

（中国科学院半导体研究所，北京，100083）

摘要 本文利用多种实验手段，包括 OTCS、DLTS、低温 PL 等，对影响 LEC SI-GaAs 单晶电学性质热不稳定的可能因素进行了系统研究。在仔细分析对比文献发表的实验结果的基础上，提出了普遍成立的多能级电学补偿模型。这个模型不但能成功地对 SI-GaAs 单晶热不稳定的本质进行合理解释，而且还为研制高热稳定的 GaAs 材料提供了科学依据。

Study on thermal stability and electrical compensation mechanism of SI-GaAs

Wang Zhanguo, Dai Yuanjun, Xu Shouding, Yang Xiquan
Wan Shouke, Sun Hong, Lin Lanying

(Institute of Semiconductors, Chinese Academy of Sciences, Beijing 100083, China)

Abstract The effects of all possible factors on thermal unstability of LEC SI-GaAs materials subject to heat-treatment have been systematically studied by using various techniques such as OTCS, DLTS and low temperature PL etc. On the basis of carefully analysing our and published experimental results, a new compensation modes is presented in which five-level has been taken into account. This model can be used not only to successfully explain all phenomena related to conductivity of GaAs crystals, but also to provide a clue of improving the quality of this material.

1 引言

LEC SI-GaAs 作为超高速微波器件和集成电路的基础材料，受到广泛的重视，多年来一直是国内外的重点研究课题之一。

与元素半导体相比，GaAs 材料的纯度远不如 Si 和 Ge，特别是高温生长过程中存在的非正化学配比的问题，不但会导致本征点缺陷及其络合物产生，而使材料的质量

原载于：固体电子学研究与进展，1991，11（3）：216-224.

难以控制；而且这些本征缺陷会在器件制备工艺热处理过程中发生变化，从而严重地影响到电路性能和器件成品率。因此，对原生和热处理后 SI-GaAs 中杂质缺陷行为的研究具有特殊重要的意义。

本工作针对不同生长条件下制备的 SI-GaAs 材料，利用光注入瞬态电流谱（OTCS）、深能级瞬态电容谱（DLTS）、红外吸收、低温光致发光（PL）等手段，对影响 SI-GaAs 热稳定性的原因进行了系统的实验研究，并在此基础上提出了五能级的电学补偿模型，从理论上解释了 SI-GaAs 单晶在热处理过程中变中阻和转型问题。

2　LEC SI-GaAs 中可能的杂质、缺陷及其行为的分析

2.1　体 GaAs 中主要的电活性剩余杂质

业已知道，GaAs 中的浅施主杂质主要是占 As 位的Ⅵ族元素（如 O、S、Se、Te 等）和占 Ga 位的Ⅳ族元素（C、Si、Ge、Sn、Pb 等）；而浅受主杂质则主要由占 Ga 位的Ⅱ族元素（Be、Mg、Zn 等）和占 As 位的Ⅳ族元素所贡献。其中Ⅳ族元素是两性杂质，既可占 Ga 位起施主作用，也可以占 As 位而充当受主。

上述元素在体 GaAs 中的行为不但可从理论上加以推测，而且可通过制作有意掺杂样品，进行物理测量来证实。红外局域模（LVM）吸收研究表明[1]，Si 在体 GaAs 中主要占 Ga 位（Si_{Ga}），而 C 却主要占 As 位（C_{As}）。图 1 是高纯 GaAs 的低温高分辨光致发光谱，可以清楚地看到，除了带边的自由激子发光峰（FE1. 516eV）外，还有束缚在中性受主和施主上的激子发光峰（A^0X、D^0X）、自由电子到中性受主的发光峰（eA^0）、施主–受主对发光峰（D^0A^0）以及它们的声子伴线等。其中与 C、Si、Ge 相应

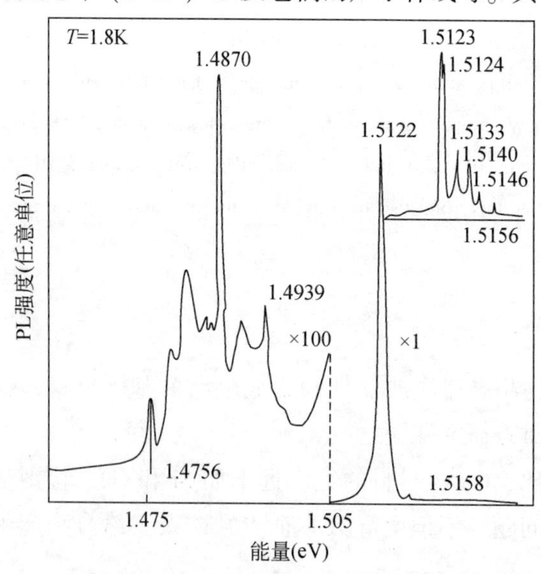

图 1　LPE-GaAs 的低温高分辨 PL 谱

T=1. 8K。右上角为主峰 1. 5122eV 的精细谱

的（eA^0）PL 峰是经常出现的。对于一些激活能很浅（几个 meV）、彼此又很靠近的浅施主，则只能用光热电离谱来加以区分。

综合各种测试结果，现已知道在 LEC SI-GaAs 中对电学补偿起主要作用的浅杂质是 C_{As}、Si_{Ga}和 S_{As}，它们的含量在 $1\sim5\times10^{15}cm^{-3}$。此外，在体 GaAs 中很难排除 O 的沾污，其含量往往较高，但 O 在 GaAs 中的行为究竟是浅施主还是深施主，或者是非电活性杂质的问题，至今仍不很清楚。

2.2 GaAs 中的本征点缺陷及其行为

作为二元化合物，GaAs 中可能的本征点缺陷有六种：V_{As}，V_{Ga}，As_i，Ga_i，Ga_{As}，As_{Ga}。其中 V_{As}，As_i，Ga_i起浅施主作用；V_{Ga}则起受主作用；反位缺陷 As_{Ga}和 Ga_{Aa}分别为双施主和双受主中心，而且能级位置也较深（几十至几百 meV）。

这些在晶体生长、处理过程中形成的缺陷及其络合物对晶体的质量有着重要影响。而它们产生的根本原因在于材料中化学配比的偏离，但正化学配比的获得十分困难。从热力学分析可知，在热平衡条件下，GaAs 中 V_{As}的浓度与 V_{Ga}空位的浓度之积可近似看做一个常数[2]，即 $[V_{As}]\cdot[V_{Ga}]=K(T)$，$K(T)$ 是温度的函数，而且随着温度的升高而增大。由此可见 GaAs 中本征点缺陷的浓度是依赖于生长条件的。

LEC SI-GaAs 通常是在富 As 条件下生长，由于非平衡 As 压的存在，相对抑制了 V_{As}的产生。由上述关系式可知，在减小 $[V_{As}]$ 的同时，将会使 $[V_{Ga}]$ 增大，而 V_{Ga}的增多也会使 As 占 Ga 位的概率加大，即 As_{Ga}的浓度增加。众所周知，As_{Ga}正是形成半绝缘 GaAs 的关键因素之一，即 *EL*2 缺陷的核心。

与此相反，如果是在富 Ga 条件下生长，那将会使得 V_{Ga}的浓度减少，而 V_{As}的浓度增加，Ga 占 As 位的反位缺馅（GaAs）就会占优势。这正是导致材料转变为 *P* 型的重要原因之一，因为 Ga_{As}在 GaAs 中是作为一个双受主中心而存在的[3,4]。

2.3 GaAs 中过渡金属元素和其他杂质缺陷中心的行为

实验已证明，过渡金属 Mn、Fe、V、Cu、Cr 等都能在 GaAs 禁带中引入受主中心，而且一般为深中心。GaAs：Mn 和 GaAs：Cu 的低温 PL 谱如图 2 和图 3 所示。

$Mn_{Ga}^{0/-}$ $E_v+0.11eV$，$Fe_{Ga}^{0/-}$[5] $E_v+0.52eV$，$V_{Ga}^{0/-}$[6] $E_c-0.55eV$，

$Cr_{Ga}^{0/-}$[7] $E_v+0.76eV$，　$Cu_{Ga}^{0/-}$ $E_v+0.15eV$，$Cu_{Ga}^{-/-}$ $E_v+0.45eV$

其中 Cr 引入的受主中心位于禁带中心附近，曾被用作获得 SI-GaAs 晶体的掺杂剂。另外几种杂质引入的受主能级都将直接或间接对 *N* 型半绝缘晶体的热反型产生影响。因而在 GaAs 晶体生长和热处理过程中应精心控制这些杂质的沾污。

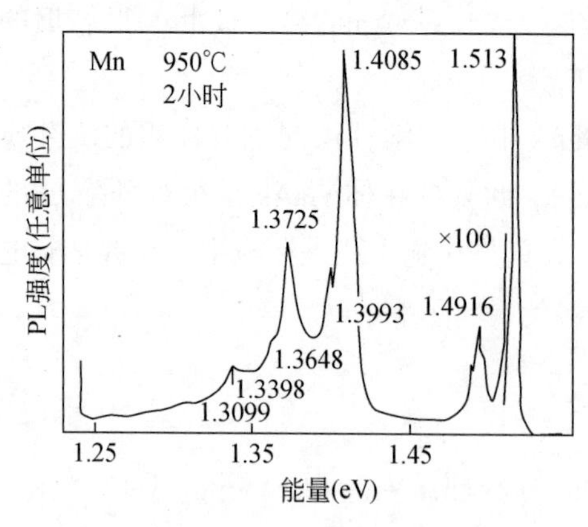

图2　掺 Mn 的 GaAs 低温高分辨 PL 谱

T=1.8K。$Mn^{0/-}_{Ga}$ 跃迁相应的光致发光峰能量位置在 1.4085eV 处

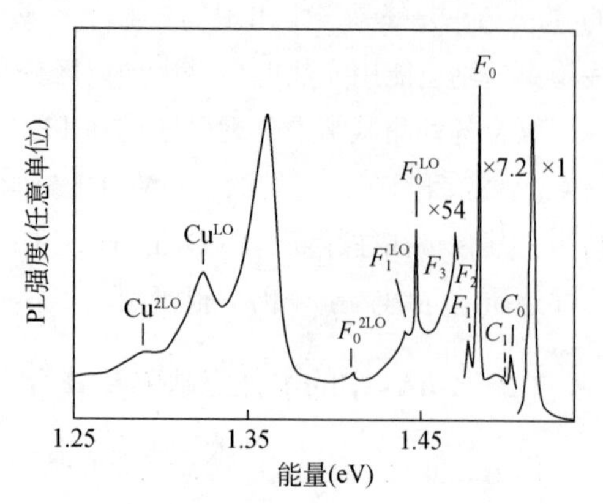

图3　掺 Cu 的 GaAs 低温高分辨率 PL 谱

T=1.8K。$C^{0/-}_{Ga}$ 跃迁相应的光致发光峰能量位置在 1.365eV 处，$Cu^{-/-}_{Ga}$ 的发光峰不在本图中

3　LEC SI-GaAs 中电学补偿机理分析

未掺 LEC SI-GaAs 之所以具有半绝缘特性是由于晶体内各种电活性杂质、缺陷相互补偿的结果，同时，这也是使材料热不稳定和电学不均匀的原因。下面对可能的电学补偿模型进行分析。

3.1　浅施主和浅受主补偿模型

假定体 GaAs 仅有浅杂质存在，浅施主的总浓度为 N_D，浅受主的总浓度为 N_A。那么，要获得半绝缘材料的条件就只能是 N_A 与 N_D 相等，或者说两者之比很接近 1（0.9999999）。这在实验中显然不可能实现。

3.2　三能级补偿模型[8]

假定除了浅杂质以外，在禁带中心附近还存在一个深能级 E_T，其浓度为 N_T。这时形成半绝缘的补偿条件可大大放宽。

如果深中心为施主中心（例如 $EL2$），只要式子

$$\begin{cases} N_D < N_A \\ N_{TD} > N_A - N_D \end{cases}$$

成立即可。

如果深中心为受主中心（例如 Cr），那么要满足的条件为

$$\begin{cases} N_A < N_D \\ N_{TA} > N_D - N_A \end{cases}$$

三能级模型原则上可以解释材料的半绝缘特性，但却无法说明晶体生长，处理过程中

常常出现中阻及转型问题。这表明模型有一定的局限性。最近，M. Baeumler 等根据 EPR 实验结果提出了四能级补偿模型[9]，解释了他们所遇到的问题。

3.3 多能级补偿模型

(1) 多能级补偿模型的实验基础

N 型和 P 型中阻 LEC SI-GaAs 的 DLTS 谱分别示于图 4(a) 和图 4(b)。从图中可以看到体 GaAs 中存在着许多电子和空穴陷阱。这些分别位于禁带中心上半部和下半部的深中心，是由 GaAs 中本征缺陷、剩余杂质及它们的络合物形成的。这类深中心在热处理过程中发生迁移、离解或重新组合，都将导致材料电学性质发生变化。

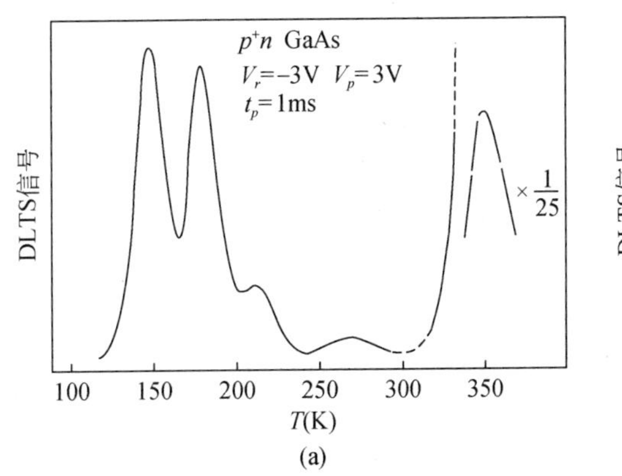

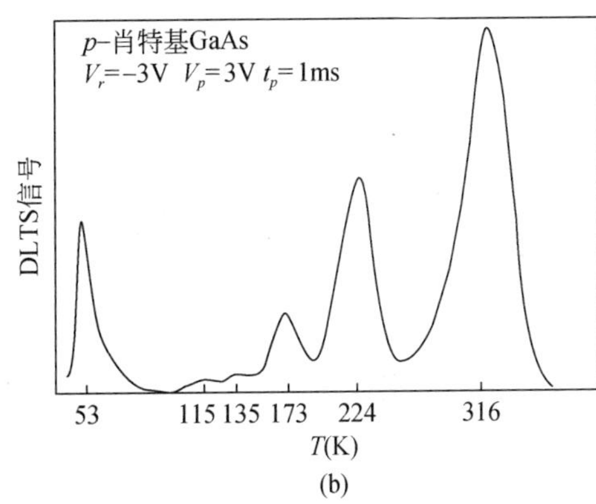

图 4 (a) p^+n GaAs 的 DLTS 谱；(b) p-肖特基 GaAs 的 DLTS 谱

(2) 多能级补偿模型的分析

在三能级模型的基础上，现分别在禁带中心的上半部和下半部引入两组深中心：上半部为施主中心 E_{TD}，总浓度为 N_{TD}；下半部为受主中心 E_{TA}，总浓度为 N_{TA}，就未掺 LEC SI-GaAs 而言，其禁带中心附近的能级为 $EL2$，其浓度用 N_{EL2} 表示。然后来分析材料出现不同导电类型和电阻率的可能成因。

如果只考虑中心的单个荷电状态（即荷 0 个电子和荷 1 个电子），且不考虑激发态的存在，电中性方程可以一般地写成[10]

$$n + \sum_{K_{受主}} n_K = p + \sum_{K_{施主}} (N_K - n_K)$$

$$n_K = \frac{N_K}{1 + g_K e^{(E_K - E_F)/KT}}。$$

式中，n_K 是 K 中心荷 1 个电子时的浓度，N_K 是 K 中心的总浓度。$g_K = g_{K0}/g_{K1}$，g_{K0}，g_{K1} 分别是中心 K 荷 0 个电子和荷 1 个电子时的简并度，g_K 是简并因子。这里均以导带底作为零能级位置。

1) N 型 SI-GaAs（$\rho > 10^5 \Omega \cdot cm$）。

此时满足关系式

$$\begin{cases} N_A > N_D + N_{TD}, \\ N_{EL2} > (N_A + N_{TA}) - (N_D + N_{TD}) \text{。} \end{cases} \tag{1}$$

电学补偿机构如图 5(a)所示。补偿后系统的费米能级 E_F 位于 $EL2$ 上面附近，材料的电导完全由 $EL2$ 激发到导带电子所决定。由于 $EL2$ 处于禁带中心附近，所以材料表现半绝缘特性。这是所希望的情况。补偿后，中心 E_A、E_{TA}、E_D、E_{TD} 全部电离，考虑此时 n，$p \ll N_A$、N_D，电中性方程写为

$$N_A + N_{TA} = N_D + N_{TD} + \frac{N_{EL2}}{1 + n/\Phi_{EL2}} \text{。} \tag{2}$$

其中，$\Phi_{EL2} = g_{EL2} N_c' T^{\frac{3}{2}} \mathrm{e}^{-\frac{EL2(T)}{KT}}$，$N_c' = \frac{2(2\pi m_n \cdot K)^{\frac{3}{2}}}{h^3}$；后面的 Φ_D、Φ_{TD}、Φ_A、Φ_{TA} 有相似的关系。

从（2）式得到

$$n = \Phi_{EL2}\left[\frac{N_{EL2}}{(N_A + N_{TA}) - (N_D + N_{TD})} - 1\right], \tag{3}$$

从此式可以看出 $\ln nT^{-\frac{3}{2}} \sim \frac{1}{T}$ 呈线性关系，其斜率为 $-EL2(0)/K$。

2）N 型中阻（$p < 10^5\Omega \cdot \mathrm{cm}$）。

此时有关系式

$$\begin{cases} N_D < N_A + N_{TA}, \\ N_{TD} > (N_A + N_{TA}) - N_D \text{。} \end{cases} \tag{4}$$

图 5(b) 即表示此种情况。系统的 E_F 位于禁带中心上半部施主中心附近，$EL2$ 填满电子呈电中性。材料电导几乎完全受 N_{TD} 控制，呈中高阻特性。由于此时中心 E_D、E_A、E_{TA} 全部电离，$p \ll n$，电中性方程写为

$$n + N_A + N_{TA} = N_D + \frac{N_{TD}}{1 + n/\Phi_{TD}} + \frac{N_{EL2}}{1 + n/\Phi_{EL2}} \text{。} \tag{5}$$

（5）式右边最后一项只在更高温度下起作用，较低温时，$EL2$ 是不电离的，呈电中性，因而略去。

因此得到

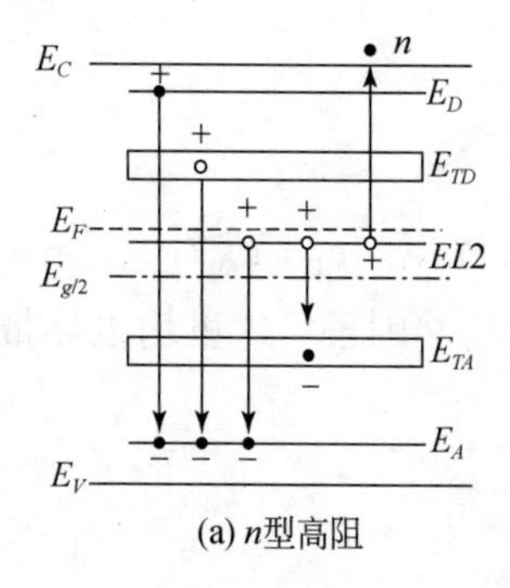

(a) n型高阻

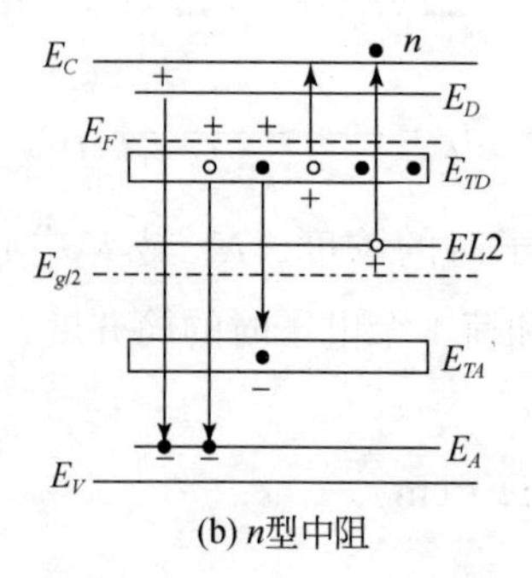

(b) n型中阻

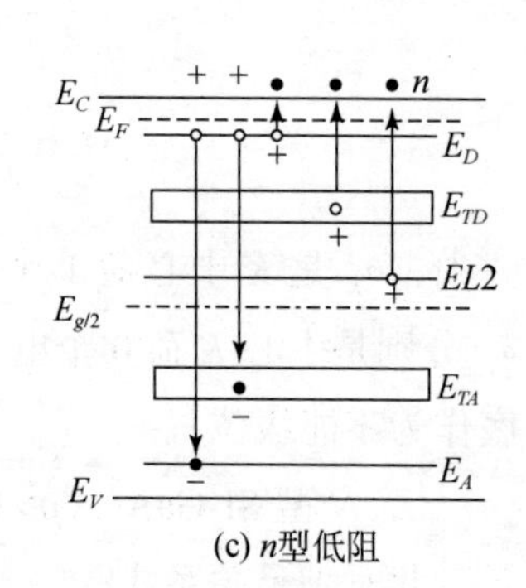

(c) n型低阻

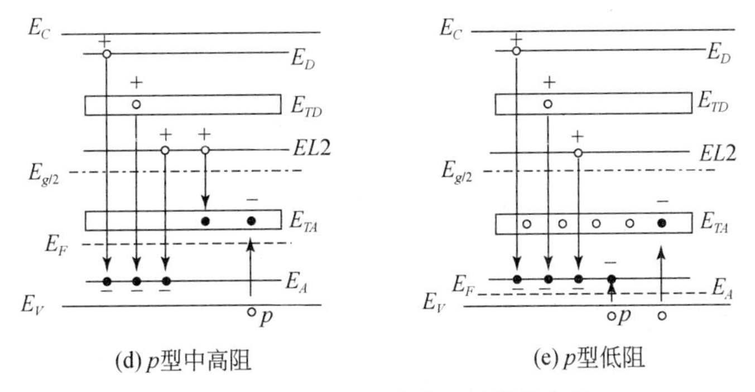

(d) p型中高阻　　(e) p型低阻

图 5　五能级电学补偿模型的补偿机构

$$n=\frac{1}{2}(N_1+\Phi_{TD})\left[\sqrt{1+\frac{4\Phi_{TD}N_2}{(N_1+\Phi_{TD})^2}}-1\right]。\tag{6}$$

式中，$N_1=(N_A+N_{TA})-N_D$，$N_2=(N_D+N_{TD})-(N_A+N_{TA})$。由于 $4\Phi_{TD}N_2/(N_1+\Phi_{TD})^2\propto T^{-\frac{3}{2}}$，所以当温度升高满足关系式 $4\Phi_{TD}N_2/(N_1+\Phi_{TD})^2\ll 1$ 时，(6)式可简化为 $n=\Phi_{TD}N_2/(N_1+\Phi_{TD})$。这是一个很简单的关系式。当温度进一步升高，使得 $\Phi_{TD}\gg N_1$，这时 $n=N_2$，表示 N_{TD}全部电离，在 $\ln nT^{\frac{-3}{2}}\sim 1/T$ 关系曲线上将出现一个平台；温度继续升高，电子浓度将来自 $EL2$ 的贡献，直到 $EL2$ 全部电离，相应地出现第二个平台为止。

3）N 型低阻（$\rho<10\Omega\cdot\text{cm}$）。

此时关系成为

$$N_D>N_{TD}>N_A+N_{TA}。\tag{7}$$

系统的 E_F 由浅施主决定，N_A、N_{TA}全由 N_D 所补偿，导带电子也由 N_D 贡献。材料表现为低阻，如图 5(c) 所示。电中性方程因 N_A、N_{TA}全部电离，且 p 可略去而写成

$$n+N_A+N_{TA}=\frac{N_D}{1+n/\Phi_D}+\frac{N_{TD}}{1+n/\Phi_{TD}}+\frac{N_{EL2}}{1+n/\Phi_{EL2}}\tag{8}$$

由于（8）式右边后两项只在更高温度时起作用，略去后得到

$$n=\frac{1}{2}(N_A+N_{TA}+\Phi_D)\left[\sqrt{1-\frac{4N_1\Phi_D}{(N_A+N_{TA}+\Phi_D)^2}}-1\right],\tag{9}$$

此时 $N_1<0$。在 $\ln nT^{-\frac{3}{2}}\sim 1/T$ 关系曲线中，在一定条件下将依次出现 3 个平台，分别对应于 N_D、N_{TD}、N_{EL2}的全部电离。一种可能的情况是，由于温度太高，本征电离的电子和空穴浓度会大于由深施主 $EL2$ 电离产生的电子浓度，从而掩盖了第三平台。

4）P 型中高阻。

这时出现关系

$$\begin{cases}N_D+N_{TD}+N_{EL2}>N_A,\\ N_A+N_{TA}>N_D+N_{TD}+N_{EL2}。\end{cases}\tag{10}$$

在低温下 N_D、N_{TD}、N_{EL2}和 N_A 全部电离，仅 N_{TA}为部分电离。系统 E_V 在 E_{TA}附近，

见图 5(d)，n 可略去，电中性方程为

$$N_A+\frac{N_{TA}}{1+p(\Phi_{TA}/n_i^2)}=p+(N_D+N_{TD}+N_{EL2}),\tag{11}$$

空穴浓度为

$$p=\frac{1}{2\Phi_{TA}}(n_i^2+N_3\Phi_{TA})\left[\sqrt{1+\frac{4\Phi_{TA}N_4}{(n_i^2+N_3\Phi_{TA})^2}}-1\right]。\tag{12}$$

式中，$N_3=(N_D+N_{TD}+N_{EL2})-N_A$；$N_4=(N_A+N_{TA})-(N_D+N_{TD}+N_{EL2})$。同 b）的情形类似，在某一温度范围下深受主全部电离，$\ln pT^{-\frac{3}{2}}\sim 1/\mathrm{T}$ 关系曲线将出现一个平台。

5）P 型低阻。

此时满足关系

$$N_A>N_{TA}>N_D+N_{TD}+N_{EL2}。\tag{13}$$

系统的 E_F 在 E_A 附近，深浅施主全由 N_A 补偿，材料的电导为 P 型低阻，见图 5(e)。电中性方程为

$$\frac{N_A}{1+p\ (\Phi_A/n_i^2)}+\frac{N_{TA}}{1+p\ (\Phi_{TA}/n_i^2)}=p+\ (N_D+N_{TD}+N_{EL2})。\tag{14}$$

左边第二项只在温度较高时起作用，因而略去

$$p=\frac{1}{2\Phi_A}(n_i^2+N_0\Phi_A)\left[\sqrt{1-\frac{4\Phi_AN_3}{(n_i^2+N_0\Phi_A)^2}}-1\right]。\tag{15}$$

式中，$N_0=N_D+N_{TD}+N_{EL2}$，此时 $N_3<0$。在 $\ln pT^{-\frac{3}{2}}\sim 1/T$ 关系曲线上将出现两个平台，一个对应于 E_A 的全部电离，另一个对应于 E_{TA} 的全部电离。

4 LEC SI-GaAs 热不稳定的可能原因

从上述讨论中可知，未掺体 GaAs 中剩余浅施主是 Si_{Ga}，S_{As} 也许还有 Sn_{Ga}；受主主要是 C_{As}，以及少量的 Si_{As} 和 Ge_{As}。这些替位浅杂质通常具有较高的热稳定性，不容易发生转化。所以在晶体生长、热处理过程中 Si_{Ga} 和 C_{As} 沾污仅仅是分别导致原生晶体为低阻 N 型和 P 型的原因，而不能对材料的热稳定起直接的决定性作用。

4.1 非化学配比问题

可以认为，LEC SI-GaAs 晶体热不稳定的根本原因是由晶体生长过程中化学配比偏离所产生的本征缺陷及其热处理过程中的相互转化造成的。热处理时的点缺陷和快扩散杂质的迁移、运动、相互作用络合、重新分布，必将改变原生晶体内的杂质缺陷行为，从而也就改变了它们的补偿机构。

以掺 In 对 GaAs 中深能级的影响为例来说明上述观点。用来降低位错的等电子杂质 In，从原则上推想是不应当影响到 GaAs 的电学性质的。可是实验结果却并非如此。

样品在未掺和掺 In 条件下测得的实验曲线，有着明显的差异，见图6(a) 和 (b)。*EL*3、*EL*6 是体 GaAs 常出现的深施主中心，分别是与 V_{As} 和 V_{Ga} 有关的络合物。In 掺入 GaAs 后择优占据 Ga 位，降低了 V_{Ga} 的浓度，根据上面讨论过的关系式，$[V_{As}]\cdot[V_{Ga}]=K(T)$，$V_{As}$ 的浓度就将增高，这正是图 6 曲线差异的原因所在。此外，V_{As}、V_{Ga} 浓度的变化还会引起 As_{Ga} 浓度的变化，从而影响到 *EL*2 的浓度，使材料的电学补偿度 $(N_A-N_D)/N_{EL2}$ 增大，这对热稳定性和均匀性都是不利的。实验结果也支持了这种看法。

可见 In 的掺入虽不直接对 GaAs 的电学性质产生影响，但它通过上述途径影响到 GaAs 中杂质缺陷的行为，也就间接地对材料的电学性质发生了作用。而本征缺陷是这一系列变化能够发生的基础。

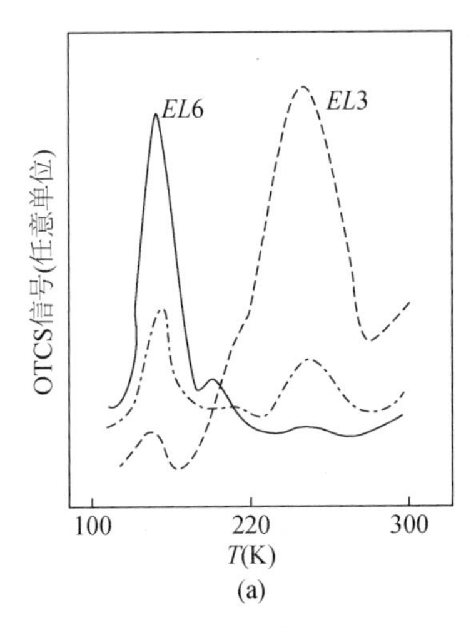

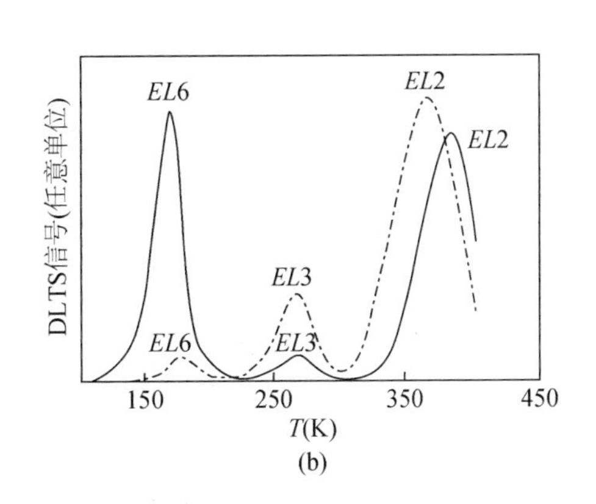

图6（a） 未掺和掺 In 后，SI-GaAs 的 OTCS 谱

——未掺 In，—·—In 含量为 0.0024，---In 含量为 0.0072

图6（b）未掺和掺 In 后，SI-GaAs 的 DLTS 谱

——未掺 In，—·—掺 In

4.2 重金属杂质沾污问题

晶体生长、热处理过程中重金属杂质沾污也将严重地影响材料的热稳定性。这是因为多数过渡金属元素（Mn、Fe、Cu）等在 GaAs 中是很不稳定的，甚至在室温下也会发生迁移。因而如何避免这些杂质的沾污是制备 SI-GaAs 晶体的又一个关键问题。

上述讨论表明，影响 SI-GaAs 热稳定因素主要有两个：一个为本征缺陷，另一个为重金属沾污。

对于减少位错密度，改善电学均匀性的要求，可通过设计合理的晶体生长热场和优化热处理条件来实现。

5　结论

综上所述，要制备热稳定性好的未掺 LEC SI-GaAs 单晶，必须满足如下 3 个条件：

1）浅受主杂质（主要是 C_{As}）的总浓度 N_A 与浅施主浓度 N_D 和禁带中心上半部较深能级浓度 N_{TD} 应满足关系式：$N_A>N_D+N_{TD}$，而且 C_{As} 应控制在 $(1-5)\times10^{15}cm^{-3}$。

2）*EL2* 中心的浓度 N_{EL2} 应满足关系式：

$N_{EL2}>(N_A+N_{TA})-(N_D+N_{TD})$，其浓度应控制在 $(1-2)\times10^{16}cm^{-3}$。

3）尽可能少的重金属杂质沾污。

前两个条件既是保证材料半绝缘特性的需要，也是提高热稳定性的要求。可以通过合理控制 C_{As}、Si_{Ga} 的浓度及合适的 As/Ga 来实现。由此也可看出，在人们减少 C 沾污的努力中，并不一定是使 C 越少越好，还应同时考虑到 Si 沾污的情况。应当使两者保持一个适当的浓度差距，以保证上述关系式的成立。

参考文献

[1] W. M. Theis et al., Appl. Phys. Lett., 41(1982), 70.

[2] Z. G. Wang, S. D. Xu, B. H. Yang, S. K. Wan and L. Y. Lin, 6th Conference on Semi-insulating Ⅲ-Ⅴ Materials, Toronto, May, 13-16th, 1990.

[3] Z. G. Wang, L-Å Ledebo and H. G. Grimmeiss, J. Phys. C: Solid State Physics, 17(1984), 259-272.

[4] P. W. Yu, W. C. Mitchel, M. G. Mier, S. S. Li and W. L. Wang, Appl. Phys. Lett., 41(1982), 532.

[5] D. V. Lang, Logan R. A., J. Electron. Mater., 4(1975), 1053.

[6] V. K. Bazhenov and N. N. Solov′ov, Fiz. Tekh. Poluprouodn., 5(1971), 1828 [Sov. Phys. Semicond., 5(1972), 1589].

[7] B. Clerjaud, A. M. Hennel, and G. Martinez, Sol. Stat. Comm., 33(1980), 983.

[8] 王占国等，半导体通讯，3(1977)，32。

[9] M. Baeumler, U. Kaufmann, P. M. Mooney and J. Wagner, 6th Conference on Semi-insulating Ⅲ-Ⅴ Materials, Toronto, May, 13-16th, 1990.

[10] D. C. Look, Electrical Charact. of GaAs Materials and Devices, (John Wiley & Sons Ltd. 1989), 244.

Interface roughness scattering in GaAs-AlGaAs modulation-doped heterostructures

Bin Yang, Yong-hai Cheng, Zhan-guo Wang, Ji-ben Liang, Qi-wei Liao, Lan-ying Lin, Zhan-ping Zhu, Bo Xu, and Wei Li

(Laboratory of Semiconductor Materials Science, Institute of Semiconductors, Chinese Academy of Sciences, Beijing 100083, China)

Abstract We have studied the influence of interface roughness scattering on the mobility of two-dimensional electron gas (2DEG) in GaAs-AlGaAs modulation-doped heterostructures (MDH) both experimentally and theoretically. When the background ionized impurity concentration in the GaAs layer is smaller than $2.5\times10^{15}\,cm^{-3}$, our investigation shows that interface roughness scattering is the dominant scattering mechanism in the high 2DEG density ($N_s \geqslant 5\times10^{11}\,cm^{-2}$) GaAs-AlGaAs MDH. We also demonstrate that interface roughness scattering is about an order of magnitude stronger than alloy disorder scattering in GaAs-AlGaAs MDH if the AlGaAs/GaAs interface fluctuation is only one monolayer of GaAs.

It is established that interface roughness scattering dominates the low-temperature mobility of two-dimensional electron gas (2DEG) in the following systems: (i) Si metal-oxide-semiconductor (MOS) inversion layers[1] with a high 2DEG sheet density N_s ($\geqslant 10^{12}\,cm^{-2}$), (ii) thin AlAs/GaAs/AlAs quantum wells,[2] and (iii) AlGaAs/GaAs/AlGaAs[3] quantum wells. However, almost no attention has been paid to interface roughness scattering in GaAs-AlGaAs modulation-doped heterostructures (MDH) since Ando[4] theoretically predicted that interface roughness scattering has only a small contribution to the electron mobility in this system. Recently, we grew GaAs-AlGaAs MDH by molecular beam epitaxy (MBE) and measured the dependence of electron mobility on 2DEG density by Van der Pauw method. The results demonstrated that electron mobility began to decrease with 2DEG density at about $N_s = 4.5\times10^{11}\,cm^{-2}$ (shown in Fig. 1). This value of N_s is much smaller than the predicted value of N_s by Walukiewicz et al.'s[5] model based on Ando's work.[4] According to Walukiewicz et al.'s[5] model, the AlGaAs/GaAs interface is "ideal", i. e., interface roughness scattering (μ_{ro}) is negligible, so electron mobility is mainly limited by the following

原载于: Appl. Phys. Lett., 1994, 65(26): 3329-3331.

major seven scattering mechanisms: (i) deformation potential scattering (μ_{de}), (ii) piezoelectric acoustic scattering (μ_{pi}), (iii) polar optical scattering (μ_{op}), (iv) remote ionized impurity scattering (μ_{re}), (v) background ionized impurity scattering (μ_{ba}), (vi) alloy disorder scattering (μ_{al}), and (vii) intersubband scattering. Following Walukiewicz et al.'s[5] model, we simulated our experimental data, but found large discrepancies between theory and experiment especially if $N_s \geqslant 5\times10^{11}\text{cm}^{-2}$. We also found Walukiewicz's calculated results[5] was not in good agreement with Hiyamizu's experimental data[6] when $N_s \geqslant 6\times10^{11}\text{cm}^{-2}$. We think these discrepancies resulted from Walukiewicz's[5] oversimplified model because the real AlGaAs/GaAs interface is not "ideal" as Walukiewicz assumed. So in this letter, we take into account interface roughness scattering in GaAs-AlGaAs MDH to calculate the density dependence of the mobility, our theoretical results are in good agreement with both our and Hiyamizu's[6] experimental data. Furthermore, from a detailed comparison between theory and experiment,[7] we determine the height and lateral width of the AlGaAs/GaAs interface roughness of both our samples and Hiyamizu's sample grown by MBE.

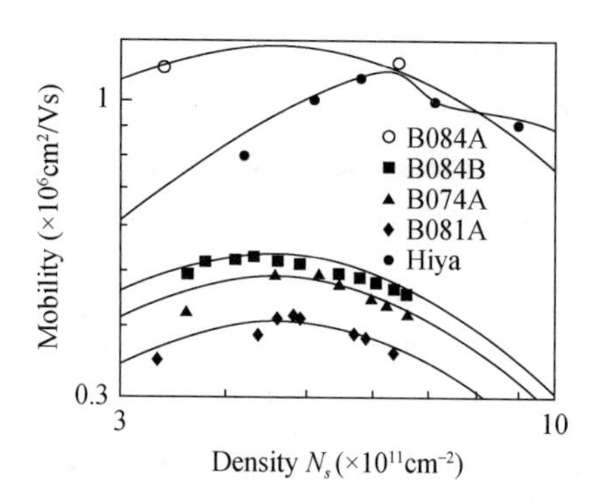

Fig. 1　Dependence of mobility on 2DEG density at $T=5\text{K}$ in GaAs-AlGaAs MDH. Points are experimental data, the curves are calculated mobilities. The parameters used in our calculation are listed below: Al composition $x=0.3$, spacer layer width $d=250\text{Å}$ for our samples, $d=150\text{Å}$ for Hiyamizu's sample, background ionized impurity concentrations are $N_b=5\times10^{14}\text{cm}^{-3}$, $1.7\times10^{15}\text{cm}^{-3}$, $2\times10^{15}\text{cm}^{-3}$, and $1\times10^{15}\text{cm}^{-3}$ for samples B084A, B084B, B074A, B081A and Hiyamizu's sample, respectively. The points labeled "Hiya" are Hiyamizu's experimental data

All of the samples used in this study were grown on semi-insulating (SI) liquid-encapsulated Czochralski (LEC) undoped GaAs (100) wafers under optimum growth conditions in a Riber 32P MBE system. The GaAs-AlGaAs MDH were prepared by growing successively a 0.5μm undoped GaAs buffer layer, a 20 periods GaAs/AlGaAs superlattice, a 0.5μm-thick undoped GaAs buffer layer, a 250Å-thick undoped $Al_{0.3}Ga_{0.7}As$ spacer layer, a 500Å-thick Si-doped $Al_{0.3}Ga_{0.7}As$ layer with doping density N_D of $6\times10^{17}\text{cm}^{-3}$, and finally a 100Å-thick undoped GaAs cap layer. Hall mobilities of the samples were measured by Van der Pauw method at 5K, a red light emitting diode (LED) with a pulse period of 3ms was used as the light source, persistent photoconductive effect is employed to gradually enhance the 2DEG density within a single sample by nearly a factor of 2. With each light pulse (3ms), the electron density increased by a certain amount. Hall measurements were performed in the dark following light exposure.

The measured dependence of mobility on 2DEG density for our samples is shown in Fig. 1. For sample B084A, its electron mobility is $1.14\times10^6 cm^2/Vs$ at 5K before illumination, $N_s=3.38\times10^{11} cm^{-2}$; when it is illuminated with red LED, N_s saturates at $N_s=6.4\times10^{11} cm^{-2}$, the mobility increases to $1.17\times10^6 cm^2/Vs$. For other three samples B084B, B074A, and B081A, their mobilities are much smaller than that of B084A, smaller than $5\times10^5 cm^2/Vs$ at 5K before illumination. In addition, with illumination, their mobilities increase initially with increasing N_s, reach the maximum value at about $N_s=4.5\times10^{11} cm^{-2}$, and then decrease with further increase in N_s.

The dependence of electron mobility on 2DEG density mentioned above cannot be explained by Walukiewicz et al.'s model.[5] According to their model, electron mobility increases with 2DEG density until the second subband begins to be occupied. In samples similar to ours, according to triangular quantum well approximation and self-consistent results as well as experiments,[5, 8, 9] the critical 2DEG density at which the second subband begins to be occupied is about $7\times10^{11} cm^{-2}$, this value of N_s is much larger than our experimental results. Furthermore, even if intersubband scattering begins to take effect at about $N_s=5\times10^{11} cm^{-2}$, our calculation demonstrated that there was still large discrepancies between theory and experiment by this model.

To explain the above experiments, we take into account of interface roughness scattering as another major scattering mechanism in GaAs-AlGaAs MDH. The relaxation time for interface roughness scattering is given by the same expression as that for Si inversion layer:[10]

$$\frac{\hbar}{\tau_{ro}} = 2\pi \sum \left(\frac{\Delta \Lambda F_{eff}}{\epsilon(q)}\right)^2 \exp\left(-\frac{1}{4}q^2\Lambda^2\right) \times (1-\cos\theta)\delta(\epsilon_k - \epsilon_{k-q}). \qquad (1)$$

Here, F_{eff} is the effective electric field defined by

$$F_{eff} = \frac{4\pi e^2}{4\pi\epsilon_0\epsilon}\left(\frac{1}{2}N_s + N_{depl}\right). \qquad (2)$$

Here, k is the 2D wave vector and $\boldsymbol{q}(=\boldsymbol{k}-\boldsymbol{k}')$ is the 2D scattering wave vector, $q=|\boldsymbol{q}|$, $\hbar$ is the reduced Plank constant, e is an electron charge, ϵ_0 and ϵ are the static dielectric constant and dielectric constant for GaAs, respectively, N_s is the 2DEG density, N_{depl} is the concentration of fixed space charges in the GaAs layer. The interface roughness is assumed to be characterized by the height Δ and lateral size Λ of the Gaussian fluctuations of the interface, and expressed as the autocorrelation function:[1]

$$\langle \Delta(\boldsymbol{r})\Delta(\boldsymbol{r}') \rangle = \Delta^2 \exp\left(-\frac{|\boldsymbol{r}-\boldsymbol{r}'|}{\Lambda^2}\right), \tag{3}$$

where $\langle\ \rangle$ means an ensemble average, $\boldsymbol{r}$ and $\boldsymbol{r}'$ are the 2D coordinate vectors along the interface. The mobility μ_{ro} limited by interface roughness scattering is then given by

$$\mu_{\text{ro}} = \frac{e}{m}\langle \tau_{\text{ro}} \rangle \propto \frac{g(\Lambda,\ N_s)}{\Delta^2 \Lambda^2 \left[\left(\frac{1}{2}\right) N_s + N_{\text{depl}}\right]^2}. \tag{4}$$

Where, $g(\Lambda,\ N_s)$ is function of Λ and N_s. Because the total 2DEG mobility is a function of both Δ and Λ, it's impossible to obtain the exact value of both Δ and Λ directly from experimental data independently. But through studying the photoluminescence spectra and cathodoluminescence spectra of GaAs/AlGaAs quantum wells, it is demonstrated that atomically flat GaAs/AlGaAs interface can be obtained by MBE,[10, 11] our recent study of the photoluminescence spectra of GaAs/AlGaAs quantum wells also proved that atomically flat GaAs/AlGaAs interface can be reproducibly obtained under optimum growth conditions in our Riber 32P MBE system.[12] Because the samples used in this letter were all grown under optimum growth conditions in our Riber 32P MBE system, so we think that the GaAs/AlGaAs interface where the 2DEG is formed is atomically flat. A reasonable way to get the value of Δ and Λ from experiment is that supposing the height of the AlGaAs/GaAs interface roughness is one monolayer of GaAs, i. e., $\Delta = 2.83$Å, then by comparing theoretical and experimental results, we can obtain the value of Λ. To demonstrate the interface roughness dependence of 2DEG mobility, Fig. 2 shows the dependence of total 2DEG mobility (μ_{total}) on the lateral size of GaAs/AlGaAs interface roughness (Λ) for different values of N_s at 5K by assuming $\Delta = 2.83$Å. Obviously, the higher the 2DEG density, the more strongly the interface roughness scatters the 2DEG. In addition, μ_{total} decreases drastically initially, reaches minimum at a certain value of Λ_c, and then increases with increasing Λ. The value of Λ_c is of the order of $\pi/2k_F$, where k_F is the Fermi wave vector. This indicates that electrons are most strongly scattered at Λ_c by interface roughness.

The calculated dependences of electron mobility on 2DEG density for sample B084B using both our model and Walukiewicz's model are shown in Fig. 3. Assuming $\Delta = 2.83$Å, $\Lambda = 18$Å, using our model, a very good agreement is obtained between theory and experiment, but Walukiewicz's model results in a large discrepancy. The simulations of samples B084A, B074A, B081A, and Hiyamizu's experimental data[6] are all carried out by our model assuming the lateral size of interface roughness to be 10, 19, 21, and 6Å, respectively. As shown in Fig. 1, the calculated results are also consistent with experiments. In the calculation, intersubband scattering is excluded from our samples, but it is includedfor Hiyamizu's sample.

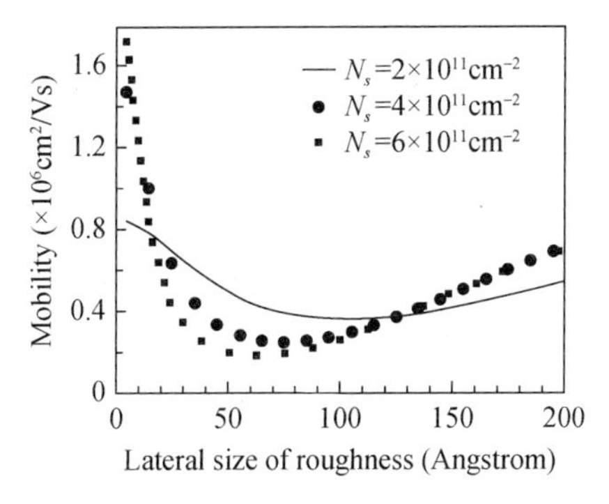

Fig. 2 Dependence of total electron mobility on lateral size of AlGaAs/GaAs interface roughness for different 2DEG densities at $T=5$K, $N_b=5\times10^{14}$cm^{-3}, $x=0.3$, $d=250$Å, and $\Delta=2.83$Å

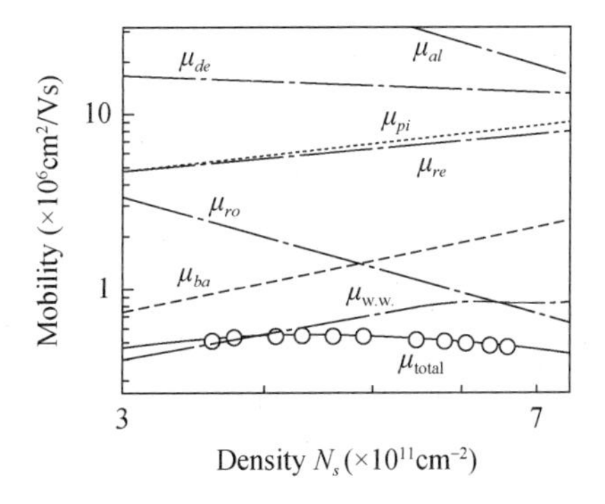

Fig. 3 Simulated dependence of electron mobility on 2DEG density for sample B084B. Points are our experimental data, curves are mobilities determined by the major eight scattering mechanisms together with the total 2DEG mobility calculated by our model(labeled "$M_{\rm total}$") or by Walukiewicz's model(labeled $\mu_{\rm W.\,W.}$). Parameters used in our calculation are $N_b=1.7\times10^{15}$cm^{-3}, $d=250$Å, $\Delta=2.83$Å, and $\Lambda=18$Å

It is shown in Fig. 3 than, for sample B084B, at smaller N_s, background ionized impurity scattering N_{ba} is the dominant scattering mechanism($N_b=1.7\times10^{15}$cm^{-3}). With increasing N_s, the screening effect of 2DEG to ionized impurity scattering increases, so the total mobility($\mu_{\rm total}$) increases with N_s initially. When $N_s\geqslant5\times10^{11}$cm^{-2}, interface roughness scattering(μ_{ro}) becomes dominant and results in the decrease of 2DEG mobility with N_s. It is also obvious to know from Fig. 3 that, for sample B084B, although the AlGaAs/GaAs interface roughness is only one monolayer of GaAs, interface roughness scattering is about an order of magnitude stronger than alloy disorder scattering, this conclusion is just opposite to Ando's[4] theoretical results. Walukiewicz[5] also indicated that Ando's calculation for alloy disorder scattering mobility was more than one order of magnitude smaller than their results, i. e., Walukiewicz's[5] result is consistent with our calculation. So, what should be pointed out here is that interface roughness scattering is a more stronger scattering process than alloy disorder scattering in high density GaAs-AlGaAs MDH. For Hiyamizu's experimental data, using Walukiewicz's model, a large discrepancy is resulted when 2DEG density is larger than $6\times10^{11}cm^{-2}$, but our model gives a better agreement when assuming $\Delta=2.83$Å, $\Lambda=6$Å. This indicates that our model is more practical (See Fig. 1).

When the background ionized impurity concentration in the GaAs layer is larger than $5\times10^{15}cm^{-3}$, our calculation demonstrated that the ionized impurity scattering becomes much stronger than interface roughness scattering even when $N_s\geqslant6\times10^{11}cm^{-2}$ (assuming $\Delta=2.83$Å, $\Lambda=20$Å). Interface roughness scattering is negligible in this case.

In conclusion, we have demonstrated that interface roughness scattering will be a dominant scattering mechanism in MBE grown high 2DEG density ($N_s\geqslant5\times10^{11}cm^{-2}$) GaAs-AlGaAs MDH if (i) background ionized impurity concentration is lower than $2.5\times10^{15}cm^{-3}$, (ii) interface roughness is large ($\Delta\geqslant2.83$Å, 10Å$\leqslant\Lambda\leqslant300$Å). We also demonstrate that interface roughness scattering is much stronger than alloy disorder scattering in GaAs-AlGaAs MDH. Comparison between theory and experiment has revealed that AlGaAs-on-GaAs interface grown by MBE has a roughness with a height of about 2.83Å and lateral size of 6–21Å for the samples studied in this letter.

References

[1] T. Ando, A. B. Fowler, and F. Stern, Rev. Mod. Phys. 54, 437 (1982).

[2] H. Sakaki, T. Noda, K. Hirakawa, M. Tanaka, and T. Matsusue, Appl. Phys. Lett. 51, 1934 (1987).

[3] R. Gottinger, A. Gold, A. Abstreiter, G. Weimann, and W. Schlapp, Europhys. Lett. 6, 183 (1988).

[4] T. Ando, J. Phys. Soc. Jpn. 51, 3900 (1982).

[5] W. Walukiewicz, H. E. Ruda, J. Lagowski, and H. C. Gatos, Phys. Rev. B 30, 4571 (1984).

[6] S. Hiyamizu, J. Saito, K. Nanbu, and T. Ishikawa, Jpn. J. Appl. Phys. 22, L609 (1983).

[7] B. Yang, Y. H. Cheng, Z. G. Wang, J. B. Liang, Q. W. Liao, and L. Y. Lin, Chin. J. Semicond. (to be published).

[8] T. Ando, J. Phys. Soc. Jpn. 51, 3893 (1982).

[9] H. L. Stormer, A. C. Gossard, and W. Wiegmann, Solid State Commun. 41, 707 (1982).

[10] J. Singh and K. K. Bajajaj, J. Appl. Phys. 57, 5433 (1985).

[11] D. Bimberg, J. Christen, T. Fukunaga, H. Nakashima, D. E. Mars, and J. N. Miller, J. Vac. Sci. Technol. B 5, 1192 (1987).

[12] J. B. Liang, Z. G. Wang, M. Y. Kong, Z. P. Zhu, B. Yang, B. Xu, W. Li, and G. K. Kuang, Proceedings of the Fourth Chinese National Conference on Thin Solid Films (unpublished), 171.

Simulation of lateral confinement in very narrow channels

Q. H. Du

(China Center for Advanced Science and Technology(World Laboratory), Beijing 100080, Laboratory of Semiconductor Materials Science, Institute of Semiconductors, Chinese Academy of Sciences, Beijing 100083, China)

Z. G. Wang

(Laboratory of Semiconductor Materials Science, Institute of Semiconductors, Chinese Academy of Sciences, Beijing 100083, China)

J. M. Mao

(China Center for Advanced Science and Technology(World Laboratory), Beijing 100080, China and nm-Structure Group Institute of Physics, Chinese Academy of Sciences, Beijing 100080, China)

Abstract A theoretical study is presented of the lateral confinement potential (CP) in the very narrow mesa channels fabricated in the conventional two-dimensional (2D) electron gas in GaAs-$Al_xGa_{1-x}As$ heterostructures. The 1D electronic structures are calculated in the framework of the confinement potential: $V(x) = m^* \omega_0^2 x^2/2$ for $|x| < d_0/2$ and $V(x) = \infty$ at $|x| = d_0/2$ where d_0 and ω_0, the parameters characterizing such a composite potential well, are the effective width and the characteristic frequency, respectively. Calculations suggest that, for wires having structural widths $W_{str} = 1500$, 550, and 500nm, the CP models consisting of flat bottoms and soft walls are more realistic than this complex potential. The experimental result for a wire of $W_{str} = 250$nm cannot be explained within the composite-well model including these two cases. Thus how to explain the experiment for this wire remains an open question.

Recent advances in microfabrication technology have made it possible to obtain narrow conducting channels or "quantum wires" by introducing a lateral confinement in a two-dimensional(2D) electron system. Such wire structures are most commonly fabricated from high-mobility, modulation-doped GaAs-$Al_xGa_{1-x}As$ heterojunction materials with use of shallow[1-4] or deep[3] mesa-etching and split-gate[5-9] techniques. Many numerical and analytic studies regarding quantum wires deal with the shape of the lateral confinement potential(CP). This is mainly due to the need to obtain an exact description of the CP,[2, 7, 10-14] a better understanding of its origin,[15] and more importantly, a qualitatively

原载于: Phys. Rev. B, 1994, 49(24): 17452-17455.

correct interpretation of some of the experimental findings observed on the narrow mesa channels.[3, 10, 16-19] The properties of the CP's for these mesa and split-gate structures should be different due to the difference in the physical origin of the confinement. This had not been realized until Sun and Kirczennow found it very recently through a theoretical investigation based on the density-function formula.[14] Here we will focus on the mesa-channel structures.[1, 3] In principle, a quantum-mechanical simulation based on self-consistent calculations of the electronic structures should give a general realistic idea about the shape of the CP.[10] The above-mentioned theoretical investigation is so far the most reliable one of this kind. However, their results need to be confirmed further due to the assumption of a simple model of the wire structure. In fact, as realized by Berggren, Roos, and van Houten previously,[10] it is not a straightforward matter to obtain a general and quantitative description of the shape of the CP in this way, due to some uncontrolled factors introduced through the fabrication process.

Information about the CP can also be extracted, in an approximate but simple manner, by fitting a selected potential model with one variable parameter to the experimental data from Shubnikov-de Haas (SdH) oscillations.[3, 10, 11] To date, two types of potentials have been assumed for this purpose: the parabolic[3, 10] and the squarewell[11] (SW) potentials. Berggren, Roos, and van Houten first applied the former to a narrow mesa channel of W_{str} = 500nm. The resulting fit to the experimental data seemed to be excellent. That is, both the linearity of the experimental sublevel index n versus inverse magnetic field in a high-B region and the nonlinearity at low magnetic fields can be explained quite well. In reality, however, the explanation of the former was rather approximate. Thus, such a model is not realistic for wires of W_{str} close to or larger than 500nm where the SdH oscillations exhibit the $1/B$ periodicity causing the above-mentioned linearity. Nevertheless, the explanation of the nonlinearity is good, which can also be seen from the fit for a wire of W_{str} = 550nm in Ref. [3]. On the contrary, the fits of the SW model to the high-field data for the 500-and 1500nm-wide wires are extremely excellent, while the fits to the low-field data were not good, especially for the narrower one.[11] The same should hold for the 550nm-wide wire, as it will turn out. Thus, it is expected that the more realistic shape of the lateral CP is intermediate between that of a harmonic oscillator and a SW for the 500-and 550nm-wide wires. In addition, we find that the "revised" $n-B^{-1}$ relation[16] was used in fitting the data for a wire of W_{str} = 250nm within the parabolic model.[3] Thus, the explanation is not convincing.

We present here a simulation of the lateral CP by assuming a potential model that consists of infinitely steep walls and a parabolic bottom. The former is described by a width

d_0 and the latter by a characteristic frequency ω_0. The main feature of this complex potential is that it can change over to a parabolic one as $d_0 \to \infty$ or/and $\omega_0 \to \infty$ and to a SW one as $d_0 \to 0$ or/and $\omega_0 \to 0$. This model has recently been employed to describe the confinement in remotely doped wide-parabolic-quantum-well structures.[20, 21] If the lateral confinement is chosen to be along the x direction, the potential can be written as $V(x) = 1/2m^* \omega_0^2 x^2$ for $|x| < 1/2\ d_0$ and ∞ for $|x| < 1/2\ d_0$, where m^* is the electron effective mass. The value of $m^* = 0.067m_e$ is adopted in this paper, with m_e the free-electron mass.

Choosing the Landau gauge $\boldsymbol{A} = (0,\ Bx,\ 0)$ and neglecting spin splitting because of the small g value of GaAs, one finds the Hamiltonian

$$\hat{H} = \frac{\hat{P}_x^2}{2m^*} + \frac{1}{2} m^* \omega^2 (x - x_0)^2 + \frac{\hat{P}_y^2}{2m^*(B)}, \tag{1}$$

where $\omega = (\omega_c^2 + \omega_0^2)^{1/2}$, $\omega_c = eB/m^*$, $x_0 = (\hat{P}_y/eB)(\omega_c/\omega)^2$, and $m^*(B) = (\omega/\omega_0)^2 m^*$ is the effective "magnetic" mass. The second term in this expression is a magnetoelectric parabola centered at x_0. This parabola will change into a pure magnetic one when $\omega_0 \to 0$. In the past, the Schrödinger equation with Hamiltonian (1) was solved for the case of $\omega_0 = 0$ by means of numerical methods[11, 22] or by using confluent hypergeometric functions (CHGF's).[17, 18] The latter is generally believed to be an approximate method due to the nature of the CHGF's.[22] Indeed, we find that the values of magnetoelectric sublevels obtained by using a conventional method in the calculations of the CHGF's are not reliable at all in the case of small magnetic fields (including $B = 0$) and also large values of n. We employ a technique in these calculations and the obtained solutions of the Schrödinger equation are absolutely reliable. Details about this technique will appear elsewhere.

For a given CP well within our model, exact dispersion relations can be obtained for every B value in the region of interest here. Then we can calculate the depopulation of 1D subbands once an area electron concentration N_e is given. In this paper, a minimum in the measured magnetoresistance[2] or magnetoresistivity[3] is assumed to be associated with the complete depopulation of a particular subband.[10] The corresponding B value B_{dep} will be referred to as depopulating field in the following, which can be given by the above-mentioned calculations. In particular, the field at which the $(n+1)$ th subband and those above this one are, at $T = 0\text{K}$, empty is defined as $B_{\text{dep},\,n}$. On the basis of this assumption, the variable parameters characterizing the electron channels that include ω_0, d_0, and N_e can be determined by a comparison of the calculated results with experiments, which is the minimization of the expression

$$F = \sum (B_{\text{dep}}^{-1} - B_{\text{exp}}^{-1})^2, \tag{2}$$

where $B_{\exp}$ is the magnetic field at which a particular minimum in R_{yy} appears. According to the assumption $B_{\exp}$ is actually an experimental depopulating field, and $B_{\mathrm{dep},\,n}$ corresponds to the nth minimum in R_{yy}.

The experimental data used here are taken from the magnetoresistance measurements in wires of $W_{\mathrm{str}}=1500$, 550, and 250nm, and from the quasi-dc magnetoconductivity in a wire with $W_{\mathrm{str}}=550$nm, which is the only one belonging to the deep mesa-etched structure. The four wires will be, in the order of their widths, referred to as wire A, B, C, and D, respectively. Wire A, the widest one, and wire C were fabricated from the same 2D sample. In Fig. 1(a), we display the experimental n versus $(1/B)$ plot for wire B. This plot is typical among those for the three wider wires, since the corresponding SdH oscillations are all periodic in $1/B$ at large magnetic fields. [2, 3] The experimental plot for wire D is shown in Fig. 1(b), which demonstrates that the SdH oscillations do not exhibit such a periodicity. For this reason the three wires and the narrowest one will be treated independently in the following. We first restrict our consideration to the case of the SW model. For the three wider wires, N_e can be directly estimated from the high-field slopes of the n versus $(1/B)$ plots according to

$$N_e = \frac{2e}{h}B_0 \tag{3}$$

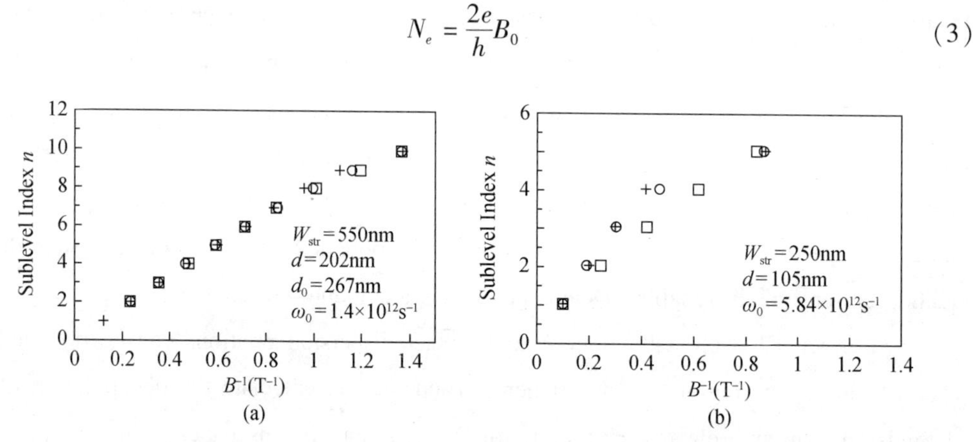

Fig. 1　Experimental and calculated sublevel index n vs inverse magnetic field $1/B$ within both the square-well and the composite-well model for wires with (a) $W_{\mathrm{str}}=550$nm and (b) $W_{\mathrm{str}}=50$nm. Squares refer to the minima in the measured magnetoresistance or magnetoresistivity. Circles and crosses denote the theoretical results within the square-well and the composite-well model, respectively

with B_0 the inverse period of the SdH oscillations as in a 2D case. Such a model is found to fit very well the high-field data for these wires, and the fits are not affected by varying d within relatively wider ranges. The values of N_e for wire A and C are 3.07×10^{15} and $2.35\times10^{15}\,\mathrm{m}^{-2}$, respectively, which are smaller than that for the original 2D sample, namely, $\sim5\times10^{15}\,\mathrm{m}^{-2}$. The N_e value for wire B is estimated to be $4.06\times10^{15}\,\mathrm{m}^{-2}$. For wire D, the narrowest one, the oscillatory magnetoresistance does not exhibit the above-mentioned periodicity and thus N_e cannot be estimated in this way. To determine it we assume that d is several times larger than the cyclotronorbit diameters[2, 10, 16] for electrons in the two lowest subbands at the magnetic field where the first minimum occurs in R_{yy}. This assumption is consistent with the d value obtained in the following. In such a condition, we find that $E_n = (2n+3/2)\hbar\omega_c$ for the electrons in these two subbands confined in a magnetic parabola with $|x| = d/2$. That is, the energy of such electrons in the lowest or the first subband equals the energy for the bottom of the second subband. Thus, when the first minimum in R_{yy} appears, the absolute values for x_0 must be less but very much close to $d/2$. According to a simple analysis,[16] the electron concentration is given by $N_e = (2e/h)B_{\mathrm{dep},1}$. This expression is much similar to that for the wider wires. From it we find for wire D, $N_e = 4.86 \times 10^{15}\,\mathrm{m}^{-2}$. The remaining parameter d can then be obtained through the minimization procedure as stated previously. The values are 384, 202, 162, and 105nm for wires A, B, C, and D, respectively. The deduced lateral depletion width for wire A is amazingly large, which turns out to be 553nm. A typical fit for wire B is shown in Fig. 1(a). It is excellent in the linear region of the experimental n versus$(1/B)$plot, because the calculated plot is strictly linear in this region. In the low-B region where the experimental plot is nonlinear, the calculated points deviate leftwards from the corresponding experimental ones, except the last one of the former, which deviates slightly rightward. For wire D, the leftward deviations are much wider and they actually exist in a whole B region, as shown in Fig. 1(b). It is striking that the calculated B_{dep}^{-1}, increases linearly with n at high magnetic fields. This indicates that the SW model cannot fit the experimental data for this wire in any rate.

We now turn to the case of the composite-well model. In this case, we find that the CP's for wires A, B, and C are best described by wells with $d_0 = 385$, 234, and 164nm, $\omega_0 = 0.09 \times 10^{12}$, 1.4×10^{12}, and $0.9 \times 10^{12}\,\mathrm{s}^{-1}$ for wires A, B, and C, respectively. For comparison, the calculated n versus$(1/B)$plots for wire B are displayed in the same figure for both the present and the SW model cases [see Fig. 1(a)]. The smallest F values are found to be, in units of $10^{-4}\,\mathrm{T}^2$, 1.61, 18.41, and 3.26 for wires A, B, and C, respectively. For the SW model the corresponding value are, in thc same units, 10.00, 103.42, and 6.98, respectively. A key feature of the best fit for wire B within the composite-well model is that the calculated n versus$(1/B)$plot deviates slightly from linearity in the original linear region, while in the nonlinear region, the leftward deviations, as mentioned above, are considerably diminished. We find that the same holds for wires A and C. Thus, the composite-well model improves the fits for the three wider wires, which can also be seen clearly from the change in the F values. This improvement actually means that the real CP's are soft in general for these wires.

The CP for wire *D* is found to be best described by a composite well with $\omega_0 = 5.84 \times 10^{12}\,s^{-1}$. Such a value is so large that the composite well can entirely be substituted with a parabolic one. The fit to the parabolic model is also shown in Fig. 1(b). As is clear from this figure, the above-mentioned leftward derivations are considerably diminished, and thus this model improves the fit of the SW one. The improvement is much more pronounced than for the cases of the three wider wires. Nevertheless, such deviations are still rather large.

According to the definition of $B_{dep,\,n}$, it is the magnetic field at which Fermi level lies below but approaches the bottom of the $(n+1)$th subband. As *B* decreases, the bottoms for the subbands above Fermi energy will go down and pass through the Fermi level in turn. Thus, if an E_f versus *B* plot and an $E_{n,\,0}$ versus *B* plot are displayed in the same figure, $B_{dep,\,n}$'s are actually the magnetic fields for their intersections, where $E_{n,\,0} = E_{n,\,k_y}(B)\,|_{k_y=0}$ is the energy for the bottom of the *n*th subband. In particular, $B_{dep,\,n}$ is the magnetic field for the intersection of a Fermi-energy curve with the plot of $E_{(n+1),\,0}$ versus *B*. The solid curves in Fig. 2(a) and 2(b) show the *B* dependence of $E_{n,\,0}$ corresponding to the situations, in Fig. 1(a), for the SW and the composite-well models, respectively. The variation of Fermi energy for the former case is given by the dashed curve in Fig. 2(a). The dotted straight lines in the two figures are for a 2D case. As is clear from Fig. 2(a), the $E_{n,\,0}$ versus *B* curve for the case of the SW model coincides with the corresponding 2D one above a critical magnetic field B_{crit} depending on *n*, and deviates upward below such a field. Apparently, the overwhelming majority of the electrons confined in a wire by a SW potential do not experience the confinement at sufficiently high magnetic fields. Such a system can then be considered to be a 2D one. It follows that the E_f versus *B* curve should coincide with that for the 2D case with the same area electron density in a high-*B* region. This was confirmed by the calculation finding that such a curve, in Fig. 2(a), holds for the *d* values larger than the one adopted in this figure, in the high-*B* region where the above-mentioned upward deviation does not occur. Consequently, $B_{dep,\,n}$ for the SW case behaves just as in the 2D case in this B region, which leads to B_{dep}^{-1} increasing linearly with *n*. For the case of the composite-well model, the $E_{n,\,0}$ oversus *B* curves do not coincide with the corresponding 2D ones in such a *B* region due to the nonflat bottom [see Fig. 2(b)], and B_{dep}^{-1} is almost impossible to increase with *n* linearly for this reason only. The same should hold for CP models with any other types of nonflat bottoms unless they contribute very little to the total confinements. This suggests that the real CP's have flat regions for the wires of this type with sufficiently large W_{dep}'s such that the SdH oscillations are periodic in $1/B$. Since the CP's are soft in general for the three wider wires, the CP models that consist of flat bottoms and soft kinds of walls are more realistic than both the SW and the composite-well ones in these cases. This is consistent with the theoretical result obtained on the basis of the self-consistent calculations in Ref. [14].

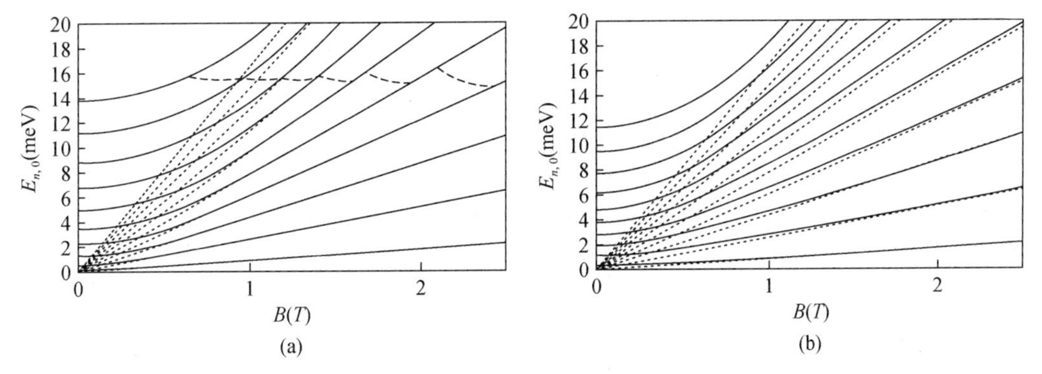

Fig. 2 Variation of $E_{n,0}$ with B for the case, in Fig. 1(a), of (a) the square-mell model and (b) the composite-well model. The dotted lines are for a 2D case. Due to the parabolic bottom the $E_{n,0}$ vs B curves for the case of the composite-well model deviate upward from the corresponding 2D ones at high magnetic fields, which is different from the case in(a). The variation of E_f with B for this case is given by the dashed curve

In brief, we present a simulation of the lateral confinement potential in the narrow mesa etched channels. For the three wires with $W_{str} \geqslant 500$nm, the calculated n versus $(1/B)$ plots are found to be strictly linear in high-B regions, which agrees well with the experimental results. The fits are not so good in the nonlinear regions. Within the composite model, the theoretical plots are only quasilinear in the original linear regions, which suggests flat bottoms in the real CP wells for the three wires and those of this type with sufficient larger W_{str}'s such that SdH oscillations are periodic in $1/B$ at high magnetic fields. The composite-well model improves the fits of the SW one in general. This actually indicates that the walls of the potential wells are, in general, somewhat soft rather than infinitely steep for the three wider wires. Therefore, the CP models, which consist of flat regions and soft kinds of walls, are more realistic than both the SW and the composite-well ones for these wires. This is consistent with the theoretical result obtained from self-consistent calculations. For the narrow wire of $W_{str} = 250$nm, the SW model cannot fit the experimental data in any rate. The composite-well model, which turns out to be a parabolic one in the case of this wire due to the relatively large ω_0 value obtained, improves the fit considerably. Nevertheless, the departure of the calculated data from the experimental results is still rather large. Thus, how to explain the experiment still remains an open question.

We would like to acknowledge Professor Kun Huang, Professor B. Y. Gu, and Dr. G. B. Ren for helpful discussions. This work was supported by a grant from the Youth

Foundation of the National"863"High Technology in China, and was partially financed by the Department of Solid State Physics, University of Lund.

References

[1] H. van Houten, B. J. Wees, M. G. J. Heyman, and J. P. Andre, Appl. Phys. Lett. 49, 1789 (1986).

[2] H. van Houten, B. J. Wees, G. Roos, and K. F. Berggren, Superlatt. Microstruct. 3, 497(1987).

[3] T. Demel, D. Heitmann, P. Grambow, and K. Ploog, Appl. Phys. Lett. 53, 2177(1988).

[4] R. G. Mani and K. I. Klitzing, Phys. Rev. B 46, 9877(1992).

[5] H. Z. Zheng, H. P. Wei, D. C. Tsui, and G. Weimann, Phys. Rev. B 34, 5635(1986).

[6] T. J. Thornton, M. Pepper, H. Ahmed, D. Andrews, and G. G. Davies, Phys. Rev. Lett. 56, 1198(1986).

[7] K. F. Berggren, T. J. Thornton, D. J. Newson, and M. Pepper, Phys. Rev. Lett. 57, 1769 (1986).

[8] A. C. Warren, D. A. Antoniads, and H. I. Smith, Phys. Rev. Lett. 56, 1858(1986).

[9] T. L. Cheeks, M. L. Roukes, A. Scherer, and H. G. Craighead, Appl. Phys. Lett. 53, 1964 (1988).

[10] K. F. Berggren, G. Roos, and H. van Houten, Phys. Rev. B 37, 10 118(1988).

[11] J. H. Rundquist, Semicond. Sci. Technol. 4, 455(1989).

[12] J. F. Weisz and K. F. Berggren, Phys. Rev. B 40, 1325(1989).

[13] K. F. Berggren, Int. J. Quantum Chem. 33, 217(1988).

[14] Y. Xinlong Sun and George Kirczennow, Phys. Rev. B 47, 4413(1993).

[15] S. E. Laux, D. J. Frank, and F. Stern, Surf. Sci. 196, 101(1988).

[16] Q. H. Du, T. H. Wang, J. M. Mao, W. Q. Cheng, J. M. Zhou, and Q. Huang, Phys. Rev. B 46, 4992(1992).

[17] F. M. Peeters, Phys. Rev. Lett. 61, 589(1986).

[18] F. M. Peeters, Superlatt. Microstruct. 6, 217(1989).

[19] T. Geisel, R. Ketzmerick, and O. Schedletzky, Phys. Rev. B 69, 1680(1992).

[20] L. Brey, N. F. Johnson, and Jed Dempsey, Phys. Rev. B 42, 1240(1990);42, 2886(1990).

[21] Jed Dempsey and B. I. Halperin, Phys. Rev. B 47, 4662(1993);47, 4674(1993).

[22] S. Chaudhari and S. Bandyopadhyay, J. Appl. Phys. 71, 3027(1992).

Theoretical investigation of the dynamic process of the illumination of GaAs

Ren Guang-bao
(China Center of Advanced Science and Technology(World Laboratory), Beijing 100080, Laboratory of Semiconductor Materials Science, Institute of Semiconductors, Chinese Academy of Sciences, Beijing 100083, China)
Wang Zhan-guo and Xu Bo
(Laboratory of Semiconductor Materials Science, Institute of Semiconductors, Chinese Academy of Sciences, Beijing 100083, China)
Zhou Bing
(Physical Department, Guangxi Normal University, Guilin 541000, Guangxi, China)

Abstract The dynamic process of light illumination of GaAs is studied numerically in this paper to understand the photoquenching characteristics of the material. This peculiar behavior of GaAs is usually ascribed to the existence of *EL2* states and their photodriven metastable states. To understand the conductivity quenching, we have introduced nonlinear terms describing the recombination of the nonequilibrium free electrons and holes into the calculation. Though some photoquenching such as photocapacitance, infrared absorption, and electron-paramagnetic-resonance quenching can be explained qualitatively by only considering the internal transfer between the *EL2* state and its metastability, it is essential to take the recombination into consideration for a clear understanding of the photoquenching process. The numerical results and approximate analytical approach are presented in this paper for the first time to our knowledge. The calculation gives quite a reasonable explanation for *n*-type semiconducting GaAs to have infrared absorption quenching while lacking photoconductance quenching. Also, the calculation results have allowed us to interpret the enhanced photoconductance phenomenon following the conductance quenching in typical semi-insulating GaAs and have shown the expected thermal recovery temperature of about 120K. The numerical results are in agreement with the reported experiments and have diminished some ambiguities in previous works.

1 Introduction

EL2 is one of the most important native centers appearing in GaAs. It is located near the

原载于：Phys. Rev. B, 1994, 50(8): 5189–5195.

midgap(E_c−0.76) and it has been intensively studied for decades now for the following two reasons. On the one hand, the *EL*2 center is important technologically for its role in producing semi-insulating GaAs. On the other hand, it exhibits an optically driven metastability at low temperature ($T<120$K). Thus GaAs shows peculiar photoquenching behavior. Though there are some arguments about it (such as the electric charge transformation model), this characteristic is generally ascribed to the transition of *EL*2 to its metastable state($EL2^*$) on illumination with white or infrared light(about 1.13eV) and the thermal or optical recovery associated with the reverse transformation, $EL2^* \rightarrow EL2$. The metastable state is electrically inactive and not accessible by most experimental methods. Therefore very little is known about it, and only the transition between the two states has been investigated. Techniques like electron paramagnetic resonance (EPR), [1-3] photoconductance (PC), [4, 5] infrared absorption (IA), [6, 7] photocapacitance, [7, 8] and internal friction[9] have been applied to this study. It is generally accepted that the *EL*2 defect contains As_{Ga}, and this is confirmed by EPR and electron-nuclear double-resonance (ENDOR) measurements.[10-12] The configuration of *EL*2, however, still remains uncertain; it may be an isolated As_{Ga} or a complex defect such as As_{Ga}-V_{As}, As_{Ga}-As_i, or As_{Ga}-V_{Ga}-V_{As}. The same uncertainties exist, of course, for the atomic configuration of $EL2^*$. In this paper we will present a theoretical study on the photoquenching process for further understanding of the *EL*2 center.

We focus our study mainly on the dynamic process of photoconductivity quenching. Semi-insulating (SI) GaAs materials show this peculiar behavior at low temperature under infrared light illumination, and some of them also present enhanced photoconductance characteristics.[23] Although there have already been some theoretical investigations on the dynamic processes of photoquenching,[13-16] such as infrared absorption, EPR, and photocapacitance quenching, no one has yet given the correct and satisfactory calculated result for photoconductivity quenching to our knowledge. The EPR and absorption quenching features are determined by *EL*2 and $EL2^*$ themselves and may be qualitatively explained when only the internal transition between the normal state *EL*2 and the metastable state $EL2^*$ is considered. Photocapacitance quenching can easily be interpreted since the measurements are usually made in *p-n* junctions in which the recapture of free carriers is normally negligible. However, the recombination of free carriers must be considered in PC measurements, since the photoconductivity is directly associated with the free electrons and holes emitted from deep centers. It is known that the concentration of photoinduced free carriers in SI GaAs is of the order of 10^8-10^{11} cm^{-3}, much smaller than the *EL*2 concentration (about 10^{16}cm^{-3}). In fact, it is the rapid recombination that leads to the low concentration of free carriers. To describe these direct and indirect recombinations with the deep center

*EL*2, nonlinear terms should be introduced in the general rate equations for the overall quenching process. As a result, the equation set needs numerical calculation. Even worse, the equation set we faced here is very stiff and unstable. Correct and convergent calculation results cannot be obtained by simple methods such as the Euler and Runge-Kutta methods. We have dealt with it carefully by combining the Newton iterative method and the Gear method, and obtained quite a good result.

In the next section, we will discuss the calculation model in detail, give the general rate equation set describing the overall photoquenching process, and present an approximate analytical approach under certain conditions. Section Ⅲ analyzes and discusses the numerical results in different initial conditions corresponding to different type of GaAs. Finally, Sec. Ⅳ summarizes this paper and presents a conclusion.

2 Calculation model

2.1 General rate equation set

It is generally accepted that the metastable state ($EL2^*$) can only be accessed by an internal transition when the *EL*2 is filled with electrons, i. e. , in the neutral state of the defect, $EL2^0$. The photoquenching behavior of GaAs at low temperature can then be interpreted by considering this internal transition, the free-carrier emission from *EL*2, and the free-carrier recombination. The optical transition involved is depicted in Fig. 1. The free electrons and holes emitted by *EL*2 occur with the rates $\sigma_p^0\phi$ and $\sigma_p^0\phi$, respectively, while $\sigma^*\phi$ is the rate of emission to the metastable state. Here, ϕ represents the photon flux, σ_n^0, and σ_p^0 are the electron and hole photoionization cross sections of the *EL*2 center, and σ^* is the optical cross section describing the emission $EL2 \rightarrow EL2^*$. Let us denote the concentrations of the neutral state $EL2^0$, ionized state $EL2^+$, and metastable state $EL2^*$ as N, N^+, and N^*, respectively. Then the general rate equation set for the overall photoquenching process, can be written as

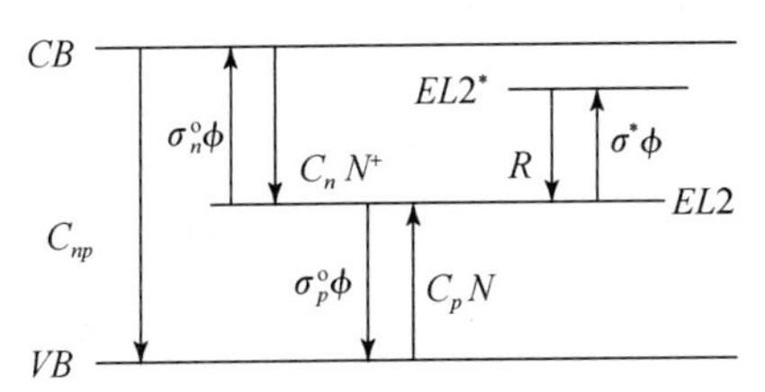

Fig. 1 Optical transitions of the *EL*2 center

$$
\begin{aligned}
dn/dt &= N\sigma_n^0\phi - C_n nN^+ - Cnp, \\
dp/dt &= N^+ \sigma_p^0\phi - C_p pN - Cnp, \\
dN/dt &= -(\sigma_n^0 + \sigma^*)\phi N + \phi_p^0\phi N^+ + RN^* + C_n nN^+ - C_p pN, \\
dN^*/dt &= N\sigma^*\phi - RN^*, \\
N + N^* + N^+ &= N_T,
\end{aligned} \tag{1}
$$

where t is the illumination time, n and p are the concentrations of free electrons and holes emitted from the $EL2$ centers, and N_T is the total concentration of $EL2$ defects which is usually of the order of $10^{16}\,cm^{-3}$ in SI GaAs. R represents the recovery rate for the reverse transition $EL2^* \rightarrow EL2$, which consists of two parts, the thermal recovery R_{th} and the optical regeneration R_{op}(σ_r^* is the optical cross section for $EL2^0$ regeneration):

$$R = R_{th} + R_{op} = R_{th} + \sigma_r^*\phi.$$

C is the direct recombination coefficient of the free carriers which takes the value[17]

$$C \cong 2 \times 10^{-10}\,cm^3 s^{-1}.$$

C_n and C_p denote the free-electron and-hole indirect recombination rates with the $EL2$ center, which are defined by the usual relations

$$C_n = \sigma_n V_n, \quad C_p = \sigma_p V_p,$$

where $\sigma_n(\sigma_p)$ is the electron(hole) capture cross section and $V_n(V_p)$ is the thermal velocity of the electrons(holes).

Equation (1) is a set of nonlinear differential equations for there are nonlinear recombination terms, which need numerical calculation. Since the nondimensional coefficients $C_n N_T/(\sigma_p^0\phi)$, $C_p N_T/(\sigma_p^0\phi)$, and $CN_T/(\sigma_p^0\phi)$ are about $10^6 - 10^9$ (much greater than 1), this differential equation set is very stiff and unstable, as we mentioned in the previous section, and must be dealt with carefully. However, we managed to obtain numerical results by combining the Gear method and the Newton iterative method. The calculation results will be shown and discussed in the next section. Now we are going on to discuss an approximate analytical solution and a steady-state solution under certain conditions.

2.2 Analytical approach for *n*-type SI GaAs

Suppose the initial concentrations of the neutral state $EL2^0$ and the ionized state $EL2^+$ before illumination are N_0 and N_p respectively in the semi-insulating GaAs. It is known that N^+ remains nearly unchanged during the whole quenching process, according to the numerical calculation shown in the next section. The free holes can be neglected ($p \cong 0$) when the illumination time is not long. Therefore we can obtain a simplified differential equation set for this situation:

$$\begin{aligned} &\mathrm{d}n/\mathrm{d}t = N\sigma_n^0\phi - C_n nN^+, \\ &\mathrm{d}N/\mathrm{d}t = -(\sigma_n^0 + \sigma^*)\phi N + \sigma_p^0\phi N^+ + RN^* + C_n nN^+, \\ &N + N^* + N^+ = N_T, \\ &N^+ = N_p + n - p \cong N_p, \end{aligned} \tag{2}$$

since n, $p \ll N_p$. From the above equation set, one has

$$\mathrm{d}n^2/\mathrm{d}t^2 + 2A\,\mathrm{d}n/\mathrm{d}t + Bn = N_0 R\sigma^*\phi, \tag{3}$$

where

$$A = 0.5[(\sigma_n^0 + \sigma^*)\phi + R + C_n N_p],$$
$$B = C_n N_p(\sigma^*\phi + R).$$

Using the initial conditions

$$n\,|_{t=0} = 0 \text{ and } \mathrm{d}n/\mathrm{d}t\,|_{t=0} = N_0\sigma_n^0\phi,$$

it is given that

$$n = 0.5N_0\sigma_n^0\phi[\exp(-At - \sqrt{A^2 - B}t) - \exp(-At + \sqrt{A^2 - B}t)]/\sqrt{A^2 - B}. \tag{4}$$

Since $C_n N_p \gg (\sigma_n^0 + \sigma^*)\phi + R$ with normal light intensity at low temperature, we can further simplify the expression to

$$n \cong N_0\sigma_n^0\phi\{[R + \sigma^*\phi\exp(-\sigma^*\phi t - Rt)]/(\sigma^*\phi + R) - \exp(-C_n N_p t)\}/(C_n N_p). \tag{5}$$

Finally, it can be written as

$$n \cong N_0\sigma_n^0\phi[\exp(-\sigma^*\phi t) - \exp(-C_n N_p t)]/(C_n N_p), \tag{6}$$

when $R \ll \sigma^*\phi$, which is satisfied at low temperature. Based on Eq.(6), we can deduce that

$$\begin{aligned} t = t_m &\cong \ln[C_n N_p/(\sigma^*\phi)]/(C_n N_p - \sigma^*\phi) \\ &\cong \ln(C_n N_p/\sigma^*\phi)/(C_n N_p) \cong 1\,\mathrm{ms}, \end{aligned}$$

and the photoinduced free-electron concentration will take its maximum value

$$n = n_m \cong N_0\sigma_n^0\phi/(C_n N_p) \ll N_0.$$

Thus it is known that the photoconductance reaches its maximum value at a very early time t_m, and the conductance decay time $\tau \cong 1/\sigma^*\phi$. The above result agrees with previous experiments. It was reported[5, 18] that the photoconductivity of SI GaAs at low temperature reached its maximum point at the very beginning, the peak conductance was proportional to photon flux ϕ, and its quenching time was inversely proportional to ϕ.

Under the same conditions, we can further obtain

$$\begin{aligned} N^+ \cong N_p &+ N_0\sigma_n^0\phi\{[R + \sigma^*\phi\exp(-\sigma^*\phi t - Rt)]/(\sigma^*\phi + R) \\ &- \exp(-C_n N_p t)\}/(C_n N_p) \cong N_p, \end{aligned} \tag{7}$$

$$N \cong N_0\{[R+\sigma^*\phi\exp(-\sigma^*\phi t-Rt)]/(\sigma^*\phi+R)-\sigma_n^0\phi$$
$$[\exp(-\sigma^*\phi t-Rt)]-\exp(-C_nN_pt)]/(C_nN_p)\}$$
$$\cong N_0\exp(-\sigma^*\phi t). \tag{8}$$

Clearly, the $EL2^0$ concentration N decays nearly exponentially and is dominated by $\sigma^*\phi$.

The above results are quite different from the results obtained from a simple model excluding the free-carrier recombination. According to the simple model calculation,

$$N = N_T[C_1\exp(-\lambda_1 t)+C_2\exp(-\lambda_2 t)]+N_T\sigma_p^0\phi R/(\lambda_1\lambda_2), \tag{9}$$

$$n = N_T\sigma_n^0\phi\ \{\{C_1[1-\exp(-\lambda_1 t)]/\lambda_1+C_2[1-\exp(-\lambda_2 t)]/\lambda_2\}+t\sigma_p^0\phi R/(\lambda_1\lambda_2)\}, \tag{10}$$

where C_1, C_2, λ_1 and λ_2 are defined as

$$C_1 = (N_0/N_T-\sigma_p^0\phi/\lambda_1)(R-\lambda_1)/(\lambda_2-\lambda_1),$$
$$C_2 = (N_0/N_T-\sigma_p^0\phi/\lambda_2)(R-\lambda_2)/(\lambda_1-\lambda_2),$$
$$\lambda_{1,2} = 0.5\{e\pm\sqrt{e^2-4[R(\sigma_n^0+\sigma_p^0)\phi+\sigma^*\sigma_p^0\phi^2]}\},$$
$$e = (\sigma_n^0+\sigma_p^0+\sigma^*)\phi+R.$$

In the simple model calculation, N and N^+ are somewhat complicated and mainly controlled by σ_n^0[15], and the concentration of emitted free electrons n may be greater than N_T. Eq.(10) is obviously an incorrect result since it can be simplified as $n\cong N_T(1+Rt)\sigma_n^0/\sigma^*$, if $R\sim 0$ and $\sigma_n^0\phi t\gg 1$. Thus the simple model without the free-carrier recombination is of doubtful validity.

Our calculated result for N can be used to fit the infrared absorption experimental data [see Fig. 6(c) below]. The absorption $\alpha(t)$ after irradiation can be expressed as

$$\alpha(t) \cong \text{const}+\alpha_{EL2^0}N_0\exp(-\sigma^*\phi t),$$

assuming $\alpha_{EL2^*}\ll\alpha_{EL2^0}$, where α_{EL2^*} and α_{EL2^0} are the optical cross sections of $EL2^*$ and $EL2^0$ defects.

2.3 Steady-state solution for SI GaAs

In this part, we continue to discuss the steady-state situation following a long duration of light illumination. After the $EL2^0$ defects are transformed to the electrically inactive metastable state by light, the photoinduced holes will dominate the electrical conductance instead of electrons. Thus an enhanced photoconductivity appears following the photoconductance quenching, and at last a steady photocurrent will occur. For this stationary state, we obtain

$$N\sigma_n^0\phi - C_n nN^+ - Cnp = 0,$$
$$N^+\sigma_p^0\phi - C_p pN - Cnp = 0,$$
$$N\sigma^*\phi - RN^* = 0, \quad (11)$$
$$N + N^* + N^+ = N_T,$$
$$N^+ = N_p + n - p.$$

Since $C\sigma_n^0\phi(C_pC_nN_p) \ll 1$ at low temperature, $n \cong 0$ can be simply eliminated. Substituting this into the above equation set, we obtain

$$p \cong -0.5(D + N_0) + 0.5\sqrt{(D + N_0)^2 + 4N_0D} \cong N_p/(1 + N_0/D), \quad (12)$$

where $D = (1+\sigma^*\phi/R)\sigma_p^0\phi/C_p$. We see that $D/N_0 \ll 1$ is usually satisfied, and therefore

$$p \cong N_p(\sigma_p^0\phi/C_pN_0)\sigma^*\phi/R \cong N_p(\sigma_p^0\phi/C_pN_0)\sigma^*\phi/(R_{th} + R_{op}), \quad (13)$$

where $\sigma^*\phi \gg R$ is required, which is satisfied at low temperature as mentioned above. It is known that the optical regeneration dominates at low temperature.[19, 20] Consequently, we can predict that the stationary photocurrent is proportional to $\phi N_p/N_0$ at low temperature ($T <$ 100K). The above result may be applied to investigate the compensation of the *EL*2 center in SI GaAs, which will be discussed in detail elsewhere.

3 Numerical results and discussion

Numerical values of the parameters used for the calculation are listed in table 1 unless otherwise specified. The results of the calculation under different conditions are shown in the figures below. In table 1, the photoionization cross sections are defined under infrared light of 1.17eV, and $\sigma_r^*/\sigma^* = 0.005$ and $\mu_n/\mu_p = 15$, are fitting parameters, where μ_n and μ_p are the electron and hole mobility. The total concentration of *EL*2, N_T, is set to be $10^{16}\,cm^{-3}$, and n, p, N, and N^+ in the figures are normalized by N_T.

Table 1 Typical values of the calculation parameters

$\sigma_n^o = 1.2\times10^{-16}\,cm^2$ [a]	$\sigma^*/\sigma_n^0 = 0.08$ [b]	$\mu_n = 4.5\times10^4\,cm^2/Vs$
$\sigma_p^0 = 1.3\times10^{-17}\,cm^2$ [a]	$\sigma_r^*/\sigma^* = 0.005$	$\mu_p = 3\times10^3\,cm^2/Vs$
$\phi/N_T = 0.05\,cm/s$	$N_p/N_T = 0.15$	$T = 77K$
$R_{th} = 2\times10^{11}\exp[(-0.3ev)/kT]\,s^{-1}$ [c]		$C = 2\times10^{-10}\,cm^{-3}s^{-1}$ [d]
$\sigma_n = 5\times10^{-19} + 6\times10^{-15}\exp[(-56.6meV)/kT]\,cm^2$ [e]		$\sigma_p = 2\times10^{-18}\,cm^2$ [e]

[a] Refs. [8, 15].

[b] Ref. [15].

[c] Ref. [20].

[d] Ref. [17].

[e] Ref. [19].

In Fig. 2, we show the numerical results for typical n-type SI GaAs. The calculation indicates that the photoconductivity reaches a maximum at the very beginning, and then decreases gradually down to a minimum point. As light illumination goes on, the concentration of photoinduced electrons will be diminished and the holes will become dominant. As a result, the conductance increases again, the enhanced photoconductivity. Finally it reaches a stationary state. PC quenching is a well-known feature of SI GaAs. Besides the photoconductivity quenching, some authors have also reported[5, 18, 21] enhanced photoconductance (EPC) in SI GaAs, and photo-Hall measurement has confirmed that the photocurrent has converted from n type to p type after $EL2$ is completely quenched.[5] Our calculation results are in good agreement with these experiments, and predict a necessary condition of EPC, that is, $N_p/N_T>0.01$. However, if the nonlinear recombination is neglected, as some authors have done in their calculation,[13-16] the concentration of emitted electrons and holes will never decrease. Therefore PC quenching and the EPC phenomenon cannot be deduced theoretically from a simple model calculation.

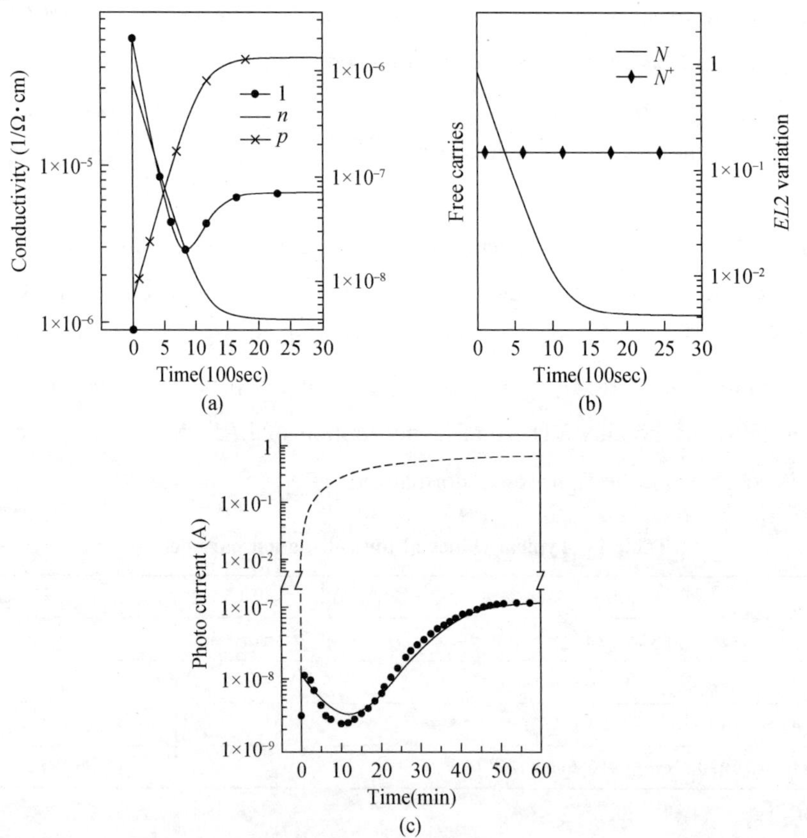

Fig. 2　(a) Variation of conductivity (1), free electrons (n), and free holes (p) under light. (b) Variation of concentrations of $EL2^0$ (N) and $EL2^+$ (N^+) under light. (c) Comparison of the photocurrent calculation (solid curve) with experiment data (solid points) from Ref. [23]. The dashed line shows the I_{ph} calculation from the simple model without free-carrier recombination; it is much higher than experiment and never decreases

The concentration variations of the $EL2^+$ and $EL2^0$ are also given in Fig. 2(b). As we mentioned, $N^+(EL2^+)$ could hardly be changed by illumination while the neutral-defect($EL2^0$) concentration N keeps decaying exponentially. The calculation results are drastically different quantitatively from the simple model calculation excluding free-carrier recombination[15] [see Eqs. (9) and (10) in Sec. Ⅱ B]. In Fig. 2(c), we give a comparison of the calculated photocurrent I_{ph} with experiment.[23] The sample size is $5\times2\times0.3\text{mm}^3$, $\sigma_n^0=4\times10^{-17}\text{cm}^2$, and $\sigma_p^0=3\times10^{-17}\text{ cm}^2$ at a photon energy of 1.12eV.[8] We set the applied voltage at 1V, $\phi=5.5\times10^{13}\text{cm}^{-2}\text{s}^{-1}$, $\sigma^*=5\times10^{-17}\text{cm}^2$, $N_p/N_T=0.3$, $\sigma_T^*/\sigma^*=0.002$, and the other parameters are as listed in table 1. The calculation agrees well with experiment. From Fig. 2(c), we also see that the simple model calculation cannot give a reasonable result for photoconductance quenching, for the calculated result for I_{ph} (dashed line) is much higher than expected and never decreases. Therefore we can conclude that the free-carrier recombination is very important in photoquenching, especially in photoconductance quenching, and should not be neglected in the calculation.

Fig. 3(a) and 3(b) give the calculation results for different photon flux ϕ and σ^*. They show that the maximum photoconductivity is proportional to ϕ, the conductance decay time is proportional to $1/\phi\sigma^*$, and the stationary photocurrent intensity is proportional to ϕ. The results agree with experiments[5] and also confirm the analytical approach in the previous section.

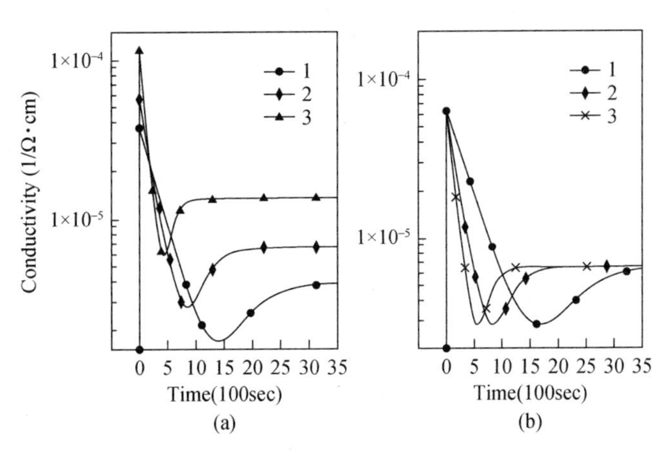

Fig. 3 (a) Conductivity under different light intensities $\phi/N_T=0.03,\ 0.05,\ 0.1$ (lines 1, 2, 3). (b) Photoconductivity for different $\sigma^*/\sigma_n^0=0.04,\ 0.08,\ 0.12$ (lines 1, 2, 3)

Fig. 4(a) and 4(b) show the corresponding plots for different initial fractions of ionized defects ($EL2^+$) $k=N_p/N_T$. Here N_p is equal to the concentration of residual acceptors(N_A-N_D) in SI GaAs, where N_A, N_D are the total shallow acceptors and donors. We see that the stationary photoconductance varies greatly with N_p/N_T. The higher the compensation, the

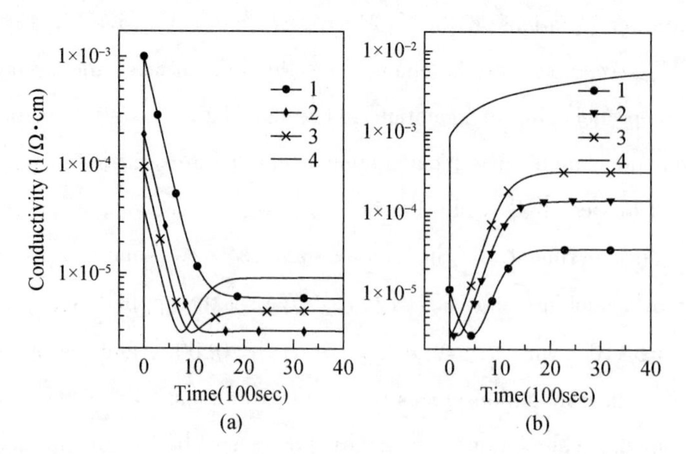

Fig. 4 Photoconductivity for different compensation ratios $k=(N_D-N_A)/N_T$.
(a) $k=0.01, 0.05, 0.1, 0.2$ (lines 1, 2, 3, 4). (b) $k=0.5, 0.8, 0.9, 1.0$ (lines 1, 2, 3, 4)

higher the current intensity. According to the calculated figures, the EPC feature may not appear if the compensation is very low, e. g., N_A-N_D much less than $0.05\ N_T$. However, with high compensation, for example, $k=N_p/N_T>0.8$, the EPC may only be observed while the photoconductance does not show obvious quenching behavior.

Now we are going to discuss the photoquenching behavior at different temperatures. The numerical results shown in Fig. 5 indicate that the thermal recovery from $EL2^*$ to $EL2^0$ plays a major role at higher temperature as the recovery rate R_{th} increases exponentially. The calculated recovery temperature of 120K, which will be slightly shifted by other parameters, is consistent with the measurements.[19-21]

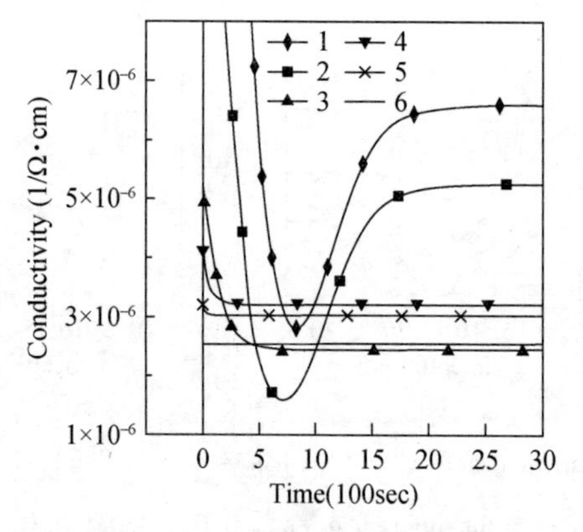

Fig. 5 Photoconductivity under different temperatures $T=77, 90, 110, 115, 120, 125$K (lines 1, 2, 3, 4, 5, 6)

Finally, we come to discussing the calculation for low-resistance material, doped n-type and p-type semiconducting (SC) GaAs. $EL2$ usually has two different charge states, $EL2^0$ and

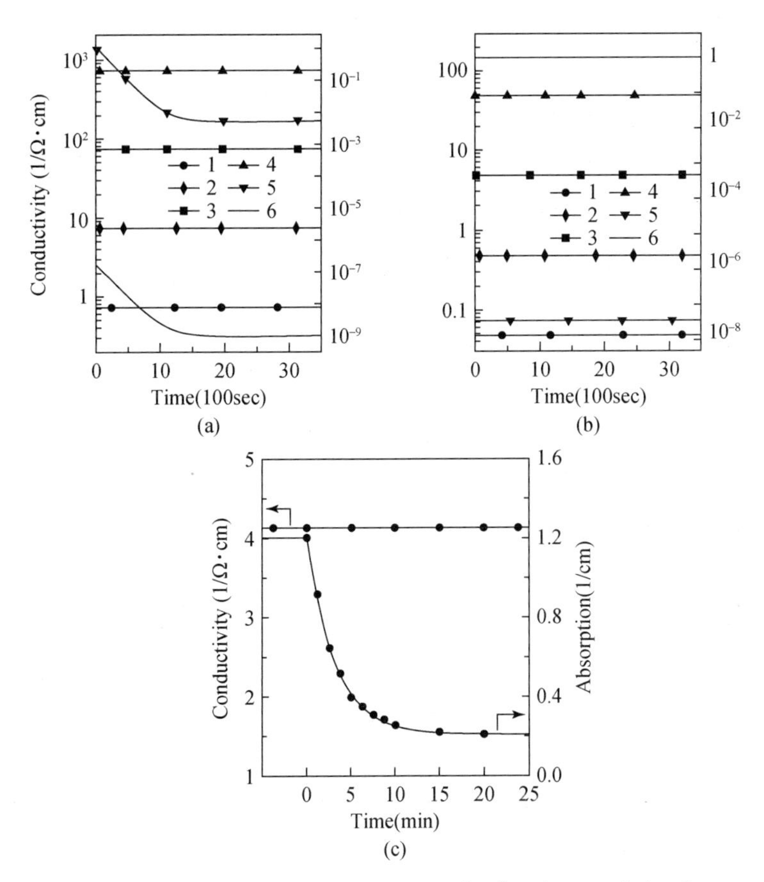

Fig. 6 Photoconductivity and $EL2$ concentration variation for doped n-type (a) and p-type (b) GaAs. n_0/N_T or $p_0/N_T=0.01$, 0.1, 1.0, 10 (lines 1, 2, 3, 4). Lines 5 and 6 correspond to $EL2^0$ and $EL2^+$ variation with n_0/N_T or $p_0/N_T=0.1$. (c) Comparison of the theoretical calculation with the experiment for the photoconductivity and optical absorption of n-type SC GaAs at 1.17eV. The solid points are the experiment data from Ref. [6]; the fitting photon flux ϕ is $5.4\times10^{14}\text{cm}^{-2}\text{s}^{-1}$

$EL2^+$. In n-type SC GaAs, $EL2$ defects are all in the neutral state $EL2^0$, since N_D-$N_A>0$ (thermal emission from the deep level is neglected). In p-type SC GaAs, the $EL2$ defects are all ionized since N_A-$N_D>N_T$. According to the compensation situation we can determine that $N_p/N_T=0$ and $n|_{t=0}=n_0$ for n-type GaAs, $N_p/N_T=1$ and $p|_{t=0}=p_0$ for p type, where n_0 and p_0 are the corresponding residual donors and acceptors. The plots in Figs. 6(a) and (b) indicate that the photoconductance remains nearly unchanged in both n-type and p-type SC GaAs even if it is slightly doped, i. e., n_0/N_T, $p_0/N_T=0.01$. Moreover, the $EL2$ concentration will not be obviously changed by illumination in p-type GaAs since $EL2^+$ cannot be directly transformed into $EL2^*$. In contrast to p-type GaAs, the concentration of the neutral state $EL2^0$ in n-type GaAs can be decreased significantly by light. Thus n-type GaAs may present EPR and infrared absorption quenching while lacking quenching of the photoconductivity. The above calculation results have interpreted some experiment measurements which were not clearly explained in previous works. [6, 15, 22]

Fig. 6(c) shows the comparison with experiment[6] of the calculation results for the photoconductance and infrared absorption for n-type SC GaAs. Our calculated results are in good agreement with the experiments and present a quite natural explanation for n-type SC GaAs to have infrared absorption quenching while lacking PC quenching. Based on the general rate equation, we can also deduce the conditions for the doped GaAs to lack additional photoconductivity: $(N_T\phi\sigma_n^0)/(C_n n_0^2) \ll 1$ for n type, $(N_T\phi\sigma_p^0)/(C_p p_0^2) \ll 1$ in p type, which are often satisfied with normal light intensity. We should mention that the Auger-like regeneration of the $EL2$ which is dominant in n-type GaAs material[17] is neglected in our calculation. Consideration of the Auger-like regeneration does not change the above conclusion except to result in a lower recovery temperature.

4 Conclusion

We have studied the dynamic process of photoquenching in GaAs based on the metastability model. Our calculation shows that considering free-carrier recombination is essential to understanding the photoquenching characteristics of GaAs, otherwise some ambiguities will arise. The numerical calculation and analytical approach give the following results.

(a) Typical SI n-type GaAs. Besides EPR and infrared absorption quenching, it will present both photoconductance quenching and the EPC phenomenon. However, EPC does not appear in very slightly compensated GaAs, while highly compensated GaAs only presents the EPC phenomenon. Our numerical results also give an expected thermal recovery temperature of about 120K and predict that the steady-state current intensity is proportional to $\phi N_p/N_0$ (N_p, N_0 are the concentrations of $EL2^+$ and $EL2^0$) at low temperature, which indicates a potential application in investigating $EL2$ compensation in GaAs.

(b) Doped GaAs. It will not present photoconductance quenching. However, n-type semiconducting GaAs may show other photoquenching behaviors such as EPR and IA quenching. p-type GaAs does not show any photoquenching behavior in contrast to n type.

The above numerical calculation results agree well with experiments and favor the metastability model.

Acknowledgments

One of the authors (Ren) would like to thank Professor Nan-xian Chen at Beijing

University of Science and Technology for his encouragement and helpful discussion. This work was partly supported by the National Foundation of Science in China.

References

[1] U. Kaufmann, W. Wilkening, and M. Baeumber, Phys. Rev. B 36, 7726(1987).

[2] Noriaki Tsukada, Jpn. J. Appl. Phys. 24, L689(1985).

[3] M. Baeumber, U. Kaufmann, and J. Windscheif, Appl. Phys. Lett. 46, 781(1985).

[4] Noriaki Tsukada, Toshio Kikuta, and Koichi Ishida, Jpn. J. Appl. Phys. 24, L302(1985).

[5] Z. Q. Fang and D. C. Look, Appl. Phys. Lett. 59, 48(1991).

[6] J. C. Parker and Ralph Bray, Phys. Rev. B 38, 3610(1988).

[7] B. Dischler and U. Kaufmann, Rev. Phys. Appl. 23, 779(1988).

[8] P. Silverberg, P. Omling, and L. Samuelson, Appl. Phys. Lett. 52, 1689(1988).

[9] J. Ertel, H. G. Brion, and P. Haasen, Acta Phys. Pol. A 83, 11(1993).

[10] B. K. Meyer et al. , Phys. Rev. B 36, 1332(1987).

[11] B. K. Meyer, J. M. Spaeth, and M. Swcheffler, Phys. Rev. Lett. 52, 885(1984).

[12] J. M. Spaeth, D. M. Hofmann, and B. K. Meyer, in Microceopic Identification of Electronic Defects in Semiconductors, edited by N. M. Johnson, S. G. Bishop, and G. D. Watkins, MRS Symposia Proceedings No. 46(Materials Research Society, Pittsburgh, 1985), 185.

[13] G. Vincent, D. Bois, and A. Chantre, J. Appl. Phys. 53, 3643(1982).

[14] G. Vincent, D. Bois, and A. Chantre, Phys. Rev. B 23, 5335(1981).

[15] T. Benchigues et al. , Appl. Surf. Sci. 50, 277(1991).

[16] G. R. Baraff and M. A. Schluter, Phys. Rev. B 45, 8300(1992).

[17] U. Strauss, W. W. Ruhle, and K. Kohler, Appl. Phys. Lett. 62, 55(1993).

[18] J. Jimenez, P. Hernandez, J. A. de Saja, and J. Bonnafe, Solid State Commun. 55, 459(1985).

[19] J. C. Bourgoin, H. J. V. Bardeleben, and D. Stievenand, J. Appl. Phys. 64, R65(1988).

[20] H. J. Von Bardeleben, N. T. Bagraev, and J. C. Bourgoin, Appl. Phys. Lett. 51, 1451(1987).

[21] U. V. Desinca, D. I. Desnica, and B. Santic, Appl. Phys. Lett. 58, 278(1991).

[22] Ralph K. Wan and J. C. Parker, Phys. Rev. Lett. 57, 2434(1986).

[23] J. Jimenez et al. , Jpn. J. Appl. Phys. 27, 1841(1988), and Refs. 6 and 7 therein.

Effect of image forces on electrons confined in low-dimensional structures under a magnetic field

V L Dostov[1] and Zhanguo Wang[2]

([1] Ioffe Institute, Politekhnicheskaya 26, St Petersburg, 194021, Russia)

([2] Laboratory of Semiconductor Materials Science, Institute of Semiconductors, Chinese Academy of Sciences, Beijing 100083, China)

Abstract　We consider the effect of image forces, arising due to a difference in dielectric permeabilities of the well layer and barrier layers, on the energy spectrum of an electron confined in a rectangular potential well under a magnetic field. Depending on the value and the sign of the dielectric mismatch, image forces can localize electrons near the interfaces of the well or in the well centre and change the direct intersubband gaps into indirect ones. These effects can be controlled by variation of the magnetic field, offering possibilities for exact tuning of electronic devices.

1 Introduction

The model of 2D electrons confined in a potential well $U(x)$ under a magnetic field $\boldsymbol{B} = [0, 0, B_z]$ parallel to potential's 'interfaces' was first used by Zawadzki[1] to describe the energy spectrum of electrons in a quantum well. Introducing the Landau gauge

$$A = [-B_z y, 0, 0],$$

and (quasi)-2D wavefunctions

$$\Psi_n = \phi_n(x)\exp(ik_y y + ik_z z),$$

one can write for the energy

$$E = E_n(k_y, B_z) + \hbar^2 k_z^2/2m^*.$$

where E_n can be found from the 1D Schrödinger-like equation with a confinement potential $U(x)$

$$-\frac{\hbar^2}{2m^*}\frac{d^2\phi_n}{dx^2} + \frac{m^*\omega_c^2}{2}(x + x_0)^2\phi_n + U(x)\phi_n = E_n\phi_n. \tag{1}$$

where $\omega_c = Be/2m^*$ is the cyclotron frequency, $x_0 = k_y\hbar/eB$ and a spin splitting is neglected.

原载于：Semicond. Sci. Technol., 1994, 9(10): 1781-1786.

More recently Berggren et al. [2] pointed out that the same formalism can be used to describe electrons confined in quasi-1D structures in which an additional confinement potential $U_c(z)$ is applied along with $U(x)$. The only modification is the replacement of $\hbar^2 k_z^2/2m^*$ by the discrete spectrum E_m^c corresponding to $U_c(z)$. Usually this z confinement is considerably stronger than the x confinement and 1D electrons occupy only one z subband. Under this assumption the z confinement results in a constant energy shift

$$E = E_n(k_y, B_z) + E_0^z. \tag{2}$$

which vanishes from further consideration[2-6]. Numerous publications have followed[2] and this model was successfully used to describe experimental data on magnetoconductivity and quasi-DC transport in split-gate heterojunctions[3], deep and shallow mesa-structures[4, 6] and other low-dimensional structures[5].

The reliability of this approach depends on a proper choice of the confinement potential $U(x)$. In[1] Zawadzki chose the rectangular potential

$$\begin{aligned} U_0 &= 0 \quad \text{if} \quad |x| < a/2, \\ U_0 &= \infty \quad \text{otherwise}. \end{aligned} \tag{3}$$

to represent the confinement in a quantum well of width a. At the same time numerous publications[7] have demonstrated that carriers in quantum wells and other low-dimensional structures may be greatly influenced by image forces arising due to a difference in a dielectric permeability between the confinement region and the barrier layers. In the present work we modify the model of Zawadzki to take into account the image potential. To the best of our knowledge, this is the first detailed treatment of an interplay of image forces with the magnetic field in low-dimensional structures. The model and basic properties of the image potential are discussed in section 2. Dimensionless equations and the method of calculation are presented in section 3. Section 4 discusses results of numerical calculations demonstrating quantitative effects produced by image forces. Analytical results for some limiting cases are given in the appendix.

2 The model

Let us consider electrons confined in a quantum well of width a and a static dielectric permeability ε_1 sandwiched between two thick barrier layers with a static dielectric permeability ε_2 (see Fig.1). It is known from electrostatics[8, 9] that in addition to confinement(3) arising due to a difference in hand structure of the layers an electron at a point x inside such a structure experiences the electrostatic image force with a potential $U_e(x)$. This potential can be calculated using the method of images. The principle of this method is to find fictitious charges e' which, together with given charge e, provide an electric field satisfying boundary conditions on interfaces[9]. For a single interface at, say, $x=a/2$, the boundary conditions are satisfied by placing the image charge

$$e' = \frac{\varepsilon_1 - \varepsilon_2}{\varepsilon_1 + \varepsilon_2} e,$$

at the point $x' = a - x$. The energy of the given charge is

$$U_c = \frac{e^2(\varepsilon_1 - \varepsilon_2)}{4\varepsilon_1(\varepsilon_1 + \varepsilon_2)} \frac{1}{a/2 - x}. \tag{4}$$

Corresponding expressions for a single dielectric/metal interface read

$$e' = -e,$$

$$x' = a - x,$$

$$U_e = -\frac{e^2}{4\varepsilon_1} \frac{1}{a/2 - x}.$$

It has been demonstrated in electrostatics[9] that one can obtain the solution for an arbitrary dielectric/metal interface from the solution for a corresponding dielectric/dielectric interface via the formal substitution $\varepsilon \to \infty$ for the metal medium.

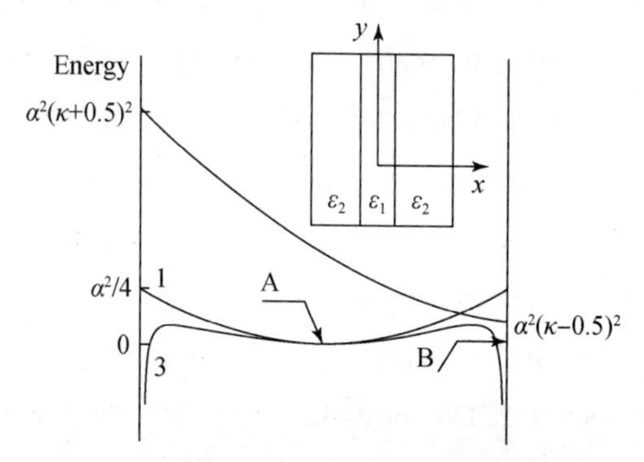

Fig. 1　Sketch of the effective potential in equation (9) at $\beta=0$, $\kappa=0$ (1); $\beta=0$, $\kappa>0.5$ (2) and $\beta<0$, $\kappa=0$ (3)

For the double interface of the model under consideration, boundary conditions can be satisfied by an infinite series of image charges, yielding the following expression for U_e

$$U_e(x) = \frac{e^2\gamma}{4\varepsilon_1 a}\tilde{U}(\gamma, x), \qquad |x| < a/2 \tag{5}$$

$$\tilde{U}_e = \frac{1}{1/2 + x/a} + \frac{1}{1/2 - x/a} + \Phi(\gamma, 1/2 + x/a) + \Phi(\gamma, 1/2 - x/a),$$

where

$$\Phi(\gamma, l) = \sum_{n=1}^{\infty} \frac{\gamma^{2n}}{n + l} \frac{\gamma^{2n-1}}{n}$$

and

$$\gamma = \frac{\varepsilon_1 - \varepsilon_2}{\varepsilon_1 + \varepsilon_2}.$$

The same results have been obtained by Gabovich and Rozenhaum using a more general method[10]. The solution for a metal/dielectric/metal sandwich is given in[8] and coincides with(5) after the substitution $\varepsilon_2 \to \infty$, namely, with $U_e(\gamma = -1, x)$. For this particular case, series(5) can be rewritten using digamma functions[8]. For a qualitative interpretation we will use the first two singular terms in (5), keeping accurate summarizing for quantitative numerical calculations. Here and hereafter we follow the common assumption and neglect the interaction of electrons inside the well[1-6].

The first two terms in(5) are singular near interfaces $|x| \to a/2$

$$U_e \to \frac{e^2\gamma}{4\varepsilon_1 a} \frac{1}{1/2 - |x/a|}, \tag{6}$$

coinciding with the solution for the single interface. More detailed calculations for dielectric/metal interfaces taking into account dispersion of ε_2 due to plasmons in the metal lead to a saturation of U_e at small distance d_0 from the interface. The distance is about a few angstroms according to Sols and Ritchie[11]. This distance is usually considerably smaller than other characteristic sizes of the problem and gives negligible amendment for the infinitely high barrier model(see also the appendix). Similar results for semiconductor interfaces have been obtained in[12].

The potential(5) is the even function of x and contains the energy shift

$$U_e(0) = -\frac{e^2}{\varepsilon_1 a}\ln(1 - \gamma). \tag{7}$$

For a potential well of finite height this results in the classical shift of the electron's binding energy, but it is not important while one considers the states inside the infinite well. To emphasize quantum effects we omit this shift in numerical calculations and on figures having zero of the image potential at $x = 0$ (point A on Fig. 1). At small x equation (5) can be expanded in series as

$$U_e \approx \frac{e^2\gamma}{4\varepsilon_1 a}\left[\zeta\left(\frac{x}{a}\right)^2 + \cdots\right], \tag{8}$$

where

$$\zeta(\gamma) = \frac{a^2}{2}\frac{\partial^2 \tilde{U}}{\partial x^2}\bigg|_{x=0} = 16\left(1 + \frac{\gamma^2}{27} + \frac{\gamma^4}{125} + \cdots\right).$$

The static dielectric permeability of the well layer depends on its composition and is about I2 for A^3B^5-based structures. Typical values of γ for semiconductor/semiconductor interfaces are relatively small, namely about 0.04 for the GaAs/$Al_{0.3}Ga_{0.7}As$ interface or 0.15 for the Ge/Si interface. Due to a positive correlation between bandgap and $1/\varepsilon$, values of γ are mostly positive for the structures based on type-Ⅰ heterostructures, but they may be negative for some kinds of type-Ⅱ heterostructures.

The value of γ is considerably larger ior deep mesa structures, i. e. thin slabs etched from a semiconductor layer[4]. For semiconductor/vacuum interfaces one has $\varepsilon_2 = 1$ and $\gamma \approx 0.9$. Due to strong additional z confinement only a few 1D subbands (2) are populated and quantum effects as 1D Shubnikov-de Haas oscillations can be clearly observed. These oscillations are sensitive to the shape of the confinement potential[4,6]. These factors make deep mesa structures convenient objects for experimental observations of image force effects.

Negative values of γ close to -1 are typical for semiconductor/metal and oxide/metal interfaces[13]. and gaps between metal or dielectric plates[11]. The model also approximates a semiconductor layer sandwiched between oxide-metal barriers (MOS structure) when oxide layers are thin, say a few monolayers. Such thin oxide layers provide high potential barriers for electrons. but their contributions in U_e can be neglected. Then expression (5) with $\gamma \rightarrow -1$ combined with (3) seems to be a good appoximation for the confinement potential.

3 Dimensionless equations

We introduce dimensionless variables measuring a length in the units of a and an energy in the units of characteristic confinement energy $\hbar^2/2ma^2$. In these variables equation (1) reads

$$\phi''_n - (\alpha^2(\tilde{x} + \kappa)^2 + \beta\tilde{U}(\gamma, \tilde{x}) - \tilde{E}_n)\phi_n = 0 \quad \text{if} \quad |\tilde{x}| < \frac{1}{2}$$

$$\phi_n = 0. \qquad \text{otherwise} \tag{9}$$

where

$$\tilde{x} = x/a,$$

$$\kappa = \frac{\hbar k_y}{m^* \omega_c a},$$

$$\tilde{E}_n = 2m^* a^2 E_n/\hbar^2,$$

α is the characteristic magnetic energy.

$$\alpha = \frac{m^* \omega a^2}{\hbar} \sim B_z,$$

and β is the characteristic electrostatic energy

$$\beta = \frac{e^2 m^* a}{2\varepsilon_1 \hbar^2}\gamma.$$

The confinement potential (3) results in boundary conditions

$$\phi_n = 0, \quad \text{at} \quad |x| = \frac{1}{2}. \tag{10}$$

The values of these parameters are displayed in the example of experimental conditions for deep mesa structures[4]. For $a\approx 200\text{mm}$, $B\approx 1\text{T}$, $\varepsilon_1\approx 12$, $\varepsilon_2\approx 1$ and $\gamma\approx 0.9$. one can get $\alpha\approx 80$ and $\beta\approx 20$. In such structures under reported experimental conditions the electrons occupy the lowest states with $|\kappa|\sim 0\cdots 1$. The number of occupied states varies from 1 to a maximum number depending on the applied magnetic field. This maximum number corresponds to zero magnetic field and it is from 6 to 16 for specific structures[4, 6]. For narrower structures α and β decrease with a, reflecting the increasing contribution of spatial confinement(3). For deep mesa structures $\beta\approx 1$ at $\alpha\approx 10\text{nm}$. This value is the characteristic width below which electrostatic effects are dominated by the spatial confinement in our model. For AlGaAs/GaAs quantum wells this width is about 200nm due to the small value of γ. The same parameters except negative β can be obtained for structures with a negative γ. Aforementioned values give the range of parameters to study.

Singular terms(6) do not contain γ explicitly. Therefore, while β is formally fixed the value of γ does not qualitatively influence the obtained results. In the numerical calculations we take(arbitrarily) $|\gamma|=0.8$.

The energy spectrum can be found by solving equation(9) with boundary conditions (10): The effective potential in equations(9) and(10) is a combination of the parabolic magnetic term, the infinite rectangular potential and the image potential term that is similar to the 1D Coulomb potential in the vicinity of the edges. The problem is symmetric with respect to κ: $E_n(\kappa)=E_n(-\kappa)$. To solve equations(6) and(7) at arbitrary α, β, γ and κ we use a numerical method. We obtain the best convergence for a wide range of parameters by applying Jacobi diagonalization[14] to the matrix that is a Fourier presentation of Hamiltonian (9). The perturbation approach can also be used for some limiting cases. The corresponding explicit expressions(17)-(19) for the electrostatic amendment are given in the appendix.

4 Discussion

First we consider the energy spectrum at $\kappa=0$. This problem is relevant for inter-and intra-subband transitions[1]. At $\beta=0$ the problem reduces to the problem treated by Zawadzki. It allows two types of energy levels as limiting cases. For the lowest energy levels $E_n\ll\alpha^2/4$ the effective potential in(9) is parabolic. Its wavefunctions can be approximated as

$$\phi_n=\left(\frac{\alpha^{1/2}}{2^n n!\ \pi^{1/2}}\right)^{1/2}\exp(-\alpha\tilde{x}^2/2)H_n(\alpha^{1/2}\tilde{x}),\tag{11}$$

$$\tilde{E}_n=2\alpha\left(n+\frac{1}{2}\right).\quad n=0,1,\cdots\tag{12}$$

where $H_n(z)$ are Hermitian polynomials.

The highest levels with the energy $E_n \gg \alpha^2/4$ are confined between infinite potential walls. Their wavefunctions can be approximated as

$$\phi_n = 2^{1/2}\cos\pi n\tilde{x}, \tag{13}$$

$$\tilde{E}_n = \pi^2(n+1)^2 + \alpha^2\kappa^2. \quad n = 0, 1, \cdots \tag{14}$$

In dimensional units the last term in expression (14) is the kinetic energy $\hbar^2 k_y^2/2m^*$. Finally, there are hybrid levels with $E_n \approx \alpha^2/4$.

The effect of the image potential on these levels is different for negative and positive β. Introducing the image potential with $\beta < 0$ one changes the one-valley effective confinement potential in (9) to a three-valley one, as shown, in Fig. 1. While β is small, the modification only slightly decreases E_n and interlevel distances in agreement with explicit equations (17), (18). New effects appear when the absolute value of β is large enough to lower E_n deeper than the bottom of the parabolic valley (point A on Fig. 1). The further increase of $|\beta|$ leads to localization of wavefunctions in image potential valleys B near the edges of the structure. At small distances from the edges the image potential is given by equation (6) similar to the 1D Coulomb potential. The eigenfunctions of the 1D Coulomb potential consistent with boundary conditions (10), i. e. with nodes at $|\tilde{x}| = \frac{1}{2}$, can be expressed through radial functions of a 3D hydrogen atom with $l=0$ as $\rho\psi_{n,0}(\rho)$[15]. At sufficiently large $|\beta|$ this eigenfunction can be chosen as

$$\varphi_n = \exp\left(-\frac{\beta(1/2-|x|)}{2(n+1)}\right)(1/2-|x|)U_n^{(1)}\left(\frac{\beta(1/2-|x|)}{2(n+1)}\right), \tag{15}$$

$$\tilde{E}_n = \beta^2/4(n+1)^2. \quad n = 0, 1, \cdots \tag{16}$$

where $U_n^{(1)}$ are normalized associated Laguerre polynomials as defined in[14]. The parabolic dependences (16) can be seen in Fig. 2 at large negative β. For a given n th level the localization occurs at $\beta < \beta_0(\alpha)$ where $\tilde{E}_n(\beta_0) = 0$. The absolute value of $\beta_0(\alpha)$ increases with magnetic field (with α). It can be roughly estimated by equating the characteristic electrostatic energy with the characteristic energy at $\beta=0$.

$$-\beta_0 \approx \pi^2(n+1)^2 + 2\alpha\left(n+\frac{1}{2}\right).$$

Numerical calculations give $\beta_0(0) \approx -9$ and $\beta_0(30) \approx -30$ for $n=0$ which is consistent with the estimation.

Let us consider the structure with $\beta < \beta_0(0)$ and compare the shape of wavefunctions without a magnetic field and under a strong magnetic field α such as $\beta > \beta_0(\alpha)$ (see Fig. 3). Comparison of (11) and (15) shows that the strong magnetic field can completely change the localization of carrier wavefunctions, compressing them from the edges to the centre of the

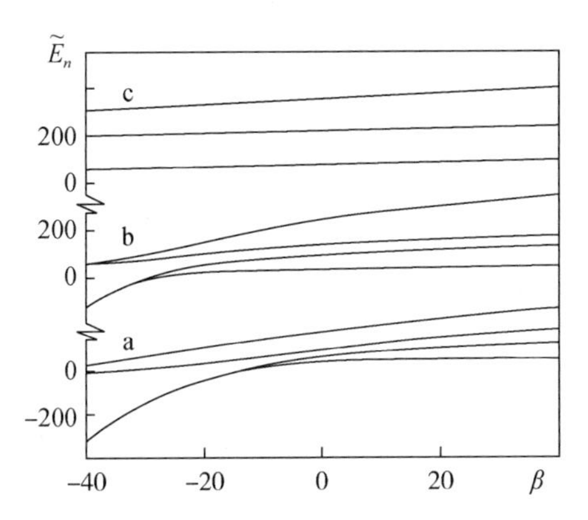

Fig. 2 Lowest energy levels versus β at $\alpha=0$(a), $\alpha=30$(b) and $\alpha=70$(c)

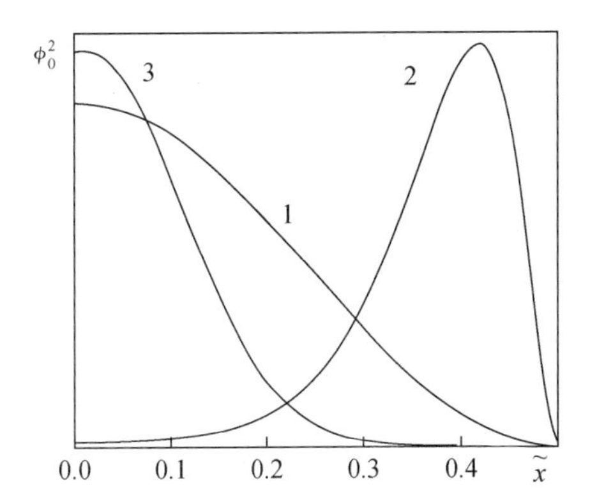

Fig. 3 The shape of the wavefunction of the ground state at $\alpha=0$, $\beta=0$(1), $\alpha=0$, $\beta=-25$(2) and $\alpha=70$, $\beta=-25$(3)

structure. If some of surface-related states have small characteristic energies $\tilde{E}<\alpha$, β, then one can control effectively their contribution to electronic processes by variation of the magnetic field. Evidently, the choice of composition of the structure's layers controls the localization of electrons as well. The necessary edge localization of carriers can be achieved at sufficiently high negative β, which are expected from the structures with $\gamma\to-1$ listed in section 2.

When β is positive, the shape of the confinement potential remains one-valley. The image potential decreases the effective well width oppositely to the case of a negative β. As a result, E_n and interlevel distances increase in agreement with equation (17). The image potential tends to localize wavefunctions further from edges, thus decreasing the contribution of other surface-related factors.

Often one needs to calculate E_n as functions of κ at a given magnetic field, for instance to describe Shubnikov-de Haas oscillations or, more generally, to calculate Fermi energy.

The $E_n(\kappa)$ dependences are presented in Fig. 4 by numerical calculations at $\alpha=70$. At large κ, $E_n(\kappa)$ increase rapidly at any β, but at small κ their behaviour changes qualitatively with β. At $\beta>0$. $E_n(\kappa)$ increase, at $\beta=0$ they are almost constant and at $\beta<0$ new minima of $E_n(\kappa)$ arise at $\kappa\neq 0$.

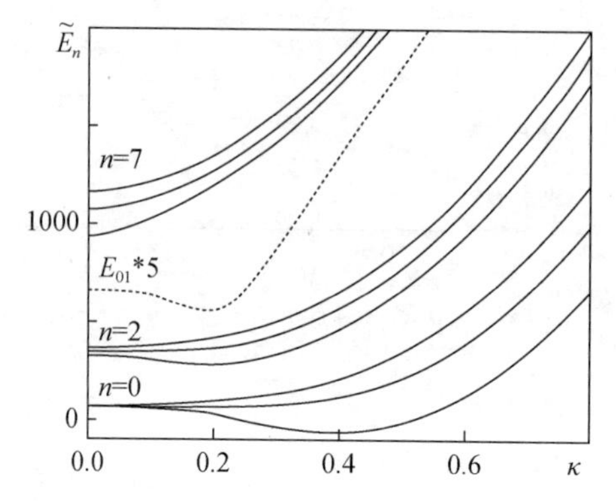

Fig. 4　Energy-momentum dispersion $E_n(\kappa)$ at $\alpha=70$ (full curves). Numbers designate the levels. For every level the data at $\beta=-25$ (lowest curve). $\beta=0$ (curve in the middle) and $\beta=25$ (highest curve) are given. The dotted curve represents the energy gap between zero and first levels for $\beta=-25$. For clarity the other levels are not indicated

To explain this behaviour it is convenient to consider a positive κ and sufficiently large α, i. e. strongly localized states near $\tilde{x}=-\kappa$. The value of κ affects the energy spectrum because it determines a position of the parabolic magnetic term in (9). When κ increases, the parabola shifts towards an edge. While $\kappa<1/2$ and $\beta=0$, the shift does not affect significantly the shape of wavefunctions of the lowest levels (11) but it changes their localization: $\varphi_n(\tilde{x})\to\varphi_n(\tilde{x}-\kappa)$. Therefore. the initial part of $E_n(\kappa,\beta=0)$ is almost dispersionless. When $\beta\neq 0$, the increase of κ shifts the wavefunctions nearer to an edge where image forces are stronger. Therefore, the absolute value of the contribution of the image potential increases with κ, and total E_n increase or decrease depending on the sign of β. At $\kappa>1/2$ and $\beta=0$ the position of the minimum of the effective potential increases as $\alpha^2(\kappa-1/2)^2$, so, approximately, do $E_n(\kappa)$. At sufficiently large κ this effect dominates and $E_n(\kappa)$ increase notwithstanding the value of β. At $\beta<0$ the combination of the increase with the decreasing initial part of $E_n(\kappa)$ gives two minima at some $0<|\kappa_n|<1/2$. The intersubband energy gaps also have such minima at some $\kappa\neq\kappa_n$. At $\beta>0$ one obtains a parabola-like curve with its minimum at $\kappa=0$. This argument can be illustrated by explicit formulae (18) and (19) in the appendix. For the conditions under consideration the electrostatic amendment is comparable with interlevel distances and may cause an observable shift of Shubnikov-de Haas oscillations.

The highest levels(13) are practically insensitive to the magnetic term, and at moderate κ the electrostatic contribution $\delta E_n(\kappa) \approx \delta E_n(\kappa=0)$ is given by equation(17).

Obtained results can be qualitatively applied to the lowest energy levels in the potential well of finite height $E_W \gg E_n$. In this case it is more convenient to add the classical shift(7) to previous results everywhere

$$E_n \rightarrow E_n - \frac{e^2}{\varepsilon_1 a}\ln(1-\gamma)$$

counting energies from the bottom of the well(3). A more accurate consideration should replace(10) by proper boundary conditions and consider the Schrödinger-like equation(9) with an effective potential for an electron in barrier layers as well as inside the well. The corresponding wavefunctions should have finite amplitudes on interfaces and the singular potential(5) leads to a divergence in calculations of the energy of states. For example. an analogue of integral(17) for a finite-height well is no longer finite. This difficulty can be bypassed by taking into account the saturation of the image potential near interfaces discussed in section 2. In such calculations the electrostatic amendment to E_n due to final well height very likely depends on d_0 and increases with $1/E_w$, because of the increase in the amplitude of the wavefunctions near interfaces. The specific behaviour of this amendment is a subject for future work.

Summarizing, if the dielectric permeability of the confinement region is higher than the dielectric permeability of barrier layers, image forces quantitatively affect energy-momentum dispersion and tend to localize wavefunctions in the centre of the structure. In the opposite case the image forces may result qualitatively in the additional localization of wavefunctions near the edges of the structure, giving rise to new minima on dispersion curves and the transformation of direct intersubband energy gaps into indirect ones. One can also get asymmetric dependences $E_n(\kappa)$ by placing the confinement region between two layers with different dielectric permeability.

5 Conclusions

In conclusion, we have demonstrated that the interplay of magnetic field and image forces may influence significantly the energy spectrum and the shape of wavefunctions of electrons confined in low-dimensional structures. By choosing materials for the structure and applying a magnetic field one can change considerably the shape of the effective confinement potential, control energy-momentum dispersions, the density of states and the spatial localization of carriers. Thus one can accurately fit parameters of electronic devices. The effects of image forces on Shubnikov-de Haas oscillations and intersubband transitions can be observed experimentally. Such treatment may also shed light on the nature of the confinement in low-dimensional structures.

Acknowledgments

The authors are very grateful to V Rossin, S Karpov and Jia-Jiong Xiong for helpful discussions.

Appendix: perturbation theory calculations

The case when image forces are small enough to be taken into account via first-order perturbation theory is interesting from both practical and theoretical points of view. Unfortunately, explicit expressions for basis wavefunctions $\varphi_n^0(\tilde{x},\beta=0)$ can be obtained only in some limiting cases.

Keeping for U_e two first singular terms (8) one can get the following electrostatic amendment for highest levels (13)

$$\delta\tilde{E}_n = \frac{\langle \varphi_n^0 | U_e | \varphi_n^0 \rangle}{\langle \varphi_n^0 | \varphi_n^0 \rangle} = 2s_n\beta. \tag{17}$$

where

$$s_n = \int_0^{2\pi(n+1)} \frac{1 - \cos z}{z} \mathrm{d}z$$

is the positive slowly increasing function of n : $s_0 \approx 2.4$, $s_8 \approx 4.5$.

At small $\kappa<1/2$ and high α the lowest levels are strongly localized near $x=0$. For such states it is more convenient to substitute directly the approximate expression (8) in equation (9) obtaining

$$\tilde{E}_n = 2(\alpha^2 + \zeta\beta)^{1/2}\left(n + \frac{1}{2}\right) + \zeta\beta\frac{\alpha^2\kappa^2}{\alpha^2 + \zeta\beta}. \tag{18}$$

or, at $\zeta\beta \ll \alpha^2$

$$\delta\tilde{E}_n \approx \zeta\beta\left(\frac{n + \frac{1}{2}}{\alpha} + \kappa^2\right).$$

where κ should be small. Equation (18) coincides with the energy spectrum of an electron confined in a parabolic potential well under a magnetic field obtained by Berggren et al. [2]

At $\kappa=0.5$ and $\beta=0$ the lowest states $\tilde{E}_n<\alpha^2$ are confined in a half-parabolic potential with an infinitely high wall in its middle. Correspondingly, the wavefunctions can be approximated by odd eigenfunctions of the parabolic oscillator (13), (14), $n=2m+1$, with corresponding energy levels[1]. Perturbation theory gives

$$\delta\tilde{E}_n = b_n \alpha^{1/2} \beta \tag{19}$$

where b_n decreases with n : $b_1 = 2\pi^{-1/2}$, $b_2 = \frac{5}{3}\pi^{-1/2}$, $b_3 = \frac{89}{60}\pi^{-1/2}$.

The same calculations can be done taking into account the saturation of U_e at the small distance d_0 from interfaces discussed in section 2. Simple estimations give the relative amendment of the order d_0/a to expression(17) and of the order $d_0\alpha^{1/2}/a$ to expression(19). The states(19) are more sensitive to U_e behaviour near interfaces where they are localized. On the contrary. the states(18) localized far from interfaces are practically insensitive to U_e behaviour near interfaces.

References

[1] Zawadzki W 1987 Semicond. Sci. Technol. 2 550.

[2] Berggren K-F, Roos G and van Houten H 1988 Phys. Rev. B 37 10118.

[3] Rundquist J H 1989 Semicond. Sci. Technol. 4 455.

[4] Demel T, Heitmann D. Grambow P and Ploog K 1988 Appl. Phys. Lett. 53 2176.

[5] Motohisa J and Sakdki H 1993 Appl. Phys. Lett. 63 1786.

[6] Dostov V L. Du Q H and WaiiZ G. 1994 Solid State Commun. 87 833.

[7] The quantum effects related with image forces at zero magnetic field $B=0$ have been widely discussed. See, for instance, Andreani L C and Pasquarello A 1990 Phys. Rev. B 42 8928; Deng Z Y, Yang X L and Gu S W 1993 Solid State Commun. 86 399 and [10, 11, 13, 15].

[8] Sommerfeld A and Bethe H 1933 Handbirch der Physik vol XXN/2 ed H Geiger and K Shell(Berlin: Springer)450.

[9] Landau L D and Lifshitz E M 1960 Electrodynamics of Continuous Media(Oxford Pergamon)40.

[10] Gabovich A M and Rozenbaum V M 1984 Sou. Phys. Semicond. 18 308.

[11] Sols F and Ritchie R H 1987 Phys. Rev. B 35 9314.

[12] Lang N D and Kohn W 1973 Phys. Rev. B 7 3541.

[13] Simmons J G 1969 Tunneling Phenomena in Solids ed E Burstein and S Lundquist (New York: Plenum) 135.

[14] Korn G and Korn T 1968 Mathemarical Handhook(New Yorn: McGraw-Hill) ch 20.

[15] Forstmann F 1993 Progr. Surf. Sci. 42 21.

Photoluminescence studies of single submonolayer InAs structures grown on GaAs(001) matrix

Wei Li, Zhanguo Wang, Jiben Liang, Bo Xu, and Zhanping Zhu
(Laboratory of Semiconductor Materials Science, Institute of Semiconductors, Chinese Academy of Sciences, Beijing 100083, China)
Zhiliang Yuan
(National Semiconductor Superlattice Laboratory, Institute of Semiconductors, Chinese Academy of Sciences, Beijing 100083, China)
Jian Li
(National Integrated Optoelectronics Laboratory, Institute of Semiconductors, Chinese Academy of Sciences, Beijing 100083, China)

Abstract Optical properties of single submonolayer InAs structures grown on GaAs(001) matrix are systematically investigated by means of photoluminescence and time-resolved photoluminescence. It is shown that the formation of InAs dots with 1 ML height leads to localization of excitons under certain submonolayer InAs coverages, which play a key role in the highly improved luminescence efficiency of the submonolayer InAs/GaAs structures.

The growth of highly strained InAs/GaAs heterostructures has attracted much interest from a fundamental point of view due to their unique electronic and optical properties as well as for potential applications in electronic and optoelectronic devices.[1-3] Strong modification of electronic structures and surface morphology due to high strain(7% lattice mismatch) has been reported.[4] More recently, it has been discovered that an InAs/GaAs superlattice(SL) composed of submonolayer InAs exhibited greatly improved luminescence efficiency at room temperature compared to either the $In_xGa_{1-x}As$ alloy or monolayer InAs/GaAs SL with the same average In composition.[5] However, the underlying physical origin of this striking phenomenon has so far not been understood as there are few data available on such a system. In this letter, we investigate systematically the optical properties of single submonolayer InAs structures grown on GaAs(001) matrix, and present our experimental findings on the origin of highly improved luminescence efficiency of submonolayer InAs/GaAs structures in relation to the localization of excitons.

原载于：Appl. Phys. Lett., 1995, 67(13): 1874-1876.

The samples investigated were grown by molecular beam epitaxy (MBE, Riber 32P). Growth rates were nominally 1.0ML/s for GaAs and 0.2ML/s for InAs. All samples consisted of a 0.3μm thick GaAs buffer layer grown at 600℃, followed by a 20 period SL grown by depositing alternating layers of 2nm GaAs and 2nm $Al_{0.3}Ga_{0.7}As$ to trap impurities and smooth the growth front. Then a single fractional layer of InAs was deposited at 400℃. After the ultrathin InAs layer was completed the growth was interrupted for 30sec and the arsenic beam equivalent pressure was 1×10^{-5} Torr. Finally, a 100nm thick GaAs capping layer was grown at 580℃. During InAs layer deposition reflection high energy electron diffraction (RHEED) was used to monitor the growth process in situ. The RHEED patterns observed indicated a two-dimensional growth behavior. Photoluminescence (PL) measurements were made between 10 and 300K, with the 514.5nm line of an Ar^+ laser serving as excitation source. The PL signal was analyzed by a 1m double monochromator set in all cases to a resolution better than 0.5meV, and detected by a cooled photomultiplier with a GaAs photocathode. Time-resolved PL (TRPL) measurements were carried out with excitation pulses of 5ps FWHM at an 80MHz repetition rate from a synchronously mode locked Styrl 9 dye laser. The TRPL signal is spectrally resolved by a streak camera with the time resolution of 10ps. The ultrathin InAs layer thickness determined according to RHEED oscillations was in good agreement with characterization by high-resolution double-crystal X-ray diffraction (HRDXD) based on the simulation of the measured curves using the dynamical theory of X-ray diffraction.

PL spectra of samples with different InAs layer coverages ($\Theta = 1/3$, 1/2, and 1ML) were investigated thoroughly. In Fig. 1 we show the typical PL spectra under cw excitation of the samples. The GaAs free-exciton emission can be seen on the high-energy side of the spectra and the InAs layer-related PL line at lower energies due to $n=1$ heavy hole exciton recombination. With reduction of InAs coverage deposition the InAs-related peaks are systematically blueshifted, accompanied by a strongly asymmetric broadening to higher energies. For samples with $\Theta = 1/2$ML, the InAs-related peak exhibits the highest luminescence efficiency and narrower linewidth (FWHM = 3.5meV) although only half of the InAs coverage is deposited as compared with the 1ML sample. For the $\Theta = 1/3$ML sample the GaAs free exciton emission is comparable to that of the attached InAs layer (FWHM = 3.7meV), and the weak peak (labeled "C") around 1.495eV is identified as carbon-related transitions in GaAs and, accordingly, its intensity saturates when the excitation density is increased. Fig. 2 displays the typical spectra of samples with $\Theta = 1/2$ML excited with different power densities. Under low-power excitation (0.05mW) and relatively low temperatures, several peaks appear on the high-energy side of the line (1e–1hh). Another striking feature is the high-energy tail of GaAs free exciton emission in the submonolayer InAs structures. Such phenomena have never been observed under similar conditions in quantum wells as well as in the 1ML thick InAs sample. [6] In addition, from excitation-intensity-dependent PL experiments it can be seen that the integrated intensity of these InAs-related peaks exhibits a strictly linear dependence on excitation in the range from 0.05 to 100mW, which shows that a purely radiative decay channel is introduced by the InAs insertions. From the change in the intensity ratio between InAs-related and GaAs free-excitons peaks (Fig. 3), it is found that with increase in excitation of submonolayer samples the InAs-related peaks tends to saturate, while that of the 1ML thick InAs sample is nearly proportional to the excitation.

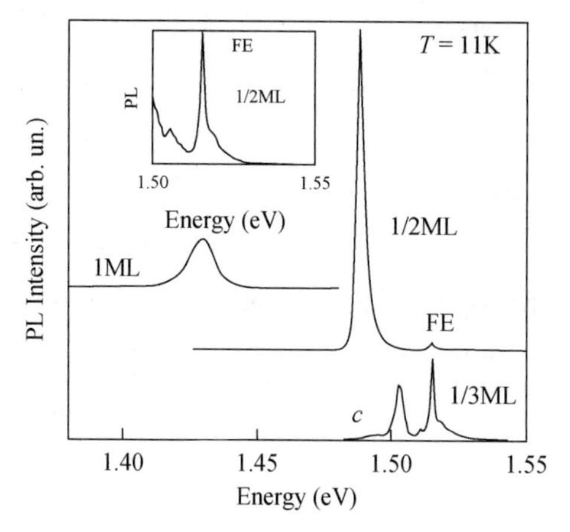

Fig. 1　cw PL spectra at 11K of single InAs layers of different coverages (Θ = 1/3, 1/2, 1ML) grown on (001) GaAs surfaces under an excitation of 5mW. In the inset, the high-energy side of the PL spectrum for a 1/2ML sample is shown

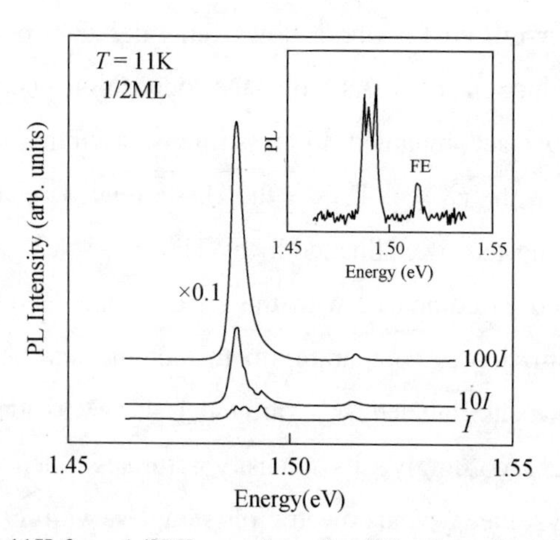

Fig. 2　PL spectra taken at 11K for a 1/2ML sample under different excitation intensities (I = 0.05mW). In the inset, the PL spectrum of a 1/2ML sample excited with 0.05mW is shown

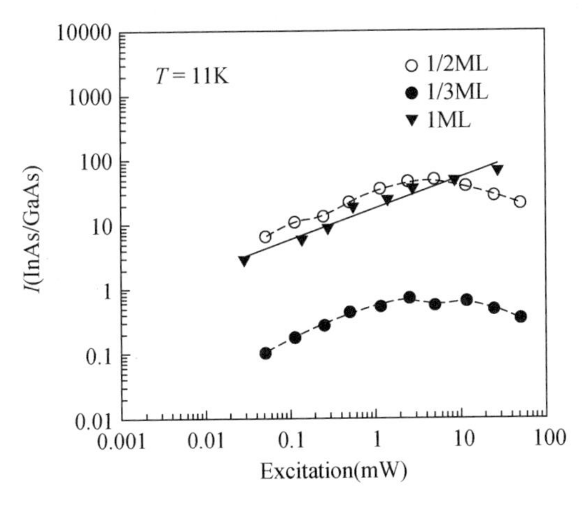

Fig. 3 Excitation dependence of the change in the intensity ratio between InAs-related and GaAs free exciton peaks for samples with $\Theta = 1/3$, $1/2$, and 1ML

These unusual features can be fully understood by considering the structural configuration of the system. Based on recent scanning tunnel microscopy (STM) studies in the submonolayer InAs coverage deposition on GaAs matrix,[7] we can draw a physics picture as follows: the initial growth process of the InAs layer on a GaAs surface involves adatom nucleation, as well as InAs island (1ML height) growth and coalescence. It is clear that upon going from one-monolayer to submonolayer coverages, the excitons become localized as a result of the potential of InAs islands (dots). The formation of InAs dots breaks the in-plane symmetry, resulting in a band of localized states. The lower the energy of the state, the more the state is localized. There is a critical InAs coverage (Θ_c) deposition beyond which the distance between neighboring InAs dots is so close that lateral transport of excitons is optional [i. e., the band of localized states is replaced by the two-dimensional (2D) plane-wave-like states], which exhibits quasi-2D exciton behavior. This simple picture can explain our observations. At low temperatures, once electrons and holes are photoexcited in the GaAs barrier, they quickly relax into the localized states with InAs dots by phonon emission. The closely localized electrons and holes pair to form excitons followed by radiative recombination. So, the low-temperature PL shape can be viewed as a rough estimate of the density of states of the localized band weighed by the carrier distribution in the band, which can account for the asymmetric broadening to higher energies of the InAs-related peak in the submonolayer InAs structures. Furthermore, localization of excitons in the submonolayer InAs structures can increase the radiative efficiency because of the prevention of nonradiative recombination at impurities or defects outside the InAs dots. The reason for poor luminescence efficiency of the 1/3ML sample is probably due to lower binding energy and less localized states as compared with the 1/2ML sample. The asymmetric broadening of

GaAs exciton emission in the submonolayer samples is attributed to the breakdown of the in-plane k-conservation selection rule as a result of GaAs insertion.

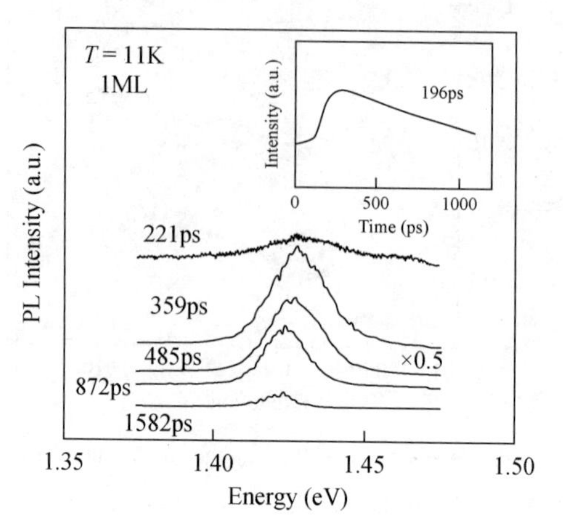

Fig. 4　Transient PL spectra taken at 11K after pulsed excitation for a 1ML sample. The spectra are taken at different delay times after the arrival of the exciting pulse, as denoted in the figure. In the inset, the time dependence of the spectrally integrated emission is shown on a semilogarithmic scale

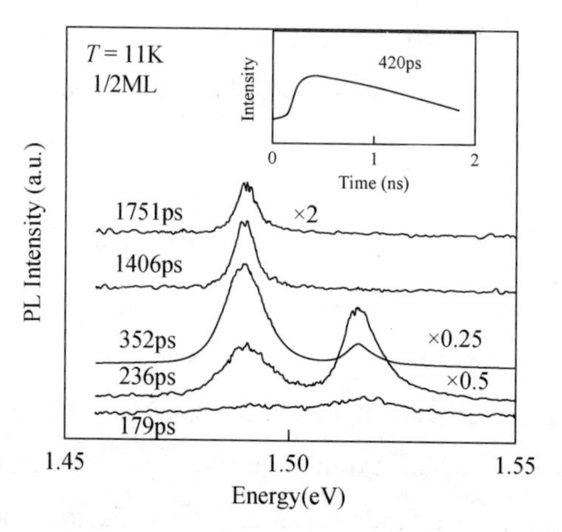

Fig. 5　Transient PL spectra taken at 11K after pulsed excitation for 1/2ML sample. The spectra are taken at different delay times after the arrival of the exciting pulse, as denoted in the figure. In the inset, the time dependence of the spectrally integrated emission is shown on a semilogarithmic scale

To further explore the carrier dynamics in the submonolayer InAs structures, TRPL measurements were carried out and a simple three-level model was used to analyze the temporal profiles. As Figs. 4 and 5 show, the emission of the 1ML sample is characterized by a rapid decay of the luminescence intensity. In contrast, the emission intensity of a submonolayer (1/2ML) sample exhibits a much slower decay. The spectrally integrated emission exhibits an exponential decay for both samples, consistent with the expected monomolecular decay of excitons. The carrier lifetimes are determined by fitting the luminescence decay with a single exponential of 196ps for the 1ML sample, 168ps for the 1/3ML sample, and 420ps for the 1/2ML sample. This can be explained in relation to the localization of exciton as follows: it is well-known that the coherent area of a localized exciton is determined by the localization length, which is smaller than that of a free exciton (the spatial extent of the localized state is always smaller than that of the extended state). Thus the radiative decay time of localized excitons is longer than that of the extended state). Thus the radiative decay time of localized excitons is longer than that of free excitons. The shorter decay time in the 1/3ML sample could be due to the less localized excitons and more existing nonradiative recombination centers. In addition, in the real 1ML sample localized excitons at the rough interface of GaAs/InAs could be contributing to the measured decay time at low temperature. From Figs. 4 and 5 it is obvious that the emission band in the 1ML sample spectrally red-shifts in the course of time (equivalent to the Stokes shift), which is not the case for submonolayer InAs samples. This result provides extra evidence for the localization of excitons in the submonolayer InAs structures grown on a GaAs matrix.

In conclusion, in submonolayer InAs samples we have observed strongly asymmetric broadening to higher energies of the PL peaks. With increase in the excitation of the submonolayer InAs samples the InAs-related peaks tend to saturate. In TRPL observations the submonolayer InAs structures exhibit no red-shift of exciton emission in the course of time in contrast to the 1ML sample, but they do exhibit an increase of the carrier lifetime and improved luminescence efficiency in a 1/2ML sample as compared to the 1ML sample. These findings allow us to deduce that the InAs dots in situ formed after submonolayer InAs coverage deposition leads to localization of excitons, which plays a key role in the highly improved luminescence efficiency of submonolayer InAs/GaAs structures.

References

[1] J. Y. Marzin and J. M. Gerard, Phys. Rev. Lett. 62, 2172(1989).

[2] O. Brandt, L. Tapfer, R. Cingolani, K. Ploog, M. Hohenstein, and F. Phillipp, Phys. Rev. B 41, 12 599(1990).

[3] J. Lee, K. Kudo, S. Kuniyoshi, K. Tanaka, Y. Makita, and A. Yamada, J. Cryst. Growth 115, 164(1991).

[4] M. Ilg, R. Nötzel, K. Ploog, and M. Hohenstein, Appl. Phys. Lett. 62, 1472(1993).

[5] P. D. Wang, N. N. Ledentsov, C. M. Sotomayer Torres, P. S. Kop'ev, and V. M. Ustinov, Appl. Phys. Lett. 64, 1526(1994).

[6] O. Brandt, G. C. LaRocca, A. Heberle, A. Ruiz, and K. Ploog, Phys. Rev. B 45, 3803(1992).

[7] V. Bressler-Hill, A. Lorke, S. Varma, P. M. Petroff, K. Pond, and W. H. Weinberg, Phys. Rev. B 50, 8479(1994).

Influence of DX centers in the $Al_xGa_{1-x}As$ barrier on the low-temperature density and mobility of the two-dimensional electron gas in GaAs/AlGaAs modulation-doped heterostructure

Bin Yang, Zhan-guo Wang, Yong-hai Cheng, Ji-ben Liang, Lan-ying Lin, Zhan-ping Zhu, Bo Xu, and Wei Li

(Laboratory of Semiconductor Materials Science, Institute of Semiconductors, Chinese Academy of Sciences, Beijing 100083, China)

Abstract In GaAs/AlGaAs modulation-doped heterostructure, adopting triangular quantum well approximation and including the seven major scattering mechanisms, we considered the existence of the DX centers in the $Al_xGa_{1-x}As$ barrier and calculated the dependence of low-temperature two-dimensional electron gas (2DEG) density and mobility on spacer layer thickness, Al composition and Si-doping concentration of the $Al_xGa_{1-x}As$ barrier. The calculated results explained the experimental results that cannot be explained by the previous studies. Our calculations demonstrate that DX centers in the $Al_xGa_{1-x}As$ barrier play an important role in determining low-temperature 2DEG density and mobility.

It is well known that extremely high low-temperature electron mobility can be achieved in GaAs/AlGaAs modulation-doped heterostructures (MDH).[1,2] Such superior low-temperature transport properties attracted great attention in the fields of both fundamental physical studies[3] and high-speed device implementations.[4] To approach maximum mobility in GaAs/AlGaAs MDH, besides making extensive experimental studies, people have expended great efforts to theoretically calculate the dependence of low-temperature two-dimensional electron gas (2DEG) density and mobility on spacer-layer thickness(d),[5,6] Al composition(x),[6] Si-doping concentration(N_D),[7] and doping profile,[8] etc. But up to this date, the experimental dependence of low-temperature 2DEG density on Al composition and Si-doping concentration has not yet been successfully explained by the calculated results.[6,7,9] As for the optimum Al composition for maximum mobility, several

原载于：Appl. Phys. Lett., 1995, 66(11): 1406-1408.

authors[1, 2, 10] published their experimental results, but the mechanism is not clear so far.

Our recent study discovered that the large discrepancies between calculation and experiment in the previous studies[6, 7] mainly resulted from one thing: the authors had not taken into account the existence of the DX centers in the $Al_xGa_{1-x}As$ barrier and their influence on low-temperature 2DEG density and mobility. Because these deep centers[11] hardly ionize at low temperature in the dark, neglecting their existence and their influence on 2DEG density will result in large discrepancies. In this letter, we took into account the existence of these DX centers. Adopting a triangular quantum well approximation and including the seven major scattering mechanisms in GaAs/AlGaAs MDH, we systematically calculated the dependence of low-temperature 2DEG density and mobility on Al composition (x), Si doping concentration (N_D), and spacer layer thickness. Our calculated results successfully explained the experimental results[1, 2, 10] that cannot be explained by the previously calculated results.[6, 7]

Through experiments, Schubert and Ploog[11] concluded that when Si is doped into $Al_xGa_{1-x}As$, in the range of $0<x\leqslant 0.2$, only a shallow donor level is introduced ($E_D \approx$ 6meV); but when $0.2<x\leqslant 0.4$, a deep donor level (the level of the DX centers) is also introduced ($E_{dd}=140\pm 10$meV), i. e., the shallow donors and the DX centers coexist in $Al_xGa_{1-x}As$ in this case. Because these deep donors result in persistent photoconductivity (PPC) effect in both GaAs/AlGaAs MDH and Si-doped $Al_xGa_{1-x}As$ epilayer ($x>0.2$),[1, 11, 12] so we think that in GaAs/AlGaAs MDH, the DX centers play an important role in determining low-temperature 2DEG density. We also noticed that Schubert[8] observed the influence of DX centers on the low-temperature 2DEG density in GaAs/AlGaAs MDH in his experiment. So in this letter, we use N_l and N_d to denote the shallow donor and the deep donor concentration, respectively, then $N_D=N_l+N_d$, N_D is the total Si-doping concentration in the $Al_xGa_{1-x}As$ barrier.

Approximately, we use the following equation to calculate the shallow donor and the deep donor concentration in the Si-doped $Al_xGa_{1-x}As$ barrier according to the experimental results reported in Ref. [11]:

$$\frac{N_l}{N_D}=1-4.75(x-0.2),\quad 0.2\leqslant x\leqslant 0.4. \tag{1}$$

In this letter, the low-temperature ($T=5$K) 2DEG density and mobility are discussed. Because at 5K in the dark, the shallow donors do not freeze out, but the deep donors hardly ionize,[11] approximately, we suppose only shallow donors contribute to 2DEG density. Let E_F denote the Fermi energy level of the GaAs/AlGaAs 2DEG system: in our calculation, when $T=5$K in the dark, we suppose E_F is pinned at the shallow donor level ($E_D=6$meV);

N_r denotes the remote ionized impurity concentration, then $N_r \leqslant N_I$.

Adopting a triangular quantum well approximation, Fig. 1 shows the energy band diagram of the GaAs/AlGaAs MDH. Let V_b and x denote the barrier height and Al composition, respectively, then

$$V_b = \frac{70}{100} \times 1.247x = 872.9x. \tag{2}$$

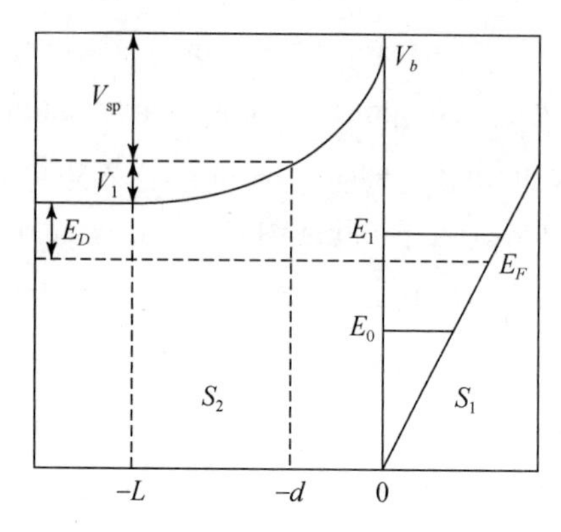

Fig. 1　Energy band diagram of GaAs/AlGaAs MDH. S_1 is GaAs layer, S_2 is $Al_xGa_{1-x}As$ layer, E_0, E_1, and E_F are the ground state subband energy level, first excited state subband energy level, and the Fermi energy level, respectively

The potential energy drop on the spacer layer(d) is denoted as V_{sp},

$$V_{sp} = \frac{e^2 d(N_s + N_{depl})}{\epsilon_0 \epsilon_2}. \tag{3}$$

where ϵ_2 is the dielectric constant of $Al_xGa_{1-x}As$($\epsilon_2 = 12.1$), N_s is 2DEG density, N_{depl} is the fixed space charge in the GaAs layer, and e is electron charge. Let $N_{depl} = 5\times10^{10}\text{cm}^{-2}$ in our calculation.[13] The potential energy drop within the Si-doped $Al_xGa_{1-x}As$ layer is denoted as V_1,

$$V_1 = \frac{e^2(N_s + N_{depl})^2}{2\epsilon_0 \epsilon_2 N_r}. \tag{4}$$

When the 2DEG system is in equilibrium with the donors in the $Al_xGa_{1-x}As$ layer, we have the equation:

$$V_b = E_0 + \frac{\pi \hbar^2 N_s}{m} + E_D + V_1 + V_{sp}, \tag{5}$$

where E_0 (Ref. [14]) is the 2DEG ground state subband energy level calculated by the triangular quantum well approximation, m is the electron effective mass, $m = 0.067m_0$, E_D is the binding energy of the shallow donors measured from the bottom of the $Al_xGa_{1-x}As$ conduction band. From Eqs. (1)-(5) and E_0,[14] we have the relation of x, d, N_s, N_r, and N_D, thus the dependence of 2DEG density and mobility on d, x, N_s, and N_D can be calculated.[13] When only the ground state subband is occupied by electrons, 2DEG mobility[5, 13] is limited by deformation potential, piezoelectric acoustic, polar optical, remote ionized impurity, background ionized impurity, alloy disorder, and interface roughness scattering.[6, 12] Including these seven major scattering mechanisms, both our calculated temperature and 2DEG density dependence of 2DEG mobility[12, 13] are in good agreement with our experimental data. In order to eliminate any ambiguities, AlGaAs/GaAs interface roughness is supposed to be the same, $\Delta = 2.83$Å, $\Lambda = 6$Å, in this letter. Influence of interface roughness scattering on 2DEG mobility is reported in another paper.[12]

Fig. 2 shows the calculated dependence of 2DEG density on spacer layer thickness for $0.2 \leqslant x \leqslant 0.4$. Obviously, 2DEG density decreases with increasing spacer layer thickness. When $x = 0.4$, N_s has the smallest value. This is because when $x = 0.4$, most of the Si atoms become deep donors, only the small amount of shallow donors contribute to low-temperature 2DEG density. For larger spacer layer thickness ($d > 200$Å), when $x \sim 0.35$, 2DEG density is higher than the others. We grew GaAs/AlGaAs MDH by MBE, with $N_D = 5 \times 10^{17} \text{cm}^{-3}$, $x = 0.3$; when $d = 150$Å and $d = 250$Å, respectively, the measured 2DEG densities at 5K are $5 \times 10^{11} \text{cm}^{-2}$ and $3.4 \times 10^{11} \text{cm}^{-2}$, respectively. Obviously, our calculated results for $x = 0.3$ are in agreement with these experimental data. The results in Fig. 2 are also in agreement with Hiyamizu's experimental results.[15] The previous authors did not consider the existence of DX centers. They roughly supposed that the Fermi level was pinned at the donor level whose binding energy was estimated to be 50meV (Ref. [6]) or 100meV,[16] thus resulting in large discrepancies. At room temperature or when 2DEG density is saturated with illumination at low temperatures, we suppose E_F is pinned at the deep donor level ($E_{dd} \approx$ 140meV). Then we obtained approximately the same results reported in Ref. [9].

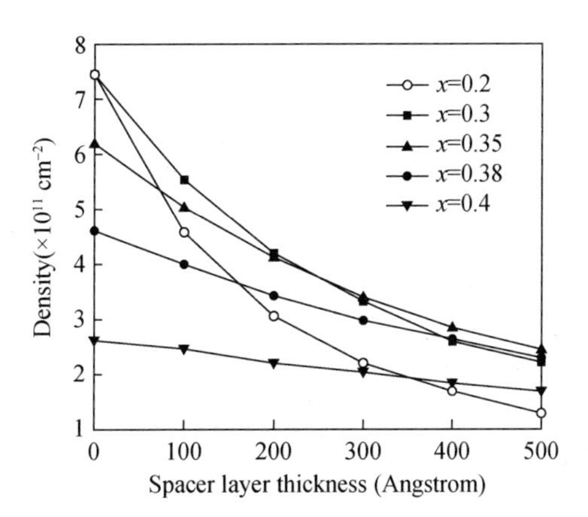

Fig. 2 Calculated dependence of 2DEG density on spacer layer thickness, parameters used in our calculation are background ionized impurity concentration $N_b = 2.5 \times 10^{14}$ cm^{-3}, $N_D = 5 \times 10^{17} \text{cm}^{-3}$, $T = 5$K

The calculated dependence of 2DEG mobility on spacer layer thickness for $0.2 \leqslant x \leqslant 0.4$ is shown in Fig. 3. Obviously, for $d<200$Å, when $x \sim 0.38$, 2DEG mobility is larger than the others; this result is in agreement with the experimental results by Drummond et al. [10] Drummond et al. fabricated a series of GaAs/AlGaAs MDH by MBE. The spacer layer thickness of their samples was 75Å. Their results demonstrated that when $x = 0.38$, the maximum 2DEG mobility was obtained. Fig. 3 also shows that for $d>200$Å, when $x \sim 0.35$, 2DEG mobility is larger than the others. This result is in agreement with experimental results of Pfeifier[1] and Saku[2] In Fig. 3, when $x = 0.2$ or $x = 0.4$, 2DEG mobility is much smaller. This is because when $x = 0.2$, the barrier height is smaller ($V_b \approx 200$meV) which results in strong alloy disorder scattering; [12] 2DEG mobility is small. When $x = 0.4$, because of the influence of the DX centers, 2DEG density is much smaller (Fig. 2), which results in very strong ionized impurity scattering, and 2DEG mobility is also small. So Al composition (x) is a very important parameter in GaAs/AlGaAs MDH. In the range of $0.2 \leqslant x < 0.4$, larger values of x enhance the barrier height, which is good for obtaining higher low-temperature 2DEG density and mobility; but larger values of x also enhance the amount of DX centers, which are unfavorable for obtaining high low-temperature 2DEG density and mobility. Both these two factors should be considered in choosing the proper value of x. The previous authors did not consider the existence of the DX centers in their calculation, so their calculated results cannot explain the experimental results.

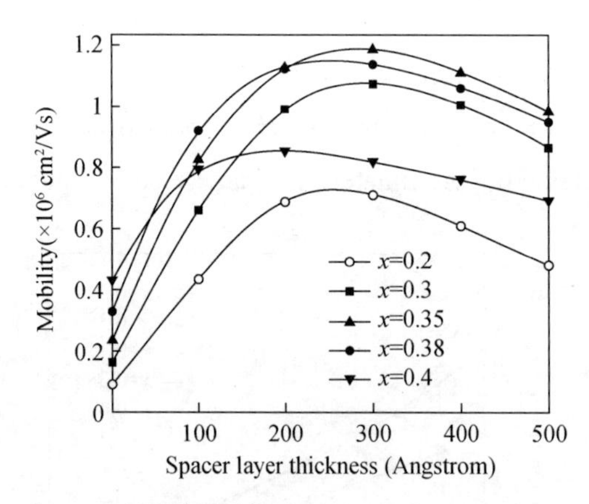

Fig. 3 Calculated dependence of 2DEG mobility on spacer layer thickness. Parameters used in our calculation are background ionized impurity concentration $N_b = 2.5 \times 10^{14}\text{cm}^{-3}$, $N_D = 5 \times 10^{17}\text{cm}^{-3}$, $T = 5$K

It is known from Fig. 4 that for $0.2 \leqslant x \leqslant 0.38$, when $1 \times 10^{16}\text{cm}^{-3} < N_D < 1 \times 10^{18}\text{cm}^{-3}$, 2DEG mobility increases steeply with increasing N_D. But when $N_D > 1 \times 10^{18}\text{cm}^{-3}$, 2DEG mobility levels off with increasing N_D. These results show the same dependence of 2DEG mobility on Si-doping concentration as Saku's experimental results. [2] Our calculation demonstrated that when 1×10^{16} cm^{-3}<N_D<1×10^{18}cm^{-3}, 2DEG density increases steeply with increasing N_D,[13] so the screening effect of 2DEG to ionized impurity scattering increases with increasing N_D, which results in increasing 2DEG mobility. But when N_D>1×10^{18}cm^{-3}, 2DEG density is high enough(N_s>3.5×10^{11}cm^{-2}) to make interface roughness scattering and alloy disorder scattering begin to effectively scatter the 2DEG,[6, 12] which limits the further increase in 2DEG mobility. It was reported[2] when N_D>1×10^{18} cm^{-3}, parallel conduction becomes noticeable, so we should choose $N_D\leqslant1\times10^{18}$cm^{-3}.

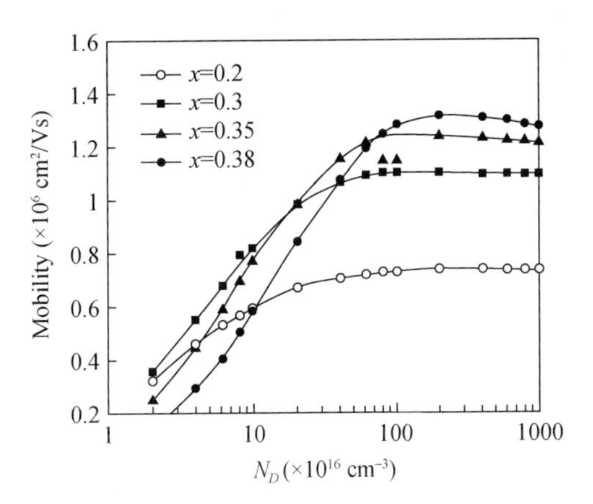

Fig. 4 Calculated dependence of 2DEG mobility on Si-doping concentration in $Al_xGa_{1-x}As$ barrier. Parameters used in our calculation are background ionized impurity concentration $N_b=2.5\times10^{14}$ cm^{-3}, $d=300$Å, $T=5$K

In conclusion, the influence of the DX centers on low-temperature 2DEG density and mobility is studied and the dependence of low-temperature 2DEG density and mobility on 2DEG structure parameters is calculated in this letter. Our calculated results are in good agreement with experimental results and demonstrate that the deep donors in the AlGaAs barrier play an important role in determining the low-temperature 2DEG density and mobility in GaAs/AlGaAs MDH.

References

[1] L. Pfeiffer, K. W. West, H. L. Stormer, and K. W. Baldwin, Mater. Res. Soc. Symp. Proc. 45, 3(1989).

[2] T. Saku, Y. Hirayama, and Y. Horikoshi, Jpn. J. Appl. Phys. 30, 902(1991).

[3] D. C. Tsui and H. L. Stormer, IEEE J. Quantum Electron. QE-22, 1711(1989).

[4] T. Minura, Surf. Sci. 113, 454(1982).

[5] W. Walukiewicz, H. E. Ruda, J. Lagowski, and H. C. Gatos, Phys. Rev. B 30, 4571(1984).

[6] T. Ando, J. Phys. Soc. Jpn. 51, 3900(1982).

[7] K. Lee, M. Shur, T. J. Drummond, and H. Morkoc, J. Appl. Phys. 54, 2093(1983).

[8] E. F. Schubert, L. Pfeiffer, K. W. West, and A. Izabelle, Appl. Phys. Lett. 54, 1350(1989).
[9] E. F. Schubert, Doping in Ⅲ-Ⅴ Semiconductors(Cambridge University Press, New York, 1993), 398.
[10] T. J. Drummond, W. Kopp, R. Fischer, and H. Morkoc, J. Appl. Phys. 52, 1028(1982).
[11] E. F. Schubert and K. Ploog, Phys. Rev. B 30, 7021(1984).
[12] B. Yang, Y. h. Cheng, Z. g. Wang, J. b. Liang, Q. w. Liao, L. y. Lin, Z. p. Zhu, B. Xu, W. Li, Appl. Phys. Lett. 65, 3329(1994).
[13] B. Yang, Y. h. Cheng, Z. g. Wang, J. b. Liang, Q. w. Liao, L. y. Lin(unpublished).
[14] T. Ando, A. B. Fowler, and F. Stern, Rev. Mod. Phys. 54, 437(1982).
[15] S. Hiyamizu, J. Saito, K. Nanbu, and T. Ishikawa, Jpn. J. Appl. Phys. 22, L609(1983).
[16] F. Stern, Appl. Phys. Lett. 43, 974(1983).

Photoluminescence studies on very high-density quasi-two-dimensional electron gases in pseudomorphic modulation-doped quantum wells

Wei Li, Zhanguo Wang, Aimin Song, Jiben Liang, Bo Xu, Zhanping Zhu, Wanhua Zheng, Qiwei Liao, and Bin Yang

(Laboratory of Semiconductor Materials Science, Institute of Semiconductors, Chinese Academy of Sciences, Beijing 100083, China)

Abstract Photoluminescence studies on highly dense quasi-two-dimensional electron gases (2DEGs) in selectively Si δ-doped GaAs/$In_{0.15}Ga_{0.85}As$/$Al_{0.25}Ca_{0.75}As$ quantum wells ($N_s = 4.24 \times 10^{12} cm^{-2}$) are presented. Five well-resolved photoluminescence lines centered at 1.4194, 1.4506, 1.4609, 1.4695, and 1.4808eV were observed, which are attributed to the recombinations of 2DEG with holes in the $n=1$ heavy-hole sub-band. The sub-band separations clearly exhibit the feature of the investigated quantum wells structure. The linewidths of these peaks are in the range of 2.2–3.4meV, indicating the high quality of the structures. The reason for discrepancy between the calculated and observed linewidth is given. Their dependence on the excitation intensity and temperatures are also discussed.

Modulation-doped structures are of great interest in the study of the quasi-two-dimensional electron gas (2DEG) in both transport and optical experiments as well as device applications, because of the high mobility achieved by separating the electrons from the ionized donors. [1] Moreover, these structures provide a nearly ideal system for the study of many-body effects. The 2DEG at the interface of GaAs/$Al_xGa_{1-x}As$, for example, made the discovery of the fractional quantum Hall effect possible and has been used to observe new quantum phenomena. [2] Studies of the electron-electron interaction effects can be extended to three-dimensional electron-gas systems.

Photoluminescence (PL) spectroscopy is a useful nondestructive method which can provide valuable information on the Coulomb interactions of the photoexcited holes with 2DEG as well as crystalline quality of the structures, etc. The optical hole created by the

原载于：J. Appl. Phys. , 1995, 78(1): 593–595.

photoexcitation is used as a probe of the physical properties of the equilibrium electron system, which is of interest both in theory and experiment. Fritze et al. investigated wide, parabolic (Al, Ga)As quantum well(QWs) structures with low electron-gas density($N_s = 2.5\times10^{11}cm^{-2}$) and observed a broad PL band attributed to the sub-band exciton emission. [3] In order to further study many-electron-one-hole exciton effects, in this paper we designed a modulation-doped asymmetric quantum well structure to realize high-density quasi-2DEG ($N_s = 4.24\times 10^{12}cm^{-2}$) and strong electron-hole overlap. Five well-resolved PL lines originating from 2DEG have been observed. We further provide a linewidth analysis of these lines and discuss their origin.

The samples employed in this study were grown in a Riber 32P molecular-beam epitaxy system on semi-insulating (100) GaAs substrates. The epitaxial layer structure typically consists of a 1μm undoped GaAs buffer, a 15nm $In_{0.15}Ga_{0.85}As$ channel, a 2nm undoped $Al_{0.25}Ga_{0.75}As$ spacer, Si δ-doping of $6.0\times10^{12}cm^{-2}$, a 30nm undoped $Al_{0.25}Ga_{0.75}As$ barrier, and a Si-doped n^+-GaAs cap layer. A reference sample with the same structure but undoped was also grown for comparison. The Van Der Pauw method was employed to measure the electrical properties of the samples at 77 and 300K. Photoluminescence measurements were made between 10 and 77K using excitation from the 514.5nm line of an Ar-ion laser. The luminescence was analyzed by a grating monochromator, detected using a lock-in technique with a cooled GaAs photocathode photomultiplier.

Fig. 1 shows the PL spectra from the sample with a sheet carrier concentration of $4.24\times 10^{12}cm^{-2}$ at various excitation intensities. Under excitation of $0.6W/cm^2$, several weak luminescence lines appear between the two intensive peaks centered at 1.4197 and 1.4936eV, respectively. These lines become steadily stronger and discernible with enhancement of excitation power density, while the PL intensities of the two lines centered at 1.4868 and 1.4936eV reach the saturation values soon. As the sample temperature is raised (see Fig. 3), these two peaks disappear in the PL spectra. This is the typical behavior of extrinsic luminescence from the transition related to impurities or defects. The PL lines centered at 1.4868 and 1.4936eV are emitted from the GaAs buffer layer, which are probably carbon impurity related to the free to bound(FB) and DAP transitions, respectively.

In Fig. 1 it can be seen that the PL intensities of the lines at 1.4194, 1.4506, 1.4609, 1.4695, and 1.4808eV become steadily stronger (nearly linearly) with increasing excitation intensity, exhibiting the striking free excitonlike feature despite the presence of an electron Fermi sea. For comparison, the PL spectra measured for the undoped GaAs/InGaAs/AlGaAs quantum well structures with same structure only show three peaks attributed to (e, A^0), (DAP), and excitons emission. Therefore, it can be concluded that these intrinsic

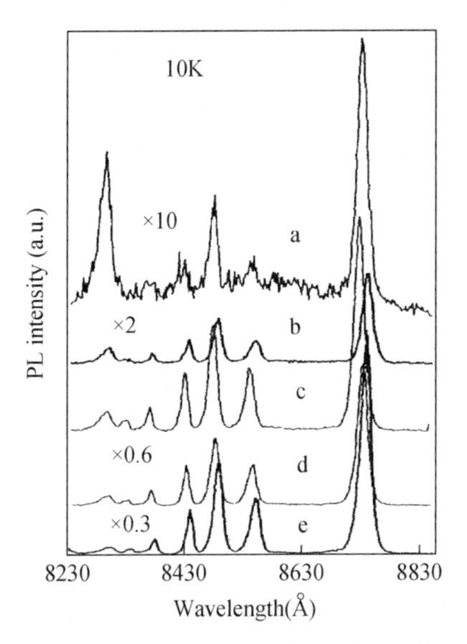

Fig. 1 PL spectra measured at 10K for the sample under various excitation intensities: I=(a)0.6,(b)0.9,(c)2.0 and 3.6 and (d)8.0W/cm^2

PL signals originate from the radiative recombination of high-density electron gas in the quantum well with the photoexcited holes.

We have performed self-consistent calculations (solving the Schrödinger and Possion equations simultaneously) of the sub-band structure to determine the energy levels and the results are shown in Fig. 2. The sub-band separations are as follows: $E_2-E_1=41.2$, $E_3-E_2=14.5$, $E_4-E_3=10.3$, and $E_5-E_4=11.6$meV. Because of the self-consistent electric field, photoexcited hole wave functions indeed are predominantly located at the GaAs/InGaAs interface. At weak excitation the recombination process only involves the photocreated holes in the 1Hh sub-band. The sharp and intense peak at 1.4194eV is assigned to E_1H_1 transition, and the others are 1.4506(E_2H_1), 1.4609(E_3H_1), 1.4695(E_4H_1), and 1.4808eV(E_5H_1) peaks, respectively. From the positions of the E_1H_1, E_2H_1, E_3H_1, E_4H_1, and E_5H_1, we can give the optically determined energy separations in terms of these sub-band energies as $E_2-E_1=31.20$, $E_3-E_2=10.30$, $E_4-E_3=8.60$, and $E_5-E_4=11.30$meV, which clearly exhibit the feature of the quantum well structure investigated. To our knowledge, it is the first time such quantum effects in PL spectra have been demonstrated. While the general agreement of these energies with the calculated values justifies the overall spectroscopic interpretation, the finite discrepancies between the two sets of energies are unsurprising due to the influence of electron-hole interaction on the sub-band energies as well as many-body effects.

From Fig. 1 the PL spectra show that the intensity of the E_3H_1 peak is higher than that of E_2H_1. This can be explained by the larger overlap between E_3 and H_1 wave functions. In

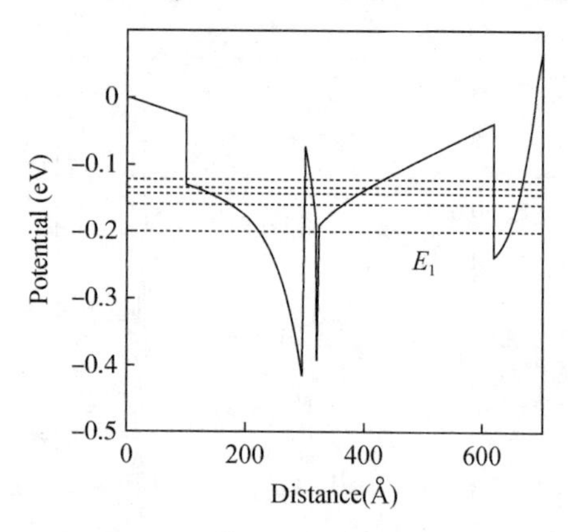

Fig. 2　Calculated conduction-band diagram of the structures and sub-band energy levels (dotted lines)

addition, it is noteworthy that the photoluminescence signals exhibit surprising narrow linewidths and symmetric line shapes. The full widths at half maximum (FWHMs) are in the range of 2.2–3.4meV. Feng and Spector have theoretically calculated the linewidth due to scattering of the screened electron-hole pairs by the electron gas.[4] Their results can be expressed as FWHM (meV) = $0.0611 + 7.43 \times 10^{-12} N_s$, which predicts a much larger linewidth than observed. The reason for the discrepancy is that they have not considered the effect of carriers populated in higher sub-bands on the exciton linewidth. Generally, the linewidth broadening mechanisms of the low-temperature PL spectra are of various kinds: homogeneous, thermal, and inhomogeneous. The homogeneous broadening depends on the 2DEG density, since the *e-h* pairs are scattered by the 2DEG. Interface roughness, fluctuations in the alloy composition, crystalline defects, etc. cause inhomogeneous broadening of the luminescence line shape in the quantum wells.[5] These effects lead to hole localization, resulting in the breakdown of the *k*-conservation selection rule and broad luminescence lines. Therefore it can be reasonably deduced that narrower linewidths of the luminescence peaks from recombination between 2DEG and photoexcited holes are the result of less hole localization in real space (i. e., the *k* values of the electrons and holes involved in *e-h* pairs emission are near $k = 0$). In other words, a narrower linewidth of the PL line attributed to 2DEG indicates better alloy uniformity and lower interface related effects.

The PL spectra at various temperatures of the sample are shown in Figs. 3 and 4. As the temperature is raised from 10K, only the intrinsic luminescence lines exist in the PL spectra, and the luminescence intensities of these peaks show substantial decay, demonstrating a strong temperature sensitivity of the many-electron-one-hole excitons. At 77K the luminescence line of E_1H_1 is still observable. When enhancing excitation intensity, the PL signals exhibit a broadband. This can be explained in consideration of the scattering of the screened *e-h* pairs by the electron gases.

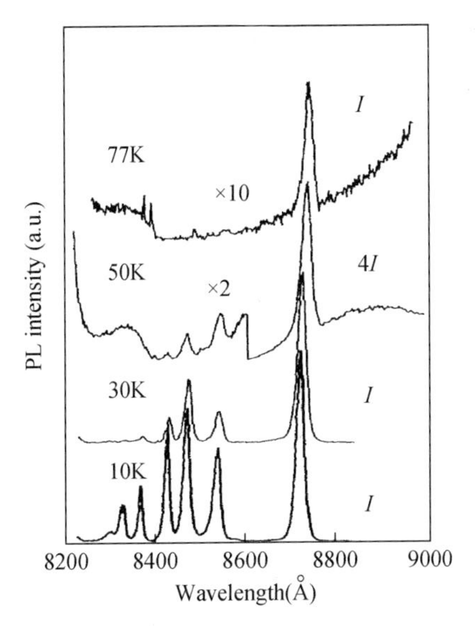

Fig. 3 Temperature-dependent PL spectra for the sample, excitation density $I=3.6\mathrm{W/cm^2}$

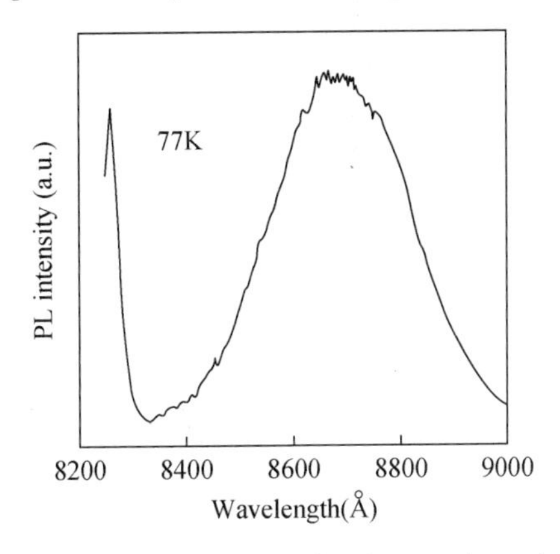

Fig. 4 PL spectrum measured at 77 K for the sample excited at $38\mathrm{W/cm^2}$

In conclusion, PL spectroscopy is a useful technique to investigate the optical properties of 2DEG. The five well-resolved PL peaks originating from 2DEG have been observed. The sub-band separations clearly exhibit the feature of the investigated QWs structure. The very sharp PL lines with linewidths in the range of 2.2–3.4 meV have been achieved, indicating the high crystalline quality of our samples.

References

[1] H. Peric, B. Jusserand, D. Richards, and B. Etinne, Phys. Rev. B 47, 12722(1989).

[2] D. C. Tsui and H. L. Störmer, IEEE J. Quantum Electron. 22, 1711(1986).

[3] M. Fritze, W. Chen, A. V. Nurmikko, J. Jo, M. Santos, and M. Shayegan, Phys. Rev. B 45, 8408(1985).

[4] Y. P. Feng and H. N. Spector, IEEE J. Quantum Electron. 24, 1677(1988).

[5] R. Cingolani, Y. H. Zhang, R. Rinaldi, M. Ferrara, and K. Ploog, Surf. Sci. 267, 457(1992).

Ordering along ⟨111⟩ and ⟨100⟩ directions in GaInP demonstrated by photoluminescence under hydrostatic pressure

Jianrong Dong, Zhanguo Wang, Dacheng Lu, Xianglin Liu,
Xiaobing Li, Dianzhao Sun, Zhijie Wang, and Meiying Kong
(Laboratory of Semiconductor Materials Science, Institute of Semiconductors, Chinese Academy of Sciences, P. O. Box 912, Beijing 100083, China)
Guohua Li
(National Laboratory for Superlattices and Microstructures, Institute of Semiconductors, Chinese Academy of Sciences, P. O. Box 912, Beijing 100083, China)

Abstract Photoluminescence of GaInP epilayers under hydrostatic pressure is investigated. The Γ valley of disordered GaInP shifts sublinearly upwards with respect to the top of the valence band with increasing pressure and this sublinearity is caused by the nonlinear dependence of lattice constant on the hydrostatic pressure. The Γ valleys of ordered GaInP epilayers rise slower than that of the disordered one. Considering the interactions between the Γ valley and folded L and X valleys, the pressure dependence of the band gap of ordered GaInP is calculated and fitted. The results demonstrate that not only ordering along ⟨111⟩ directions but also sometimes simultaneous ordering along ⟨111⟩ and ⟨100⟩ directions can occur in ordered GaInP.

GaInP is an attractive material for the applications to electronic, light-emitting devices and high-efficiency solar cells.[1] It has been found that the band gap of GaInP grown by metalorganic vapor phase epitaxy (MOVPE) sometimes is smaller than that of GaInP grown by liquid phase epitaxy (LPE) with the same composition.[2] The red-shift is due to the ordering of group Ⅲ atoms in GaInP as confirmed by transmission electron diffraction (TED) patterns.[3] In the ordered structure, the elements Ga and In occupy the alternate (111) atomic planes of group Ⅲ sublattice forming a GaP/InP monolayer natural superlattice in the direction of ⟨111⟩. The formation of this structure depends on the growth conditions and substrate orientation.[3-5] The ordering in GaInP leads to changes in the band structure and a reduction of band gap.[6, 7] Hydrostatic pressure can produce large variations in the electronic

---

原载于：Appl. Phys. Lett., 1996, 68(12): 1711-1713.

band structure preserving the crystal symmetry. Thus photoluminescence (PL) under hydrostatic pressure provides a powerful means to study the band structure of ordered GaInP. Kobayashi et al. found that the band gap of ordered GaInP had a nonlinear dependence on hydrostatic pressure and they interpreted the nonlinearity to be a result of repulsion between Γ and L valleys.[8] Uchida et al. found that the X valleys were lowered by the same amount as the Γ valley, indicating the existence of ordering in GaInP along the ⟨100⟩directions.[9]

In this letter, the photoluminescence of GaInP under hydrostatic pressure is investigated. The relationships between the band gaps of ordered GaInP and pressure are calculated and fitted using first-order perturbation theory, and the results show that ⟨100⟩ ordering in ordered GaInP sometimes occurs, as found by Uchida et al.[9]

The undoped GaInP epilayers were grown on semi-insulating GaAs substrates. Samples A and B were grown by atmospheric pressure MOVPE on(001)GaAs at temperatures of 650 and 610℃, respectively, and at a V/Ⅲ ratio of 70. Sample C was grown using low-pressure MOVPE on(001)GaAs at a temperature of 705℃ and a V/Ⅲ ratio of 200. The sample D was grown by gas-source molecular beam epitaxy (GSMBE) on (111) GaAs at the temperature of 500℃. In all samples, a GaAs buffer layer of some 200nm in thickness was grown prior to the growth of GaInP epilayers. The GaInP epilayer thicknesses are in the range of 1–2μm. The lattice mismatch between GaInP epilayer and the GaAs substrate was confirmed by double-crystal X-ray diffraction(DCXD)to be less than 2×10^{-3} for all samples. The Ga compositions of $Ga_xIn_{1-x}P$ epilayers for samples A, B, C, and D were determined to be 0.496, 0.508, 0.525, and 0.514, respectively. PL measurements were carried out at room temperature with a diamond-anvil high-pressure cell in combination with a micro-optical system. The samples were thinned down to about 20μm thick, cleaved into pieces of about $100\times100\mu m^2$ in size and then a piece of sample was loaded together with a small ruby chip as the pressure sensor into the gasket hole. A 4 : 1 methanol-ethanol mixture was used as the pressure-transmitting medium. The 488 nm line of an Ar ion laser was employed as the excitation source with an incident power of 200mW. The emissions were dispersed by an HRD-2 double-grating monochromator and detected by a cooled GaAs photocathode photomultiplier tube.

At room temperature and atmospheric pressure, the PL peak energies of samples A, B, C, and D are predicted to be 1.892, 1.905, 1.918, and 1.911eV, respectively, according to the relationship between PL peak energy and Ga composition.[10] The measured PL peak energies of samples A, B, C, and D at atmospheric pressure are 1.788, 1.811, 1.860, and 1.903eV, less than the corresponding calculated values by 104, 94, 58, and 8meV, respectively. Therefore, ordering occurs in samples A, B, and C. In contrast to them, sample D is in disordered state. For convenience, hereafter we will take the PL peak energies as the corresponding band gaps without any correction.

Shown in Fig. 1 are the PL spectra of sample D at different pressures. It can be seen that the PL peak energy shifts to higher energy with increasing pressure. Evident asymmetry of the PL line shape can be observed when the pressure is increased to 3.9GPa, indicating that the PL peak is a superposition of two peaks. In most Ⅲ – Ⅴ compound semiconductors, increasing pressure primarily causes an increase in the Γ and L conduction bands while inducing a small lowering in the X band. In Fig. 1, the weak negative pressure coefficient of lower energy peak (labeled X) suggests that it is due to an indirect transition between the X conduction band minima and the valence band states. Since the Γ valley is lower than the L valleys in energy, the higher energy peak (labeled Γ) comes from the recombination of electrons in the Γ valley and the holes at the top of the valence band. The Γ and X valleys are expected to cross when the pressure is increased to a certain pressure because they have positive and negative pressure coefficients, respectively. The decrease in intensity of peak Γ is due to the transfer of electrons from Γ valley to X valleys. PL spectra of samples A, B, and C under high pressures are similar to those of sample D. The pressure dependence of PL peak energy for the four samples are shown in Fig. 2. It is noticed that the energy of peak Γ has a sublinear dependence on pressure whether GaInP epilayers are ordered or disordered. The relationship between the energy of peak Γ and pressure can be fitted using a quadratic dependence. However, the dependence of the energy of peak X on pressure is often fitted by a linear relation. We summarize in Tab. 1 the energies at atmospheric pressure $E(0)$, linear and quadratic pressure coefficients (a) and (b) obtained by a least-squares fit to the data.

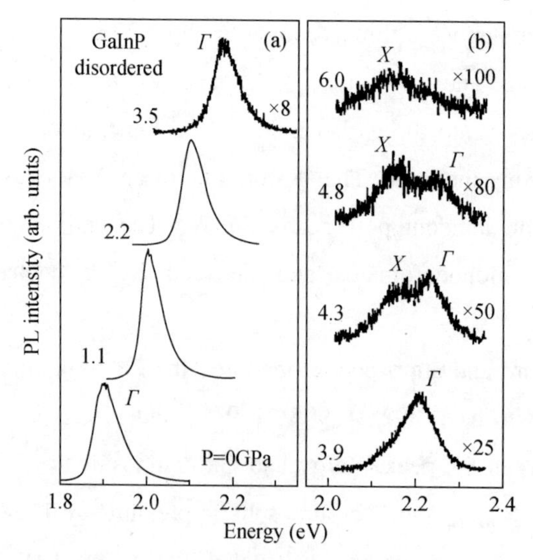

Fig. 1 PL spectra of disordered GaInP at room temperature at different pressures

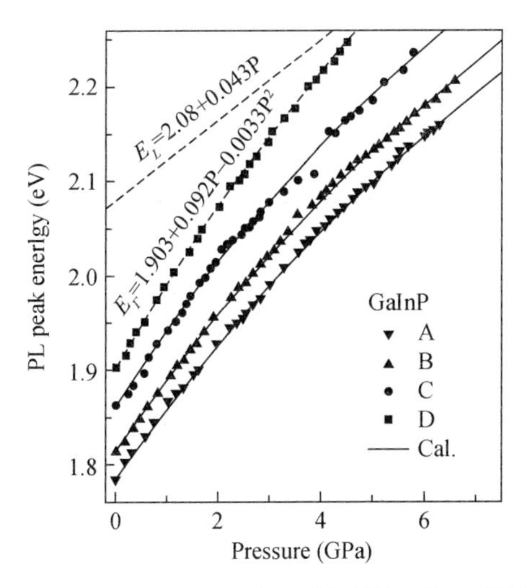

Fig. 2 Pressure dependence of PL peak energies of GaInP. The solid lines are the calculated pressure dependence of the energies of peak Γ for samples A, B, and C, and the dotted lines are the E_Γ and E_L of disordered GaInP

The sublinear pressure dependence of the energy of peak Γ likely arises from the nonlinear compression of the lattice constant under high pressure. The pressure dependence of lattice constant can be expressed by the Murnaghan's equation: [11]

$$P = (B_0/B'_0)[(a/a_0)^{3B'_0} - 1], \quad (1)$$

where B_0 is the bulk modulus with the same units as P, B'_0 is the derivative of B_0 with respect to P, a_0 the lattice constant of crystal at atmospheric pressure, and a the lattice constant at high pressures. Here, the B_0 of GaInP is determined by interpolating between the values of GaP [112. 7GPa (Ref. [12])] and InP [72. 5 GPa (Ref. [13])]. No measurements of B'_0 seem to be reported in the literature. In view of the near constancy of the parameter B'_0 for semiconductor materials, we take the B'_0 of GaP(4.79) (Ref. [12]) as that of GaInP. For sample D, we calculated the $\Delta a/a_0$ according to the Murnaghan's equation and plotted the $E_\Gamma \sim \Delta a/a_0$ curve. A straight line was obtained, indicating that the sublinear dependence of E_Γ on P completely results from the nonlinear relation between P and $\Delta a/a_0$. However, for the other three samples the $E_\Gamma \sim \Delta a/a_0$ relationship is still sublinear. Since ordering occurs in these samples, the nonlinear dependence of band gap on $\Delta a/a_0$ is probably related to the modified band structure of ordered GaInP.

In ordered GaInP epilayers, a GaP/InP monolayer superlattice is formed and the folding of L valleys to the Γ point in the Brillouin zone is induced by ordering. Here, we denote the folded L valley by L_{c1}. The repulsion between the Γ and L_{c1} valleys results in the down shift of the Γ valley and, hence, the reduction in band gap of ordered GaInP. [14] This situation can be treated by first-order perturbation theory, and the energies of the Γ and L_{c1} valleys can be expressed as follows:

Table 1　Data related to the band gap of GaInP at high pressures; band gaps E_Γ and E_X at atmospheric pressure, pressure coefficients obtained by a least-squares fit to the experimental data using $E=E(0)+\alpha P+\beta P^2$, and the interaction potentials V_L and V_X

Sample	$E_\Gamma(0)$ (eV)	α_Γ (meV/GPa)	β_Γ (meV/GPa^2)	$E_X(0)$ (eV)	α_X (meV/GPa)	V_L (eV)	V_X (eV)
A	1.788	75	-2.5	2.131	-8	0.191	—
B	1.811	81	-3.2	2.135	-8	0.151	—
C	1.860	83	-3.4	2.145	-8	0.101	0.011
D	1.903	92	-3.3	2.203	-10	—	—

$$E_\pm = 1/2\{(E_L + E_\Gamma) \pm [(E_L - E_\Gamma)^2 + 4V_L^2]^{1/2}\}, \tag{2}$$

where E_Γ and E_L are the energies of Γ and L conduction band minima from the top of the valence band, respectively, in disordered GaInP; E_+ and E_- are the energies of the L_{c1} and Γ conduction band minima, respectively, when considering the interaction between Γ and L_{c1} valleys; and V_L is the interaction potential between Γ and L_{c1}. E_L of disordered $Ga_{0.5}In_{0.5}P$ is 2.08eV.[15] No pressure dependence of E_L is reported, and we obtain the pressure coefficient α_L to be 43meV/GPa under the assumption that the ratio of α_L/α_Γ in GaInP is the same as that in GaAs(50/107).[16] V_L can be determined using the E_- of ordered GaInP, and the E_Γ(1.903eV) and E_L of disordered GaInP at atmospheric pressures by Eq. (2), and the results are given in Tab. 1.

Assuming that the V_L does not vary with pressure, then we can calculate the pressure dependence of E_- according to the E_Γ and E_L of disordered GaInP. The calculated results are given in Fig. 2 (solid lines). It is found that the calculated results are in good agreement with the experimental values for samples A and B. This confirms that the folding of L valleys to the Γ point indeed occurs. For sample C, there are discrepancies between the calculated and experimental results, especially in the pressure range of 3–4 GPa. In the calculations, only the Γ–L interaction is considered. The band gap reduction of 58meV in sample C is much smaller than the theoretically predicted value of 260meV,[6] thus the ordering in sample C along the ⟨111⟩ directions is only partial. On the other hand, Uchida et al.[9] found the evidence of ordering in the ⟨100⟩ directions in ordered GaInP. As a consequence, ordering along the ⟨100⟩ directions in sample C is possible. Considering this, we fit the $E \sim P$ relation of sample C by the similar method as mentioned above, using the Γ–X interaction potential V_X as the fitting parameter. The Γ–X interaction potential V_X = 0.011eV is obtained and the result is shown in Fig. 3. The good agreement between fitting curves and experimental data indicates that ordering in sample C is not only along ⟨111⟩ but also along the ⟨100⟩ directions. Since V_L =

0.101eV and V_X=0.011eV, in sample C the ordering in the ⟨100⟩ directions is weak, but it cannot be neglected in explaining the pressure dependence of band gap of ordered GaInP.

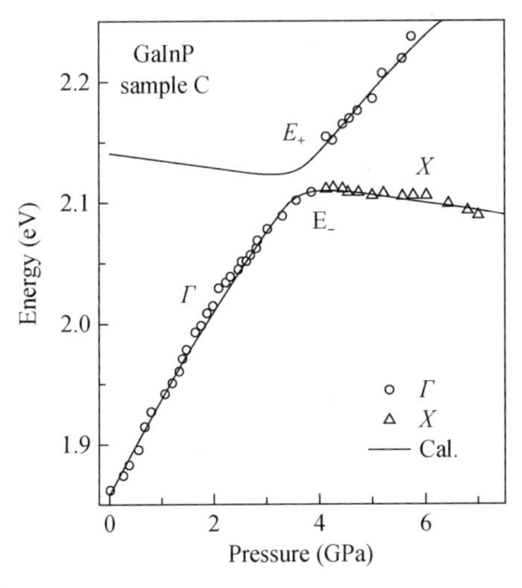

Fig. 3 Pressure dependence of PL peak energies of sample C. The solid lines are the fitted curves considering the Γ–X interaction

The last point worth noting is that even in the ordered GaInP with only ⟨111⟩ ordering, the corresponding X valleys are lower than those of the disordered GaInP, evidencing that the position of X valleys is also affected by the folding of L valleys to the Γ point.

In summary, we have investigated the photoluminescence of GaInP under high pressure. Considering the interactions between the Γ valley and the folded L and X valleys, the pressure dependence of band gap of ordered GaInP are calculated and fitted using simple theory. The results demonstrate that ordering in ordered GaInP along the ⟨111⟩ and ⟨100⟩ directions can simultaneously occur.

The authors are grateful to Du Wang, Xiaohui Wang, Jianping Li, Lingxiao Li, and Shirong Zhu for assistance in the growth of samples. The authors also thank Weibin Gao for DCXD measurements. This work is funded by National Natural Science Foundation of China.

References

[1] K. A. Bertness, S. R. Kurtz, D. J. Friedman, A. E. Kibbler, C. Kramer, and J. M. Olson, Appl. Phys. Lett. 65, 989(1994).

[2] A. Gomyo, T. Suzuki, K. Kobayashi, S. Kawata, and I. Hino, Appl. Phys. Lett. 50, 673(1987).

[3] M. Kondow, H. Kakibayashi, and S. Minagawa, J. Cryst. Growth 88, 291(1988).

[4] A. Gomyo, K. Kobayashi, S. Kawata, I. Hino, T. Suzuki, and T. Yuasa, J. Cryst. Growth 77, 367(1986).

[5] S. Minagawa and M. Kondow, Electron. Lett. 25, 758(1989).

[6] S. -H. Wei and A. Zunger, Appl. Phys. Lett. 56, 662(1990).
[7] L. C. Su, I. H. Ho, N. Kobayashi, and G. B. Stringfellow, J. Cryst. Growth 145, 140(1994).
[8] T. Kobayashi, M. Ohtsji, and R. S. Deol, J. Appl. Phys. 74, 2752(1993).
[9] K. Uchida, P. Y. Yu, N. Noto, and E. R. Weber, Appl. Phys. Lett. 64, 2858(1994).
[10] G. B. Stringfellow, J. Appl. Phys. 43, 3455(1972).
[11] F. D. Murnaghan, Proc. Natl. Acad. Sci. USA 30, 244(1944).
[12] Numerical Data and Functional Relationship in Science and Technology, edited by O. Madelung, H. Weiss, and M. Schulz, Landolt-Börnstein, New Series(Springer, Heidelberg, 1982), Vol. 17a.
[13] R. Trommer, H. Müller, M. Cardona, and P. Vogl, Phys. Rev. B 21, 4869(1980).
[14] S. R. Kurtz, J. Appl. Phys. 74, 4130(1993).
[15] G. D. Pitt, M. K. R. Vyas, and A. W. Mabbitt, Solid State Commun. 14, 621(1974).
[16] D. J. Wolford and J. A. Bradley, Solid State Commun. 53, 1069(1985).

Influence of the semi-insulating GaAs Schottky pad on the Schottky barrier in the active layer

J. Wu, Z. G. Wang, and L. Y. Lin

(Laboratory of Semiconductor Materials Science, Institute of Semiconductors, Chinese Academy of Sciences, Beijing 100083, China)

C. B. Han

(General Research Institute of Non-ferrous Metals, Beijing 100088, China)

M. Zhang and S. W. Bai

(13th Research Institute of MMEI, Shijiazhuang 050051, China)

Abstract The influence of the sidegate voltage on the Schottky barrier in the ion-implanted active layer via the Schottky pad on the semi-insulating GaAs substrate was observed, and the mechanism for such an influence was proposed.

The semi-insulating GaAs substrate in the GaAs technology, based on metal-semiconductor field-effect transistors (MESFETs), plays a role in providing isolation between devices. However, as the semi-insulating property of the substrate is achieved by impurity compensation by deep traps and their occupation can be altered electrically or optically, some problems arise. The crosstalk between GaAs MESFETs through the substrate is a major obstacle to the use of the densely packed GaAs digital integrated circuits (ICs). The essential phenomenon of the crosstalk between devices is the so called sidegating effect: the drain current of the MESFET is modulated by applying a voltage to an adjacent device. The sidegating effect is caused by the modulation of the space charge region at the substrate-active channel interface by the bias voltage applied to the substrate.[1] However, the influence of the Schottky pad lying directly on the substrate of the MESFET is recently recognized.[2-6] Miller and Bujatti[2] proposed that the oscillations in substrate leakage current can induce periodic voltage fluctuation on the MESFET gate via the gate pad contact on the substrate. Inokuchi et al.[3] found a large leakage current between the Schottky contact and sidegate. Wu et al.[5] observed that the negative voltage applied to the substrate modulated the Schottky barrier on the *n* channel. In this letter, the experimental results are presented to

原载于: Appl. Phys. Lett., 1996, 68(18): 2550-2552.

show that the negative voltage applied to the substrate modulates the Schottky barrier of the gate via the gate pad on the substrate, and a simple mechanism for this modulation is proposed.

The sidegating effect and C-V profiling were measured in the configuration shown in Fig. 1. The active layer was formed by the Si ion-implantation at the energy of 140keV with a dose of $4\times10^{12}\ cm^{-2}$. The n^+ region for the ohmic contacts were formed by the ion implantation at 100keV with does of $1\times10^{13}cm^{-2}$. The FET processing consisted of Au/Ge/Ni contacts for the drain and source, the selective ion implantation with O^+ and B^+ for electrical isolation, a constant gate recess by chemical etching, and Al gate deposited by evaporation. The FEE had a gate length of 1μm and gate width of 80μm. The Al Schottky contact for C-V profiling is $90\times70\mu m^2$ in size. Three sidegates (SGs) of ohmic contact were made. Both of SG1 and SG2 were separated from the n channel with a distance of 20μm. However, SG2 was placed near the Schottky pad on the substrate and the gap between them is just 5μm. During the conventional C-V profiling of the Schottky contact on the n channel, 1MHz capacitance was measured and a variable negative voltage from 0 to −24V was applied to SG1 or SG2. In the measurement of the sidegating effect, the gate was grounded or floating, the source was grounded and the drain was biased at a constant voltage $V_D=3V$. The sidegate voltage was applied by biasing SG2 or SG3 with a voltage ramp from 0 to −15 V at a scanning speed of 0.13/s. The drain current I_{DSS} of the FET was measured as a function of V_{SG}. Fig. 2(a) and 2(b) show the Schottky barrier capacitance versus the bias voltage V_{Sch} applied between the Schottky and ohmic contacts at various V_{SG} applied to SG1 and SG2, respectively. It can be seen from Fig. 2(a), as V_{SG} applied at SG1 increases, C decreases, as reported earlier.[5] At first, the C-V curve shifts downwards very slightly, and as V_{SG} exceeds about 16V, the downward shift is remarkable with V_{SG} at SG1 increasing. In the case of applying V_{SG} to SG2, which is much closer to the Schottky pad on the substrate than SG1, the C-V curve shifts downwards much more rapidly with increasing V_{SG} than the case of applying V_{SG} to SG1.

Fig. 3 shows the variations in I_{DSS} with V_{SG} when the gate was grounded and floating, respectively. Curve 1, usually observed in experiments on sidegating, represents I_{DSS} as a function of V_{SG} applied to the SG3 with the gate grounded and it can be seen that I_{DSS} begins to decrease rapidly as V_{SG} exceeds about 9V, which is the threshold voltage for the onset of the sidegating. However, I_{DSS} begin to decrease at $V_{SG}=0$ as V_{SG} increases when the gate is floating, as shown in curve 2. There is a turning point in curve 2 at about 9V, as V_{SG} exceeds the turning point, I_{DSS} decreases more rapidly with increasing V_{SG}. If V_{SG} is applied to SG2 which is closer to the Schottky pad and much further away from the FET than SG3, the

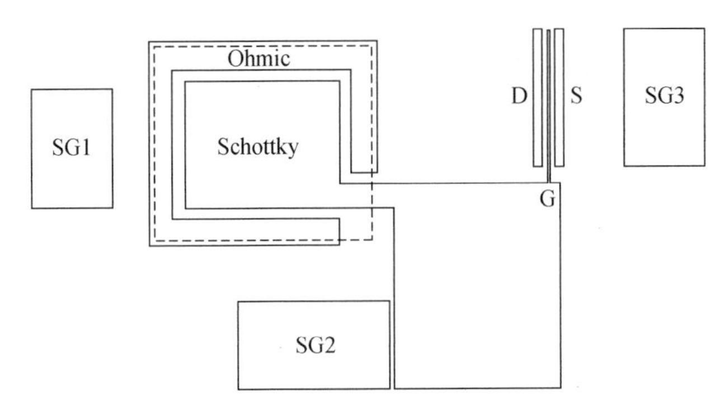

Fig. 1 The test structure for the C-V profiling and the sidegating effect measurement of MESFET. The area within the dash line is the n region

sidegate voltage has no effect on I_{DSS} when the gate was grounded even when it exceeds 15V. However, when the gate is floating, I_{DSS} decreases much more rapidly with V_{SG} applied to SG2 increasing than the case of applying V_{SG} to SG3.

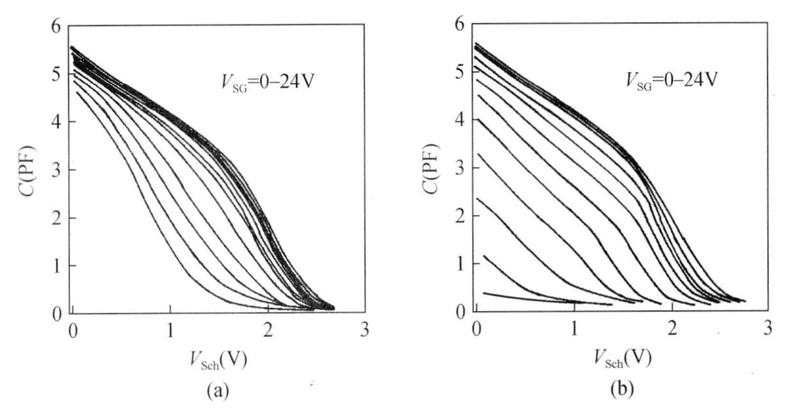

Fig. 2 The Schottky capacitance C vs the bias voltage V_{Sch} at various negative substrate voltage V_{SG}, (a) V_{SG} is applied at SG1, (b) V_{SG} at SG2. The increasing step of V_{SG} is 2V

As described above, the C-V curve shifts downwards and, when the gate is floating, I_{DSS} begin to decrease even at $V_{SG}=0$ with V_{SG} increasing. These phenomenon imply that the negative voltage V_{SG} applied to the substrate via SGs has the influence on the Schottky barrier, as if V_{SG} were applied at the Schottky contact through the substrate. As shown in Figs. 2 and 3, the closer to the Schottky pad on the substrate is SG, the more obvious is the influence, as the C-V curve shifts downwards and I_{DSS} decreases much more rapidly with increasing V_{SG} when it is applied to SG2 than to SG1 and SG3, respectively. Therefore, it can be concluded that it is the Schottky pad on the substrate through which V_{SG} influences the Schottky barrier on the active layer. This conclusion is consistent with what Miller and Bujatti observed in their experiments on low-frequency oscillations in GaAs FETs.[2] They

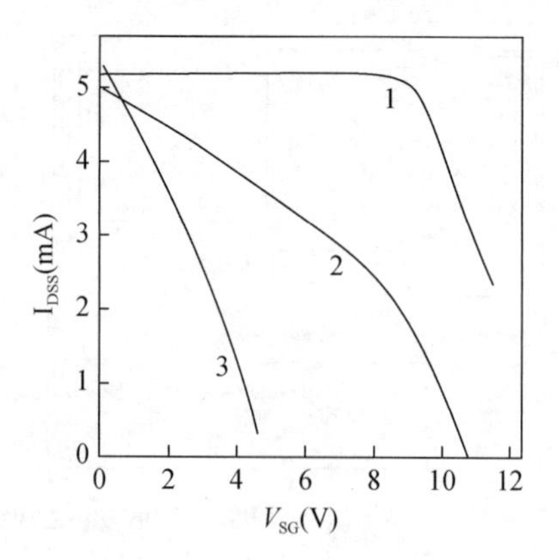

Fig. 3 The drain current I_{DSS} as a function of the sidegate voltage V_{SG}. The gate is grounded and V_{SG} is applied at SG3 for curve 1, the gate floating and V_{SG} at SG3 for curve 2, and the gate floating and V_{SG} at SG2 for curve 3

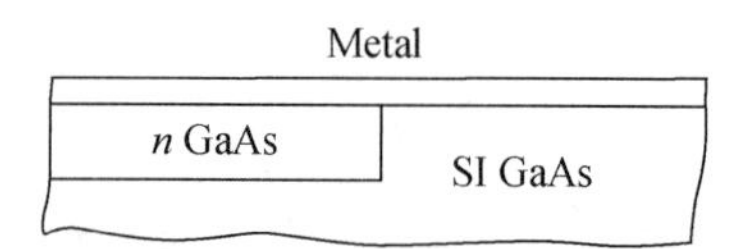

Fig. 4 The Schottky metal on the active layer and the substrate

found that the oscillation in the substrate leakage current caused the oscillation in the FET through the Schottky pad on the substrate and the substrate-active layer interface.

It should be mentioned that when the gate was grounded, the voltage between the source and gate was set at zero and, therefore, V_{SG} has no effect on the Schottky barrier of the gate.

As shown in Fig. 4, when the gate metal lies directly on the SI substrate, a Schottky barrier forms in the substrate as in the active layer,[7] and the two Schottky barriers are connected through the gate metal. If the influence of the substrate-active layer interface is ignored, the energy band diagram of the system of the two Schottky barriers together with gate metal at thermal equilibrium can be simplified to the one as shown in Fig. 5Φ, the barrier height for electrons enter into the semiconductor from the gate metal, is fixed to be approximately 0.8eV with the presence of a high density of interface states.[8, 9] As shown

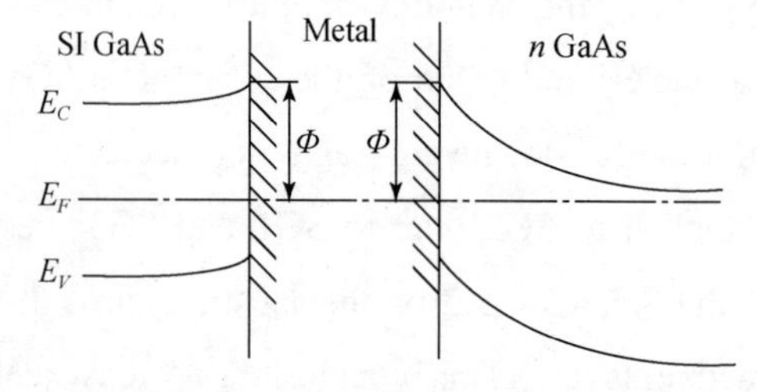

Fig. 5 The energy band diagram of the Schottky barriers in the active layer and the substrate

by Horio et al., [10] when a negative voltage V_{SG} is applying to SG, electrons are injected from it into the substrate and they will be captured by the deep donor levels, such as *EL2*. As V_{SG} increases and the electron injection proceeds, the quasi Fermi level in the substrate rises up remarkably. If such an effect is far reaching in distance, the energy band diagram shown in Fig. 5 should be influenced by V_{SG} and not be in thermal equilibrium any more. As the quasi Fermi level in the substrate is raised, more electrons will be injected from the substrate to the metal and its electrical potential becomes more negative with respect to the active layer, and, therefore, the depletion region in the active layer will be increased.

The system of the two Schottky barriers with the gate metal could be equivalent to two Schottky diodes. The application of V_{SG} makes the substrate Schottky diode forward biased and the active-layer Schottky diode reversely biased. In such a way, V_{SG} modulates the Schottky barrier on the active layer.

References

[1] C. Kocot and C. A. Stolte, IEEE Trans. Electron Devices ED-29, 1059(1982).

[2] D. Miller and Bujatti, IEEE Trans. Electron Devices ED-34, 1239(1987).

[3] K. Inokuchi, Y. S. Itoh, and Y. Sano, J. Electrochem. Soc. 137, 464C(1990).

[4] S. J. Chang and C. P. Lee, IEEE Trans. Electron Devices ED-40, 698(1993).

[5] J. Wu, Z. G. Wang, T. W. Fan, L. Y. Lin, and M. Zhang, J. Appl. Phys. (to be published).

[6] Y. Liu and R. W. Dutton, IEEE Electron Device Lett. 11, 505(1990).

[7] J. F. Wager and A. J. Mccamant, IEEE Trans. Electron Devices ED-34, 1001(1987).

[8] L. J. Brillson, J. Phys. Chem. Solids 44, 703(1983).

[9] S. M. Sze, Physics of Semiconductor Devices(Wiley, New York, 1981).

[10] K. Horio, T. Ikoma, and H. Yanai, IEEE Trans. Electron Devices ED-33, 1242(1986).

Electrical properties of semi-insulating GaAs grown from the melt under microgravity conditions

Wang Zhan-guo, Lin Lan-ying, Li Cheng-ji, Zhong Xing-ru, Li Yun-yan, Wan Shou-ke, Sun Hong

(Laboratory of Semiconductor Materials Science, Institute of Semiconductors, Chinese Academy of Sciences, Beijing 100083, China)

Abstract A technologically important undoped semi-insulating(SI) GaAs single crystal was successfully grown in the Chinese recoverable satellite as far as we know for the first time by using a similar growth configuration described previously. The experimental results proved that the space SI GaAs crystals have a lower density of defects and defect-impurity complexes as well as a better uniformity.

Microgravity conductions provide a unique utilization potential for various areas of fundamental and applied research in key technologies. Growth-rate fluctuations, especially stoichiometric variations of Ⅲ-Ⅴ compound semiconductor materials caused by gravitational convections can be effectively suppressed under a reduced-gravity environment. Therefore, the reduction of native point defects, in turn, their complexes and enhancing the homogeneity of the crystal are expected which have been primarily confirmed by our early publications.

In this paper, we present a detailed study on electrical and optical properties of both the low resistivity(LR) GaAs and semi-insulating(SI) GaAs samples grown from the melt under microgravity conditions using various techniques. The behaviors of native defects, including a technologically important mid-gap deep level *EL2*, in the space-grown GaAs crystal are also reported.

Single crystal growth configuration and growth method of SI GaAs in space were the same as those used in the first flight experiment.[1] A ground-grown SI GaAs single crystal by liquid encapsulated Czochralski method with ⟨100⟩ orientation was used as a seed and charge ingot. Unfortunately, however, during the space crystal growth, a small part of the molten-zone was occasionally contacted with the quartz well which leads to a lower resistivity

原载于：Chin. Phys. lett. ,1996, 13(7): 553-556.

portion formation of the space GaAs crystal because of the silicon contamination.

The space-grown and the ground-seed GaAs single crystals were cut perpendicular to the growth direction and the wafers with the (100) face were mechanically first and then chemically polished for characterization. Optical transient current spectroscopy (OTCS) was used to characterize the behavior of deep centers in both the space-grown and the ground-grown SI GaAs samples. The uniformity of the space GaAs wafers was determined by using cathodeluminescence (CL). On the other hand, deep level transient spectroscopy (DLTS) was employed to measure the concentration of the *EL2* centers in the space-grown LR GaAs samples. The microwave field effect transistor (FET) devices were also fabricated on the Si[28] implanted spacegrown SI GaAs single crystal wafer for further examining the quality of this material.

Fig. 1 shows the typical OTCS spectra of the space-grown and Earth-grown SI GaAs samples. It can be seen that the general shapes of the OTCS spectra for both samples are similar, but the defect densities of the most deep levels for Earth-grown seed SI GaAs are higher than that of the space samples, which is in good agreement with our first space experiment. Again, the better stoichiometry of the space-grown SI GaAs crystal has been obtained. Typical CL images (CLI) of undoped SI GaAs grown in space and on Earth are shown in Figs. 2 (a) and 2 (b), respectively. It is interested to note that the cell structures are existing in both samples, however, the cell structure is significantly changed. For the space crystal, the short range fluctuations are dramatically reduced. This means that the uniformity of the space material is really improved. The fact that high gain and low noise microwave FET devices made on the Si[28] implanted space SI GaAs wafer are comparable to the best result obtained from the SI GaAs grown on Earth further confirms the better quality of the space material.

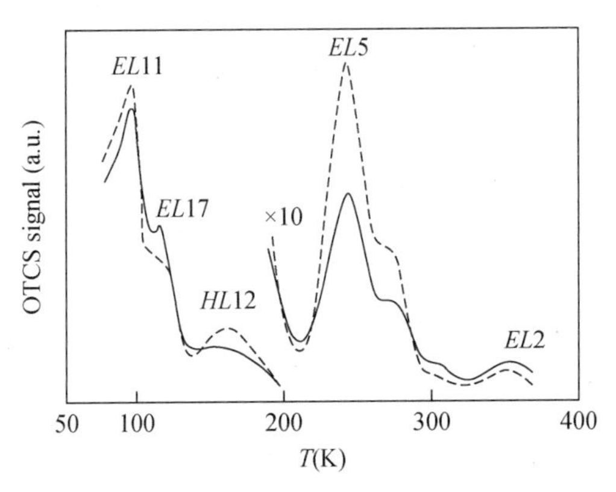

Fig. 1 Typical OTCS spectra of the spacegrown and Earth-grown SI GaAs samples. Dashed line: the ground-seed SI GaAs sample; solid line: the space-grown SI GaAs ample, rate window t_1/t_2 = 5ms/50ms

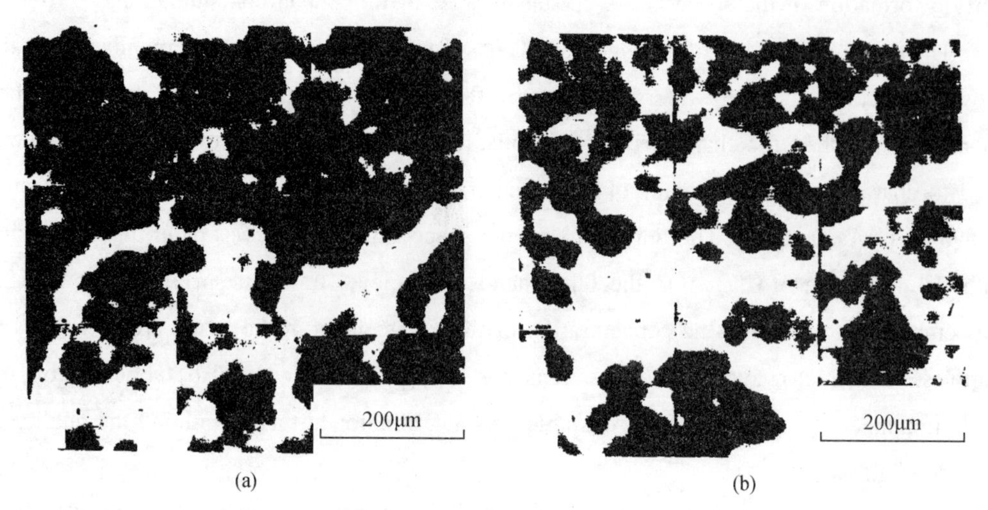

Fig. 2　CLI of SI GaAs grown in space and on Earth. (a) Undoped SI GaAs sample grown in space; (b) undoped SI GaAs seed grown on Earth

Typical DLTS spectrum performed on the space-grown LR GaAs portion is shown in Fig. 3, in which, at least, four electron trap sare observed. The energy positions in the band gap are found to be very close to the wellknown centers *EL*6, *EL*5, *EL*3 and *EL*2, and the densities of defects, including the dominating deep donor *EL*2, are a little bit smaller than that of the ground-GaAs single crystal grown by the horizontal Bridgman method from a quartz boat. However, thermal stability of the space LR GaAs material was found to be unstable. For an example, a heat treatment at 600℃ in vacuum for 30min, the space-grown LR GaAs was completely changed into an inhomogeneous high resistivity material. The origin of thermal unstability of the space-grown LR GaAs material will be published elsewhere.

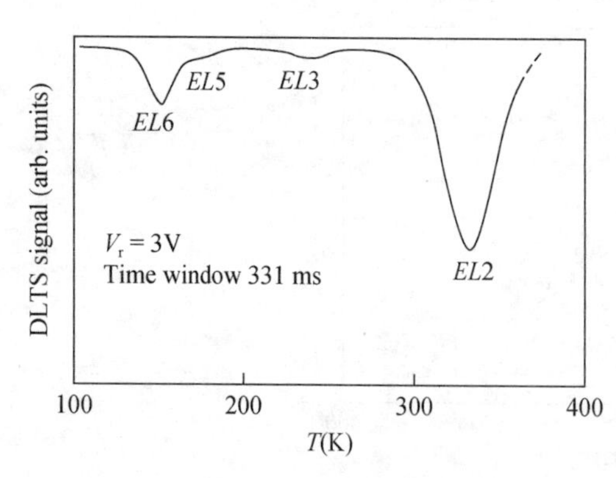

Fig. 3　Typical DLTS spectrum of the spacegrown LR GaAs sample

Is it possible to grow undoped SI GaAs in space? As is known that the native defectrelated *EL*2 center has the ability to lock the Fermi level in the mid-gap position of the band gap of GaAs, thus resulting in SI materials. This is doubted by our earlier studies on the DLTS spectrum of Te-doped GaAs crystal grown in space in which the native defect-related deep centers and their densities are reduced remarkably in space crystal. This is because the stoichiometry of the GaAs crystal can be more precisely controlled under a reduced gravity environment. The present results further confirmed this finding. Nevertheless, the density of the *EL*2 centers for the space GaAs crystal grown under a suitable arsenic vapour pressure, for an example, a slightly arsenic-rich, could be still kept high enough for completely compensating the rest shallow acceptor states, thus resulting in a high resistivity of the crystal.

This also strongly supports that the *EL*2 center is actually the simple arsenic antisite defect As_{Ga}. If the nature of the *EL*2 is a complex of the arsenic antisite defect As_{Ga} and the vacancies V_{As} or/and V_{Ga} as suggested by many authors,[2, 3] we would not be able to easily grow undoped SI GaAs crystal in space.

Typical PL spectra taken from different parts of the space GaAs wafer measured at 13K are shown in Fig. 4. Near the intrinsic region of the PL spectrum emission lines are observedat 1.512(*A*), 1.492(*B'*), 1.4827(*B*), 1.4529(*C*) and 1.406eV(*D*), which are attributed to AX, DX, FB and DAP transitions. The deep PL bands located at 1.215(*E*) and 0.942eV(*F*), which only appear in the LR GaAs sample will be discussed in detail. The spatial distribution of the PL intensity for both bands *E* and *F* across the wafer(see Fig. 4 and insert(a) in it) can be well understood according to the following arguments. As mentioned above, during the space GaAs crystal growth, a small part of the molten-zone was occasionally contacted with the quartz well, in this case silicon is obviously incorporation with GaAs. The silicon atoms incorporated in the GaAs melt near the quartz well migrate into the bulk by a pure diffusion process during the space crystal growth, this is because thermal and solutal convections are illuminated under microgravity conditions. It is well known that, silicon in bulk GaAs growth condition preferentially occupies the Ga site as a shallow donor, and it tends to reduce the concentration of Ga vacancy, $[V_{Ga}]$. The reduction of $[V_{Ga}]$ consequently results in enhancing the concentration of arsenic vacancy, $[V_{As}]$, because of $[V_{Ga}] \cdot [V_{As}] = K(T)$, where $K(T)$ is the mass action constant for the formation of Schottky pairs at a certain temperature. This indicates that $[V_{Ga}]$ and $[V_{As}]$, also Si_{Ga}, have opposite profiles in the direction from LR GaAs to SI GaAs(see insert in Fig. 4). This means that the spatial distribution tendency for Si_{Ga} and V_{As} across the wafer should be in a similar manner, i. e. higher concentration at the edge(*A*) of the LR GaAs and decreasing dramatically as the measured point reaching the transition region(*B*). On the other hand, one should keep in mind that the vacancy V_{Ga} is a dominant defect in As-rich space GaAs growth condition.

On the basis of the above information, we tentatively consider that the natures of the *E* and *F* PL bands are the complexes of $Si_{Ga}V_{Ga}$ and V_{As}, $Si_{Ga}V_{Ga}$ respectively, and the spatial distribution of the PL intensity for both bands *E* and *F* are then well explained(see insert(a)

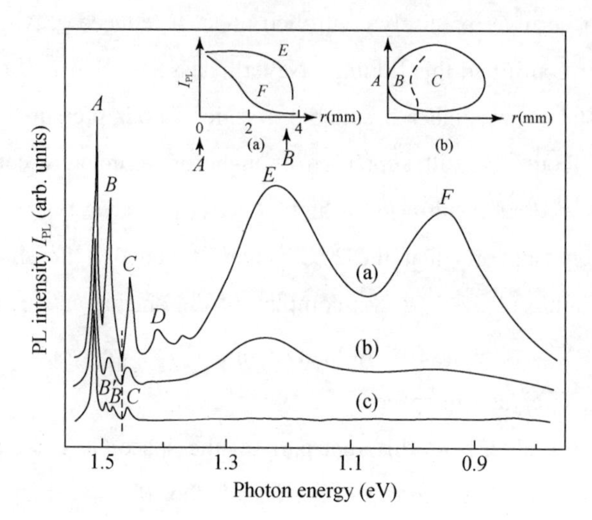

Fig. 4 Spatial distribution of PL centers across the wafer (from the LR GaAs to SI GaAs) cut perpendicular to the space GaAs crystal growth axis. (a) PL spectrum for the LR GaAs portion, (b) transition region between the LR GaAs and SI GaAs, and (c) SI GaAs portion. Upright insert (a) shows the profiles of the PL intensity of deep PL centers E and F across the wafer from *A* to *B* as shown in insert (b)

in Fig. 4). These results also directly confirm the structural models of these two PL bands proposed by Williams[4] and Wang et al.,[5] in which the 1.2eV band is associated with a localized Ga vacancy-donor complex, whereas the PL band around 1.0eV is a complex of a Ga vacancydonor-As vacancy.

In summary, we can conclude that if the crystal growth conditions are well designed in microgravity environments, they would be in favor of high quality SI GaAs crystal growth because in space thermal convection in the melt is suppressed, and the stoichiometry of GaAs single crystal can be more precisely controlled.

References

[1] Lin Lanying group and Da Daoan group, Mater. Sci. Forum, 50(1989)183.

[2] Zou Yuanxi, Zhou Jicheng, Mo Peigen, Lu Fengzhen, Li Liansheng, Shao Jiuan and Huang Lei, Inst. Phys. Conf., Ser. No. 65(1983)49.

[3] J. Lagowski, M. Kaminska, J. M. Pavsey, H. C. Gatos and W. Walukiewicz, Inst. Phys. Conf., Ser. No. 65. (1983) 41.

[4] E. W. Williams, Phys. Rev. 168(1968)992.

[5] Z. G. Wang, C. J. Li, S. K. Wan and L. Y. Lin, J. Cryst. Growth, 103(1990)38.

808nm high-power laser grown by MBE through the control of Be diffusion and use of superlattice

Donghai Zhu, Zhanguo Wang, Jiben Liang, Bo Xu, Zhanping Zhu, Jun Zhang, Qian Gong, Shengying Li

(Laboratory of Semiconductor Materials Science, Institute of Semiconductors, Chinese Academy of Sciences, Beijing 100083, China)

Abstract 808nm high-power laser diodes are grown by MBE. In the laser structure, the combination of Si-doped GRIN (graded-index) region adjacent to *n*-AlGaAs cladding layer with reduced Be doping concentration near the active region has been used to diminish Be diffusion and oxygen incorporation. As compared with the laser structure which has undoped GRIN region and uniform doping concentration for Si and Be, rsepectively, in the cladding layers, the slope efficiency has increased by about 8%. Typical threshold current density of $300A/cm^2$ and the minimum threshold current density of $220A/cm^2$ for lasers with 500μm cavity length are obtained. A high slope efficiency of 1.3W/A for coated lasers with 1000μm cavity length is also demonstrated. Recorded CW output power at room temperature has reached 2.3W.

1 Introduction

High-power laser diodes for the light source of solid-state laser pumping have been extensively investigated[1,2]. To realize high-power operation, it is needed to achieve a lower operating current and higher quantum efficiency. The performance characteristics of the laser diodes are determined by the quality of the epitaxial layer, laser structure design and the control of growth condition. Beryllium (Be) diffusion plays an important role in the performance of GaAs/AlGaAs laser diodes grown by MBE. Both increased oxygen incorporation and *p-n* junction displacement are related to Be diffusion[3]. In addition, the Be is believed to be responsible for the failure of laser diode[4]. Although MBE technique can provide better reproducible control over composition, thickness and doping profile in the direction of growth, careful treatments are needed for Be doping. In this paper, we report the MBE growth of GRIN-SCH SQW 808nm high-power laser diodes through the control of

原载于：Journal of Crystal Growth, 1997, 175/176(Part 2): 1004-1008.

Be diffusion. The application of the superlattice buffer is also included.

2 Laser structure and epitaxial growth

The MBE machine used in this experiment is a Riber 32p system. Because the high-power laser diodes operate under crude conditions, it is required to improve the quality of the laser materials. It is believed that the growth of high-quality laser structure by MBE requires a more sophisticated methods than that for MOCVD-grown lasers[5]. In our laser structure, superlattices in GaAs buffer and n-AlGaAs cladding layers were incorporated. The superlattice can bring some benefits such as the reduction in the propagation of defects from substrate, a relief of strain at the interface between high aluminum concentration cladding layer and the substrate[5]. The superlattice also offers additional defense against oxygen incorporation, makes the epitaxial layer flat and improves the quality of the lower interface of the QW[6]. The advantage of superlattice may improve the uniformity of the laser materials which is beneficial to the highpower or broad-area laser diodes. Low aluminum composition AlGaAs layer is introduced as a getter buffer to remove the transient oxygen because impurities including oxygen are released as the Mo holder is heated to higher temperatures.

Fig. 1 is the schematic diagram of GRINSCH AlGaAs SQW laser composition profiles. The laser was grown by MBE on an n^+-GaAs substrate. The growth sequence is described as follows: a 1μm thick n-GaAs buffer layer containing a five-period n-GaAs/$Al_{0.6}Ga_{0.4}As$ (10nm/10nm) superlattice buffer layer, 0.5μm thick n-$Al_{0.1}Ga_{0.9}As$ layer, 1.3μm thick n-$Al_{0.6}Ga_{0.4}As$ cladding layer including a five-period GaAs/$Al_{0.6}Ga_{0.4}As$ (10nm/10nm) superlattice; 0.2μm thick slightly Si doped $Al_xGa_{1-x}As$ confining layer (linearly graded from $x=0.6$ to 0.3), undoped 10nm thick $Al_{0.07}Ga_{0.93}As$ active layer, undoped 0.2μm thick $Al_xGa_{1-x}As$ confining layer (linearly graded from $x=0.3$ to 0.6), 1.3μm thick p-$Al_{0.6}Ga_{0.4}As$ cladding layer, and 0.3μm thick p^+-GaAs contact layer. Silicon was used as the n-type and beryllium as the p-type dopant. Growth rate was 1.2μm/h for $Al_{0.6}Ga_{0.4}As$.

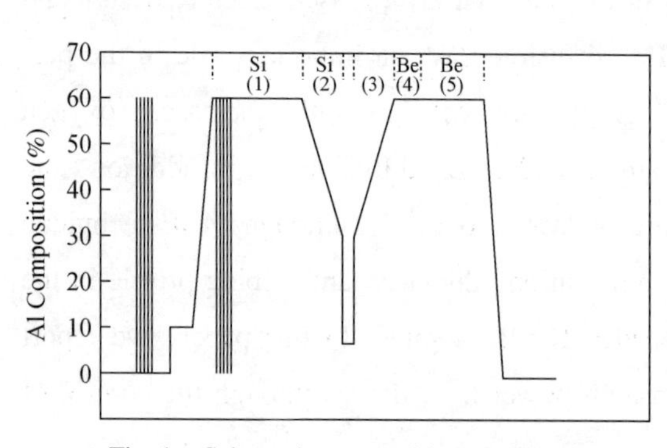

Fig. 1 Schematic composition profiles for the laser structure

Beryllium has a high diffusivity in AlGaAs/GaAs structure[7]. The unstable behavior of this dopant during crystal growth, as well as during operation of laser diodes make the control of $p-n$

junction difficult and affect the stability of lasers[8,4]. On the other hand, Naresh Chand et al.[3] reported that compared to undoped AlGaAs layer, the oxygen content is lower in the Si-doped AlGaAs layer and higher in the Be-doped one. Oxygen which forms nonradiative recombination centers degrades the performance and reliability of the laser diodes. The above research results indicate that beryllium doping and diffusion have important effects on the performance of laser diodes. Experimental analysis[3] also shows that the diffusion of Be can be retarded by increasing the doping of n-layer and doping the n-side of the GRIN region. Considering the Be, Si and undoping-related oxygen incorporation, it is reasonable to dope the GRIN region adjacent to the n-AlGaAs cladding layer with Si in order to reduce oxygen content in the active region nd avoid the p-n junction displacement (a separation of p-n junction from the active layer heterojunction).

We adopted tailoring doping profiles in the epitaxial growth. In the laser structure (type Ⅰ) as shown in Fig. 1, the Si doping concentration in the GaAs buffer and n-AlGaAs cladding layer (1) are 2×10^{18} and 1×10^{18} cm^{-3}, respectively. The GRIN region (2) adjacent to the n-AlGaAs cladding layer was doped with Si to 10^{16} cm^{-3}. GRIN region (3) was undoped. The low Be-doped region (4) (5×10^{17} cm^{-3}) in the p-AlGaAs cladding layer near the GRIN region at a range of 0.1μm was adopted. Except for the doping concentration in this region, the Be doping in the p-AlGaAs cladding layer (5) is 8×10^{17} cm^{-3}. The reason for the lower Be doping concentration near the GRIN region is as follows: First, it is found that undoped AlGaAs spacer layer (as thick as 0.2μm) is ineffective to prevent Be diffusion[13]. The Be atoms piled up in the middle of the n-side GRIN region by ~60nm from the QW region (the experimental condition is similar to ours). The distance from the beginning of the diffusion to the pile-up of the Be atoms is about 0.3μm. Our GRIN region is 0.2μm for each side. We expected that the lower Be doping region (0.1μm) plus the 0.2μm GRIN region might be helpful to prevent Be out-diffusion. Second, the reduced doping concentration proximate to the active region can enhance the external quantum efficiency in a proper laser structure[9]. In order to show the influence of the tailoring doping profiles on the laser performance, another type of laser structure (type Ⅱ) was grown. In the type Ⅱ laser structure, all the GRIN region and active layer are undoped. The n-and p-AlGaAs cladding layers are uniformly doped with Si (1×10^{18} cm^{-3}) and Be (8×10^{17} cm^{-3}). The laser structures of type Ⅰ and type Ⅱ differ only by the doping profiles and the rest of the structures is identical. More than ten growths for each structure were carried out.

3　Device fabrication and laser performance

First, 100μm wide and 500μm long uncoated lasers were fabricated in order to obtain the parameters reflecting the quality of epitaxial wafer. There is no obvious difference in the threshold current density for the two structures. The typical threshold current density is $300A/cm^2$. The minimum J_{th} is $220A/cm^2$. These values are satisfactory when compared with lasers of similar structure[10, 11], which indicates the high-quality epitaxial materials grown by MBE and fairly low-concentration of nonradiative recombination centers. However, a difference in the slope efficiency exists. Statistical data show that the slope efficiency is 0. 84 and 0. 78W/A per facet for type Ⅰ and type Ⅱ structures, respectively. Enhancement in the slope efficiency is obtained by about 8% in type Ⅰ as compared to type Ⅱ. It was reported that doping modifications had no measurable effect on external quantum efficiency for a structure in which the optical field is tightly confined[9] because there were less changes in the free-carrier absorption. If this is true, for our laser structure(type Ⅰ) in which the optical field is tightly confined[9], we consider the improvement of slope efficiency as the result of decreased concentration of nonradiative recombination center or the minimizing *p-n* junction displacement, or both, in type Ⅰ structure. The displacement allows majority carrier current flow out of the active layer. The current does not contribute to lasing[14].

Second, dielectric(SiO_2) defined metal-stripe lasers with 100, 200μm wide stripes and 1000μm cavity length were also fabricated for type Ⅰ structure. High-and low-reflectivity facet coatings were applied to each facet using SiO_2 and Si, which gives the reflectivity of 10% and 90% at the front and rear facets, respectively. Zn diffusion was performed to reduce the contact resistance. Ti/Pt/Au was used for the *p*-side contact, the *n*-side contact was formed with AuGe/Ni/Au. Laser chips were mounted on Cu heat sinks with In solder in the junction down configuration.

Typical slope efficiency of these lasers is 1W/A. Output power versus current characteristics at room temperature under CW operation are shown in Fig. 2. These lasers are different in stripe width and same in the cavity length(1000μm). Curve(a) represents one of coated lasers with 100μm stripe width. Curve(b) represents one of coated-laser with 200μm stripe width. In curve(a), laser can be operated up to 2W. In curve(b), a high slope efficiency of 1. 3W/A for the coated-front facet alone is demonstrated, recorded CW output power reached 2. 3W. The typical emission wavelength of 100μm wide stripe laser diodes under 1W output power was measured to be 808nm, as shown in Fig. 3.

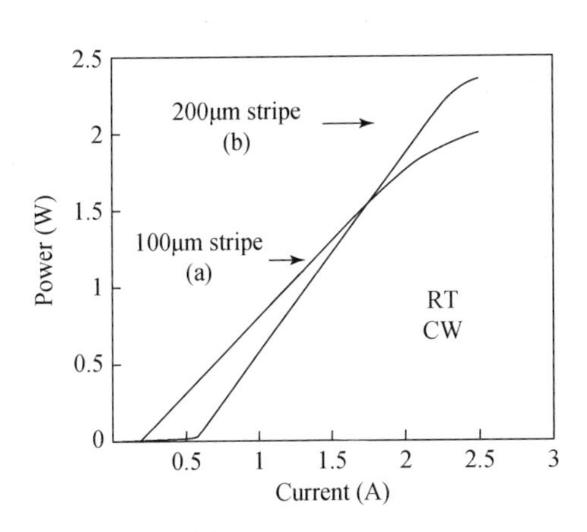

Fig. 2 Light output power vs current characteristics of the lasers with 100 and 200μm stripe width

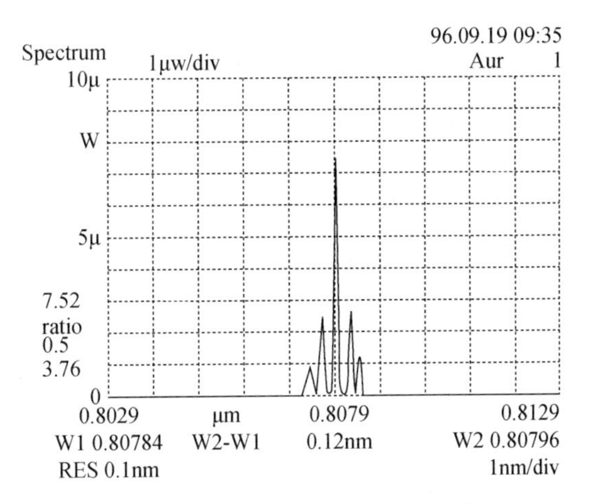

Fig. 3 Typical emission spectrum for the 100μm×1000μm laser devices under output power 1W at room temperature

The temperature dependence of CW light output versus current characteristics for 100μm stripe laser is depicted in Fig. 4. The temperature dependence of I_{th} is given by

$$I_{th} = I_0 \exp(T/T_0),$$

where I_{th} is the threshold current at the temperature T, I_0 is a constant and T_0 is the characteristic temperature. The laser has been operated up to 95℃ at 0.5W under CW operation. The characteristic temperature T_0 is as high as 185K between 35–85℃ and 163K between 85 – 95℃, which indicates the good thermal characteristic of the laser structure. Because shorter wavelength AlGaAs/GaAs lasers tend to have a small T_0 [12], this value is relatively high.

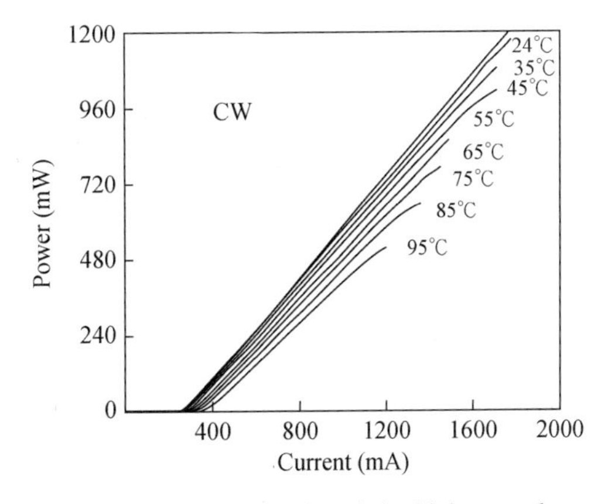

Fig. 4 Light output power vs current characteristics of the highpower laser at different temperatures under CW conditions

In summary, high-power 808nm-AlGaAs quantum-well lasers are grown by MBE. Both reduced Be doping concentration near the active region and the Si doping in the *n*-side GRIN region are used to reduce the Be diffusion and oxygen incorporation near the active region. Compared with nontailoring-doping laser structure, the enhancement of the slope efficiency is observed. The minimum threshold-current density of 220A/cm^2 is obtained. High slope efficiency of 1.3W/A is demonstrated. Recorded CW output power at room temperature has reached 2.3W.

Acknowledgements

The authors are grateful for the help from Zujie Fang, Bin Liu, Xiongwei Hu, Caizheng Jin, Xiaojie Wang and Qin Han throughout this work.

References

[1] V. P. Chaly, D. M. Demidov, G. A. Folin, S. Yu. Karpov, V. E. Myachin, Yu. V. Pogorelsdy, I. Yu. Rusanovieh, A. P. Shkurku and A. L. Ter-Martirosyan, J. Crystal Growth 150(1995)1350.
[2] I. Eliashevich, J. Diaz, H. Yi, L. Wang and M. Razeghi, Appl. Phys. Lett. 66(1995)3087.
[3] N. Chand, S. N. G. Chu, A. S. Jordan, M. Geva and V. Swaminathan, J. Vac. Sci. Technol. B 10 (1992)807.
[4] A. Jakubowicz, A. Oosenbrug and Th. Forster, Appl. Phys. Lett. 63(1993)1185.
[5] M. E. Givens, L. J. Mawst, C. A. Zmudzinski, M. A. Emanuel and J. J. Coleman, Appl. Phys. Lett. 50(1987)301.
[6] H. Imamoto, F. Sato, K. Imanaka and M. Shimura, Appl. Phys. Lett. 54(1989)1388.
[7] P. Enquist, G. W. Wisks and L. F. Eastman, J. Appl. Phys. 58(1985)4130.
[8] G. E. Kohnke, M. W. Koch, C. E. C. Wood and G. W. Wicks, Appl. Phys. Lett. 66 (1995)2786.
[9] R. G. Waters, D. S. Hill and S. L. Yellen, Appl. Phys. Lett. 52(1988)2017.
[10] Y. Nagai, K. Shigihara, A. Takami, S. Karakida, Y. Kokubo and A. Tada, IEEE Photon. Technol. Lett. 3(1991)97.
[11] C. A. Wang, H. K. Choi and M. K. Connors, IEEE Photon. Technol. Lett. 1(1989)351.
[12] L. A. Coldren and S. W. Corzine, in: Diode Lasers and Photonic Integrated Circuits(Wiley, New York, 1995)ch. 2, 58.
[13] V. Swaminathan, N. Chand, M. Geva, P. J. Anthony and A. S. Jordan, J. Appl. Phys. 72 (1992)4648.
[14] P. J. Anthony, J. R. Pawlik, V. Swaminathan and W. T. Wang, IEEE J. Quantum. Electron. QE-19(1983)1030.

Reflectance-difference spectroscopy study of the Fermi-level position of low-temperature-grown GaAs

Y. H. Chen, Z. Yang, R. G. Li, and Y. Q. Wang

(Department of Physics, Hong Kong University of Science and Technology, Clearwater Bay, Kowloon, Hong Kong, China)

Z. G. Wang

(Laboratory of Semiconductor Materials Science, Institute of Semiconductors, Chinese Academy of Sciences, Beijing 100083, China)

Abstract The results of a reflectance-difference spectroscopy study of GaAs grown on (100) GaAs substrates by low-temperature molecular-beam epitaxy (LT-GaAs) are presented. In-plane optical anisotropy resonances which come from the linear electro-optic effect produced by the surface electric field are observed. The RDS line shape of the resonances clearly shows that the depletion region of LT-GaAs is indeed extremely narrow ($\ll$200Å). The surface potential is obtained from the RDS resonance amplitude without the knowledge of space-charge density. The change of the surface potential with post-growth annealing temperatures reflects a complicated movement of the Fermi level in LT-GaAs. The Fermi level still moves for samples annealed at above 600℃, instead of being pinned to the As precipitates. This behavior can be explained by the dynamic properties of defects in the annealing process.

GaAs grown by low-temperature molecular-beam epitaxy (LT-GaAs) has recently attracted much interest due to its unique properties and potential electronic and optoelectronic applications.[1-5] This material is crystalline, with a high content of excess arsenic,[4] which agglomerates into small metallic clusters after post-growth annealing.[5,6] With the formation of arsenic precipitates the electrical and optical properties of LT-GaAs change significantly.[3-6] For example, annealing at temperatures above 600℃ usually results in LT-GaAs with extremely high resistivity. However, the mechanism of such high resistivity is still in debate. Some believe that it was the result of depletion of carriers by the Schottky barrier at the surface of the metallic As precipitates,[5,6] while others suggested a compensation model of defects similar to the conventional semi-insulating (SI) GaAs.[4] The

原载于：Phys. Rev. B, 1997, 55(12): 7379–7382.

change of carrier and defect densities inevitably changes the Fermi level, and the determination of which is therefore important in the study of defect-state evolution in post-growth annealing.

Photoreflectance (PR) is extensively used to study the surface electric field and the related Fermi level of semiconductors. However, so far no Franz-Keldysh Oscillation (FKO) produced by the electric field inside LT-GaAs was ever detected except for samples annealed at above 700℃.[7,8] We believe that this is due to the highly nonuniform electric field and the extremely narrow surface depletion region which are the results of high density of defects and traps in LT-GaAs samples.[4] As a result, the Fermi level of LT-GaAs could only be determined indirectly by PR measurements of the electric field in the adjacent normal GaAs region near a LT-GaAs/ GaAs interface.[9]

Here we report a direct measurement of the surface potential of LT-GaAs by reflectance-difference spectroscopy (RDS). Recent RDS studies of doped GaAs showed strong anisotropy resonances at the critical points E_1 and $E_1+\Delta_1$ due to the electro-optic (LEO) effect generated by the surface electric field.[10-12] In this paper we present our RDS study of a series of LT-GaAs samples annealed at different temperatures from 300 to 850℃. Similar resonances are observed for these samples at the same GaAs critical energies. The line shape of the resonances evidently show that the sign of the surface potential is the same as that of n-type conventional GaAs, and the surface space-charge region is indeed very narrow (≪200Å). This enables us to determine the value of the surface potential without a knowledge of the actual space-charge density, and the effects of post-growth annealing on the Fermi level in LT-GaAs.

The LT-GaAs samples used in this study were grown by a Riber-32p molecular-beam-epitaxy (MBE) system at 250℃ using a Ga-to-As beam flux ratio of 10 under arsenic-stable growth condition. A 1000-Å GaAs buffer layer was grown first on (001) liquid-encapsulated Czochralski SI-GaAs substrates at 580℃. The substrate temperature was then lowered to 250℃ and a 2μm thick LT-GaAs layer was grown at a growth rate of 1μm per hour. No dopants of any kind were intentionally introduced. After growth, the wafer was cut into pieces. Some of them were annealed at 300, 350, 400, 450, 500, and 600℃ for 30min in vacuum or nitrogen environment, while others were subjected to rapid thermal annealing (RTA) at 700, 800, and 850℃ for 10s in arsenic environment. All samples studied were of a high degree of crystalline, as was confirmed by double-crystal X-ray-diffraction measurements. Arsenic precipitates were observed in samples annealed at 600℃ by transmission electron microscopy. In addition, some doped conventional MBE GaAs samples of different carrier densities were grown in a separate system for comparison.

The RDS setup was essentially the same as the one employed earlier.[13] It was set in a configuration in which only the in-plane anisotropy was measured. Regular photoreflectance was also measured for all the samples. However, only the *n*-type GaAs sample grown by regular MBE showed FKO, which provided another way to determine the surface electric field.

Fig. 1 shows the in-plane anisotropy spectra of a semi-insulating (001)-oriented GaAs substrate, an *n*-type GaAs epilayer, and three LT-GaAs samples. The three LT-GaAs samples are the as-grown sample, the sample annealed at 500℃, and the sample annealed at 850℃, respectively. The principal axes of the anisotropy are along the [110] and the $[1\bar{1}0]$ directions. The optical anisotropy of the SI-GaAs sample is typical and very similar to that found in other SI-GaAs substrates with even unpolished surfaces, and undoped MBE GaAs samples. It was suggested that this optical anisotropy is due to the structure anisotropy of (001) GaAs surface,[12] even when the surface is covered by an oxide layer. Compared to that of the SI-GaAs substrate, the anisotropy spectra of the *n*-type GaAs shows an additional positive resonance near $E_1 = 2.95\text{eV}$, and a negative resonance near $E_1 + \Delta_1 = 3.15\text{eV}$. This is typical of the *n*-type GaAs samples reported previously,[12] and is resulted from the LEO effect produced by the surface electric field.[10, 11] The structures in the vicinity of E_1 and $E_1 + \Delta_1$ for the as-grown LT-GaAs samples are similar to that of the *n*-type GaAs sample. This suggests that they are also from the LEO effect. The amplitude of the LEO structures for the LT-GaAs sample annealed at 500℃ is reduced as compared to that of the as-grown one,

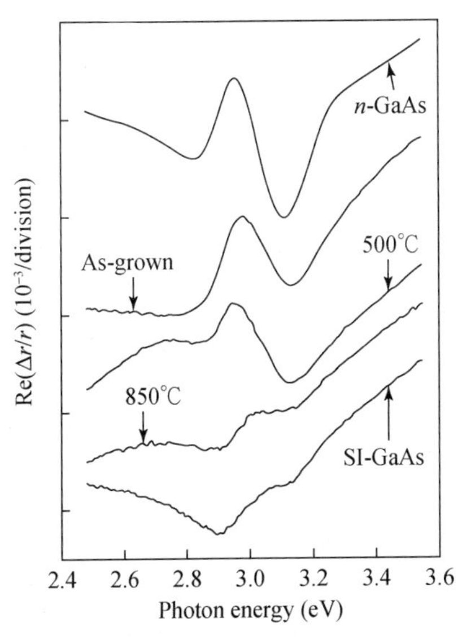

Fig. 1 The real part of the reflectance difference spectra of a semi-insulating GaAs, an *n*-type GaAs, and three LT-GaAs grown on a (001) substrate and annealed at different temperatures

and the LEO structures almost disappear for the samples after 850℃ RTA. This indicates that the surface electric field of LT-GaAs can be greatly altered by the post-growth annealing. For doped normal GaAs, the surface electric field is determined by the surface potential and the density of the space charges. To find the surface potential, the doping density must be separately measured. However, as the following discussions will show, for LT-GaAs the surface potential can be directly determined from the LEO resonance amplitude without the knowledge of the net space-charge density.

In order to see clearly the difference of the LEO part of RDS between the *n*-type GaAs and the LT-GaAs samples, we plot only their LEO component, which are obtained by subtracting the SI-GaAs RDS spectra from their RDS spectra. Typical results are shown in Fig. 2. For the *n*-type GaAs sample, the real part of RDS consists of a positive peak at E_1 and a negative peak at $E_1+\Delta_1$, and the imaginary part consists of a positive peak between the E_1 and $E_1+\Delta_1$ energies. For LT-GaAs, the real part of its RDS is similar to the imaginary part of the *n*-doped GaAs, while its imaginary part is similar to the negative real part of *n*-doped GaAs. The apparent difference between the two is a factor $-i$ which can be explained as follows. All other LT-GaAs samples show the same resonances but with different amplitudes. The resonance positions of the LT-GaAs are slightly redshifted as compared to that of the *n*-type GaAs. This is due to the lattice expansion by As interstitials.

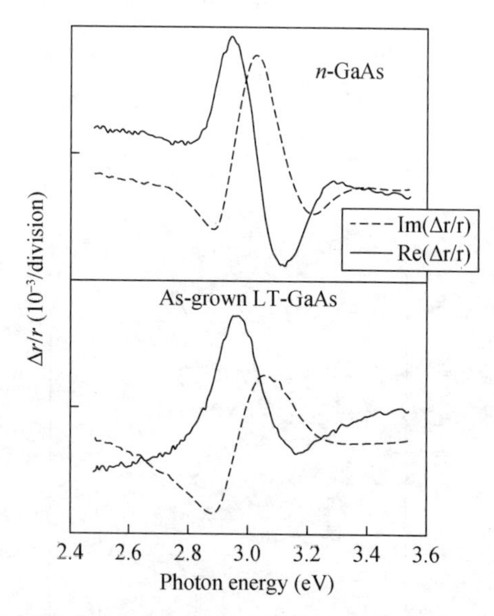

Fig. 2　The linear electro-optic part of the reflectance difference spectra of an *n*-type GaAs and the as-grown LT-GaAs sample

The LEO part of RDS measures the difference of the dielectric function along the two perpendicular directions in the surface plain. When the electric field varies with the depth,

RDS measures an averaged difference of the dielectric function in the two directions over the light penetration length $\langle \Delta\varepsilon \rangle$, which is given by[12]

$$\langle \Delta\varepsilon \rangle = -2ik\int_{-d}^{0} e^{-2ikz}\Delta\varepsilon(z)\,dz\,, \tag{1}$$

where $\Delta\varepsilon(z)$ is the LEO-induced difference of the dielectric function proportional to the electric field $F(z)$, d is the space-charge region depth, and k is the complex wave vector of light. Because $\Delta\varepsilon(z)$ is zero outside the space-charge region, and the exponential term exp $(-2ikz)$ approaches zero beyond the penetration depth, it is apparent that $\langle \Delta\varepsilon \rangle$ depends on the magnitude of the penetration depth or the depth of the space-charge region, whichever is smaller. For a photon energy of 3eV, the penetration depth is about 170Å for GaAs. For our n-type GaAs sample, the depth of the spacecharge region is about 1500Å, with the density of space charges in the range of 10^{17} cm^{-3} and the surface potential $V_b = 0.7$eV, as separately measured from the PR spectrum and the Hall data. In this case $\Delta\varepsilon(z)$ can be approximated by $\Delta\varepsilon(0)$ in Eq. (1), which leaves

$$\langle \Delta\varepsilon \rangle \cong \Delta\varepsilon(0) = \beta F_s\,, \tag{2}$$

where F_s is the surface electric field, and β is the LEO coefficient.

For the LT-GaAs the surface potential is about 0.1eV, and the space-charge density is about 10^{20} cm^{-3}.[4] This produces a narrow space-charge region with a depth on the order of 10Å, much smaller than the light penetration depth. Under this condition, the exponential term in the integral of Eq. (1) can be neglected. Equ. (1) is then converted into an integral of the electric field, since $\Delta\varepsilon(z)$ is proportional to the electric field at z based on the LEO effect. The final result can be simplified as

$$\langle \Delta\varepsilon \rangle = -2ik\beta V_b. \tag{3}$$

It is clear that the different line shape of the LEO structures of the n-type GaAs and the LT-GaAs samples is simply due to the factor $-i$. This will change the imaginary (real) part of the RDS for the n-type GaAs to the real (negative imaginary) part for the LT-GaAs, as is clearly shown in Fig. 2, since the wave vector k is mostly real. Moreover, Eq. (3) shows that the LEO resonance of LT-GaAs is determined only by the surface potential V_b, and is independent of the space-charge density.

The LEO coefficient β of (100) GaAs was obtained by comparing the surface field obtained from the PR data and the RDS peak-to-peak amplitude of the n-type sample at the E_1 and $E_1 + \Delta_1$ energies. Our results are consistent with those reported in Ref. [12]. Assuming that the LEO coefficient of the LT-GaAs is the same as normal GaAs, the effective surface electric field of the LT-GaAs samples, $2kV_b$, can then be obtained. Finally the surface potential is calculated according to Eq. (3). Fig. 3 shows the surface potential versus the annealing temperature. It is clear that the surface potential depends strongly on the annealing temperature. The surface potential decreases from 89mV in the as-grown sample, to about 67mV in the sample annealed at 350℃. For higher annealing temperatures it begins to increase, and reaches a maximum value of 90mV at 450℃, and then drops down to about 30mV for the sample annealed at 600℃. RTA at 850℃ further reduces the potential to less than 10mV. It is clear that the Fermi level is not yet pinned at 600℃. These results are consistent with the earlier work by Look et al.[4] Furthermore, RTA at 700℃ increases the surface potential at 115meV.

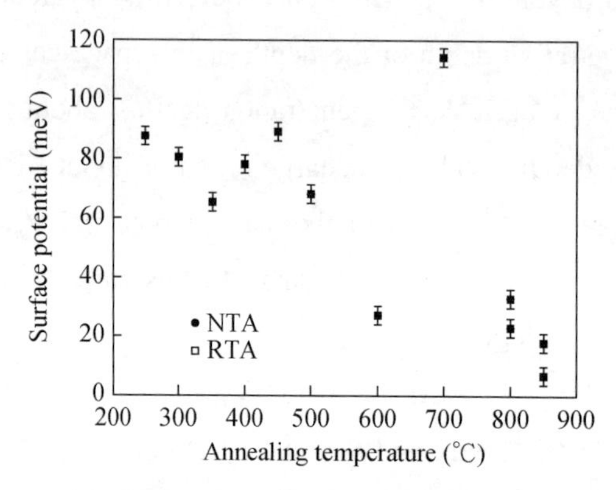

Fig. 3 The surface potential of LT-GaAs as a function of postgrowth annealing temperatures. The closed points are for thermal annealing, while the open squares are for rapid thermal annealing

Assuming that post-growth annealing does not affect the Fermi level at the surface, the change of the surface potential would then result from the shift of the bulk Fermi level in the forbidden gap of LT-GaAs. The nonmonotonic change of the Fermi level with the annealing temperature indicates that there are competing mechanisms during the annealing process, and a simple model of Fermi-level pinning to As precipitates is not adequate. It is well known that most of the excess As in LT-GaAs is in the form of defects such as As interstitials(As_i), As antisites(Aa_{Ga}), Ga vacancies(V_{Ga}), or small complexes of these defects. As_{Ga} serves as a deep donor near the midgap, V_{Ga} is believed to be a deep acceptor at 0.28eV above the valence band,[14] while As_i may not be electrically active but expands the lattice of LT-GaAs. The Fermi level is usually several kT above the position of As_{Ga} depending on the ratio of the density of As_{Ga} to the net density of acceptors below the midgap. When the ratio decreases, more As_{Ga} are ionized, leaving a lower Fermi level and a lower surface potential. During the annealing at 300–350℃, only As_i is mobile.[15] It moves to form As precipitates and release the strain. The processes($As_{Ga}\rightarrow As_i+V_{Ga}$), ($As_i-V_{Ga}$ complex$\rightarrow As_i+V_{Ga}$) and (As_i-As_{Ga} complex$\rightarrow As_i+As_{Ga}$) take place to balance the decrease of As_i. These processes will change the As_{Ga} concentration slightly, while increasing that of V_{Ga}, thus increasing the V_{Ga}/As_{Ga} ratio. As a result, the Fermi level moves away from the conduction band, and leaves a lower surface potential. At temperatures of 400–450℃, V_{Ga} becomes mobil. There are several possibilities for it. The first is that V_{Ga} assists the diffuse of As_{Ga} to form As precipitates and finally being absorbed by the Ga atoms expelled by the precipitates.[16] The second possibility is that V_{Ga} collides with As, and forms the complex $V_{As}-As_{Ga}$, which is probably a donor of 0.17eV below the conduction band.[14] The third is just to form vacancy clusters.[17] These processes will all cause the V_{Ga} density to decrease, which causes the Fermi level to shift back to the conduction band, and increases the height of the surface potential. When the annealing temperature is higher than 500℃, most of As_{Ga} moves to form the As precipitates, which moves the Fermi level to the midgap again. The change of the surface potential therefore comes from the increase and decrease of V_{Ga} and As_{Ga} at different annealing temperatures. It was observed previously that the ionized As_{Ga} would increase from 4×10^{18} to $1\times10^{19}cm^{-3}$ after annealing at 350℃, and then decreases to $1\times10^{18}cm^{-3}$ after 400℃ annealing.[4] Noting that the density of the ionized As_{Ga} is approximately equal to that of V_{Ga}, this result supports our discussion above. The same discussions also apply to the RTA samples. The relative variation of the deep donors and acceptor usually shift the Fermi level by several kT which is consistent with the data in Fig. 3 that show a maximum shift of about 100mV.

In summary, we have shown that RDS is an effective tool for the study of surface potential of LT-GaAs, and that the surface electric field of LT-GaAs is limited to an extremely narrow space-charge region. Unlike the case of doped normal GaAs, the amplitude of the LEO resonances of LT-GaAs is determined by the surface potential instead of the surface electric field. The change of the surface potential, and therefore the Fermi level, can be determined to a precision of better than 5meV. Such precision is very difficult to achieve in conventional PR measurements of the electric field at LTGaAs/ GaAs interfaces, since the potential involved is of the order of 0.5eV. The effect of the post-growth annealing on the surface potential of LT-GaAs shows a complicated movement of the Fermi level in LT-GaAs. The evolution of the Fermi level as a function of annealing temperature shows that the Fermi level is not determined by pinning to As precipitates alone. At low annealing temperatures it is determined by a process of various defect formation and annihilation, while at high annealing temperatures the effect of surface states at the As precipitates may also be taken into account. The Fermi level is not pinned to the surface state of As precipitates which are present after annealing at 600℃ and above, but still moves at higher annealing temperatures.

This work was supported by the Research Grant Council Grant 609/95P from the Hong Kong Government, and most of the experiments were performed in the William Mong's Semiconductor Cluster Laboratory at Hong Kong University of Science and Technology.

References

[1] F. W. Smith, A. R. Calawa, C. L. Chen, M. J. Manfra, and L. J. Mahony, IEEE Electron Device Lett. 9, 77(1988).

[2] T. Motet, J. Nees, S. Williamson, and G. Mourou, Appl. Phys. Lett. 59, 1455(1991).

[3] S. U. Dankowski, D. Streb, M. Ruff, P. Kissel, M. Kneissl, B. Knapfer, G. H. Dohler, U. D. Keil, C. B. Sorenson, and A. K. Verma, Appl. Phys. Lett. 68, 37(1996).

[4] D. C. Look, D. C. Walters, G. D. Robinson, J. R. Sizelove, M. G. Mier, and C. E. Stutz, J. Appl. Phys. 74, 306(1993).

[5] A. C. Warren, J. M. Woodall, J. L. Freeouf, D. Grischkowski, D. T. McInturff, M. R. Mellochl, and N. Otsuka, Appl. Phys. Lett. 57, 1331(1990).

[6] A. C. Warren, J. M. Woodall, P. D. Kirchner, X. Yin, F. Pollak, M. R. Melloch, N. Otsuka, and K. Mahalingam, Phys. Rev. B 48, 4617(1992).

[7] A. Giordana, O. J. Glembocki, E. R. Glasker, D. K. Gaskill, C. S. Kyono, M. E. Twigg, M. Fatemi, and B. Tadayon, J. Electron. Mater. 22, 1391(1993).

[8] T. M. Cheng, C. Y. Chang, T. M. Hsu, W. C. Lee, and J. H. Huang, J. Appl. Phys. 77, 2124 (1995).

[9] H. Shen, F. C. Rong, R. Lux, J. Pamulapati, M. Taysing-Lara, M. Dutta, E. H. Poidexter, L. Calderon, and Y. Lu, Appl. Phys. Lett. 61, 1585(1992).

[10] S. E. Acosta-Ortiz and A. Lastras-Martinez, Phys. Rev. B 40, 1426(1989).

[11] S. E. Acosta-Ortiz, J. Appl. Phys. 70, 3239(1991).

[12] H. Tanaka, E. Colas, I. Kamiya, D. E. Aspnes, and R. Bhat, Appl. Phys. Lett. 59, 3443 (1991).

[13] Z. Yang, I. K. Sou, Y. H. Yeung, G. K. L. Wong, Jie Wang, Cai-xia Jin, and Xiao-Yuan Hou, J. Vac. Sci. Technol. B 14, 2973(1996).

[14] Z. Q. Fang and D. C. Look, Appl. Phys. Lett. 63, 219(1993).

[15] M. Uematsu, P. Werner, M. Schultz, T. Y. Yan, and U. Gosele, Appl. Phys. Lett. 67, 2863 (1995).

[16] D. E. Bliss, W. Walukiewicz, J. W. Ager III, E. E. Haller, K. T. Chan, and S. Tanigawa, J. Appl. Phys. 71, 1699(1992).

[17] J. Stormer, W. Triftschauser, N. Hozhabri, and K. Alavi, Appl. Phys. Lett. 69, 1867(1996).

半导体材料的现状和发展趋势

王占国

（中国科学院半导体研究所，北京，100083）

Status and development of semiconductor materials

Wang Zhanguo

（Laboratory of Semiconductor Materials Science，Institute of Semiconductors，
Chinese Academy of Sciences，Beijing 100083，China）

本文共分为三个部分。

第一部分，从材料是人类社会发展的物质基础与先导着手，对作为现代信息社会两大支柱高技术产业（微电子和光电子）的核心和基础的半导体材料，在国民经济建设、社会可持续发展以及国家安全中的战略地位和作用进行了分析。

第二部分，首先分析了几种重要半导体材料目前达到的水平，器件、电路应用及其发展趋势。认为：直拉硅单晶在现代微电子技术中所起的主导地位在本世纪末和下世纪初都是无可质疑的；然而，理论分析表明，0.05–0.07μm 将是硅 MOS 集成电路线宽的极限尺寸，这不仅是指器件缩小到深亚微米，纳米量级时的量子尺寸效应对器件特性的影响所带来的物理和技术问题，更重要的是将受 Si 和 SiO_2 自身性质的限制（如直拉硅中高浓度过饱和间隙氧和 SiO_2 的介电击穿等），也就是说 Si 将最终无法满足人类对更大信息量的需求。以 GaAs、InP 为代表的Ⅲ–Ⅴ族化合物半导体材料和以它们为基的超晶格量子阱材料以及与 Si 平面工艺相容的 GeSi 合金材料等则是最有希望的替代材料。

GaAs 等Ⅲ–Ⅴ族化合物半导体材料（包括体材料、薄层和超薄层微结构材料）大多是直接带隙材料，与 Si 相比，具有光学跃迁几率高、电子饱和漂移速度快，耐高温，抗辐照等特点，在超高速、超高频、低功耗、低噪声器件和集成电路，特别是在光电子器件，光电集成和光计算应用方面占有独特的地位，受到广泛的重视。它们的发展将会使全球通信，高速计算（包括光计算），大容量信息处理，空间防御，电子对抗以

原载于：人工晶体学报，1997，26（3–4）：284.

及武器装备的微型化、智能化等这些对于国民经济和国家安全都至关重要的领域产生巨大技术进步，这是目前国际上发展最快而又接近规模化生产的一个重要领域，我国虽有一定的基础，但差距仍然很大。加强研究开发中心（基地）建设和必要的经费的投入，以缩小差距，进而迎头赶上，是当务之急。

此外，还特别对近年来兴起的材料科学前沿研究领域——低维（一维、零维）半导体材料所特有的基本性质，低维材料自组装生长和纳米加工制备技术，低维材料的检测和评价方法以及基于一维、零维半导体异质结构材料的固态量子器件研制所取得的进展和存在问题及其发展趋势作了评述。该领域的研究属超前性的探索，虽有一定风险，但极为重要，一旦突破，将带动全面发展，甚至触发新的产业革命，应给予重视。

本文最后一部分，结合国情和目前我国在该领域的发展水平，提出了发展我国半导体材料的战略设想。

Effects of annealing on self-organized InAs quantum islands on GaAs(100)

Q. W. Mo, T. W. Fan, Q. Gong, J. Wu, and Z. G. Wang

(Laboratory of Semiconductor Materials Science, Institute of Semiconductors, Chinese Academy of Sciences, Beijing 100083, China)

Y. Q. Bai

(College of Materials Science and Technology, Beijing University of Science and Technology, Beijing 100083, China)

Abstract Self-organized InAs islands on (001) GaAs grown by molecular beam epitaxy were annealed and characterized with photoluminescence (PL) and transmission electron microscopy (TEM). The PL spectra from the InAs islands demonstrated that annealing resulted in a blueshift in peak energy, a reduction in intensity, and a narrower linewidth in the PL peak. In addition, the TEM analysis revealed the relaxation of strain in some InAs islands with the introduction of the network of 90° dislocations. The correlation between the changes in the PL spectra and the relaxation of strain in InAs islands was discussed.

"Self-assembled" nanostructures grown on planar substrates, such as InAs/GaAs and GeSi/Si, have attracted considerable attention due to their high potential for optical and electronic applications.[1-3] Dense arrays of InAs/GaAs quantum islands are usually fabricated at low substrate temperature (~500℃ or lower).[4] However, the temperature must be increased to grow high-quality GaAs caps and AlGaAs layers, which are essential for light-emitting devices. Therefore, the effects of high-temperature annealing on quantum islands are important, and many works have been carried out on this subject.[5-7] Previous works indicated that high-temperature annealing could introduce severe changes in the island shape, size, and composition. In this work, we demonstrate that annealing resulted in the reduction in the linewidth as well as the intensity of the photoluminescent (PL) peak from InAs islands. In addition, transmission electron microscope (TEM) analysis shows that annealing introduced a dislocation network in InAs islands and induced a relaxation of strain. These results suggest that the changes in PL spectra and structures are correlated.

原载于：Appl. Phys. Lett., 1998, 73(24): 3518-3520.

The structure investigated in this work consists of a 300nm thick GaAs buffer layer grown at 580℃ on a semi-insulating (001) GaAs substrate, followed by five periods of bilayer structure consisting of an InAs dot sheet of 2.5ML and a 9nm GaAs spacer layer grown at 490℃, and finally, a 80nm thick GaAs capping layer grown at 580℃. The growth rates for InAs and GaAs are 0.11 ML/s and 0.7μm/h, respectively, and a 4s growth interruption was introduced before and after the InAs island growth. The PL measurement was performed at 15K on an IFS120HR Fourier transform infrared spectrometer with an InGaAs detector. After the growth, some samples are annealed at 700℃ for 60min under As pressure. Plan-view TEM samples were prepared by mechanical polishing the resulting wafer to ~70μm thick, and the final thinning to the perforation was performed with Ar-ion milling from the substrate side. For a better image quality, the annealed specimens were also sputtered for about 40nm from the topside by Ar-ion at a large angle. Bright-field and weak-beam images were taken using a JEOL2000EX electron microscope.

Fig. 1 shows the PL spectra of InAs islands before and after annealing. It can be seen from Fig. 1 that annealing induces some changes in PL spectra in three respects: (1) a blueshift of the peak energy by 77meV; (2) the full width at half maximum (FWHM) is narrowed by about 30meV; and (3) a reduction in PL intensity by 70%. The blueshift is believed to be caused by the intermixture of In and Ga atoms via diffusion,[8,9] and the reduction of FWHM and intensity may result from internal structure changes which will be discussed later.

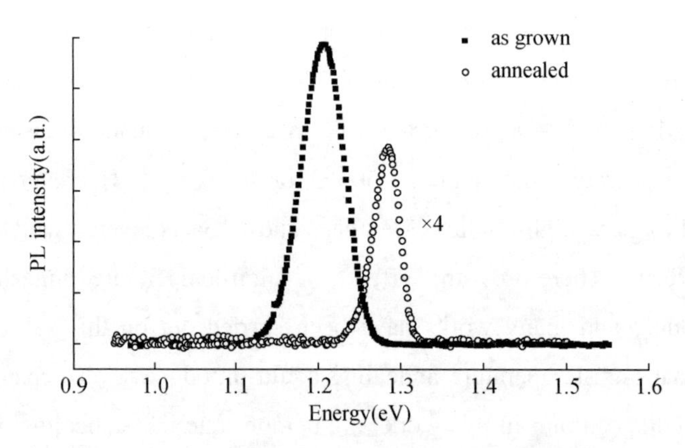

Fig. 1　PL spectra (15K) from as-grown InAs islands on GaAs substrates and InAs islands annealed at 700℃ for 60min

Fig. 2(a) shows the plan-view bright-field images of the InAs quantum islands in the as-grown sample under the [022] diffraction condition, and it can be seen that InAs islands can be divided into two groups according to sizes—small islands and large ones. The small islands in the first group are uniform in size, of about 10nm, and their density is approximately 600/μm^2. While the bigger ones in the second group, with sizes ranging from 20–80nm, have a density of about 30/μm^2. It should be mentioned that these sizes may be bigger than the real ones as the contrast is formed by the strain in the matrix. However, this discrepancy does not influence our following qualitative results. The contrast of InAs islands demonstrates that they are mostly coherent strain precipitates in a matrix.[10] It can be seen from Fig. 2(a) that there is a black-white lobe in the contrast of small islands in the first group, while no such contrast is observed in big islands of the second group. This difference in contrast between the two groups is believed to be caused by the island size.[10] Fig. 2(b) is the image of the annealed sample and it can be seen that both the shape and density of the InAs islands are almost the same as in the as-grown sample shown in Fig. 2(a). However, the contrasts of the large islands are now dominated by a network of moiré fringes running along the [011] or $[01\bar{1}]$ direction perpendicular to the diffraction vector. Figs. 3(a) and 3(b) are the weak-beam dark-field images of the annealed samples, and parallel dislocation lines can be seen perpendicular and parallel to the [011] direction, respectively, in the large islands. The ***g*** · ***b*** analysis shows the existence of two perpendicular sets of dislocations, and they are supposed to be a network of 90° Lomer dislocations introduced during strain relaxation.[11, 12] Such a network of misfit dislocations was observed in Ge/Si, (In, Ga) As/GaAs, and other highly misfit systems.[11-14] In Figs. 3(a) and 3(b), besides the large islands, a few small islands can be observed as indicated by the arrows. The contrasts of white-white lobes with a no-contrast line in the weak-beam image manifest the incoherence of the small islands. Such islands can also be found in samples before annealing (not shown)

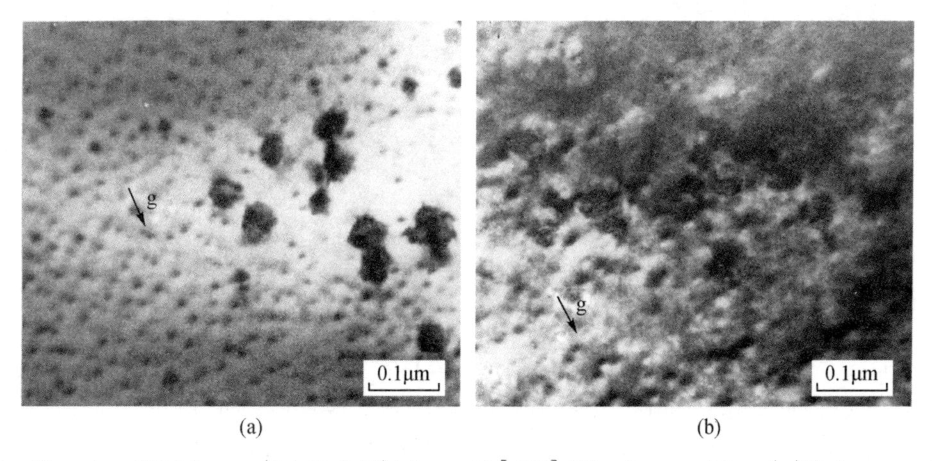

Fig. 2 Plan-view TEM images (bright field) taken with [022] diffraction condition. (a) Before annealing, both the small and large InAs islands show strain contrast. (b) After annealing, the large InAs islands contain moiré fringes perpendicular to the [022] direction, which are caused by misfit dislocations

and their density is about 5/μm². It is clear now that three types of InAs islands have been observed in our experiment: small coherent islands with overwhelming quantities (600/μm²), a few incoherent small islands (5/μm²), and some coherent large islands (30/μm²), which undergo relaxation to some extent during annealing with the introduction of a network of misfit dislocations.

From Fig. 3, the residual strain in the large islands after annealing can be estimated by[15]

$$\bar{\boldsymbol{\varepsilon}}=\boldsymbol{\varepsilon}-\boldsymbol{b}/\boldsymbol{d},$$

where $\bar{\boldsymbol{\varepsilon}}$ is the residual strain in the InAs islands, $\boldsymbol{\varepsilon}$ is the mismatch between InAs and GaAs (i.e., the strain of InAs islands without dislocations), and $\boldsymbol{b}$, $\boldsymbol{d}$ are Burgers vector and dislocation spacing, respectively. The result shows that the dislocation network with a spacing of 7nm can result in almost complete relaxation of the large islands. Taking the dislocation network as a sign of relaxation, Fig. 4 shows the size distribution of unrelaxed islands before and after annealing. It indicates that before annealing the size of the unrelaxed islands has a wide range, from 10 to 80nm. However, almost all the islands with a size bigger than 30nm have become relaxed completely after annealing.

The changes in the PL spectra demonstrated in the Fig. 1 should be correlated with the relaxation of strain in large islands during annealing. As it is well known, a network of dislocations acts as a nonradiative center to drastically reduce the luminescence efficiency, and those relaxed large islands cannot emit luminescence any more. We can explain the observed

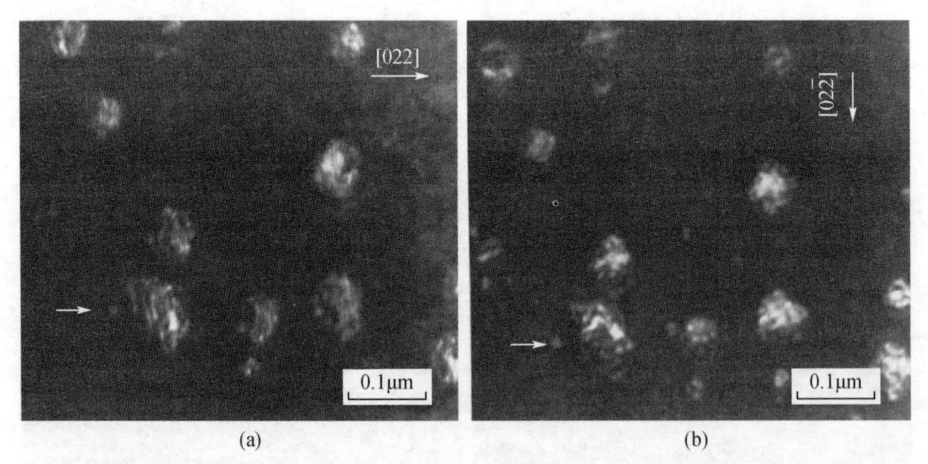

Fig. 3　Weak-beam dark-field images of annealed InAs islands in the same area taken with two $\boldsymbol{g}$ vectors perpendicular to each other. Both images show dislocation lines perpendicular to the $\boldsymbol{g}$ vector used. A typical small incoherent island is indicated by the arrow, which shows contrasts of white-white lobes with a line of no-contrast in the center. (a) Vertical lines on the InAs islands are dislocations with the [022] diffraction. (b) Using the $[02\bar{2}]$ diffraction condition, only horizontal dislocation lines are now visible

PL spectra: (1) after annealing, a narrower size range of radiative islands leads to the narrower linewidth; and (2) a smaller population of active islands results in the intensity reduction. Here, annealing plays a role like a "filter" to reduce the size dispersion of "active" InAs islands.

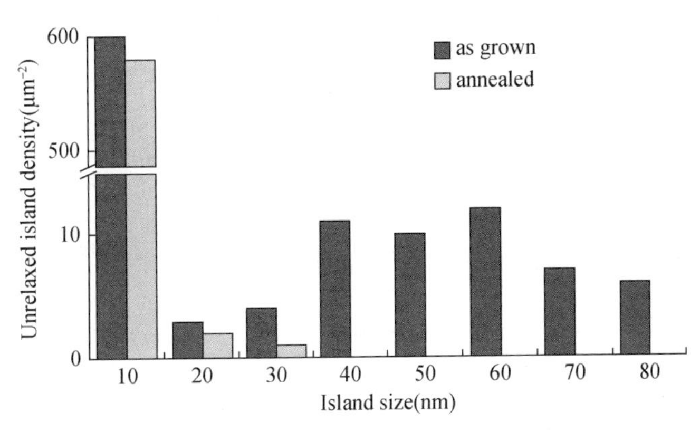

Fig. 4 The size distribution of unrelaxed InAs islands before and after annealing at 700℃ under As pressure for 60min

We attribute such a dramatic effect of annealing to the special stage of InAs islands growth here. According to the valence force field (VFF) calculation,[16] 2.5ML is about the "critical thickness" of InAs islands grown on GaAs, exceeding which dislocations begin to be introduced in much the same way as the two-dimensional peudomorphic layer. Thus, when 2.5ML of InAs are deposited, some incoherent small islands appear among the previous coherent ones and some grow much faster because of their lower strain energy,[17] which leads to the coexistence of these three types of islands in our samples. However, the large islands do not continue to relax with the dislocation introduction although they "should". Therefore, the kinetic limitation of dislocation generation should be taken into account here at such a low growth temperature. When the sample is heated at 700℃, the nucleation and motion of enough dislocations can be activated and the full relaxation can be completed.

However, while we can explain the narrowed linewidth of InAs PL peak using the shrinking of the size range of "active" islands, we may meet difficulty when ascribing the intensity reduction solely to the decrease of the active InAs island amounts. Some other factors should be considered, such as the quality of the capping layer. Our Raman spectra indicate the existence of As clusters in the capping layer, which implies some problems exist in our annealing experiment, and it may affect the intensity of the light emission. Further work is in progress to understand the detailed correlation.

In conclusion, we have shown that high-temperature annealing of InAs islands results in a reduction of the FWHM and intensity as well as a blueshift of energy in PL spectra. The full relaxation of the large strained islands is confirmed and believed to be the cause of the changes in PL spectra.

This work was supported by Chinese National Natural Science Foundation Contract No. 69576028.

References

[1] L. Goldstein, F. Glas, J. Y. Marzin, M. N. Charasse, and G. Le Roux, Appl. Phys. Lett. 47, 1099(1985).

[2] D. Leonard, M. Krishnamurthy, C. M. Reaves, S. P. Denabaars, and P. M. Petroff, Appl. Phys. Lett. 63, 3203(1993).

[3] J. M. Moison, F. Houzay, F. Barthe, L. Leprice, E. Andre, and O. Vatel, Appl. Phys. Lett. 64, 196(1994).

[4] N. N. Ledentsov, M. Grundmann, N. Kirstaedter, O. Schmidt, R. Heitz, J. Böhrer, D. Bimberg, V. M. Ustinov, V. A. Shchukin, P. S. Kop'ev, Zh. I. Alferov, S. S. Ruvimov, A. O. Kosogov, P. Werner, U. Richter, U. Gösele, and J. Heydenreich, Solid-State Electron. 40, 785(1996).

[5] A. O. Kosogov, P. Werner, and U. Gösele, Appl. Phys. Lett. 69, 3072(1996).

[6] D. G. Deppe and N. Holonyak, J. Appl. Phys. 64, R93(1988).

[7] G. P. Kothiyal and P. Bhattacharya, J. Appl. Phys. 63, 2760(1988).

[8] B. Elman, E. S. Koteles, P. Melman, C. Jagannath, C. A. Armiento, and P. Rothman, J. Appl. Phys. 68, 1351(1990).

[9] S. Bürkner, M. Baeumler, J. Wagner, E. C. Larkins, W. Rothemund, and J. D. Ralston, J. Appl. Phys. 79, 6818(1996).

[10] P. B. Hirsch, A. Howie, R. B. Nicholson, and D. W. Pashley, Electron Microscopy of Thin Crystal(Butterworths, London, 1965).

[11] K. Tillmann, D. Gerthsen, P. Pfundstein, A. Förster, and K. Urban, J. Appl. Phys. 78, 3824(1995).

[12] F. K. Legoues, J. Tersoff, M. C. Reuter, M. Hammar, and R. Tromp, Appl. Phys. Lett. 67, 2317(1995).

[13] D. J. Eaglesham, R. T. Tung, J. P. Sullivan, and J. P. Shrey, J. Appl. Phys. 73, 4064(1993).

[14] A. Sakai and T. Tatsumi, Phys. Rev. Lett. 71, 4007(1993).

[15] J. Tersoff, Appl. Phys. Lett. 62, 693(1993).

[16] A. Sasaki, J. Cryst. Growth 160, 27(1996).

[17] J. Drucker, Phys. Rev. B 48, 18203(1993).

Wurtzite GaN epitaxial growth on a Si(001) substrate using γ-Al_2O_3 as an intermediate layer

Lianshan Wang, Xianglin Liu, Yude Zan, Jun Wang, Du Wang, Da-cheng Lu, and Zhanguo Wang

(Laboratory of Semiconductor Materials Science, Institute of Semiconductors, The Chinese Academy of Sciences, Beijing 100083, China)

Abstract Wurtzite GaN films have been grown on(001) Si substrates using γ-Al_2O_3 as an intermediate layer by low pressure(~76Torr) metalorganic chemical vapor deposition. Reflection high energy electron diffraction and double crystal *x*-ray diffraction measurements revealed that the thin γ-Al_2O_3 layer of "compliant" character was an effective intermediate layer for the GaN film grown epitaxially on Si. The narrowest linewidth of the X-ray rocking curve for (0002) diffraction of the 1.3μm GaN sample was 54 arcmin. The orientation relationship of GaN/γ-Al_2O_3/Si was (0001)GaN ‖ (001)γ-Al_2O_3 ‖ (001)Si, [11-20]GaN ‖ [110]γ-Al_2O_3 ‖ [110]Si. The photoluminescence measurement for GaN at room temperature exhibited a near band-edge peak of 365nm(3.4eV).

Gallium nitride(GaN) is a wide gap semiconductor of great promise for application to blue, violet, and ultraviolet(UV) light emitting devices and high temperature/highpower devices.[1-4] GaN films have been fabricated on a number of substrates such as Si,[5,6] GaAs,[7] $MgAl_2O_4$,[8] 6H-SiC,[9] ZnO,[10] as well as various crystallographic orientations of sapphire.[11,12] Recently, much improvement has been made in the high quality of GaN films by the predeposition of a thin AlN or GaN buffer layer.[13,14] *P*-type conductivity has been achieved in wurtzite GaN films doped with Mg exposed to low energy electron beam irradiation[15] or N_2 gas ambient thermal annealing[16] after their growth. High efficiency GaN-based blue light emitting diodes have been commercially available.[17]

Silicon is viewed, because of its high quality, large size, and low cost, as one of the most promising substrates for the growth of GaN. However, due to the large difference in lattice constant, crystal structure, and thermal expansion coefficient, it is rather difficult to grow epitaxial GaN on Si. Many attempts have led to amorphous or polycrystalline

原载于: Appl, Phys. Lett., 1998, 72(1): 109-111.

GaN.[10, 18] Lei et al. have successfully grown zinc blende and wurtzite GaN on Si by molecular beam epitaxy,[5, 6] whereas Takeuchi et al. also reported the growth of GaN on Si (111) substrates using SiC as an intermediate layer.[19] AlN is another material for an intermediate layer in GaN films on Si.[20, 21] We chose the γ-Al_2O_3 material on Si to replace the sapphire substrate for growth of GaN because almost all high quality GaN films are grown on sapphire(α-Al_2O_3), and γ-Al_2O_3 may serve as an insulator layer between the GaN and the Si. In particular, epitaxial growth of insulator layers on Si is of great importance in achieving Si on insulator(SOI) structures and for the long-range goal of three-dimensional integrated circuits. If GaN can be prepared on γ-Al_2O_3/Si, it offers very attractive potential to harmonically incorporate GaN optoelectronic devices in silicon-based very large scale integrated circuits. However, this structure for GaN on γ-Al_2O_3/Si has not been studied yet. In the present letter, we report wurtzite GaN thin films grown on Si(001) substrates using γ-Al_2O_3 as an intermediate layer by low pressure metalorganic chemical vapor deposition (LPMOCVD).

Deposition of our Al_2O_3 films on Si substrates was carried out by the LPCVD method using trimethylaluminium (TMA) and N_2O for source materials, and like the procedure reported elsewhere.[22, 23] A horizontal-type low pressure (~76 Torr) MOCVD system was used for growth of GaN and the sources were trimethylgallium(TMG) and ammonia(NH_3). Polished Si(001) substrates of slightly doped *p* type were used. Prior to the growth of the GaN films, an Al_2O_3 layer (~60nm) was predeposited on (001) Si, then the substrate was transferred to the horizontal reactor to grow the GaN films via a two-step growth process.[14] GaN buffer layers of 20 – 80nm were prepared between 450 and 600℃ and subsequent epilayers were grown at temperature of over 1000℃.

The as-deposited Al_2O_3 layer on Si (001) had a mirror-like featureless surface. The reflection high energy electron diffraction (RHEED) pattern has both spots and rings, implying a highly oriented polycrystal. The result of analysis showed that our Al_2O_3 was of a γ phase(γ-Al_2O_3), namely, a tetragonal distortion of the spinel arrangement with a = 7.95Å and c = 7.79Å.[24] Additionally, cross-sectional transmission electron microscopy showed that our γ-Al_2O_3 possessed preferential orientation by (001) γ-Al_2O_3 ∥ (001) Si and [110] γ-Al_2O_3 ∥ [110] Si.

An OPTON CSM 950 scanning electron microscope (SEM) inspected the surface morphology of the GaN films grown on (001) Si substrates covered by the γ-Al_2O_3 intermediate layers. The surfaces of single crystal GaN films were roughened with many well-oriented mosaics. For an investigation of a series of growth runs by SEM, the surfaces of a few samples had cracks arising from the stress produced by thermal mismatches between the films and substrates (25%) when the samples were cooled from the growth temperature to room temperature. The cracks could be seen only in high magnification of over 10000 by the scanning electron microscope. All GaN films to the substrates were of very good adhesion quality and no films were observed to peel off from the substrates although there were mosaic structures on the surfaces of the films. The films investigated here are free of any cracks.

Fig. 1 plots the X-ray diffraction profile and its rocking curve from the GaN film 1.3μm thick. The (0001) plane of GaN is parallel to the (001) plane of Si. A strong peak at $2\theta = 34.4°$ originated from the (0002) reflection of the wurtzite GaN, so we obtain a *c*-axis lattice constant of 5.208Å. The narrowest full width at half-maximum (FWHM) obtained for the (0002) diffraction from the 1.3μm GaN sample is 54 arcmin (see the inset of Fig. 1). This value is much greater than that of wurtzite GaN on (0001) sapphire,[13] but it approaches and is comparable to that for GaN on Si using an AlN buffer as reported by Kung et al.[21]. We also measured the rocking curves of the (0002) diffraction with a third crystal in the ω mode and the $\omega/2\theta$ mode, respectively, to evaluate the influence of the mosaic structure and the variation of the spacing lattice. The experimental results for the above sample demonstrated that the FWHM ($\Delta\theta = 47$arcmin) due to the orientation fluctuation of the crystallites (the mosaic effect) was much larger than that ($2\Delta\theta = 3.4$arcmin) due to the variation of the lattice spacing. The serious mosaic effect was presumably associated with the crystallographic quality of γ-Al_2O_3 since the oxide was only a preferentially oriented polycrystal and could not serve as a good template for the GaN epitaxy. Moreover, it should be under a strong stress from the silicon substrate and would work at a manner which resembled a "compliant" substrate. In fact, assuming that the γ-Al_2O_3 layer was a fully single crystal, we may speculate that the quality of the GaN films is remarkably improved, especially by greatly eliminating the mosaic effect. In addition, we should note that the linewidth of X-ray

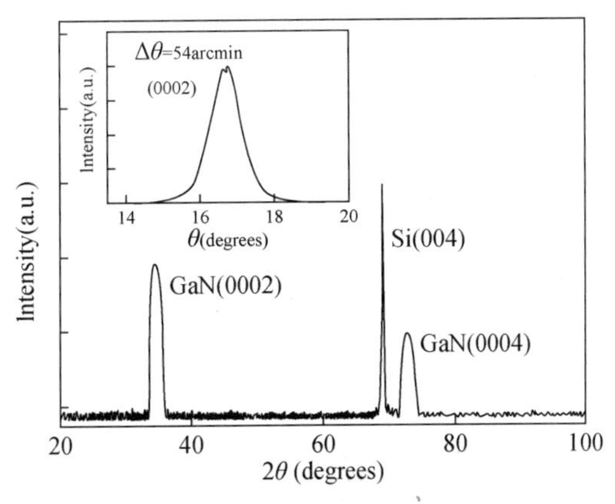

Fig. 1 X-ray diffraction profile and its rocking curve of the (0002) reflection for the 1.3μm thick GaN film grown on Si covered by a thin γ-Al_2O_3 layer

diffraction also depends on the thickness of films and on the optimized condition of buffer layers.[25]

The RHEED pattern recorded along [10−10] azimuth in Fig. 2 also demonstrates that the GaN film was epitaxially grown on the γ-Al_2O_3/Si(001) substrate. Clearly, the zero order and high order Laue spots were superposed and the spotty pattern was associated with a rough surface of the film. By measuring the spacings of the diffraction spots, the lattice constants were found to be $a = 3.185 \pm 0.008$ Å and $c = 5.196 \pm 0.005$ Å, consistent with the X-ray diffraction measurement. Our observation for the GaN/γ-Al_2O_3/Si sample by cross-sectional transmission electron microscopy showed that the orientation relationship between them was (0001) GaN ∥ (001) γ-Al_2O_3 ∥ (001) Si, [11−20] GaN ∥ [110] γ-Al_2O_3 ∥ [110] Si. Fig. 3 illustrates the schematic lattice mismatching model of GaN and γ-Al_2O_3. As the [−1100] axis of GaN lies along the [1−10] axis of γ-Al_2O_3, we can consider the lattice mismatching relation, $\sqrt{2}a_0/2(\gamma\text{-}Al_2O_3) \approx \sqrt{3}a_0(\text{GaN})$, to be 1.73%; on another orientation, if the [11−20] axis of GaN lies along the [110] axis of γ-Al_2O_3, we can assume the lattice mismatching relation, $\sqrt{2}a_0/2(\gamma\text{-}Al_2O_3) \approx 2a_0(\text{GaN})$, to be 13.5%. To some extent, these values are still smaller than that for GaN on Si(17%).

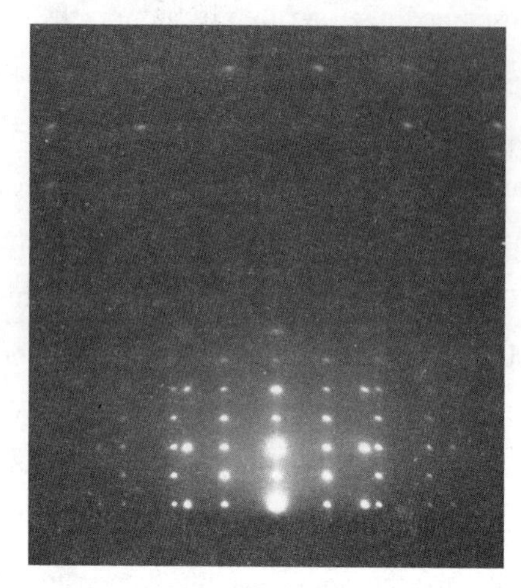

Fig. 2 RHEED pattern recorded along [10−10] azimuth for the GaN film grown on Si(001) covered by a γ-Al_2O_3 layer

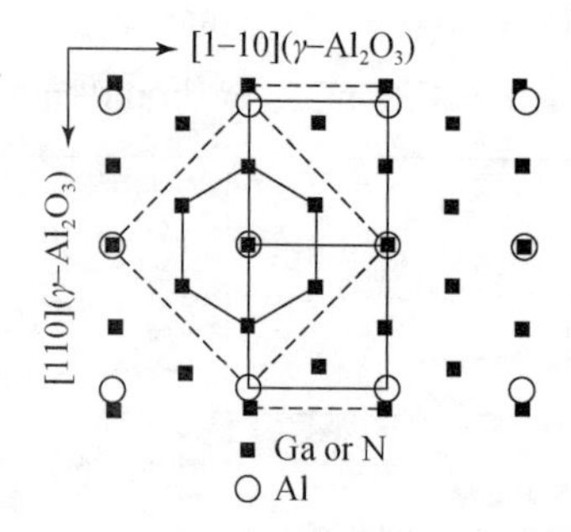

Fig. 3 Schematic model of the lattice mismatching between the (0001) GaN and the (001) γ-Al_2O_3. Open circles indicate the underlying Al atoms on the (001) plane of γ-Al_2O_3 while closed squares denote the Ga or N atoms on the (0001) plane of GaN. The square connected by dotted lines constitutes a unit cell for the (001) plane of γ-Al_2O_3, whereas the hexagon linked by the dot-dashed lines is a unit cell for the (0001) plane of GaN. The nearly coincident repeated units of Ga or N and Al are marked by the parallelograms constructed by the bold lines and heavy dashed lines, respectively

More interesting, why does GaN grown on this thin γ-Al_2O_3 have pseudocubic symmetry in the wurtzite modification? It has been established that the crystal structure of the epitaxial GaN films is most strongly influenced by the substrate material, orientation, and growth temperature. The equilibrium crystal structure for GaN is wurtzite while GaN has been known to have zinc blende when grown as a cubic substrate. The lattice mismatch should not be a primary factor, but the growth temperature is an important reason for our GaN being grown in the wurtzite structure because it is thermodynamically stable under our growth condition of up to 1000℃. Although we cannot rule out the possible coexistence of two variants, we cannot yet attain evidence from our observation in detectable limitation.

Indeed, the γ-Al_2O_3 layer on Si plays another important role as does the sapphire in the fabrication of GaN epitaxy besides acting as a "compliant" substrate and having a lower lattice mismatch between GaN and γ-Al_2O_3(compared to the GaN grown directly on Si). It is strange to us why GaN could be grown on the preferentially oriented oxide. The following are possible factors therefore: (1) the AlN formation on the γ-Al_2O_3 surface by the nitridation is one of the reasons because the lattice mismatch between GaN and AlN is as small as 2.5%; (2) the crystallite nuclei in γ-Al_2O_3 must be taken in consideration. It is the crystallite nuclei that may provide the template for the GaN buffer layer deposited at low temperature. Again, because the γ-Al_2O_3 layer remained stable even at high temperature, the thermal fluctuation of crystallite nuclei in γ-Al_2O_3 was thus relatively small and the GaN buffer layer was easily crystallized to achieve solid phase epitaxy on these crystallite nuclei during the elevated temperature annealing stage. In our experiment, when GaN was grown directly on Si, we observed that the GaN buffer layer could not nucleate on Si. This suggests that γ-Al_2O_3 is favorable for nucleation of the GaN buffer layer. Cross-sectional transmission electron microscopy also confirmed that γ-Al_2O_3 was not yet converted to a fully single crystal but that the GaN buffer layer on γ-Al_2O_3 achieved solid phase epitaxy.

Our Hall effect measurements were performed for the GaN films on Si covered by the γ-Al_2O_3 layers at room temperature because γ-Al_2O_3 has good insulator performance.[22] The unintentionally doped films were normally *n* type with a typical carrier concentration of 2×10^{19} cm^{-3} and an electron mobility of $17cm^2/Vs$. This high carrier concentration was associated with the native defects in the films because of their poor quality and the oxygen impurity incorporation thereof detected by Auger electron spectroscopy.

At room temperature, we measured the photoluminescence (PL) spectra of GaN on γ-Al_2O_3/Si(001) substrates using a He-Cd laser(20mW, 325nm) as the excitation source. A typical spectrum is shown in Fig. 4. Besides the broad yellow band occurring with a center wavelength of 550nm (2.25eV) which is in connection with the deep level states in this material, a band edge emission is situated at 365nm(3.4eV).

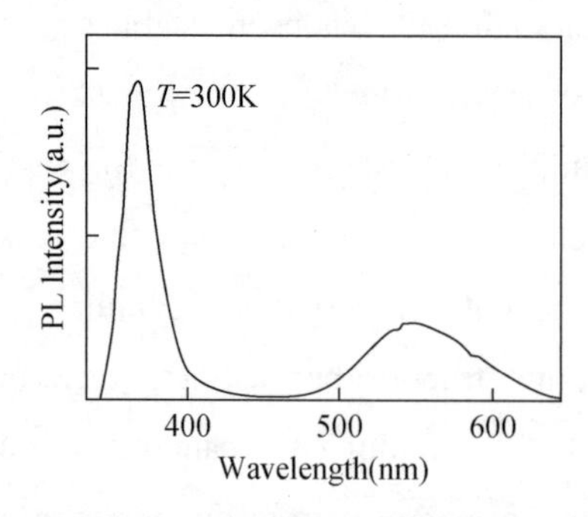

Fig. 4　A typical photoluminescence spectrum at room temperature for GaN on(001)Si covered by a thin γ-Al_2O_3 layer

In summary, wurtzite GaN films have been grown epitaxially on Si(001) substrates using a γ-Al_2O_3 intermediate layer of compliant character by low pressure MOCVD. The narrowest FWHM for(0002) reflection from the 1.3μm thick GaN film was 54arcmin. The orientation relationship of GaN/γ-Al_2O_3/Si was (0001) GaN ‖ (001) γ-Al_2O_3 ‖ (001) Si, [11-20] GaN ‖ [110] γ-Al_2O_3 ‖ [110] Si. A near band edge emission peak of 3.4eV at room temperature was observed. Work for obtaining high quality GaN on γ-Al_2O_3/Si under the optimized conditions is still underway.

The authors wish to thank Professor Weibin Gao, Professor Yutian Wang, and Dr. Xiaojun Wang. They are also grateful to senior engineer Yuping Wang of Peking University and to engineer Wenyan Yang of Tsinghua University. This study was financially supported by the National Advanced Materials Committee of China and by the Planning Commission of China under Contract No. 857010502.

References

[1] H. Morkoç, S. Strite, G. B. Gao, M. E. Lin, B. Sverdlov, and M. Burns, J. Appl. Phys. 76, 1363(1994).

[2] S. D. Lester, F. A. Ponce, M. G. Craford, and D. A. Steigerwald, Appl. Phys. Lett. 66, 1249 (1995).

[3] S. Nakamura, M. Senoh, S. I. Nagahama, N. Iwasa, T. Yamada, T. Matsushita, H. Kiyoku, and Y. Sugimoto, Appl. Phys. Lett. 68, 2105(1996).

[4] Ö. Aktas, W. Kim, Z. Fan, A. Bothkarev, A. Salvador, S. N. Mohammad, B. Sverdlov, and H. Morkoç, Electron. Lett. 31, 1389(1995).

[5] T. Lei, M. Fanciulli, R. J. Molnar, T. D. Moustakas, R. J. Graham, and J. Scanlon, Appl. Phys. Lett. 59, 944(1991).

[6] T. Lei, T. D. Moustakas, R. J. Graham, Y. He, and S. J. Berkowitz, J. Appl. Phys. 71, 4933

(1992); T. Lei, K. F. Ludwig, Jr., and T. D. Moustakas, ibid. 74, 4430(1993); S. N. Basu, T. Lei, and T. D. Moustakas, J. Mater. Res. 9, 2370(1994).

[7] S. Fujieda and Y. Matsumoto, Jpn. J. Appl. Phys., Part 2 30, L1665(1991).

[8] A. Kuramata, K. Hirino, K. Domen, K. Shinohara, and T. Tanahashi, Appl. Phys. Lett. 67, 2521 (1995).

[9] B. N. Sverdlov, G. A. Martin, H. Morkoç, and D. J. Smith, Appl. Phys. Lett. 67, 2063(1995).

[10] Z. Sitar, M. J. Paisley, B. Yan, and R. F. Davis, Mater. Res. Soc. Symp. Proc. 162, 537 (1990).

[11] T. Sasaki and S. Zembutsu, J. Appl. Phys. 61, 2533(1987).

[12] C. J. Sun and M. Ragzeghi, Appl. Phys. Lett. 63, 973(1993).

[13] H. Amano, N. Sawaki, I. Akasaki, and Y. Toyoda, Appl. Phys. Lett. 48, 353(1986).

[14] S. Nakamura, Jpn. J. Appl. Phys., Part 1 30, 1705(1991).

[15] H. Amano, M. Kito, K. Hiramatsu, and I. Akasaki, Jpn. J. Appl. Phys., Part 2 28, L2112 (1989).

[16] S. Nakamura, T. Mukai, M. Senoh, and N. Iwasa, Jpn. J. Appl. Phys., Part 2 31, L139(1992).

[17] S. Nakamura, T. Mukai, and M. Senoh, Jpn. J. Appl. Phys., Part 2 30, L1998(1991).

[18] Z. J. Yu, B. S. Sywe, A. U. Ahmed, and J. H. Edgar, J. Electron. Mater. 21, 383(1992).

[19] T. Takeuchi, H. Amano, K. Hiramatsu, N. Sawaki, and I. Akasaki, J. Cryst. Growth 115, 634 (1991).

[20] A. Watanabe, T. Takeuchi, K. Hirosawa, H. Amano, K. Hiramatsu, and I. Akasaki, J. Cryst. Growth 128, 391(1993).

[21] P. Kung, A. Saxler, X. Zhang, D. Walker, T. C. Wang, I. Ferguson, and M. Razeghi, Appl. Phys. Lett. 66, 2958(1995).

[22] M. Ishida, I. Katakabe, N. Ohtake, and T. Nakamura, Mater. Res. Soc. Symp. Proc. 116, 375 (1988).

[23] M. Ishida, I. Katakabe, T. Nakamura, and N. Ohtake, Appl. Phys. Lett. 52, 1326(1988).

[24] R. W. G. Wyckoff, Crystal Structures, 2nd ed. (Interscience, New York, 1965), Vol. 3, p. 84.

[25] J. N. Kuznia, M. Asif Khan, D. T. Olson, R. Kaplan, J. Freitas, J. Appl. Phys. 73, 4700 (1993).

High-density InAs nanowires realized in situ on(100)InP

Hanxuan Li

(Laboratory of Semiconductor Materials Science, Institute of Semiconductors, Chinese Academy of Sciences, Beijing 100083, China; Department of Electrical and Computer Engineering, Duke University, Durham, North Carolina 27708-0291)

Ju Wu and Zhanguo Wang

(Laboratory of Semiconductor Materials Science, Institute of Semiconductors, Chinese Academy of Sciences, Beijing 100083, China)

Theda Daniels-Race

(Department of Electrical and Computer Engineering, Duke University, Durham, North Carolina 27708-0291)

Abstract High-density InAs nanowires embedded in an $In_{0.52}Al_{0.48}As$ matrix are fabricated in situ by molecular beam epitaxy on(100)InP. The average cross section of the nanowires is $4.5\times10nm^2$. The linear density is as high as 70wires/μm. The spatial alignment of the multilayer arrays exhibit strong anticorrelation in the growth direction. Large polarization anisotropic effect is observed in polarized photoluminescence measurements.

Considerable effort has been made towards the goal of producing nanostructures such as quantum dots and quantum wires, owing to their unique physical properties and potential device applications. Fabrication of nanostructures with uniform dimensions and defect-free interfaces will be essential in fully exploiting their anticipated merits of multidimensional quantum confinement. For practical device applications, it is also desirable to increase the active volume by stacking the nanostructures into multilayer arrays.

The InAs/InP system is expected to play an important role in the technologically important 1.3-1.55μm wavelength range.[1] Recently, InGaAs quantum wires on vicinal (100) InP have been investigated.[2] Wire-like one dimensional patterns have also been reported in several systems such as InAs/AlAs(Ref. [3]) and $In_xAl_{1-x}As/In_yAl_{1-y}As$ (Ref. [4]) superlattices on singular(100)InP. These reports have in common the appearance of

原载于: Appl. Phys. Lett., 1999, 75(8): 1173-1175.

lateral composition modulations within the as-designed heterostructures. In this letter, we report the self-organized growth of InAs nanowires. For the multilayer arrays grown, we also find that the nanowires in successive layers exhibit strong anticorrelation in the growth direction.

The samples were grown on (100) InP by a Riber MBE system. Both single and multilayer samples were prepared. The multilayer samples consist of a 200nm $In_{0.52}Al_{0.48}As$ buffer layer, six periods of InAs/$In_{0.52}Al_{0.48}As$ superlattices, and a 50nm $In_{0.52}Al_{0.48}As$ cap layer. For the single-layer sample, the strained InAs layer was grown under three different temperatures(470, 500, and 530℃) to investigate the temperature effects on the formation of nanostructures. The growth rates of $In_{0.52}Al_{0.48}As$ and InAs were 0.7 and 0.2μm/h, respectively. Ex situ double crystal X-ray diffraction (DCXD) was used to examine the degree of lattice matching of the $In_{0.52}Al_{0.48}As$ buffer layer to the InP. Growth was carried out under As stabilization with a Ⅴ/Ⅲ flux ratio of 15-20. No growth interruptions were introduced during the process.

The samples were characterized by atomic force microscopy (AFM), transmission electron microscopy (TEM), and polarized photoluminescence (PL) measurements. Specimens for TEM were prepared by mechanical thinning followed by ion milling. 15K polarized PL measurements were performed using a variable-temperature cryostat. The excitation source was a linearly polarized 514.5nm line of an Ar-ion laser. The incident laser beam was polarized parallel to either the[110]or[$\bar{1}$10]direction.

Fig. 1 is an AFM image of 6ML InAs deposited at 500℃. Dense nanowires along the [$\bar{1}$10] direction form continuously over the entire surface. Most were straight over 100nm, and some were almost ~0.5μm in length. The average spacing, base width, and height of the nanowires are 15, 10, and 4.5nm, respectively. Under our growth conditions, quasiperiodic wire-like nanostructures are observed in a wide growth temperature range near 500℃ (470-530℃). However, it is found that both high and low temperatures result in inhomogeneity of the nanowires. The shapes of wires inside these samples are almost irregular and show a weaker periodicity.

An exact explanation for the wire-like nanostructure formation is still lacking, and is usually believed to be related to the surface anisotropy.[5-7] Our in situ reflection high-energy electronic diffraction (RHEED) patterns clearly indicate surface anisotropy due to As-dimer reconstruction. Thus it costs less energy to create steps along the [$\bar{1}$10] direction than along [110]. Moreover, such a surface reconstruction also allows easier diffusion in the [$\bar{1}$10] direction (i.e., parallel to As dimers).[5-7] It thus appears that surface anisotropy can favor the formation of wire-like structures. Note that our results are different from those in Ref. 8.

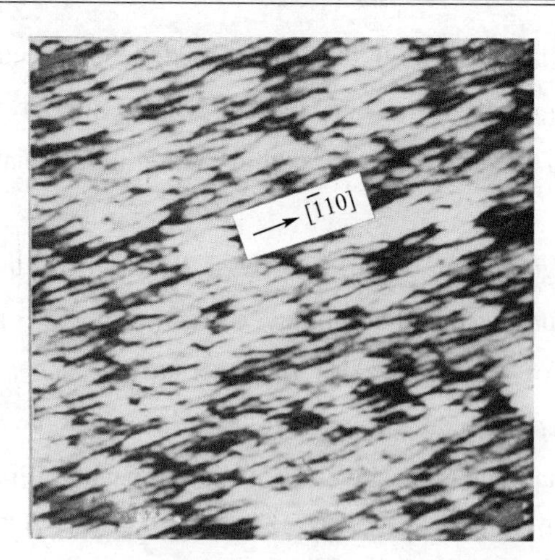

Fig. 1 1μm size AFM image of the uncapped 6ML sample

In that work, more isotropic quantum dots were formed. We believe that the observed different self-organized behavior should be attributed to the different growth conditions used. A number of previous studies have also shown that the growth mode of InAs on InP is more complex and sensitive to the deposition conditions.[1,5,9] In particular, the nominal growth rate used here is much higher than that used in Ref. [8]. Higher growth rates result in less time for adatom surface migration, and have been shown to favor the formation of elongated structures for InAs/InP.[5]

Fig. 2(a) shows the [$\bar{1}$10] cross-sectional TEM image of the six-period InAs (6.5ML)/ $In_{0.52}Al_{0.48}As$ (20nm) multilayer array. The nanowires appear as the thicker bright regions connected with the thinner quantum wells in between; they are embedded in the $In_{0.52}Al_{0.48}As$ layers which show darker contrast. Thus, the InAs nanowires are surrounded by a higher band gap $In_{0.52}Al_{0.48}As$ region in both [110] and [100] directions. This two-dimensional band gap confinement leads to the formation of the nanowire arrays. For this sample, the wire dimensions in the lowest layers are much larger, and many neighboring wires are merged together, resulting in a rough surface. With subsequent InAs/InAlAs layer deposition, the nanowires become smaller and well separated. The TEM images demonstrate that it is possible to fabricate nanowire arrays without creating any extended defects.

It is found that by varying the thickness of InAs, the uniformity of nanowires can be improved. Fig. 2(b) shows the cross-sectional TEM image of the sample with six-period InAs (6ML)/$In_{0.52}Al_{0.48}As$(10nm). A uniform spatial distribution is observed. Because of the small lateral periodicity(~15nm), we can estimate a linear density as large as 70 wires/μm, much higher than the ever reported values of 2.5 wires/μm (MBE-grown quantum wires on GaAs(311)*A* patterned with sub-μm-pitch gratings),[10] 15wires/μm [CBE-grown InGaAs wires on vicinal (100) GaAs],[11] and 33 wires/μm [MBE-grown InGaAs wires on (100)

InP]. [3] A high density will be extremely important for future device applications. Fig. 2(c) presents the plan-view TEM images of this sample, quite consistent with the morphology profile of single-layer samples(Fig. 1).

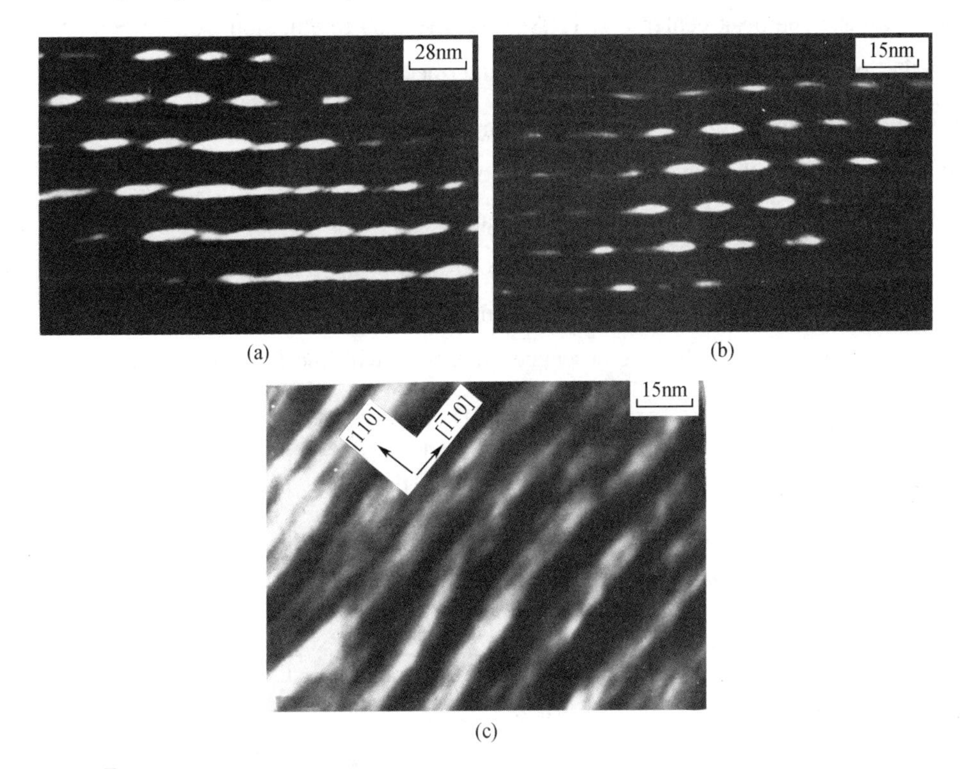

Fig. 2 [$\bar{1}$10] cross-sectional TEM image of stacked nanowire arrays with: (a) six-period InAs (6.5ML)/$In_{0.52}Al_{0.48}As$(20nm); (b) six-period InAs(6ML)/$In_{0.52}Al_{0.48}As$(10nm); (c) Is the planview TEM image of the sample in(a)

The surprising observation of the multilayer arrays is that the nanowires of one layer are positioned in the interstices of the previous layer, exhibiting strong anticorrelation in the growth direction, thus providing further experimental evidence[12] for the vertical anticorrelation predicted by Shchukin et al. [13] According to their theory, if the spacer thickness of multilayer islands is decreased, a transition may occur from vertical anticorrelation to vertical correlation. However, in a series of multilayer samples with different $In_{0.52}Al_{0.48}As$ spacer thicknesses l ($5nm \leq l \leq 20nm$), we observe the same anticorrelation. If l is small enough ($< 5nm$), we might be able to observe vertical correlation. Thus samples with $l < 5nm$ and different InAs coverage are needed to fully understand the phenomenon observed here. Quantitative explanation for the formation of the anticorrelated array should examine the energetics of the array, taking into account the material parameters and shape(facet, height, diameter, etc.) of InAs/InAlAs nanowires.

The representative 15K polarized PL spectra of the samples with six-period InAs (6.5ML)/$In_{0.52}Al_{0.48}As$(10nm) is shown in Fig. 3. The spectra are dominated by emission from the nanowires at 0.65eV, revealing effective carrier capture from the adjacent InAs thin quantum wells. The peak centered at 1.08eV is attributed to emission from the quantum well due to the localization of carriers at random interface fluctuations. As shown in Fig. 3, the peak associated with nanowires exhibits a large anisotropy for the orthogonal [$\bar{1}$10] and [110] polarization with a peak intensity ratio of 3. This polarization anisotropy in PL spectra is compatible with the one-dimensional nature of nanowires. The large polarization anisotropic effect may be related to the enhanced nonlinear optical properties associated with the additional degree of quantum confinement and/or strain.[14] We did not observe any significant polarization for the luminescence associated with the InAs quantum well.

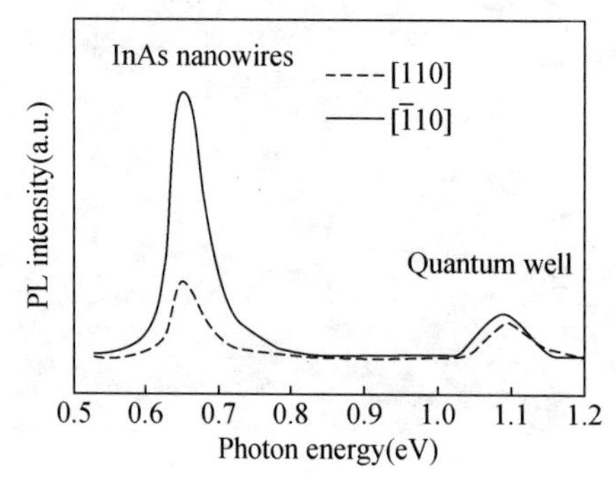

Fig. 3 15K polarized PL spectra of the multilayer sample with six-period InAs(6.5ML)/$In_{0.52}Al_{0.48}As$(20nm). The incident laser beam was polarized in the [110] or [$\bar{1}$10] direction

In conclusion, we have studied the growth behavior of InAs on InAlAs/InP. We find that dense InAs nanowire arrays can be formed spontaneously. A large polarization anisotropic effect is observed in polarization PL measurements. The spatial alignment of the multilayer array exhibits typical anticorrelation in the growth direction.

This work is supported partially by the National Natural Science Foundation of China and the National Advanced Material Committee of China. Daniels-Race and Li acknowledge the partial support of this work by NSF Grant No. NSFECS-95-33780 and the U. S. Department of Energy Grant No. DE-FG02-97ER45648.

References

[1] H. Marchand, P. Desjardins, S. Guillon, J. -E. Paultre, and Z. Bougrioua, Appl. Phys. Lett. 71, 527(1997); A. Ponchet, A. Le Corre, H. L'Haridon, B. Lambert, and S. Salaün, ibid. 67, 1850 (1995), and references therein.

[2] M. Brasil, A. Bernussi, M. A. Cotta, and M. Marquezini, Appl. Phys. Lett. 65, 857(1994).

[3] S. T. Chou, K. Y. Cheng, L. J. Chou, and K. C. Hsieh, J. Appl. Phys. 78, 6270(1995).

[4] A. G. Norman, S. P. Ahrenkiel, H. Moutinho, and M. M. Al-Jassim, Appl. Phys. Lett. 73, 1844 (1998).

[5] M. A. Cotta, R. A. Hamm, T. W. Staley, S. N. Chu, L. R. Harriott, M. B. Panish, and H. Tempkin, Phys. Rev. Lett. 70, 4106(1993).

[6] M. Kasu and N. Kobayashi, J. Cryst. Growth 170, 246(1997).

[7] J. Sudijono, M. D. Johnson, C. W. Snyder, M. B. Elowitz, and B. G. Orr, Phys. Rev. Lett. 69, 2811(1992).

[8] S. Fafard, Z. Wasilewski, J. McCaffrey, S. Raymond, and S. Raymond, Appl. Phys. Lett. 68, 991(1996).

[9] J. Brault, M. Gendry, G. Grenet, G. Hollinger, Y. Désieres, and T. Benyattou, Appl. Phys. Lett. 73, 2932 (1998); A. Weber, O. Gauthier-Lafaye, F. H. Julien, J. Brault, M. Gendry, Y. Désieres, and T. Benyattou, ibid. 74, 413(1999); M. Phaner-Goutorbe, Y. Robach, P. Krapf, A. Solére, and L. Porte, Surf. Sci. 402-404, 268(1998), and references therein.

[10] R. Nötzel, U. Jahn, Z. Niu, A. Trampert, J. Fricke, H. Schönherr, T. Kurth, D. Heitmann, L. Däweritz, and K. Ploog, Appl. Phys. Lett. 72, 2002(1998).

[11] S. Hara, J. Motohisa, and T. Fukui, J. Cryst. Growth 170, 579(1997).

[12] M. Strassburg, V. Kutzer, U. W. Pohl, A. Hoffmann, I. Broser, N. N. Ledentsov, D. Bimberg, A. Rosenauer, U. Fischer, D. Gerthsen, I. L. Krestnikov, M. V. Maximov, P. S. Kop'ev, and Zh. I. Alferov, Appl. Phys. Lett. 72, 942(1998).

[13] V. A. Shchukin, D. Bimberg, V. G. Malyshkin, and N. N. Ledenstov, Phys. Rev. B 57, 12262 (1998).

[14] P. J. Pearah, A. C. Chen, A. M. Moy, K. C. Hsieh, and K. Y. Cheng, IEEE J. Quantum Electron. QE-30, 608(1994).

High power continuous-wave operation of self-organized In(Ga)As/GaAs quantum dot lasers

Z. G. Wang, J. B. Liang, Gong Qian and B. Xu

(Laboratory of Semiconductor Materials Science, Institute of Semiconductors, Chinese Academy of Sciences. Beijing 100083, China)

Abstract Quantum dot (QD) lasers are expected to have superior properties over conventional quantum well lasers due to a delta-function like density of states resulting from three dimensional quantum confinements. QD lasers can only be realized till significant improvements in uniformity of QDs with free of defects and increasing QD density as well in recent years. In this paper, we first briefly give a review on the techniques for preparing QDs, and emphasis on strain induced self-organized quantum dot growth. Secondly, self-organized In(Ga)As/GaAs, InAlAs/GaAlAs and InAs/InAlAs QDs grown on both GaAs and InP substrates with different orientations by using MBE and the Stranski-Krastanow (SK) growth mode at our labs are presented. Under optimizing the growth conditions such as growth temperature, Ⅴ/Ⅲ ratio, the amount of InAs, $In_xGa_{1-x}As$, $In_xAl_{1-x}As$ coverage, the composition x etc., controlling the thickness of the strained layers, for example, just slightly larger than the critical thickness and choosing the substrate orientation or patterned substrates as well, the sheet density of QDs can reach as high as $10^{11}cm^{-2}$, and the dot size distribution is controlled to be less than 10% (see Fig. 1). Those are very important to obtain the lower threshold current density (J_{th}) of the QD Laser. How to improve the dot lateral ordering and the dot vertical alignment for realizing lasing from the ground states of the QDs and further reducing the J_{th} of the QD lasers are also described in detail. Thirdly, based on the optimization of the band engineering design for QD laser and the structure geometry and growth conditions of QDs, a 1W continuous-wave (cw) laser operation of a single composite sheet or vertically coupled In(Ga)As, quantum dots in a GaAs matrix (see Fig. 2) and a larger than 10W semiconductor laser module consisted nineteen QD laser diodes are demonstrated. The lifetime of the QD laser with an emitting wavelength around 960nm and 0.614W cw operation at room temperature is over than 3000hrs, at this point the output power was only reduced to 0.83db. This is the best result as we know at moment. Finally the future trends and perspectives of the QD laser are also discussed.

原载于:1999 IEEE Hong Kong Electron Devices Meeting,1999,2-3.

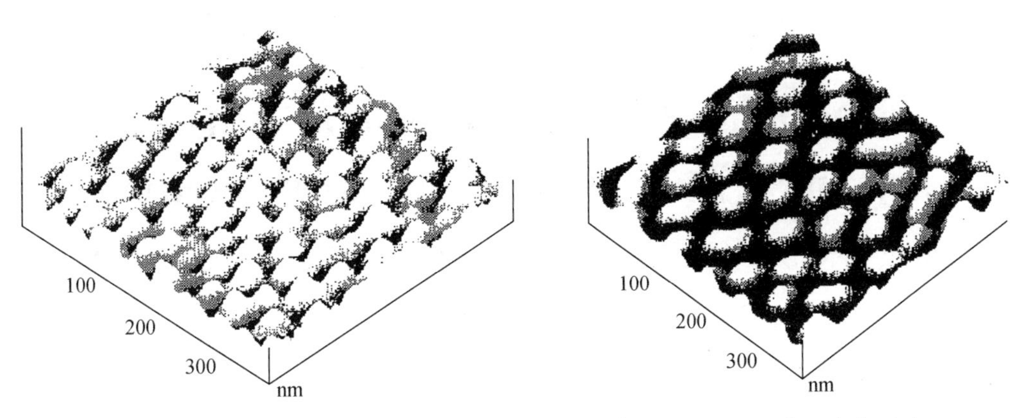

Fig. 1 AFM images of 9ML $In_{0.4}Ga_{0.6}As$ QDs(left) and 13ML $In_{0.3}Ga_{0.7}As$(right) Grown on(311)B GaAs Substrate

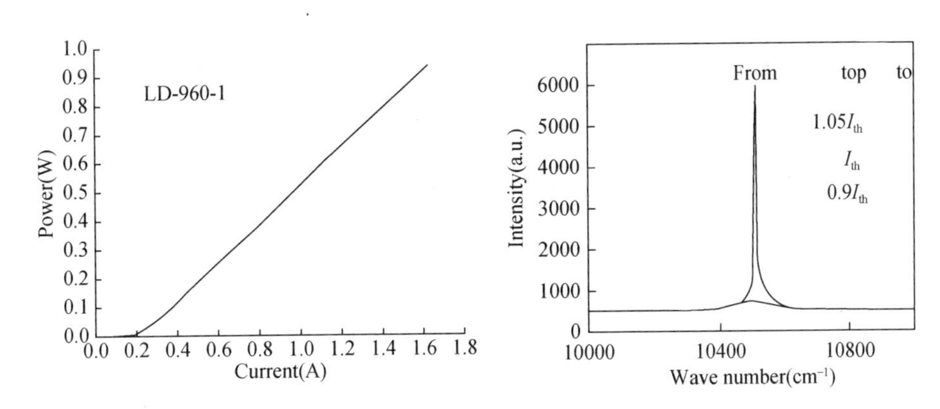

Fig. 2 CW Light Output Power Versus Current Characteristics(left), and Lasing spectra of InAs/GaAs QD Laser(right)

Quantum-dot superluminescent diode: A proposal for an ultra-wide output spectrum

Zhong-Zhe Sun, Ding Ding, Qian Gong, Wei Zhou, Bo Xu and Zhan-Guo Wang

(Laboratory of Semiconductor Materials Science, Institute of Semiconductors, Chinese Academy of Sciences, Beijing, 100083, China)

Abstract We propose a novel superluminescent diode (SLD) with a quantum dot (QD) active layer, which should give a wider output spectrum than a conventional quantum well SLD. The device makes use of inhomogeneous broadness of gain spectrum resulting from size inhomogeneity of self-assembled quantum dots grown by Stranski-Krastanow mode. Taking a design made out in the $In_xGa_{1-x}As/GaAs$ system for example, the spectrum characteristics of the device are simulated realistically, 100–200nm full width of half maximum of output spectrum can be obtained. The dependence of the output spectrum on In composition, size distribution and injection current of the dots active region is also elaborated.

1 Introduction

Superluminescent diodes (SLDs) are the most promising light sources for application in optical measurement systems such as optical gyroscopes and sensors (Burns et al., 1983), optical time domain reflectometry (Takada et al., 1987), and short and medium distance optical communication systems (Friebele and Kersey, 1994). Large spectral width is one of the most important features of SLDs, since the broadband characteristics and consequent short coherence length of SLDs can significantly improve the system performance.

Light emitted from SLDs consists of amplified spontaneous radiation. This causes the spectral width and shape of the emission spectrum to be directly determined by the gain spectrum. According to this principle, some efforts have been devoted to broadening the spectral width of SLD by proper design of device structure and operation condition. At first, introduction of quantum well (QW) structure as an active region of the SLDs was proved to be an effective and simple way to broaden the spectrum. SLDs with QW structure, which

原载于：Optical and Quantum Electronics, 1999, 31(12): 1235–1246.

have spectra two to three times wider than those employing a conventional bulk double-heterostructure (DH) structure, had been obtained in the initial experiments (Chen et al., 1990). Then, some further improvements on QW-SLD bring about wider spectra. For example, (1) large current operation SLDs were designed (Kondo et al., 1992; Semenov et al., 1993, 1996), which introduce the amplification of intrinsic spontaneous emission by $n=1$ and $n=2$ transitions simultaneously; Spectral widths of 68 and 170nm were obtained for the 0.8μm AlGaAs SLDs and 1.5μm InGaAsP SLDs respectively. However, there were some limits in the means, e.g., the wide spectrum is obtained only under certain current density, when gains at $n=1$ and $n=2$ transitions are equal. And a dip will appear in the spectrum if the separation of $n=1$ and $n=2$ transitions is too large. (2) Mikami et al. (1990) designed the SLDs having a stacked active layer structure with different band-gaps. Spectral widths of 80 and 140nm were obtained for 1.3 and 1.5μm SLDs respectively; Lin et al. (1996, 1997) designed SLDs with multi-quantum wells of different widths, which introduces different $n=1$ transition energies. Maximal spectral width of 91.5nm was obtained for 0.8μm SLDs. All the above values of spectral width significantly exceed those of conventional SLDs without the improved factors.

2 Design idea and theory basis

In this paper, we propose a new SLD aimed at wide spectrum-quantum dot superluminescent diode (QD-SLD), which is characterized by the introduction of a self-assembled quantum dots (QDs) active region into conventional SLD structure (Fig. 1(a)). In recent several years, the formation of strained self-assembled quantum dots (Leonard et al., 1993, 1994; Leon et al., 1995) by heteroepitaxial growth in Stranski-Krastanow (S-K) mode has been studied extensively for their fundamental properties and applications in optoelectronics. Such QDs have been successfully fabricated with many different material combination on GaAs substrates, such as, InAs/GaAs, InGaAs/GaAs, and GaSb/GaAs of near-infrared emission, and AlInAs/AlGaAs, InP/InGaP of visible emission. In current self-assembled quantum dot technology, certain size inhomogeneity is common where range of dot size is evidenced in many experiments by spectrally broad photoluminescence (PL) signals, and atomic force microscopy (AFM), transmission electron microscopy (TEM) photographs, typically of nonuniformity of not less than 10%. Such inhomogeneity of QDs is formed naturally in the S-K mode growth, which is essential and not accidental (Ebiko et al., 1998). In general, such inhomogeneous size distribution of QDs in the active region is disadvantageous for achieving lasing of QD laser. However, for the designed QD-SLD, it becomes an effective, intrinsic advantage for broadening gain spectrum, as described below.

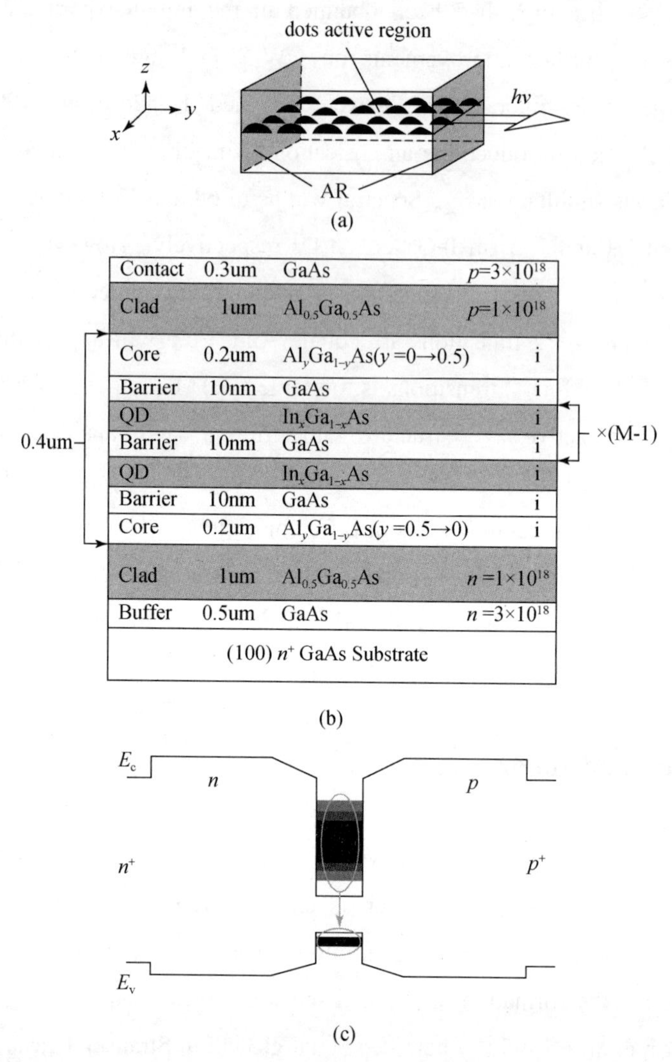

Fig. 1 (a) Schematic view of the geometry of a QD-SLD; (b) GRIN-SCH structure with $In_xGa_{1-x}As$ quantum dot layers and GaAs barrier layers where M is the number of dot layers; (c) the diagram of the conduction band and valence band of a QD-SLD; The occupation of levels in QDs is indicated by shading

We know, the width of the gain spectrum is determined by all of the shapes of the density-of-state (DOS), the relaxation broadening and the thermal distribution. For an ideal QDs ensemble of uniform size or a single QD of a given size, the DOS is given by the δ function; so the gain spectrum will be very sharp because the linewidth of gain is determined only by relaxation broadening. However, for the actual QDs ensemble, size inhomogeneity will smear the sharpness out of the DOS function of the QD structure. For illustration purposes, we assume the QDs are some cubic boxes in an infinite depth potential well, and their sizes follow Gaussian distribution:

$$f(D) = \frac{1}{\sqrt{2\pi}\Delta D}\exp\left[-\frac{(D - D_0)^2}{2(\Delta D)^2}\right],$$

where, D is the side length of a single QD, D_0 is the average side length of QDs; size distribution function $f(D)$ represents the ratio of the density of QD with side length D to the density of total QDs; ΔD is standard deviation, the values of $\Delta D/D_0$ represent the degree of size inhomogeneous distribution. Then, for a single QD, DOS is given by δ function:

$$\rho_{QD}(E) = \frac{1}{V_{QD}}\sum_{l,m,n} 2 \cdot \delta(E - E_{l,m,n}),$$

where, the factor of two is spin degeneracy, V_{QD} the volume of a single QD, l, m, n are the labels of the quantized levels in the dots. While for the actual QDs ensemble with size fluctuation, the DOS is given by:

$$\rho_{QDs}(E) = \frac{\text{the sum of state density per unit energy spacer in total dots}}{\text{the volume sum of total dots}}$$

$$= \frac{\int V_{QD} \cdot \rho_{QD}(E) \cdot N \cdot f(D) \cdot \mathrm{d}D}{\int V_{QD} \cdot N \cdot f(D) \cdot \mathrm{d}D},$$

where, N is the total number of dots in the ensemble: In Fig. 2 is shown schematically the variation of the DOS of QDs ensemble with increase of the degree of size inhomogeneous distribution. It can be seen obviously that the effects of the inhomogeneity are quite dramatic. Only slight size inhomogeneity will make DOS change from ideal δ function to a broadened peaks distribution around some central energies. When the degree of size inhomogeneous distribution increases, the peak value of DOS distribution decreases and the degree of broadening increases, until obvious overlap occurs between the adjacent distribution peaks. The shape of DOS will change from peaks, to waves, up to a flattened slope. Finally, the effect of the quantum confine in producing an abrupt DOS function will be lost entirely.

In a word, the DOS of an actual QDs ensemble is not a sharp δ function but a flattened distribution determined by the degree of size inhomogeneity. So, the shape of gain spectrum will also not be sharp, but be closely related to the shape of DOS and the site of Fermi level, that is, an actual QDs ensemble will result in an inhomogeneously broadened gain spectrum. Moreover, the optical gain spectrum of the QD structure will be much wider than that of QW structure devices at the same injection current. The reasons are twofold: (1) the active region volume of the QD structure is much smaller than that of a conventional QW structure. If we assume similar carrier lifetimes, then, at a given current density, the carrier density in the QDs is much larger than that of the conventional QW active region. This increased density causes the quasi-Fermi level penetrations into the conduction and valence band to increase, so

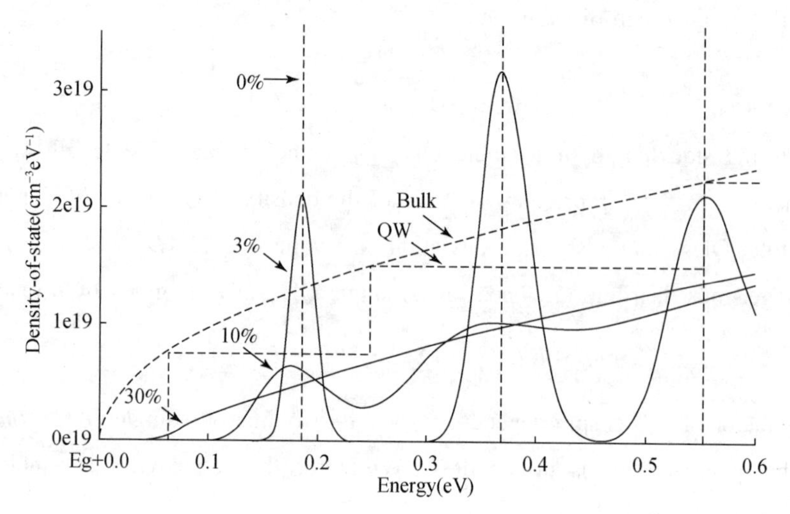

Fig. 2　Variation of density-of-state of QDs ensemble with the increase of the size inhomogeneous distribution; the ratio of standard deviation to the average size of QDs ensemble is signed by percent number. The DOS curves of QW (thickness is equal to D_0) and bulk material are also plotted for comparison

that the gain condition $\hbar\omega < E_{fc} - E_{fv}$ is satisfied over a larger range of ω. (2) Another factor contributing to the increased penetration of the quasi-Fermi level is that the DOS of the QDs ensemble is usually smaller at a given E than the QW value as shown in Fig. 2 (unless ideal size uniformity of the QDs was achieved). This necessitates a larger Fermi energy at a given carrier density and leads to a broader gain spectrum. So, the design SLD with the QDs structure is suitable for producing a wider output spectrum than QW-SLD. As mentioned at the beginning, such superiority on broadening spectrum of QW-SLD to DH-SLD had been proved experimentally (Chen et al., 1996).

It should be indicated that, in the above discussion, an implicit assumption is that a common quasi-Fermi level lies on the QDs ensemble i. e. a quasi-equilibrium distribution can be established. Such an assumption means the rates of carrier relaxation from higher levels to lower levels and carrier transfer between different dots are fast enough as compared to the radiative lifetime of dots. Since relatively fast relaxation with characteristic times of the order of 10 – 100ps has been observed experimentally for $In_xGa_{1-x}As/GaAs$ QDs (Ohnesorge, 1996), our assumption will not cause any significant difference in carriers distribution. So the simplified model is valid for qualitative study on inhomogeneous broadness of gain.

As far as the practical fabrication of a QD-SLD is concerned, SLD is generally achieved from an LD chip with some structure for suppressing lasing, such as antireflection coating (AR) etc. For the past several years, much progress has been made in the growth of self-assembled QDs by S-K mode, and the QD lasers of various material systems have been achieved (Kirstaedtev et al., 1994; Fafard et al., 1996; Kamath et al., 1996; Mirin et al.,

1996; Shoji et al., 1996; Ustinov et al., 1997, 1998), which provides a good basis for the fabrication of QD-SLD. At the same time, we noticed that for the reported QD lasers, there is usually a broadened electroluminescence (EL) spectrum caused by the size fluctuation of QDs for e. g., the full width at high maximum (FWHM) is ~50nm for $In_{0.3}Ga_{0.7}As/GaAs$ (Mirin et al., 1996), ~60nm for $In_{0.4}Ga_{0.6}As/GaAs$ (Kamath et al., 1996), ~70nm for $In_{0.8}Ga_{0.2}As/GaAs$ (Ustinov et al., 1997), ~100nm for (In, Ga) As/GaAs (Shoji et al., 1996), ~20nm for $Al_{0.36}In_{0.64}As/Al_{0.25}Ga_{0.75}As$ [21], and ~50nm for $InAs/In_{0.53}Ga_{0.47}As/InP$ dot lasers (Ustinov et al., 1998). These experimental proofs are indicative of the feasibility of wide spectrum QDSLD.

3 Simulations results and discussion

Taking the well studied $In_xGa_{1-x}As/GaAs$ material system as an example, we give a practical design of QD-SLD with the graded index separate confinement heterostructure (GRIN-SCH) as shown in Figs. 1 (b) and (c); The QD-SLD structure has one or several $In_xGa_{1-x}As$ self-assembled quantum dot layers and GaAs barriers in the center of a ~0.4μm $Al_y Ga_{1-y}As$ ($y=0\to0.5$) core with 1μm $Al_{0.5}Ga_{0.5}$ As clad on each side. The 5μm width current injection stripe and 400μm cavity length is adopted in the preliminary design; Antireflection coating on both facets should provide $<10^{-3}$ facet reflectivity for suppressing lasing.

To illustrate the performance characteristics of designed QD-SLD, we present realistic simulations of gain spectra and output optical spectra of the device. According to most experimental results reported, we assume that: the shape of QD is a flat box with base width D, height h and a fixed aspect ratio $h/D=0.2$ [13], which is closer to the actual shape of the QD than the assumption of cubic box; the areal density of QDs in the active layer is $6\times10^{10}cm^{-2}$ and the average base width D_0 is 20nm. For the QD-SLD with both facets AR coated, the output spectrum of the device is given by (Dutta and Deimel, 1983):

$$P=\frac{\beta\cdot R_{sp}(\hbar\omega)}{G(\hbar\omega)}(e^{G(\hbar\omega)\cdot L}-1),$$

where, $R_{sp}(\hbar\omega)$ denotes the total spontaneous emission rate per volume and $\beta=0.005$ is the fraction of spontaneous emission that is directed in the appropriate solid angle and gets amplified; $G(\hbar\omega)$ is the net gain given by:

$$G(\hbar\omega)=\Gamma g_{QDs}(\hbar\omega)-\alpha,$$

where, $\alpha=10cm^{-1}$ is the absorption, $\Gamma=\Gamma_x\cdot\Gamma_y\cdot\Gamma_z=0.5\times0.5\times0.037$ is the optical confinement factor [24], $L=400\mu m$ is the length of the device, $g_{QDs}(\hbar\omega)$ is the total gain of the QDs active region and is given by:

$$g_{QDs}(\hbar\omega) = \int f(D) \cdot g_{QD}(\omega, D) \cdot dD,$$

where, $g_{QD}(\omega, D)$ is the linear gain of QD of size D and is calculated as in Asada et al. (1986) with strain correcting for the band edge of $In_xGa_{1-x}As$ material.

Following are the calculated results on gain spectra and output spectra of QD-SLD under different In compositions, different size distributions, and different injection currents, which provide some basic evidences for device feasibility and useful guides for device fabrication.

(1) The size distribution is the key factor for achievement of wide spectrum QD-SLD. As shown in Fig. 3, for an $In_{0.7}Ga_{0.3}As$/GaAs QD-SLD under $8\times10^{18}cm^{-3}$ injection carriers density, when $\Delta D/D_0 = 10\%$, FWHM of gain spectra and output spectra achieve 155meV and 14nm respectively. The corresponding coherence length given by $L_c = \lambda_0^2/\Delta\lambda$ is as small as ~8μm. And the spectral widths increase with the increase of the degree of size variation. According to current studies on QDs, the ratio of standard deviation to the average size of the QDs ensemble is usually about 10% or more, so the size distribution required by QD-SLD can be readily satisfied. And, if necessary, increasing appropriately the size variation of QDs in the active region is available in technique (Leon et al., 1995). Furthermore, relatively

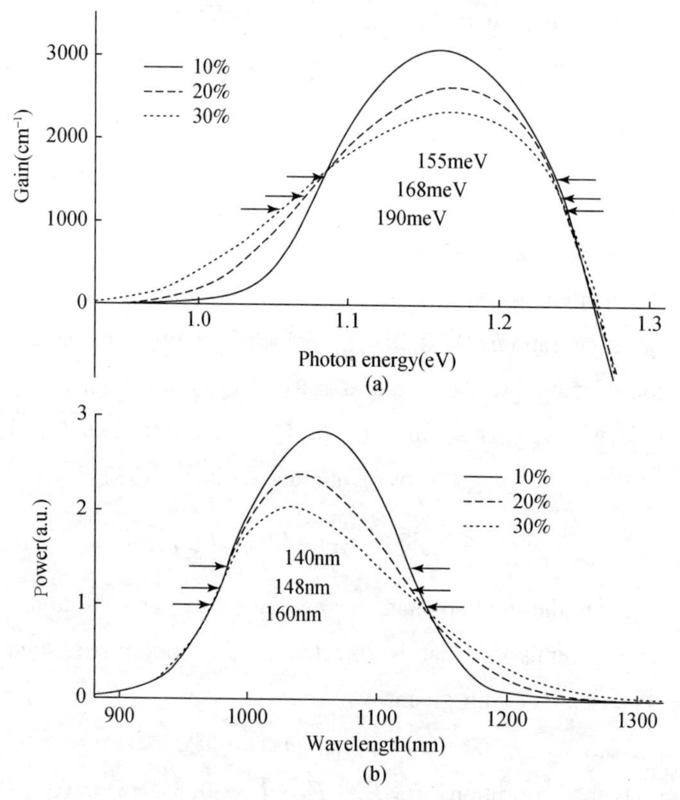

Fig. 3 The gain spectra (a) and power spectra (b) of $In_{0.7}Ga_{0.3}As$/GaAs QD-SLDs with different size variation under injection carrier density ($8\times10^{18}cm^{-3}$)

smaller average size of dots ensemble is more favorable for QD-SLD, because the size non-uniformities of the smaller dots will manifest themselves in the gain spectra more strongly as compared to the larger dots. To obtain smaller dots, higher In composition should be applied in the dots layer according to the study on $In_xGa_{1-x}As/GaAs$ QDs growth (Leonard et al., 1994).

(2) The confinement potential between dots and barriers is another important factor for modifying spectral width. As shown in Fig. 4, for the $In_xGa_{1-x}As/GaAs$ QD-SLDs with same size variation (10%) and under same injection carriers density ($8\times10^{18}\ cm^{-3}$), with increase of In composition, on the one hand, the decrease of band gap will lead to a spectral red-shift, and on the other hand, a larger band offset will lead to carrier distribution in a larger energy range with the size variation of QDs, so spectral width increase; Such effects of the composition on PL have been observed experimentally (Lobo et al., 1998). So, in view of the band offset, higher In composition in dots layer is also more favorable for achieving wide spectrum output.

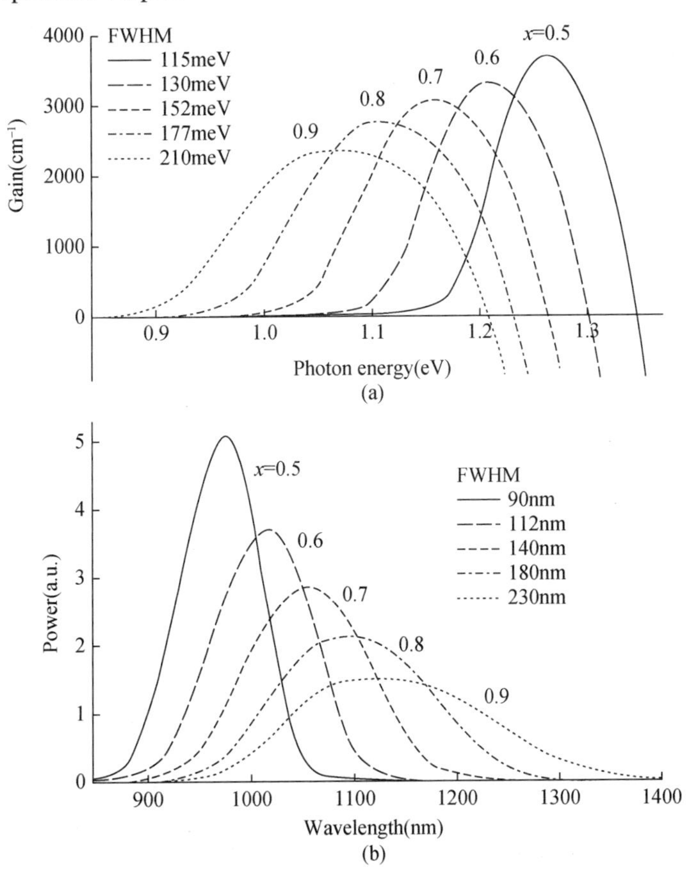

Fig. 4 The gain spectra (a) and power spectra (b) for $In_xGa_{1-x}As/GaAs$ QD-SLDs with different In composition under injection carriers density ($8\times10^{18}\ cm^{-3}$) and with size variation (10%)

(3) Higher power is another request for SLD and usually higher operation current is needed; Fig. 5 shows the gain and power spectra under different injection carrier densities for an $In_{0.7}Ga_{0.3}As/GaAs$ QD-SLD with 10% size variation. It can be seen that, for the QD-SLD with enough size variation, since the shape of gain spectra is smooth, wide spectra can be obtained in a large current range, and no dip structures appear in the power spectrum. The spectral width increases obviously with the increase of current, due to the carrier filling at higher energy levels.

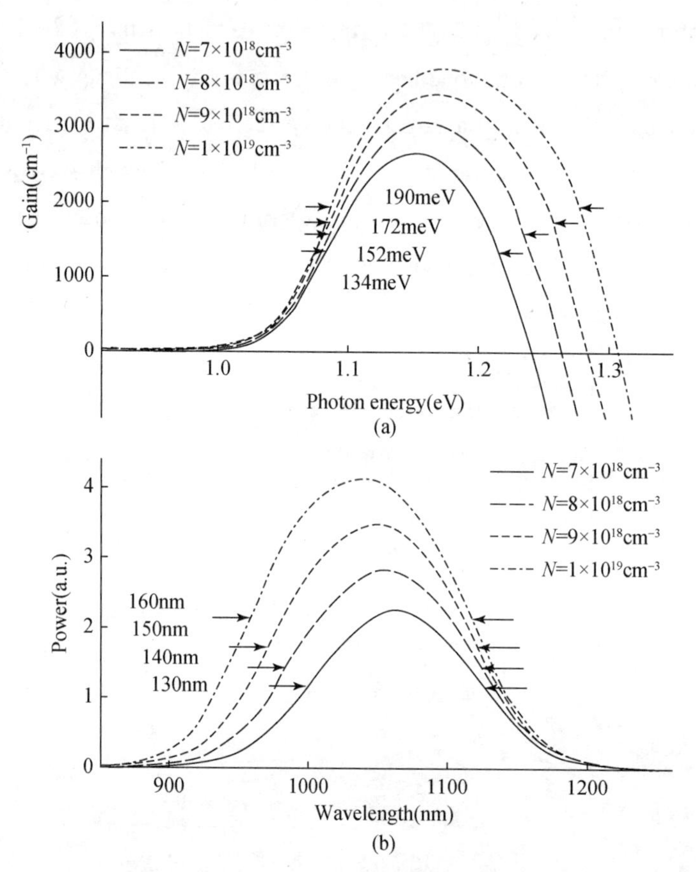

Fig. 5　The gain spectra (a) and power spectra (b) under different injection carrier density for an $In_{0.7}Ga_{0.3}As/GaAs$ QD-SLD with 10% size variation

Furthermore, it is worth noting that, unlike QD lasers, where size inhomogeneity is a factor which decreases the output power (Fafard, 1996), emission from a QD-SLD is contributed by all QDs of each size. And, introduction of multilayer structure of dots will be an effective method to obtain higher power from QD-SLD. In the multilayer structure of quantum dots, to prevent the coupling between dots from decreasing inhomogeneous broadness of gain spectrum (Solomon et al., 1996), the spacer layers should be relatively thick (e. g. 10nm for InAs dots as shown in Fig. 1 (b)).

(4) The above results have shown the basic characteristics of $In_xGa_{1-x}As/GaAs$ QD-SLD. It can be concluded that QD-SLD are promising structures for broadening the spectrum. Moreover, the design of QD-SLD is not restricted to In_xGa_{1-x} GaAs material system. Introduction of various QD material systems (such as InAlAs/AlGaAs, $InAs/In_{0.52}Al_{0.48}As/InP$) will produce more extensive emission in the technologically interesting wavelength region.

4 Conclusions

In summary, we have proposed a new type of SLD, the quantum dot superluminescent diode, which is more suitable for producing wide spectrum than QW-SLD. The key merit of the designed device is that the effects of size inhomogeneity formed naturally in S-K growth of QDs on broadening gain spectrum is unutilized, and the superluminescent diode with a wider spectrum than QW-SLD can be achieved with the current technique.

Acknowledgements

This work was supported by the National Natural Science Foundation of China. The authors would like to thank Y. W. Liu for stimulating discussions.

References

Asada, M., Y. Miyamato and Y. Suematsu. IEEE J. Quantum Electronics QE-22 1915, 1986.

Burns, W. K., C. L. Chen and R. P. Moeller. IEEE/OSA J. Lightwave Technol. LT-1 98, 1983.

Chen, T. R., L. Eng, Y. H. Zhuang, A. Yariv, N. S. Kwong and P. C. Chen. Appl. Phys. Lett. 56 1345, 1990.

Dutta, N. K. and P. P. Deimel. IEEE J. Quantum Electronics QE-19 496, 1983.

Ebiko, Y., S. Muto, D. Suzuki, S. Itoh, K. Shiramine, T. Haga, Y. Nakata and N. Yokoyama. Phys. Rev. Lett. 80 2650, 1998.

Fafard, S., K. Hinzer, S. Raymond, M. Dion, J. McCaffrey, Y. Feng, S. Charbonneau. Science 274 1350, 1996.

Friebele. E. J. and A. D. Kersey. Fiberoptic sensors measure up for smart structure, Laser Focus World 30 (5) 165-171, 1994.

Kamath, K., P. Bhattacharya, T. Sosnowski, T. Norris and J. Phillips. Electron. Lett. 32 1374, 1996.

Kirstaedtev, N., N. N. Ledeustov, M. Grundmann, D. Bimberg, V. M. Stinov, S. S. Ruvimov, M. V. Maximov, P. S. Kop'ev, Zh. I. ALferov, U. Richter, P. Werner, U. Gosele and J. Heydenreich. Electron. Lett. 30 1416, 1994.

Kondo, S., H. Yasaka, Y. Noguchi, K. Magari, M. Sugo and O. Mikami. Electron. Lett. 28

132, 1992.

Leon, R., S. Farfard, D. Leonard, J. L. Merz and P. M. Petroff. Appl. Phys. Lett. 67 521, 1995.

Leonard, D., M. Krishnamurthy, C. M. Reaves, S. P. Denbaars and P. M. Petroff. Appl. Phys. Lett. 63 3203, 1993.

Leonard, D., M. Krishnamurthy, S. Fafard, J. M. Merz and P. M. Petroff. J. Vac. Sci. Technol. B. 12 1063, 1994.

Leonard, D., K. Pond and P. M. Petroff. Phys. Rev. B. 50 11687, 1994.

Lin, C. F., B. L. Lee and P. -C. Lin. IEEE, Photon. Technol. Lett. 8 1456, 1996.

Lin, C. F., B. L. Lee. Appl. Phys. Lett. 71 1598, 1997.

Lobo, C., R. Leon, S. Fafard and P. G. Piva. Appl. Phys. Lett. 72 2850, 1998.

Mikami, O., H. Yasaka and Y. Noguchi. Appl. Phys. Lett. 56 987, 1990.

Mirin, R., A. Gossard and J. Bowers. Electron. Lett. 32 1732, 1996.

Ohnesorge, B., M. Albrecht, J. Oshinowo, A. Forchel and Y. Arakawa. Phys. Rev. B. 54 11532, 1996.

Semenov, A. T., V. R. Shidlovski and S. A. Safin. Electron. Lett. 29 854, 1993.

Semenov, A. T., V. R. Shidlovski, D. A. Jackson, R. Willsch and W. Ecke. Electron. Lett. 32 255, 1996.

Shoji, H., Y. Nakata, K. Mukai, Y. Sugiyama, M. Sugawara, N. Yokayama and H. Ishikawa. Electron. Lett. 32 2023, 1996.

Solomon, G. S., J. A. Trezza, A. F. Marshall and J. S. Harris. Phys. Rev. Lett. 76 952, 1996.

Takada, K., I. Yokohama, K. Chida and J. Noda. Appl. Opt. 26 1603, 1987.

Ustinov, V. M., A. E. Zhukov, A. Yu. Egorov, A. R. Kovsh, S. V. Zaitsev, N. Yu. Gordeev, V. I. Kopchatov, N. N. Ledentsov, A. F. Tsatsul'nikov, B. V. Volovik, P. S. Kop'ev, Z. I. Alferov, S. S. Ruvimov, Z. Liliental-Weber and D. Bimberg. Electron. Lett. 34 670, 1998.

Ustinov, V. M., A. Yu. Egorov, A. R. Kovsh, A. E. Zhukov, M. V. Maximov, A. F. Tsatsul'nikov, N. Yu. Gordeev, S. V. Zaitsev, Yu. M. Shernyakov, N. A. Bert, P. S. Kop'ev, Zh. I. Alferov, N. N. Ledentsov, J. Böhrer, D. Bimberg, A. O. Kosogov, P. Werner and U. Gösele. J. Cryst. Growth. 175/176, 689, 1997.

Optical properties of InAs self-organized quantum dots in $n-i-p-i$ GaAs superlattices

J. Z. Wang, Z. M. Wang, and Z. G. Wang
(Laboratory of Semiconductor Materials Science, Institute of Semiconductors, Chinese Academy of Sciences, Beijing, China)
Z. Yang
(Department of Physics and the Advanced Materials Research Institute, The Hong Kong University of Science and Technology, Clearwater Bay, Kowloon, Hong Kong, China)
S. L. Feng
(National Laboratory for Superlattices and Microstructures, Institute of Semiconductors, Chinese Academy of Sciences, Beijing, China)

Abstract The optical properties of InAs quantum dots in $n-i-p-i$ GaAs superlattices are investigated by photoluminescence(PL) characterization. We have observed an anomalously large blueshift of the PL peak and increase of the PL linewidth with increasing excitation intensity, much smaller PL intensity decrease, and faster PL peak redshift with increasing temperature as compared to conventional InAs quantum dots embedded in intrinsic GaAs barriers. The observed phenomena can all be attributed to the filling effects of the spatially separated photogenerated carriers.

Self-assembled InAs quantum dots(QDs) are a system for both basic physical study and practical applications due to their properties of zero-dimension and the resulting atomic-like, discrete energy states. The optical properties of self-organized QDs have been intensively studied to elucidate the quantum states and their potential applications in QD lasers and photodetectors.[1,2] Many basic physical properties, such as many-body effects, electric transport, and thermal effects in QDs have been widely studied.[3-7] Recently, QD structures that could separately store photogenerated electrons and holes were designed and studied, both for their fundamental properties and potential memory device applications. So far there have been only a few reports in this field,[8,9] and the detailed carrier separation effects in QDs have still not been investigated. In this letter, we report the unusual dependence of photoluminescence(PL) properties on pumping intensity and temperature of InAs self-assembled QDs in $n-i-p-i$ GaAs superlattices, which proved to be a structure that can effectively

原载于：Appl. Phys. Lett., 2000, 76(15): 2035-2037.

separate photogenerated carriers,[10-12] and show that such PL properties are the result of separation of photogenerated electrons and holes.

The samples were grown by molecular-beam epitaxy on a (001) N^+ GaAs substrate. After the deposition of a 1μm semi-insulating GaAs buffer layer, ten periods of *n*-GaAs (40nm)/*p*-GaAs(40nm) were grown. The *n* and *p* dopants are Si and Be, respectively, and the *n* and *p* doping concentration is $1\times10^{18}cm^{-3}$. In the middle of each *n*-and *p*-GaAs layer, a 2ML InAs QD layer was inserted. For comparison, a reference sample with twenty 2ML InAs QDs separated by 40nm of intrinsic GaAs was also grown. The growth temperature was 520℃ for the entire structure. Here, we name QDs in the *n*-GaAs, *p*-GaAs, and intrinsic GaAs layers *n*–*n*, *p*–*p*, and *i*–*i* QDs, respectively. The PL measurements were performed using a Fourier transform spectrometer equipped with an InGaAs detector. The spectra were obtained under excitation from the 514.5nm line of an argon ion laser.

Fig. 1 shows the schematic band diagram of the *n*–*i*–*p*–*i* sample along the growth direction. Most of the photogenerated electrons and holes will be spatially separated by the doping potential into *n*–*n* and *p*–*p* QDs, respectively, resulting in the accumulation of electrons in *n*–*n* QDs and holes in *p*–*p* QDs. Possible optical transitions are marked by the solid arrows in Fig. 1. Due to the large spatial separation (40nm of the GaAs barrier layer) and the strong localization of these separated electrons and holes, the possibilities of nonvertical transitions, which would produce PL at lower photoenergies, should be very small. Vertical transitions in *n*–*n* and *p*–*p* QDs are expected to completely dominate the optical properties of the *n*–*i*–*p*–*i* sample. Thus, we can simply expect that the PL of the *n*–*i*–*p*–*i* sample should be very weak, and filling effects of electrons to the conduction band of *n*–*n* QDs and holes to the valence band of *p*–*p* QDs should be very obvious.

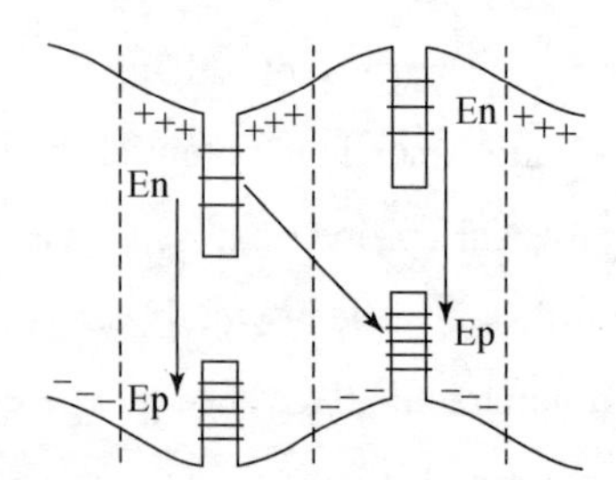

Fig. 1 Band diagram of the *n*–*i*–*p*–*i* sample along the growth direction. The arrows indicate the possible optical transitions

Figs. 2(a) and 2(b) show the PL spectra of the *n*–*i*–*p*–*i* sample and the *i*–*i* sample at 15K under excitation intensities of 0.5, 2, 8, and 32W/cm². It is seen that under the same excitation intensity, the PL peak energy of the *n*–*i*–*p*–*i* sample is higher than that of the *i*–*i* sample and the PL intensity of the *n*–*i*–*p*–*i* sample is about ten times weaker than that of the *i–i* sample. With increasing excitation intensity from 0.5 to 32W/cm^2, for the *n–i–p–i* sample, a large blueshift of the PL peak (by 18meV) and an increase of the PL linewidth (by 20meV) are observed [inset of Fig. 2(a)]. Whereas for the *i–i* sample, both the PL peak and linewidth remain almost the same (both increased by about 3meV). The high-energy PL peak is an indication that the PL originates from the vertical transitions. The large blueshift of the PL peak and increase of the PL linewidth with increasing excitation intensity result from the filling effects of the spatially separated photogenerated carriers. Note that under a low excitation intensity of 0.5W/cm^2, when the filling effects of the photogenerated carriers are relatively weak, the PL peak energy of the *n–i–p–i* sample is about 30meV larger than that of the *i–i* sample, this is expected and can be attributed to the filling effects of carriers from dopants.

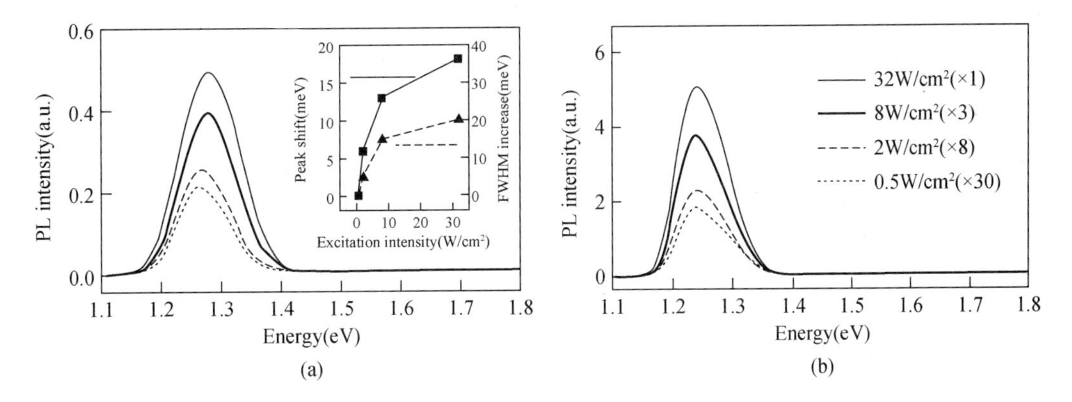

Fig. 2 PL spectra of (a) the *n–i–p–i* sample and (b) the *i–i* sample at 15K under excitation intensities of 0.5, 2, 8, and 32W/cm^2. For clarity, these spectra of both samples are multiplied, which are shown in (b). The inset in (a) shows the PL peak blueshift and the increase of the PL peak full width at half maximum of the *n–i–p–i* sample with increasing excitation intensity; the lines are guides for the eyes

Fig. 3(a) shows the temperature dependence of the integrated PL intensity of both the *n–i–p–i* sample and the *i–i* sample under an excitation intensity of 10W/cm^2. It is seen that at high temperatures, the intensity decrease of the *n–i–p–i* sample (ten times from 15 to 250K) is much smaller than the *i–i* sample (300times). The rapid decrease in PL intensity of the *i–i* sample is consistent with the reported results of similar *i–i* InAs/GaAs QDs and can be attributed to the thermal activation of charge carriers from the QDs into the GaAs barriers followed by nonradiative recombination.[5-7] For the *n–i–p–i* sample, when the temperature is increased, some of the electrons in the *n–n* QDs will be thermally activated out and fall into the *p–p* QDs, and give off PL by recombining with the large amount of trapped holes there. Likewise, some of the holes in the *p–p* QDs can go to the *n–n* QDs and recombine with the electrons there. This additional increase of PL intensity with increasing temperature will slow down the overall PL intensity decrease in the $n-i-p-i$ QD sample. The temperature-dependent separation effect of photogenerated carriers is similar to the case in GaAs $n-i-p-i$ superlattices and type-I hetero-$n-i-p-i$ superlattices.[11,12]

Fig. 3(b) shows the PL peak energy as a function of temperature of the $n-i-p-i$ QD sample and the $i-i$ QD sample. The solid line is for the InAs band gap calculated according to the Varshni law, and are vertically shifted for clarity (the GaAs band gap is also shown by the dotted line). It is seen that the PL peak redshift with increasing temperature of both the $i-i$ and $n-i-p-i$ samples is faster than that of the InAs band-gap shrinkage at T<200K. The faster redshift of QDs compared to their bulk material band-gap shrinkage is known to be caused by the carrier redistribution between coupled QDs which favors carrier population into QDs with lower ground-energy states.[5-7] However, from 15 to 250K, the redshift of the $n-i-p-i$ sample is 97meV as compared to 70meV for the $i-i$ sample. The larger redshift of the $n-i-p-i$ sample is the result of the loss of high-energy electrons in the $n-n$ QDs and holes in the $p-p$ QDs due to thermal activation. The redshift of the two samples eventually follows the Varshni law of InAs, as seen in Fig. 3(b). For the $i-i$ sample, the redshift is similar to the shrinkage of the InAs band gap when the temperature is above 170K. For the $n-i-p-i$ sample, the redshift tends to follow the band-gap shrinkage of InAs when the temperature is increased to 225K.

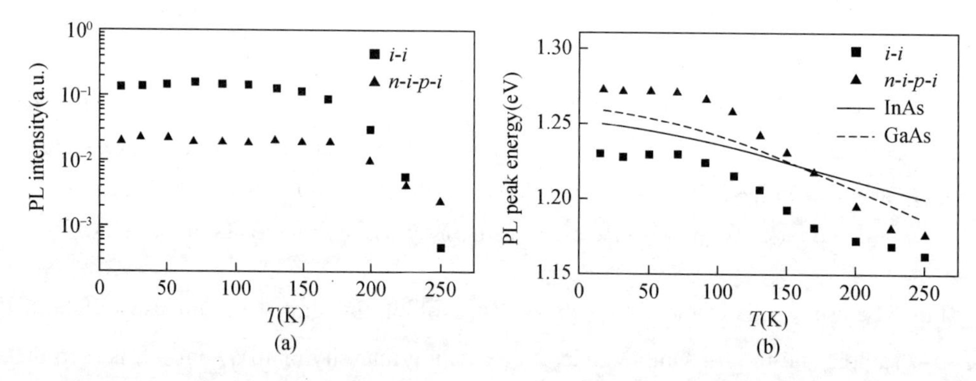

Fig. 3 (a) Integrated PL intensity and (b) PL peak energy as a function of temperature under an excitation intensity of 10W/cm^2. The solid line is for the InAs band-gap calculated according to the Varshni law using the parameters of InAs and are shifted along the energy axis (the GaAs band gap is also shown as the dotted line and are vertically shifted). The temperatures are 15, 30, 50, 70, 90, 110, 130, 150, 170, 200, 225, and 250K

In summary, we have observed the anomalous PL properties of InAs QDs in an $n-i-p-i$ GaAs superlattice. The observed large blueshift of the PL energy peak and linewidth with increasing the excitation intensity, the much smaller PL intensity decrease, and the faster PL peak redshift with increasing temperature can all be attributed to the filling effects of the spatially separated photogenerated carriers.

This work is supported by the National Natural Science Foundation of China.

References

[1] J. Y. Marzin, J. M. Gérard, A. Izraë, D. Barrier, and G. Bastard, Phys. Rev. Lett. 73, 716 (1994).

[2] S. Sauvage, P. Boucaud, F. Glotin, R. Prazeres, J. M. Ortega, A. Lemaître, J. M. Gérard, and V. Thierry-Flieg, Appl. Phys. Lett. 73, 3818(1998).

[3] A. Wojs and P. Hawrylak, Phys. Rev. B 55, 13066(1997).

[4] W. Heller, U. Bockelmann, and G. Abstreiter, Phys. Rev. B 57, 6270(1998).

[5] U. H. Lee, D. Lee, H. G. Lee, S. K. Noh, J. Y. Leem, and H. J. Lee, Appl. Phys. Lett. 74, 1597(1999).

[6] D. I. Lubyshev, P. P. González-Borrero, E. Marega, Jr., E. Petitprez, N. La Scala, Jr., and P. Basmaji, Appl. Phys. Lett. 68, 205(1996).

[7] A. Polimeni, A. Patané, M. Henini, L. Eaves, and P. C. Main, Phys. Rev. B 59, 5064(1999).

[8] G. Yusa and H. Sakaki, Appl. Phys. Lett. 70, 345(1997).

[9] W. V. Schoenfeld, T. Lundstrom, P. M. Petroff, and D. Gershoni, Appl. Phys. Lett. 74, 2194 (1999).

[10] G. H. Döhler, H. Künzel, D. Olego, K. Ploog, P. Ruden, H. J. Stolz, and G. Abstreiter, Phys. Rev. Lett. 47, 864(1981).

[11] P. Ruden and G. H. Döhler, Surf. Sci. 132, 540(1983).

[12] K. Köhler, G. H. Döhler, J. N. Miller, and K. Ploog, Superlattices Microstruct. 2, 339(1986).

High-performance strain-compensated InGaAs/InAlAs quantum cascade lasers

Feng-Qi Liu[1], Yong-Zhao Zhang[1], Quan-Sheng Zhang[1], Ding Ding[1], Bo Xu[1], Zhan-Guo Wang[1], De-Sheng Jiang[2] and Bao-Quan Sun[2]

([1]Laboratory of Semiconductor Materials Science, Institute of Semiconductors, Chinese Academy of Sciences, Beijing 100083, China)

([2]National Laboratory for Superlattices and Microstructures, Institute of Semiconductors, Chinese Academy of Sciences, Beijing 100083, China)

Abstract We report on the realization of quantum cascade (QC) lasers based on strain-compensated $In_xGa_{1-x}As/In_yAl_{1-y}As$ grown on InP substrates using molecular beam epitaxy. X-ray diffraction and cross section transmission electron microscopy have been used to ascertain the quality of the QC laser materials. Quasi-continuous wave lasing at $\lambda \approx 3.54-3.7\mu m$ at room temperature was achieved. For a laser with 1.6mm cavity length and 20μm ridge-waveguide width, quasi-continuous wave lasing at 34℃ persists for more than 30min, with a maximum power of 11.4mW and threshold current density of $1.2kA/cm^2$, both record values for QC lasers of comparable wavelength.

Quantum cascade (QC) lasers are a fundamentally new semiconductor laser source. They are not only renewing the field of mid-infrared injector lasers, but also represent a source of novel unconventional ideas for semiconductor lasers in general. QC lasers are based on electronic transitions between quantized conduction band states of a multiple quantum well structure and are grown by molecular beam epitaxy[1-3]. The wavelength of QC laser is essentially determined by the layer thickness of the active region rather than by the bandgap of the material. As such, it can be tailored over a wide range using the same heterostructure material[4-6]. QC lasers for the first atmospheric window (3–5μm) are important for a variety of commercial and military applications. However, in intersubband QC lasers, the short-wavelength operation is limited by the size of the conduction band discontinuity ΔE_c, which exists between the two semiconductor materials. To achieve a QC laser with wavelength shorter than 4μm, it is required to use strain-compensated $In_xGa_{1-x}As/In_yAl_{1-y}As$ ($x>53\%$,

原载于：Semicond. Sci. Technol, 2000, 15(12): 44–46.

$y<52\%$) materials, which operate in the active region of the QC laser, because this strain-compensated material system gives enlarged conduction band discontinuity[6]. This approach adds flexibility in the QC laser design by allowing a selection of the desired discontinuity but also adds the constraint that the tensile strain balances the compressive strain in the structure. A strain-compensated InGaAs/InAlAs QC laser grown on InP was first demonstrated in 1998 to achieve a laser operating at a wavelength of 3.4μm. However, the maximum operating temperature and output power of these strain-compensated InGaAs/InAlAs devices is limited relative to the corresponding lattice-matched devices. The most arduous problem is the difficulty of fabricating high-quality laser material. Reducing the threshold current density and enhancing the output power of a QC laser and its operating temperature are important for device applications. Here we demonstrate our results obtained on strain-compensated $In_xGa_{1-x}As$/$In_yAl_{1-y}As$ QC lasers, operating at a wavelength which is as short as 3.54–3.7μm. Very low threshold current density of $1.2kA/cm^2$ at 34℃ is realized. The breakthrough is that quasi-continuous wave operation at 34℃ with output power 11.4mW persists for more than 30min without obvious degradation.

The strain-compensated InGaAs/InAlAs laser structures are grown by molecular beam epitaxy (MBE) on n-doped InP (Si, $1\times10^{18}cm^{-3}$) substrates in a Riber 32p MBE system. The active region of each laser structure consists of 25 superlattice periods, which are alternating n-doped injector regions and undoped triple-quantum-well (wafer B1143) or double-quantum-well (wafer B1146) active regions. The laser structures are similar to that described in[4, 7, 8]. The complete laser structures are schematized in Fig. 1. The active regions of QC lasers are designed as partially strain compensated, while the remaining strain is compensated in the cladding layers. The net strain in the complete structure tends to zero. The active region of sample B1143 is based on the $In_{0.55}Ga_{0.45}As$/$In_{0.51}Al_{0.49}As$ alloy pair, and the layer sequence of one period of the structure, in nanometres starting from the injection barrier, is **4.5**/0.7/**1.4**/4.2/**2.0**/3.6/**2.7**/2.1/**2.1**/2.0/**2.0**/1.8/**1.8**/1.8/**2.1**/1.7/**2.5**/1.7nm. $In_{0.51}Al_{0.49}As$ layers are in bold, $In_{0.55}Ga_{0.45}As$ layers are in roman and n-doped layers (Si, $2\times10^{17}cm^{-3}$) are underlined. The active region of sample B1146 is based on the $In_{0.55}Ga_{0.45}As$/$In_{0.5}Al_{0.5}As$ alloy pair; the layer sequence is **6.1**/4.3/**2.5**/3.5/**2.4**/2.0/**2.0**/1.9/**1.9**/1.8/**1.8**/1.8/**1.8**/1.6/**1.6**/1.5/**1.8**/1.4/**2.0**/1.4/**2.2**/1.3nm, and the n-doping level is 3×10^{17}.

The control of the laser materials and the interface quality are of vital importance in improving the device performance and reproducibility. In particular, problems arise from factors such as lack of control (during growth) of the composition of complex ternary systems, the quality of interfaces and layer thickness control and ineffective carrier doping and distribution. The device characteristics are very much influenced by the material

InGaAs	20nm	$8\times10^{18}cm^{-3}$
InGaAs/InAlAs graded	25	7×10^{18}
InAlAs	2000	$2\times10^{17}\rightarrow7\times10^{18}$
InGaAs/InAlAs graded	30	2×10^{17}
InGaAs	300	1×10^{17}
(Active+Injector)25×		
InGaAs	300	1×10^{17}
InGaAs/InAlAs graded	25	3×10^{17}
InP substrate		*n*-doped

Fig. 1　Schematic cross section of the complete $In_xGa_{1-x}As/In_yAl_{1-y}As$ laser structures grown by MBE. Indicated are the *n*-type doping levels and the layer thicknesses in nanometres. The thicknesses of 25 period active regions are 1018 and 1220 nm for sample B1143 and sample B1146, respectively

quality[9]. After optimizing the growth condition, high-quality QC laser materials are grown. The structures were grown without interruption. An optimum growth temperature of 520℃ was used. Fig. 2 shows a typical X-ray diffraction (XRD) spectrum for a 25-period $In_{0.55}Ga_{0.45}As/In_{0.5}Al_{0.5}As$ QC laser waveguide core (sample B1146) and the complete $In_{0.55}Ga_{0.45}As/In_{0.51}Al_{0.49}As$ QC laser structure (sample B1146) grown on InP substrates. The satellite peaks have excellent periodicity and narrow linewidths. This demonstrates that compositional gradients in the growth direction are negligible, otherwise the linewidths would not be as narrow. A transmission electron microscope (TEM) micrograph of laser structure B1146 is displayed in Fig. 3. The InGaAs material is the lighter and the InAlAs has the darker shade. This picture clearly shows that the interfaces are very sharp and flat. As an example, the 1.3nm well is clearly visible and its thickness is constant across the three periods of the structure that are displayed. In general, the measured thicknesses agree with the XRD results. The TEM and the XRD results clearly demonstrate that the strain-compensated laser materials are of high quality.

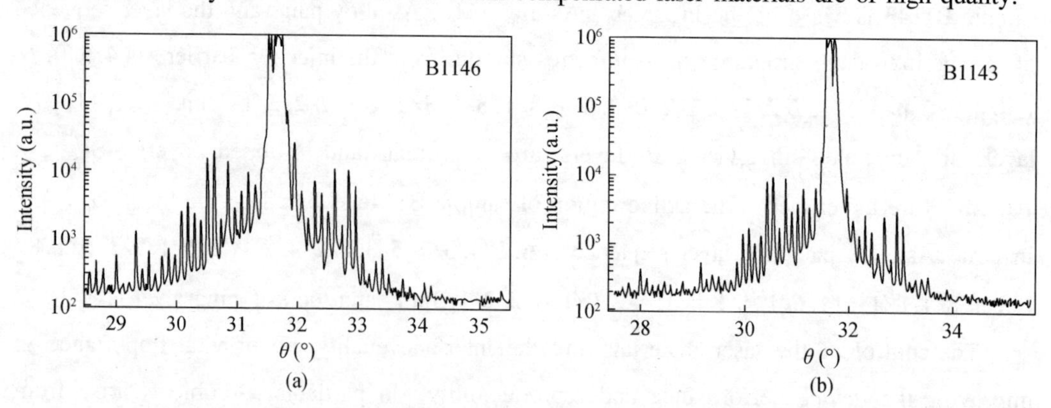

Fig. 2　XRD spectra: (a) 25 period $In_{0.55}Ga_{0.45}As/In_{0.5}Al_{0.5}As$ QC laser waveguide core (sample B1146). (b) The complete $In_{0.55}Ga_{0.45}As/In_{0.51}Al_{0.49}As$ QC laser structure (sample B1143). The satellite peaks have excellent periodicity and narrow linewidths

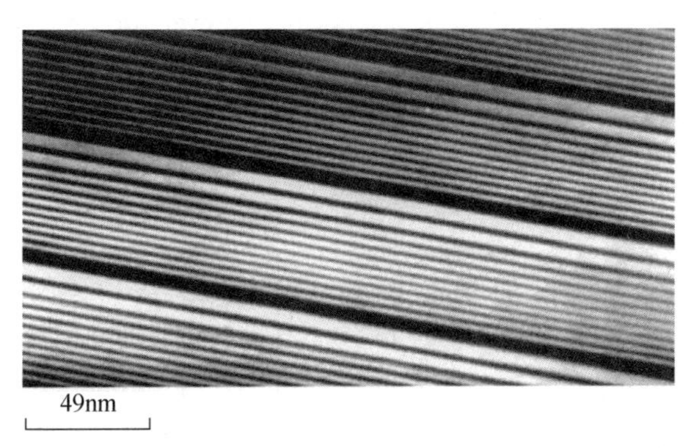

Fig. 3 Transmission electron micrograph of a portion of the cleaved cross section of a QC laser (sample B1146). Three periods of the 25-stage structure are shown. The white bands are the InGaAs alloy layers, the dark ones the InAlAs alloy layers. The largest dark band is the InAlAs injection barrier

Devices fabricated from the sample wafers have been processed into ridge mesas of various widths (10–30μm) by photolithography and wet chemical etching in a HBr : HNO_3 : H_2O solution. A SiN_xO_y layer coverage thickness of 400nm was deposited around the mesa for electrical isolation. AuGeNi/Au was used for top and bottom contacts. No annealing of the contacts was performed. The lasers were cleaved into cavities of various lengths, with the facets left uncoated for testing. All devices were tested with epitaxial-layer-side down bonded to Cu heat sinks in ambient conditions. The spectral characteristics of the devices were measured with a Fourier transform infrared spectrometer in continuous scan. The laser emission was collected using a Ge lens. A liquid nitrogen-cooled HgCdTe detector was used for detection of the laser radiation. The lasers were driven with 7.2μs pulses at a 6kHz repetition rate at room temperature (32℃). Typical laser spectra are shown in Fig. 4. The emission wavelength of the lasers, centred at 3.54μm for B1143 and 3.7μm for B1146, is in very good agreement with the theoretical prediction. This fact and the results of XRD measurements on the structure show agreement between MBE growth and the appropriate modelling of the band structure of the strained $In_{0.55}Ga_{0.45}As$ in the active region. To evaluate accurately the optical power, light-current curves are measured using a black-body radiation optical power meter in ambient conditions. Measured in pulsed operation, the lasers operate up to 34℃ and exhibit the lowest reported threshold current density ($1.2kA/cm^2$) for a QC laser at 34℃. Fig. 5 shows the quasi-CW output power versus current for a typical device at RT. The threshold current densities for the two samples differ. The threshold current densities are 1.2 and $2.7kA/cm^2$ for B1143 and B1146, respectively; while the slope efficiencies are about $dP/dI = 12.4$ and $19mWA^{-1}$ for B1143 and B1146, respectively. For sample B1143, slightly lower threshold current densities are expected because of the shorter laser wavelength (and the shorter injector) and thus smaller waveguide losses (and larger gain) compared to sample B1146. These lasers also showed impressive results, with quasi-CW peak power 11.4mW at 34℃, and persist in lasing for more than 30min without obvious degradation. This is further evidence of the material quality; these are the highest values reported for a QC laser.

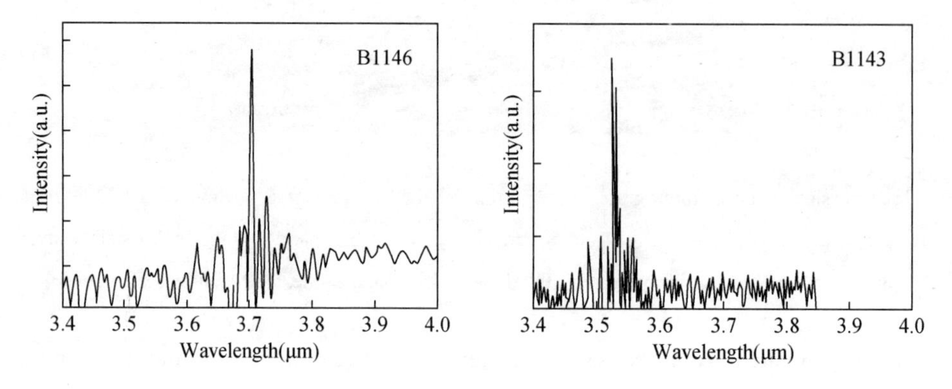

Fig. 4　Lasing spectra of the QC lasers measured at 32℃ in ambient conditions. The two lasers are 1.6mm long and 20μm wide. Pulses are 7.2μs long at a repetition rate of 6kHz

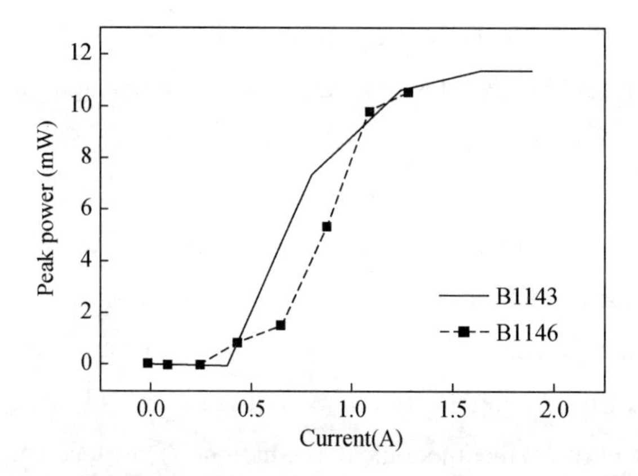

Fig. 5　Measured peak power from a single facet versus drive current at 34℃ for two QC lasers. The two lasers are 1.6mm long and 20μm wide. Pulses are 7.2μs long at a repetition rate of 6kHz

In summary, our paper is intended mainly to result in improvement in the QC laser characteristics and average output power available at wavelengths shorter than 4μm and ambient operating temperature. The main goal was to achieve, by improved multi-quantum-well design, higher gain. As a result, we obtain laser sources at the technologically important wavelength of 3.54–3.7μm operating at 32–34℃. We report on the realization of a QC laser based on strain-compensated $In_xGa_{1-x}As/In_yAl_{1-y}As$ growth on InP substrate using molecular beam epitaxy. Quasi-CW lasing at $\lambda \approx 3.54$–$3.7\mu m$ above room temperature was achieved. This strain-compensated QC laser displays threshold current density as low as

1.2kA/cm^2 at 34℃. The pronounced feature is that quasi-CW operation at 34℃ with a maximum power of 11.4mW persists for more than 30min without obvious degradation. This fascinating feature of QC lasers may promote their potential applications.

This work was supported by the National Natural Science Foundation and National Advanced Materials Committee of China (contract numbers 69786002 and 715-001-0111, respectively). XRD was measured using the Beijing Synchrotron Radiation Facility. The authors would like to acknowledge the help of Chang-Zhi Guo, Shui-Lian Chen, Quan-Jie Jia, Wen-Li Zheng and Zhong-Jun Chen.

References

[1] Faist J, Capasso F, Sivco D L, Sirtori C, Hutchinson A L and Cho A Y 1994 Science 264 553.

[2] Faist J, Capasso F, Sirtori C, Sivco D L, Baillargeon J N, Hutchinson A L and Cho A Y 1996 Appl. Phys. Lett. 68 3680.

[3] Slivken S, Jelen C, Rybaltowski A, Diaz J and Razeghi M 1997 Appl. Phys. Lett. 71 2593.

[4] Faist J, Capasso F, Sivco D L, Hutchinson A L, Chu S N G and Cho A Y 1998 Appl. Phys. Lett. 72 680.

[5] Strasser G, Gianordoli S, Hvozdara L, Schrenk W, Unterrainer K and Gornik E 1999 Appl. Phys. Lett. 75 1345.

[6] Tredicucci A, Gmachl C, Capasso F, Sivco D L, Hutchinson A L and Cho A Y 1999 Appl. Phys. Lett. 74 638.

[7] Faist J, Tredicucci A, Capasso F, Sirtori C, Sivco D L, Baillargeon J N, Hutchinson A L and Cho A Y 1998 IEEE J. Quantum Electron. 34 336.

[8] Faist J, Capasso F, Sirtori C, Sivco D L, Hutchinson A L, Chu S N G and Cho A Y 1996 Superlatt. Microstruct. 19 337.

[9] Razeghi M 1999 Microelectron. J. 30 1019.

Research and development of electronic and optoelectronic materials in China

WANG Zhan-guo

(Laboratory of Semiconductor Materials Science, Institute of Semiconductors, The Chinese Academy of Sciences, Beijing 100083, China)

Abstract A review on the research and development of electronic and optoelectronic materials in China, including the main scientific activities in this field, is presented. The state-of-the-arts and prospects of the electronic and optoelectronic materials in China are briefly introduced, such as those of silicon crystals, compound semiconductors, synthetic crystals, especially nonlinear optical crystals and rare-earth permanent magnets materials, etc., with a greater emphasis on Chinese scientist's contributions to the frontier area of nanomaterials and nanostructures in the past few years. A new concept of the trip chemistry proposed by Dr. Liu Zhongfan from Peking University has also been described. Finally the possible research grants and the national policy to support the scientific research have been discussed.

1 Bulk materials

1.1 Silicon materials

Electronic materials play more and more important role in making electronic devices and integrated circuits. It was predicted that up to FY2000 the world electronics market would reach about USD $200 billion, in which 95% is silicon-related electronic devices and systems. Unfortunately, China only has a productive capability of 80tons polycrystal silicon and 300tons silicon single crystals a year, and epi-ready silicon wafers are 0.5% of the total wafers output in the world (4 billion square inch). However, a polycrystal silicon factory with annual produce of 1000 tons and two 200mm wafer mass production lines are under construction, and in the meantime, fabrication of 300mm silicon crystal is in its experimental phase. Hopefully, we would reach 5% of the total world silicon production by the year of 2005.

原载于:Chinese Journal of Semiconductors, 2000, 21(11): 1041-1049.

1.2 Semi-insulating GaAs crystals

Undoped Semi-Insulating (SI) GaAs crystals are widely used for high-speed devices, MMIC and GaAs IC's, which are important to the highspeed computation, mobile optical fiber communications and the military purpose as well. Compared with silicon, the situation of SI-GaAs in China is relatively good. SI-GaAs crystal wafer of 50 – 75mm can meet the commercial requirements in domestic market, and the pulling SI-GaAs crystals with a diameter 100mm is still in a short production run. SI-GaAs crystal with a diameter of 150mm is prepared in the laboratory. By the end of year of 2005. China is expected to have the annual capability of SI-GaAs crystals of 3tons, which is about 6% of the world annual yield.

1.3 Synthetic crystals

Growth and study of the synthetic crystals, especially of the NonLinear Optical (NLO) crystals belong to high-tech fields, in which China is in the leading position. Synthetic crystals can be applied to some important industries, military, scientific and technological fields and daily life, such as the generation and frequency-tuning of lasers, optical communication, optical storage, optical calculations and laser processing and surgery etc.

The crystal research in China has gradually entered into "crystal engineering" stage. According to requirements of the applications, we are now able to design, explore and grow new crystals. A number of new crystals with Chinese characteristic such as Cesium triborate ($C_3B_3O_5$, CBO), BBO, Lithium triborate (LBO), KDP (KH_2PO_4), Ce: $BaTiO_3$, etc. have been discovered during the past two decades, during which China occupies the leading position in the world. The NLO crystals are potentially useful materials for the harmonic of high power density in laser systems, especially the tunable UV generation to 185nm by Sum Frequency Generation (SFG).

Techniques for growing of large-sized Ti: Al_2O_3 (Ti: sapphire) crystals by temperature gradient method have been developed at the Shanghai Institute of Optics and Fine Mechanics in early 90's. High perfection and excellent optical homogeneity $\Delta n \leqslant 10^{-5}$/inch for the laser rods have been obtained, as well as the Ti doping concentration of 0.6wt%, high FOM (Figure of Merit) value of 200 – 300 and the high optical damage threshold of 15J/cm^2. Combining the semiconductor laser diodes or diode arrays, some kinds of all solid-state tunable lasers are obtained, i. e. a prototype of fs laser with a power of 120mW and a pulse width of 28fs has been produced.

Besides various kinds of artificial crystal, such as Cr : Tm : Ho : YAG laser crystals, a new type of photorefractive crystal, KNSBE, Fe : Mg : $LiNbO_3$ crystals and tunable laser materials of Cr^{3+}LiCAF, Cr^{3+}LiSAF, etc. are also synthesized in labs.

1.4 Rare-earth permanent magnet material

China has an abundance of rich rare-earth, with the reserves being about 80% of the total in the world. The verified and the prospective reserves of rare-earth elements in China are 48million tons and over 100 million tons respectively, which are distributed over Baotou of Inner Mongolia and Jiangxi province. In 1999, the total output of rare-earth is 6×10^4 tons and that of pure oxides and metals are 2.5×10^4 tons, as shares about 70% of the global market.

The rare-earth permanent magnet materials, especially NdFeB magnets, have been explored widely in recent years both in China and out of China. The global output of sintered NdFeB was 285tons in 1987 and 9600tons in 1998, while the output of NdFeB magnets in China was 4000tons in 1998 and 5180tons in 1999. However, the grade of the production yielded in China is still very low.

On the other hand, the rare-earth giant magnet stricture materials, Tb-Dy-Fe with<110> oriented, and 20mm in diameter, has been developed and of the international advanced standard. Research on the rare-earth element related to the electroluminescence materials is also in progress.

1.5 Ceramics powder with high dielectric constant for multilayer ceramic capacitors

More than 200tons of ceramics powder with high dielectric constant and low firing temperature have been produced in China, ten percent of which was exported in 1998. About 5 billion pieces of Multilayer Ceramics Capacitors (MLCC) were fabricated.

Two kinds of $BaTiO_3$, based and relaxed the ferroelectric composite materials, are studied systematically, and the dielectric constant, in the range of 4600–5200, has been obtained.

2 Superlattice and quantum well structures

In early 80's, Chinese scientists and engineers began to work with the GaAlAs/GaAs (InGaAsP/InP etc.) superlattice and quantum well materials using the home-made MBE (MOCVD) system. After their solving many unexpected problems caused by the homemade machines, the stability of the system and growth conditions were greatly improved. The electron mobility of GaAlAs/GaAs two dimensional electron gas(2DEG), of $1.14\times10^6 cm^2/(V\cdot s)$, was obtained at 4.8K in 1993, which demonstrates that the quality of 2DEG materials reached the international advanced state. The experimental results, as shown in Fig. 1, demonstrate that the roughness of 2DEG GaAlAs/GaAs interface grown by us is less than one atomic layer.

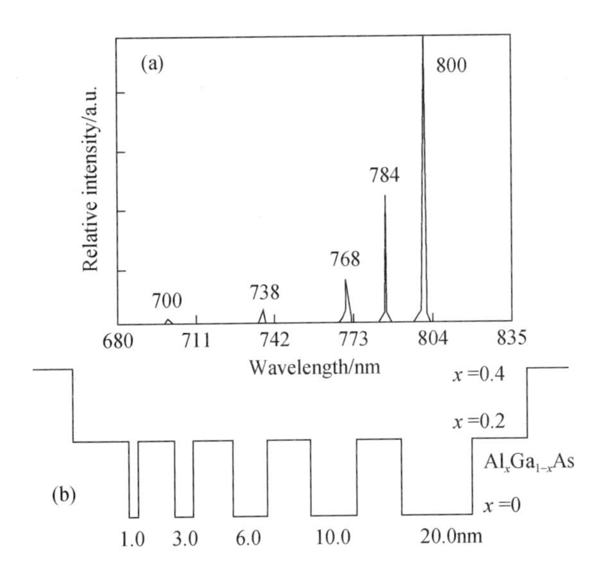

Fig. 1 AlGaAs/GaAs Quantum Well Structure(a) and Corresponding PL Spectra(b)

On the basis of high quality 2DEG materials, HEMT and PHEMT devices with excellent performance (for an example, GaAs based PHEMT, $g_m = 690mS/mm$, $f_T = 82GHz$) are realized at labs. In the meantime, high power quantum well lasers in pilot production and the module of 1.55μm Distributed Feed Back(DFB) lasers/EA modulator for optical fiber communication are successfully fabricated. In addition, the heterostructures have also been prepared for the vertical cavity surface emission laser and Self-Electronic-Optic Effect Devices(SEED), etc. in some institutes and universities in China.

High performance semiconductor laser in the mid-infrared region are desired to be applied to the remote chemical sensing, infrared countermeasures, biomedical diagnosis and free space optical communication, etc. In recent years, the Quantum Cascade (QC) laser based on both intersubband[1] and interband[2] transitions have been demonstrated successfully in InGaAs/InAlAs and type Ⅱ InAs/InGaSb/AlSb quantum well structures, respectively, with the wavelength ranging from 3.4 to 13μm. The intersubband QC lasers based on InGaAs/InAlAs quantum well structures have been fabricated at Institute of Shanghai Metallurgy, Chinese Academy of Sciences(CAS) and Institute of Semiconductors, CAS as well. The mid-infrared QC laser emitting at the wavelength of 5μm and 8μm respectively were operated in pulsed mode at the temperature up to 220K. Fig. 2(left) is the schematic band diagram of the active and the injective regions for InGaAs/InAlAs intersubband QC laser, while the Fig. 2 (right) is the high-resolution cross-section TEM image for the QC laser structure. They indicate that the high quality structure has been obained with dislocaton-feee and smooth interfaces.

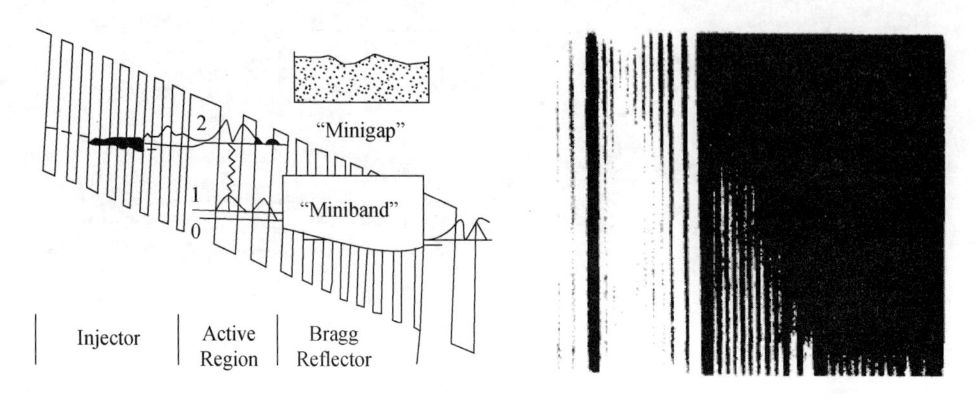

Fig. 2　Schematic Band Diagram of InGaAs/InAlAs Intersubband QC Laser(left), High-Resolution Cross-Section TEM Image of Structure(right)

Similar idea has been proposed in early 70s by a Chinese scientist, about the high power laser structure based on the interband quantum tunneling and the photon coupling between the multi quantum well active regions. A 5-W Continuous-Wave (CW) laser operation of three active regions coupling at room temperature with wavelength of 980nm has been demonstrated recently, which indicates its potential application in the optical pumping source of optical fiber amplifier.

3　Studies on nanomaterials and nanostructures in China

The preferable properties of nanomaterials and nanostructures (or low dimensional materials and quantum devices), which are quite different to those of bulk materials, have attracted much attention during the past decades. To these materials, the effects of quantum's size, quantum tunneling, quantum interference, surface and interface, as well as the non-linear optical effect will all lead to their distinctive properties. A new type of the device with these materials has been proposed and demonstrated. The important contributions to this new field made by Chinese scientists in past few years are reported briefly in this paper.

3.1　Carbon nanotubes and gallium nitride nanorods

Many groups in China are working on the study of carbon nanotubes. Prof. Xie Sishen et al. from Institute of Physics, Chinese Academy of Sciences, have succeeded in preparing "very long" (~1cm) well-aligned carbon nanotube arrays, which makes it easier to study the mechanical and electrical properties of the carbon nanotubes. And a single well carbon nanotube with a diameter of 0.5nm was fabricated recently by his group, which is very close to the theoretical limitation of 0.4nm. Prof. Fan Shoushan's group, from Tsinghua University, on the other hand, has made GaN nanorods via a carbon nanotubes-confined reaction[3], in which Ga_2O vapor reacts with NH_3 gas in the presence of carbon nanotubes to form GaN nanorods. The nanorods have a diameter of 4–50 nanometers and a length up to 25 micrometers. It is proposed using the carbon nanotubes as a template to confine the reaction, as results in the free-standing hexagonal wurtzite GaN nanorods, which have a similar diameter to that of the original nanotubes after the evaporation of the carbon nanotubes via heat treatment at a higher temperature. Fan Shoushan et al.[4] have succeeded in the synthesis of massive arrays of monodispersed carbon nanotubes on patterned porous silicon substrates and the plain ones. The well-ordered self-oriented nanotubes have been tested, which are uniformly on typical 2cm×2cm silicon substrates and acting as electron field emitter arrays with the results shown in Fig. 3.

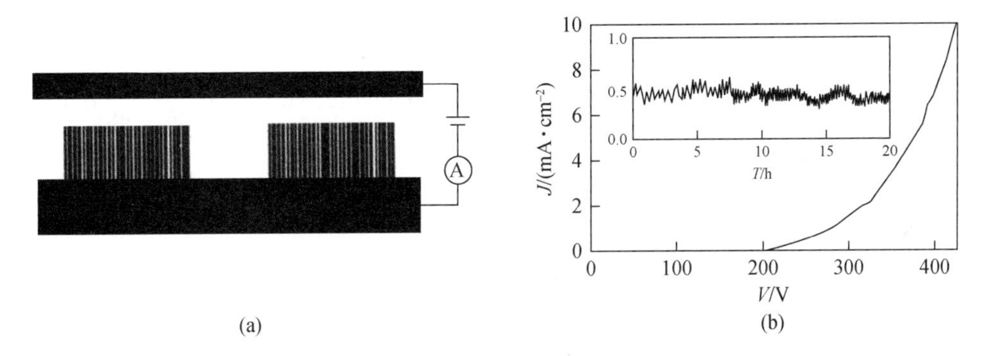

Fig. 3 Self-Oriented Nanotube Arrays as Electron Field Emitter Array. (a) Experimental Setup. The cathode consists of a n-type porous silicon substrate with 250μm by 250μm nanotube blocks. The height of the blocks is 130μm. Aluminum-coated silicon substrate serves as the anode and is kept 200μm a way from the sample by a mica spacer containing a hole in the center. (b) Current Density Versus Voltage Characteristics

In addition, Chinese scientists have also made important contributions to the discovery of nanobalance for the following purpose of: (1) measuring the mass of a single nanoparticle, (2) preparation of large-sized C_{60} single crystal (~13mm long and ~6mm in diameter), (3) hydrogen storage with carbon nanotubes, (4) identification of Buckyball (a C_{60} molecular) orientation on silicon surface and (5) preparation of nanocables, such as SiC core with SiO_2 sheathing, TaC superconductivity core with graphite or SiO_2 sheathing, etc.

3.2 Strain-induced self-assembled quantum dots and quantum wires[5-7]

The strain-induced self-assembled technique in lattice mismatched system, which is on the basis of Stranski-Krastonow(SK) growth mode, has been widely applied in fabrication of density-packed defect-free QDs, which might lead the great developments of nanoscale opto-electronic devices in future. However, the uniformity and density of QDs, especially the quantum wires(QWRs), are extremely difficult to control, as makes it impossible to obtain the favorable performance of nanoscale devices as theoretical predictions. Recently, it is found by Zhanguo Wang's group in Institute of Semiconductors, CAS that by adjusting the growth conditions, the density and ordering of self-assembled InGaAs/GaAs, InAs/InAlAs/InP and InAs/InGaAs/InP QDs and QWRs can be controlled, which are grown by SK growth mode using MBE technique. At the same time, we observed the surprising alignment of dots and their remarkable elongation under specific condition. Spontaneous formation of InAs QWRs in InAlAs/InP(001) via sequential chain-like coalescence of QDs along[1-10] is realized. Theoretical calculation based on the energetic of interacting steps proves the experimental results. Sequential coalescence of initial isolate dots reduces the total free energy greatly. Thus the wire-like structure is rather favorable. This fascinating crossover from QDs to QWRs might become a new approach to the fabricate high-density QWRs. Multi-stacked bilayer InAs/InAlAs has been grown with the purpose of proving the uniformity and shapes of QWRs. Contrary to the result observed in vertically coupled InAs/GaAs QDs array, the anti-correlation in the growth direction of InAs/InAlAs multilayer array was firstly discovered in our lab(see Fig. 4).

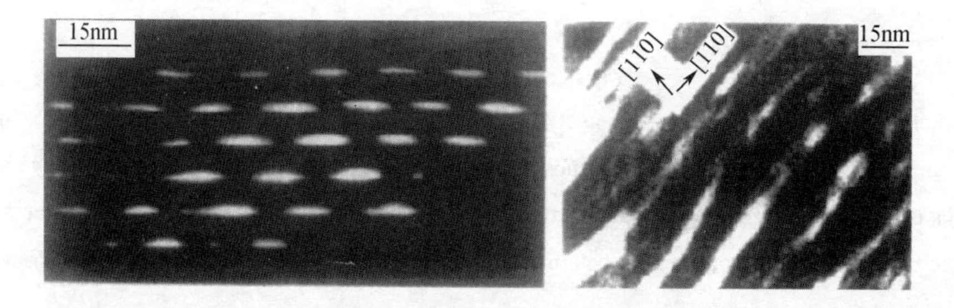

Fig. 4　Plan-View TEM Image of Six-period 6.5ML InAs/20nm InAlAs(left) and [1-10] Cross Sectional TEM Image of Six-Period 6ML InAs/10nm InAlAs(right)

To explain this abnormal feature, the strains of two-dimensional(2D) QWR arrays in infinite isotropic materials are calculated on the basis of principle of minimum energy. Fig. 5 shows the calculated strains of 2D QWR arrays. If the vertical distance between QWR arrays is larger enough than the critical value, the materials above the QWRs are laterally compressed while the materials above interstice of QWRs are laterally expanded. The vertical anti-correlation can be obtained when the spatial distance between QWR arrays is larger than the critical value, which is mainly determined by the width of QWRs. The observation is well explained by our model.

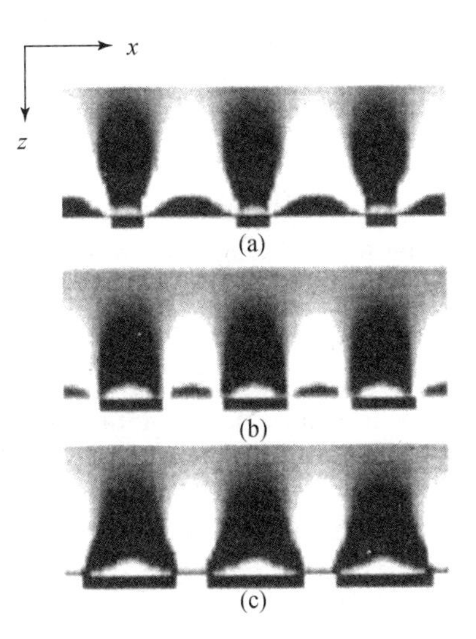

Fig. 5 Calculated Strain Component ε_x of Two-Dimensional QWR Arrays in $x-z$ Plane for (a) $\omega=d/4$, (b) $\omega=d/2$ and (c) $\omega=3d/4$. Here w is the width of the QWRs and d is the in-plane period of the QWR arrays. The black rectangular represents the QWRs

Comparing the self-assembled QDs and QWRs, we find their potential applications in quantum electronic or optoelectronic devices. Consequently, based on the optimization of band gap engineering design for QD lasers, structure geometry and the growth conditions of QDs, 3.62W (two facets) CW laser operation of three-stacked vertically coupled InAs/GaAs QDs at Room Temperature (RT) has been demonstrated. The lifetime of the QD laser is over 3000hr, which is of an emitting wavelength about 960nm with CW operating at 0.614W at RT. At this point the output power was reduced to 0.83db, which is the best result as far as we know.

3.3 Concept of tip chemistry[6]

The concept, "tip chemistry", was proposed by Dr. Zhongfan Liu, Peking University. The essence is to study various physicochemical properties of supramolecular assemblies by chemically modified SPM tips with special nanoparticles, which play a leading role in serving as a nanoprobe, nanolens and nanobeaker of a chemical reaction. The contributions made by his group is as follows: (1) AFM tip is used a force probe to examine the local chemical reaction properties. It has been modified with an organic monolayer to have a specific chemical interaction with the sample surface, so that the surface reaction processes can be monitored by measuring the specific tip-sample interaction. The dissociation properties of acid-base SAMs and the electrochemical reduction-oxidation of azobenzene SAMs are also studied in this way, as indicates that it is really an effective approach to explore the local reaction properties of inhomogenous surfaces, (2) Should tip be used as a nanolens to induce spatially-confined chemical reactions, the highly localized surface chemical reaction for nano-structure can be induced due to the highly-localized fields between the SPM tip and the sample surface, which is created by electric field, etc. The nanooxidation of Si and GaAs using conductive AFM tips and the nanolithgraphy have been studied, based on an organized colloidal nanoparticles mask. (3) To study the single electron phenomena, SPM tip is modified, with an organic SAM and colloidal nanoparticles serving as the tunneling barrier and Coulomb island, respectively. Positioning SPM tips on a SAM-modified surface can form a double-barrier tunneling structure with extremely low capacitance. Single electron tunneling phenomena have been investigated in such self-assembled nanostructures at room temperature with Au, CdS and PbS as Coulomb islands.

And besides, they have succeeded in the chemical treatment on single walled carbon nanotubes and the fabrication of various nanotube assemblies on silicon or gold surfaces with nanotubes on them regularly.

3.4 Solvothermal preparation of non-oxide nanomaterials

It is worthless for Prof. Yitai Qian's group at the University of Sciences and Technology of China to make a great progress with the solvothermal preparation of non-oxide nanomaterials. As we know, those non-oxide materials, such as Nitride, Carbides and Boride, are prepared conventionally by direct combination of elements at high temperatures. Some non-oxides can be obtained by the complex organometallic precursors in organic solvents at certain temperatures, however, the post-treatment at a higher temperature is needed, as makes it difficult to obtain the nanocrystalline materials. Solvothermal progress is a non-aqueous thermal process, which is relatively simple and easy to control. And the sealed system can effectively prevent the contamination by the toxic and air-sensitive starting materials, too.

GaN with a diameter of 30nm was successfully prepared by the reaction of gallium halide on Lithium nitride at 300℃ in Benzene. Nanocrystalline InAs was also obtained through the co-reduction of $AsCl_3$ and $InCl_3$ by using metallic Zinc at 180℃. In addition, nanomaterials such as InP, FeAs, SnS_2 and ternary semiconductors of $CuInSe_2$, $CuSbS_2$ etc. were obtained in temperatures of 200℃ below. The morphology of nanomaterials is controlled by the solvents. The nanorods, such as CdSe, InAs, $CoFe_2$ etc., were successfully prepared in corresponding solvents. Recently, PbSe and SnSe nanorods have been fabricated in organic solvent at room temperature under standard atmospheric pressure.

4 Research grants and national policy

There are many grants to support various research programs in China, such as National Natural Science Foundation of China (NSFC), National Basic Research Project (NBRP), National High Technology R&D Program (HTR&D, also called as "863" program), National Five-Year Planning Project (FYPP), Foundation of the Education Ministry and Foundation of the Chinese Academy of Sciences, etc. The NSFC and NBRP are mainly for the fundamental studies, while the HTR&D and FYPP especially for the R&D projects. It is in nation's consideration to encourage the combination of institutes and/or universities and the industry or companies for more founding. The total founding is about RMB 10 billion (~USD $1.2 billion) per annum, not including that from industry and companies nor for some special projects. For example, the founding for No. 909 project (0.35μm IC's production line in Shanghai) amounts to more than RMB 10 billion, about 10% of which is used for the research on materials and engineering. The founding for the study concerned with electronic and optoelectronic materials are less than 5% in total. China is a developing country and relatively poor. To catch up with the developed countries, we have to make more efforts and have a long way to go.

Nearly 15000 Chinese scientists and engineers are working in the fields of materials science and engineering distributed over more than 30 national and ministerial key laboratories and about 300 groups in China. One third of them are engaged in the study on electronic and optoelectronic materials. At least ten or more national conferences in this field have been held in China, and hundreds of researchers take part in various international conferences annually abroad. More than 2000 referred papers per year are published in international authoritative journals.

Acknowledgements

With acknowledgements to Professors Fan Shoushan, Professor Liu Zhongfan, Professor Qian Yitai, Professor Wang Zhenxi, Professor Zhang Lide, Professor Li Aizheng, Professor Sheng Guangdi and my colleagues at Laboratory of Semiconductor Materials Sciences, CAS for their providing the useful materials and figures for this paper. Special thanks to Dr. Chen

Yonghai, who spent a lot of time helping me to draw figures. This work is partly supported by the NSFC and the NAMCC(National Advanced Materials Committee of China).

References

[1] J. Faist, F. Capasso, D. L. Sivco et al., Science, 1994, 264: 553.

[2] R. Q. Yang, Superlattices Microstruct., 1995, 17: 77.

[3] Weiqiang Han, Shoushan Fan, Qunqing Li and Yongdan Hu, Science, 1997, 277: 1287.

[4] Shoushan Fan et al., Science, 1999, 283: 512.

[5] F. Q. Liu, Z. G. Wang et al., Phys. Lett., 1998, A249: 555.

[6] J. Wu. Z. G. Wang et al., J. Crystal Growth, 1999, 197: 95.

[7] H. X. Li, Z. G. Wang et al., Appl. Phys. Lett., 1999, 75: 1173.

[8] Z. F. Liu, Universitatis Pekinensis, 1998, 34: 161.

半导体量子点激光器研究进展

王占国

（中国科学院半导体研究所 半导体材料科学实验室，北京）

摘要 首先简要地回顾了半导体激光器发展的历史和量子点激光器所特有的优异性能，进而介绍半导体量子点及其三维量子点阵列的制备技术。然后分别讨论了量子点激光器（能带）结构设计思想，实现基态激射所必须具备的条件和近年来国内外半导体量子点激光器的研究进展。最后分析讨论了量子点激光器研制中存在的问题和发展趋势。

Semiconductor quantum dot lasers

Wang zhan-guo

(Laboratory of semiconductor Materials Science, Institute of Semiconductors, Chinese Academy of Scienes, Beijing 100083, China)

Abstract The history of semiconductor laser diodes, and especially quantum dot (QD) lasers with their superior lasing characteristics, are briefly reviewed. Typical techniques for fabricating semiconductor QDs and three-dimensional QD arrays are described. Various ideas for the band energy structure design and the conditions for realizing lasing via the ground states of QD lasers are presented, and recent exciting research advances are summarized in detail. Finally, the key problems for improving the properties of QD lasers, future trends and prospects for this new type of laser diode are discussed.

1 引言[1]

1962 年，第一个砷化镓（GaAs）同质 *PN* 结激光器在 4K 实现受激辐射，开辟了半导体激光器（LD）的新时代。它与几乎同时发明的光导纤维技术一起，奠定了今天光纤通信技术的基础。1963 年，Alferov，Kazarinov[2] 和 Kroemer[3] 提出的双异质结概念，是继半导体激光器发明以来的一个重要里程碑，使 GaAlAs/GaAs 激光器的性能得

原载于：物理，2000，29（11）：643-648.

到显著改善；虽然与起源于衬底的穿透位错和缺陷相关的暗线缺陷导致的激光器早期退化问题，曾一度使激光器实用化的步伐进展缓慢，但随着低位错 GaAs 衬底研制成功和外延层质量的提高，半导体激光器在 20 世纪 70 年代末 80 年代初已有商品出售基于 1978 年 Casey 和 Punish[4] 等以及稍后的 Tsang（1981）[5] 和 Hersee（1982）[6] 等提出的电子、光子分别限制的概念和分子束外延（MBE）及金属有机物化学气相淀积（MOCVD）技术的不断发展与完善，导致了 1980 年 GaAlAs/GaAs 量子阱激光器（QWLD）的研制成功，这是半导体激光器发展历史上又一个里程碑，具有更重要的意义。载流子受到一维量子限制的 QWLD 与双异质结激光器相比，具有更低的阈值电流密度（J_{th}）和 J_{th} 对温度依赖的不敏感以及窄的增益谱等，因而受到广泛重视，大大加快了半导体激光器，特别是大功率激光器及其阵列的实用化步伐，并逐步形成了高技术产业。

理论分析表明，当材料的特征尺寸在 3 个维度上都与电子的德布罗意波长或电子平均自由程相比拟或更小时，电子在材料中的运动受到三维限制，也就是说电子的能量在 3 个维度上都是量子化的，称电子在 3 个维度上都受限制的材料为量子点。图 1 是不同维度材料和相应的态密度函数。如果量子点的最低两个分立量子能级能量差大于几倍的 kT（室温约 25meV），那么就不会出现增益函数的热依赖或激光依赖的展宽。这种具有类原子的态密度函数的量子点激光器（QDLD），可望具有比 QWLD、量子线激光器更加优异的性质，如超低阈值电流密度（$J_{th} \leqslant 2A/cm^2$，目前最好的 QWLD 的 $J_{th}=50A/cm^2$）、极高的阈值电流温度稳定性（理论上 $T_0=\infty$）、超高的微分增益（至少为 QWLD 的一个量级以上）和极高的调制带宽以及在直流电流调制下无啁啾工作等。量子点激光器已显示出从大功率、光计算到光纤数字传输用高速光源以及红外探测器等方面的极重要的应用前景，是目前国际上最前沿的重点研究方向之一。

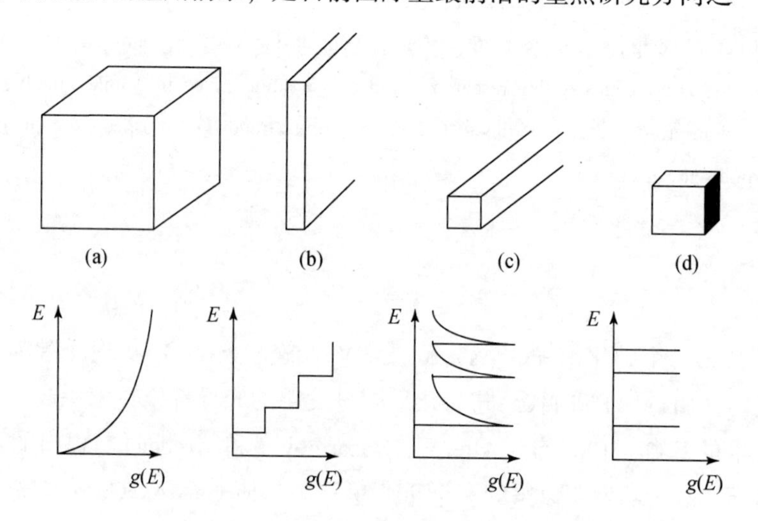

图 1　不同维度材料（上）和相应的态密度函数（下）

（a）三维体材料；（b）二维量子阱材料；（c）一维量子线材料；（d）零维量子点材料

2 量子点的制备技术

量子点的制备方法很多，但常用的有两种。一是微结构材料生长与微细加工技术相结合的方法[7]，即采用 MBE 或 MOCVD 技术在图形化衬底上进行选择性外延生长或高质量的外延材料生长，结合高空间分辨电子束直写、干法或湿法刻蚀，然后再进行外延生长。这种方法的优点是 QD 的尺寸、形状和密度可控，但由于加工带来的界面损伤和工艺过程引入的杂质污染等，使其器件性能与理论的预言值相差甚远。1994 年，Hirayama 等[8]用上述方法制备的 InGaAs/InGaAsP 量子点激光器，在 77K 时，J_{th}仍高达 7. 6kA/cm^2！

为此，近年来人们利用 SK 生长模式，又发展了应变自组装制备 QD 的新技术[9]。SK 生长模式适用于晶格失配较大，但表面、界面能不是很大的异质结材料体系。SK 外延生长初始阶段是二维层状生长，通常只有几个原子层厚，称之为浸润层［图 2（a）］随层厚的增加，应变能不断积累，当达到某临界厚度 t_c 时，外延生长则由维层状生长过渡到三维岛状生长［图 2（b）］，以降低系统的能量。三维岛生长初期形成的纳米量级尺寸的小岛周围是无位错的。若用禁带宽度大的材料将其包围起来，小岛中的载流子受到三维限制，称之为量子点［图 2（c）］。在生长的单层量子点基础上，重复上述的生长过程，可获得量子点超晶格结构。量子点超晶格结构究竟是垂直对准还是斜对准，依赖于隔离势垒层的厚度和量子点顶层应力的分布［图 2（d）为垂直对准，图 2（e）为斜对准］。在 MBE 生长量子点时，可用高能电子衍射仪通过衍射斑点形状的变化直接控制量子点的形成。这种方法的缺点是其尺寸、形状、分布均匀性和密度较难控制。图 3 是我们实验室利用 SK 生长模式分别在 InP 基和 GaAs 基衬底上生长的三维量子线点阵的 TEM 截面像和 InAs 量子点的 AFM 图。

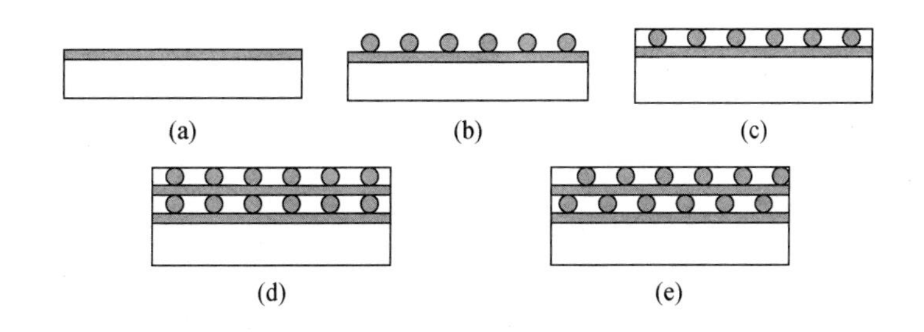

图 2　利用 SK 生长模式制备量子点的示意图

［白色方框为 GaAs 材料，灰色的小岛和长条为 In(Ga)As，灰长条为浸润层。(a) 浸润层；(b) In(Ga)As 岛；(c) 量子点；(d) 量子点的垂直对准；(e) 量子点的斜对准］

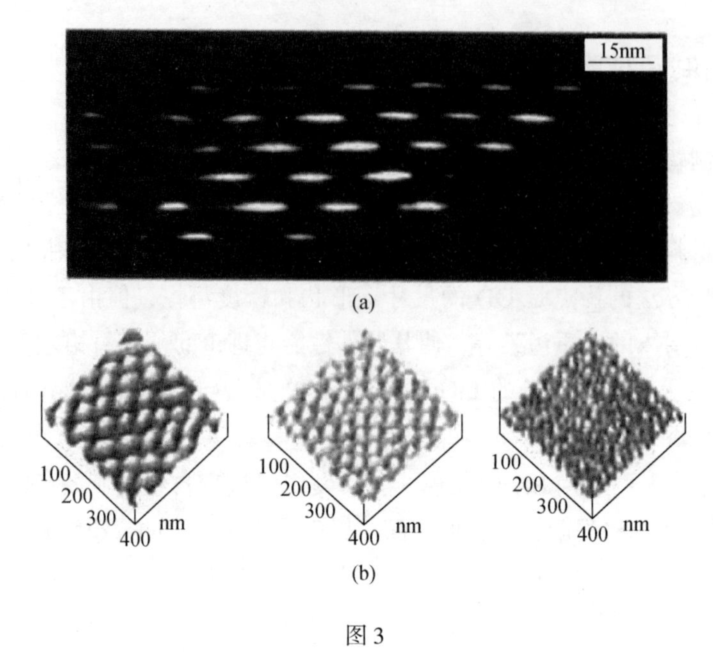

图 3

（a）InAs/InAlAs/InP 三排量子线阵列的 TEM 截面像；（b）$In_xGa_{1-x}As$/GaAs（311）*B* 量子点的 AFM 图（左 $x=0.3$，中 $x=0.4$，右 $x=0.5$）

3　量子点激光器设计与制造

简单地说，量子点激光器是由一个激光母体材料和组装在其中的量子点以及一个激发并使量子点中粒子数反转的泵源所构成。一个实际的量子点激光器的能带结构和生长结构示意图如图 4 所示。

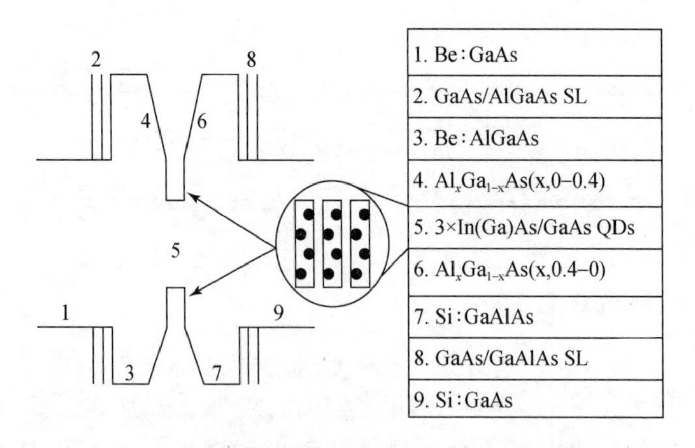

图 4　量子点激光器能带结构和生长结构示意图

（1，9 为上下欧姆电极接触层；2，8 为超晶格缓冲层；3，7 为上下包层；4，6 为上下折射率梯度改变分别限制区；5 为量子点有源区）

一个理想的量子点激光器，首先，要求量子点的尺寸、形状相同，其变化范围应小于10%，即量子点应只有单一电子能级和一个空穴能级，以利于QDLD的基态激射.量子点的多能态存在，浸润层和势垒能态将引起激射波长蓝移，无法实现QDLD的无啁啾工作。无啁啾工作是量子点LD最重要的优点之一。所谓啁啾是指激光器在直流电流的调制下激光器发射波长的改变（$\Delta\lambda$），而波长改变是远距离光纤通信中一个亟待解决的关键问题。它的物理起因是来自激光介质中复数磁导率的实部与虚部之间的耦合。理想的量子点LD是无啁啾工作的，实际上由于量子点激发态等贡献使增益曲线不再对称，导致反映耦合强度的线宽增强因子不再为零，而增大到0.5，但这仍要比最好QWLD的值（1-2）优2-4倍。第二，要求有尽量高的量子点面密度和体密度，以保证QD材料有尽可能大的增益和防止增益饱和，这将有利于QDLD的低阈值基态工作。第三，要正确选择量子点的尺寸，因为量子点的临界尺寸同选用材料体系导带带阶（ΔE_c）紧密相关. 若选用QD的尺寸不适当（如太小），量子点中第一电子能级（基态）与势垒层连续能量差很小，那么在有限的温度下，量子点中的载流子的热激发将使量子点中载流子耗尽，无法实现基态激射。若量子点中能级差与kT相比拟时，高能级的热填充难以避免。简单的估算表明，InAs/GaAs球形量子点的尺寸下限大约为20nm。第四，量子点激光器工作波长可通过选择材料体系，控制量子点形状、尺寸等实现，目前已报道的QDLD的波长已覆盖了从红光到近红外波段。此外，激光腔面制作、镀膜质量、腔长选择等对QDLD的工作模式也产生重要影响。

量子点激光器材料通常是采用MBE技术生长（也有用MOCVD技术的），生长顺序依次是（见图4）：在重掺硅的n^+(001）面GaAs衬底上，先生长AlGaAs/GaAs短周期超晶格缓冲层，以屏蔽来自衬底的缺陷和使生长平面平整，进而生长出厚度为1-1.5μm的掺硅N型$Al_{0.4}Ga_{0.6}As$下包层；折射率梯度改变$As_xGa_{1-x}As$（$x=0.4\to0$）下光学限制层；包含单层或多层In(Ga)As/GaAs或InGaAs/AlGaAs量子点有源区，折射率梯度改变$Al_xGa_{1-x}As$（$x=0\to0.4$）上光学限制层；掺Be P型$Al_{0.4}Ga_{0.6}As$上包层；短周期超晶格层和重掺Be的p^+ GaAs电极接触层。GaAs，AlGaAs和量子点的生长温度分别为600℃，700℃和490℃。In(Ga)As/GaAs应变自组装量子点生长是采用SK生长模式，二维平面生长向三维岛状（量子点）生长过渡，从高能电子衍射图像由线状向点状改变来进行原位监控。若单层量子点的面密度为$4\times10^{10}cm^{-2}$，厚度为200nm，包括有三层垂直耦合量子点的复合辐射体积内总的量子点密度可达$6\times10^{15}cm^{-3}$。激光器结构生长完成后，可按需要制成不同器件结构（宽条或窄条脊形等）、不同工作模式的量子点激光器。

4 量子点激光器的研究进展

1980年，Arakawa和Sakaki[10]首先尝试了利用磁场对量子阱中载流子进行三维限制，结果仅发现激光器的 T_0 增加，但未使 J_{th} 降低。直到 1994 年，第一个基于应变自组装量子点的 Fabry-Perot 注入激光器由俄罗斯-德国联合实验小组的 Kirstaedter 等[11]研制成功。它是在 AlGaAs/GaAs 折射率梯度改变分别限制结构中加入 InGaAs/GaAs 单层量子点有源区，器件制成带有无镀膜解理腔面的浅台面条形几何结构，腔长 1mm。77K 时，工作波长 $\lambda=0.95\mu m$ 的 $J_{th}=120A/cm^2$，$T_0=350K$（50-120K），明显地超过了量子阱激光器的 T_0；室温工作时，J_{th}增加到 $950A/cm^2$，波长向短波长移动，很接近浸润层发光能量。分析表明，这是由于量子点中载流子激发到势垒层中和增益饱和所致。1995 年，日本的一个研究小组[12]，采用 MBE 技术和 SK 生长模式，制备了 $In_{0.4}Ga_{0.6}As$/GaAs单层量子点激光器，85K 时，$J_{th}=800A/cm^2$，输出波长为 $0.92\mu m$，激射跃迁发生在浸润层。1996-1997 年是量子点激光器研制迅速发展的两年，除俄罗斯-德国联合实验小组之外，日本、英国、美国、法国、加拿大和中国等的一些研究小组也加入了量子点激光器的研制行列。为了实现量子点激光器的基态激射，研究人员在优化生长工艺条件下，使量子点的尺寸、形状的均匀性得到了明显改善，量子点尺寸均匀性可控制在10%以下，基本上满足了基态激射的要求。为了降低室温阈值电流密度，多个实验小组尝试了增加有源区垂直耦合量子点的层数的方法，取得了理想的效果。1996 年，Ledantsov 等[13]采用10 层 $In_{0.5}Ga_{0.5}As$/GaAs 量子点超晶格结构为量子点激光器的有源区，使室温下 J_{th}降低到 $90A/cm^2$；采用3 层 $In_{0.5}Ga_{0.5}As/Al_{0.15}Ga_{0.85}As$ 量子点超晶格结构，室温下，四侧解理面几何构形的量子点激光器，在工作波长为 $1\mu m$ 时，J_{th}仅为 $62A/cm^2$！创下了量子点激光器的最好记录。1997 年，Heinrichsdorff 等[14]报道了腔长为 1.5mm 的量子点激光器，77K 时，$J_{th}=12.7A/cm^2$，室温下 $J_{th}=110A/cm^2$ 的好结果。1997 年，我们实验室的 MBE 组[15]也实现了腔长为 $500\mu m$ 的 InAs/GaAs 单层量子点激光器的研制，室温下在工作波长为 $0.98\mu m$ 时，J_{th}为 $590A/cm^2$，达到了当时国际先进水平。1998 年中国[16]和美国[17]等研制成功的大功率量子点激光器（多模）的室温 J_{th}分别为 $218A/cm^2$ 和 $270A/cm^2$。1999 年美国的一个实验小组[18]，研制成功室温 J_{th}仅 $26A/cm^2$ 的 $InAs/In_{0.15}Ga_{0.85}As$ 量子点激光器。它是将一个密度高达 $7\times10^{10}cm^{-2}$ 的单层 InAs 量子点置入应变 $In_{0.15}Ga_{0.85}As$ 量子阱中作为有源区来实现的。器件工作波长为 $1.25\mu m$，但这个双侧解理面激光器的腔长长达 $7800\mu m$，以增加有源区的增益。同年 Gyoungwon park[19]等又报道了室温单模 CW 工作波长为 $1.3\mu m$ 的量子点激光器，J_{th}为 $45A/cm^2$，脉冲工作 $J_{th}=25A/cm^2$，向光通信应用又跨进了一大步。

理想的量子点激光器的另一个突出优点是阈值电流密度（J_{th}）不依赖于温度，即特征温度 $T_0=\infty$。实际的量子点激光器的 T_0 与量子点尺寸、势垒层材料及其质量密切相关。通过优化量子点尺寸和提高包层生长温度（改善势垒层质量），可使量子点激光器的 T_0 明显提高。1996 年，Bimberg 等[20]和 Alferov[21]等报道了他们研制的量子点激光器的 T_0 分别高达 425K（<100K）和 530K（80-220K）。1997 年，Maximov 等[22]将量子点置入 GaAs/AlGaAs 量子阱中，使量子点中载流子的逃逸势垒高度增加，大大降

低了载流子的逃逸概率，减小了漏电流，从而使他们研制的量子点激光器的 T_0 在 80-330K 仍能保持高达 385K 的值，这是量子阱激光器无法做到的！

典型的单层量子点的增益谱半宽约 50meV，这远大于满足 0.5mm 腔长的条形激光器单模工作要求的 1meV。然而在实验上确已观察到 QDLD 在稍大于 J_{th} 时（$<1.1J_{th}$）就出现单纵模工作模式，并且随注入电流增大，多个可明显分辨的纵模工作模式在短波长方向相继出现，可能的原因有两个：一是由于量子点分布不均匀，其增益在接近增益极大值附近可能存在局域最小值，这就增加了另外一些纵向模式的阈值电流当注入电流增大时，最初被抑制的模式开始激射；二是尺寸不同的量子点，对载流子的俘获时间常数不同，只有那些能迅速充满载流子的量子点才对激射有贡献。因此，要实现量子点激光器单纵模工作，对量子点的尺寸均匀性的控制要求是很严格的，要实现这一点，还需做更多的工作。

量子点激光器所特有的低 J_{th}、高 T_0 工作和灾难性光学镜面损伤（COMD）以及衬底缺陷影响被减小等优异性能，使量子点激光器更有利于高功率工作。大功率量子点激光器的研制工作在过去的 3 年里取得了很大进展。1997 年，Shernyakov 等首先报道了室温（287K）高达 1W 连续波（CW）工作的量子点激光器研制成功的消息。量子点激光器的有源区为 10 层垂直耦合 $In_{0.5}Ga_{0.5}As/Al_{0.15}Ga_{0.85}As$ 量子点超晶格，激光器结构为 114μm 的浅台面，微分效率 40%，J_{th} 在 1100μm 腔长时为 290A/cm^2，工作波率长为 ~0.99μm，当热沉温度为 70-75℃时仍能观察到激射。1998 年 5 月，俄罗斯-德国联合实验小组的 Maximov 等[23]也报道了与 Shernyakov 等类似的结果。他们还发现，经 800℃ 0.5h 的退火，量子点激光器的主要结构性能参数 J_{th} 和量子效率未见结改变或退化，表明量子点激光器工作是稳定的。就在这篇文章的后记中，他们又公布了在增加量子点有源区的密度后，获得了室温 CW 工作 1.5W 的光输出功率的好结果。同年，我国的大功率量子点激光器的研制也取得了突破性进展，中国科学院半导体材外料科学实验室的 MBE 小组，采用 3 层垂直耦合 In(Ga)As/GaAs 量子点超晶格为有源区，制作成条宽为 100μm、腔长 800μm 的宽接触激光器结构，腔面镀膜，连续波工作单面光输出功率大于 1W。图 5 是该实验室研制成功的大功率量子点激光器的光输出功率与注入电流（a）和激射特性曲线（b）。$T_0=333K$（20-180K），$T_0=157K$（180-300K），室温 $J_{th}=218A/cm^2$，工作波长 0.96μm，器件在 0.61W 和 0.54W 条件下室温工作 3000h，功率分别仅下降 0.81db 和 0.49db，为目前国际上报道的最好结果之一。将 19 路量子点激光器通过光纤制成耦合模块，室温连续输出光功率可高达 10W 以上。

1999 年，俄罗斯-德国联合实验小组的 Kovsh[24]等在分析了限制量子点激光器输出最大功率的可能原因后，通过增加有源区量子点的面密度（$1.5\times10^{11}cm^{-2}$）和采用 3 层垂直耦合 $InAs/Al_{0.15}Ga_{0.85}As$ 量子点超晶格有源区，制备成功工作波长为 0.87μm、室温连续双面（镀膜）最大输出功率为 3.5W 的最好成绩。激光器为 100μm×920μm 宽条结构，

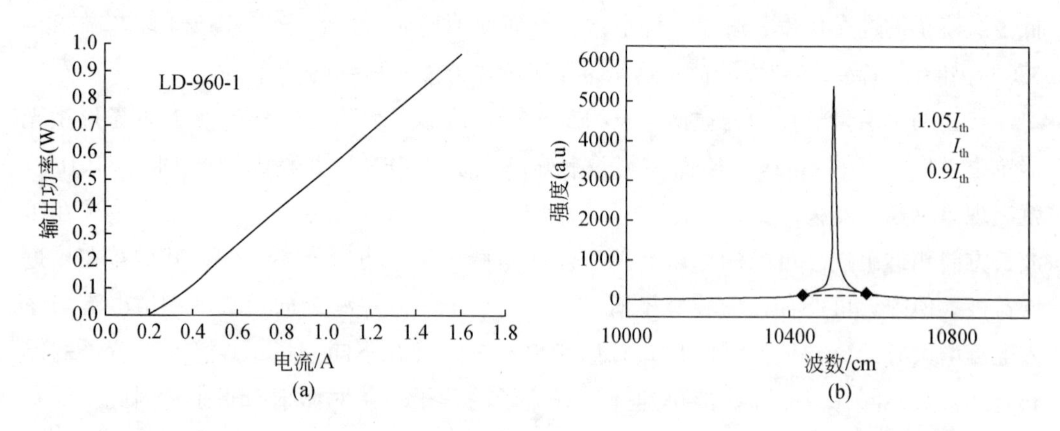

图5　我们实验室研制成功的大功率量子点激光器的光输出功率与注入电流（a）和激射特性（b）曲线

J_{th}=400A/cm^2，峰值转换效率达45%。他们相信，进一步优化激光器的设计和制作工作，量子点激光器的输出功率还会进一步增加。

合理选择$In_xGa_yAl_{1-x-y}As$材料体系的组分x和y，量子点激光器的工作波长可覆盖红光到近红外光纤通信的两个重要窗口。1999年，美国得克萨斯大学[17]和日本富士通的实验小组[25]分别报道了试制成功波长为1.30μm室温单模连续工作的量子点激光器的信息。目前，InP基量子点的光致发光谱峰值波长已扩展到光纤通信的另一个窗口1.55μm附近，显示了它作为光纤通信光源的重要应用前景。此外，早在1996年美国-加拿大联合实验组的Fafard等[26]研制成功以$In_{0.64}Al_{0.36}As$单层量子点为有源区的宽接触（400×600μm）红光激光器，77K工作波长为0.707μm，J_{th}=730A/cm^2，脉冲输出功率为200mW。红光量子点激光器的研制进展虽然不大，但也不断有文章报道，我们实验室也开展了$In_{0.35}Al_{0.65}As/Ga_{0.5}Al_{0.5}As$红光量子点的研制工作，在80K已观察到了红光激射。

垂直腔面发射激光器（VCSEL）由于具有光束质量好、低电流工作和面发射、便于器件平面集成和光耦合应用等特点，是半导体激光器研制的另一个热点。若将VCSEL的量子阱有源区用单层或多层量子点代替，就可制成量子点VCSEL。人们预料，它会有更优越的性能。VCSEL的关键是微腔的设计与制造。微腔实际上是由一个带宽为几十毫电子伏（meV）的抑制频带和在抑制频带中另加一个约1meV的通频带组成，若使尺寸均匀量子点的基态能量与微腔的通频带（pass band）匹配，则可实现对光子和电子的优化控制。由于Fabry-Perot腔模式间距离很大，微腔又很短，故量子点VCSEL的单模运转可自动满足。量子点VCSEL激光器的输出波长只受通频带的限制，而不受量子点尺寸不均匀性的影响，若量子点最大增益与通频带不匹配，则仅导致J_{th}增加。

第一个室温电注入工作的量子点VCSEL是1996年Saito等[27]研制成功的，量子点有源区是由面密度为2×10^{10}cm^{-2}的10层垂直耦合量子点超晶格组成，发射面为25μm×

25μm，激射波长为960.4nm，处于基态和激发态之间，显然是两者共同的贡献。Lott等[28]于1997年又实现了电注入室温基态激射的量子点VCSEL，台面直径为8μm时，阈值电流为200μA(J_{th}=400A/cm^2)。当台面直径缩小到1μm时，阈值电流低到68μA，这与目前报道最好的量子阱VCSEL的结果相当[29]。

5 量子点激光器研制存在的问题和发展趋势

在短短的几年里，虽然量子点激光器的研制已取得了令人瞩目的成绩，但其性能与理论预测相比，仍有较大差距！进一步提高量子点激光器的性能，必须解决下述问题：一是如何生长尺寸均匀的量子点。尽管单个量子点的光增益很大，但由于它的尺寸不均匀，会导致发光峰的非均匀展宽，现有的量子点发光峰的半高宽最好的也只是20meV左右，远大于量子阱材料（~meV）；所以只有很小一部分量子点对激光器发光有贡献，限制了光增益，影响了J_{th}的进一步减小，生长尺寸高度均匀的量子点是面临的提高量子点激光器质量的一个严峻挑战！二是如何增加有源区量子点面密度和体密度（增加耦合层数），保证有源区高的光增益，也是要进一步解决的重要问题之一。三是如何优化量子点激光器的结构设计，使其有利于量子点对载流子的俘获和束缚。通常是采用增大量子点受限势垒高度和提高势垒区材料生长质量来实现。另外，通过控制量子点尺寸和选择材料体系的组分，可在较大范围内改变激光器的波长，但在实验上，要获得可精确控制的工作波长，仍有较长的路要走。

综上所述，研制用于光纤通信光源的1.3μm和1.5μm单模、低J_{th}和无啁啾工作的可实用化的量子点激光器，仍是今后相当长一段时间里人们追求的目标；研制用于光纤放大器和全固态激光器泵浦源的大功率量子点激光器是另一个重要方向，大功率量子点激光器研制有可能在近几年内取得突破，并首先走向实用化。

（后记：在本文定稿时，中国科学院半导体材料科学实验室又研制成功双面连续波工作输出功率为3.618W的大功率量子点激光器）

致谢

感谢梁基本、姜卫红等提供的有益资料以及在打印、制图等方面的帮助。

参考文献

[1] Bimbery D, Grunclmann M, Ledentsov N N. Quantum Dot Heterostructures. Chichester-New York etc: John Wiley & Sons, 1999. 279.

[2] Afferov Zh I, Kazaronov R F. Semiconductor laser with electric pumping. Author's Certificate. N181737.

[3] Kroemer H. Proc. IEEE, 1963, 51: 1782.

[4] Casey H C, Panish M B. Heterostructure Lasers(Part A). New York: Acade mic. 1978.

[5] Tsang W T. Appl. Phys. Lett., 1981, 38: 835.
[6] Hersee S D, Baldy M, Assenat P et al. Electron. Lett., 1982, 18: 870.
[7] Forchel A, Leier H, Maile E et al. Festkörperproble me(Advances in solid state physics), 1988, 28: 99
[8] Hirayama H, Matsunaga K, Asada M et al. Electron Lett., 1994, 30: 142.
[9] Seifert W, Carlsson N, Miller M et al. Prog. Crystal Growth Charact, 1996, 33: 423.
[10] Arakawa Y, Sakaki H. Appl. Phys. Lett.,1982,40:939.
[11] Kirstaedter N, Ledentsov N N, Grundmann M et al. Electron. Lett., 1994, 30: 1416.
[12] Shoji H, Mukai K, Ohtsuka N et al. Photon. Tech. Lett., 1995, 12: 1385.
[13] Ledentsov N N, Shchukin V A, Grundmann M et al. Phys. Rev. B, 1996, 54: 8743.
[14] Heinrichsdorff F, Mao M H, Kirstadter N et al. Appl. Phys. Lett., 1997, 71: 22.
[15] 王占国. 第四届全国 MBE 会议文集. 江苏无锡, 1997 年 9 月[WANG Zhan-Guo. Proceeding of the 4th National Conference on MBE. Wuxi, Jiangsu, 1997, Sept. (in Chinese)].
[16] Wang Z G, Gong Q, Zhou W et al. 3th Rim-pacific international conference on advanced materials and processing, edited by I mam M A et al. Hosted by TMS, 1998, 2097.
[17] Huffaker D L, Park G, Zhou A et al. Appl. Phys. Lett., 1998, 73:2564.
[18] Liu G T, Stintz A, Li H et al. Electron. Lett., 1999, 35: 1163.
[19] Gyoungwon park, Oleg B Shcheking, Sebastion et al. RT continuous ware operation of a single-layered 1.3μm qumtun dot laser; Appl. 1999, 75: 3267.
[20] Bimberg D, Kirstaetsov N, Grundmann M et al. Phys. Stat. Sol. (b), 1996, 194: 159.
[21] Aferov Z I, Gardeev Yu N, Zaitsev S V et al. Semicond., 1996, 30: 197.
[22] Maximov M, Gardeev N Yu, Zaitsev S V et al. Semicond., 1996, 31: 124.
[23] Maximov M V, Shernyakov Yu M, Tsatsul' nikov A F et al. J. App. Phys., 1998, 83: 5561
[24] Kowsh A R, Zhukov A E, Licshits D A et al. Electron. Lett., 1993, 35: 1161.
[25] Mukai K, Nakata Y, Otsubo K et al. IEEE Photonics Technology. Lett., 1999, 11: 1205.
[26] Fafard S, Hinzer K, Ray mond S et al. Science, 1996, 274: 1350.
[27] Sairo H, Vishi K, Ogura I et al. Appl. Phys. Lett., 1996, 69: 3140.
[28] Lout J A, Ledentsov N N, Vstinov V M et al. Electron. Lett., 1997, 33: 1150.
[29] Huffaker P L, Deppe D G. Appl. Phys. Lett., 1997, 70: 1781.

High-power and long-lifetime InAs/GaAs quantum-dot laser at 1080nm

Hui-Yun Liu, Bo Xu, Yong-Qiang Wei, Ding Ding, Jia-Jun Qian, Qin Han, Ji-Ben Liang, and Zhan-Guo Wang

(Laboratory of Semiconductor Materials Science, Institute of Semiconductors, Chinese Academy of Sciences, Beijing 100083, China)

Abstract High power and long lifetime have been demonstrated for a semiconductor quantum-dot(QD) laser with five-stacked InAs/GaAs QDs separated by an InGaAs strain-reducing layer (SRL) and a GaAs spacer layer as an active medium. The QD lasers exhibit a peak power of 3.6W at 1080nm, a quantum slope efficiency of 84.6%, and an output-power degradation rate of 5.6%/1000h with continuous-wave constant-current operation at room temperature. A comparative reliability investigation indicates that the lifetime of the InAs/GaAs QD laser with the InGaAs SRL is much longer than that of a QD laser without the InGaAs SRL. This improved lifetime of the QD laser could be explained by the reduction of strain in and around InAs QDs induced by the InGaAs SRL.

Semiconductor quantum-dot (QD) lasers have attracted considerable attention.[1-9] QD lasers are expected to attain high power, less temperature-sensitive operation, and a remarkable reduction in threshold current, due to the discrete atom-like states in QDs. Recently, QD lasers have progressed rapidly and have been demonstrated at room temperature with high output power of 4W with continuous-wave (cw) operation and 4.5W with pulsed operation,[1,2] high characteristic temperature of 120K near 1.3μm[3], and extremely low threshold current density of $19A/cm^2$,[4] which are comparable to those of a conventional semiconductor quantum-well laser. The biggest remaining issue concerning the practical application of QD lasers is their reliability. The investigations on the reliability of QD lasers are rarely reported. In particular, many doubt whether InAs/GaAs QD lasers can survive during an aging test, due to the larger lattice misfit between the self-assembled InAs QDs and the surrounding GaAs barrier layers in the laser active region.

In this letter, an InGaAs overgrowth layer directly deposited on InAs QDs in the active

原载于：Appl. Phys. Lett., 2001, 79(18): 2868-2870.

region has been designed to improve the lifetime of the InAs/GaAs QD laser.[5] We demonstrate that the InAs/GaAs QD lasers with the InGaAs strain-reducing layer(SRL) have a long lifetime of approximately 9000h with cw room-temperature(RT) operation, which is much longer than that of a laser without the InGaAs SRL. In addition, the InAs/GaAs QD lasers with the InGaAs SRL have a high power of 3.6W and a wavelength of 1080nm. Such lasers are important for the application of solid-state laser and fiber laser pumping.[2]

The QD laser structures were grown by solid source molecular beam epitaxy(Riber 32P) on n^+-doped GaAs(100) substrates. The active medium in the lasers consists of five layers of 2.0 monolayer(ML) InAs QDs separated by 5ML $In_{0.2}Ga_{0.8}As$ SRL and 15ML GaAs spacer layer, with 10nm GaAs below and above the 5 QD layer. The active region is grown at the center of an undoped 200 nm GaAs/AlGaAs waveguide clad by p and n AlGaAs layers of 1μm thickness. The evolution of the surface morphology during the growth process was monitored by reflection high-energy electron diffraction. The InAs quantum dots, the InGaAs strain-reducing layer, and the GaAs spacer layer were grown at 480℃, the GaAs layers at 580℃, and the AlGaAs layers at 700℃, as measured by an optical pyrometer. The sample is processed into 100μm broad-stripe lasers with uncoated cleaved facets to form cavities with a cavity length of $L = 800\mu m$. The laser diodes were mounted for characterization in a variable-temperature liquid-helium optical cryostat. The luminescence spectra were detected with a Fourier transform infrared spectrometer operating with an $In_xGa_{1-x}As$ photodetector.

The inset of Fig. 1 shows a transmission electron microscopy(TEM) image for vertically stacked InAs/GaAs QDs separated by $In_{0.2}Ga_{0.8}As$ SRL and a GaAs spacer layer, which corresponds to what was used as active material in the device here. The image was taken under two-beam, (200) dark-field conditions to maximize the contrast due to chemical composition. The QDs from each layer have a base diameter of ~20nm and thickness of ~3.5nm. The distinct increment of QD dimensions in the upper layer suggested by previous investigation[6] is not observed here. The light output power from both uncoated facets against driving current for this laser diode operating in cw mode at 25℃ heatsink temperature is present in Fig. 1. We measured a maximum output power of 3.6W at 4A without power saturation and the quantum slope efficiency η_d is nearly constant and equals 84.6%. These results could compare with the best data reported in the literature. By comparison, the previously highest reported cw powers of QD lasers are 3.5W with $\eta_d = 73\%$ from a 100μm stripe 920μm cavity InAs/GaAs QD laser and 4W with $\eta_d = 80\%$ from a 100μm stripe 2mm cavity InGaAs/GaAs QD laser.[1,7] Extrapolation of the linear regime yields a threshold current density of 221A/cm^2.

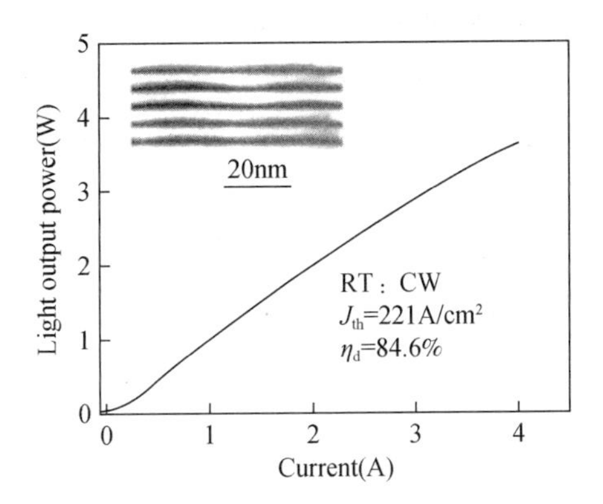

Fig. 1 Light output power from both uncoated facets as a function of current for an 800μm long cavity at room temperature under the continuous wave condition. The inset shows a cross-sectional TEM image of a laser active region, where the InAs/GaAs QDs are separated by an InGaAs SRL and a GaAs spacer layer

The electroluminescence (EL) spectra of the QD laser below and above threshold at RT are shown in Fig. 2 The EL spectrum below threshold is multiplied by 5 and the ground state energy of InAs QDs at 1.153eV with a full width at half maximum (FWHM) of 23meV, which agrees with the photoluminescence (PL) peak position (1.160eV) of the laser wafer at room temperature (not shown here). Above threshold, a narrow laser line at 1.152eV with FWHM of 2.8meV is observed. We assume that lasing action is dominated by ground-state transitions in the InAs/GaAs QDs, because the energy of the laser line is only 1meV smaller than subthreshold EL and 8meV smaller than PL measured at room temperature.

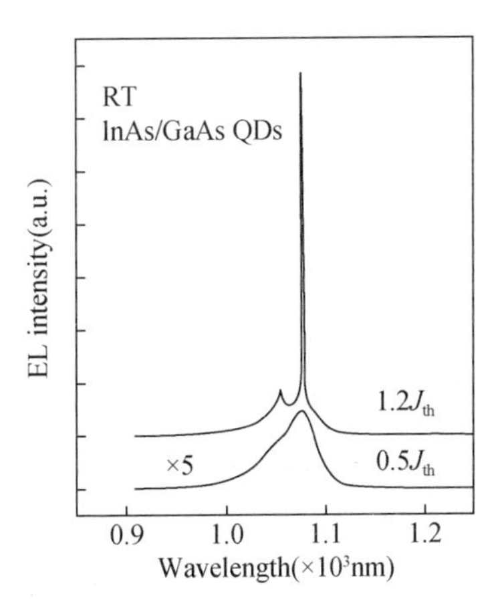

Fig. 2 Electroluminescence spectra of a QD laser below and above threshold

The temperature T dependence of J_{th} is shown in Fig. 3. The lines are a fit to the data using the following formula: $J_{th} = J_0 \exp(T/T_0)$. The characteristic temperature T_0 is 258K for temperatures between 150 and 250K, and it decreases to $T_0 = 84$K for T>250K. Thermal excitation of electrons and holes into upper discrete QD levels probably increases the threshold for $T>250$K. [8,9] The inset in Fig. 3 shows the EL spectra above the threshold current at different temperatures.

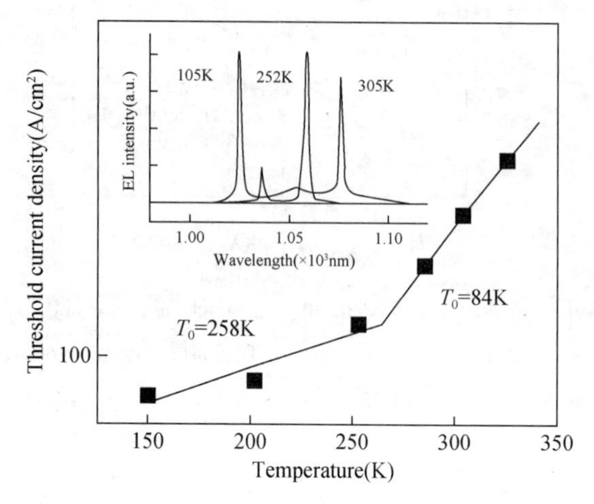

Fig. 3 Temperature dependence of J_{th} for the InAs/GaAs QD laser. The values of the characteristic temperature T_0 are given. The inset shows the laser spectra recorded at different temperatures ($J=1.2J_{th}$)

The aging and reliability studies of the InAs/GaAs QD lasers under the constant current (corresponding to an operating power of 613mW at t = 0h) were performed at room temperature. We also prepared reference lasers where five layers of InAs QDs separated by a 20ML GaAs spacer layer were used to supersede the InAs QDs with the InGaAs SRL and the GaAs spacer layer in the active region. The change in the output power was measured over time to determine the degradation rate of these lasers.

Fig. 4 shows the results of measured output power versus time. For the laser with InAs/GaAs QDs covered by the InGaAs SRL in the active region, the reliability study was continued for 3760h of operation at room temperature. After 3760h of cw operation, a degradation rate of 5.6%/1000h for output power has been obtained as shown in Fig. 4. Based on the degradation speed, the lifetime was estimated to be as long as approximately 9000h. The lifetime was defined as the time when the operating output power drops with 3dB. For the reference laser, the output power decreases by a factor of 24.8% during the initial 1041h and then the power decreases rapidly. On the basis of the results of the lifetime test, it is evident that using a thin InGaAs SRL instead of part of a GaAs spacer layer directly deposited on InAs QDs will increase the reliability of the InAs/GaAs QD laser.

In general, the rapid degradation of the GaAs-based semiconductor laser is related to the dark line defect (DLD) in the EL topograph of the active region. [10] To obtain a long lifetime GaAs-based semiconductor laser, much effort has been focused on the suppression or

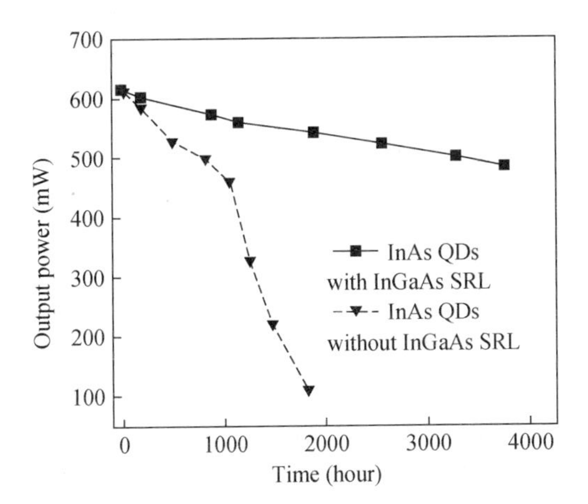

Fig. 4 The aging-time dependence of light output power of the InAs/GaAs QD lasers with the QDs covered with and without the InGaAs strain-reducing layer in an active region under the cw constant-current operation at room temperature

elimination of DLD, which is related to the minimization of residual stress during the crystal growth process and the decrease of stress of chip bonding introduced in the fabrication process.[10] Due to the same fabrication progress after crystal growth of the two types of QD lasers shown in Fig. 4, the improvement of the reliability for the InAs QD laser with InGaAs SRL should result from the minimization of residual stress in the active region of the QD laser.[5] Here, the reduction of stress in and around InAs islands induced by InGaAs SRL will lead to elimination of DLD formation during InAs QDs' capping process and suppression of DLD growth during the aging test progress. Therefore, the long lifetime QD laser would be realized using a thin InGaAs overgrowth layer instead of partial GaAs to directly deposit onto InAs QDs in the laser active region.

In conclusion, we have demonstrated a 1080nm emitting InAs/GaAs QD laser with high cw output power of 3.6W, reliable operation at a power of 0.613W, and record high quantum slope efficiency of 84%. The long laser lifetime of about 9000h for the InAs/GaAs QD laser is achieved by directly depositing a thin InGaAs strain-reducing layer on InAs islands in the active region. These results suggest that the high-power InAs/GaAs QD laser with a practical lifetime may soon be realized.

This work was supported by the National Advanced Materials Committee of China and Special Funds for Major State Basic Research Project No. G20000683 and by the National Natural Science Foundation of China under Contract No. 60076024.

References

[1] F. Klopf, J. P. Reithmaier, A. Forchel, P. Collot, M. Krakowski, and M. Calligaro, Electron.

Lett. 37,353(2001).

[2] F. Heinrichsdorff, Ch. Ribbat, M. Grundmann, and D. Bimberg, Appl. Phys. Lett. 76, 556 (2000).

[3] K. Mukai, Y. Nakata, K. Otsubo, M. Sugawara, N. Yokoyama, and H. Ishikawa, Appl. Phys. Lett. 76, 3349(2000).

[4] G. Park, O. Shchekin, D. L. Huffaker, and D. G. Deppe, IEEE Photonics Technol. Lett. 12, 230 (2000).

[5] H. Y. Liu, X. D. Wang, J. Wu, B. Xu, Y. Q. Wei, W. H. Jiang, D. Ding, X. L. Ye, F. Lin, J. F. Zhang, J. B. Liang, and Z. G. Wang, J. Appl. Phys. 88, 3392(2000).

[6] Q. Xie, A. Madhukar, P. Chen, and N. P. Kobayashi, Phys. Rev. Lett. 75, 2542(1995).

[7] A. R. Kovsh, A. E. Zhukov, D. A. Livshits, A. Yu. Egorov, V. M. Ustinov, M. V. Maximov, Yu. G. Musikhin, N. N. Ledentsov, P. S. Kop'ev, Zh. I. Alfeerov, and D. Bimberg, Electron. Lett. 35, 1161(1999).

[8] G. Park, O. B. Shchekin, and D. G. Deppe, IEEE J. Quantum Electron. 36, 1065(2000).

[9] O. B. Shchekin, G. Park, D. L. Huffaker, and D. G. Deppe, Appl. Phys. Lett. 77, 466(2000).

[10] M. Fukuda, Reliability and Degradation of Semiconductor Lasers and LEDs (London, Boston, 1991).

Self-assembled quantum dots, wires and quantum-dot lasers

Wang Zhan-Guo, Chen Yong-Hai, Liu Feng-Qi, Xu Bo

(Laboratory of semiconductor materials science, Institute of semiconductors, Chinese Academy of Sciences, Beijing 100083, China)

Abstract Molecular beam epitaxy-grown self-assembled In(Ga)As/GaAs and InAs/InAlAs/InP quantum dots(QDs) and quantum wires(QWRs) have been studied. By adjusting growth conditions, surprising alignment, preferential elongation, and pronounced sequential coalescence of dots and wires under specific condition are realized. The lateral ordering of QDs and the vertical anti-correlation of QWRs are theoretically discussed. Room-temperature (RT) continuous-wave (CW) lasing at the wavelength of 960nm with output power of 3.6W from both uncoated facets is achieved from vertical coupled InAs/GaAs QDs ensemble. The RT threshold current density is $218A/cm^2$. A RT CW output power of 0.6W/facet ensures at least 3570h lasing (only drops 0.83dB).

1 Introduction

Much effort has been devoted to the formation of self-assembled quantum dots(QDs) in lattice mismatched systems by the Stranski-Krastanov (S-K) growth mode due to their potential application in nanoscale optoelectronic devices[1-3]. Theoretically, QD laser structures have lower threshold current density, higher gain, and a weak temperature dependence etc, compared with quantum well lasers[4]. The recent developed QD lasers based on self-assembled In(Ga)As/(Al)GaAs QD system have already demonstrated that the room temperature (RT) threshold current density reached at as low as $\sim 60A/cm^2$ for lasing via QD states[2,3]. However, there is still a long way to go to achieve high output power and long lifetime of QD lasers. Here we demonstrate our research on molecular beam epitaxy-grown self-assembled In(Ga)As/GaAs and InAs/InAlAs/InP QDs and quantum wires (QWRs), and the high power InAs/GaAs QD lasers. We first describe the growth procedures and characterization of self-assembled QDs formed in the S-K growth mode. Then we focus

原载于:Journal of Crystal Growth, 2001, 227-228:1132-1139.

our attention on the mechanisms of lateral and vertical ordering. Self-assembled InAs/GaAs QD lasers with a continuous-wave(CW) output power of 3.6W at RT is given finally.

2 Self-assembled growth and characterization of QDs

All the GaAs-and InP-based QD structures in sandwich form were grown in a Riber 32p MBE system. For GaAs-based QD structures, the growth temperatures and growth rates of GaAs, InAs, $Al_{0.5}Ga_{0.5}As$, and AlInAs are (600℃, 0.67μm/h), (500℃, 0.16μm/h), (620℃, 0.8μm/h), and (530℃, 0.2μm/h), respectively. For the In(Ga)As/$In_{0.52}Al_{0.48}As$/InP QD structures, the growth temperatures and growth rates of $In_{0.52}Al_{0.48}As$ and In(Ga)As are (500℃, 0.6–0.7μm/h), (470–500℃, 0.18–0.3μm/h), respectively. No cap layer was deposited for samples designed for ex situ AFM measurement: after the strained In(Ga, Al)As were deposited, the samples were cooled to 200℃ while arsenic pressure was maintained in order to reduce surface reorganization.

For device application, it is necessary to improve the density of defect-free QDs. Towards this purpose, we have investigated the optimized deposition thickness besides the optimized growth temperature and growth rate as illustrated in the former section. Table 1 summarizes the statistical results from InAs/GaAs(001) QDs grown with different nominal InAs deposition thickness. We find that the optimized InAs thickness is 1.8ML. From 1.6 to 1.8ML, the self-limiting growth of QDs is predominant[5,6]. The dot size increases very slowly, while the dot density increases abruptly. Samples with 2.0ML InAs thickness give rise to dot density smaller than that of 1.8ML and bimodal size distributions (one branch group presents very large dot size), indicating obvious dot-dot coalescence.

Table 1　Statistics of InAs/GaAs(001) QDs grown with different nominal InAs deposition thickness

Nominal InAs thickness	1.6ML	1.8ML	2.0ML	
Average QD height(nm)	2.33	2.95	2.78	6.0
Average lateral size(nm)	16.3	19.2	17.9	28
Dot density(μm^{-2})	289	647	136	

InAs/GaAs QDs exhibit a high emission efficiency, and the emission wavelengths can be controlled in the range of 0.93–1.3μm. For applications requiring shorter wavelengths, the red light emission at 0.72μm can be obtained by use of material such as InAlAs and AlGaAs that have band gap wider than GaAs. On the other hand, the study of the self-assembled QDs on an InP substrate permits shifting of the emission wavelength into the technologically important 1.3–2.0μm range by using the InAs as the narrow band-gap material and the InGaAs or InAlAs as the wide band-gap materials.

3 Lateral ordering of S-K QDs

Uniformity in size and shape is a prerequisite for their potential use in QD electronic or optoelectronic devices. Ordering of QDs, which is helpful for improving dot uniformity, may be achieved if advantage is taken of the elastic interaction between QDs. However, ordering along the plane of the dot formation presents greater difficulties, even though some progress has been made in this regard with different substrate orientations. To improve the performances of QD devices, the fundamental and most arduous question on lateral ordering and uniformity of QDs must be overcome. Fig. 1 shows AFM and TEM images of InGaAs/GaAs, InAs/ $In_{0.52}Al_{0.48}As$/InP, and $In_{0.9}Ga_{0.1}As/In_{0.53}Ga_{0.47}As$/InP samples. Conspicuous self-organized QDs are realized. $In_xGa_{1-x}As$/GaAs QD samples in Figs. 1 (a) and (b) show preferential ordering. This novel feature is strongly correlated with the indium content. The lower indium content, the more obvious lateral ordering. This preferential ordering can also be revealed in $In_{0.9}Ga_{0.1}As/ In_{0.53}Ga_{0.47}As$/InP, and InAs/$In_{0.52}Al_{0.48}As$/InP QD structures. What is the nature of this lateral ordering? We think three significant factors should be particularly noted: (i) surface migration of indium, (ii) anisotropic stress, and (iii) indium segregation related alloying or mixing between S-K layer and buffer (cap) layer.

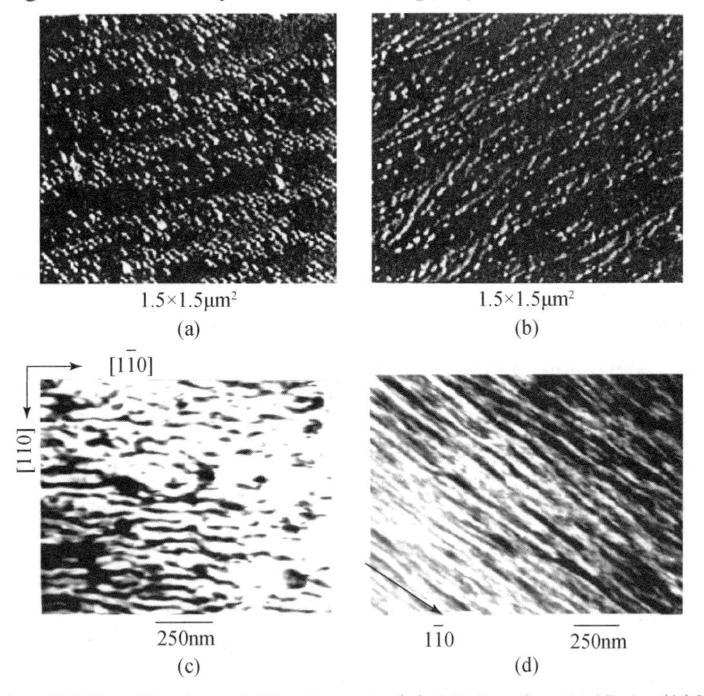

Fig. 1 AFM (a, b) or TEM (c, d) images of QDs structures: (a) 4ML $In_{0.6}Ga_{0.4}As$/GaAs, (b) 8ML $In_{0.4}Ga_{0.6}As$/GaAs, (c) Five stacked structure of 4ML InAs capped by 2ML $In_{0.53}Ga_{0.47}As$ and separated by 5nm $In_{0.52}Al_{0.48}As$ spacer layer, which was grown on $In_{0.52}Al_{0.48}As$/InP. (d) 4ML $In_{0.9}Ga_{0.1}As/In_{0.53}Ga_{0.47}As$/InP

Now, we discuss the above factors sequentially. (i) Thermal roughening provides nucleation sites for “islanding”, which can be supported for the undulated growth front after strained layer overgrowth[5,6]. Because of the high surface mobility of In, the higher In content, the greater possibility for finding more nucleation sites. The driving force for the mass transport toward the islands is assumed to be a reduction of surface strain energy of the (strained) islands. Consequently, higher In content results higher dot density. On the other hand, lower In content indicates relative smaller strain, and the dot is relative sparse and flat. For the reason of statistical “triggering” or fluctuation of growth front and strain, some stepped facets will preferentially appear. The adatoms should get to these step edges, and eventually form the embryo of elongated dot. The elongated dot will induce anisotropic strain, providing a prerequisite for another dot to form, and so on, allows the sequential alignment of QDs. (ii) Since the spatial variation of stress associated with the step instability can dramatically affect the activation barrier for 3D dot nucleation, the anisotropy in step creation energies governs the equilibrium shape of dots. This behavior is governed by the energetics of the strain field and the surface, subject to the kinetics of adatom transport that regulate nucleation and evolution of dots, in competition with dislocation introduction. We simply assume the ratio of step creation energies running [110] and $[1\bar{1}0]$ directions is α. The energy change ΔE due to coalescence of N isolate dots is[5]

$$\Delta E=\left\{-7(N-1)+5\alpha(N-1)-2(1+\alpha)\times\ln\left[\frac{1}{3}(4N-1)\right]\right\}\frac{64(1+\alpha)^2}{675\sqrt{5}}\times\frac{\alpha\lambda_{AO}^2\sqrt{\lambda_{AO}\lambda_{Ad}}}{h^4c^2}. \quad (1)$$

We obtain $\Delta E<0$ on condition that

$$1<\alpha<\frac{6}{2.5-(N-1)^{-1}\ln\left[\frac{1}{3}(4N-1)\right]}-1. \quad (2)$$

As a matter of fact the condition (2) is automatically satisfied, meaning that QWR can be spontaneously formed. The rather fascinating result of this study is that anisotropic stress in combination with interacting steps governs the metamorphosis of QDs into QWRs. This fascinating crossover from QDs to QWRs may point the way toward a new approach to the fabrication of such structures. (iii) Intuitively, alloying between the InAs deposit and the buffer is thought to be one reason for the observed differences in nanostructure size, shape, and distribution. The smaller the relevant mixing enthalpy is, the greater the alloying will be. On the other hand, strong In surface segregation in the buffer can reduce this InAs overlayer/buffer mixing and produce an increase of the roughness responsible for the nucleation sites. The composite formed by the buffer plus the strained overlayer can relax its excess energy via the 2D/3D surface morphology change and/or via some chemical exchange of the Ⅲ or Ⅴ elements in their respective sublattices[6]. Observed difference between InAs/

$In_{0.52}Al_{0.48}As/InP$ and $In_{0.9}Ga_{0.1}As/In_{0.53}Ga_{0.47}As/InP$ are mainly associated with the role played by In surface segregation during the growth of the two alloy buffers and distinctive anisotropy.

4 Vertical alignment of S-K QDs and wires

Multilayering can improve dot uniformity apparently due to the elastic interaction between dots: with increasing number of bilayers (spacer layer plus QD layer), the 3D dots become more uniform in size, shape, and spacing[7,8]. Typical cross section TEM images of vertically coupled QDs are displayed in Fig. 2(a). Five sets of $In_{0.65}Al_{0.35}As$ dots separated by different thickness of $Al_{0.5}Ga_{0.5}As$ spacers are clearly evident. The subsequent dots locate right above the buried ones, leading to the vertical correlation of QDs that has been studied by many authors[7,8].

Recently, the vertically anti-correlated QDs have been obtained[9-11], which gives a new possibility to control effectively the electronic spectrum of 2D QDs. The vertical anti-correlation of InAs QWRs in InAlAs/InP has also been observed in our laboratory[12,13]. Fig. 2(b) shows TEM images of 6 stacked structures of 6.5ML InAs in 5 nm InAlAs space layers. The details were discussed in Refs[12,13]. Clearly, the subsequent wires locate in between two neighboring wires. Theoretical studies reveal that the vertical anti-correlation can be related to the elastic anisotropy of the surrounding medium[9,10,14]. However, we find that the vertical anti-correlation can also occur in isotropic medium.

Now we come to our calculations. The self-assembled InAs QWRs in Fig. 2(b) can be treated as 2D QWRs (the height of QWRs approaches to zero), since their thickness is much smaller than their width[12,13]. Consider a periodic array of such 2D QWRs buried in an infinite medium. The growth direction is along the z-axis, and the 2D QWR array lies in the $z=0$ plane with the y-axis lying at the center of one QWR. From the symmetry of the problem that the strain components are independent of y, and specially, $\varepsilon_y = \varepsilon_{xy} = \varepsilon_{yz} = 0$ outside the 2D QWRs. This plane problem can be solved using an Airy stress function Φ; a biharmonic function of x and z[15]. Due to the periodicity in the x direction, the Airy stress function Φ for the QWRs in an isotropic medium can be analyzed into cosine series

$$\Phi = \frac{E}{1+v}\sum \frac{\cos(\alpha_n x)}{\alpha_n^2}(c_n + d_n\alpha_n z)\mathrm{e}^{-\alpha_n z}, \tag{3}$$

where E and v are the Young's modulus and the Poisson ration of the medium, respectively, α_n is given by $\alpha_n = 2n\pi/d$ with d denoting the inplane period of the 2D QWR array, c_n and d_n are the unknown parameters. Based on the Airy stress function, one directly obtains the stress and strain components[15]. Then, the elastic energy outside and inside the QWRs can

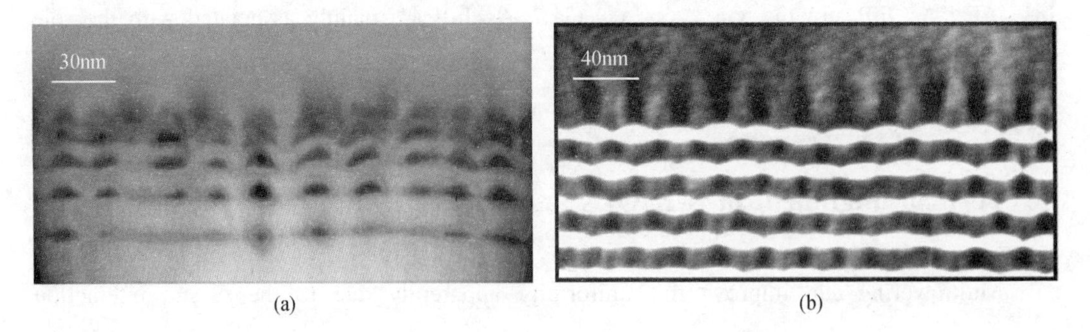

(a)　　(b)

Fig. 2　Cross-section TEM images of vertically coupled QD structures: (a) 5-period 10ML $In_{0.65}Al_{0.35}As$/$Al_{0.5}Ga_{0.5}As$, the thickness' of $Al_{0.5}Ga_{0.5}As$ spacers are 15, 12, 10, and 7nm from bottom to top, and (b) 5-period of 6ML InAs/20nm $In_{0.52}Al_{0.48}As$ QWR arrays

be calculated. According to the minimum-energy principle, the unknown parameters c_n and d_n are determined finally by minimizing the total elastic energy of the QWRs. The same method can be applied for a 2D QWR array in anisotropic medium, where the Airy stress function F satisfies a fourth-order partial differential equation instead of the biharmonic equation (Ref. [16] shows a similar example).

Fig. 3 shows ε_x of a 2D QWR array with $w = d/4$ in isotropic and anisotropic mediums. The parameters $\varepsilon_0 = 0.03$ and $v = 0.36$ and $C_{11} \sim 2C_{12} \sim 2C_{44}$ are adopted for isotropic and anisotropic medium, respectively. We first discuss the results from the isotropic model (Fig. 3(a)). With the increase in the distance away from the QWR array, the material in the region A is first laterally stretched ($\varepsilon_x > 0$) and then becomes laterally compressed ($\varepsilon_x < 0$). There exists a critical value z_{c1} at which the strain ε_x changes from tensile to compressive. In contrast, the material in the region B is first laterally compressed ($\varepsilon_x < 0$) and then becomes laterally stretched ($\varepsilon_x > 0$). Another critical value z_{c2} can be defined in this region. The calculated strain distribution implies that: when it is buried deeply enough, the 2D QWR array will produce compressive surface regions right above the QWRs and stretched surface regions above the interstices between QWRs, resulting in the vertical anti-correlation. Our calculation shows that the maximum of z_{c1} and z_{c2} is about $0.2d$, which means that the critical thickness for 2D InAs/$In_{0.48}Al_{0.52}As$ QWR arrays equals $0.2d$ approximately. As to the results from the anisotropic model (Fig. 3(b)), the alternation of tensile and compressive regions in the vertical direction is very clear, which indicates the alternation of vertical correlation and anti-correlation with the distance between the 2D QWR arrays. Such conclusion has also been obtained by Schukin et al. based on the studies on the elastic interaction between 2D QWR arrays[14]. Except the oscillatory behavior, the distribution of ε_x from the anisotropic model is very similar to that from the isotropic model, especially in the space near the QWR array. The above conclusion is right even if the strain relaxation due to the surface is taken into account. The detailed discussion will be given elsewhere.

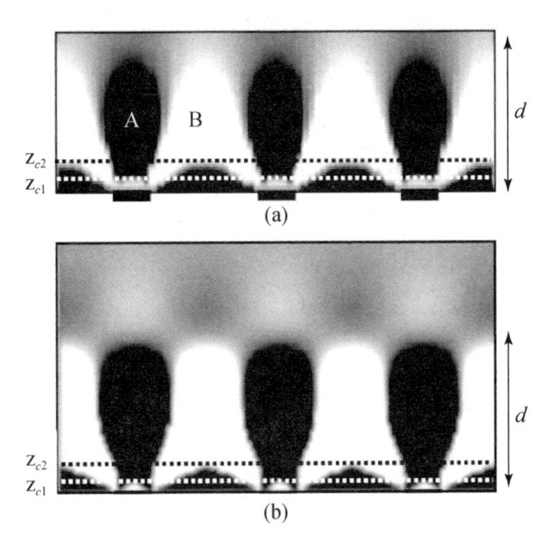

Fig. 3 The ε_x distribution of a two-dimensional(2D) QWR array with $w=d/4$ in isotropic medium(a), and anisotropic medium(b). The dark denotes compressive region while the white denotes tensile region. The rectangles represent the 2D QWRs schematically

The calculated strain distribution is consistent with the TEM observation in Fig. 2(b). Note that the contrast in the $In_{0.48}Al_{0.52}As$ spacer layer in Fig. 2(b) indicates modulation in the chemical composition. This TEM result directly proves that the regions above the interstices of QWR arrays are laterally stretched and can attract more In atoms than the regions right above the QWRs during the growth of $In_{0.48}Al_{0.52}As$. The same results have been obtained by the finite element technique[12].

5 InAs/GaAs QD lasers

QD laser structures were grown on(001) Si doped GaAs substrates. A N^+ GaAs buffer, a N type 1μm thick AlGaAs cladding layer, a 0.2μm gradual refractive index (GRIN) waveguide, an active layer consisted of three-period 1.8ML InAs/5nm GaAs vertical coupled QD array, a 0.2μm GRIN wave guide, a P type 1μm thick AlGaAs cladding layer are sequentially deposited, a 0.3μm P^+-cap layer terminates the structure growth. The growth temperatures are 500℃, 600℃, and 700℃ for InAs, GaAs, and AlGaAs, respectively. Single layer InAs QDs sample grown with identical condition to active region was used for AFM measurement. The average height, average lateral radius, and dot density are 2.95, 19.2nm, and $6.47\times10^{10}cm^2$, respectively.

The grown structure was processed into a broad area laser with one and two facet-coated. The strip width and cavity length are 300 and 800μm, respectively. Room temperature CW lasing at 950–960nm has been achieved. The cross-section TEM image of QD active layer is presented in Fig. 4 (a). Vertical coupled InAs QDs can be identified clearly. Fig. 4(b) shows the power output versus CW drive current. The CW output power is about 3.6W from both uncoated facets, and the threshold current density is 218 A/cm^2. From the relationship between the threshold current density and temperature, the characteristic temperature T_0 is 333K in the range of 20–180K, which is much higher than that of quantum well (QW) lasers. Between 180 and 300K, the measured characteristic temperature T_0 is about 157K. In order to assess reliability, 100×800μm^2 devices were life-tested at a heatsink temperature of 30℃. Typical life-tested results is shown in Fig. 5 An output power of 0.6W ensures at least 3670h and only drops 0.83dB. This is one of the best results ever reported. What is more, fiber coupling module diodes with RT CW output power above 10W have been successfully fabricated.

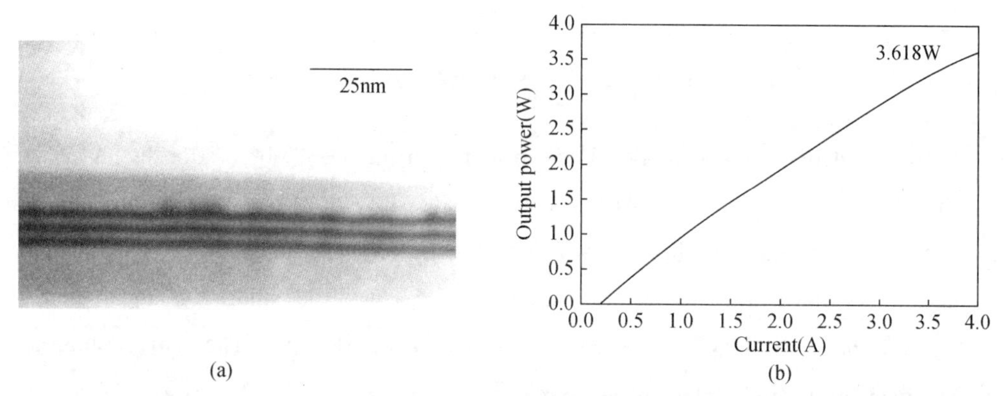

Fig. 4 (a) TEM cross-sectional view of the active region, the InAs QDs can be observed clearly. (b) RT output versus current characteristics of InAs/GaAs QD lasers

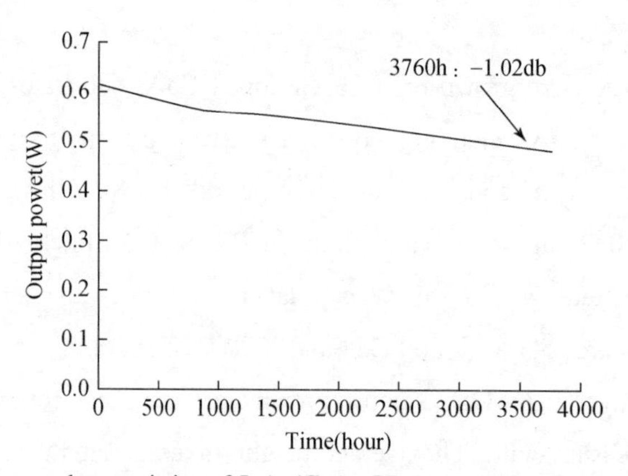

Fig. 5 Life test characteristics of InAs/GaAs QD laser diodes operating at 960nm

The realization of high power QD lasers indicates better localization of carriers in QDs. Deduced from the long lifetime QD lasers, the point defects concentration is effectively suppressed. The high-temperature AlGaAs growth (700℃) serves as post-growth annealing to the QD active region. This ensures high crystalline quality of QD active region. The improvements in laser characteristics by post-growth annealing may be attributed to a reduction of nonradiative recombination centers not only in the QDs region but also in the GRIN and cladding barriers[17]. As a result, fewer carriers are captured by the nonradiative centers before they recombine radiatively in the active region, and the internal quantum efficiency of the laser increases. Besides this important factor, other critical reasons should be disclosed. (i) Good performance of QD lasers correlates strongly with the InGaAs alloy-driven lateral ordering in the active region. Post-growth annealing induces InAs S-K layers mixing with GaAs spacer layers to form InGaAs alloy layer to some extent. This will facilitate lateral ordering of the QDs. The lateral ordering may be one fatal factor in improving the laser's performances. (ii) Just like strained QW lasers providing lower threshold current density and superior reliability over conventional QW lasers, it is necessary to maintain a desired strain level for potential improvements in device properties, thus the sufficient In content is essential. The propagation of defects in the active layer is retarded because the In atom is larger than the Ga, Al, and As atoms, which are almost the same size. (iii) Although atomic layer epitaxy grown vertical coupled QDs can reduce threshold current density of QD laser strikingly[3], it is of no use to enhance the output power of QD lasers. The distinctive merit of atomic layer epitaxy is the high QD uniformity in growth direction, but, it is of no use to improve the lateral ordering. On the other hand, this method results in insufficient In content in the QDs active region and the lower QD density. To increase the dot density, adding the stacked layers is necessary. However, just like multi-QW laser this does not necessarily enhance the output power, too many stacked QD layers may give a bad result on output power[18]. (iv) We adopt the optimized triple-stacked QD layer as the active region for two aspects: too many stacked layers will decrease the output power for one aspect, and multi-stacked structure contribute to the volume of active region, and consequently facilitate to the ground state lasing of QD laser for another aspect.

6 Conclusion

In summary, our aim of this work is to optimize growth conditions to achieve high quality self-assembled QDs and QWRs grown on GaAs and InP substrates. For the reason of fabricating high power QD lasers, vertical ordering, and especially lateral ordering of QDs are systematically investigated. The vertical anti-correlation of QWRs has been explained. A RT CW lasing at the wavelength of 960nm with output power of 3. 6W is achieved from vertical coupled InAs/GaAs QDs ensemble. The RT threshold current density is 218A/cm^2. A RT CW output power of 0. 6W/facet ensures at least 3670h lasing(only drops 0. 83dB). Fiber-coupling module consisted 19 QD lasers with RT CW output power above 10W has been successfully fabricated. This is one of the best results ever reported.

References

[1] Z. G. Wang, Q. Gong, W. Zhou, in: M. A. Imam, R. DeNale, S. Hanada (Eds.), The Third Pacific Rim International Conference on Advanced Materials and Processing (PRICM 3), Australia, 1998, 2097.

[2] V. M. Ustinov, A. Yu Egorov, A. R. Kovsh, J. Crystal Growth 175/176(1997)689.

[3] H. Ishikawa, H. Shoji, J. Vac. Sci. Technol. A 16(1997)794.

[4] N. N. Ledentsov, N. Kirstaedter, M. Grundmann, Microelectron. J. 28(1996)915.

[5] F. Q. Liu, Z. G. Wang, B. Xu, Phys. Lett. A 249(1998)555.

[6] R. Nötzel, J. Temmyo, A. Kozen, Appl. Phys. Lett. 66(1995)2525.

[7] Q. Xie, A. Madhukar, P. Chen, Phys. Rev. Lett. 75(1995)2542.

[8] F. Liu, S. E. Davenport, H. M. Evans, Phys. Rev. Lett. 82(1999)2528.

[9] G. Springholz, V. Holy, M. Pinzolits, G. Bauer, Science 282(1998)734.

[10] V. Holy, G. Springholz, M. Pinzolits, G. Bauer, Phys. Rev. Lett. 83(1999)356.

[11] I. L. Krestnikov, M. Strassburg, M. Caesar, in: David Gershoni (Ed.), Physics of Semiconductors, World Scientific, Singapore, 1999, 70.

[12] J. Wu, Z. G. Wang, J. Crystal Growth 197(1999)95.

[13] H. X. Li, J. Wu, Z. G. Wang, T. Daniels-Race, Appl. Phys. Lett. 75(1999)1173.

[14] V. A. Schukin, D. Bimberg, V. G. Malyshkin, N. N. Ledentsov, Phys. Rew. B 57(1998)12 262.

[15] J. P. Hirth, J. Lothe, Theory of Dislocations, Wiley, New York, 1982, 43.

[16] D. A. Faux, J. Haigh, J. Phys.: Condens. Matter 2(1990)10 289.

[17] J. Ko, C. H. Chen, L. A. Coldren, Electron. Lett. 32(1996)2099.

[18] P. W. A. McIlroy, A. Kurobe, Y. Uemastu, IEEE J. Quantum Electron. QE-21(1985)1958.

Controllable growth of semiconductor nanometer structures

Z. G. Wang, J. Wu

(Key Laboratory of Semiconductor Materials Science, Institute of Semiconductors, Chinese Academy of Sciences, P. O. Box 912, Beijing 100083, China)

Abstract Self-assembled quantum dots and wires were obtained in the $In_xGa_{1-x}As/GaAs$ and $InAs/In_{0.52}Al_{0.48}As/InP$ systems, respectively, using molecular beam epitaxy (MBE). Uniformity in the distribution, density, and spatial ordering of the nanostructures can be controlled to some extent by adjusting and optimizing the MBE growth parameters. In addition, some interesting observation on the InAs wire alignment on InP(001) is discussed.

1 Introduction

Molecular beam epitaxy (MBE) self-assembled nanostructures in the mismatched system may be used in devices application as quantum dots (QD) and wires (QWR). The self-assembling of the dots and wires during MBE involves random atomic processes, and there are various structural nonuniformities in these structures which cover their intrinsic properties and hamper their applications. In this work, the efforts are made to control the growth of the self-assembled nanostructures in the In(Ga)As/ GaAs and InAs/InAlAs/InP systems by adjusting and optimizing the growth parameters, and improvements in structural properties are achieved to some extent. In addition, some significant phenomenon is observed in the symmetry of spatial ordering of InAs wires on InP(001), which may result from some kinetic process during the self-assembling of the InAs wires.

2 Experimental

2.1 $In_xGa_{1-x}As/GaAs$

In the fabrication of $In_xGa_{1-x}As$ dots, GaAs substrate orientation, In composition x, and

原载于：Microelectronics Journal, 2003, 34(5-8): 379-382.

the annealing temperature are used as variable experimental parameters to improve the structural property in island size, shape and density. In addition, $In_{0.5}Al_{0.15}Ga_{0.35}As$ layer is introduced between GaAs substrate and the InGaAs dot layer. The effects of these factors on the structural properties of the $In_xGa_{1-x}As$/GaAs QDs are investigated by atomic force microscope(AFM) and photoluminescence(PL).

2.1.1 Substrate orientation and In composition

Both the substrate orientation and In composition x[1-5] have influential effects on the structural properties of InGaAs dots. The InAs layer of 1.8 monolayers(ML) is deposited on GaAs(100), (311)A, (311)B, (511)A, and(511)B, respectively. Fig. 1 shows the AFM images of the InAs dots with the density of a few of $10^{10}/cm^2$ on the substrate of different orientations. The uniformity of the dots on(113)B with density of $2.3\times10^{10}/cm^2$ is apparently the best of all. This is consistent with the PL spectra shown in Fig. 2, and the peak from the dots on GaAs(113)B, 22eV, is narrowest.

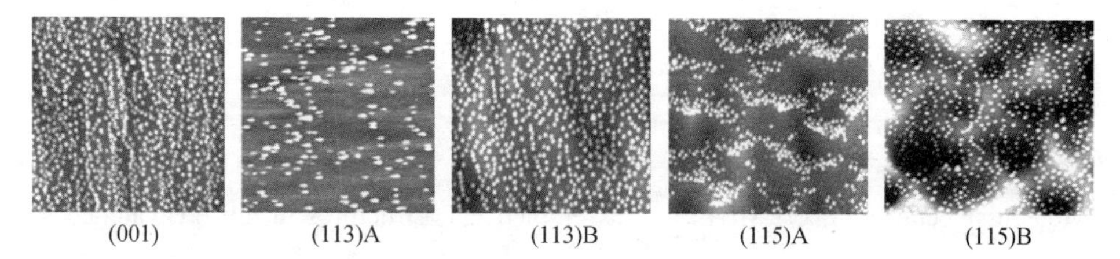

Fig. 1 AFM images of InAs dots on GaAs(100),(311)A,(311)B,(511)A,(511)B

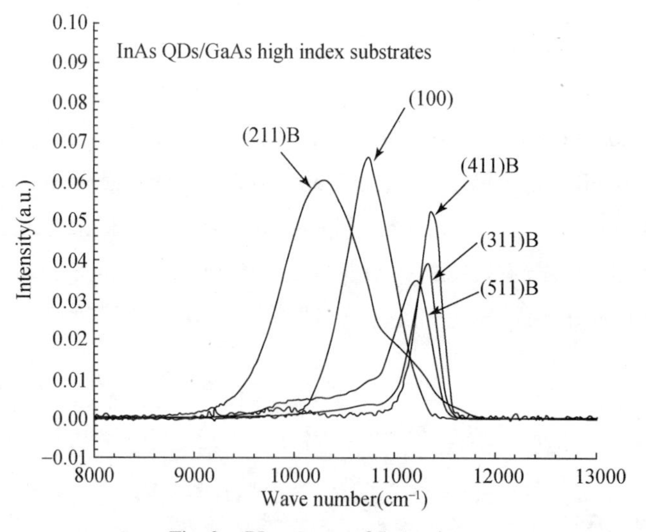

Fig. 2 PL spectra of InAs dots

In composition x is varied from 0.3 to 1.0 in the $In_xGa_{1-x}As$ on GaAs(311)B. The AFM image shows that with $x=0.4$ the dots are most uniform in size and shape, as shown in Fig. 3.

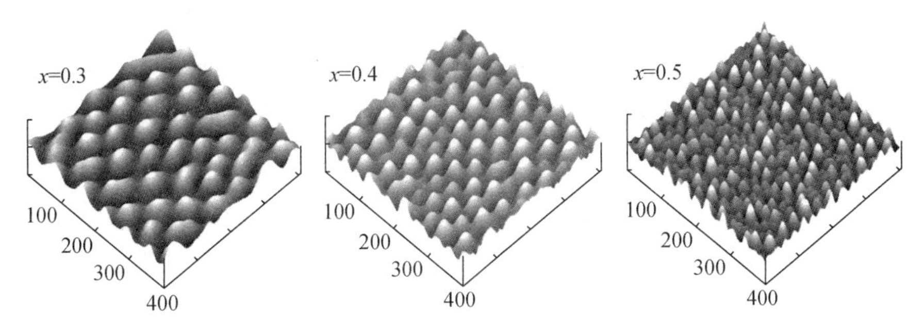

Fig. 3 $In_xGa_{1-x}As$ dots on GaAs(311)

2.1.2 $In_{0.5}Al_{0.15}Ga_{0.35}As$ buried layer

The InGaAs dots can be improved by a buried InAlGaAs layer below[6-9]. If an $In_{0.5}Al_{0.15}Ga_{0.35}As$ layer of 4.5nm is buried 5nm below, the dots are decreased in both height and lateral size besides that the dot density is increased. The results on GaAs(001) and (115)B are shown in Fig. 4 and table 1. In addition, the dots with the buried layer on GaAs(115)B tend to be aligned along the $[\bar{1}10]$ direction.

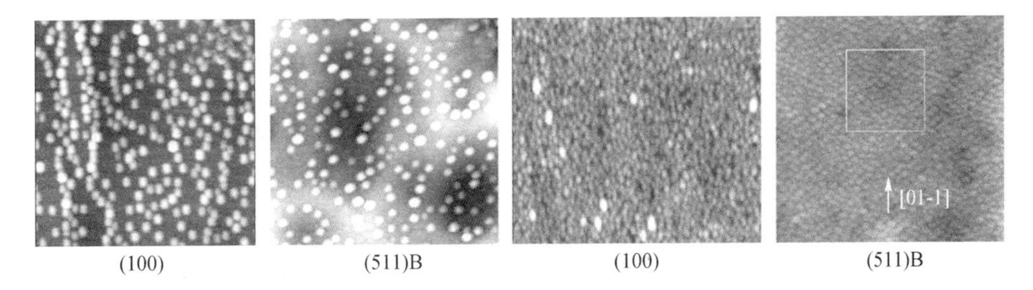

Fig. 4 AFM images of InGaAs dots: (a) and (b) are on GaAs(100), and (c) and (d) on GaAs(511)B without and with the InGaAlAs buried layer, respectively ($1\times1\mu m^2$)

Table 1 The density and size of $In_{0.4}Ga_{0.6}As$ dots with and without the buried $In_{0.5}Al_{0.15}Ga_{0.35}As$ layer

Sample density (cm^{-2})		Lateral size (nm)	Height (nm)
Without buried layer			
(100)	2.5×10^{10}	35	9.6
(115)B	1.7×10^{10}	28	8.7
With buried layer			
(100)	1.1×10^{11}	20	2.6
(115)B	1.2×10^{11}	21	2.1

2.1.3 Annealing[10,11]

The effects on the $In_{0.5}Ga_{0.5}As$/GaAs(001) dots of annealing in the temperature range

650–900℃ for 1min are investigated with PL. As shown in Fig. 5, the emission wavelength and half width from dots are almost unaltered by 650℃ annealing. At 700℃ annealing, the peak is blue shifted and its half width is reduced. At 800℃, the peak position is blue shifted by 116meV relative to the as grown and the half width is reduced from 62.5 to 28.6meV. At 850℃, the surface of the sample roughens and the PL emission from the dots is deteriorated.

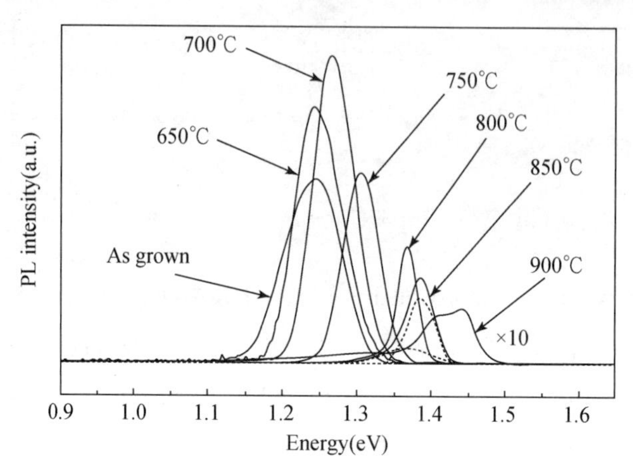

Fig. 5 PL spectra of InAs/GaAs dots

2.2 InAs/InAlAs/InP

The wires along the [$\bar{1}$10] direction are formed in the InAs/$In_{0.52}Al_{0.48}As$ multilayer system on InP(001) and they are diagonally aligned relative to the [001] growth direction[12–14]. In this work, the InAs strained layers in the InAs/InAlAs system are deposited by both the conventional MBE and MEE[15] methods, and the effect of the growth modes on the symmetry in the spacial alignment of InAs wires is observed with transmission electron microscope.

Fig. 6 shows the cross-sectional TEM images of the InAs wires array viewed from the [$\bar{1}$10] direction. Fig. 6(a) is the 8 ML-InAs wire array on InP(001). The InAs wires in the array are stacked along the direction away from the [001] direction on either side by about 33°, and the array is symmetrical about the [001] direction. Such symmetry in the MBE wire array on InP(001) is disrupted when the material is grown with the MEE. Fig. 6(b) is the unsymmetrical MBE InAs wire array on InP substrate 6°-misoriented from (001), and the angles between the growth direction and wire alignment are 26 and 43°, respectively. Fig. 6(c) is the unsymmetrical MEE InAs wire array on InP(001), and the angles between (001) and wire alignment are 47 and 38°. Fig. 6(d) is the unsymmetrical MEE 10 ML-InAs array, and the slanting angles are 30 and 70°, respectively. In addition, the QWR cross-section shape in the Figure is apparently distorted from a symmetrical one about the [001] growth direction.

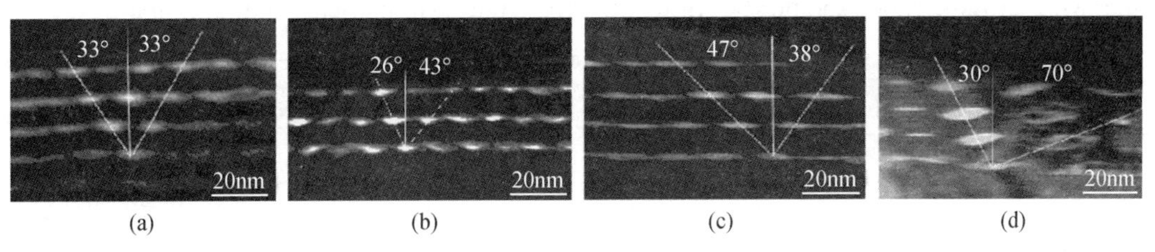

Fig. 6 XTEM images of InAs wires: (a) MBE, 8 ML InAs on InP(001); (b) MBE, 8 ML InAs on InP mis-oriented from (001) by 6°; (c) MEE, 8 ML InAs on InP(001); (d) MEE, 10 ML InAs on InP(001)

3 Discussion on the symmetry of the alignment of InAs wires

The [001] direction is symmetrical in crystallography and it is easy to understand that the MBE InAs QWR array on the InP(001) is symmetrical about the growth direction and the symmetry is disrupted when the material is grown on the misoriented substrate. However, the asymmetry in the alignment of the MEE InAs QWR array on InP(001) requires some explanation.

The main difference between the MBE and MEE growth modes is that the surface atomic diffusion during growth is significantly enhanced in the MEE mode[15]. The asymmetry in the MEE InAs QWR array may be related to the enhanced diffusion, and the mechanism is proposed as following.

Fig. 7 shows a schematic cross section of an InAs wire under self-assembling during epitaxy growth. The hills at sides A and B are both built up with terraces and steps in atomic scale. Uphill migration of randomly deposited adatoms is necessary for the wire to self-assembly and the driving force for adatom ascending is the strain energy, as an atom incorporated on a top site would be in a more relaxed state. If the terraces and steps on sides A and B are different from each other in atomic structure, the energy barriers for ascending adatoms are different and the upward motion is biased with more adatoms migrating up the hill from the side with the smaller energy barrier. This biased uphill motion of adatoms may result in the asymmetry in the MEE InAs QWR array on the (001)-oriented substrate.

Fig. 8 shows a structural model of a GaAs (2×4) surface in α-phase[16]. The reconstructed InAs surface should be similar to the case of GaAs as the (2×4) reconstruction is common in Ⅲ-Ⅴ semiconductors. A long terrace of 1ML height along the $[\bar{1}10]$ direction should be formed with the complete (2×4) reconstructed unite to keep the charge neutral. Therefore, the two edges of a terrace should be different from each other in atomic structure, resulting in the different energy barriers for ascending adatoms up the hills A and B, as shown in Fig. 2. With more atoms ascending up the hill with smaller energy barrier, the wire cross-section should be distorted from the symmetrical one and the resulting distribution of mismatch strain is not symmetrical about the growth direction. Eventually, the asymmetrical strain distribution produces the asymmetry in the QWR array on the (001)-oriented substrate.

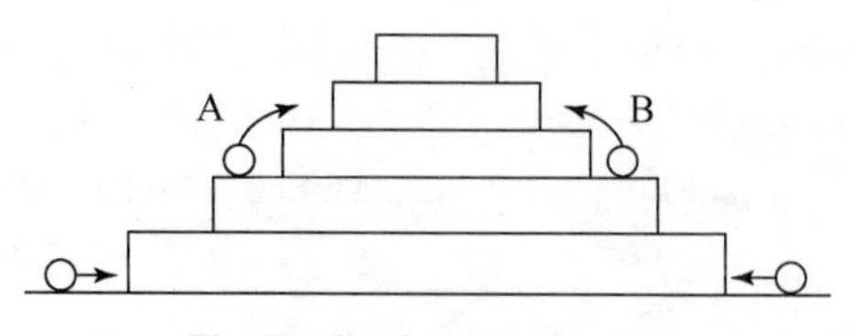

Fig. 7 Sectional InAs wire

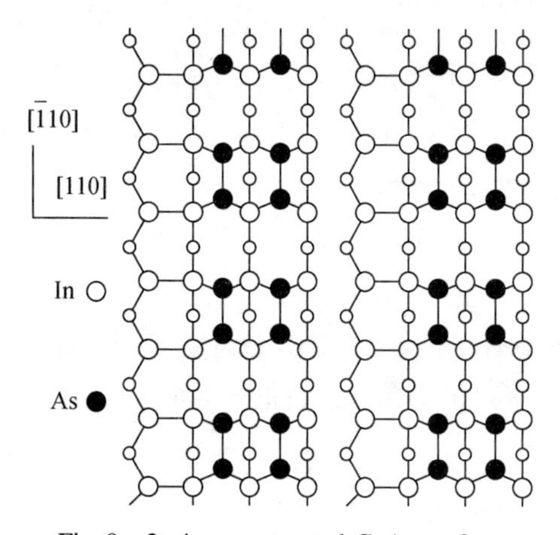

Fig. 8 2×4 reconstructed GaAs surface

For the MBE growth mode, the atomic diffusion is greatly reduced in comparison to the MEE mode and the effect proposed above may be neglected, and the MBE InAs QWR array seems symmetrical about the [001] growth direction.

Acknowledgements

This work is supported by the Special Funds for Major State Basic Research Project G20000683 and National Natural Science Foundation under the contract no 69976027.

References

[1] G. S. Solomon, J. A. Trezza, J. S. Harris Jr., Appl. Phys. Lett. 66(1995)991.

[2] G. S. Solomon, J. A. Trezza, J. S. Harris Jr., Appl. Phys. Lett. 66(1995)3161.

[3] R. Notzel, J. Temmyo, A. Kozen, T. Tamamura, T. Fukui, H. Hasegawa, Appl. Phys. Lett. 66 (1995)2525.

[4] M. Kawabe, K. Akahane, S. Ian, K. Okino, Y. Okada, H. Koyama, Jpn. J. Appl. Phys. 38

(1999)491.

[5] P. P. Gonzalez-Borrero, E. Jr. Mareyaa, D. I. Lubysheva, E. Petitpreza, P. Basmavia, J. Cryst. Growth 175/176(1997)765.

[6] S. Y. Shiryaev, F. Jensen, J. L. Hanseenm, J. L. Hanseni, J. W. Peterson, A. N. Larsen, Phys. Rev. Lett. 78(1997)503.

[7] G. S. Solomon, J. A. Trezza, A. F. Marshall, J. S. Hams, Jr., Phys. Rev. Lett. 76(1996)952.

[8] I. Mukhametzhanov, R. Heiz, J. Zeng, P. Chen, A. Madhuka, Appl. Phys. Lett. 73(1997)1841.

[9] R. Heiz, I. Mukhametzhanov, P. Chen, A. Madhukar, Phys. Rev. B 58(1998)10151.

[10] R. Leon, Y. Kim, C. Jagadish, M. Gal, J. Zou, D. J. H. Cockayne, Appl. Phys. Lett. 69 (1996)1888.

[11] S. Malik, Ch. Robert, R. Murray, M. Pate, Appl. Phys. Lett. 71(1997)1987.

[12] J. Wu, B. Xu, H. X. Li, Q. W. Mo, Z. G. Wang, M. Zhao, D. Wu, J. Cryst. Growth 197 (1999)95.

[13] H. X. Li, J. Wu, Z. G. Wang, T. Daniel-Race, Appl. Phys. Lett. 75(1999)1173.

[14] J. Brault, M. Gendry, O. Marty, M. Pitaval, J. Olivares, G. Grenet, G. Hollinger, Appl. Surf. Sci. 162/163(2000)584.

[15] Y. Horikoshi, M. Kawashima, H. Yamaguchi, Jpn. J. Appl. Phys. 25(1986)L868.

[16] H. H. Farrell, C. J. palmstrom, J. Vac. Sci. Technol. B 8(1990)903.

Effect of $In_{0.2}Ga_{0.8}As$ and $In_{0.2}Al_{0.8}As$ combination layer on band offsets of InAs quantum dots

J. He, B. Xu, and Z. G. Wang

(Key Laboratory of Semiconductor Materials Sciences, Institute of Semiconductors, Chinese Academy of Sciences, Beijing 100083, China)

Abstract We demonstrate the self-organized InAs quantum dots capped with thin and $In_{0.2}Al_{0.8}As$ and $In_{0.2}Ga_{0.8}As$ combination layers with a large ground and first excited energy separation emission at 1.35μm at room temperature. Deep level transient spectroscopy is used to obtain quantitative information on emission activation energies and capture barriers for electrons and holes. For this system, the emission activation energies are larger than those for InAs/GaAs quantum dots. With the properties of wide energy separation and deep emission activation energies, self-organized InAs quantum dots capped with $In_{0.2}Al_{0.8}As$ and $In_{0.2}Ga_{0.8}As$ combination layers are one of the promising epitaxial structures of 1.3μm quantum dot devices.

Since quantum dots (QDs) realized by self-organization during strained layer heteroepitaxy show good optical quality, they have recently found many applications in optoelectronics.[1-4] GaAs-based QDs laser can now be made to operate at 1.3μm using InAs or InGaAs quantum dot.[5-9] As a result, these devices may offer a low cost alternative to InP-based lasers which operate at the same wavelength. To get higher performance GaAs-based QD lasers, numerous works have been done to extend the wavelength longer with narrow linewidth and large ground to first excited state transition energy separation, which is an important parameter to improve the T_0 of QD lasers.[10-12] Of all the works, studies on energy-control for InAs QDs capped with thin AlAs ($In_{0.2}Al_{0.8}As$ or $In_{0.2}Ga_{0.8}As$) layer(s) have been carried out and the results are promising. High intensity Photoluminescence (PL) emission around 1.3μm has been obtained together with narrow linewidth of 21meV and energy separation as large as 108 meV.[13,14] In these letters, the authors attribute these merits either to the suppression of In and Ga atoms intermixing between the InAs QDs and the overgrowth GaAs barrier layer or to the reduction of the strain of the dots by the introduction of the thin capped layer(s), but the experimental evidence for the above-

原载于: Appl. Phys. Lett., 2004, 84(25): 5237-5239.

mentioned conclusions is not sufficient, furthermore, the quantitative effect of the thin capped layer(s) on the band offsets is still unclear, which is crucial for the design and understanding of the high performance GaAs-based QD lasers.

In this letter, we report the band offsets of self-organized InAs QDs capped with $In_{0.2}Al_{0.8}As$ and $In_{0.2}Ga_{0.8}As$ combination layers by performing deep level transient spectroscopy (DLTS) measurement on Schottky diode and p–n junction structure.

The samples under investigation were fabricated by solid source molecular beam epitaxy on n^+-GaAs (100) substrate in a Riber 32P apparatus. The growth is initiated with appropriate n^+-GaAs substrate. For sample A, 500nm Si-doped ($2\times10^{18}cm^{-3}$) GaAs buffer was grown at 580℃, then five layers of 2.7 monolayer InAs capped with 5nm $In_{0.2}Ga_{0.8}As$ and 2nm $In_{0.2}Al_{0.8}As$ with 30nm GaAs space layer were grown at 500℃. Finally, a 150nm layer was deposited. Sample A was homogeneously doped with Si of $2\times10^{16}cm^{-3}$. Sample B is doped with Be of $2\times10^{16}cm^{-3}$ with the same structure except for the top 50nm GaAs layer with concentration of $3\times10^{18}cm^{-3}$ Ohmic contact. The growth process was monitored by a 10keV reflection highenergy electron diffraction system. Substrate rotation and As_4 partial pressure of 7.3×10^{-6} Torr were used during growth.

Atomic force microscopy (AFM) measurements were carried out using a Park Scientific Instruction Microscope, model VP, in contact mode. PL measurements were performed under the excitation of 632.8nm line of a He-Ne laser. The luminescence spectra were detected by a liquid nitrogen cooled Ge detector.

The DLTS measurements were performed using the Innovance AB-type deep level transient spectroscopy. Schottky diodes were fabricated from sample A by evaporating Au onto the top GaAs layer. The back Ohmic contacts were formed by alloying In onto the n^+-GaAs substrate. For sample B, we alloy AuAgZn onto the top GaAs layer and AuGeNi onto the n^+-GaAs substrate to form Ohmic contact. Mesa diodes were fabricated for DLTS measurement.

An AFM image of the crystal surface with the growth halted following the 2.7 monolayer deposition of InAs is shown in Fig. 1. The typical lateral size was found to be about 41nm and height about 7nm. Size uniformity was found in the measurements. Dots density was about $3.9\times10^{10}cm^{-2}$

Fig. 2 displays the PL spectra of sample A at room temperature with excitation power of 20 and 100mW, respectively. The PL wavelength is at about 1.35μm and the full width at half maximum is about 37meV. With the increase of excitation power, a high energy peak emerges, which originates from the excitation states. The emission of the excited states indicates a more efficient quantum confinement of the carriers in the InAs QDs capped with

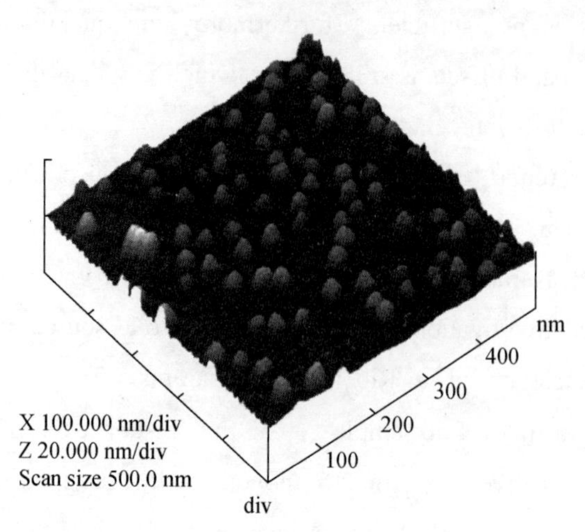

Fig. 1　500nm×500nm atomic force microscope image of InAs quantum dots. The nominal grown thickness was 2.7 monolayers

InAlAs and InGaAs combination layers. The energy separation between the ground and the first excited radiative transition is 102meV. For the previous studies of In(Ga)As/GaAs QDs, the energy separation between ground and first excited radiative transition range from 82meV for QD lasers operating at 1.28μm to 66meV for QDs laser operating at 1.33μm. [15,16] This relatively large energy separation, which is of great significance to decrease the temperature sensitivity of lasing threshold of QD lasers, obviously, is attributed to the combination $In_{0.2}Al_{0.8}As$ and $In_{0.2}Ga_{0.8}As$ capped layer.

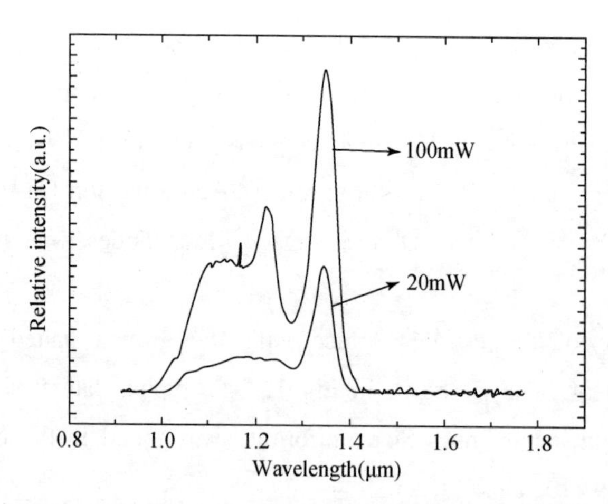

Fig. 2　PL spectra of (a) sample A at room temperature with excitation power of 20 and 100mW, respectively

Drawing an analogy between multielectron traps and quantum dots and from a detailed balance between thermal capture and emission rates of electrons from quantum dots, the emission rate of electrons or holes is given by

$$\tau_c = \frac{\exp(\Delta E/kT)}{\sigma v_{th} N_c},$$

where ΔE is the electron activation energy, σ is the capture cross section V_{th}, is the thermal velocity, and N_c is the conduction band effective density of states.

The DLTS measurements were made under dark condition. The sample biasing conditions were obtained from the C–V measurements. Fig. 3 displays the typical DLTS spectra of the two samples under majority injection. A significant peak is observed with the maximum position at 219K. From the activation plots of these signatures, the emission activation energy is determined to be 0.57 and 0.40eV. The peaks are conspicuously absent in the scan excluding the quantum dots by changing the reverse bias, which demonstrated the dominating emission signals in our DLTS measurements correspond to the energy level of the quantum dots.

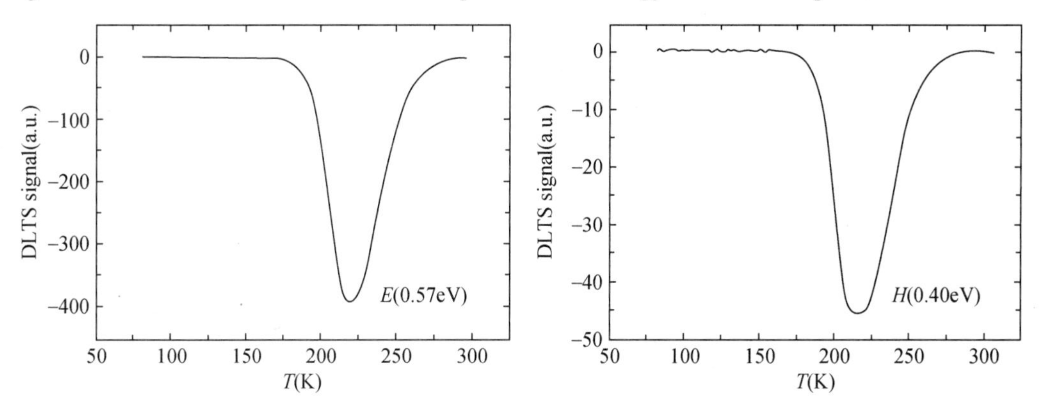

Fig. 3 The typical DLTS spectra of the two samples under majority injection. All the spectra were recorded for the rate window of 8.28ms, the filling pulse was 1ms, the reverse bias was −1V, and the pulse height was 0V

The DLTS peak height changes distinctly with the variation of the rate windows, indicating that large capture barriers exist. As the capture cross section, σ can be determined independently by the pulse-filling method. The capture barrier E_b can be evaluated from the temperature dependence of σ_n. The amplitude of the DLTS signal by the filling pulse t_p is expressed as

$$S(t_p)_{\max} = S(\infty)[1-\exp(-t_p/\tau_c)],$$

where $S(t_p)_{\max}$ is the peak height, $S\infty$ is the DLTS peak height when all quantum dots are fully filled by carriers, t_p is the pulse width, and τ_c is a time constant of capture.

The capture time constant τ_c and consequently, the capture rate c_n which can be measured with variation of the rate window, were obtained from the initial slope of the $\ln[1-S(t_p)/S(\infty)]$ versus t_p plot shown in Fig. 4. The curves reflect that the capture process is an exponential one. Due to a little fluctuation of QD sizes, the DLTS spectra of QD are broadened. The energy level broadening can cause the nonexponential transient. So, we see a little deviation from an exponential transient occurring in the curves in Fig. 4.

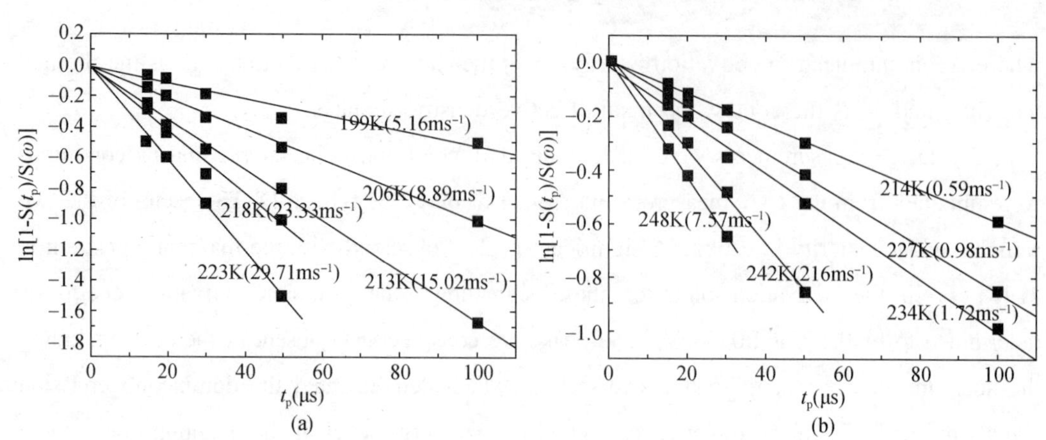

Fig. 4 DLTS signal amplitudes as a function of the activation pulse width at various temperatures for samples A and B, respectively. The values in brackets are the capture rate obtained from the slope of related curve

The capture cross section σ_n is connected with the capture c_n by

$$\sigma_n = \frac{c_n}{v_{\text{th}} n},$$

with v_{th} being the mean thermal velocity of the carriers during the capture process and n the free-carrier concentration, which is assumed to be a constant in the temperature range considered. Using the following relationship,

$$\sigma(T) = \sigma(\infty)\exp\left(-\frac{E_b}{k_B T}\right),$$

where σ is the capture barrier, k_B is the Boltzmann constant, and $\sigma(\infty)$ is a constant independent of temperature. The capture barrier of σ is plotted against the inverse of temperature in Fig. 5, from which we got the capture barrier $E_{eB} = 0.29$ eV and $E_{hB} = 0.24$eV for electron and hole, respectively. From DLTS measurement, we get the intrinsic emission activation energy (the binding energy) by combination of capture barriers with the apparent emission activation energies of InAs QDs capped with $In_{0.2}Al_{0.8}As$ and $In_{0.2}Ga_{0.8}As$ combination layers. $E_e^{\text{QD}\to\text{GaAs}} = 0.28$eV, $E_h^{\text{QD}\to\text{GaAs}} = 0.16$eV for electron and hole, respectively. The theoretical values for the electrons' and holes' emission activation energies of InAs/GaAs QDs are about 0.12 and 0.09eV for electrons and holes,[17] respectively, and moreover, previous experimental studies for self-assembled InAs/GaAs QDs show that the emission activation energies is only 0.12eV for electrons.[18] In our system, we got deeper emission activation energies for the electrons and holes together with a large ground to first excited state transition energy separation. Obviously, with the introduction of the combination layer of and $In_{0.2}Al_{0.8}As$ and $In_{0.2}Ga_{0.8}As$ capped layer, $In_{0.2}Ga_{0.8}As$ capped layer mainly serves as a strained-reduced layer which can lower the ground level of the QDs, Meanwhile, for

$In_{0.2}Al_{0.8}As$ layer, Al atoms are immobile at 500℃ and the chemical bonding in this layer is expected to be stronger. The Al accumulation on the InAs QDs prevents the In atoms from diffusing to the wetting layer. Therefore, the QD decomposition is reduced and the QD height is larger than that in the case with GaAs as a cap layer. And $In_{0.2}Al_{0.8}As$ and $In_{0.2}Ga_{0.8}As$ combination layer can both enhance the potential barrier surrounding InAs QDs[19] and reduce the strain of InAs QDs. Therefore, large energy separation and deeper emission activation energies are obtained, which leads to significant improvement in the temperature sensitivity of threshold. A combination of the PL and DLTS measurements, the schematic band structure of InAs QDs capped with $In_{0.2}Al_{0.8}As$ and $In_{0.2}Ga_{0.8}As$ combination layers is shown in Fig. 6. The apexes in the interface are caused by both the reduced strain and enhanced barrier with the introduction of $In_{0.2}Al_{0.8}As$ and $In_{0.2}Ga_{0.8}As$ combination layers.

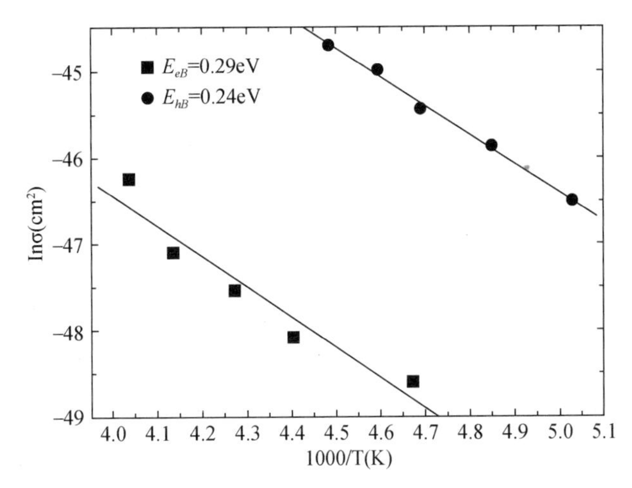

Fig. 5 The capture cross section as a function of the inverse of the temperature. The rate window varies from 8.28 to 165.6ms

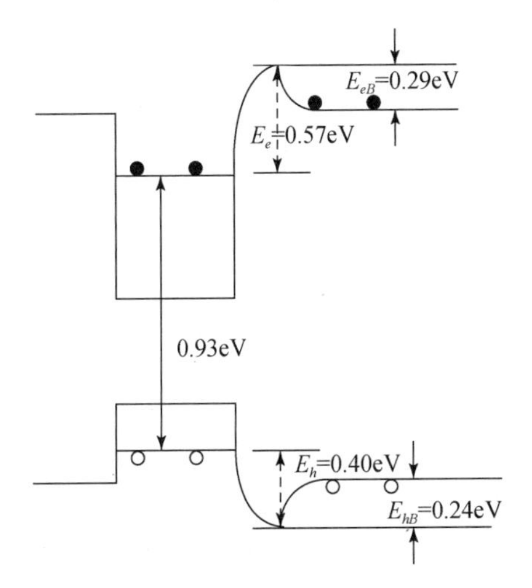

Fig. 6 The schematic band structure of of InAs QDS capped with InAlAs and InGaAs combination layers

In conclusion, we have reported investigations of carrier emission and capture from InAs QDs capped with $In_{0.2}Al_{0.8}As$ and $In_{0.2}Ga_{0.8}As$ combination layers by DLTS technique. From DLTS measurements, we determine emission activation energies of and 280 and 160 meV for the electron and hole ground states of the dot, which is larger than that of InAs/GaAs system, and these parameters are of great significance for the design and understanding of the high performance GaAs-based QDs lasers.

This work was supported by The National High Technology Research and Development Program (No. 2002AA302107), National Natural Science Foundation of China (under Contract Nos. 60176006, 60076024, 60006001, and 90101002).

References

[1] L. Brus, IEEE J. Quantum Electron. 22, 1909(1993).

[2] K. Mukai, Y. Nakata, K. Otsubo, M. Sugawara, N. Yokoyama, and H. Ishikawa, IEEE Photonics Technol. Lett. 11, 1205(1999).

[3] M. A. Migliorato, L. R. Wilson, D. J. Mowbray, M. S. Skolnick, M. Al-Khafaji, A. G. Cullis, and M. Hopkinson, Appl. Phys. Lett. 90, 6374(2001).

[4] J. He, Y. C. Zhang, B. Xu, and Z. G. Wang, J. Appl. Phys. 93, 8898(2003).

[5] S. Fafard, Z. Wasilewski, J. McCaffrey, S. Rmound, and S. Charbonneau, Appl. Phys. Lett. 68, 991(1996).

[6] J. Brault, M. Gendry, G. Grrent, G. Hollinger, Y. Desieres, and T. Benyattou, Appl. Phys. Lett. 73, 2923(1998).

[7] H. Y. Liu, B. Xu, Y. Q. Wei, D. Ding, J. J. Qian, and Z. G. Wang, Appl. Phys. Lett. 79, 2868(2001).

[8] Y. Qiu, P. Gogna, S. Forouhar, A. Stintz, L. F. Lester, and T. Benyattou, Appl. Phys. Lett. 79, 3570(2001).

[9] O. B. Shchekin and D. G. Deppe, Appl. Phys. Lett. 80, 3277(2002).

[10] M. Arzberger, U. Kasberger, G. Bohm, and G. Abstreiter, Appl. Phys. Lett. 75, 3968(1999).

[11] O. B. Shchekin, G. Park, D. L. Huffuker, and D. G. Deppe, Appl. Phys. Lett. 77, 466(2000).

[12] J. He, X. D. Wang, B. Xu, Z. G. Wang, and S. C. Qu, Jpn. J. Appl. Phys., Part 1 42, 1154 (2003).

[13] K. Nishi, H. Saito, and S. Sugou, Appl. Phys. Lett. 74, 1111(1999).

[14] Y. Q. Wei, S. M. Wang, F. Ferdos, J. Vukusic, and A. Larsson, Appl. Phys. Lett. 81, 1621 (1998).

[15] G. Park, O. B. Shchekin, S. Csutak, and D. G. Deppe, Appl. Phys. Lett. 75, 3267(1999).

[16] Yu. M. Shernyakov, D. A. Bedarev, E. Yu. Kondrat'eva, P. S. Kop'ev, A. R. Kovsh, N. A. Maleev, M. V. Maximov, S. S. Mikhrin, A. F. Tsatsul'nikov, V. M. Usintov, B. V. Volovik, A. E. Zhukov, Zh. I. Alferov, N. N. Ledentsov, and D. Bimberg, Electron. Lett. 35, 898 (1999).

[17] M. A. Cusack, P. R. Briddon, and M. Jaros, Phys. Rev. B 54, R2300(1996).

[18] S. Ghosh, B. Kochman, J. Singh, and P. Bhattacharya, Appl. Phys. Lett. 76, 2571(2000).

[19] H. Y. Liu, I. R. Sellers, M. Hopkinson, C. N. Harrison, D. J. Mowbray, and M. S. Skolnick, Appl. Phys. Lett. 83, 3716(2003).

信息功能材料的研究现状和发展趋势

王占国

（中国科学院半导体研究所，半导体材料科学重点实验室，北京，100083）

摘要　介绍了国内外信息功能材料目前的研发水平、器件应用概况和发展趋势，主要介绍了半导体材料，也涉及信息存储材料、有机光电子材料和人工晶体材料等。

关键词　信息技术，半导体微结构材料，超晶格、量子阱材料，量子线、量子点材料，量子比特

Current status and trend of research and development for functional information materials

Wang Zhanguo

(Key Laboratory of Semiconductor Materials Science, Institute of Semiconductors, Chinese Academy of Sciences, Beijing 100083, China)

Abstract　The current status and future prospects of research and development for functional information materials and its related devices are introduced in this paper, in which stress mainly on semiconductor materials, but information storage materials, organic opto-electronic materials and artificial crystal materials etc are also briefly discussed.

21 世纪是以信息产业为核心的知识经济时代。随着信息技术向数字化、网络化的迅速发展，超大容量信息传输、超快实时信息处理和超高密度信息存储已成为信息技术追求的目标。信息的载体正由电子向光电子结合和光子方向发展。与此相应，信息材料也从体材料发展到薄层、超薄层微结构材料，并正向光电信息功能集成芯片和有机/无机复合材料以及纳米结构材料方向发展。历史发展表明，信息功能材料是信息技术发展的基础和先导。没有硅材料和硅集成芯片的问世，就不会有今天微电子技术；没有光学纤维材料的发明、砷化镓材料的突破、超晶格和量子阱材料的研制成功以及半导体激光器和超高速器件的发展，就不会有今天先进的光通信、移动通信和数字化

原载于：化工进展，2004，23（2）：117-126.

高速信息网络技术。可以预料，基于量子效应的纳米信息功能材料的发展和应用，必将触发新的技术革命，并将深刻地影响着世界的政治、经济、军事对抗格局和彻底地改变人类的生产和生活方式。

信息功能材料作为21世纪信息社会高技术产业发展的基础材料，涉及信息产生、发射、传输、接收、处理、存储和显示等各个方面，并涉及半导体材料、光存储材料及有机、无机、有机/无机复合发光与显示材料和人工晶体材料等。从材料体系上看，除硅材料作为当代微电子技术的基础在21世纪中叶不会改变外，化合物半导体微结构材料以其优异的光电性质在高速、低功耗、低噪声器件和电路，特别是光电子器件、光电集成和光子集成等方面发挥着越来越重要的作用。GaN基紫、蓝、绿异质结构发光材料和器件的研制成功将引起照明光源的革命，社会经济效益巨大。航空、航天以及国防建设的要求推动了宽带隙高温微电子材料和中远红外激光材料的发展。探索低维结构材料的量子效应及其在未来纳米电子学和光子学方面的应用已成为材料科学目前最活跃的研究领域[1-7]。

1 半导体硅材料

硅(Si)是当前微电子技术的基础材料，预计其统治地位至少到21世纪中叶都不会改变。从提高硅器件、集成电路（ICs）成品率，提高性能和降低成本来看，增大直拉硅单晶的直径、解决硅片直径增大导致的缺陷密度增加和均匀性变差等问题仍是今后硅单晶发展的大趋势。8in(1in=0.0254m）硅片已普遍用于集成电路的生产，硅ICs工艺由8in向12in的过渡也将在近年内完成。预计2016年前后，18in的硅片将投入生产，直径27in硅单晶研制也正在积极筹划中。目前，300mm、线宽为0.13μm的硅超大规模集成电路（ULSI）生产线将投入大规模生产，300mm、90nm工艺已被一些大公司逐渐采用，32nm工艺技术也在实验室研制成功。从进一步缩小器件的特征尺寸、提高硅ICs的速度和集成度看，研制适合于硅深亚微米乃至纳米工艺所需的大直径硅外延片将会成为硅材料发展的另一个主要方向。直径8-12in的硅外延片已成功地用于生产，更大尺寸的外延片也在开发中。另外，以低功耗、高速和抗辐照为其特点的绝缘体上半导体（SOI）材料，如智能剥离（smart-cut）材料和SIMOX材料等的研制也取得重要进展，国际上已有直径为8in的SOI商品材料出售，SOI材料很可能成为180nm及以下的存储器电路的优先选用材料。

理论分析表明，20-30nm将是硅MOS集成电路线宽的“极限”尺寸。这不仅是指量子尺寸效应对现有器件特性影响所带来的“物理”限制和光刻技术的限制问题，更重要的是将受硅、SiO_2自身性质的限制。尽管人们正在积极寻找高介电绝缘材料(如用Si_3N_4等来替代SiO_2)、低介电互连材料（用Cu代替Al引线）以及采用系统集成芯片（system on a chip）技术等来提高超大规模集成电路（ULSI）的集成度、运算速度和功能，但硅将最终难以满足人类不断增长的对更大信息量的需求。为此，除积极探索基于全新原理的量子计算、光计算、分子计算和 DNA 生物计算等之外，应把更多的希望寄托在发展新材料、新效应和新技术上，如 Si 基半导体异质结构材料、Ⅲ－Ⅴ族化合物半导体材料，特别是低维（纳米）半导体结构材料等。

2　硅基异质结构材料

硅基光、电器件集成一直是人们所追求的目标。但由于硅是间接带隙，如何提高硅基材料发光效率就成为一个亟待解决的问题。经过多年研究，近年来在硅基Ⅲ－Ⅴ族化合物半导体材料、GeSi 合金和硅基高效发光材料研究等方面取得了重大进展。另外，人们还在硅基纳米材料（纳米 Si/SiO_2）、Ge/Si 量子点和量子点超晶格材料以及纳米硅的受激放大等研究方面也取得了令人振奋的成绩。

2.1　硅衬底上沉积高质量 GaAs 单晶薄膜

尽管 GaAs/Si 和 InP/Si 是实现光电子混合集成理想的材料体系，但由于晶格失配和热膨胀系数等不同造成的高密度失配位错而导致器件性能退化和失效，使其难以实用化。最近，Motolora 等公司宣称，他们在大尺寸的硅衬底上用钛酸锶作为协变层（柔性层）成功地生长了器件级的 GaAs 外延薄膜。研究表明，在硅衬底上生长薄层钛酸锶时，氧分子扩散到硅与钛酸锶界面处，并与下层的硅原子键合在硅与钛酸锶之间产生一个非晶界面层，这个非晶层使钛酸锶晶格常数弛豫，并与砷化镓的晶格匹配得很好。目前，美国的一些公司已经开发了 12in 硅衬底上的砷化镓生长技术。此外，他们还正在致力于开发硅衬底上生长 InP 和 GaN 的技术。大直径 GaAs/Si 复合片材的研制成功不仅给以 GaAs、InP 为代表的化合物半导体产业带来挑战，而且以其廉价、可克服 GaAs、InP 大晶片易碎和导热性能差等缺点以及与目前标准的半导体工艺兼容等优点受到关注，其最大的一个潜在应用是为实现人们长期以来的梦想——光、电集成电路（芯片）提供技术基础。但是，GaAs/Si 等复合材料能否真正获得实际应用还有待时间的考验。

2.2　GeSi/Si 应变层超晶格材料

GeSi/Si 应变层超晶格材料因其在新一代移动通信上的重要应用前景，而成为目前硅基材料研究的另一个重要方向。Si/GeSi MODFET 和 MOSFET 的最高截止频率已达 200GHz，HBT 最高振荡频率为 160GHz，噪声在 10GHz 下为 0.9dB，其性能可与 GaAs 器件相媲美，已有在手机中应用的报道。GeSi 材料生长方法主要有硅分子束外延、化学束外延和超高真空化学气相淀积（UHV/CVD）3 种，从发展趋势看，UHV/CVD 方法有较大优势。目前，8in 的 GeSi 外延片已研制成功，更大尺寸的外延设备也在筹划中。一方面，GeSi 材料以其器件、电路的工作频率高及功耗小等特点优于硅材料；另一方面，又以其价廉而胜于 GaAs 等化合物半导体材料。因而可以预料，GeSi 材料将在下一代移动通信的应用中占有一席之地。

2.3 硅基发光材料与器件

英国 Surrey 大学的 Wei Lek Ny 等在 *Nature* 杂志报道了一种“位错工程”的新方法，即将硼离子注入硅中，既是 *P* 型掺杂剂，又可与 *N* 型硅形成 *PN* 结，同时又在硅中引入位错环；位错环形成的局域场调制硅的能带结构，使荷电载流子空间受限，从而使硅发光二极管器件的量子效率提高到 0.1%。澳大利亚新南威尔士大学的 Martin A Green 等利用光发射和光吸收互易的原理设计制备的倒金字塔结构，不但减少了反射，而且通过将光俘获在电池中而增加光吸收，又将硅基 LED 的近室温功率转换效率提高到 1%。意大利卡特尼亚的 ST 微电子公司的研究人员称他们将稀土金属离子，如铒（Er）、铈（Ce）等，注入包含有直径为 1–2nm 的硅纳米晶的富硅二氧化硅中，由于量子受限效应，抑制了非辐射复合过程发生，创造了外量子效率高达 10% 的硅基发光管的世界纪录。发光管的发光波长依赖于稀土掺杂剂的选择，如掺铒（Er）发 1.54μm 标准光通信光波，掺铽（Tb）发绿光，掺铈（Ce）发蓝光。

硅基高效发光器件的研制成功，为硅基光电子集成和密集波分复用光纤通信应用提供了技术基础，具有深远的影响。

2.4 硅基氮化镓发光材料和器件

以蓝宝石和 SiC 为衬底的高亮度蓝、绿光材料和发光器件已经商业化，但由于加工困难和价格昂贵等原因导致其成本难以下降。利用硅衬底有很多优点，如尺寸大、热导率高、低成本、易加工和可与硅微电子集成等，但由于 GaN 和 Si 之间大的晶格和热失配，而导致的外延层龟裂、高密度的穿透位错和表面形貌差等问题，使其难以得到实际应用。日本 Nagoya 技术研究所的 T Egawa 等报道了在硅（111）衬底上，应用 MOVVD 生长技术制备的 InGaN 基蓝、绿发光管性能得到明显改善的结果。他们采用 AlN/AlGaN 缓冲层和 AlN/GaN 多层结构，在 2in 的硅衬底上，生长出高结晶质量的、无龟裂的 GaN 基发光管。蓝光发光管在 20mA 时的工作电压为 4.1V，串联电阻 30Ω，输出功率为蓝宝石衬底的一半。从总体来看，其特性可与蓝宝石衬底的结果相比。这个结果说明，采用硅衬底制造 InGaN 基蓝、绿光发光器件是一个很有应用前景的方法，随着材料质量进一步提高和改进器件设计，光输出功率也将会得到进一步改善。

3 Ⅲ–Ⅴ族化合物半导体材料

与硅相比，Ⅲ–Ⅴ化合物材料以其优异的光电性质在高速、大功率、低功耗、低噪声器件和电路、光纤通信、激光光源、太阳能电池和显示等方面得到了广泛的应用。GaAs、InP 和 GaN 及其微结构材料是目前最重要、应用最广泛的Ⅲ－Ⅴ族化合物半导体材料。

3.1　GaAs 和 InP 单晶材料

GaAs 和 InP 是微电子和光电子的基础材料，为直接带隙，具有电子饱和漂移速度高、耐高温、抗辐照等特点，在超高速、超高频、低功耗、低噪声器件和电路，特别在光电子器件和光电集成方面占有独特的优势。目前，世界 GaAs 单晶的总年产量已超过 200t（日本 1999 年的 GaAs 单晶的生产量为 94t，GaP 为 27t），其中以低位错密度的 VGF 和 HB 方法生长的 2–3in 的导电 GaAs 衬底材料为主。近年来，为满足高速移动通信的迫切需求，大直径（6–8in）的 SI-GaAs 发展很快，4in 70cm 长及 6in 35cm 长和 8in 的半绝缘砷化镓（SI-GaAs）也在日本研制成功，美国 Motolora 公司正在筹建 6in 的 SI-GaAs 集成电路生产线。InP 具有比 GaAs 更优越的高频性能，发展的速度更快，但研制直径 4in 以上大直径的 InP 单晶的关键技术尚未完全突破，价格居高不下。

GaAs 和 InP 单晶的发展趋势是：①增大晶体直径，目前 4in 的 SI-GaAs 已用于大生产，预计直径为 6in 的 SI-GaAs 在 21 世纪初也将投入工业应用；②提高材料的电学和光学微区均匀性；③降低单晶的缺陷密度，特别是位错；④GaAs 和 InP 单晶的 VGF 生长技术发展很快，很有可能成为主流技术。

3.2　GaAs、InP 基超晶格、量子阱材料

半导体超薄层微结构材料是基于先进生长技术（MBE，MOCVD）的新一代人工构造材料。它以全新的概念改变着光电子和微电子器件的设计思想，即从过去的所谓“杂质工程”发展到“能带工程”，出现了“电学和光学特性可剪裁”为特征的新范畴，是新一代固态量子器件的基础材料。

GaAlAs/GaAs、GaInAs/GaAs、AlGaInP/GaAs、GaInAs/InP、AlInAs/InP、InGaAsP/InP 等 GaAs、InP 基晶格匹配和应变补偿材料体系已发展得相当成熟，已成功地用来制造超高速、超高频微电子器件和单片集成电路。高电子迁移率晶体管（HEMT）、赝高电子迁移率晶体管（P-HEMT）器件最好水平已达 f_{max} = 600GHz，输出功率 58mW，功率增益 6.4dB；双异质结晶体管（HBT）的最高频率 f_{max} 也已高达 500GHz，HEMT 逻辑大规模集成电路研制也达很高水平。基于上述材料体系的光通信用 1.3μm 和 1.5μm 的量子阱激光器和探测器，红、黄、橙光发光二极管和红光激光器以及大功率半导体量子阱激光器已商品化，表面光发射器件和光双稳器件等也已达到或接近达到实用化水平。目前，研制高质量的 1.5μm 分布反馈（DFB）激光器和电吸收（EA）调制器单片集成 InP 基多量子阱材料和超高速驱动电路所需的低维结构材料是解决光纤通信瓶颈问题的关键。西门子公司已完成了 80×40Gbps 传输 40km 的实验。另外，用于制造准连续兆

瓦级大功率激光阵列的高质量量子阱材料也受到人们的重视。

虽然常规量子阱结构端面发射激光器是目前光电子领域占统治地位的有源器件，但由于其有源区极薄（约0.01μm）端面光电灾变损伤、大电流电热烧毁和光束质量差一直是此类激光器的性能改善和功率提高的难题。采用多有源区量子级联耦合是解决此难题的有效途径之一。法国汤姆逊公司研制出了三有源区带间级联量子阱激光器，2000年年初在美国SPIE（The Society of Photo-optical Instrumentation Engineer）会议上，报道了单个激光器准连续输出功率超过10W的结果。中国早在20世纪70年代就提出了这种设想，随后又从理论上证明了多有源区带间隧穿级联、光子耦合激光器与中远红外探测器，与通常的量子阱激光器相比，具有更优越的性能，并从1993年开始了此类新型红外探测器和激光器的实验研究。1999年年初，980nm InGaAs新型激光器输出功率达5W以上，包括量子效率、斜率效率等均达当时国际最好水平。最近，又提出并开展了多有源区纵向光耦合垂直腔面发射激光器研究，这是一种具有高增益、极低阈值、高功率和高光束质量的新型激光器，在未来光通信、光互联与光电信息处理方面有着良好的应用前景。

为克服*PN*结半导体激光器的能隙对激光器波长范围的限制，基于能带设计和对半导体微结构子带能级的研究，1994年美国贝尔实验室发明了基于量子阱内子带跃迁和阱间共振隧穿的量子级联激光器（QCLs），突破了半导体能隙对波长的限制，成功地获得3.5–17μm波长可调的红外激光器，为半导体激光器向中红外波段的发展以及在光通信、超高分辨光谱、超高灵敏气体传感器、高速调制器、无线光学连接和红外对抗等应用方面开辟了一个新领域。自1994年以来，QCLs在向大功率、高温和单膜工作等研究方面取得了显著进展。2001年瑞士Neuchatel大学物理研究所的Faist等采用双声子共振和三量子阱有源区结构使波长为9.1μm的QCLs工作温度高达312K，连续输出功率3mW。目前，量子级联激光器的工作波长已覆盖近红外到中、远红外波段（3–70μm）。中国科学院上海微系统和信息技术研究所于1999年研制成功120K、5μm和250K、8μm的量子级联激光器，中国科学院半导体研究所于2000年又研制成功3.7μm室温准连续应变补偿量子级联激光器，使中国成为能研制这类高质量激光器材料为数不多的几个国家之一。

目前，Ⅲ–Ⅴ族超晶格、量子阱材料作为超薄层微结构材料发展的主流方向，正从直径4in向6in过渡，生产型的MBE（如Riber的MBE6000和VG Semicon的V150 MBE系统，每炉可生产9片4in，4片6in或45片2in；每炉装片能力分别为80片6in，180片4in和64片6in，144片4in；Applied EPI MBE的GEN2000 MBE系统，每炉可生产7片6in，每炉装片能力为182片6in）和MOCVD设备（如AIX 2600G3，5片6in或9片4in，每台年生产能力为3.75×10^4片4in或1.5×10^4片6in；AIX 3000，5片10in或25片4in，95片2in也正在研制中）已研制成功，并已投入使用。EPI MBE研制的生产型设备中，已有50kg的砷和10kg的镓源炉出售，设备每年可工作300天。英国卡迪夫的 MOCVD 中心、法国的 Picogiga MBE 基地、美国的 QED 公司、Motorola 公司、日本的富士通、NTT、索尼等都有这种外延材料出售。生产型 MBE 和 MOCVD 设备的使用，必然促进衬底材料和材料评价设备的发展。

3.3　一维量子线、零维量子点材料

基于量子尺寸效应、量子干涉效应，量子隧穿效应和库仑阻效应以及非线性光学效应等的低维半导体材料是一种人工构造（通过能带工程实施）的新型半导体材料，是新一代量子器件的基础，其应用极有可能触发新的技术革命。这类固态量子器件以其固有的超高速、超高频（1000GHz）、高集成度（10^{10} 电子器件/cm^2）、高效低功耗和极低阈值电流（亚微安）、极高量子效率、极高增益、极高调制带宽、极窄线宽和高的特征温度以及微焦耳功耗等特点在未来的纳米电子学、光子学和新一代超大规模集成电路（VLSI）等方面有着极其重要的应用背景，受到各国科学家和有远见高技术企业家的高度重视。

目前低维半导体材料生长与制备主要集中在几个比较成熟的材料体系上如 GaAlAs/GaAs、In（Ga）As/GaAs、InGaAs/InAlAs/GaAs、InGaAs/InP、In（Ga）As/InAlAs/InP、InGaAsP/InAlAs/InP 以及 GeSi/Si 等，并在量子点激光器、量子线共振隧穿、量子线场效应晶体管和单电子晶体管和存储器研制方面，特别是量子点激光器研制上取得了重大进展。应变自组装量子点材料与量子点激光器的研制已成为近年来国际研究热点。1994 年俄罗斯和德国联合小组首先研制成功 InAs/GaAs 量子点材料，1996 年量子点激光器室温连续输出功率达 1W，阈值电流密度为 290A/cm^2，1998 年达 1.5W，1999 年 InAlAs/InAs 量子点激光器在 283K 温度下最大连续输出功率（双面）高达 3.5W。中国科学院半导体研究所在继 1996 年研制成功量子点材料、1997 年研制成功量子点激光器后，1998 年年初研制成功 3 层垂直耦合 InAs/GaAs 量子点有源区的量子点激光器室温连续输出功率超过 1W，阈值电流密度仅为 218A/cm^2，工作寿命超过 3000h。2000 年以来，量子点激光器的研制又取得很大进展，俄罗斯约飞技术物理所 MBE 小组、柏林的俄罗斯和德国联合研制小组及中国科学院半导体研究所半导体材料科学重点实验室的 MBE 小组等研制成功的 In(Ga)As/GaAs 高功率量子点激光器，工作波长 1μm 左右，单管室温连续输出功率高达 3.6-4W。中国科学院半导体研究所半导体材料科学重点实验室的 MBE 小组于 2001 年通过在高功率量子点激光器的有源区材料结构中引入应力缓解层，抑制了缺陷和位错的产生，提高了量子点激光器的工作寿命，室温下连续输出功率为 1W 时工作寿命超过 5000h。2001 年，这个小组在 InAlAs/AlGaAs/GaAs 红光量子点激光器的研制方面也取得了显著进展，其性能为目前国际报道的最好水平。俄罗斯约飞技术物理所和德国柏林技术大学联合实验组于 2002 年在大功率亚单层量子点激光器研制方面取得突破进展。亚单层量子点激光器（200μm 条宽、腔长 1040μm）的有源区是由被 12nm 的 GaAs 空间隔离层隔开的两组亚

单层 InAs 量子点层组成，每一个亚单层量子点由 12 个周期、0.3ML 的 InAs/2.4ML GaAs 组成，置于激光器结构的 2 个波导层中心。量子点激光器的工作波长 0.94μm，阈值电流密度 290A/cm^2，单管室温连续输出功率高达 6W，特征温度 150K；器件输出功率在 0.8-6W，总转换效率高于 50%，为目前国际报道的最高水平。中国科学院半导体材料科学重点实验室 MBE 组的研究人员于 2002 年利用自组织量子点所固有的尺寸分布宽的特点，在国际上首次研制成功自组织量子点超辐射发光管。超辐射量子点发光管采用通常的分别限制结构和特殊的倾斜条型电流注入结构（抑制 F-P 模式激光振荡），有源区由 5 层非耦合 InGaAs/InAs 量子点堆叠构成。在腔长为 1600μm，注入电流 1.4A 时，室温连续波工作（中心波长 1μm）的光输出功率大于 200mW，光谱半宽 60nm，为目前国际已报道的超辐射发光管的最好结果。超辐射发光管在光纤传感器、密集波分复用光纤通信和细胞组织干涉层析成像技术等方面有广泛的应用。

2002 年美国康奈尔大学和哈佛大学的科学家在 *Nature* 杂志上发表论文声称，他们成功地将大小相当于单个分子的原子团结构置于相距仅 1nm 的电极之间，由原子团包裹的单个过渡族金属原子传送或中断电流，就其特性相当于一个晶体二极管，这是人们用单个原子或分子组装纳米机器研制方面取得的新进展。

中国科学院物理研究所表面物理实验室的谢其坤和贾金锋等[8]利用他们提出的“幻数团簇+模板”方法和分子束外延技术，成功地在硅衬底上制备出了尺寸相同、空间排列严格有序、面密度高达 10^{13}/cm^2 的金属纳米团簇阵列。这种结构具有良好的热稳定性，并与硅平面工艺兼容，有望在超高密度存储方面得到应用。

与半导体超晶格、量子阱和量子点材料相比，高度有序的半导体量子线的制备技术难度更大。近年以来，量子线的生长制备和性质研究也取得了长足的进步。中国科学院半导体研究所半导体材料科学重点实验室的 MBE 小组利用 MBE 技术和 SK 生长模式，继 2000 年成功制备出空间高度有序的 InAs/InAl(Ga)As/InP 的量子线和量子线超晶格结构的基础上，又对 InAs/InAlAS 量子线超晶格的空间自对准（垂直或斜对准）的物理起因和生长控制进行了研究，并取得了较大进展。2001 年，王中林领导的佐治亚理工大学的材料科学与工程系和化学与生物化学系的研究小组基于无催化剂、控制生长条件的氧化物粉末的热蒸发技术，成功地合成了诸如 ZnO、SnO_2、In_2O_3 和 Ga_2O_3 等一系列半导体氧化物纳米带，它们与具有圆柱对称截面的中空纳米管或纳米线不同。这些原生的纳米带呈现出高纯、结构均匀和单晶体，几乎无缺陷和位错；纳米线呈矩形截面，典型的宽度为 20-300nm，宽厚比为 5-10，长度可达数毫米。这种半导体氧化物纳米带是一个理想的材料体系，可以用来研究载流子维度受限的输运现象和基于它的功能器件制造。同年，中国的香港城市大学材料科学与工程系的李述汤[9]和瑞典隆德大学固体物理系纳米中心的 Lars Samuelson 领导的小组[10]，分别在 SiO_2/Si 和 InAs/InP 半导体量子线超晶格结构的生长制备方面也取得了有意义的结果。

2002 年采用汽-液-固相反应（V-L-S）生长制备半导体纳米线和纳米线超晶格的

工作又取得重要进展。美国哈佛大学的 Gudiksen 等[11]分别利用激光协助催化方法和应用金纳米团簇催化剂结合化学气相沉积技术，成功生长 2–21 层的组分调制纳米线超晶格结构 GaAs/GaP 和 *P*-Si/*N*-Si，*P*-InP/*N*-InP 调制掺杂纳米线超晶格结构。纳米线的直径和异质结或 *PN* 结界面组分与掺杂的陡度依赖于催化剂金等纳米团簇的大小，纳米线超晶格的直径从几个纳米到数十纳米不等，长度可达几十微米。发光和输运性质测量表明，这种纳米超晶格结构具有优异的光电性质，其潜在应用可覆盖从纳米条形码到纳米尺度偏振发光二极管的整个范围。美国加利福尼亚大学伯克利的 Johnson 等利用镍催化剂和 V-L-S 方法，通过金属镓和氨在 900℃蓝宝石衬底上直接反应，合成了直径在几十到几百纳米之间、长达数十微米的 GaN 纳米量子线，X 射线衍射证实纳米线具有钎锌矿晶体结构。四倍频光参量放大器（波长 290–400nm，平均功率 5–10mW）用作泵浦激光器，在泵的单个 GaN 单晶纳米线（直径约 300nm，长约 40μm）的两端观察到了蓝、紫激光发射。激射波长随泵浦功率增加的红移，支持了高温下电子–空穴等离子体是 GaN 主要的激射机制观点。上述研究结果将有力地促进实现基于纳米线的电注入蓝–紫相干光源的研制步伐。

美国加利福尼亚大学和劳伦兹伯克利国家实验室的科学家杨培东（Peidong Yang）等在 2001 年 *Science* 杂志报道了他们成功研制了 ZnO 纳米线紫外激光器。他们认为紫外激光器将在信息存储和微分析等芯片实验室器件（lab on a chip device）上有应用前景。单晶 ZnO 纳米线结构是在镀金的蓝宝石衬底上，以金作为催化剂，沿垂直于衬底方向生长出来的。纳米线长 2–10μm，直径为 20–150nm。ZnO 纳米线和衬底之间的界面形成激光共振腔的一个镜面，纳米线的另一端的六方理想解理面为另一个镜面。在 266nm 光的激发下，由纳米线阵列发出波长在 370–400nm 的激光。单个纳米线激射也曾观察到。

低温工作的单电子晶体管早在 1987 年就已研制成功。1994 年，日本 NTT 就研制成功沟道长度为 30nm 的纳米单电子晶体管，并在 150K 观察到栅控源–漏电流振荡。1997 年，Zhuang 等又报道了室温工作的单电子晶体管开关。近年来，中国科学院物理研究所王太宏小组在单电子晶体管研制方面也取得了很好成绩。利用单电子晶体管的电导对岛区电荷极为敏感的性质，可制成超快和超灵敏的静电计，分辨率高达 $1.2\times10^{-5}e/Hz^{1/2}$，比目前最好的商用半导体静电计分辨率高六七个数量级，可用来检测小于万分之一电子电荷的电量。按照目前的技术水平，制备室温工作的单个 SET 已无不可克服的困难，但由于所需要的不仅是单个器件，而是每个 MPU 芯片可集成数量为 10^9–10^{10}功能完全相同的 SET，以满足超高速运算要求。1998 年 Yano 等[12]采用 0.25μm 工艺技术实现了 128MB 的单电子存储器原型样机的制造，这是单电子器件在高密度存储电路的应用方面迈出的重要的一步，但要实现单电子器件的大规模集成，还有很长的路要走。目前，基于量子点的自适应网络计算机业已取得进展，其他方面的研究正在深入地进行中。

半导体量子点、量子线材料的制备方法虽然很多，但总体来看不外乎自上而下（top down）、自下而上（bottom up）和两者相结合的方法。细分起来主要有：微结构材料生长和精细加工工艺相结合的方法，应变自组装量子线、量子点材料生长技术，图形化衬底和不同取向晶面选择生长技术，单原子操纵和加工技术，纳米结构的辐照制备技术，以及在沸石的笼子中、纳米碳管和溶液中等通过物理或化学方法制备量子点和量子线的技术。目前发展的主要趋势是寻找原子级无损伤加工方法和应变自组装生长技术，以求获得无缺陷的、空间高度有序和大小、形状均匀、密度可控的量子线和量子点材料。

4 宽带隙半导体材料

宽带隙半导体材料主要指的是金刚石、Ⅲ族氮化物、碳化硅、立方氮化硼以及Ⅱ-Ⅵ族硫、锡碲化物、氧化物（ZnO 等）及固溶体等，特别是 SiC、GaN 和金刚石薄膜等材料，因具有高热导率、高电子饱和漂移速度和大临界击穿电压等特点，成为研制高频大功率、耐高温、抗辐照半导体微电子器件和电路的理想材料，在通信、汽车、航空、航天、石油开采以及国防等方面有着广泛的应用前景[13]。另外，Ⅲ族氮化物等也是优良的光电子材料，在蓝、绿光发光二极管（LED）和紫、蓝、绿光激光器（LD）以及紫外探测器等应用方面也显示了广泛的应用前景。随着 1993 年 GaN 材料的 *P* 型掺杂突破，GaN 基材料成为蓝绿光发光材料的研究热点。1994 年日本日亚公司研制成功 GaN 基蓝光 LED，1996 年实现室温脉冲电注入 InGaN 量子阱紫光 LD，次年采用横向外延生长技术降低了 GaN 基外延材料中的位错，使蓝光 LD 室温连续工作寿命达到 10000h 以上。目前，GaN 基蓝、绿 LED 已实现规模生产，年销售额已达数十亿美元。近年来，功率达瓦级（最大为 5W）的 GaN 基蓝、紫光发光二极管的研制成功，使人们看到了固态白光照明诱人前景。固态照明与目前常用的白炽灯相比，不仅发光效率高，节约能源 2/3，而且工作寿命可提高 10 倍以上，加之工作电压低、安全可靠和无污染等，是当前国内外研发的热点。国际上许多大公司，如 GE、Philips 和 Osram 等，都投入巨资从事固态白光光源的开发，希望能在这一具有巨大潜在商业利益的高技术领域占据优势地位。GaN 基激光器的研制也取得进展，工作波长在 400-450nm，最大室温连续输出光功率业已达 0. 5W 以上。在微电子器件研制方面，GaN 基 FET 的最高工作频率（f_{max}）已达 140GHz，f_T=67GHz，跨导为 260mS/mm。HEMT 器件也相继问世，发展很快。另外，在 2001 年，基于 InGaAlN 材料体系，波长短达 280nm 的紫外发光二极管和 256×256 太阳盲 AlGaN 焦平面阵列探测器的研制成功，使其在军事上有着广泛的应用前景。

众所周知，以 GaN 为代表的Ⅲ族氮化物因为没有同质衬底材料，而只能生长在与其晶格失配很大的蓝宝石、碳化硅、硅或砷化镓等衬底上，大的晶格失配导致的高缺陷密度，严重地影响着器件性能和它进一步的应用。目前，GaN 基衬底材料的研制包含两方面的工作：一个方法是采用各种生长技术制备块状 GaN 晶体，但进展不大，最大尺寸约 1cm；另一个方法是采用氢化物汽相外延（HVPE）技术，首先在蓝宝石或 GaAs 等衬底上长厚 0.5-1mm 的 GaN 外延薄膜，然后通过激光剥离技术将其与衬底分开并经表面加工，形成所谓的自支撑 GaN 衬底。经过多年的努力，日本的 Sumitomo 公司于 2000 年年底宣称“2in 自支撑 GaN 衬底制备获得突破，2001 年将有商品出售”。遗憾的是，至今尚未广泛地被采用，原因可能与价格昂贵或质量尚需提高等问题有关。尽管如此，自支撑 GaN 衬底制备成功与应用将对 GaN 基激光器和高温微电子器件和电路研制起着重要的推动作用。

Ⅲ族氮化物系统与传统半导体系统的显著差别之一是Ⅲ族氮化物表现出很强的压电效应，其中 AlN 具有已知半导体中最大的压电系数。Ⅲ族氮化物系统，特别是 AlGaN/GaN 系统的这一特征使得其对材料中的应变及所处的电学环境异常敏感。在由Ⅲ族氮化物材料组成的异质结构中，晶格失配将引起应变，从而显著影响材料的能带结构，引起简并能量状态的分裂，同时应变导致的压电效应能进一步调制能带，改变系统电子能级分布和态密度分布，表现出特异的、其他系统中不常见的效应。与传统半导体器件中通过掺杂改变材料中载流子浓度从而调制电导率不同，压电诱导能带工程主要通过调节材料中的应变（由衬底、晶向、材料组分和厚度决定）和压电系数（由材料组分决定）来改变材料中的极化电场，从而实现对材料能带的调制，改变材料的导电能力。在这种结构中，材料导电能力的提高将不受杂质浓度、散射和复合增强作用的限制。因此，采用这种结构的器件能够较容易地通过改变极化场的方向实现电子或空穴的积累，因而能有效避开目前在Ⅲ族氮化物材料中普遍存在的 *P* 型掺杂困难。这不仅有助于改善现有的场效应器件（FET）、二维电子（空穴）器件等新型器件的性能，而且大大有助于发展出目前难以实现的Ⅲ族氮化物双极型器件。目前科学家们对Ⅲ族氮化物压电诱导能带工程的机理和方法了解和很不全面，急需加以解决。

近年来具有反常带隙弯曲的窄禁带 InAsN、InGaAsN、GaNP 和 GaNAsP 材料的研制也受到了重视，这是因为它们在长波长光通信和太阳能电池等方面显示了重要应用前景。2002 年，1300nm 垂直腔面发射激光器（VCSELs）材料与器件研制方面取得了长足的进步。德国慕尼黑的信息技术所的 H Riechert 等应用 MBE 和 MOCVD 技术，分别以 $In_{0.35}Ga_{0.632}AsN_{0.018}$ 双量子阱和三量子阱为 VCSEL 的有源区，量子阱厚 6nm，垒层 20-25nm；上镜面和下镜面分别由 28 对 $Al_{0.8}Ga_{0.2}As$/GaAs 和 32-34 对 AlAs/GaAs 组成。氧化孔径为 $4\times6\mu m^2$ 的 MBE 生长器件，室温连续工作波长为 1306nm，阈值电流 2.2mA，边模抑制比优于 30dB（传输速率 2.5Gb/s，典型驱动电流 5mA），输出功率大于 1mW，器件直到 80℃仍保持激射。采用氧化电流孔径为 5μm，单模发射功率 700μW 的样品，传输速率 2.5Gb/s，传输距离超过 20.5km 时，比特误码率低于 10^{-11}。该小组应用 MOCVD 技术研制的 InGaAsN 基 VCSELs，也取得了数据传输率为 10Gb/s、背对背运用比特误码率低于 10^{-11} 的好结果。

以 Cree 公司为代表的体 SiC 单晶的研制业已取得突破性进展，2in 的 4H 和 6H-SiC 单晶与外延片以及 3in 的 4H-SiC 单晶已有商品出售；以 SiC 为 GaN 基材料衬底的蓝绿光 LED 业已上市，参与以蓝宝石为衬底的 GaN 基发光器件的竞争；其他 SiC 相关高温器件的研制也取得了长足的进步。目前存在的主要问题是材料中的缺陷密度高，且价格昂贵。

宽带隙半导体异质结构材料往往也是典型的大失配异质结构材料。所谓大失配异质结构材料是指晶格常数、热膨胀系数或晶体的对称性等物理参数有较大差异的材料体系，如 GaN/蓝宝石（Sapphire）、SiC/Si 和 GaN/Si 等。大晶格失配引发界面处大量位错和缺陷的产生极大地影响着微结构材料的光电性能及其器件应用，如何避免和消除这一负面影响是目前材料制备中的一个迫切要解决的关键科学问题。20 世纪 90 年代以来，国际上提出了多种解决方法，虽有进展，但未能取得重大突破。近年来，中国科学院半导体研究所的王占国、陈涌海和汪连山等基于缺陷工程、晶面特征与表面再构、晶体结构对称性和生长动力学等方面的考虑，提出了柔性衬底的概念，并在 ZnO/Si、γ-Al_2O_3/Si、SiC/Si 和 GaN/Si 等异质结构材料准备方面取得了进展。这个问题的解决，必将极大地拓宽材料的可选择余地，开辟新的应用领域。

目前，除 SiC 单晶衬底材料、GaN 基蓝光 LED 材料和器件已有商品出售外，大多数高温半导体材料仍处在实验室研制阶段。制约这些材料实用化的关键问题有：GaN、ZnO 等体单晶材料、宽带隙 *P* 型掺杂和欧姆电极接触，单晶金刚石薄膜生长与 N 型掺杂，Ⅱ-Ⅳ族材料的退化机理等。国内外虽已做了大量的研究，至今仍未取得重大突破。

5 固态量子构筑和量子计算

随着微电子技术的发展，计算机芯片集成度不断增高，器件尺寸越来越小（纳米尺度）并最终将受到器件工作原理和工艺技术限制，而无法满足人类对更大信息量的需求。为此，发展基于全新原理和结构的功能强大的计算机是 21 世纪人类面临的巨大挑战之一。1994 年 Shor[14] 基于量子态叠加性提出的量子并行算法并证明可轻而易举地破译目前广泛使用的公开密钥 Rivest、Shamir 和 Adlman（RSA）体系，引起了人们的广泛重视。

所谓量子计算机是应用量子力学原理进行计算的装置，它的基本信息单元叫作量子比特（qubit），是实现量子计算的关键。根据量子理论，电子可以同时处于两个位置，原子的能级在某一时刻既可以处于激发态，也可以处于基态。这意味着以这些系统构造出的基本计算单位——比特，不仅能在相应于传统计算机位的逻辑状态 0 和 1 稳定存在，而且也能在相应于这些传统位的混合态或叠加态存在，因此称为量子比特。也就是说，量子比特能作为单个的 0 或 1 存在，也可以同时既作为 0 也作为 1，而且用数字系数代表了每种状态的可能性。

构筑量子比特是实现量子计算的基础，文献报道了很多物理系统都可以用于构筑量子比特，如液态核磁共振、施主杂质核自旋、超导体和半导体量子点中的电子自旋等。在这些系统中，可能最有前途的是半导体量子点，因为现在已经有了生产半导体材料的成熟工艺，而且人们对于半导体量子点，特别是自组装量子点的研究无论在理论上还是实验上也趋于完善。

1998 年，Loss 和 Divincenzo[15] 利用耦合单电子量子点上的自旋态来构造量子比特，实现信息传递。电子自旋有上、下两个方向，所以一个量子点就相当于传统计算机中的一个晶体管开关，形成了一个单量子比特，每个点都可表示 0 或 1。对于有 3 个量子点“Q1、Q2、Q3”的物理体系，假设在 Q1 和 Q2 上各有一个多余电子，自旋为 1/2；用铁磁材料量子点控制 Q1 上的自旋态，用外加电压控制 Q1、Q2 两个量子点之间的耦合。若电压高，则量子点间的隧穿势垒增高，电子隧穿被禁止；若电压低，则势垒低，Q1 和 Q2 上的电子自旋会发生海森伯交换耦合，电子发生隧穿。电子隧穿到达顺磁（PM）点，可作为一个观察窗口，电子隧穿进入 Q3，则可通过静电计对自旋进行测量。量子计算机工作时，信息就是在这样的量子比特对之间相互交换。2001 年，普渡大学的 Jeong 和 Chang[16] 首次探测到连在一起的一对量子点中每个量子点上电子的自旋方向，这无疑使量子计算机的实现又向前迈进了一步。

实现量子比特构造和量子计算机的设想方案很多，其中最引人注目的是 Kane[17] 最近提出的一个实现大规模量子计算的方案。其核心是利用硅纳米电子器件（这个器件是由高纯硅、掺杂原子磷和绝缘层以及金属栅的重复结构组成，栅宽和栅距为 10nm，施主磷精确地掺在设计规定的每一个栅下的硅晶体中）中磷施主核自旋进行信息编码，通过外加电场控制核自旋间相互作用实现其逻辑运算，自旋测量是由自旋极化电子电流来完成，计算机要工作在 mK 的低温下。这种量子计算机的最终实现依赖于与硅平面工艺兼容的硅纳米电子技术的发展。除此之外，为了避免杂质对磷核自旋的干扰，必须使用高纯（无杂质）和不存在核自旋不等于零的硅同位素（^{29}Si）的硅单晶；减小 SiO_2 绝缘层的无序涨落以及如何在硅里掺入规则的磷原子阵列等是实现量子计算的关键。然而，阻碍量子计算机实现的另一个难题就是量子态在传输、处理和存储过程中可能因环境的耦合（干扰）而从量子叠加态演化成经典的混合态，即所谓消相干。因此，特别是在大规模计算中能否始终保持量子态间的相干是量子计算机走向实用化前所必须克服的难题。

6　信息存储材料

目前磁记录材料仍是最重要的信息存储材料。通过技术革新和巨磁阻材料的利用，磁性材料的存储密度仍有大幅度提高的空间。但是，2001–2006 年，磁材料中磁记录单元（磁晶）的尺寸将达到其记录状态的物理极限，相应的存储密度为 10–100G/in^2。在这种背景下，从 20 世纪 80 年代末以来，光存储技术得到了十分迅速的发展，光存储的市场不断扩大。目前，一般的光存储技术已经成熟，一次性和可擦写的光盘都已商业化。由于20世纪90年代末GaN蓝色激光的出现，光存储密度由于使用光波波长的变短而得到成倍的增长。下一步的发展方向是研究和开发适合蓝紫激光波长的光盘材料。此外，由于光存储技术的面密度已接近光学衍射极限，国际上正在寻找下一代的光存储技术，如三维光存储技术、全息存储技术和近场光存储等。在这些新的存储技术中，关键还是可实用的光存储材料的研究和开发。

7 有机光电子材料

有机发光材料以其低廉的成本和良好的柔性，已成为全色高亮度发光材料研究的又一大热点。目前有机电致发光材料LED的发光亮度最高已达10^5Cd/m^2，发光效率151m/W，工作寿命超过20000h，并实现了红、绿、蓝及全色发光，并有16in彩屏研制成功的报道，商业化前景看好。中国从20世纪80年代末和90年代初就开始这方面的研究，现已制备出高发光效率的PPV有机材料，其绿色高分子材料的发光效率约为88%，是目前国际上报道的最高值，但有机发光材料的稳定性尚未得到彻底解决。

折射率渐变新型塑料光纤由于其优良的机械性能以及抗干扰、易连接、低成本等特点，成为短途光纤通信和光纤入户的关键材料。1994年，国际开始研制渐变型塑料光纤技术。目前美国第六舰队“小鹰号”航空母舰已利用塑料光纤建立了局部电话通信系统，并有小批量的渐变型塑料光纤生产和应用。日本庆应大学实验室中已经研制出带宽为2GHz/100m、衰减仅为100dB/km的折射率渐变型塑料光纤。中国在20世纪90年代中期也开展了相应研究，并在短距离数据、资料传输、局域网互联等方面进行了示范演示，取得了较大进展。

8 人工晶体材料

近年以来，高光学质量、大尺寸激光晶体材料制备等方面取得了长足进步，并在可调谐、大功率和复合功能3个应用方面也取得了重大的进展。掺钛蓝宝石等可调谐激光晶体已实用化；Nd：YAG和铝酸镁镧等新型大功率激光基质正向千瓦级器件发展。中国通过对非线性光学晶体微观结构与宏观性能相互关系的研究，建立了相关的理论模型，经过工艺优化和反复的实验筛选，相继研制成功偏硼酸钡（BBO）和三硼酸锂（LBO）等有着重要应用价值的多种新型倍频晶体，在国际上享有盛誉。此外，中国还在特大尺寸KDP晶体和以铌酸锂为代表的三维光存储材料研制和应用方面做了很好的工作。

9 结语

从总体上看，在几个重要材料体系上，中国信息功能材料的研究水平已达到或接近国际先进水平，个别材料体系则处于领先地位，例如：非线性光学晶体，用于主干线通信网络的 1. 55μm InP 基 DFB 激光器材料，MOCVD 一次生长横向集成 DFB LD/EA 调制器发射模块芯片材料、GSMBE 一次生长纵向集成 InP 基 MSM PD/P-HEMT 接收模块芯片材料，用于接入网、硅平面波导和 CATV 的 1. 3μm 无制冷多量子阱激光器材料，高纯度 GaAs 外延材料，AlGaAs/GaAs 调制掺杂二维电子气材料（2DEG），高性能 InP、GaAs 基 HEMT、PHEMT 和 HBT 微结构材料，应变自组装量子点激光器材料，中远红外量子级联激光器材料，大功率量子阱激光器材料，超高亮度黄橙光材料，大晶格失配 GaN/γ-Al_2O_3/Si、Si/γ-Al_2O_3/Si 柔性衬底材料研究，以及带间量子隧穿激光器材料和器件研制等方面都做出了国际先进水平的工作。但从总体上看，多属跟踪研究，创新特别是原创性的工作不多；而在产业化方面，差距则更大。

致谢

本文部分资料引自国家重点基础研究发展规划“信息功能材料相关基础问题”项目课题组专家研讨形成的内容，特此声明，并表示感谢。

参考文献

[1] 王大中，杨叔子. 技术科学发展与展望——院士论技术科学[M]. 济南：山东教育出版社，2002. 317.

[2] 王占国. [J]. 世界科技研究与发展，1998，25(5)：51.

[3] Faist J, Capasso F, Sivco D L, et al. [J]. Science, 1994, 264 : 553.

[4] Seifert W, Carlsson N, Miller M, et al. [J]. Prog. Crystal Growth Charact., 1996, 33 : 423.

[5] Burroughes J H, Bradley D D C, Brown A R, et al. [J]. Nature, 1990, 347 : 539.

[6] Zah C E, Bhat R, Pathak B H, et al. [J]. IEEE J. Quantum Electron, 1994, 30(2) : 511.

[7] Service R F. [J]. Science, 2000, 287(5453): 561a.

[8] 贾金锋，谢其坤. 科学发展报告[M]. 北京：科学出版社，2003. 98.

[9] Zhang Rui Qin. Lifshitz Yeshayahu, Lee Shuit-Tong. [J]. Adv. Mater, 2003, 15 : 635.

[10] Lars Samuelson. [J]. Materials Today, 2003, 22(6): 10.

[11] Gudiksen M S, Lauhon L J, Wang J F, et al. [J]. Nature, 2002, 415: 617.

[12] Yano K, Ishii T, Sane T, et al. [C]. San Francisco: ISSCC Digest of Technical Papers, 1998, 344.

[13] Alan Mills. [J]. Ⅲ-Ⅴs Review, 2000, 13 : 23.

[14] Shor P W. Proceeding of the 35th Annual Symposium on Foundation of Computer Science. [C]. New York : IEEE Computer Society Press, 1994, 124 ~ 134.

[15] Loss D, Divincenzo D P. [J]. Phys. Rev. A, 1998, 57 : 120.

[16] Jeong H, Chang A M, Melloch M R, [J]. Science, 2001, 293 : 2221.

[17] Kane B E. [J]. Nature, 1998, 393: 133.

High-performance quantum-dot superluminescent diodes

Z. Y. Zhang, Z. G. Wang, B. Xu, P. Jin, Z. Z. Sun, and F. Q. Liu

(Key Laboratory of Semiconductor Materials Science, Institute of Semiconductors, Chinese Academy of Sciences, Beijing 100083, China)

Abstract By inclining the injection stripe of a multiple layer stacked self-assembled InAs quantum dot (SAQD) laser diode structure of 6° with respect to the facets, high-power and broad-band superluminescent diodes (SLDs) have been fabricated. It indicates that high-performance SLD could be easily realized by using SAQD as the active region.

Superluminescent diodes (SLDs) are required as light sources for many applications including optical gyroscopes and sensors, optical time-domain reflectometers (OTDRs), and wavelength-division-multiplexing (WDM) system testing. High output power and large spectral bandwidth are key features for SLDs. High output power has been attempted by many efforts, including antireflection coating of the facet[1] and utilizing the tapered active region[2],[3]. The high output power of continuous wave (CW) above 200 mW at room temperature was obtained at 5000mA injection current[2]. In addition, as the spectral width broadened, the coherence length was reduced. The short coherence length can reduce the Rayleigh backscattering in fiber gyroscope systems and can improve the spatial resolution in an OTDR application. Therefore, broadening spectral width of SLDs offers an advantage for obtaining the ultimate sensitivity in these applications. Until now, several different methods have been used to broaden the output spectrum width of SLDs, such as using the quantum wells of different widths as the active region of SLDs[4] or increasing the pumping current of SLDs[5]. The former is intent on introducing different $n=1$ energy transition, and the latter is focused on introducing the $n=1$ and $n=2$ energy transition synchronously. Both ways are effective in broadening the spectrum width, however, they also have some disadvantageous factors. Although the different quantum-well widths can afford the different energy transition, the changing of energy is not successive, which will lead to the irregular shape of spectrum; the large pumping current for SLD will easily result in the saturation of the output power.

原载于：IEEE Photonics Technology Letters, 2004, 16(1):27-29.

In recent years, quasi-zero dimensional systems, especially self-assembled semiconductor quantum dots (QDs) grown by the Stranski-Krastanow mode, have been investigated from both fundamental studies and potential device applications such as laser diodes[6]. But in current self-assembled QD technology, the size of inhomogeneity is common and the typical nonuniformity is not less than 10%. Such inhomogeneity of QDs is formed naturally in the S-K growth mode, which is essential and not accidental[7]. In general, such inhomogeneous size distribution of SAQD in the active region is disadvantageous for achieving lasing of the QD laser. The coherence length L_{coh} of SLD is given by $L_{coh} = \lambda^2/\Delta\lambda$, where $\Delta\lambda$ is the emission spectral width of the SLD and λ is its central wavelength; so for the designed QD superluminescent diodes (QD-SLDs), it becomes an effective intrinsic advantage for broadening spectrum, which we have reported in our previous work[8]. Moreover, there are also two other advantages when making the SAQD as the active region of SLD. One is that the efficiency of luminescence of QD materials is higher than that of quantum well and other low-dimensional materials due to the atomic-like density of state in the QD system. The other is that the changing of energy of the SAQD is continuous, the flat spectrum can be realized, and it will be good for WDM and other applications of SLD.

In this letter, the tilted-stripe SLDs were fabricated on the molecular beam epitaxy (MBE)-grown wafer with multiple layer stacked QD structure. Because the tilted-stripe structure could well eliminate the Fabry-Pérot resonance[9], and the multiple layer stacked QD active region could not only benefit from the broadened spectral width but also enhance the efficiency of luminescence, the high-power and broad-band SLD has been easily realized.

The self-assembled QD superluminecent diode (SAQD-SLD) structures were grown by solid source MBE (Riber 32P) on n^+-doped GaAs(100) substrate. As seen in Fig. 1(a), the two graded index-separating confinement heterostructures (GRINSCH) were formed in connection with the multiple stacked QD active layers. The active medium in the SLDs consisted of five layers of two monolayer (ML) InAs SAQDs, which were separated by five ML $In_{0.2}Ga_{0.8}As$ strain-reducing layer (SRL) and 15 ML GaAs spacer layer, with 10nm GaAs below and above the five QD layers [in Fig. 1(b)]. The waveguide layers and the active region were claded by p and n $Al_{0.5}Ga_{0.5}As$ layers of 1μm thickness. The epilayer was followed by a SiO_2 layer on which a stripe window was etched out at the 6° angle. The stripe widths of all the devices were 10μm, and the lengths varied from 800 to 1600μm. No facet coatings were applied to the devices.

Fig. 2 shows the light output power versus current characteristics of the devices of different lengths at room temperature under continuous operation. From Fig. 2, we could find that all the devices with different lengths had high output power of above 200mW and no heat saturation phenomena occured. Among them, the former three curves [Fig. 2(a)-(c)] had a knee that corresponds to a threshold for lasing, whereas the last one[Fig. 2(d)] had no knee. To gain the same output power, the larger pumping currents were required for the longer length of devices, which indicated that quantum efficiency was reduced along with the increasing of length of devices. Additionally, it could be seen clearly that the longer the length of devices, the more obvious the superluminescent phenomena were. We thought that to avoid the reflection feedback modes, we must lengthen the devices adequately due to the insufficient lateral confinement in our gain-guided structure[9]. As a matter of fact, it was the same reason for the different quantum efficiency among the four devices. Because the former three devices had different extent of reflection feedback modes respectively, the quantum efficiency was higher than that of the last one, naturally. In a word, among them, the device with the stripe length of 1600μm showed ideal superluminescent *p-i* characteristics, just as Fig. 2(d) demonstrated. In addition, the CW output power of above 200mW was obtained at only 1400mA injection current. As an SLD, it is a very high efficiency. Further increasing the injection current will lead to the lasing of the device.

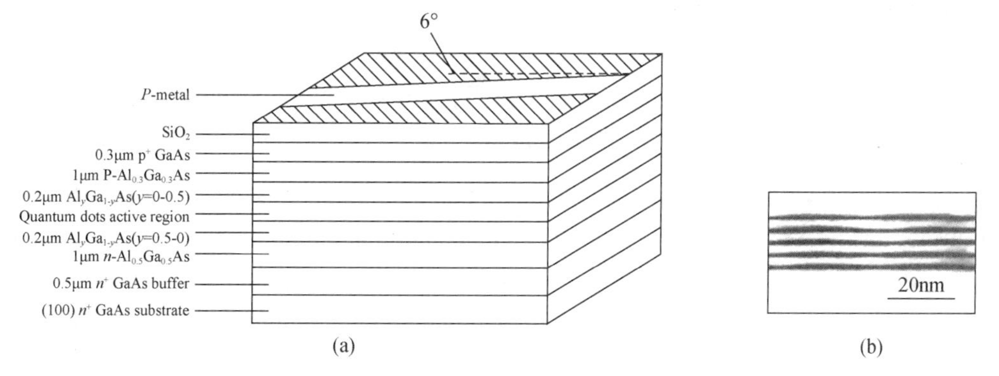

Fig. 1 (a) Layer structure and composition of inclined stripe SAQD-SLD. (b) The cross-sectional transparency electron microscope image of the active region of the multiple layer stacked QDs of the SAQD-SLD

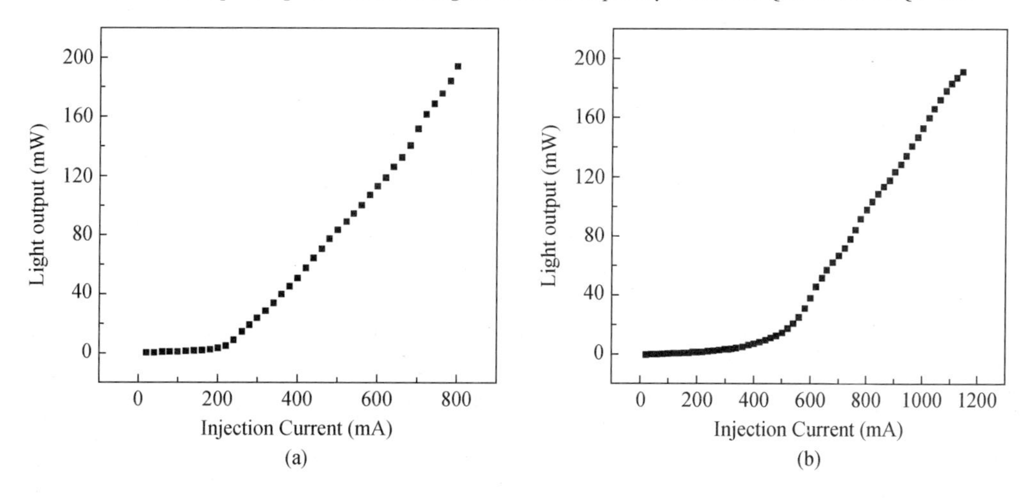

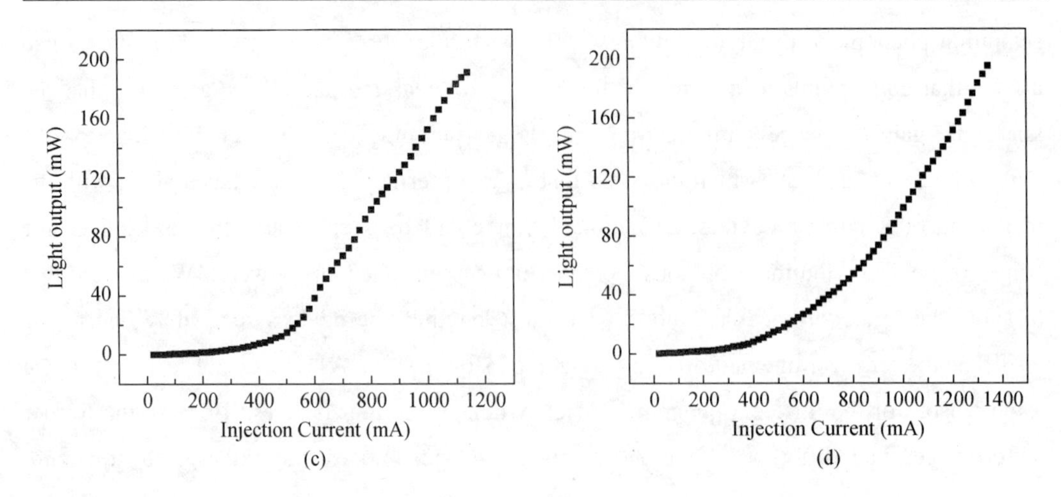

Fig. 2 Light output power of SAQD-SLD of different active stripe length [(a) 800, (b) 1000, (c) 1200, and (d) 1600μm] under continuous operation at room temperature

The corresponding spectra of the SAQD-SLD (the device with the stripe length of 1600μm) were also measured in Fig. 3. We could find that the spectrum was flat and it also exhibited an almost broad uniform shape at different power levels. The full-width at half-maximum of the spectra was above 60nm and the spectral modulation, defined as $m=(P_{max}-P_{min})/(P_{max}+P_{min})$, was less than 0.2dB in the vicinity of the spectrum peak with a 200mW output power, where P_{min} and P_{max} are the minimum and the maximum intensities in the spectrum. These characteristics of SAQD-SLD are effective to be applied in the spectrum-slicing technique and WDM system, so the continuous changing of energies of SAOD could be another favorable factor to improve the performance of SLD.

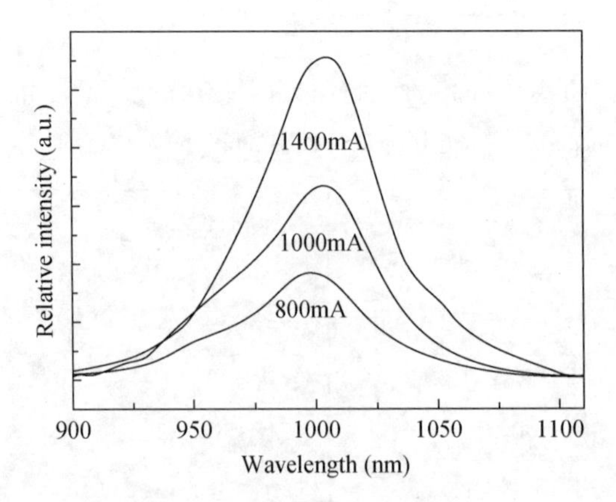

Fig. 3 Output spectrum of SAQD-SLD under different CW pumping current at room temperature

The high performance of SLD was realized with such a simple structure, so we believe that if we make use of the appropriate and complicated structure, we can obtain the higher performance of SLD. For example, because the emitted photon energy is affected by the size of the QDs, we can introduce the multiple stacked layers with different deposition amount of InAs QDs into the active region of SLD. By stacking QDs with different amounts in different layers, the extremely wide spectrum will be acquired. Furthermore, if the tapered active region and facet coating are applied to the SAQD-SLD, the output powers will be considerably increased.

In conclusion, the CW output power of above 200mW with the spectral bandwidth of above 60nm of the SAQD-SLD has been obtained at room temperature. It is also approved that the introduction of the SAQDs structure as the active region of SLD is an effective way to improve their performance.

References

[1] T. L. Paoli, R. L. Thornton, R. D. Burnham, and D. L. Smith, "Highpower multiple-emitter AlGaAs superluminescent diodes," Appl. Phys. Lett., vol. 47, 450, 1985.

[2] T. Yamatoya, S. Sekiguchi, F. Koyama, and K. Iga, "High-power CW operation of GaInAsP/InP superluminescent light-emitting diode with tapered active region," Jpn. J. Appl. Phys., vol. 40, 678, 2001.

[3] I. Middlemast, I. Sarma, and S. Yunus, "High power tapered superluminescent diodes using novel etched deflectors," Electron. Lett., vol. 33, 903, 1997.

[4] B. Wu, C. F. Lin, L. W. Laih, and T. T. Shih, "Extremely broadband InGaAsP/InP superluminescent diodes," Electron. Lett., vol. 36, 945, 2000.

[5] A. T. Semenov, V. R. Shidlovski, D. A. Jackson, R. Willsch, and W. Ecke, "Spectral control in multisection AlGaAs SQW superluminescent diodes at 800 nm," Electron. Lett., vol. 32, 255, 1996.

[6] H. Chen, Z. Zou, O. B. Shchekin, and D. G. Deppe, "InAs quantum-dot lasers operating near 1.3μm with high characteristic temperature for continuous-wave operation," Electron. Lett., vol. 36, 1703, 2000.

[7] Y. Ebiko, S. Muto, D. Suzuki, S. Itoh, K. Shiramine, T. Haga, Y. Nakata, and N. Yokoyama, "Island size scaling in InAs/GaAs self-assembled quantum dots," Phys. Rev. Lett., vol. 80, 2650, 1998.

[8] Z. Z. Sun, D. Ding, Q. Gong, W. Zhou, B. Xu, and Z. G. Wang, "Quantum-dot superluminescent diodes: a proposal for an ultra-wide output spectrum," Opt. Quantum Electron., vol. 31, 1235, 1999.

[9] G. A. Alphonse, D. B. Gilbert, M. G. Harvey, and M. Ettenberg, "Highpower superluminescent diodes," IEEE J. Quantum Electron., vol. 24, 2454-2457, Dec. 1988.

Time dependence of wet oxidized AlGaAs/GaAs distributed Bragg reflectors

R. Y. Li, Z. G. Wang, B. Xu, and P. Jin,

(Key Laboratory of Semiconductor Materials Science, Institute of Semiconductors, Chinese Academy of Sciences, Beijing 100083, China)

X. Guo and M. Chen

(Laboratory of Photoelectronics Technology, Beijing University of Technology, Beijing 100022, China)

Abstract The time dependence of wet oxidized AlGaAs/GaAs in a distributed Bragg reflector (DBR) structure has been studied by mean of transmission electron microscopy and Raman spectroscopy. The wet oxidized AlGaAs transforms from an initial amorphous hydroxide phase to the polycrystalline γ-Al_2O_3 phase with the extension of oxidation time. The thickness of oxide layers will contract due to the different volume per Al atom in AlGaAs and in the oxides. In the samples oxidized for 10 and 20min, there are some fissures along the AlGaAs/GaAs interfaces. In the samples oxidized longer, although no such fissures are present along the interfaces, the whole oxidized DBR delaminates from the buffer.

1 Introduction

The vertical cavity surface-emitting laser (VCSEL) offers various advantages over other light sources, such as a symmetrical output beam and ease of manufacturing in the form of two-dimensional arrays. In fact, it is now replacing the standard edge-emitting laser and light emitting diode in many optical data communications applications. In recent years, advances in the VCSEL, such as ultralow threshold current and high efficiencies take place,[1, 2] which are attributed to the wet oxidation of high Al content Ⅲ–Ⅴ compounds.[3] The wet oxidation process can proceed laterally (along AlGaAs layers) to create buried apertures that not only guide electron and hole currents but also provide a lateral refractive-index contrast in active layers of the VCSEL. Furthermore, wet oxidation also forms higher refractive index contrast Al_xO_y/GaAs Bragg reflectors than usual low contrast AlGaAs/GaAs distributed Bragg reflectors (DBRs). Compared with a conventional AlGaAs/GaAs DBR, the oxidized Al_xO_y/

原载于：J. Vac. Sci. Technol. B, 2005, 23(5): 2137-2140.

GaAs DBR can increase the mirror band width, elevate the reflectivity of DBR, and reduce the number of pairs of the mirror which is useful to decrease the growth thickness and growth time of the VCSEL.[4-6] As a result, the VCSEL utilizing an oxidized aperture and oxidized DBR demonstrates improved performance, such as high-power conversion efficiency, low threshold current, and low threshold voltage.[3, 5, 7]

In this article, by means of transmission electron microscopy (TEM) and Raman spectroscopy, we investigate the effect of oxidation time on the microstructure and interface of an oxidized DBR.

2 Experiment

The samples used in this work were grown on semiinsulated(100) GaAs substrates in a RIBER molecular-beam epitaxy system. First, a 200nm thick undoped GaAs buffer layer was grown. Then, a seven-period undoped DBR was grown, each period including an $Al_{0.97}Ga_{0.03}As$ layer (153nm thick) and a GaAs layer (70nm thick). After photolithography and wet etching ($H_2SO_4:H_2O_2:H_2O=5:1:1$), 136μm wide trenches spaced 64μm apart were formed. Then, the samples were oxidized in a quartz furnace at above 400℃ through water vapor with nitrogen flow for 10, 20, 40, and 50min, as Samples A, B, C, and D, respectively. Oxidation, in such case as is well known, proceeds laterally in the AlGaAs layers of DBR, beginning at the edge defined by chemical etching and proceeding inwards. The oxidized samples for TEM observation were mechanically polished and ion thinned by Ar^+ in standard fashion. Raman spectra were measured using an excitation wavelength of 514.5nm light. The probe beam was focused to apply less than $85W/cm^2$ on the samples. Scattering was measured in the $x(y', y'+z')\bar{x}$ backscattering configuration with y' and z' parallel to (110) planes.

3 Experimental results

Fig. 1(a)-1(c) show the electron diffraction patterns from cross-sectional TEM of Samples B, C, and D, respectively. The diffraction speckle includes a region of the Al oxide layer and GaAs layer. In each electron diffraction pattern of AlGaAs layer oxidized for 20min, 40min, and 50min, a ring pattern superimposed on a single-crystal spot pattern can be seen. The single-crystal pattern corresponds to the unoxidized GaAs, while the ring pattern corresponds to the polycrystalline Al oxide formed by wet oxidation. During wet oxidation, Al atoms in AlGaAs layers react with water vapor to form mainly γ-Al_2O_3 phase.[8-10]

From Figs. 1(a)-1(c), we can observe the three brightest rings of the oxide in the diffraction pattern. They can be matched to the three lattice planes, (311), (400), and (440) with the strongest diffraction intensity in γ-Al_2O_3 which has a cubic structure and space group Fd3m. Consequently, the diffraction pattern can further prove the formation of polycrystalline γ-Al_2O_3 after AlGaAs is wet oxidized. However, in the sample oxidized for 10min, only a single-crystal spot pattern from the unoxidized GaAs layer appears, no ring was present.

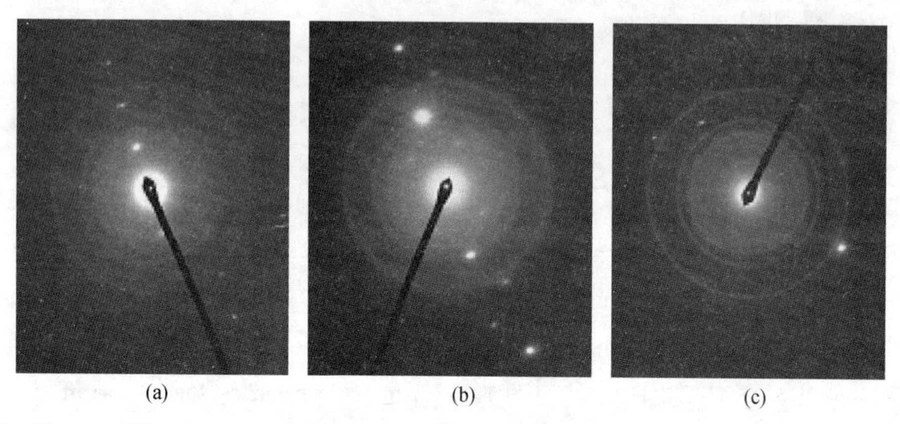

(a) (b) (c)

Fig. 1 Electron diffraction patterns from cross-sectional, TEM of the wet oxidized AlGaAs/GaAs DBR samples with different oxidation times (a) 20min, (b) 40min, and (c) 50min

The microstructure of the sample oxidized for 10min is amorphous while that of the samples oxidized for 20, 40, and 50min is polycrystalline γ-Al_2O_3. The Al_2O_3-H_2O phase diagrams may contain pure oxide and hydrated phases. The amorphous phase could be a pure Al_2O_3 phase, but this is unlikely because the crystallization of amorphous Al_2O_3 to γ-Al_2O_3 requires 800℃ annealing.[11] It is more likely the amorphous phase is a hydrated phase, such as gibbsite phase of $Al(OH)_3$ or boehmite phase of AlO(OH), since they are unstable and dehydrate to form spinel γ-Al_2O_3 at 400℃.[11] Consequently, we infer that wet oxidation forms an initial amorphous hydroxide phase and then the amorphous phase dehydrates to create the polycrystalline γ-Al_2O_3 with the extension of oxidation time. In addition, from Figs. 1(a)-1(c), we can see that with the continued heating the electron diffraction rings are clearer, which suggests that more polycrystalline γ-Al_2O_3 phase is produced.

The spinel γ-Al_2O_3 is significantly more dense than zinc blende AlAs. The volume per Al atom is (2.85Å) (Ref. [3]) γ-Al_2O_3, (3.49Å) (Ref. [3]) in $Al(OH)_3$, while (3.57Å) (Ref. [3]) in AlAs, which as almost the same as in $Al_{0.97}Ga_{0.03}As$. As a result, the thickness of the oxidized AlGaAs layer will contract in the growth direction. According to TEM images, we measured the contraction of the oxide layer in Samples A, B, C, and D to be about 4%, 8%, 10%, and 10%, respectively, compared with unoxidized AlGaAs layer.

The last two are nearly identical. When wet oxidized for 10min, AlGaAs is mainly transformed to amorphous $Al(OH)_3$. The thickness shrinkage is small due to little difference of the volume per Al atom in $Al_{0.97}Ga_{0.03}As$ and in $Al(OH)_3$. Through further oxidation, more and more $Al(OH)_3$ hydrates and form γ-Al_2O_3, which has a much smaller volume per Al atom than $Al_{0.97}Ga_{0.03}As$. Accordingly, the thickness contraction is more serious. When oxidized for 40 and 50min, AlGaAs is completely transformed to γ-Al_2O_3, leading to about 10% contraction. A shrinkage of this magnitude would produce stress, which greatly affects the mechanical and optical properties of a DBR mirror stack with γ-Al_2O_3 layers. It deserves our consideration when we design DBR and VCSEL.

Fig. 2 is the Raman spectra of Samples A, B, C, and D oxidized for 10min, 20min, 40min, and 50min, respectively. Both peaks at 292 cm^{-1} and 269cm^{-1} are from GaAs. The $Al_{0.97}Ga_{0.03}As$ related peaks at near 360cm^{-1} and 400cm^{-1} are missing in the spectra, suggesting that AlGaAs has completely reacted with water vapor. The peak of As_2O_3 is so weak that it is below the detection limit. A broad feature between 200 and 250 cm^{-1} centered at 227cm^{-1} is attributed to amorphous As. Not like most other work,[8, 11, 12] element As intensity is not at a relatively constant level. The amount of As increases (Figs. 2(a) and 2(b)) and then decreases (Figs. 2(c) and 2(d)) with the continued oxidation.

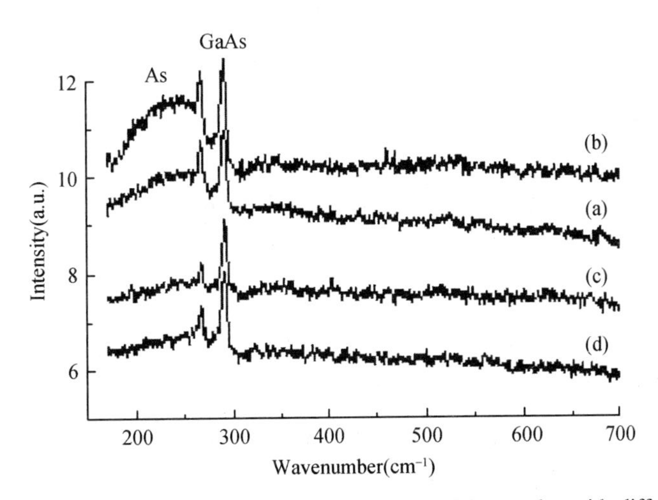

Fig. 2 Raman spectra of the wet oxidized AlGaAs/GaAs DBR samples with different oxidation times (a) 10min, (b) 20min, (c) 40min, and (d) 50min

In the wet oxidation process, AsH_3, As_2O_3, and As serve as intermediates. Their presence can be explained by the following reactions:[11, 12]

$$2AlAs+3H_2O=Al_2O_3+2AsH_3(\Delta G^{698}=-451kJ/mol),$$

$$2AsH_3=2As+3H_2(\Delta G^{698}=-153kJ/mol),$$

$$2AsH_3+3H_2O=As_2O_3+6H_2(\Delta G^{698}=-22kJ/mol),$$

$$As_2O_3+3H_2=2As+3H_2O(\Delta G^{698}=-131kJ/mol),$$

$$As_2O_3+6H=2As+3H_2O(\Delta G^{698}=-1226kJ/mol).$$

We can see that both AsH_3 and As_2O_3 can be readily converted to elemental As for the removal from the oxidized films by the reactions. Therefore, As is the dominant volatile product during wet oxidation. The TEM observations of the laterally wet oxidized AlGaAs layers show that the oxide layers are far from fully dense. [10] The voids among the grains provide natural channels for the escape of volatile reaction products. The amount of volatile products depends on the processing time. When the samples are oxidized for 10min and 20min, on the one hand, As converted from AsH_3 and As_2O_3 is continually produced by the above equations. On the other hand, it is removed through the voids in the oxidized layers. However, the loss of As cannot keep up with the formation. As a result, As gradually accumulates in the oxides, as shown in Figs. 2 (a) and 2 (b). Once the oxidation is terminated, the sample temperature will rapidly drop down to room temperature, and the volatilization process will be frozen. A large amount of volatile products still remains in the oxides since there is not enough time to remove them completely, which can induce stress. The stress cannot be released effectively and ultimately compels the oxidized AlGaAs/GaAs interfaces to fissure. As pointed by the arrows in Figs. 3 (a) and 3 (b), the interfaces are porous. If the oxidation time is long enough, for example, 40min and 50min, the conversion from AsH_3 and As_2O_3 to As is finished. There is adequate time for As to continually transport out from the oxidized layers until As is free. When the oxidation ends, little residual As exists in the oxides, as shown in Figs. 2(c) and 2(d). Accordingly, we do not observe the fissures along the oxidized AlGaAs/GaAs interfaces, as shown in Figs. 3(c) and 3(d). Figs. 2 and 3 prove that the presence of the fissures along the oxide/GaAs interfaces is not due to the thickness shrinkage of oxide layer indicated in other work, [10] but is a result of the accumulation of volatile reaction products in the oxidized layers.

Although there are no such fissures along the interfaces in Samples C and D, it does not mean they have an excellent thermal stability. The arrows in Figs. 3 (c) and 3 (d) show the whole oxidized DBR delaminates from the buffer. In the above discussion, the thickness shrinkage of oxidized AlGaAs layer would produce the tensile stress. In Samples A and B, the fissures along the interfaces can release the stress, while in Samples C and D, there are no such fissures. The stress induced by the thickness shrinkage accumulates and ultimately contributes to the delamination of the whole DBR from the buffer. The thermal stability of wet oxidized AlGaAs/GaAs DBR should be improved by in situ annealing, which needs our further research.

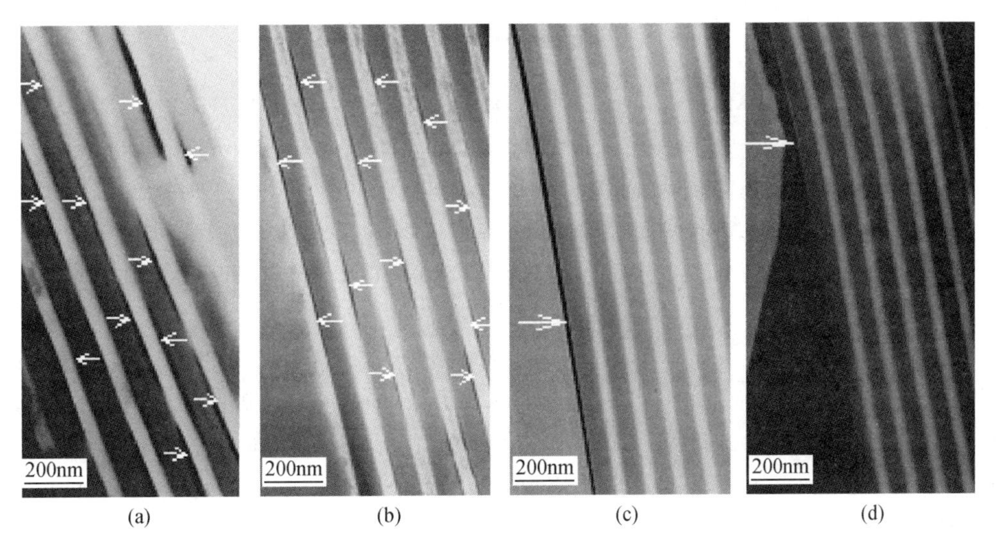

(a) (b) (c) (d)

Fig. 3 Cross-sectional TEM images of the wet oxidized AlGaAs/GaAs DBR samples with different oxidation times (a) 10min, (b) 20min, (c) 40min, and (d) 50min

4 Conclusion

In summary, the wet oxidized AlGaAs transforms from an initial amorphous hydroxide phase to the polycrystalline γ-Al_2O_3 phase with the extension of oxidation time. The thickness of the oxide layers will contract as a result of the different volume per Al atom in AlAs, in $Al(OH)_3$ and in γ-Al_2O_3. With the proceeding of wet oxidation, the amount of element As increases and then decreases because the loss of element As cannot keep up with the formation. In the samples oxidized for 10 and 20min, there are some fissures along the oxidized AlGaAs/GaAs interfaces due to the stress induced by a large amount of residual As in the oxides. In the samples oxidized for longer time, although there are no such fissures along the interfaces, the whole oxidized DBR delaminates from the buffer due to the stress accumulation induced by the thickness shrinkage. How to improve the thermal stability of oxidized DBR needs our further research.

Acknowledgments

The above work was supported by Special Funds for Major State Basic Research Project of China (No. G2000068303), National Natural Science Foundation of China (Nos. 60390071, 60276014, 90101004, and 90201033, and National High Technology Research and Development Program of China (No. 2002AA311070).

References

[1] G. M. Yang, M. H. MacDougal, and P. D. Dapkus, Electron. Lett. 31, 886(1995).

[2] B. Weigl, M. Grabherr, C. Jung, et al., IEEE J. Sel. Top. Quantum Electron. 3, 409(1997).

[3] D. L. Huffaker, D. G. Deppe, and K. Kumar, Appl. Phys. Lett. 65, 97(1994).

[4] M. H. MacDougal, H. Zhao, P. D. Dapkus et al., Electron. Lett. 30, 1147(1994).

[5] M. H. MacDougal, P. D. Dapkus, V. P. Hanmin Zhao, et al., IEEE Photonics Technol. Lett. 7, 229(1995).

[6] T. Takamore, K. Takemasa, and T. Kamijoh, Appl. Phys. Lett. 69, 659(1996).

[7] K. L. Lear, K. D. Choquette, R. P. Schneider, et al., Electron. Lett. 31, 208(1995).

[8] C. l. H. Ashby, J. P. Sullivan, and P. P. Newcomer, Appl. Phys. Lett. 70, 2443(1997).

[9] S. Guha, F. Agahi, B. Pezeshki, J. A. Kash, D. W. Kisker, and N. A. Bojarczuk, Appl. Phys. Lett. 68, 906(1996).

[10] R. D. Twesten, D. M. Follstaedt, K. D. Choquette, and R. P. Schneider, Appl. Phys. Lett. 69, 19(1996).

[11] C. l. H. Ashby, M. M. Bridges, A. A. Allerman, and B. E. Hammons, Appl. Phys. Lett. 75, 73(1999).

[12] C. l. H. Ashby, J. P. Sullivan, K. D. Choquette, K. M. Geib, and H. Q. Hou, J. Appl. Phys. 82, 3134(1997).

Materials science in semiconductor processing

Preface

The 11th conference on Defects: Recognition, Imaging and Physics in Semiconductors (DRIP) is a biennial series that started in La Grande Motte (France) in 1985. The recent DRIP conferences were held in Rimini (Italy) in 2001 and BATZ-sur-MER (France) in 2003. It is the first time to hold this conference in China. The DRIP-XI conference was organized by Institute of Semiconductors, Chinese Academy of Sciences (ISCAS) and Zhejiang University. The conference was mainly supported by ISCAS, National Natural Science Foundation of China and Industrial Companies such as SEMILAB, Struers, etc.

The purpose of the conference is to provide a forum for scientists and engineers from both of universities, institutes and industry to meet together on the topics related to the methods and techniques used for the recognition and imaging of defects in semiconductor materials (Si, Ⅲ-Ⅴ's including nitrides, SiC, Ⅳ-Ⅳ's, Ⅱ-Ⅵ's, organic compounds, etc.) and in semiconductor devices at any stage of their life. This conference was arranged into 10 oral sessions, including 13 invited talks and 50 oral presentations, and 64 poster presentations. About 100 Scientists and researchers from 23 countries gathered and 126 contributed papers were submitted. The primary concern of the conference is the methodology and the physics of measurement procedures, together with specific developments in instrumentation, and on their relationship with the structural, optical and electrical properties of defects. It is believed that this conference will play an important role in advancing the understanding of defects in semiconductor materials and devices.

We would like to thank all those involved in the Organizing Committee, the International Advisory Committee and the Scientific Program Committee. They devoted their expertise, experience and time to this conference, making the organization of DRIPXI smooth and effective. All participants, speakers, invited speakers, and session chairmen and advertisers are also greatly appreciated for their contribution to the success of the conference. Special thanks should be given to Professor Deren Yang (Scientific Program Committee Chairman), Dr. Shengchun Qu (conference secretary), Professor Tomoya Ogawa, my colleagues and graduate students for their contribution to the success of the conference.

I wish Professor Jens W. Tomm, the chair of DRIP-Ⅻ, a grand success in organizing the next conference in Germany.

Chair of DRIP-XI-(2005)
Zhanguo Wang
Institute of Semiconductors,
Chinese Academy of Sciences P. O. Box 912,
Beijing 100083, China
E-mail address: zgwang@ red. semi. ac. cn

半导体照明将触发照明光源的革命

王占国

（中国科学院半导体研究所，半导体材料科学重点实验室，北京，100083）

摘要　虽然半导体白光照明相对于传统照明光源具有体积小、寿命长、节能、环保和安全等优势，但是由于生产成本较高，尚未进入到实际应用。随着能源的紧缺和对环保的日益关注，世界各国都制定了相关的研究规划开展此方面的研究。如果能够解决突破半导体照明的核心技术，提高发光效率，降低生产成本，未来的半导体白光照明应用市场前景广阔。

1　发展半导体白光照明具有重要的战略意义

自从1879年爱迪生发明白炽灯开始，随着科学技术的不断进步，新的照明光源相继问世。现在电光源的品种已超过3000种，规格有5万种以上，但大致可分为两大类，白炽灯和荧光灯。白炽灯的发光效率目前可达15lm/W，其特点是价格便宜，但效率低，寿命短。荧光灯具有发光效率高（70lm/W），寿命长（1万多小时）的特点，已成为室内最主要的照明光源，缺点是易碎、不安全，内充汞气造成环境污染等。20世纪80年代GaAlAs超高亮度红光发光管的研制成功，特别是90年代初中期，GaN基蓝、紫光料质量提高和*P*型掺杂技术的突破，为全固态全色显示和白光照明技术的迅速发展打下了科学与技术基础。

半导体白光照明就是利用红、绿和蓝三基色的半导体发光二极管芯片的组合，或在超高亮度GaN基蓝、紫和紫外发光管管芯上涂敷相应组分的荧光物质所形成的白光光源。与传统照明光源相比，具有很多显著特点：耗电量小，仅为同亮度白炽灯泡的10%–20%；低电压、低电流启动，响应速度快（仅60ns），体积小，工作安全可靠，寿命可长达10万小时；它是冷光源，无红外辐射，便于隐蔽并且环保无污染（无电磁干扰、无汞污染、无碎碴等）。半导体照明光源不仅可以节省能源，有利于可持续发展，而且有利于环保，被认为是最有可能进入普通照明领域的一种“绿色照明光源”。目前正在孕育着人类照明史上继白炽灯、荧光灯之后的又一次革命，一场抢占半导体照明新兴产业制高点和世界市场的争夺战已经在全球打响。

我国是仅次于美国的第二发电大国，2002年度我国发电总量为1.65万亿kW·h，

原载于：新材料产业，2006，2：37–40.

其中80%为火力发电，燃烧大量的原煤和石油；产生大量的粉尘和CO_2、SO_2和NO_2等有害气体，环境污染严重。我国照明用电量约占总发电量的12%（约2000亿kW·h），相当于目前三峡水电站装机容量840亿kW的2.4倍，并且随着我国农村城市化比例的增加和人民生活质量的提高，正以每年5%以上的速度增长，照明节电的潜力很大。如采用“半导体灯”替代传统光源，按节能45%计，不但每年可节电约900亿kW·h，相当少建装机容量为220万kW火力发电站10多座，可节省电力建设资金千亿元以上，而且每年可减少排放近8000万t CO_2，64.8万t SO_2和31.7万t NO_2！可见，作为“绿色照明光源”的半导体灯不仅可节约大量能源，而且还可减少火力发电带来的对环境的严重污染，符合国家可持续发展战略。

我国2001年白炽灯的产量就高达35亿只，荧光灯13.5亿只，已成为世界照明电器的第一生产大国和出口大国。据预测，我国在未来7年，半导体照明光源的市场规模（销售产值）将达到763亿元，其中仅白光LED照明灯具就高达640亿元，半导体照明光源将成为我国照明电器行业的支柱产业之一。半导体照明产业具有技术密集和劳动密集型双重特点，发展半导体照明技术，不仅对提升我国传统照明产业的技术水平，提高其国际竞争力，实现由世界照明电器生产大国到生产强国的转变，而且还将为保持社会稳定提供可观的就业机会，从而具有显著的社会效益和巨大的经济效益。

综上所述，半导体白光照明具有重要的经济和社会意义，已成为国际上新一轮高技术竞争的一个焦点。我们应该抓住机遇，加快研究开发和产业化的步骤，确保在激烈的国际竞争中占有一席之地。

2　世界各国都制定了相关的研究计划，并取得了重大技术突破

由于半导体照明关系到节能和改善全球环境，又存在巨大的潜在市场，受到了世界各国政府的重视，并相继制定了国家级的研究计划。例如日本的21世纪光计划（1998-2002年），政府耗资50亿日元，推广白光照明。预计2006年完成用白光LED替代50%的传统照明光源；美国的国家半导体照明计划（2000-2010年），计划投入5亿美元。美国光电工业发展协会（OIDA）预计，半导体灯将于2007年代替白炽灯，2012年替代荧光灯，市场和社会效益不可估量；欧盟在2000年7月也启动了用于推动白光照明应用的彩虹计划；韩国产业资源部也早在1999年就启动了为时5年的白光LED照明技术计划。

在各国政府的大力支持下，1999年美国GE、荷兰Philips和德国Osram三大世界照明生产巨头纷纷与半导体公司合作，组建了GElcore、Lumileds和Osram Opto Semiconductors公司。还有日本的Nichia和美国的Cree公司等都正在集中全力开发半导体照明光源，并通过申请专利等设置技术壁垒，企图在这一高技术产业领域占有主导地位。

实现半导体白光照明的技术路线主要有三条：蓝色 LED（440-480nm）、紫色及紫外 LED（360-420nm）单芯片加荧光粉合成白光技术；红、绿、蓝（RGB）多芯片组合白光技术；以及采用 MOCVD 等技术直接生长多有源区的白光 LED 技术。目前已经商品化的白光 LED 利用的是前两种方法，第三条技术路线还处于研究的初始阶段。

图1 作为“绿色照明光源”的半导体灯不仅可节约大量能源，而且还可减少火力发电带来的对环境的严重污染

目前国外 GaN 基 LED 正在向大功率、高亮度、高电光转换效率、低成本方向发展。功率型 LED 外延芯片的研制是实现半导体照明的关键，世界上主要研发单位和生产公司都在努力提高发光芯片模块的光通量和发光效率。美国 Cree 公司 2003 年 10 月份展出了光通量为 1200lm 的白光 LED 集成灯，发光效率 32lm/W。该公司还介绍了他们在光通量为 4.7lm 的小芯片上获得发光效率达 74lm/W 好结果。美国 Gelcore 公司研制出光通量 32lm、发光效率 22lm/W 的世界最亮的紫外激发的白光 LED；日本日亚化学工业（Nichia）在 2003 年日本电子展会上，展示了光通量为 1000lm、发光效率 33lm/W 的白光 LED 模块，据称此亮度在白光发光 LED 模块中为当时世界最高水平，模块安装面积为 40mm×40mm，厚度为 10mm。同时，日本日亚化学工业还首次向公众公开了 50lm/W 的白光 LED 模块。此外，日本松下电工、丰田合成和韩国等的一些公司也在 LED 照明灯、高亮度白光 LED 模块等研发方面做出了成绩。这些最新的进展预示着半导体照明技术和产业已达到一个快速上升和发展的阶段。

3 国内半导体照明规划及研究进展

我国在“863”等国家级计划的支持下，从 1994 年开始 GaN 基蓝绿光 LED 的研究，1999 年“863”计划新材料领域立项开展白光 LED 研究。经过近十年的发展，国内 GaN 基 LED 研究和生产已有一定基础。成功地研制出 GaN 基蓝光、绿光和白光 LED；基本掌握了拥有自主知识产权的 GaN 基发光二极管制备的关键技术；初步形成了 GaN 基 LED 外延片生产、LED 芯片制备、LED 封装和应用的工业生产链和相应的研发体系；分别在北京、上海、江西、福建等地相继建立了研发中心和年产亿只 GaN 基 LED 管芯的基地。但是，总体上说，我国大陆 LED 产业的规模和水平和日本、美国、中国台湾等国家和地区相比，还有较大差距。

面对半导体白光照明新兴产业的迅速崛起和世界照明工业转型的重大时机，为了抓住机遇，抢占世界半导体照明产业的制高点，在对国内外情况进行充分调研论证的基础上，科学技术部于2003年10月，正式启动了“国家半导体照明工程”。加强了对关键技术攻关、应用技术开发和推进产业化方面的支持。

4　实现半导体白光照明，需要攻克以下核心技术

根据国际半导体白光照明路标要求（表1），半导体照明最核心的技术是大功率、高亮度半导体LED外延片、芯片的制备和LED封装技术。提高半导体发光效率，降低生产成本是研发的主攻方向。

要实现上述路标，必须解决如下主要关键技术：

（1）功率型LED外延片制备技术；

（2）高电光转换效率的大功率LED芯片结构设计与制造技术；

（3）大功率、长寿命白光LED荧光粉及其制造技术；

（4）功率型LED衬底制备技术；

（5）功率型LED封装技术；

（6）LED光源、灯具设计和应用技术；

（7）OLED光源设计、制造和应用技术；

（8）高效、高可靠、智能化LED光源；

（9）半导体照明光源标准建立。

表1　国际半导体白光照明路标要求

年份	2002	2007	2012	2020
发光效率（lm/W）	25	75	150	200
寿命（khr）	20	>20	>100	>100
光通量（lm/灯）	25	200	1000	1500
输入功率（W/灯）	1	2.7	6.7	7.5
流明成本（美元/klm）	200	20	<5	<2
灯成本（美元/灯）	5	4	<5	<3
彩色重现指数	75	80	>80	>80
市场占领	低光通量	白炽灯	荧光灯	全部

5　半导体照明市场应用前景广阔

半导体照明光源因具有高效、节能、环保、寿命长、易维护和经济利益巨大等特

点，受到国内外的高度重视。目前已成为研发的热点。它的进一步发展，极有可能触发照明光源的革命。

从半导体 LED 转换效率的理论分析和近年来大功率白光 LED 研制进展看，到 2020 年半导体 LED 转换效率突破 50%，实现 LED 白光发光效率达到 200lm/W 的技术指标，已无原则上不可克服的困难（据预测，2025 年半导体 LED 的转换效率将达 75%，发光效率则可高达 300lm/W）。当然，要实现上述目标，关键是如何进一步提高发光芯片的质量，包括降低外延片的缺陷密度、提高其均匀性和优化器件结构设计等，其中探索新型衬底材料，探索基于新的材料体系（如 OLED 等）和新型器件结构与工作模式的白光 LED 发光等是主要研究方向。

影响国际半导体白光照明路标能否如期实现的另一个关键因素是其生产成本和销售价格，届时能否下降到用户可以接受的水平。据估计，到 2005 年，LED 的初始投资性价比达到 50 美元/klm，那时用白光 LED 光源代替一个 100W、1500lm 的白炽灯泡照明，点燃 1 万小时的费用比较如表 2（假定电费为：1 美分/kW · h，我国的电价要比 1 美分/kW · h高很多）。

两种光源在点 1 万小时后，费用即达到平衡。LED 寿命约为 10 万小时，因而从长远考虑，使用 LED 是更经济的。随着技术的进步和半导体白光照明性能价格比的不断提高，半导体照明光源将逐渐得到广泛应用。据业内人士预测，到 2008 年，全球 LED 的产值将从 2004 年的 125 亿美元提高到 500 亿美元，市场潜力巨大，发展前景看好。

表 2　白光 LED 与 100W 白炽灯点燃 1 万小时费用比较

100W 白炽灯	电费 100（美元）	初始投资 5（美元）	合计 105（美元）
白光 LED	30	75	105

6　我国半导体白光照明的发展战略

抓住照明产业革命的历史机遇，突破半导体白光照明的关键技术，形成有国际竞争力的中国半导体照明新型产业，实现由照明光源大国向照明光源强国的转变。具体说来，要瞄准成熟的低成本产业化技术，以近期解决市场应用和产品的性价比，中远期培育新兴的大功率普通白光照明产业为目标，形成从外延片、芯片到封装及产品应用的完整产业链的角度出发，解决产业化的关键技术和原创性核心技术，形成一批专利、培育一批企业、建设一批特色产业基地，建成有自主知识产权的标准体系，使半导体照明从目前的特种照明应用领域（指示灯、交通信号灯、大屏幕显示、LCD 背光照明和汽车灯等）进入普通照明领域，形成有国际竞争力的中国半导体照明新型产业，实现由照明光源大国向照明光源强国的历史转变。具体目标是：

2012 年，进入特种照明领域，替代 80% 以上的交通信号灯和 LCD 背光照明；在景观照明上替代 80% 以上的白炽灯，普通照明领域替代 50% 的白炽灯。

2020 年，形成具有自主知识产权的比较完整配套的半导体照明产业，全面进入特种照明领域，在普通照明领域替代 80% 以上的白炽灯和 50% 的荧光灯，节能 50% 以上。

7 我国半导体白光照明的发展对策

根据国际半导体白光照明发展路标，到 2020 年，“半导体灯”的发光流明效率要达到 200lm/W，为 2002 年水平的 8 倍；价格由 2002 年的 200 美元/klm 降低到 2 美元/klm！要实现这个极具挑战性的、代表人类光源的又一次革命的雄伟目标，难度很大。为此，这就需要我们从现在起，集中产、学、研优势力量，在国家的统一部署和领导下，协同攻关，突破高质量 GaN 基异质外延材料（包括衬底材料），高效、长寿命荧光材料和高效、稳定的有机/高分子三线态发光材料产业化制备，大功率白光照明器件结构设计与实现以及有效散热和封装等关键技术，形成具有我国自己的知识产权的高新技术产业，参与国际竞争。具体建议的对策如下：

第一，鉴于国际专利和技术壁垒已严重地制约了我国半导体照明产业的发展，开展有关半导体照明专利战略研究十分必要，并在此基础上选择优先发展方向，制定符合我国国情的、半导体照明中长期产业发展规划。

第二，实施研究、开发和产业化统筹规划，协调发展策略。成立“国家半导体照明工程”领导小组，发挥政府引导作用，并对其实施进行宏观指导和协调，避免一哄而上，无序竞争。

第三，建立国家级研发创新平台，突破半导体照明产业发展的共性关键技术，形成自主知识产权，为产业发展提供技术支撑。实施体制创新和人才战略，突破部门、行业之间壁垒，加大引进海内外高级人才和知识产权保护的力度，形成创新群体；鼓励企业前期介入，加快创新成果向产业转化的步伐，占领该高新技术产业领域的制高点。

第四，按照合理布局，统筹规划，协调发展的原则，加快产业化基地和示范工程建设步伐。企业应成为示范工程实施和技术创新的主体，以项目管理的方式、市场推动的策略开展工作，使研发目标始终围绕产业化来进行。建立市场化运作机制，引导和提高民间资本参与基地建设的主动性，达到政府资本与民间资本的联动，推动项目的有效实施和产业化。

第五，建立国家半导体照明工程网站。半导体照明光源的研发与产业化是一场全新的产业革命，通过网站进行宣传、技术与人才培训、交流信息和市场培育等，力求调动全社会各类相关人员的积极性，关注和推动这场革命，为专项的实施创造一个良好的外部环境。

Study of the wetting layer of InAs/GaAs nanorings grown by droplet epitaxy

C. Zhao,[1] Y. H. Chen,[1] B. Xu,[1] C. G. Tang,[1] Z. G. Wang,[1] and F. Ding[2]

([1] Key Laboratory of Semiconductor Materials Science, Institute of Semiconductors, Chinese Academy of Sciences, Beijing 100083, China)

([2] Institute for Integrative Nanosciences, IFW Dresden, Helmholtzstrasse 20, 01069 Dresden, Germany)

Abstract The properties of the wetting layer (WL) of InAs nanorings grown by droplet epitaxy have been studied. The heavy-hole (HH) and light-hole (LH) related transitions of the In(Ga)As WL were observed by reflectance difference spectroscopy. From the temperature dependent photoluminescence behavior of InAs rings, the channel for carriers to redistribute was found to be the compressed GaAs instead of the In(Ga)As layer, which strongly indicated that the wetting layer was depleted around the rings. Furthermore, a complex evolution of the WL with In deposition amount has been observed.

Low-dimensional structures, such as quantum dots (QDs) and quantum rings, have attracted much attention because they are the basis of future electronic and optoelectronic devices.[1] The most commonly used growth method for these structures is the Stranski-Krastanov (SK) mode, in which the nanostructures can be formed to relax the mismatched strain of the system. Recently, several groups have grown lattice-matched GaAs/AlGaAs structures by droplet epitaxy, which is a nonconventional growth technique for the self-assembly of high-quality nanostructures.[2] In this method, group Ⅲ element droplets first form and then react with a group V element for crystallization.[3]

Unlike SK growth, in which the presence of wetting layer (WL) is unavoidable, the WL thickness in GaAs QD structures can be tuned in droplet epitaxy.[4] There are also reports on the growth of InAs QDs with or without WLs by droplet epitaxy.[5,6] It is clear that under some conditions, there surely exists an In(Ga)As layer under the In droplets.[6,7] However, for InAs nanostructures, the details of the WL are not clear, and there is still no report on the WLs of InAs/GaAs nanostructures, especially InAs rings grown by droplet epitaxy. The optical and electrical properties of the self-assembled nanostructures grown by SK growth

原载于：Appl. Phys. lett, 2008, 92(6): 063122.

mode or by droplet epitaxy can be affected greatly by the surrounding WLs. Therefore, it is desirable to know the electronic states and structures of WLs to understand the properties of the nanostructures and their devices. [8]

In previous studies, reflectance difference spectroscopy (RDS) has been successfully adopted to characterize In(Ga)As WLs in InAs/GaAs nanostructures grown by SK mode. [8, 9] Since the WLs often act as the carrier channel in a normal InAs/GaAs system, by studying the carrier channel from the temperature dependent photoluminescence (PL) behavior of the InAs structures, we can also obtain information on the WLs. [10] In this letter, we studied the properties of the WL of InAs rings grown on GaAs by droplet epitaxy by RDS and photoluminescence (PL). The channel for carriers to be redistributed among the InAs nanostructures was found to be the compressed GaAs instead of the existing In(Ga)As layer, which strongly indicated that the In(Ga)As WL was depleted around the rings. In addition, a complex evolution of the In(Ga)As WL with the In deposition amount has been observed.

All the samples were prepared on a 2in. n-doped GaAs(100) substrate by a Riber 32P molecular beam epitaxy (MBE) system. A GaAs buffer layer with a thickness of about 200nm was first grown at 600℃ after the removal of the native oxide by heating the substrate to 580℃ under an As_4 molecular beam flux. Then, the substrate temperature was decreased to 120℃ with the valve fully closed. The reflection high-energy electron diffraction (RHEED) pattern showed a (4×2) Ga-stabilized reconstruction at this point. A nominal amount of 2.5 ML In was deposited on the GaAs buffer layer at a rate of 0.15 ML/s, and the valve was fully opened to allow an As_4 flux with a beam equivalent pressure of 1.35×10^{-5} Torr to be supplied to the sample surface for several minutes. After annealing at 450℃ for a relatively long time until the RHEED pattern showed spotty patterns, the GaAs capping layer was grown with a thickness of 100nm for PL measurements. Finally, InAs nanostructures were grown on the capping layer for atomic force microscopy (AFM) measurements. In addition, we also prepared another sample to study the dependence of the WLs on the In deposition amount. The growth procedures for this sample were the same as in our previous report[11] except that a nominal amount of 2.5ML In was deposited on the GaAs buffer layer. The two samples are called A and B, respectively. After growth, sample B was cut into 16 pieces along the direction where the In deposition amount varies gradually. The 16 pieces were numbered from 1 for the least to 16 for the most. An almost linear variation of In deposition amount with the samples can be predicted on the basis of the cosine law for the MBE source beam. [12]

The surface morphological properties of the sample were characterized by AFM in contact mode in air. Before the PL and RDS measurements, a rapid thermal annealing process was performed at 850℃ for 30s in a N_2 atmosphere to improve the crystal quality. In PL measurements, the sample was excited by a solid-state laser with a wavelength of 532nm and a power of 100mW. The luminescence spectra were detected with a Fourier transform infrared spectrometer equipped with In(Ga)As photodetectors.

The AFM image of InAs structures for the as-grown sample is shown in the inset of Fig. 1. Because the deposition amount exceeded the critical value, which is estimated to be 1.4ML,[11] only InAs mounds and rings with a density of $1.9\times10^8/cm^2$ were observed. The average outer diameter, top diameter, and height of the rings are 260, 144, and 11nm, respectively. The shape is the same as previously reported.[11]

Fig. 1 shows the RD spectra measured at room temperature for the as-grown sample and the sample after annealing, in which the WL related transitions are observed. Similar to that observed in InAs/GaAs QD system grown by SK mode,[8, 9] the features at 1.35, 1.38, and ~1.42 eV are assigned to the heavy-hole(HH) and light-hole(LH) related transitions of the In(Ga)As WL and GaAs band edge, respectively. The RDS result demonstrates that there exists an In(Ga)As layer as a WL after the rings formed. The observed RDS signal, which measures the in-plane optical anisotropy of the In(Ga)As WL, probably comes from the segregation effect of indium atoms and anisotropic strain in the In(Ga)As WL, as discussed in Refs. [8] and [9]. In this paper, the RDS intensities are ignored.

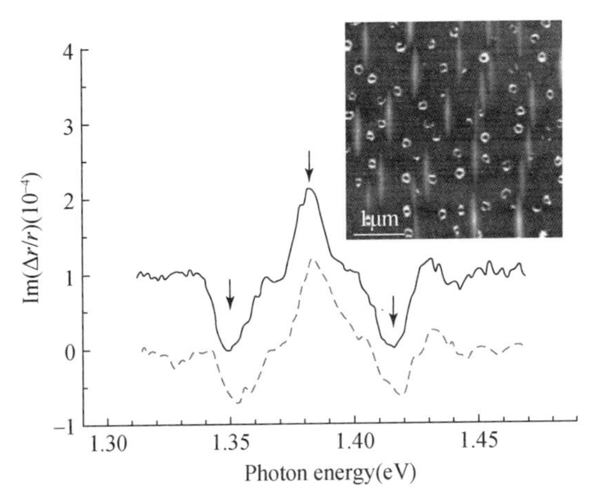

Fig. 1 (Color online) The RD spectra measured at room temperature for the as-grown sample (solid) and the sample after annealing (dashed). The HH, LH, and GaAs features are indicated by arrows from left to right. The inset shows the AFM image of InAs structures for the as-grown samples

Fig. 2 shows the PL spectra of the annealed sample at different temperatures below 180K. At temperatures higher than 180K, the PL of the InAs rings is almost quenched. The peaks at 1.30–1.32eV originate from the InAs rings, and the broad peaks at the low-energy side are related to the impurities in the substrate. The peaks' variation with the temperature from the analysis of PL results is summarized in the inset of Fig. 2, which deviates from Varshni's law for InAs bulk materials. The temperature dependence of PL peaks of nanostructures may be quite different from that of corresponding bulk materials due to carrier redistribution among the nanostructures. We use the model in Ref. [10] to study the carrier channel of the InAs rings, which proved to be suitable to describe carrier channels of QD structures. The solid line in the inset of Fig. 2 is the fitting result according to Eqs. (3a) and (3b) in Ref. [10], in which the value of the full width at half maximum of the PL spectra Γ is 95.5 meV and the carrier-to-channel activation energy E_t is 0.11eV. According to Ref. [10], the transition energy E_{ch} related to the carrier channel of the rings at the temperature of 15K is given by $E_{ch} = (1+a)E_t + E_{QR} = 1.575\text{eV}$ with $a = 1.3$, where E_{QR} is defined as the transition energy of InAs nanorings.

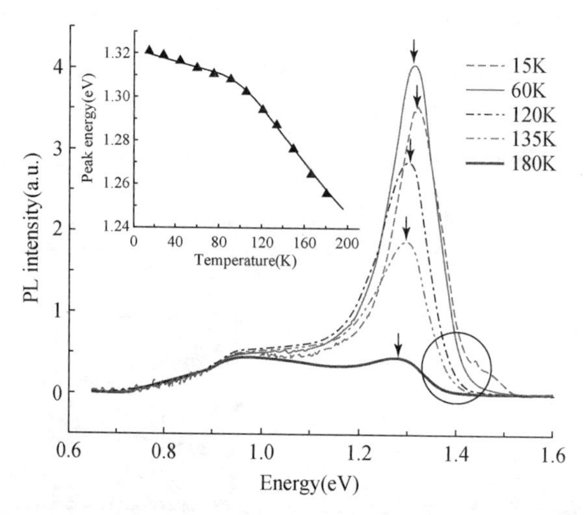

Fig. 2　(Color online) The PL spectra for the sample after annealing at different temperatures; the peaks are indicated by arrows. The lines in the circle from right to left are the results for 15, 60, 120, 135, and 180K. The inset shows the temperature dependent PL behavior of the sample after annealing. The solid symbols are experimental data, and the solid line is theoretical simulation based on the model in Ref. [10]

The obtained E_{ch} is clearly larger than the energy gap of bulk GaAs, strongly suggesting that the carrier channel is not the existing In(Ga)As WL but rather the compressed GaAs bulk material. Therefore, we believe the WL depleted around the InAs rings, which is different from other reports in which there is no WL or the WL is directly connected to the nanostructures.[3,4] Fig. 3 shows the schematic of an InAs/GaAs nanoring and the corresponding band structure according to the results above. The values of E_{WL} and E_{QR} can be acquired from RDS and PL experiments directly, and the value of E_{ch} can be obtained from the analysis of the PL results.

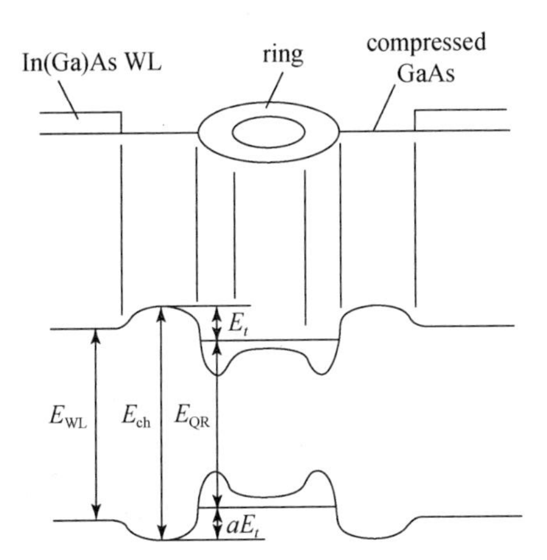

Fig. 3 The schematic of the InAs/GaAs nanorings grown by droplet epitaxy and the corresponding band structure

As mentioned in our previous report, droplets form when the deposition amount exceeds the critical value.[11] The droplets act as nucleation centers to capture not only the supplied As atoms but also the surrounding In atoms. This makes the existing In(Ga)As layer around the droplets deplete so the bulk GaAs becomes the only carrier channel of the InAs nanorings. The depletion of indium around the nanorings is also verified by the volumes of the InAs rings, which are much larger than those of the original In droplets. By averaging crystalline volumes through AFM analysis, we can estimate the quantity of In atoms inside an InAs ring to be about 3.5 times of that inside an In droplet. However, it is still not clear whether In depletion occurs during the formation of In droplets or during the crystallization of In droplets with As.

The results of AFM and RDS for the 16 pieces of sample B are given in Fig. 4. Fig. 4(a) shows the average densities of InAs rings from AFM measurement. Fig. 4(b) shows the RD spectra varying with the samples. The WL related transition is observed by RDS for all 16 samples. The shift of the LH structure actually reveals a variation of the total In amount in the buried In(Ga)As WL. As a rough estimation, the amount of In in the WL can be linearly related to the LH and HH transition energies.[8] For samples 1–6, only a portion of the deposited In atoms leads to the density increase of the rings (see Fig. 4(a)), while the rest is incorporated in the WL, leading to the redshift of LH peaks in samples 1–6. In samples 7–11, more In atoms are absorbed by InAs rings and less In atoms are incorporated in the WL, slowing the redshift rates. For samples 12–16, the density of the rings is unchanged and some large dots appear; the blueshift reveals a decrease of the amount of In in the WL, although more In is deposited. For these samples, the large dots might relax strain energy via the generation of dislocations, leading to an enhanced absorption of In atoms.

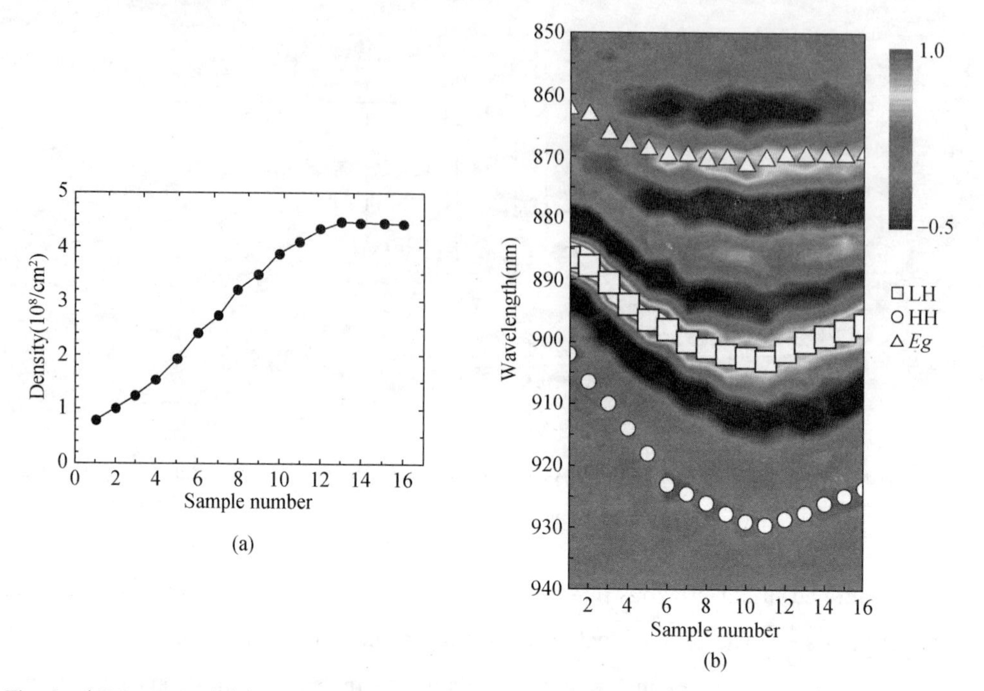

Fig. 4 (Color online) (a) Average densities of InAs rings from AFM measurement; (b) variation of the second derivative RD spectrum with respect to wavelength with the samples indicated by the color contrast. The squares, circles, and triangles indicate the wavelengths of the LH related transition, HH related transition, and GaAs band edge, respectively

The complex variation of the In(Ga)As WL with deposited In in droplet epitaxy is different from that in the SK growth mode.[9, 13] After the SK grown QDs form, the energy range of the WL transitions can be limited to a very narrow dispersion range of about 20meV, in spite of some differences in the growth conditions.[13] The wavelengths of LH and HH structures are even fixed for some conditions.[9] In the case of droplet epitaxy, the energy of the WL transitions disperse in a wide range after the rings form. This is because the WL thickness is primarily determined by the lattice mismatch in the SK mode, whereas the WL in droplet epitaxy is also affected by the droplets.

In summary, the properties of the WL of droplet epitaxy grown InAs rings have been studied by RDS and PL. It was found that the In(Ga)As WL was depleted around the rings. The formation mechanism of the WL and its complex evolution with the In deposition amount were also discussed.

This work was supported by the “973” program(2006CB604908), the National Natural Science Foundation of China (Nos. 60625402 and 60576062), and the “863” program (2006AA03Z0408).

References

[1] D. Bimberg, M. Grundmann, and N. N. Ledentsov, Quantum Dot Heterostructures (Wiley, Chichester, 1999).

[2] N. Koguchi, S. Takahashi, and T. Chikyow, J. Cryst. Growth 111, 688(1991).

[3] T. Mano, T. Kuroda, S. Sanguinetti, T. Ochiai, T. Tateno, J. S. Kim, T. Noda, M. Kawabe, K. Sakoda, G. Kido, and N. Koguchi, Nano Lett. 5, 425(2005).

[4] S. Sanguinetti, K. Watanabe, T. Tateno, M. Gurioli, P. Werner, M. Wakaki, and N. Koguchi, J. Cryst. Growth 253, 71(2003).

[5] J. S. Kim and N. Koguchi, Appl. Phys. Lett. 85, 5893(2004).

[6] J. M. Lee, D. H. Kim, H. Hong, J. C. Woo, and S.-J. Park, J. Cryst. Growth 212, 67(2000).

[7] J. H. Lee, Zh. M. Wang, and G. J. Salamo, J. Phys.: Condens. Matter 19, 176223(2007).

[8] Y. H. Chen, J. Sun, P. Jin, Z. G. Wang, and Z. Yang, Appl. Phys. Lett. 88, 071903(2006).

[9] Y. H. Chen, P. Jin, L. Y. Liang, X. L. Ye, Z. G. Wang, and A. I. Martinez, Nanotechnology 17, 2207(2006).

[10] F. Ding, Y. H. Chen, C. G. Tang, B. Xu, and Z. G. Wang, Phys. Rev. B 76, 125404(2007).

[11] C. Zhao, Y. H. Chen, B. Xu, P. Jin, and Z. G. Wang, Appl. Phys. Lett. 91, 033112(2007).

[12] M. A. Herman and H. Sitter, Molecular Beam Epitaxy: Fundamental and Current Status (Springer, Berlin, 1989), 32.

[13] G. Sek, P. Poloczek, K. Ryczko, J. Misiewicz, A. Löffler, J. P. Reithmaier, and A. Forchel, J. Appl. Phys. 100, 103529(2006).

Broadband external cavity tunable quantum dot lasers with low injection current density

X. Q. Lv, P. Jin, W. Y. Wang, and Z. G. Wang

(Key Laboratory of Semiconductor Materials Science, Institute of Semiconductors, Chinese Academy of Sciences, Beijing 100083, China)

Abstract Broadband grating-coupled external cavity laser, based on InAs/GaAs quantum dots, is achieved. The device has a wavelength tuning range from 1141.6nm to 1251.7nm under a low continuous-wave injection current density ($458A/cm^2$). The tunable bandwidth covers consecutively the light emissions from both the ground state and the 1st excited state of quantum dots. The effects of cavity length and antireflection facet coating on device performance are studied. It is shown that antireflection facet coating expands the tuning bandwidth up to ~150nm, accompanied by an evident increase in threshold current density, which is attributed to the reduced interaction between the light field and the quantum dots in the active region of the device.

1 Introduction

Grating-coupled external cavity (EC) laser is an important kind of coherent light source for the applications of spectroscopy[1], biomedical[2], interferometry[3] and so on. Combining with the fast swept frequency technique, this kind of light source can also be applied in the wavelength-division-multiplexing (WDM) system[4] and optical coherence tomography (OCT) measurement[5,6]. Wavelength tuning range is one of the most important parameters of an EC laser, as a large tuning range increases the channel amount of a WDM system and improves the spatial resolution for the OCT measurement significantly. It was proposed that the characteristic of size inhomogeneity, naturally occurring in self-assembled quantum dots (QDs) grown by Stranski-Krastanow mode, is beneficial to broadening the gain spectra and suitable for the realization of broadband emission[7]. Based on their broadband emission characteristic, QDs have successfully been used for the fabrication of superluminescent diode (SLD)[8-12] and broadband laser diode[13-15]. For 1.5μm wavelength

原载于：Optics Express, 2010, 18(9): 8916-8922.

range, broadband InP based Q-dash laser has also been achieved recently[16]. In the last few years, EC lasers with QDs as gain medium have been demonstrated[17-25]. Compared to the EC laser with quantum well as gain medium[26-29], the QDs' size inhomogeneity and relatively low ground state (GS) saturated gain make a QD-EC laser being advantageous in low injection current density, broadband and uninterrupted tuning in wavelength. Only utilizing the QD-GS emission, tuning range of 69nm[25] and 83nm[20] were realized for the QDs devices without facet coating and with single facet antireflection (AR) coating, respectively. By involving QDs' GS and excited state (ES) transitions simultaneously, a tuning wavelength range of 201nm[19] has been reported. The unique characteristics of QD-EC laser make it preferable over other broadband light source, such as QD laser and SLD. Combining with the characteristics of wide wavelength tuning and narrow linewidth, QD-EC laser can be used for the sensitive absorption spectroscopy. Besides, the QD-EC laser can also be used in Fourier-domain OCT system[5, 6] based on swept wavelength interferometry, which offers higher sensitivity and imaging speed over conventional time-domain technique.

Besides a tunable bandwidth, continuous-wave (CW) operation of an EC laser at low injection current density is also required in practical applications. Generally, in order to achieve a wide wavelength tuning range, AR coating on the device facet is needed to increase the threshold current density (J_{th}) of inner Fabry-Pérot (FP) cavity resonance. However, AR coating also increases the EC J_{th} significantly (the reason will be discussed later) and the CW operation is difficult to be realized [18, 19]. Without AR coating, the injection current density can remain at a low level, but the tuning range is restricted [17, 23]. Therefore the trade-offs between the two key features of tuning bandwidth and working current of an EC laser are very important for its practical applications.

In this paper, the effect of cavity length on the tuning bandwidth and J_{th} of a QD-EC laser has been investigated. A tuning range of 110nm (1141.6–1251.7nm) under 458A/cm^2 CW injection level has been realized for a 2mm length device without facet coating. The wavelength tuning range covers the QDs' GS and the 1st ES light emissions simultaneously. The effect of device-facet AR coating on the EC laser's performance is also studied. As compared with the EC laser employing as-cleaved facet, AR facet coating leads to enhancement of tuning bandwidth and increase of J_{th} evidently. The latter is attributed to the reduced interaction between the light field and the QD active region.

2 Experiment

The epitaxial structure of the QD gain devices used in this study was grown on n-GaAs

(001) substrate by a Riber 32P solid-source molecular beam epitaxy machine. Five layers of self-assembled InAs QDs covered by 5nm $In_{0.15}Ga_{0.85}As$ and separated from each other by 35nm GaAs spacer form the active region, which is embedded in the waveguide. 2monolayer InAs is deposited at 500℃ for the formation of QDs in each layer. The areal density is about $4\times10^{10}cm^{-2}$ obtained by atomic force microscopy for an uncapped sample, which has the same growth parameters as the device structure. Below and above the waveguide are 1.5μm *n*-and *p*-type $Al_{0.5}Ga_{0.5}As$ cladding layers grown at 620℃, respectively. Finally, a p^+-doped GaAs contact layer completes the structure. The QD epitaxial wafer was processed to fabricate gain devices of broad-area ridge structure with a stripe width of 120μm and a length of 1-3mm. They were mounted epitaxial-side down on copper heat sink.

A Littrow configuration is constructed in the EC tuning experiment[25]. The emission from one facet of a QD gain device is nearly collimated by using an aspherical lens, and then is fed back by a 1200 grooves/mm grating in its 1st diffraction order. By rotating the grating, the EC resonance wavelength is selected. The emission from the other facet of the QD gain device is used to perform emission spectra and output power measurements. The device is tested at room temperature without temperature control and with CW ($<900A/cm^2$) or pulsing ($>900A/cm^2$, 1kHz repetition rate and 3% duty cycle) injection. The spectral resolution is about 0.5nm, lying on the grating monochromator used.

3 Results and discussion

Three QD gain devices, 1mm, 2mm, and 3mm in cavity length, are fabricated. The current-injection emission spectra from the three free-running gain devices are shown in Fig. 1. As shown in Fig. 1 (c), for the device of 3mm in cavity length under $28A/cm^2$ injection, the full width at half maximum (FWHM) of the emission spectrum is 61nm, which should originate mainly from spontaneous emission of QDs' GS. This relatively wide GS emission is attributed to the size inhomogeneity naturally occurring in QDs' growth. Due to the sufficient GS gain, the 3mm device lases at GS with a J_{th} of $206A/cm^2$. As the significant increased mirror loss with the reduction of the cavity length, 2mm long device lases at the 1st ES of QDs at 1163nm. At the injection of $333A/cm^2$, simultaneous contribution of QDs' GS and the 1st ES to emission spectrum leads to a FWHM of 95nm (Fig. 1(b)). While for the device with 1mm cavity length, in addition to GS and the 1st ES, the 2nd ES can be filled before lasing, as shown in Fig. 1(a). The inset of Fig. 1(a) shows a 3peak Gaussian fitting of the emission spectrum under $833A/cm^2$ injection, presenting the GS, the 1st and the 2nd ESs transitions in QDs, respectively. Because the gain of low-energy state transition is too small to compensate for the total loss, lasings occur at the 2nd ES transitions for 1mm device.

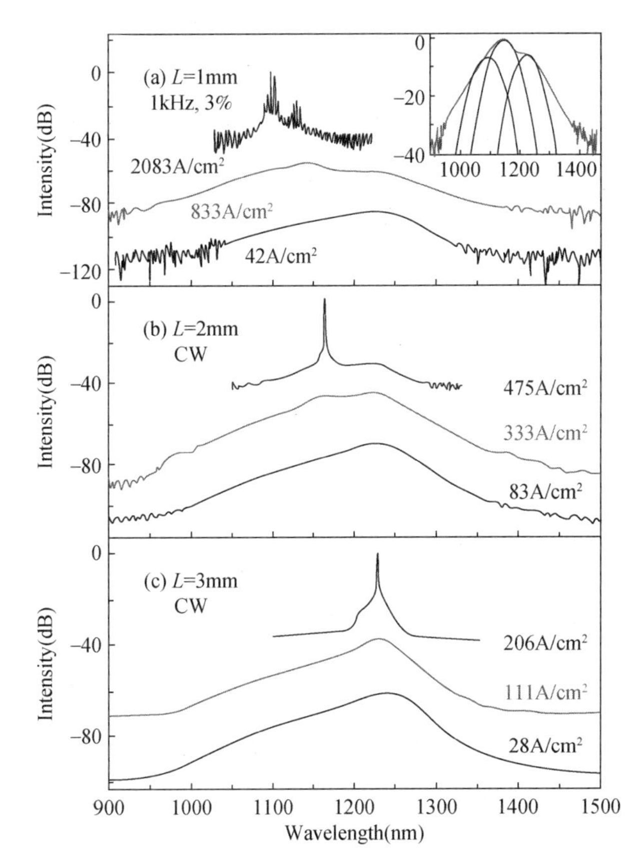

Fig. 1 Current-injection emission spectra of the free-running QD gain devices with (a) 1mm, (b) 2mm and (c) 3mm cavity length under various injection current densities. Some spectra are shifted vertically for clarity. The inset of (a) gives a 3peak Gaussian fitting of the emission spectrum at 833A/cm^2 injection, presenting the GS, the 1st and the 2nd ESs transitions simultaneously

The QD gain devices with different cavity length were put in the Littrow setup to evaluate the tuning properties. The tuning spectra of the devices are shown in Fig. 2. No facet coating was applied on the device facets. In order to avoid the inner FP resonance, the injection current density is chosen just below the J_{th} of the free-running gain device. So no inner FP resonance appears even when the wavelength is tuned to the extremes, as presented in Fig. 2. For all the tuning wavelengths, the sidemode suppression ratio is better than 25dB and the amplified spontaneous emission suppression ratio is better than 20dB. The FWHM of the lasing spectra is no more than 2nm. Because the measurements for the spectra and power are performed from one facet of gain device, the spatial distribution of the emitted optical mode should be the same as the free-running device. The gain devices with different cavity length are also different in tuning range. For the device of 3mm cavity length, there is only contribution from the QDs' GS and a tuning bandwidth of 55nm ranging from 1198. 2 to 1253. 1nm, is achieved. While for the device with 2mm cavity length, the wavelength tuning extends to 1141. 6nm at the short-wavelength side, with no sacrifice of long-wavelength tuning. Decreasing the cavity length further down to 1mm leads to a tuning bandwidth of 100nm(1073. 9 - 1173. 8nm), contributed by the 1st and the 2nd ESs. The large cavity loss makes the disappearance of the tuning across QD-GS.

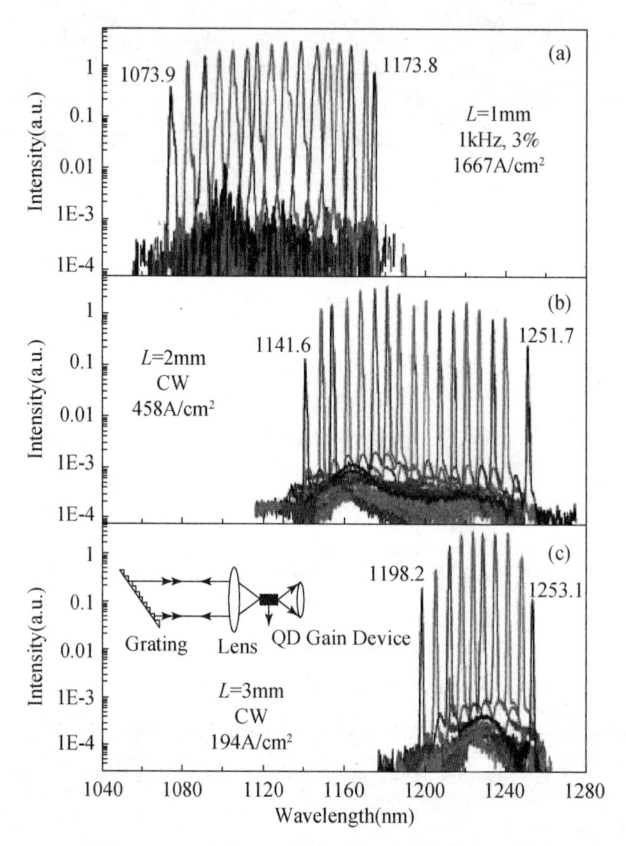

Fig. 2　Lasing spectra of grating-coupled EC lasers with InAs/GaAs QD gain devices of(a) 1mm,(b)2mm and(c)3mm cavity length. No coating was applied on the facets. The inset of(c) shows the experimental setup(Littrow configuration). The emitting directly from the gain device is used for the spectra and power measurements

Fig. 3 compares J_{th} dependence on the tuning wavelength for the QD gain devices with 1, 2 and 3mm cavity length, respectively. The J_{th} of the three free-running devices without facet coating are 206, 475, and 1708A/cm^2, respectively. The EC laser with 3mm cavity length shows the lowest J_{th}(117 - 194A/cm^2) and the narrowest tuning range. In order to avoid the inner lasing in the device, the injection current density is restricted to a relatively low value and only the GS of QDs can be populated under this injection level. The 1mm device, which is shortest in cavity length, shows broader tunability at the expense of higher J_{th}. As shown in Fig. 2(a), the gain device with 1mm cavity length allows EC laser to work under the injection level up to 1667A/cm^2. Under this injection level, the 2nd ES of QDs in the device can be populated. However, the GS saturated gain cannot compensate the external cavity loss for the 1mm length device. So the wavelength tuning range covers only the 1st and the 2nd ESs, with the GS foreclosed. The EC laser with 2mm cavity length shows better performance. Although the J_{th} increases to some extent, it still remains equivalent tuning to the long-wavelength side compared to that with 3mm cavity length device. Furthermore, the tuning on the short wavelength side can be extended to the 1st ES.

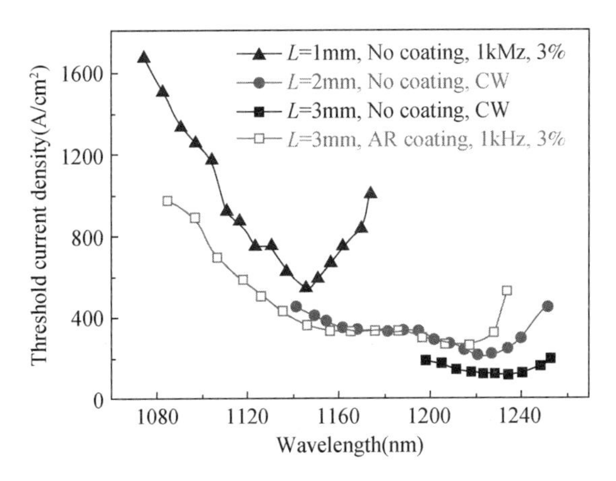

Fig. 3 Threshold current density as a function of the tuning wavelength for QD gain device with different cavity length

It can be seen from the above results that the choice of the cavity length is crucial for optimization on both J_{th} and the tuning bandwidth. On the premise that the GS saturated gain is higher than the external cavity loss, the cavity length of the gain device should be as short as possible. Thus, the tuning range can cover both the GS and the 1st ES and the tuning range above 100nm can be realized. Besides, the low QDs' GS saturated gain makes the QDs' ESs be occupied under lower injection level. This gives rise to lower J_{th} at the short wavelength side in the tuning range compared to the QW-EC laser. Generally more than 10kA/cm^2 pump level is needed for the carriers filling to the ESs in QW-EC laser[26,27]. In addition, the spectral broadening of QDs induced by its size inhomogeneity is beneficial to the consecutive tuning between two neighboring states.

Fig. 4 shows the output power as a function of tuning wavelength for the QD gain devices of 1, 2 and 3mm in cavity length. It can be seen from the figure that under the given injection level as used in Fig. 2, the maximal output power is 65, 53 and 54mW and the corresponding working current is 2, 1.1 and 0.7A, respectively. Although the output power is comparable for the three devices, the working current shows great difference and the 3mm device has the lowest power consumption. This indicates that the longer device possesses higher efficiency. From the figure, it also shows that the dependence of the output power on the wavelength represents approximately the gain characteristic of the three QD devices. A single peak can be observed in the output power vs wavelength for the 3mm cavity length device, which is attributed to the QDs' GS. However, there are two peaks in the output-power curve for both the 1 and 2mm cavity length devices, because of the simultaneous contribution of two states. For the 2mm cavity length device, the output power at GS is lower than that of the 1st ES due to finite gain of GS. Increasing the cavity length of the device slightly can eliminate the difference in output power for the two QDs' states to some extent.

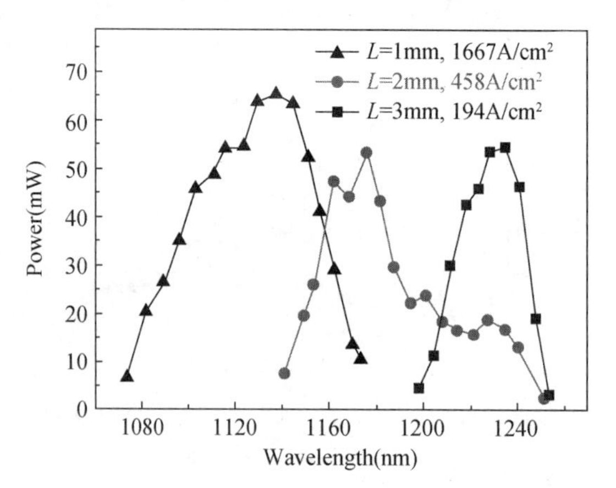

Fig. 4　Output power of the QD-EC lasers as a function of tuning wavelength for QD gain devices with different cavity length. No coating was applied on the facets. For the $L = 1$mm device, the injection is pulsed and the output power is the peak one

AR facet coating of a gain device has been proved to be effective to enlarge the tuning bandwidth[18,19]. Because the J_{th} of the inner FP resonance can be increased effectively by facet coating, the EC laser can work under a relatively high injection level and the tuning towards short wavelength can be realized easily. In order to evaluate the effect of AR coating, a single $\lambda/4$ ZrO_2 layer designed for minimum reflectivity at 1227nm was deposited on one facet (coupled with the grating in the tuning experiment) of the gain device 3mm in cavity length. After coating, the device lases at 1133nm due to the increased mirror loss. From the difference in slope efficiencies between the two facets, an effective reflectivity of approximately 2.7% is estimated at the lasing wavelength.

A ~150nm tuning bandwidth (1084.7nm – 1234.3nm), covering simultaneously the GS, the 1st and the 2nd ESs, is shown for EC tuning with the 3mm facet coating device.

The J_{th} dependence on the tuning wavelength is also presented in Fig. 3. Because of the increased facet loss, the J_{th} of the EC laser at 1228nm with AR facet coating(319.4A/cm^2) is 2.6 times as high as that without coating(122.2A/cm^2). Accordingly, the CW mode operation cannot be ensured across the whole tuning range. The increase of J_{th} can be explained as follows. With AR coating on the device facet, more radiation comes out from the active region and enters into the EC, where no active medium exists. Light field in the EC doesn't interact with the injected carriers and shows no contribution to the stimulated radiation. Namely, the optical confinement factor in the direction of light transmission decreases by the facet AR coating. As a result, the J_{th} increases. By further optimizing the structure of active region and choosing appropriate cavity length of gain device, an even lower working current density and wider tuning bandwidth should be achieved without facet coating.

4 Summary

In conclusion, the tuning characteristics of QD gain devices with different cavity length is investigated. A 110nm tuning bandwidth under the low CW injection level of 458A/cm^2, has been realized by using a device of 2mm cavity length without facet coating. The easiness of GS optical gain saturation is beneficial to the short wavelength tuning with low injection level. The AR coating is also used to expand the tuning bandwidth. A 150nm tuning band can be realized after AR coating, accompanied with an evident increase of J_{th}, which is attributed to reduced interaction between the light field and the QD active region.

Acknowledgements

The authors would like to thank Dr. J. Wu for the correction on grammar and P. Liang, H. Sun and Y. Hu for assistance on device fabrication. This work was supported by the National Basic Research Program of China (No. 2006CB604904) and the National Natural Science Foundation of China(Nos. 60976057, 60876086, 60776037 and 60676029).

References

[1] S. C. Woodworth, D. T. Cassidy, and M. J. Hamp, "Sensitive absorption spectroscopy by use of an asymmetric multiple-quantum-well diode laser in an external cavity," Appl. Opt. 40(36), 6719-6724 (2001).

[2] J. T. Olesberg, M. A. Arnold, C. Mermelstein, J. Schmitz, and J. Wagner, "Tunable laser diode system for noninvasive blood glucose measurements," Appl. Spectrosc. 59(12), 1480-1484(2005).

[3] N. Kuramoto, and K. Fujii, “Volume determination of a silicon sphere using an improved interferometer with optical frequency tuning,” IEEE Trans. Instrum. Meas. 54(2), 868-871(2005).

[4] T. Tanaka, Y. Hibino, T. Hashimoto, M. Abe, R. Kasahara, and Y. Tohmori, “100-GHz spacing 8-channel light source integrated with external cavity lasers on planar lightwave circuit platform,” J. Lightwave Technol. 22(2), 567-573(2004).

[5] S. R. Chinn, E. A. Swanson, and J. G. Fujimoto, “Optical coherence tomography using a frequency-tunable optical source,” Opt. Lett. 22(5), 340-342(1997).

[6] H. Lim, J. F. de Boer, B. H. Park, E. C. W. Lee, R. Yelin, and S. H. Yun, “Optical frequency domain imaging with a rapidly swept laser in the 815-870 nm range,” Opt. Express 14(13), 5937-5944(2006).

[7] C. K. Chia, S. J. Chua, J. R. Dong, and S. L. Teo, “Ultrawide band quantum dot light emitting device by postfabrication laser annealing,” Appl. Phys. Lett. 90(6), 061101(2007).

[8] Z. Y. Zhang, Z. G. Wang, B. Xu, P. Jin, Z. Z. Sun, and F. Q. Liu, “High-performance quantum-dot superluminescent diodes,” IEEE Photon. Technol. Lett. 16(1), 27-29(2004).

[9] L. H. Li, M. Rossetti, A. Fiore, L. Occhi, and C. Velez, “Wide emission spectrum from superluminescent diodes with chirped quantum dot multilayers,” Electron. Lett. 41(1), 41-43(2005).

[10] S. K. Ray, K. M. Groom, M. D. Beattie, H. Y. Liu, M. Hopkinson, and R. A. Hogg, “Broad-band superluminescent light-emitting diodes incorporating quantum dots in compositionally modulated quantum wells,” IEEE Photon. Technol. Lett. 18(1), 58-60(2006).

[11] X. Q. Lv, N. Liu, P. Jin, and Z. G. Wang, “Broadband Emitting Superluminescent Diodes With InAs Quantum Dots in AlGaAs Matrix,” IEEE Photon. Technol. Lett. 20(20), 1742-1744(2008).

[12] Z. Y. Zhang, R. A. Hogg, B. Xu, P. Jin, and Z. G. Wang, “Realization of extremely broadband quantum-dot superluminescent light-emitting diodes by rapid thermal-annealing process,” Opt. Lett. 33(11), 1210-1212(2008).

[13] M. Sugawara, K. Mukai, and Y. Nakata, “Light emission spectra of columnar-shaped self-assembled InGaAs/GaAs quantum-dot lasers: Effect of homogeneous broadening of the optical gain on lasing characteristics,” Appl. Phys. Lett. 74(11), 1561-1563(1999).

[14] A. Kovsh, I. Krestnikov, D. Livshits, S. Mikhrin, J. Weimert, and A. Zhukov, “Quantum dot laser with 75nm broad spectrum of emission,” Opt. Lett. 32(7), 793-795(2007).

[15] A. E. Zhukov, and A. R. Kovsh, “Quantum dot diode lasers for optical communication systems,” Quantum Electron. 38(5), 409-423(2008).

[16] C. L. Tan, H. S. Djie, Y. Wang, C. E. Dimas, V. Hongpinyo, Y. H. Ding, and B. S. Ooi, “Wavelength tuning and emission width widening of ultrabroad quantum dash interband laser,” Appl. Phys. Lett. 93(11), 111101(2008).

[17] P. Eliseev, H. Li, A. Stintz, G. T. Liu, T. C. Newell, K. J. Malloy, and L. F. Lester, “Tunable grating-coupled laser oscillation and spectral hole burning in an InAs quantum-dot laser diode,” IEEE J. Quantum Electron. 36(4), 479-485(2000).

[18] H. Li, G. T. Liu, P. M. Varangis, T. C. Newell, A. Stintz, B. Fuchs, K. J. Malloy, and L. F.

Lester, "150nm tuning range in a grating-coupled external cavity quantum-dot laser," IEEE Photon. Technol. Lett. 12(7), 759-761(2000).

[19] P. M. Varangis, H. Li, G. T. Liu, T. C. Newell, A. Stintz, B. Fuchs, K. J. Malloy, and L. F. Lester, "Low-threshold quantum dot lasers with 201nm tuning range," Electron. Lett. 36(18), 1544-1545(2000).

[20] A. Biebersdorf, C. Lingk, M. De Giorgi, J. Feldmann, J. Sacher, M. Arzberger, C. Ulbrich, G. Böhm, M. C. Amann, and G. Abstreiter, "Tunable single and dual mode operation of an external cavity quantum-dot injection laser," J. Phys. D Appl. Phys. 36(16), 1928-1930(2003).

[21] C. Ni. Allen, P. J. Poole, P. Barrios, P. Marshall, G. Pakulski, S. Raymond, and S. Fafard, "External cavity quantum dot tunable laser through 1.55μm," Physica E 26, 372-376(2005).

[22] G. Ortner, and C. Ni. Allen, C. Dion, P. Barrios, D. Poitras, D. Dalacu, G. Pakulski, J. Lapointe, P. J. Poole, W. Render, and S. Raymond, "External cavity InAs/InP quantum dot laser with a tuning range of 166nm," Appl. Phys. Lett. 88(12), 121119(2006).

[23] A. Tierno, and T. Ackemann, "Tunable, narrow-band light source in the 1.25μm region based on broad-area quantum dot lasers with feedback," Appl. Phys. B 89(4), 585-588(2007).

[24] A. Yu. Nevsky, U. Bressel, I. Ernsting, Ch. Eisele, M. Okhapkin, S. Schiller, A. Gubenko, D. Livshits, S. Mikhrin, I. Krestnikov, and A. Kovsh, "A narrow-line-width external cavity quantum dot laser for high-resolution spectroscopy in the near-infrared and yellow spectral ranges," Appl. Phys. B 92(4), 501-507(2008).

[25] X. Q. Lü, P. Jin, and Z. G. Wang, "A broadband external cavity tunable InAs/GaAs quantum dot laser by utilizing only the ground state emission," Chin. Phys. B 19(1), 018104-4(2010).

[26] A. Lidgard, T. Tanbun-Ek, R. A. Logan, H. Temkin, K. W. Wecht, and N. A. Olsson, "External-cavity InGaAs/InP graded index multiquantum well laser with a 200nm tuning range," Appl. Phys. Lett. 56(9), 816-817(1990).

[27] H. Tabuchi, and H. Ishikawa, "External grating tunable MQW laser with wide tuning range of 240nm," Electron. Lett. 26(11), 742-743(1990).

[28] X. Zhu, D. T. Cassidy, M. J. Hamp, D. A. Thompson, B. J. Robinson, Q. C. Zhao, and M. Davies, "1.4μm InGaAsP-InP strained multiple-quantum-well laser for broad-wavelength tunability," IEEE Photon. Technol. Lett. 9(9), 1202-1204(1997).

[29] S. C. Woodworth, D. T. Cassidy, and M. J. Hamp, "Experimental analysis of a broadly tunable InGaAsP laser with compositionally varied quantum wells," IEEE J. Quantum Electron. 39(3), 426-430(2003).

Experimental investigation of wavelength-selective optical feedback for a high-power quantum dot superluminescent device with two-section structure

Xinkun Li, Peng Jin, Qi An, Zuocai Wang, Xueqin Lv, Heng Wei, Jian Wu, Ju Wu, and Zhanguo Wang

(Key Laboratory of Semiconductor Materials Science, Institute of Semiconductors, Chinese Academy of Sciences, Beijing 100083, China)

Abstract In this work, a high-power and broadband quantum dot superluminescent diode (QD-SLD) is achieved by using a two-section structure. The QD-SLD device consists of a tapered titled ridge waveguide section supplying for high optical gain and a straight titled ridge waveguide section to tune optical feedback from the rear facet of the device. The key point of our design is to achieve the wavelength-selective optical feedback to the emission of the QDs' ground state (GS) and 1st excited state (ES) by tuning the current densities injected in the straight titled section. With GS-dominant optical feedback under proper current-injection of the straight titled region, a high output power of 338mW and a broad bandwidth of 65nm is obtained simultaneously by the contribution associated to the QDs' GS and 1st ES emission.

1 Introduction

Superluminescent diodes (SLDs) have attracted extensive attention for many applications including optical coherence tomography (OCT)[1,2], optical fiber based sensors[3,4], external cavity tunable lasers[5-7], optoelectronic system[8], etc. A wide emission spectrum is required for these applications, which allows to the realization of improved resolution in the systems. It has been proposed that the characteristic of size inhomogeneity, naturally occurring in self-assembled QDs grown by Stranski-Krastanow (S-K) mode, is beneficial to broaden the spectral bandwidth of the device[9]. QDs have successfully been used as the active media in several broadband light-emitting devices, such as QD-SLDs[10-18], QD semiconductor optical amplifiers (QD-SOAs)[19,20] and QD broadband laser diodes[21,22]. Till

原载于：Optics Exrpress, 2012, 20(11): 11936-11943.

now, for a typical QD-SLD with a single current-injection section inclining the waveguide at an angle with respect to the emission facet, can achieve a maximum spectrum bandwidth of about 100nm based on the balance of QDs' ground state (GS) emission and 1st excited state (ES) emission; but due to the gain saturation in the GS of QDs the output power is usually just about a few milliwatts.

Besides a wide emission spectrum, a high output power is also required in practical applications. As an example, in an OCT system, a high power is usually needed to enable great penetration depth and improve the imaging sensitivity[23]. For a regular QD-SLD device with a single current-injection section, the high output power can only be obtained at a high pumping level, where the device demonstrates a narrow spectrum emitted predominantly from the QDs' ES due to the low saturated gain of the QDs' GS. Many efforts have been made towards high-power superluminescent devices. By using an intermixed p-doped QD structure as high-gain active region, the QD-SLD exhibits a high power of 190mW with a 78-nm spectral bandwidth[18]. For geometrical designs of the high-power SLD device, A quantum-well SLD with a two-section structure which monolithically integrates an SLD with a tapered semiconductor optical amplifier (SOA) has been reported[24], which exhibits an output power one or two orders of magnitude higher than the regular SLD devices. Numerical investigation[25] and experimental evidences[26,27] have shown that the emission spectrum and output power can be tuned independently in an SLD device with the multi-section structure. Recently, we have previously demonstrated high-performance QD-SLDs with the two-section structures[28,29], which exhibits high output powers and simultaneous broad bandwidths. In [28], it exhibits a high-power QD-SLD device with two-section structure for the first time. High power (260mW) and broadband spectrum (66nm) is achieved at an optimum working point where the SOA current is 5A and the SLD current is 0.1A. But the investigation on working mechanisms of the two-section device is not presented, only the good performance of the superluminescent device is demonstrated. In [29], the high-power and broadband superluminescent device is achieved by monolithically integrating a conventional SLD with a tapered SOA. High output power is attributed to the single-pass amplification while the superluminescent light of the SLD section is propagating forward from the narrow end to the wide end of the tapered SOA. But for such a two-section SLD, the optimum working point is at high current-injection where the SOA current is about 8.5A and the SLD current is about 0.2A.

In this paper, a high performance QD-SLD device was achieved by using a two-section structure, which consists of a tapered titled ridge waveguide section and a straight titled ridge waveguide section. The working mechanisms of such a QD-SLD with the two-section structure is investigated. It is shown that double-pass gain is working in the two-section device. And wavelength-selective optical feedback to the emission of the QD' GS and 1st ES can be achieved by tuning the current densities injected in the straight titled section. With the GS-dominant optical feedback from the rear facet under proper current-injection of the straight titled section, a high output power of 338mW and a broad bandwidth of 65nm is obtained simultaneously.

2 Experiments

The epitaxial structure of the QD-SLD devices in this study was grown by a Riber 32P solidsource molecular beam epitaxy machine on *n*-GaAs(001) substrate, which is the same as our previous study[28, 29]. The active region consists of ten layers of self-assembled InAs QDs covered by 2nm $In_{0.15}Ga_{0.85}As$ and separated from each other by 35nm GaAs spacer. Each QDs layer is formed by the deposition of 1.8monolayer InAs at 480℃. The areal density is about $4\times10^{10}cm^{-2}$ obtained by atomic force microscopy for an uncapped sample, which has the same growth parameters as the epitaxial structure of the device. The whole active region included the GaAs waveguide is sandwiched between 1.5μm *n*-and *p*-type $Al_{0.5}Ga_{0.5}As$ cladding layers grown at 620℃. Finally, a p^+-doped GaAs contact layer completes the structure.

The QD-SLD devices with index-guided ridge waveguide and two-section structure were fabricated. A schematic diagram of the geometrical design (not to scale) is shown in the inset of Fig. 1. The device consists a straight titled section (S1) and a tapered titled section (S2). The straight section is 1mm long and 10μm wide. The tapered section is 2.3mm long with a full flare angle of 6°, which expands linearly from 10μm wide at the narrow end to 250μm wide at the wide end. The ridge waveguide was fabricated using standard photolithography process and wet chemical etching. The etching profile entered the bottom GaAs waveguide layer. With a deep-etched ridge waveguide, optical feedback from the rear facet of the S1 section is expected. The center axis of ridge was aligned at 6° to the facet normal to suppress Fabry-Pérot cavity resonance. Ti/Au and AuGeNi/Au ohmic contacts were evaporated on the top and back of the wafer. A 20μm wide isolating stripe between the straight region and the tapered region generated by leaving the up Ti/Au and 0.5μm semiconductor epilayers. After metallization, the device was cleaved and mounted *p*-side up on a copper heatsink using indium solder. The output facet of the S2 section were coated with a SiO_2 antireflection (AR) coating while the rear facet of the S1 section remained as-cleaved.

The QD-SLD device was characterized by optical power-injection current (*P-I*) and electroluminescence (EL) measurements at room temperature under a pulsing (1kHz repetition rate and 3% duty cycle) injection in the S2 section and a CW injection in the S1 section, respectively.

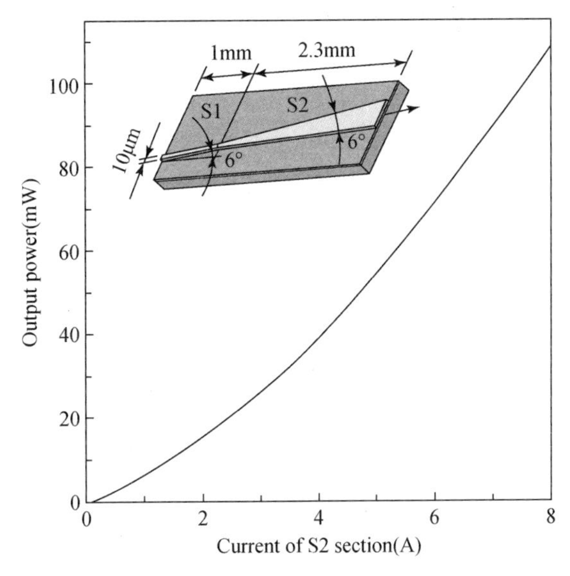

Fig. 1 *P–I* characteristic measured from the output facet of the S2 section with the S1 section un-pumped. The inset shows schematic design of the SLD device with two-section structure

3 Results

The *P–I* characteristic measured from the output facet of the S2 section while the S1 section is un-pumped as a rear optical absorption region is shown in Fig. 1. The inset depicts the schematic diagram of the device structure. As shown in Fig. 1, a superluminescent characteristic is clearly observed by the superlinear increase in optical power with the current of the S2 section (I_2). Inspection of the emission spectra (as shown in Fig. 2) indicates that lasing is successfully suppressed for such a two-section QD-SLD, and that the output optical power is due to amplified spontaneous emission. For a given I_2 of 8A, a maximum output power of 108mW was obtained.

Fig. 2 (a) shows the EL emission spectra under different injection-currents in the S2 section with the S1 section un-pumped. Fig. 2 (b) depicts the dependences of spectral bandwidth and output power by injection-currents of the S2 section. At a lower pumping level of 1A, the emission spectrum exhibits a full width at half maximum (FWHM) of 52nm with the center wavelength of 1.17μm, which corresponds to the emission from the QDs' GS. The relatively wide GS emission is attributed to the size inhomogeneity naturally occurring in self-assemble QDs. With the increase of I_2, the emission spectra are clearly broadened at the blue side, which should be attributed to the saturation of the QDs' GS and sequential carrier filling of the higher-energy ES. For a given I_2 of 4A, a wide spectrum with the maximum spectral bandwidth of 88nm is achieved, which is attributed to the balance of QDs' GS emission and 1st ES emission. However, due to the low saturated power of the GS emission, the output power is only 40mW at I_2 = 4A as shown in Fig. 2(b). It is shown that high output powers can be achieved at the high pumping levels of the S2 section, which is due to the contribution of the higher-energy ES. The 1st ES level could give out an optical gain with twice as many as the GS level due to the high angular momentum degeneracy[30]. But it is should be noticed, with an increase of injection levels from 4A to 8A, the device emits predominantly from the QDs' ES and the spectral bandwidth becomes narrow gradually. For a given I_2 of 8A, the emission spectrum exhibits a FWHM of 40nm with the output power of 108mW.

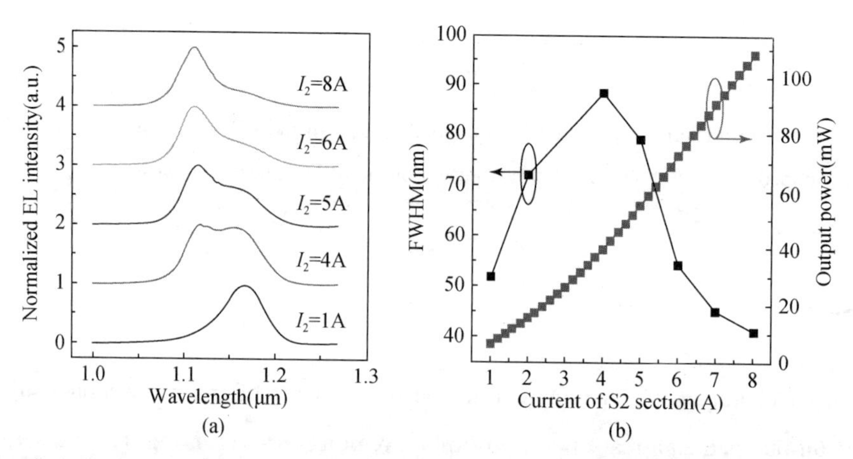

Fig. 2 (a) Normalized EL Emission spectra under different injection-currents of the S2 section. Some spectra are shifted vertically for clarity. (b) Spectral bandwidth and output power as a function of injection-current of the S2 section

The characteristics of the two-section SLD device were measured when the S1 section is under proper current-injection (I_1) to tune optical feedback from the rear facet. The output power characteristics versus I_2 under different I_1 are shown in Fig. 3. Equal power curves in the range of 100 to 800mW as function of the currents injected in the two sections of the device are shown in Fig. 4. It can be seen that the output power of the two-section SLD device increases rapidly with the increasing current-injection in the S1 section. While the S1 section is un-pumped as a rear optical absorption region, the output power of the two-section SLD device is 108mW at I_2 = 8A. At I_2 = 8A and I_1 = 200mA, the output power of the device can reach above 1.2W. Inspection of the emission spectra with various combinations of I_1

and I_2 shows that lasing appears when the output power is approximately 400mW (referring to the solid circles in Fig. 4).

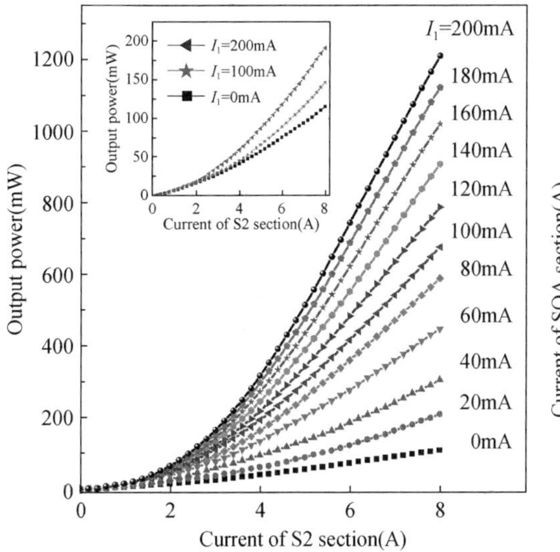

Fig. 3 Output power versus current of the S2 section under different injection-current of the S1 section

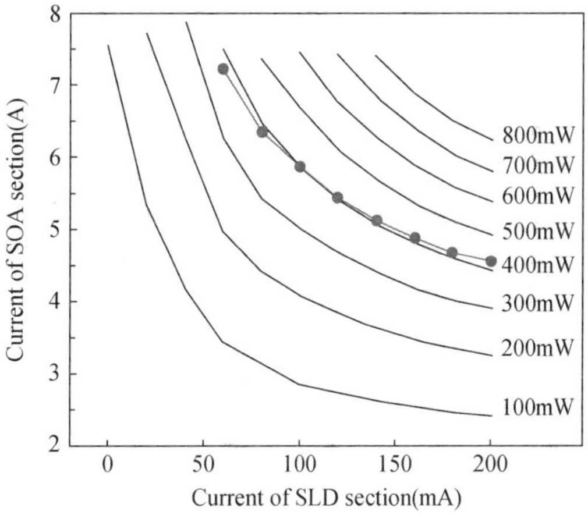

Fig. 4 Equal power curves (solid lines) as function of the currents injected in the two sections of the device. The solid circles show the combinations at which the device begins lasing

When the S1 section is pumped, part of the increasing output power of the two-section SLD device is attribute to the effect of the emission coming from the S1 section. Another reason is the double-pass amplification of the emission from the S2 section by the optical feedback from the rear facet of the S1 section. The primary increase of the output power should be attributed to the double-pass amplification by the optical feedback rather than the effect of the weak emission coming from the S1 section. The evidence as shown in the inset of Fig. 3 is that a much slighter effect on the output power by the increasing injection-current in the S1 section while both facets of the two-section SLD device are with AR coating. In addition, the high output power benefits the design of the tapered S2 section for high optical gain. With a full flare angle of 6°, the beam will expend freely to fill the full tapered region owing to diffraction[31]. The optical density will be reduced, which increases the saturated power.

Figs. 5(a) and 5(c) show the EL emission spectra under different I_1, for a given I_2 of 4A and 6A respectively. The dependences of spectral bandwidth and output power by injectioncurrents of the S1 section exhibits in Figs. 5(b) and 5(d). As we have expected, for the QDSLD device with the two-section structure, it can be found that the spectrum shape and emission bandwidth can be tuned by properly controlling the current densities injected in the S1 section.

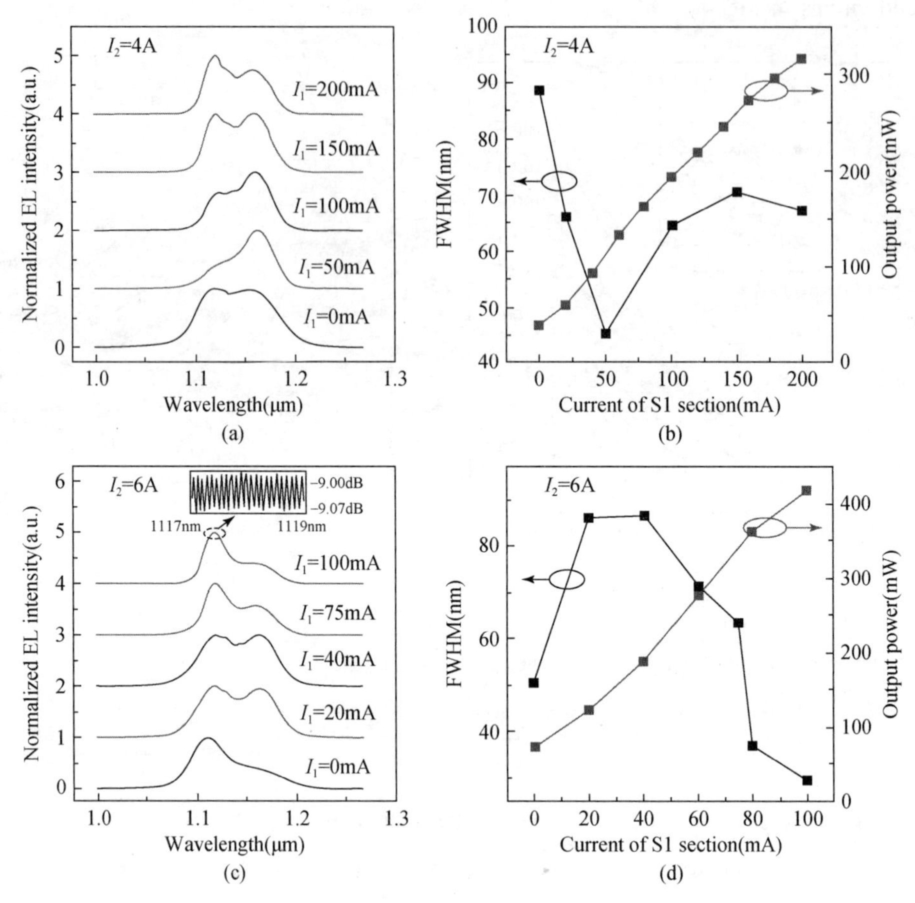

Fig. 5 Normalized EL emission spectra under different I_1, for a given I_2 of 4A(a) and 6A(c) respectively. Some spectra are shifted vertically for clarity. The dependences of spectral bandwidth and output power by injection-currents of the S1 section, for a given I_2 of 4A(b) and 6A(d) respectively. The inset of Fig. 5(c) is a segment of the high-resolution spectrum at I_2 = 6A and I_1 = 100mA

As shown in Figs. 5(a) and 5(b), under a given I_2 of 4A, the EL emission spectrum exhibits a balance of the QDs' GS emission and 1st ES emission while the S1 section is unpumped. When the S1 section is pumped, by tuning the current densities injected in the S1 section, the wavelength-selective optical feedback to the emission of the QDs' GS and 1st ES can be achieved. At I_2 = 4A and I_1 = 50mA, the GS emission provides the main contribution to the spectrum due to GS-dominant optical feedback under low pumping level of the S1 section. At I_2 = 4A and I_1 = 150mA, the balance of the QDs' GS emission and 1st ES emission appears again, which is due to the increasing optical feedback of the QDs' 1st ES. A broad emission bandwidth of 70nm and a high output power of 260mW is achieved simultaneously at I_2 = 4A and I_1 = 150mA.

For a given I_2 of 6A, the EL emission spectra under different I_1 are shown in Fig. 5(c).

When the S1 section is without pumped, a narrower spectrum with the FWHM of 51nm is obtained by the main contribution of the emission from the QDs' 1st ES. At $I_2 = 6$A and $I_1 =$ 40mA, it reaches a balance between the QDs' GS emission and 1st ES emission due to GS dominant optical feedback from the rear facet of the S1 section. A broad emission spectrum of 86nm with the output power of 186mW is obtained. At $I_2 = 6$A and $I_1 = 75$mA, a broad emission spectrum of 65nm and a high output power of 338mW is achieved simultaneously. The optical spectrum ripple of ~0.07dB is observed by a high-resolution spectral measurement at $I_2 = 6$A and $I_1 = 100$mA as shown in the inset of Fig. 5(c).

4 Conclusion

We have demonstrated a high-power InAs/GaAs QD-SLD device with broad bandwidth in the emission spectra by using a two-section structure that consists a straight titled section and a tapered titled section. The tapered section supplies for a high optical gain and the straight titled section is used to tune the optical feedback from the rear facet. It is shown that wavelength-selective optical feedback to the emission of the QDs' GS and 1st ES can be achieved by tuning the current densities injected in the straight titled section. Under proper pumping level of the straight titled section, a high output power of 338mW and a broad emission spectrum of 65nm is obtained simultaneously due to the GS-dominant optical feedback from the rear facet of the two-section QD-SLD device.

Acknowledgments

This work was supported by the National Basic Research Program of China (No. 2006CB604904) and the National Natural Science Foundation of China (Nos. 60976057, 60876086, and 60776037).

References

[1] N. Krstajic, L. E. Smith, S. J. Matcher, D. T. D. Childs, M. Bonesi, P. D. L. Greenwood, M. Hugues, K. Kennedy, M. Hopkinson, K. M. Groom, S. MacNeil, R. A. Hogg, and R. Smallwood, "Quantum dot superluminescent diodes for optical coherence tomography: skin imaging," IEEE J. Sel. Top. Quantum Electron. 16(4), 748-754(2010).

[2] S. Zotter, M. Pircher, T. Torzicky, M. Bonesi, E. Götzinger, R. A. Leitgeb, and C. K. Hitzenberger, "Visualization of microvasculature by dual-beam phase-resolved Doppler optical coherence tomography," Opt. Express 19(2), 1217-1227(2011).

[3] B. Lee, "Review of the present status of optical fiber sensors," Opt. Fiber Technol. 9(2), 57-79

(2003).

[4] N. Krstajic, D. Childs, R. Smallwood, R. Hogg, and S. J. Matcher, "Common path Michelson interferometer based on multiple reflections within the sample arm: sensor applications and imaging artifacts," Meas. Sci. Technol. 22(2), 027002(2011).

[5] X. Q. Lv, P. Jin, and Z. G. Wang, "A broadband external cavity tunable InAs/GaAs quantum dot laser by utilizing only the ground state emission," Chin. Phys. B 19(1), 018104(2010).

[6] X. Q. Lv, P. Jin, W. Y. Wang, and Z. G. Wang, "Broadband external cavity tunable quantum dot lasers with low injection current density," Opt. Express 18(9), 8916-8922(2010).

[7] K. A. Fedorova, M. A. Cataluna, I. Krestnikov, D. Livshits, and E. U. Rafailov, "Broadly tunable high-power InAs/GaAs quantum-dot external cavity diode lasers," Opt. Express 18(18), 19438-19443(2010).

[8] X. Li, A. B. Cohen, T. E. Murphy, and R. Roy, "Scalable parallel physical random number generator based on a superluminescent LED," Opt. Lett. 36(6), 1020-1022(2011).

[9] Z. Z. Sun, D. Ding, Q. Gong, W. Zhou, B. Xu, and Z. G. Wang, "Quantum-dot superluminescent diode: A proposal for an ultra-wide output spectrum," Opt. Quantum Electron. 31(12), 1235-1246(1999).

[10] N. Liu, P. Jin, and Z. G. Wang, "InAs/GaAs quantum-dot superluminescent diodes with 110nm bandwidth," Electron. Lett. 41(25), 1400-1402(2005).

[11] Y. C. Yoo, I. K. Han, and J. I. Lee, "High power broadband superluminescent diodes with chirped multiple quantum dots," Electron. Lett. 43(19), 1045-1047(2007).

[12] M. Rossetti, L. H. Li, A. Markus, A. Fiore, L. Occhi, C. Vélez, S. Mikhrin, I. Krestnikov, and A. Kovsh, "Characterization and modeling of broad spectrum InAs-GaAs quantum-dot superluminescent diodes emitting at 1.2-1.3μm," IEEE J. Quantum Electron. 43(8), 676-686(2007).

[13] Y. C. Xin, A. Martinez, T. Saiz, A. J. Moscho, Y. Li, T. A. Nilsen, A. L. Gray, and L. F. Lester, "1.3μm quantum-dot multisection superluminescent diodes with extremely broad bandwidth," IEEE Photon. Technol. Lett. 19(7), 501-503(2007).

[14] Z. Y. Zhang, Q. Jiang, M. Hopkinson, and R. A. Hogg, "Effects of intermixing on modulation *p*-doped quantum dot superluminescent light emitting diodes," Opt. Express 18(7), 7055-7063(2010).

[15] S. Haffouz, M. Rodermans, P. J. Barrios, J. Lapointe, S. Raymond, Z. Lu, and D. Poitras, "Broadband superluminescent diodes with height-engineered InAs-GaAs quantum dots," Electron. Lett. 46(16), 1144-1146(2010).

[16] H. S. Djie, C. E. Dimas, D. N. Wang, B.-S. Ooi, J. C. M. Hwang, G. T. Dang, and W. H. Chang, "InGaAs/GaAs quantum-dot superluminescent diode for optical sensor and imaging," IEEE Sens. J. 7(2), 251-257(2007).

[17] Z. Y. Zhang, R. A. Hogg, X. Q. Lv, and Z. G. Wang, "Self-assembled quantum-dot superluminescent lightemitting diodes," Adv. Opt. Photon. 2(2), 201-228(2010).

[18] Q. Jiang, Z. Y. Zhang, M. Hopkinson, and R. A. Hogg, "High performance intermixed *p*-doped

quantum dot superluminescent diodes at 1.2μm, " Electron. Lett. 46(4), 295-296(2010).

[19] Z. Bakonyi, H. Su, G. Onishchukov, L. F. Lester, A. L. Gray, T. C. Newell, and A. Tünnermann, " High-gain quantum-dot semiconductor optical amplifier for 1300nm, " IEEE J. Quantum Electron. 39(11), 1409-1414(2003).

[20] H. C. Wong, G. B. Ren, and J. M. Rorison, " Mode amplification in inhomogeneous QD semiconductor optical amplifiers, " Opt. Quantum Electron. 38(4-6), 395-409(2006).

[21] M. Sugawara, K. Mukai, and Y. Nakata, " Light emission spectra of columnar-shaped self-assembled InGaAs/GaAs quantum-dot lasers: effect of homogeneous broadening of the optical gain on lasing characteristics, " Appl. Phys. Lett. 74(11), 1561-1563(1999).

[22] A. Kovsh, I. Krestnikov, D. Livshits, S. Mikhrin, J. Weimert, and A. Zhukov, " Quantum dot laser with 75nm broad spectrum of emission, " Opt. Lett. 32(7), 793-795(2007).

[23] M. E. Brezinski and J. G. Fujimoto, " Optical coherence tomography: high-resolution imaging in nontransparent tissue, " IEEE J. Sel. Top. Quantum Electron. 5(4), 1185-1192(1999).

[24] G. T. Du, G. Devane, K. A. Stair, S. L. Wu, R. P. H. Chang, Y. S. Zhao, Z. Z. Sun, Y. Liu, X. Y. Jiang, and W. H. Han, " The monolithic integration of a superluminescent diode with a power amplifier, " IEEE Photon. Technol. Lett. 10(1), 57-59(1998).

[25] M. Rossetti, P. Bardella, and I. Montrosset, " Numerical investigation of power tenability in two-section QD superluminescent diodes, " Opt. Quantum Electron. 40(14-15), 1129-1134(2008).

[26] Y. C. Xin, A. Martinez, T. Saiz, A. J. Moscho, Y. Li, T. A. Nilsen, A. L. Gray, and L. F. Lester, " 1.3μm quantum-dot multisection superluminescent diodes with extremely broad bandwidth, " IEEE Photon. Technol. Lett. 19(7), 501-503(2007).

[27] P. D. L. Greenwood, D. T. D. Childs, K. M. Groom, B. J. Stevens, M. Hopkinson, and R. A. Hogg, " Tuning superluminescent diodes characteristics for optical coherence tomography systems by utilizing a multicontact device incorporating wavelength-modulated quantum dots, " IEEE J. Sel. Top. Quantum Electron. 15(3), 757-763(2009).

[28] Z. C. Wang, P. Jin, X. Q. Lv, X. K. Li, and Z. G. Wang, " High-power quantum dot superluminescent diode with integrated optical amplifier section, " Electron. Lett. 47(21), 1191-1193 (2011).

[29] X. K. Li, P. Jin, Q. An, Z. C. Wang, X. Q. Lv, H. Wei, J. Wu, J. Wu, and Z. G. Wang, " A high-performance quantum dot superluminescent diode with a two-section structure, " Nanoscale Res. Lett. 6(1), 625-629(2011).

[30] A. J. Williamson, L. W. Wang, and A. Zunger, " Theoretical interpretation of the experimental electronic structure of lens-shaped self-assembled InAs/GaAs quantum dots, " Phys. Rev. B 62(19), 12963-12977(2000).

[31] J. N. Walpole, " Semiconductor amplifiers and lasers with tapered gain regions, " Opt. Quantum Electron. 28(6), 623-645(1996).

19 μm quantum cascade infrared photodetectors

Shen-Qiang Zhai,[1] Jun-Qi Liu,[1] Xue-Jiao Wang,[1] Ning Zhuo,[1] Feng-Qi Liu,[1]
Zhan-Guo Wang,[1] Xi-Hui Liu,[2] Ning Li,[2] and Wei Lu[2]

([1]Key Lab of Semiconductor Materials Science, Institute of Semiconductors, Chinese Academy of Sciences; Beijing Key Laboratory of Low Dimensional Semiconductor Materials and Devices, Beijing 100083, China)

([2]National Laboratory for Infrared Physics, Shanghai Institute of Technical Physics, Chinese Academy of Sciences, Shanghai 200083, China)

Abstract Two InP based InGaAs/InAlAs photovoltaic quantum cascade detectors operating at peak wavelengths of 18μm and 19μm using different electronic transport mechanisms are reported. A longitudinal optical phonon extraction stair combined with energy mini-steps are employed for electron transport, which suppresses the leakage current and results in high device resistance. Altogether, this quantum design leads to 15K peak responsivity of 2.34mA/W and Johnson noise limited detectivity of 1×10^{11} Jones at 18μm.

Very long wave infrared (VWIR) photodetectors, covering the spectral range from 14 to 20μm, are needed for a variety of space applications, such as atmospheric temperature profiling, pollution monitoring, infrared astronomy, and satellite mapping.[1-3] One of the standard detector types in this wavelength range is the quantum well infrared photodetector (QWIP), which uses a bound to quasibound optical transition. However, QWIPs are generally designed as photoconductive devices and consequently, the applied bias results in a non-negligible dark current, which limits the operating temperature and overall performance, especially for very long wavelength detection. As an alternative solution, quantum cascade detectors (QCDs), a type of photovoltaic QWIPs, operating without bias have been presented.[4] Utilization of such detectors is theoretically possible at higher temperatures.[5] With no dark current, they are largely free from the limit of the integration time due to capacitance saturation of the read-out circuit and very promising for thermal imaging. Moreover, their thermal load is strongly reduced, which is of interest if the available cooling is limited, for example, in space borne systems. Therefore, QCD is a very competitive technology for applications in the very long wave range. To the present time, QCDs have

原载于: Appl. Physi. Lett., 2013, 102(19): 191-120.

covered a large wavelength range from the near-infrared to the terahertz region.[6-13] In this letter, the peak response wavelength is extended to 19μm in the very long wave range. Also, a quantum design is presented, which combines a longitudinal optical (LO) phonon extraction stair with a series of weakly coupled energy mini-steps for electron transport to increase device resistance and detectivity.

For a QCD, the working principle is the following: under illumination, electrons in the ground level are excited to the first excited level in the active (doped) well. They are then transferred into the ground level of the next period's active well by a carefully designed extraction cascade, which is adapted to the LO phonon energy. However, for detection wavelengths longer than 17μm (energy smaller than 73meV) and a longitudinal optical phonon energy of 34meV (InGaAs), there are no more than two phonon steps allowed in the extraction region. This implies that the extraction region includes only a few quantum well (QW)/barrier pairs, which will induce strong coupling between adjacent active well ground states and the associated large leakage current. In Ref. [12], the authors solved this problem by using miniband-based vertical transport. However, thermally assisted leakage current was still significant across the miniband, which weakened R_0A and thus the detectivity. Here, a different quantum design is presented to minimize the overlap between the active ground levels. Instead of a miniband, a series of energy mini-steps with small energy separations are fabricated for electron transport. This design can suppress the thermally assisted tunneling, which is responsible for leakage current, and increase the resistivity. The device with this design is labeled as sample A, and the structure is illustrated in table 1 in detail.

Fig. 1(a) shows the self-consistently calculated conduction band diagram of one period for sample A by Schrödinger solver. The energy between levels A1 and A2 in the active quantum well A is 67meV, which defines the QCD's peak wavelength of 18.5μm. For each period, by absorption of a photon with the energy $h\upsilon = E_{A2} - E_{A1}$, an electron is excited from A1 to A2 and then transfers to B1, which is in resonance with A2. The electron in level B1 will first undergo a fast LO phonon-assisted transition to C1 in QW C and finally relax to A'1 sequentially through five energy mini-steps. In this structure, thicker barriers are used to decrease overlap between active ground levels and the extraction levels. This helps to suppress the leakage current, which is caused by the thermally assisted tunneling, and to increase the resistivity (R_0A). On the other hand, due to a combination of acoustic-phonon scattering, electron-electron scattering, and LO-phonon scattering, efficient electron transport from C1 to A'1 through these energy mini-steps is expected.[14, 15] As a counterpart, a very long wave QCD device (labeled as sample B) with a calculated detection energy of 63 meV (19.7μm) based on miniband-based vertical transport is also presented.

The operation principle is akin to the one in Ref. [12]. The self-consistently calculated conduction band diagram and the structure for one period of sample B are shown in Fig. 1(b) and in table 1, respectively.

Table 1　Layer thicknesses for one period of the active regions of samples A and B in angstroms along with the simulated detection energy E_{12}. Bold numbers stand for $In_{0.53}Ga_{0.47}As$ wells, and roman numbers stand for $In_{0.52}Al_{0.48}As$ barriers. Underlining means doping density of 1.2×10^{17} cm^{-3}

Sample	E_{12}[meV(μm)]	A	B	C	D	E	F	G
A	67(18.5)	<u>**158**</u>/64	**65**/59	**89**/53	**96**/48	**107**/43	**121**/37	**138**/32
B	63(19.7)	<u>**163**</u>/57	**72**/27	**94**/45	**98**/42	**99**/39	**100**/29	

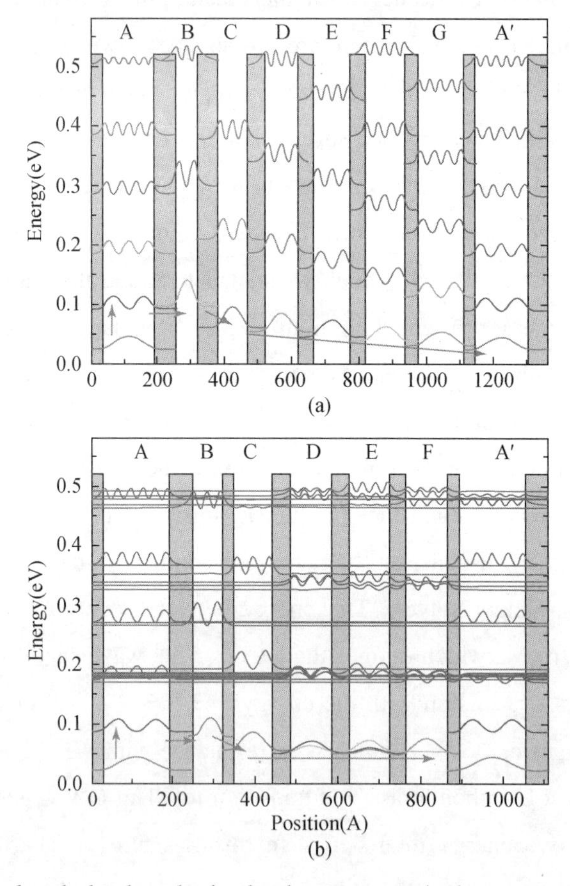

Fig. 1　Self-consistently calculated conduction band structures and relevant squared envelope functions of one period for samples A and B. (a) and (b) correspond to samples A and B, respectively

Both samples were grown by molecular beam epitaxy on semi-insulating InP substrates. Growth started with a 400nm thick Si-doped (4×10^{17} cm^{-3}) $In_{0.53}Ga_{0.47}As$ contact layer. Following that, 30 detector periods as listed in table 1 was grown. Finally, the top contacts were provided by a combination of a 200nm thick $In_{0.53}Ga_{0.47}As$ (3×10^{17} cm^{-3}) layer and a

10nm thick $In_{0.53}Ga_{0.47}As$ (1.2×10^{18} cm^{-3}) layer. X-ray diffraction (XRD) measurements confirm that the effective superlattice periods are 107nm and 83.1nm with deviations of 4.4% and 3.9% from nominal ones for samples A and B, respectively. After growth, the wafers were processed into square-shaped mesas of size $300\times300\mu m^2$ using standard photolithography and wet etching with $H_3PO_3:H_2O_2:H_2O(1:1:6)$. Electrical contacts were formed by a standard lift-off deposition of Ti/Au(10/250). To couple light into the active region efficiently, 45° multipass wedges were fabricated.

Optical and electrical characterizations were performed in a liquid He-flow cryostat equipped with ZnSe windows. The photocurrent spectrum measurement was performed using a Nicolet 6700 Fourier Transform Infrared Spectrometer(FTIR) with a KBr beam splitter and the internal glow-bar source. The photocurrent signal was then amplified with a SR560 low noise voltage amplifier and fed into the external detector port of the spectrometer. With the measured photocurrent spectrum and the incident radiation power, the peak responsivities at different temperatures were calculated using

$$R=\frac{J_m}{P_m}=\frac{T}{2}R_p\frac{\int_0^{\infty}R(\lambda)M(\lambda)\mathrm{d}\lambda}{\int_0^{\infty}M(\lambda)\mathrm{d}\lambda},\tag{1}$$

where J_m is the photocurrent of the device illuminated by a calibrated blackbody and measured with an optical chopper and a lock-in amplifier, P_m is the incident power on the detector surface calculated using the Stefan-Boltzmann law, T is the transmission coefficient across the ZnSe window of the cryostat, R_p is the peak responsivity, $R(\lambda)$ is the detector's spectral photocurrent divided by the glowbar's spectral intensity at the detector surface and normalized to unity at the peak detection wavelength, [16] and $M(\lambda)$ is normalized power density spectrum gained from the Black-body radiation law. The factor of 2 takes into account that the measurements were done with unpolarized light, whereas QCD responsivities are generally given for TM polarized light. Based on the parameters above, at 15K peak responsivity values of 2.34 and 3.31mA/W were obtained at 18 and 19μm, respectively, for samples A and B, as shown in Fig. 2. The discrepancy of the measured detection energy from the calculated one for sample A is 2.84% and for sample B is 3.65%. This can be attributed to growth deviations in period thickness and composition. For sample B, the sharp drop around the cutoff wavelength and the associated small deviation from the Lorentzian shaped spectrum are caused by the transmittance cutoff (19.5μm) of ZnSe window. For QCDs, the current responsivity can be given by $R_P=\frac{q}{h\nu}\eta\frac{p_e}{Np_c}$, where q is the elementary charge, $h\nu$ is the peak photon energy, η is the absorption efficiency, p_e is the escape probability of an excited electron in the active QW, p_c is its capture probability into the active

QW's ground state for an electron traveling down the QCD's cascade and close to unity, and N is the number of periods. To estimate the escape probability, the absorption efficiencies (η) are calculated to be 22.4% for sample A and 29.8% for sample B using Fermi's golden rule.[17, 18] The escape probabilities (p_e) are then evaluated to be 2.1% for both samples, which is rather low for QCDs. This value should be increased by a more precise alignment of the A2 and B1 states, which is essential to increase device responsivity. The escape probability p_e can be approximated using the phonon scattering lifetimes from the excited level A2 towards the ground level A1 (relaxation time, τ_{rel}) and towards the extractor state C1 (escape time, τ_{esp}),[17]

$$P_e \approx \frac{\tau_{rel}}{\tau_{rel}+\tau_{esp}}. \tag{2}$$

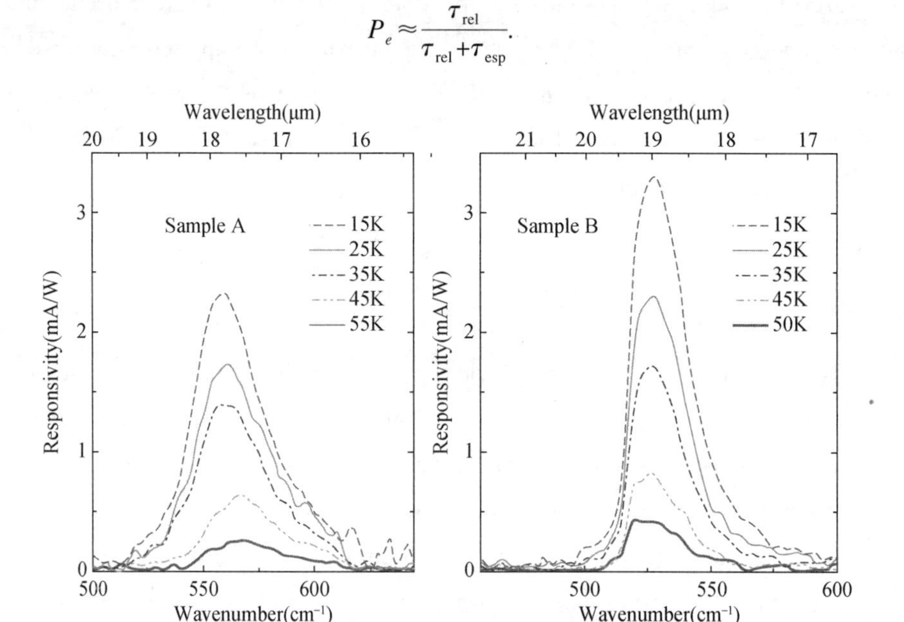

Fig. 2　Photocurrent response curves for samples A and B as a function of temperature

A more precise alignment of the A2 and B1 states will produce shorter τ_{esp} and an enhancement of escape probability. In addition, strengthening the couple between energy ministeps in a certain extent can reduce the relaxation time of the electrons through the extraction region and thus increase p_e in the case of sample A.

For both devices, a series of I–V curves at different temperatures have been measured under dark conditions using a Keithley 236 Source Meter. The dark I–V curves are shown in Fig. 3. Insets in Fig. 3 show the measured resistance-area product R_0A around 0V extracted from I–V curves versus inverse temperature. Using an Arrhenius thermal activation model, a dark current process of thermionic origin can be modeled by[19]

$$I_{dark} \propto \exp(-E_{act}/k_B T), \tag{3}$$

where E_{act} is the activation energy, k_B is the Boltzmann constant, and T is the temperature.

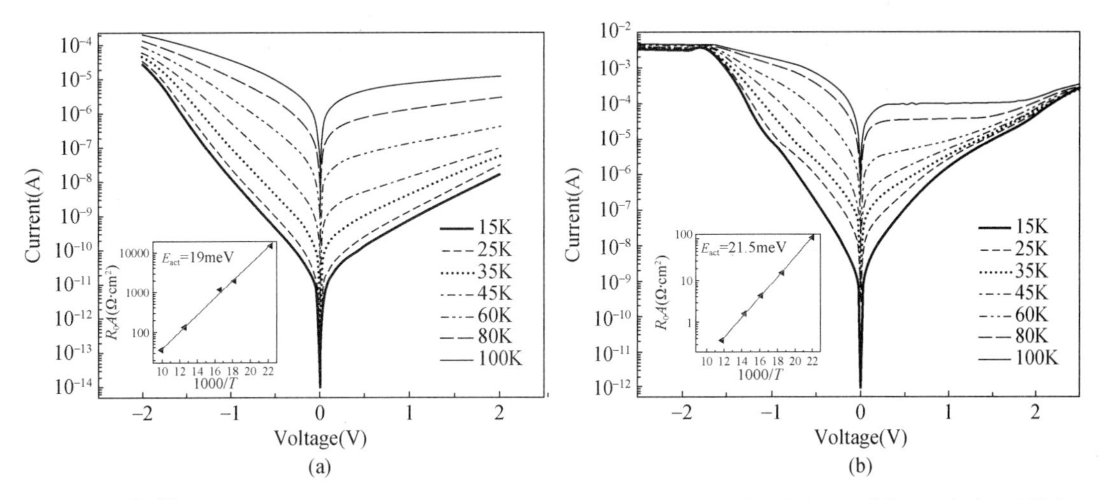

Fig. 3 $I-V$ curves for samples A and B at different temperatures under dark conditions. (a) and (b) correspond to samples A and B, respectively. Insets: R_0A extracted from $I-V$ measurements versus inverse temperature, namely Arrhenius plot

Activation energy values of 19meV and 21.5meV are calculated from the Arrhenius plot slopes for samples A and B, respectively. The activation energy of sample A is close to the simulated $\Delta E = D1 - A1 - E_{f,A1} = 18.5\text{meV}$,[19] where $E_{f,A1} = 11\text{meV}$ is the Fermi energy measured from $A1$.[20] Namely, the electron transition from QW A to the ground state in QW D is dominated for the dark current for sample A. For sample B, the activation energy is close to the calculated energy separation (20.2meV) between the Fermi level in QW $A1$ and the miniband, indicating the existence of leakage current across the miniband. For QCDs, the detectivity is Johnson noise limited at most working temperatures and can be calculated using

$$D_J^* = R_P\sqrt{\frac{R_0A}{4k_BT}}, \tag{4}$$

where R_P is the peak responsivity deduced from Eq. (1), R_0A is resistance-area product extracted from $I-V$ curves, k_B is the Boltzmann constant, and T is the temperature. The calculated results for Johnson noise limited detectivities of samples A and B are shown in Fig. 4. At 15K, peak detectivity values of 1×10^{11} and 9×10^{9} Jones were obtained for samples A and B, respectively. Compared with the 16.5μm device reported in Ref. [12], the detectivity of sample A is larger by almost two orders of magnitude. At low temperatures, the dominant contribution to the dark current is thermally assisted tunneling, so that a decrease in dark current and an increase in detectivity with increasing barrier width are expected.[21] The higher detectivity of sample B can also be explained. Although sample A has a somewhat lower responsivity value than sample B, the detectivity of sample A is higher than sample B by one order of magnitude. This implies that the concept of sample A has no significant influence on device responsivity but increases the detectivity efficiently.

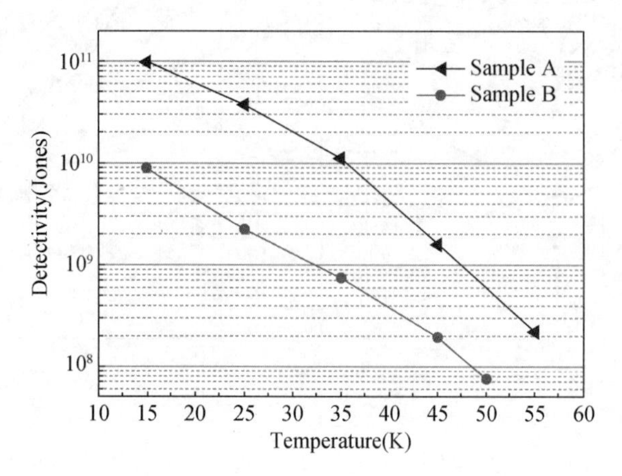

Fig. 4　Johnson noise limited detectivities D_J^* at different temperatures for samples A and B

In summary, two InP based very long wave photovoltaic quantum cascade detectors operating at peak wavelengths of 18μm and 19μm have been fabricated using energy ministeps and miniband vertical transport, respectively. Thanks to the design using an electron transport mechanism of weakly coupled energy mini-steps, high detectivity is obtained without decreasing device responsivity. Peak responsivities are 2.34 and 3.31 mA/W and Johnson noise limited detectivities are 1×10^{11} and 9×10^{9} Jones, at 15K for the 18 and the 19μm devices, respectively. Further optimization with a more precise alignment of the *A*2 and *B*1 states and an efficient extraction region will lead to a much higher device performance for both electronic transport mechanisms. This work was supported by the National Research Projects of China under Grant Nos. 10990103, 2013CB632802, 61274094, 2013CB632801, 2013CB632803, and 2011YQ13001802-04. The authors would like to thank Ping Liang and Ying Hu for their help in device processing.

References

[1] S. D. Gunapala, J. S. Park, G. Sarusi, T. L. Lin, J. K. Liu, P. D. Maker, R. E. Muller, C. A. Shott, and T. Hoelter, IEEE Trans. Electron Devices 44, 45(1997).

[2] S. D. Gunapala, K. M. S. V. Bandara, B. F. Levine, G. Sarusi, D. L. Sivco, and A. Y. Cho, Appl. Phys. Lett. 64, 2288(1994).

[3] G. Sonnabend, D. Wirtz, and R. Schieder, Appl. Opt. 44, 7170(2005).

[4] L. Gendron, M. Carras, A. Huynh, V. Ortiz, C. Koeniguer, and V. Berger, Appl. Phys. Lett. 85, 2824(2004).

[5] C. Koeniguer, L. Gendron, V. Berger, E. Belhaire, E. Costard, P. Bois, and X. Marcadet, Proc. SPIE 5957, 595704(2005).

[6] L. Gendron, C. Koeniguer, V. Berger, and X. Marcadet, Appl. Phys. Lett. 86, 121116(2005).

[7] M. Graf, N. Hoyler, M. Giovannini, J. Faist, and D. Hofstetter, Appl. Phys. Lett. 88, 241118 (2006).

[8] N. Kong, J. Q. Liu, L. Lu, F. Q. Liu, L. J. Wang, Z. G. Wang, and W. Lu, Chin. Phys. Lett. 27, 128503(2010).

[9] S. Q. Zhai, J. Q. Liu, F. Q. Liu, and Z. G. Wang, Appl. Phys. Lett. 100, 181104(2012).

[10] S. Sakr, E. Giraud, M. Tchernycheva, N. Isac, P. Quach, E. Warde, N. Grandjean, and F. H. Julien, Appl. Phys. Lett. 101, 251101(2012).

[11] A. Buffaz, M. Carras, L. Doyennette, A. Nedelcu, X. Marcadet, and V. Berger, Appl. Phys. Lett. 96, 172101(2010).

[12] F. R. Giorgetta, E. Baumann, M. Graf, L. Ajili, N. Hoyler, M. Giovannini, J. Faist, D. Hofstetter, P. Krötz, and G. Sonnabend, Appl. Phys. Lett. 90, 231111(2007).

[13] M. Graf, G. Scalari, D. Hofstetter, J. Faist, H. Beere, E. Linfield, D. Ritchie, and G. Davies, Appl. Phys. Lett. 84, 475(2004).

[14] J. H. Smet, C. G. Fonstad, and Q. Hu, J. Appl. Phys. 79, 9305(1996).

[15] B. S. Williams, B. Xu, Q. Hua, and M. R. Melloch, Appl. Phys. Lett. 75, 2927(1999).

[16] F. R. Giorgetta, E. Baumann, D. Hofstetter, C. Manz, Q. Yang, K. Köhler, and M. Graf, Appl. Phys. Lett. 91, 111115(2007).

[17] F. R. Giorgetta, E. Baumann, M. Graf, Q. K. Yang, C. Manz, K. Kohler, H. E. Beere, D. A. Ritchie, E. Linfield, A. G. Davies, Y. Fedoryshyn, H. Jackel, M. Fischer, J. Faist, and D. Hofstetter, IEEE J. Quantum Electron. 45, 1039(2009).

[18] B. F. Levine, J. Appl. Phys. 74, R1(1993).

[19] A. Majumdar, K. K. Choi, J. L. Reno, L. P. Rokhinson, and D. C. Tsui, Appl. Phys. Lett. 80, 707(2002).

[20] Y. Fu, N. Li, M. Karlsteen, M. Willander, N. Li, W. L. Xu, W. Lu, and S. C. Shen, J. Appl. Phys. 87, 511(2000).

[21] S. R. Andrews and B. A. Miller, J. Appl. Phys. 70, 993(1991).

High-performance operation of distributed feedback terahertz quantum cascade lasers

Yuanyuan Li, Junqi Liu, Fengqi Liu, Jinchuan Zhang and Zhanguo Wang

(Key Laboratory of Semiconductor Materials Science, Institute of Semiconductors, Chinese Academy of Sciences, Beijing 100083, China)

Abstract Carefully designed distributed feedback terahertz quantum cascade lasers based on surface metallic-stripe grating structure are presented. Stable single-mode emission with a side-mode suppression ratio of >20dB is obtained under all injection currents and operating temperatures. Maximum edge-emitting power of 58mW is realised in continuous-wave mode. In pulsed mode, record output powers of 286mW at 10K and 82mW at 77K are achieved at 97.1μm with a well-shaped beam pattern.

1 Introduction

Distributed feedback (DFB) terahertz (THz) quantum cascade laser (QCL) is a kind of reliable compact coherent single-mode source in THz region. For various applications such as THz spectroscopy, wireless communication, and remote sensing technology[1-3], good emission characteristics of not only stable single-mode, but also high-power and well-shaped beam pattern are desirable. Consequently, DFB THz QCLs fabricated with different structures and methods have been presented in the past years. Compared with metal-metal waveguide structure, semi-insulating surface-plasmon (SI-SP) waveguide provides favourable mode profile in the vertical dimension resulting in higher edge-emitting power and smaller beam divergence[4]. For SI-SP DFB QCLs[5-7], the surface metallic-stripe grating structure[6] is attractive due to its strong coupling effect that discontinuous periodic slits in both of the metallic layer and in the top contact layer perturb the SP mode strongly and tune the index precisely for frequency selection. To achieve high-power emission, tapered geometry is designed which combines the features of a narrow ridge section and the advantages of a wide output facet[8]. The straight ridge section is narrow enough to select the fundamental transverse mode (TM) TM_{00} for good beam pattern as well as the large tapered

原载于：Electronics Letters, 2016, 52(11): 945-947.

section is used to broaden the gain area for higher output power. In this Letter, we present high-performance operation of tapered DFB THz QCLs with surface metallic-stripe gratings based on SI-SP waveguide. Reliable and stable single-mode emission with a side-mode suppression ratio (SMSR) of 24dB is obtained. High-output power and well-shaped beam pattern are realised by adopted tapered waveguide with a 2.5° taper half-angle.

2 Device simulation and fabrication

The laser structure was grown on an Si GaAs substrate and consists of a bottom 0.5μm highly doped(Si, $2.5\times10^{18}cm^{-3}$) GaAs layer, active region, and a top 0.1μm highly doped (Si, $5\times10^{18}cm^{-3}$) GaAs contact layer. The active core includes 185 periods of $Al_{0.15}Ga_{0.85}As$/GaAs heterostructure which is similar to[9].

The grating periods of 13.4, 13.5, and 13.7μm were defined with the calculated effective index 3.6. To obtain stable single mode, the surface metallic-stripe grating structure has been simulated to optimise the duty cycle for low waveguide loss α_w and moderate grating coupling coefficient κ. The simulation was based on a finite-element electromagnetic solver (COMSOL) with periodical boundary conditions. The results are plotted in Fig. 1 with a constant grating depth 0.3μm (etching the top contact GaAs layer completely) and a variable duty cycle σ. The mode 1 which locates beneath the grating peak is always the lasing longitudinal mode due to the lower waveguide loss and the smallest value of loss for mode 1 appears at $\sigma=0.8$. The coupling coefficient κ against duty cycle σ is also presented in Fig. 1. With duty cycle increases, the coupling coefficient increases gradually and gets the maximum value at $\sigma = 0.9$. Considering a compromise between coupling coefficient and waveguide loss, 11.5μm-wide metallic-stripes were selected. Therefore, duty cycles of 85.8, 85.2, and 84% were obtained corresponding to grating periods $\Lambda =$ 13.4μm, 13.5μm, and 13.7μm, respectively.

The surface metallic-stripe gratings in the metallic layer (0.2μm) were completed by inductively coupled plasma etching after the standard fabrication process of Fabry-Perot QCLs with high-loss boundary conditions[10]. Then, the top contact GaAs layer in the slits was removed entirely by wet chemical etching. To further increase the output power, an Al_2O_3/Ti/Au/Al_2O_3 high-reflectivity layer was coated on the rear facet of cleaved lasers. The schematic cross-section through the grating structure is shown in the inset of Fig. 1.

For characterisation, the THz QCLs were mounted on the cold finger of a liquid-helium cryostat with a Winston cone in front of the emission facet. The powers were measured using an absolute THz power meter and not corrected by the transmission of the polyethylene

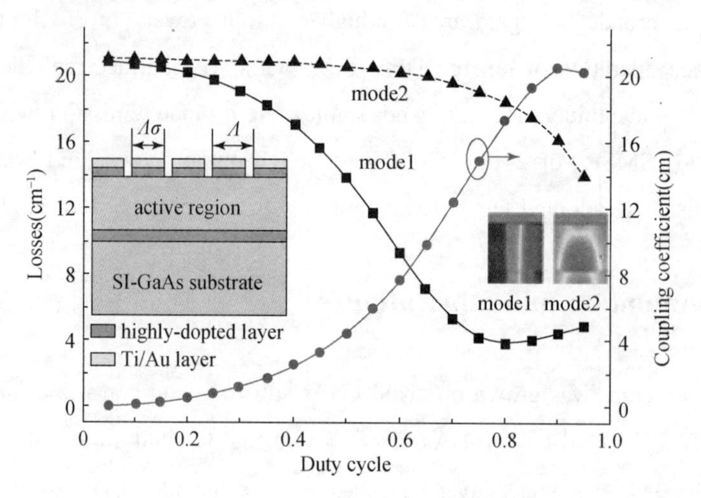

Fig. 1 Waveguide losses α_w of two modes and coupling coefficient κ against duty cycle with fixed grating etch depth(0.3μm) Inset: Sectional structure of DFB lasers

window. The measurement of lasing spectra was operated through a Fourier-transform infrared spectrometer with a resolution of 0.5cm^{-1} in rapid scan mode. Horizontal beam divergence measurement was done by placing a calibrated thermopile power meter 80mm away from the emission facet to collect lasing light with a sweep step of 1°(without Winston cone). All pulsed measurements were performed at 5kHz repetition rate with a duty cycle of 1%.

3 Results

Fig. 2 shows the power-current-voltage (P-I-V) curves of a 1.5mm-long and 200μm-wide DFB laser with 13.4μm grating period at various temperatures. For the duty cycle σ of 85.8%, coupling coefficient κ is ~19cm^{-1} as shown in Fig. 1. The coupling strength κL is calculated to be 2.85 for 1.5mm-long devices, which is appropriate to allow for a high out-coupling efficiency. In continuous-wave (CW) operation, the device realises a peak output power of 58mW at 10K and lases up to 55K. Stable single-mode is observed with an SMSR of >20dB. The characteristic temperature of 23K is extracted from the empirical relationship $J_{th} = J_0 \exp(T/T_0)$ as shown in the inset of Fig. 2.

To further enhance the output power and improve the far-field pattern, tapered DFB lasers were fabricated. The upper inset of Fig. 3 depicts the top schematic diagram of the laser. The narrow straight ridge is 0.5mm-long with a width of 200μm, which is selected to sustain fundamental TM TM_{00}. The tapered ridge is 2.5mm-long and characterised with a taper half-angle θ of 2.5°. The normalised spectra for 2.5° tapered DFB lasers for different

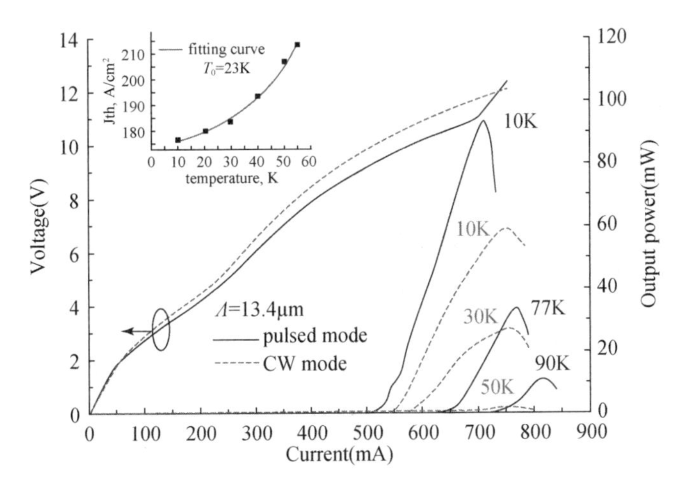

Fig. 2 Output power against driving current for 1.5mm-long DFB lasers with 13.4μm grating period both in pulsed and CW mode Inset: Empirical fitting between Jth and T in CW mode

grating periods (13.4, 13.5, 13.7μm) are displayed in Fig. 3, which were measured at nearly maximum output power. These devices lase at 96.4, 97.1, and 98.8μm, respectively, exhibiting a good linear relationship with grating periods. Stable single-mode operation is realised under all tested currents and temperatures. The lower inset of Fig. 3 shows the lasing spectrum from the 13.5μm DFB laser in logarithmic scale at 10K, an SMSR of 24dB is obtained.

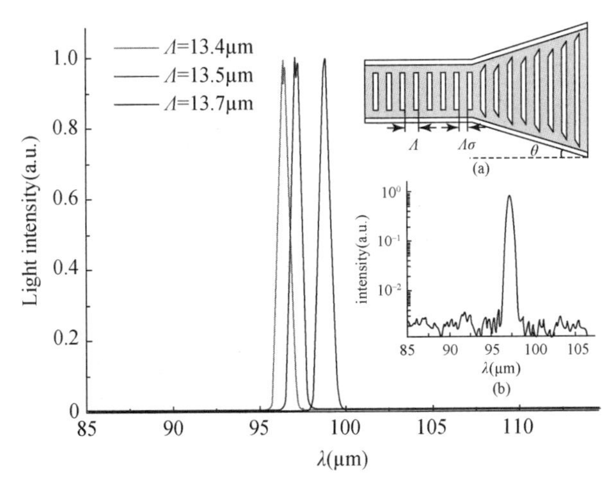

Fig. 3 Lasing spectra of tapered DFB lasers with different grating periods inset (a) Top schematic diagram of tapered DFB lasers; (b) Spectrum from Λ = 13.5μm laser in logarithmic scale

Fig. 4(a) shows the pulsed operation characteristics of the tapered lasers. The maximum emission powers are 165, 286, and 188mW at 10K for lasers of Λ = 13.4μm, 13.5μm, and 13.7μm, respectively. To our knowledge, the 286mW is the record single-mode edge-emitting power of DFB THz QCLs. The difference of peak output power among these lasers with different grating periods originates from the location of Bragg wavelength in the gain spectrum. The devices are lasing up to temperatures of ~95K. At liquid nitrogen temperature of 77K, the peak output power from the same 13.5μm laser is 82mW, which is high enough for various applications when laser is packaged in a liquid nitrogen dewar. Fig. 4(b) displays the horizontal beam patterns of the three lasers. Single-lobe beam patterns are obtained due to the fundamental TM selection effect of the narrow straight section and the existence of high-loss boundary conditions. Gaussian fitted curves of measured normalised intensity show the full-width at half-maximum divergence angles of 15.6°, 19.6°, and 20.1°. The best beam pattern is observed from the laser with 13.4μm grating period.

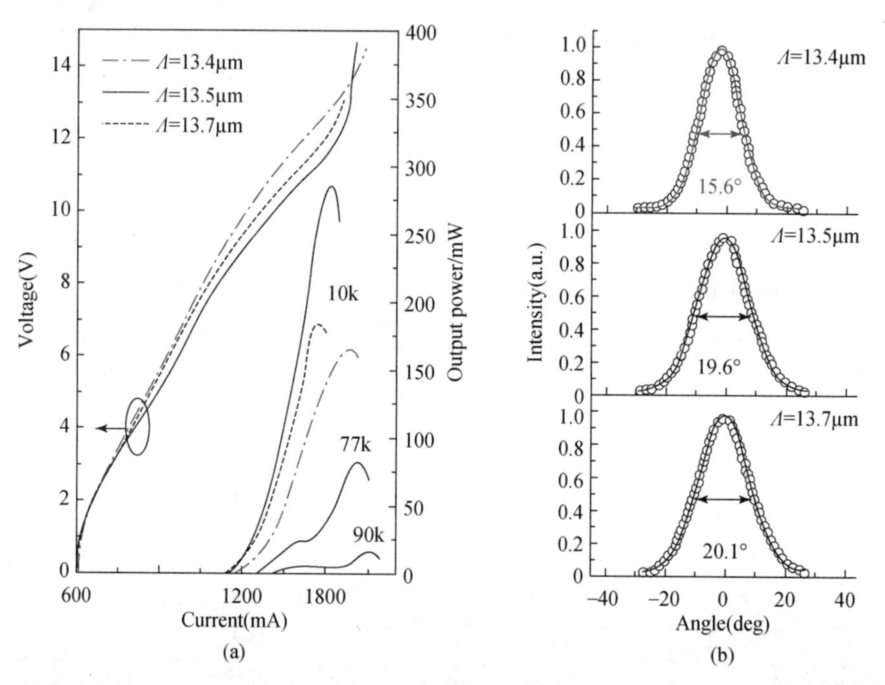

Fig. 4　Light-voltage-current curves and beam patterns (a) Emission characteristics of tapered DFB lasers with 2.5° taper half-angle in pulsed operation; (b) Single-lobe horizontal beam patterns of these lasers

4　Conclusion

Carefully designed tapered DFB THz QCLs with surface metallic-stripe gratings have been presented. Single-mode emission with an SMSR of 24dB is observed. In pulsed operation, record output powers of 286mW at 10K and 82mW at 77K are achieved with a well-shaped beam pattern.

Acknowledgment

This work is supported by the National Programs of China (grant nos. 2014CB339803, 2013CB632801, 2011YQ13001802-04, and 61376051).

References

[1] Eichholz, R., Richter, H., Wienold, M., et al.: 'Frequency modulation spectroscopy with a THz quantum-cascade laser', Opt. Express, 2013, 21, pp. 32199-32206, doi: 10.1364/OE.21.032199.

[2] Tonouchi, M.: 'Cutting-edge terahertz technology', Nat. Photonics, 2007, 1, pp. 97-105, doi: 10.1038/nphoton.2007.3.

[3] Lim, Y. L., Dean, P., Nikolic, M., et al.: 'Demonstration of a self-mixing displacement sensor based on terahertz quantum cascade lasers', Appl. Phys. Lett., 2011, 99, pp. 081108-1-081108-3, doi: 10.1063/1.3629991.

[4] Kumar, S.: 'Recent progress in terahertz quantum cascade lasers', IEEE J. Sel. Top. Quantum Electron., 2010, 17, pp. 38-47, doi: 10.1109/JSTQE.2010.2049735.

[5] Ajili, L., Faist, J., Beere, H., et al.: 'Loss-coupled distributed feedback far-infrared quantum cascade lasers', IEEE Electron. Lett., 2005, 41, pp. 419-421, doi: 10.1049/el: 20050128.

[6] Mahler, L., Tredicucci, A., Kohler, R., et al.: 'High-performance operation of single-mode terahertz quantum cascade lasers with metallic gratings', Appl. Phys. Lett., 2005, 87, pp. 181101-1-181101-3, doi: 10.1063/1.2120901.

[7] Chen, J. Y., Liu, J. Q., Liu, F. Q., et al.: 'Distributed feedback terahertz quantum cascade lasers with complex-coupled metallic gratings', IEEE Electron. Lett., 2010, 46, pp. 1340-1341, doi: 10.1049/el.2010.2223.

[8] Walpole, J. N.: 'Semiconductor amplifiers and lasers with tapered gain regions', Opt. Quantum Electron., 1996, 28, pp. 623-645, doi: 10.1007/BF00411298.

[9] Li, L. H., Chen, L., Zhu, J. X., et al.: 'Terahertz quantum cascade lasers with >1W output powers', IEEE Electron. Lett., 2014, 50, pp. 309-311, doi: 10.1049/el.2013.4035.

[10] Li, Y. Y., Liu, J. Q., Wang, T., et al.: 'High power and high efficiency operation of terahertz quantum cascade lasers at 3.3THz', Chin. Phys. Lett., 2015, 32, pp. 104203-1-104203-3, doi: 10.1088/0256-307X/32/10/104203.

Efficacious engineering on charge extraction for realizing highly efficient perovskite solar cells

Shizhong Yue,[1,2] Kong Liu,[1,2] Rui Xu,[3] Meicheng Li,[4]
Muhammad Azam,[1,2] Kuankuan Ren,[1,2] Jun Liu,[1,2] Yang Sun,[1,2]
Zhijie Wang,[1,2] Dawei Cao,[5] Xiaohong Yan,[5] Shengchun Qu,[1,2]
Yong Lei [3] and Zhanguo Wang[1,2]

([1] Key Laboratory of Semiconductor Materials Science, Beijing Key Laboratory of Low Dimensional Semiconductor Materials and Devices, Institute of Semiconductors, Chinese Academy of Sciences, Beijing 100083, China.)

([2] College of Materials Science and Opto-Electronic Technology, University of Chinese Academy of Sciences, Beijing 100049, China)

([3] Institut für Physik & IMN MacroNano(ZIK), Technische Universität Ilmenau, 98693 Ilmenau, Germany.)

([4] State Key Laboratory of Alternate Electrical Power System with Renewable Energy Sources, North China Electric Power University, Beijing 102206, China)

([5] Department of Physics, Faculty of Science, Jiangsu University, Zhenjiang 212013, China)

Abstract Efficient extraction of photogenerated charge carriers is of significance for acquiring a high efficiency for perovskite solar cells. In this paper, a systematic strategy for effectively engineering the charge extraction in inverted structured perovskite solar cells based on methylammonium lead halide perovskite($CH_3NH_3PbI_{3-x}Cl_x$) is presented. Intentionally doping the chlorine element into the perovskite structure is helpful for obtaining a high open circuit voltage. The engineering is carried out by modifying the aluminium cathode with zirconium acetylacetonate, doping the hole transport layer of nickel oxide(NiO_x) with copper and using an advanced fluorine doped tin oxide(FTO) substrate. This improves the bandgap alignment of the whole device, and thus, is of great benefit for extracting the charge carriers by promoting the transport rate and reducing the trap states. Consequently, an optimized power conversion efficiency of 20.5% is realized. Insights into how to extract charge carriers efficiently with a minimum energy loss are discussed.

1 Introduction

Recently, organic-inorganic hybrid perovskite materials have emerged as an impressive

原载于：Energy Environ. Sci. 2017, 10(12): 2570-2578.

model for fabricating highly efficient solar cells. Such a series of materials presents tunable bandgap values, a high absorption coefficient, a long lifetime of charge carriers and high charge mobility.[1-3] An attractive power conversion efficiency (PCE) of over 20% has been reported.[3,4] In the construction of perovskite solar cells (PVSCs), the active material is usually sandwiched by multiple layers to form a typical *P–I–N* configuration. The purposely designed built-in potential across the device is energetically favorable for separating excitons and capturing charge carriers by the electrodes.

The quality of the perovskite film plays a key role in the device performance, and tremendous efforts have been taken in order to acquire a good film for highly efficient photovoltaic devices.[4-6] The film quality has been intensively optimized, and thus, the extent to which the PCE can be further improved by quality optimization of the perovskite film is limited. However, the qualities of the charge transport layers and the electrodes are also of great significance and this topic deserves some extensive research attention.[7] A series of materials has been proposed to construct an effective P-I-N configuration with the perovskite film, including *p*-type semiconductors and polymers (e. g., nickel (II) oxide (NiO) and PTAA) for hole transport layers,[8,9] *n*-type semiconductors (e. g., titanium oxide and zinc oxide) and organic semiconductors {e. g., [6, 6]-phenyl C_{61} butyric acid methyl ester ($PC_{61}BM$)} for electron transport layers,[10-12] organic molecules (poly [(9, 9-bis(3′-(*N*,*N*-dimethylamino)propyl)-2,7-fluorene)-alt-2,7-(9,9-dioctylfluorene)] (PFN), BCP) as the buffer layer to enhance the charge transport to the anode or cathode.[13,14] However, the large varieties of these methodologies and materials seem to lack systematic consideration and a device design principle is needed to further enhance the device outcomes.

In this paper, a systematic strategy is provided for the construction of highly efficient PVSCs from the point of view of effective extraction of photogenerated charge carriers. The engineering is carried out using three steps. Firstly, an efficacious method was chosen to modify the aluminium (Al) cathode using zirconium acetylacetonate (ZrAcac). Such a molecule is beneficial for extracting electrons to the cathode by promoting the electron transport and reducing the corresponding trap states. Then the hole extracting capability of the nickel oxide (NiO_x) layer is optimized using a copper (Cu) dopant. This could improve the bandgap alignment of the device, thus the transport of holes here is obviously boosted with a minimum energy loss. To reduce the conductivity loss of the anode resulting from the annealing procedure of the NiO_x film, a highly conductive fluorine doped tin oxide (FTO) substrate was used. An optimized power conversion efficiency of 20.5% is realized, which could, so far, be the largest value ever obtained for the methylammonium lead halide perovskite ($CH_3NH_3PbI_{3-x}Cl_x$)-based device. The configuration principle for perovskite photovoltaic devices is given in a systematic way.

2　Results and discussion

Fig. 1(a) shows the schematic structure of the perovskite photovoltaic devices with a typical inverted structure. The perovskite film is sandwiched by NiO_x and $PC_{61}BM$ to form a *P-I-N* planar architecture, where the holes and electrons are designed to be transported to the electrodes through NiO_x and $PC_{61}BM$/ZrAcac, respectively. The corresponding cross-sectional scanning electron microscopy(SEM) image of the structure is given in Fig. 1(b), where the clear multi-layered structure can be observed. The thickness of the $CH_3NH_3PbI_{3-x}Cl_x$ is measured as 280nm, which is sufficient for absorbing the sunlight efficiently.

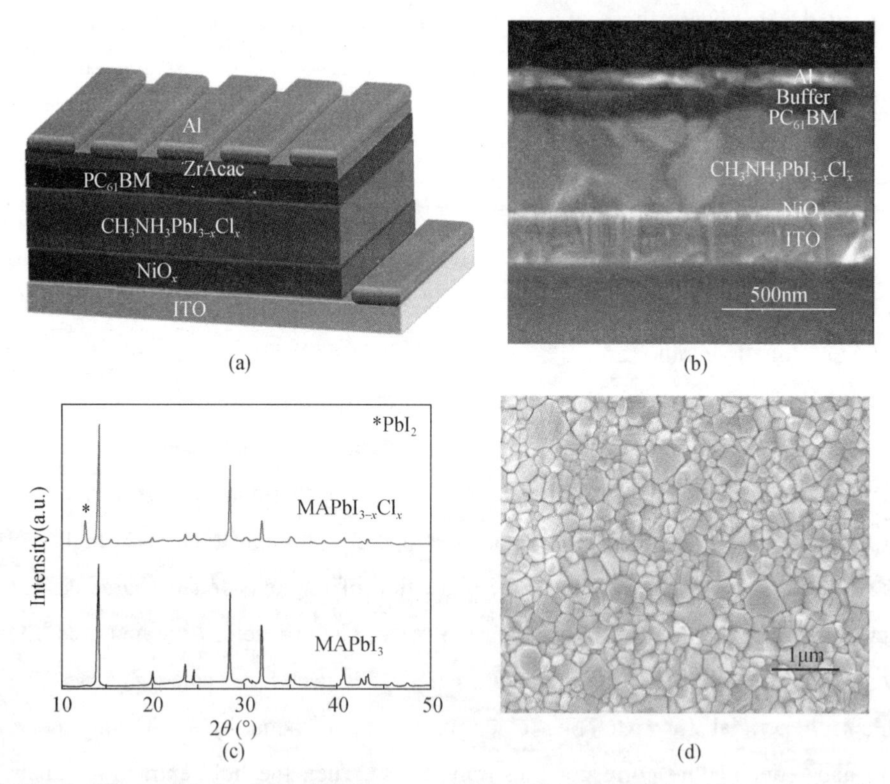

Fig. 1　(a) The schematic structure of the perovskite photovoltaic device. (b) The cross-sectional scanning electron microscopy (SEM) image of the perovskite photovoltaic devices. (c) X-ray diffraction patterns of the different perovskite films on an indium tin oxide(ITO) substrate. (d) Top view SEM image of the perovskite film on ITO/NiO_x

The perovskite film in the device was prepared using a standard one-step method with a subsequent solvent treatment procedure using toluene.[15] Fig. 1(c) displays the X-ray diffraction(XRD) pattern of the as-grown film on the ITO substrate(red curve). The series of diffraction peaks at 14°, 20.5°, 28°, and 41° corresponds to the standard diffraction from (110), (112), (220) and (224) crystal planes of $CH_3NH_3PbI_{3-x}Cl_x$, respectively.[15] Diffraction peaks of lead chloride ($PbCl_2$) were not observable and only a slight peak from lead iodide (PbI_2) can be seen. Considering the absence of PbI_2 in the film without chlorine addition (blue curve), this result indicates that the chlorine is incorporated or doped to the perovskite structure with the iodine element not being consumed well, thus leaving a tiny amount of PbI_2. The tiny amount of PbI_2 has been proven to have a somewhat positive effect on the device performance (e.g., reducing the ionic defect migrations and enhancing the lifetime of the charge carriers).[16-18] The X-ray photoelectron spectrum (XPS) of the material in Fig. S1 (ESI) also demonstrates that the material has an acceptable purity without other elemental contaminants. The stoichiometry of this film was determined using the resolved XPS spectra and the results are shown in ESI (Fig. S1 (a)) and discussed in Section S1 of the ESI. The top view SEM image of the perovskite film is shown in Fig. 1 (d), where the size of the perovskite crystal could be measured to be 100 – 500nm and no obvious pinholes were discerned. In addition, the absorption spectra given in Fig. S3 (ESI) indicate a bandgap value of 1.62eV for the $CH_3NH_3PbI_{3-x}Cl_x$ film, which is 0.04eV larger than the film prepared without chlorine (Cl) doping. The ultraviolet photoelectron spectrum (UPS) of the Cl doped film shown in Fig. S3 (c) (ESI) yields a valence band position of 5.82eV and a Fermi level of 4.70eV. These results suggest that the film already shows potential for use in highly efficient photovoltaic devices from the aspects of morphology and material.

In order to capture the photogenerated electrons in the perovskite film efficiently, a thin layer of ZrAcac was purposely inserted between the $PC_{61}BM$ and Al cathode. The Fourier infrared and Raman spectra of the ZrAcac film (Fig. S4 (a); ESI) show the specific molecular feature of ZrAcac and the corresponding XRD spectrum (Fig. S4 (b); ESI) does not demonstrate the presence of zirconium dioxide (ZrO_2), indicating the stability of the molecule in the device. Fig. 2 (a) shows the transient photoluminescence (PL) curves of structures based on ITO/$CH_3NH_3PbI_{3-x}Cl_x$/$PC_{61}BM$ and ITO/$CH_3NH_3PbI_{3-x}Cl_x$/$PC_{61}BM$/ZrAcac. Overall, the PL signals of the two structures present a fast decay, indicating a fast charge dissociation and collection.[19] It is worth noting that the structure with the buffer layer presents a faster decay than the one without it. Considering that the buffer layer is for modifying the electron transport from $PC_{61}BM$ to Al, this result demonstrates that the presence of ZrAcac could promote the transport of electrons to the cathode by reducing the possibility of recombination. The current density *versus* voltage (J–V) curves of the series of devices with various modifications on electron transport are given in Fig. 2 (b). The corresponding PCE statistical values are given in Fig. S5 (ESI) and table 1.

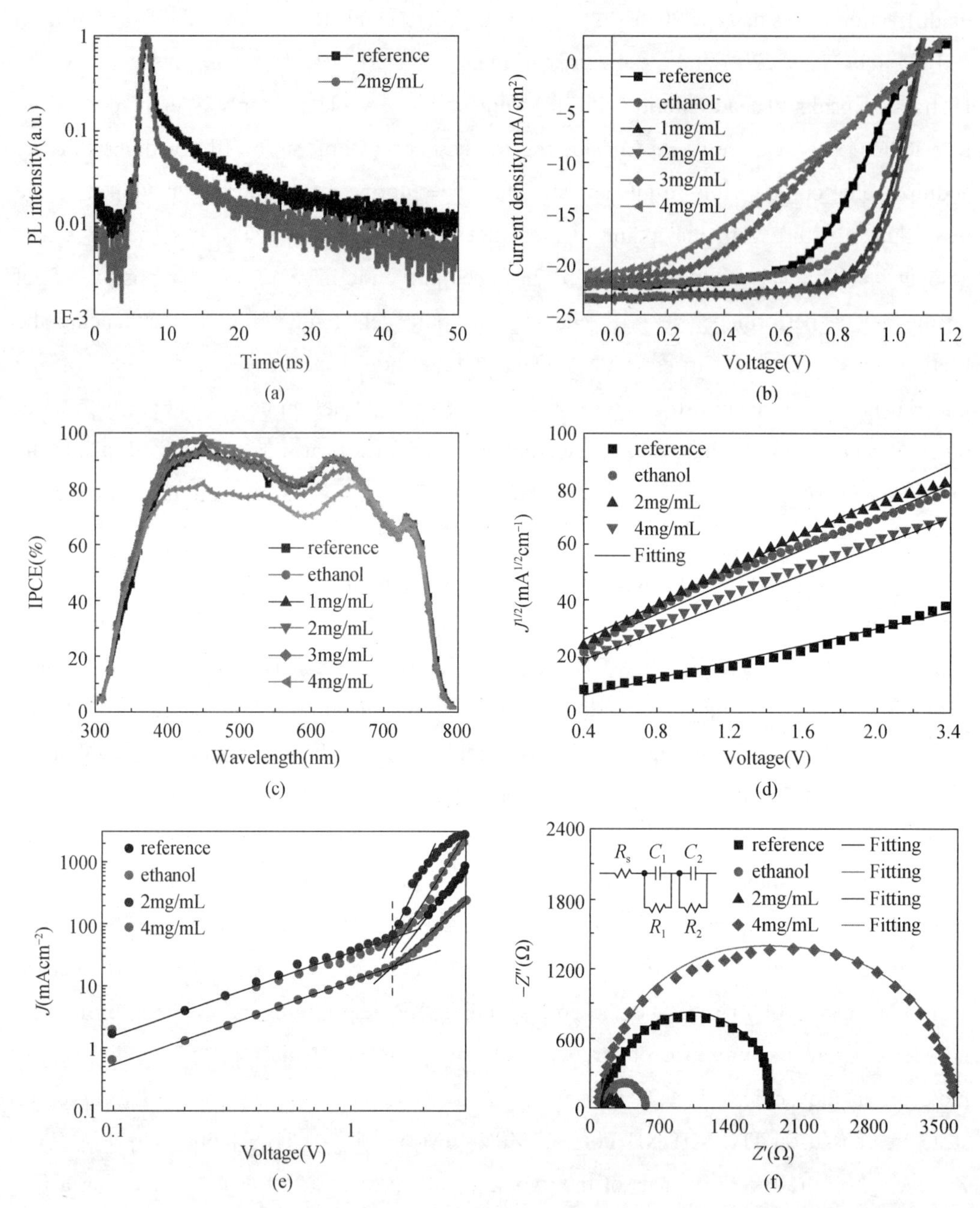

Fig. 2　(a) PL decay curves of the devices with a structure of ITO/$CH_3NH_3PbI_{3-x}Cl_x$/$PC_{61}BM$ (with and without ZrAcac). (b) The curves of current density *versus* voltage (J–V) of devices with different concentrations of stock solution for spin-coating the ZrAcac buffer layer. The curves were measured under standard solar simulated light illumination (AM 1.5). (c) The IPCE spectra of the corresponding devices. (d) $J^{1/2}$–V curves of the electron only devices with the structure of ITO/Al/$PC_{61}BM$ (or modified by ZrAcac)/Al. (e) J–V curves of the electron only devices with the structure of ITO/Al/$CH_3NH_3PbI_{3-x}Cl_x$/$PC_{61}BM$ (or modified by ZrAcac)/Al. (f) Nyquist plots of the devices in (b) measured in the dark at $V \approx V_{oc}$ (the left inset is the equivalent circuit model)

Table 1 The photovoltaic performance of the perovskite solar cells shown in Fig. 2(b)

Samples	J_{sc}(mA/cm^2)	V_{oc}(V)	FF(%)	PCE(%)
References	22.50±0.42	1.07±0.01	55.56±1.32	13.38±0.36
Ethanol	22.34±0.31	1.08±0.01	63.66±1.62	15.29±0.48
1 mg/mL	22.31±1.12	1.08±0.01	70.28±3.39	16.90±0.42
2 mg/mL	23.53±0.32	1.10±0.01	70.04±1.47	18.05±0.58
3 mg/mL	22.56±0.36	1.08±0.01	40.18±1.90	9.76±0.36
4 mg/mL	21.59±1.13	1.08±0.01	35.18±1.10	8.21±0.44

The references device without any modification delivers a short-circuit current density (J_{sc}) of 21.98mA/cm^2, an open-circuit voltage (V_{oc}) of 1.08V, fill factor (FF) of 55.48% and PCE of 13.17%. This result is comparable to previously reported values.[20] To check the modification effect of ZrAcac, the films were spin-coated with different concentrations of the ethanol stock solutions. As shown in Fig. 2(b), with the increase of the concentration from 1 to 2mg/mL, the device shows an increasing in PCE and outperforms the bare device. This indicates that the presence of ZrAcac does indeed boost the device efficiency, but the film could not be too thick because of the insulating nature of ZrAcac. As the film thickness was further increased by enhancing the concentration of the stock solution, the efficiency drops, particularly in relation to FF. These results suggest that the functionality of ZrAcac is similar to that of PFN in polymer solar cells and the positive contribution to the device performance is interface modification to reduce the electron trapping possibilities between $PC_{61}BM$ and Al.[13] It is also worth noting that the device was also fabricated by using an ethanol treatment on the $PC_{61}BM$ film before Al deposition to check the effectiveness of ZrAcac. Although the corresponding device outperforms the references device, this was attributed to the fact that ethanol could boost the crystallinity of the $PC_{61}BM$ film and improve the corresponding electron conductivity,[21, 22] and the device still delivers a lower PCE in comparison with the outperformed device in table 1 with 2mg/mL ZrAcac modification (PCE 18.6%). To obtain more information, the incident photon-to-current efficiency (IPCE) spectra of the series of devices were measured and the results are shown in Fig. 2(c). All the spectra exhibit a broad spectrum of converting photon energy into current in the range of 300–800nm, and this is consistent with the absorption spectrum shown in Fig. S3(a) (ESI). The high values of IPCE indicate a good capability for converting solar energy to electricity for the devices fabricated in this research. The integrated J_{sc} values obtained from the IPCE spectra are shown in Fig. S6 (ESI) and these values approximately agree with those from the J–V curves, although they are slightly lower than the value from J–V curve. The lower value phenomenon results from the fact the J–V curves were measured in a glove box whereas the IPCE spectra were obtained under ambient conditions. Because the perovskite solar cells are sensitive to the humidity in the air, such a slightly lowered integrated photocurrent value is reasonable.

Using the space charge limited current method on the electron only device structure, the electron mobility of the devices with different modifications at the interface of $PC_{61}BM$ and Al was investigated. Fig. 2(d) shows that the $J^{1/2}-V$ curves of the series of electron only devices(ITO/Al/$PC_{61}BM$(or modified $PC_{61}BM$)/Al) fitted with the Mott-Gurney law.[23] The device fabricated with 2mg/mL ZrAcac solution, as expected, presented the largest electron mobility, $4.20\times10^{-4}\ cm^2\ V^{-1}\ s^{-1}$, which is almost 4-fold higher than the device withoutmodification. Either a lower or higher concentration than 2mg/mL would induce a deterioration in electron mobility. This is consistent with the performance parameters of the full devices, indicating that the presence of ZrAcac with an appropriate thickness does indeed facilitate the electron transport of the device. More details on the $J^{1/2}-V$ curves in a large voltage range and on the corresponding fitted curves using the Mott-Gurney law are given in Fig. S7(ESI). In addition, the electron trap-state density of the electron only devices(ITO/Al/$CH_3NH_3PbI_{3-x}Cl_x$/$PC_{61}BM$(or modified $PC_{61}BM$)/Al) were also measured. Fig. 2(e) shows the $J-V$ curves of the corresponding devices. The linear relationship indicates an Ohmic response of the electron only devices at a low bias voltage. When the voltage goes beyond the kink point, the current quickly increases nonlinearly, demonstrating that the trap-states are completely filled. The trap-state density can be calculated using the equation:[24]

$$V_{TFL}=\frac{en_tL^2}{2\varepsilon\varepsilon_0} \tag{1}$$

where V_{TFL} is the trap-filled limit voltage, e is the elementary charge of the electron, L is the perovskite film thickness, ε is the relative dielectric constant of $CH_3NH_3PbI_{3-x}Cl_x$($\varepsilon=28.8$), ε_0 is the vacuum permittivity, and n_t is the trap-state density. The V_{TFL} values in different devices are listed in table S1(ESI). The values for the density of trap states are also included. In comparison with the references device, the modified device presents an obvious decrease in electron trap-state density and this supports the analysis described previously. To directly obtain the charge transport information of the devices, the impedance spectra were measured. As shown in Fig. 2(f), in comparison with the references device, the modified devices present a decline in the diameter of the curves as the concentration of ZrAcac stock solution increases from 0 to 2mg/mL and the diameter increases as the concentration is further increased to 4mg/mL. Considering that the diameter is inversely proportional to the charge transport resistance, this result illustrates that a suitable thickness of ZrAcac is helpful for reducing the charge transport resistance of the device. Using the equivalent circuit as shown in the inset of Fig. 2(f), the detailed parameters for charge transport are given in table S2(ESI).

Next, the hole transport of the device was optimized. Although the use of Cu doped NiO_x thin film has been reported as the hole transport layer in PVSCs,[25, 26] systemically tuning the hole conducting properties of this layer has not received much attention. The doping was performed by mixing a tiny amount of $Cu(ac)_2$ with the $Ni(ac)_2$ precursor for spin-coating the film, before annealing at 340℃ for 1h. Fig. S8 (ESI) shows the XRD patterns of the resulting NiO_x and 5% Cu doped NiO_x films. Both patterns present a series of diffraction peaks at 37°, 43° and 62° corresponding to the standard diffraction from the (111), (200) and (220) crystal planes of NiO,[27] respectively. Synthetic Ni diffraction peaks could also be observed from the two patterns, and this was consistent with a previously reported observation.[28] In addition, the 5% Cu doped film exhibits a series of diffraction peaks from copper(II)oxide(CuO)and copper(I)oxide Cu_2O, although the intensity is low because of the tiny amount used for doping. This indicates the presence of Cu related oxides in the NiO_x film. The XPS spectrum of the 5% Cu doped NiO_x film, as shown in Fig. S9(a) (ESI), also demonstrates the presence of copper oxide (CuO_x) in the NiO_x film. Additionally, resolved Ni $2p$ peaks are shown in Fig. S9 (b) (ESI) and the characteristic features including the multiplet split are in agreement with those found in the literature.[29] As shown in Fig. S9 (c) (ESI), the Cu $2p$ peaks are resolved and peaks at 932.5eV and 952.6eV correspond to the Cu $2p_{3/2}$ and Cu $2p_{1/2}$, respectively, assigned to the Cu^+.[30, 31] The shakeup satellites between 932.5eV and 952.6eV are assigned to the presence of Cu^{2+}.[32] To test the positive contribution of Cu in NiO_x, transient PL signals were measured on the structure of $ITO/NiO_x/CH_3NH_3PbI_{3-x}Cl_x$ and ITO/5% $Cu\text{-}NiO_x/CH_3NH_3PbI_{3-x}Cl_x$. As shown in Fig. 3 (a), the existence of Cu in NiO_x boosts the PL signal decay for the corresponding film, thus indicating a faster hole transport process at the interface of 5% Cu-NiO_x and $CH_3NH_3PbI_{3-x}Cl_x$ in comparison with the counterpart without doping. Using the standard bi-exponential model, the lifetime of the curves was obtained as 6.3ns and 1.4ns, respectively. Fig. S10 (ESI) shows the transmission spectra of the NiO_x film doped with different amounts of Cu. Considering that both Cu_2O and CuO have smaller bandgaps than that of NiO_x, it is expected that the increasing amount of Cu enhances the absorption capability of the corresponding films. The Tauc plots shown in Fig. S11 (ESI) demonstrate that CuO_x is able to cause a redshift of the optical bandgap values of the composite films.

Fig. 3b shows the $J-V$ curves of the devices with the NiO_x films doped with different amounts of Cu. These devices had been already optimized with modifications using 2mg/mL ZrAcac solution and the according PCE statistical values are given in Fig. S12(ESI) and table 2. As the doping amount of Cu was tuned from 0% to 5%, the devices presented a prominent increase in J_{sc}, V_{oc}, FF, and thus, PCE. The related best device in table 2(5%

Cu doped) exhibited a PCE of 18.6%. A further increase of Cu in the film would cause the device performance to deteriorate. The IPCE spectra of these devices are given in Fig. 3(c) and the values exhibit the same varying tendency of the doping amount of Cu in NiO_x as is the case in J_{sc} and PCE (With doping amount of Cu increasing from 0% to 5%, IPCE value increases and further increment of doping amount drops the IPCE value.). The photocurrent values obtained by integrating the spectra are shown in Fig. S13 (ESI) and these values are in approximate agreement with the J_{sc} given by the J–V curves.

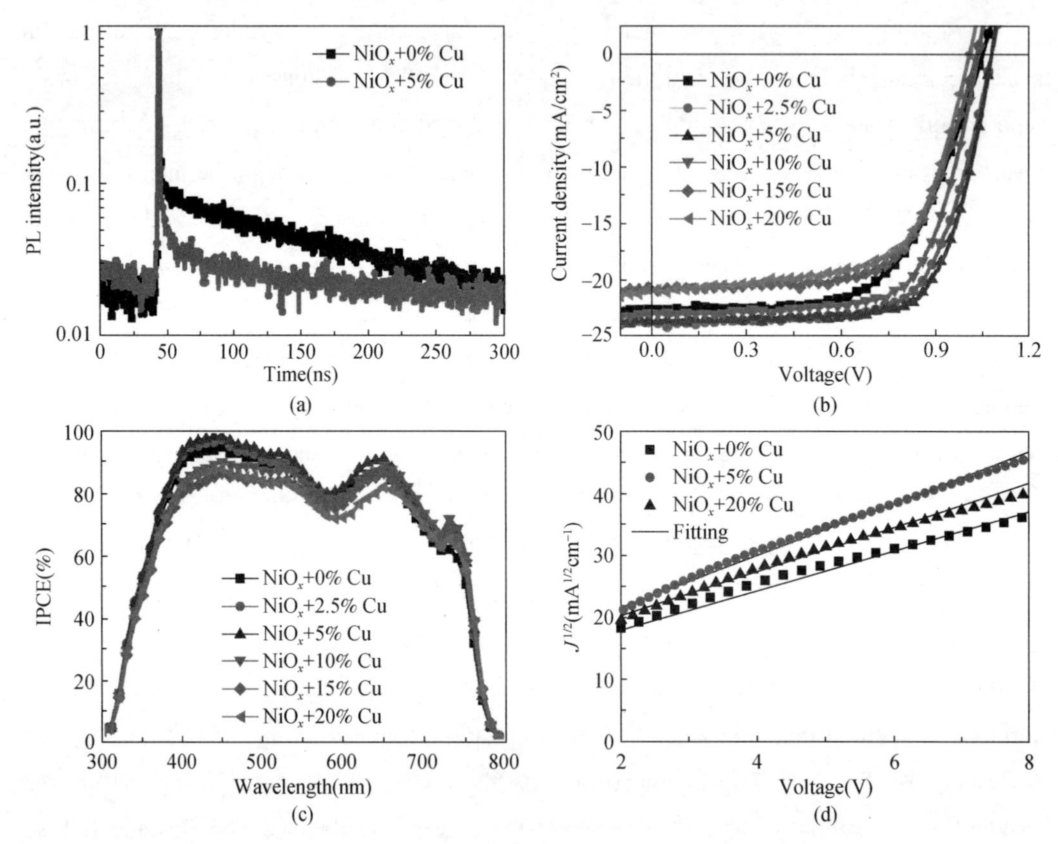

Fig. 3　(a) PL decay curves of the devices with the structure of ITO/NiO_x or 5% Cu doped NiO_x/$CH_3NH_3PbI_{3-x}Cl_x$. (b) The J–V curves of devices with different doping concentrations of Cu in NiO_x film under AM 1.5 light illumination. (c) The IPCE spectra of the corresponding devices. (d) $J^{1/2}$–V curves of the hole-only devices with the structure of ITO/NiO_x(0%, 5% and 20% of Cu doping)/$CH_3NH_3PbI_{3-x}Cl_x$/Au

To further investigate the influence of the Cu doping in the NiO_x film on the hole transport of the devices, hole-only devices based on the structures of ITO/NiO_x(0%, 5% and 20% of Cu doping)/$CH_3NH_3PbI_{3-x}Cl_x$/gold were fabricated. The relevant $J^{1/2}$ versus V curves are given in Fig. 3(d). The curves could be described by the following equation: [1]

$$J_D = \frac{9\mu\varepsilon\varepsilon_0 v^2}{8l^3} \tag{2}$$

Table 2 The photovoltaic performance of the perovskite solar cells shown in Fig. 3(b)

Samples	J_{sc}(mA/cm^2)	V_{oc}(V)	FF(%)	PCE(%)
0% Cu	22.84±0.32	1.06± 0.01	59.68±2.79	14.47±0.83
2.5% Cu	23.18±0.43	1.09±0.00	65.36±1.64	16.55±0.37
5% Cu	23.53±0.32	1.10±0.01	70.04±1.47	18.05±0.58
10% Cu	23.16±0.27	1.05±0.00	66.02±1.16	16.04±0.18
15% Cu	21.24±0.18	1.03±0.01	66.14±2.79	14.41±0.71
20% Cu	20.55±0.40	1.02±0.01	66.16±1.92	13.79±0.14

where J_D is the dark current, μ is the hole mobility, ε is the relative dielectric constant of $CH_3NH_3PbI_{3-x}Cl_x$(ε=28.8),[33] ε_0 is the vacuum permittivity, v is applied voltage and l is the thickness of the active layer. A larger slope of the curve of $J^{1/2}$ versus V definitely results in a higher value of hole mobility. The fitted values are given in table S3(ESI), where it could be concluded that the device with 5% Cu doped NiO_x film exhibits the highest mobility of holes, among the devices used in this research. More detailed $J-V$ curves in a large voltage range are provided in Fig. S14(ESI). In addition, the doping optimized device also presents a lower density of hole-trapping states when compared with other devices(Fig. S15 and table S4; ESI). Fig. S16(ESI) displays the impedance spectra of the devices using Cu doped NiO_x as the hole transport layer and illustrates that the device with 5% Cu doped NiO_x film shows an obvious superiority in the capability of its hole transport(table S5; ESI). To get more information on the positive contribution of Cu doping of NiO_x film, the UPS spectra were measured. As shown in Fig. 4, the Fermi level and valence band positions versus vacuum level could be obtained as 4.59 and 5.43eV, 4.58 and 5.51eV, 4.56 and 5.37eV for the NiO_x, 5% Cu doped NiO_x and 20% Cu doped NiO_x film, respectively. An appropriate amount of Cu doping(e. g., 5%) causes a slightly downward shift of valence band position, and this is helpful for collecting the photogenerated holes in the perovskite film with a low energy loss, because of the reduced distance between the valence band positions of 5% Cu-NiO_x and perovskite materials. Overloading of Cu in the film would make the NiO_x film show some features of CuO or Cu_2O. Because both the Cu related oxides have a valence band position that is above that of NiO_x,[34] it is understandable that the valance band position of the composite would be shifted upward by the overloading of Cu and the capability of the hole conducting of the film is worsened by the increase of energy loss for hole transport at the interface of NiO_x/$CH_3NH_3PbI_{3-x}Cl_x$. In addition, the large doping amount of Cu(e. g. , 20%) would increase the density of defects and hole traps in the NiO_x and the interface (tableS4; ESI). These are detrimental for the device performance, particularly for the V_{oc}, FF and hole mobility.

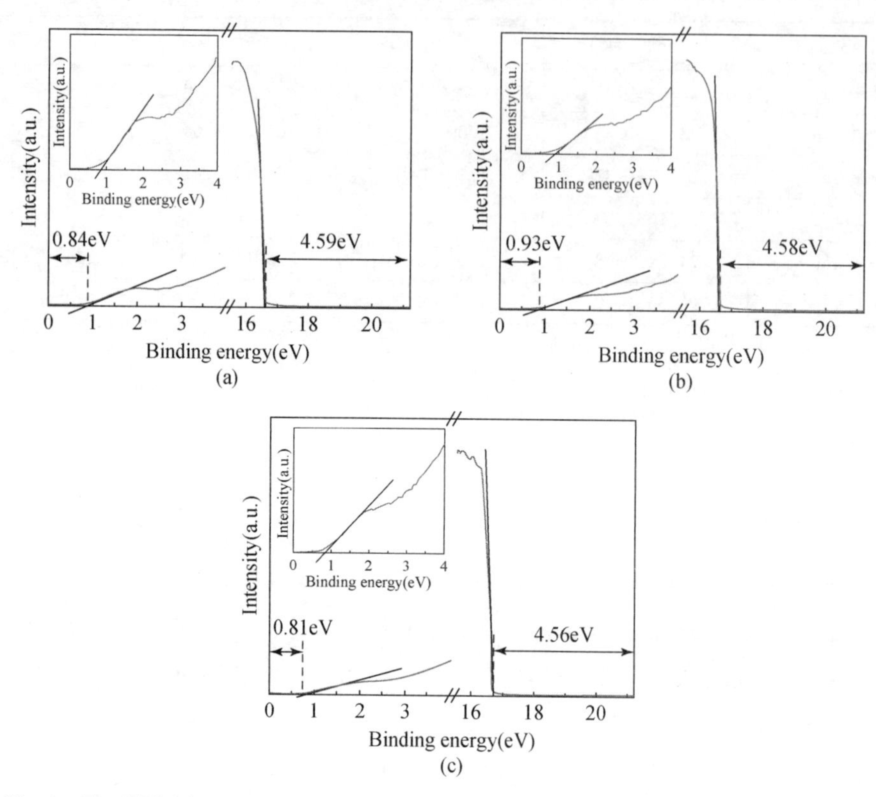

Fig. 4　The UPS of a series of Cu doped NiO_x film, the inset zooms in on the low binding energy region. (a) NiO_x, (b) NiO_x+5% Cu, (c) NiO_x+20% Cu

Table 3　The photovoltaic performance of perovskite solar cells in Fig. 5(b)

Samples	J_{sc}(mA/cm²)	V_{oc}(V)	FF(%)	PCE(%)
FTO	23.07±0.42	1.12±0.01	77.06±0.55	20.14±0.33
ITO	23.53±0.32	1.10±0.01	70.04±1.47	18.05±0.58

Considering that the NiO_x annealing process at 340℃ would cause deterioration in the conductivity of the ITO substrate, a high-conductive FTO glass was used to further improve the PCE of the devices. The advantage of the used FTO substrate is discussed in Section S2 of the ESI. As shown in Fig. 5(a), before annealing the FTO substrates exhibit a smaller resistance than the ITO substrates (FTO resistance: 8.5Ω/sq; ITO resistance: 12.8Ω/sq). After annealing, however, the resistance of the ITO substrate increases sharply from 12.8Ω/sq to 53.4Ω/sq whereas FTO only presents a small change of resistance from 8.5Ω/sq to 10.3 Ω/sq. This result shows the main advantage of using the FTO glass, because the annealing process does not obviously change the transmission capability and work function of both ITO and FTO substrates (Fig. S17; ESI). Fig. 5 (b) shows the $J-V$ curves of the FTO-based devices fabricated by using the optimized conditions of Cu-NiO_x and ZrAcac. Excitingly, the FTO device delivers a much improved PCE, compared with the ITO device, particularly in the aspect of FF. The optimized PCE is about 20.5%, which is almost the highest value obtained for the $CH_3NH_3PbI_{3-x}Cl_x$-based devices. The performance parameters with statistical values are given in table 3. The improvement in FF is mainly attributed to the satisfactory conductivity of the FTO substrates and this was also concluded from the reduction of charge transport resistance of the device as shown by the impedance spectra given in Fig. 5c. The detailed fitting parameters for charge transport are given in table S6 (ESI) and a discussion on all the fitted parameters [shown in Fig. 2 (f), Fig. S16 (ESI) and Fig. 5 (c)] corresponding to the device performance is given in Section S3 of the ESI. The forward and reverse scans for the FTO-based devices are given in Fig. S18 (ESI). In the reverse scan, FF and V_{oc} only

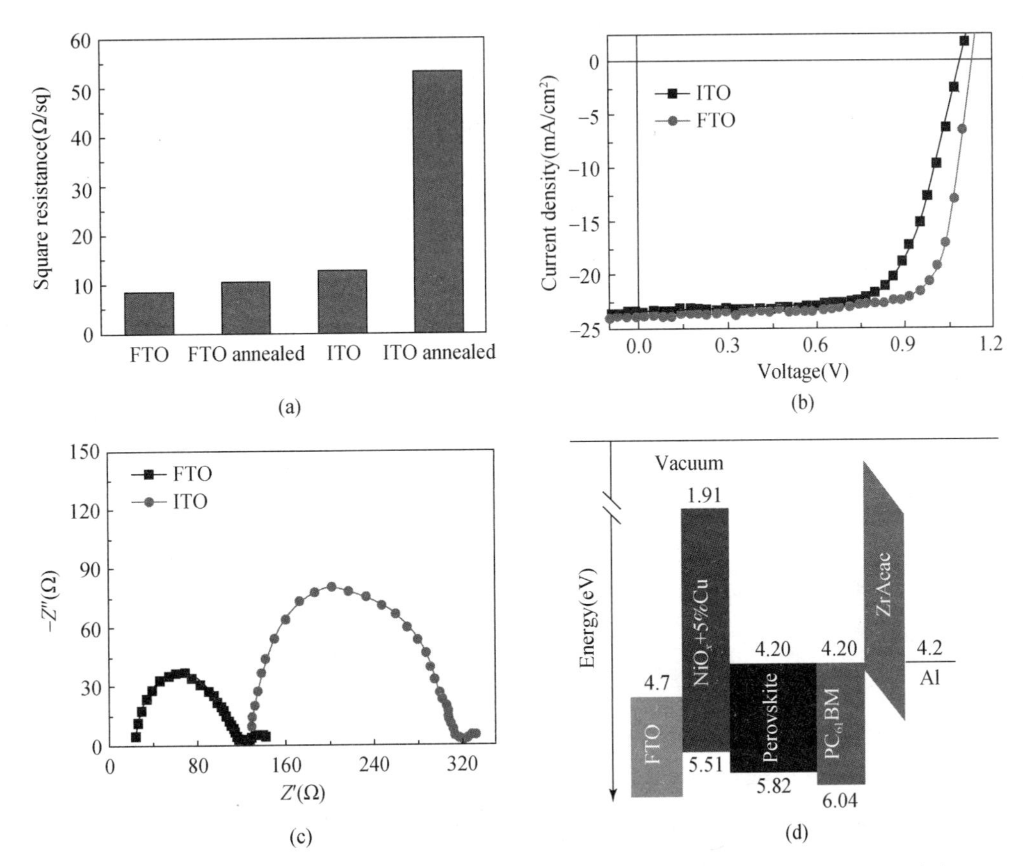

Fig. 5 (a) The variation of square resistance for the substrates before and after annealing. (b) The $J-V$ curves of devices with different substrates under AM 1.5. (c) Nyquist plots of the devices measured in dark at $V \approx V_{oc}$. (d) Schematic drawing of the band energy diagrams for the inverted structured perovskite solar cell

present a slight decrease and the PCE is 20. 1%. Such slight variation indicates that the relevant device shows a good capability against hysteresis.

Collectively, a systematic approach to optimize the performance of perovskite solar cells has been provided. The key factors governing the extraction of the charge carriers are the bandgap alignment across the whole device and the conducting nature of each component. Fig. 5(d) displays the bandgap alignment of the device. The bandgap structures of $CH_3NH_3PbI_{3-x}Cl_x$ and $PC_{61}BM$ were obtained from the relevant UPS spectra and absorption spectra(Fig. S3 and S19; ESI). The Al cathode is for collecting the electrons generated in the active layer. Because it has been systematically reported in a previous paper that lowering the work function of the cathode appropriately is beneficial for the improvement of device performance,[35] it was not focused on in this research. A purposely inserted proper thickness of ZrAcac is beneficial for promoting the electron transport from $PC_{61}BM$ to Al by reducing the corresponding transport resistance and trapping states. The formation of interfacial dipole induced by a ZrAcac thin layer at $PC_{61}BM$/Al is responsible for such an improvement because of the formation of(quasi) Ohmic contact between $PC_{61}BM$ and Al from the molecule modification.[19, 36, 37] The quality of $PC_{61}BM$ as the electron transport layer is also of importance and the corresponding electron conducting property could be ameliorated by improving the crystallinity using ethanol treatment. This is similar to the case in polymer solar cells and is the additional bonus from spin-coating the ZrAcac in ethanol solution.[21] Generally, the improvement in electron transport of the device would directly enhance the FF of the devices. Factors limiting FF are widely discussed as a field dependent competition between non-geminate recombination and charge extraction, with trap states and departure from diode ideality playing important roles.[38] As shown in Fig. S20(ESI), the contact between $PC_{61}BM$/ZrAcac/Al usually represents an efficient electron extraction. The charge transport resistance at the interfaces could be significantly reduced by the presence of ZrAcac of a suitable thickness(as discussed in Section S2; ESI). Thus the enhanced electron extraction and depressed recombination make the tradeoff positive to the FF increase. In addition, J_{sc} is also increased by the improvement of electron transport and reduction of the trap state, attributed to the depressed electron loss in the transport. However, V_{oc} is only improved slightly because of the fact that V_{oc} is mainly determined by the band gap of the perovskite material and the alignment of the energy level.[35, 39] The slightly improved V_{oc} could be attributed to the modification effect of ZrAcac, because it moves the surface potential of Al positively and would be helpful for the Al electrode to collect the electron with a relatively low energy loss(Fig. S21; ESI).

The Cl doped $CH_3NH_3PbI_3$ film shows a relatively larger bandgap than the intrinsic material and the 0.04eV widening in bandgap results in a 0.1V enchantment in V_{oc} (Fig. S22; ESI). This indicates that a large bandgap is energetically favorable for realizing a large V_{oc}.[40] The 0.1V improvement in V_{oc} particularly demonstrates that the Cl-based perovskite material exhibits a superior capability for converting the solar energy to charge carriers and transporting them to the electrodes with a low energy loss. In comparison with the band structure of $CH_3NH_3PbI_3$,[35] both the positions of valence band and conduction band of $CH_3NH_3PbI_{3-x}Cl_x$ present a downwards movement *versus* vacuum energy level. This is favorable for the NiO_x layer to collect the photogenerated holes and $PC_{61}BM$ layer to collect the electrons with low energy loss because of the match of these band energy positions. In addition, with the same device configuration, the Cl-based device presents an improvement in both electron and hole mobilities (Fig. S23; ESI). This is responsible for the remarkable improvement in FF for the Cl doped devices. The J_{sc} of the two devices is similar. Although the improvement in electron and hole mobilities could enhance the J_{sc} of the Cl-based device by reducing the charge recombination, the enlarged band gap of the $CH_3NH_3PbI_{3-x}Cl_x$ would deteriorate the generation of photo-excited charge carriers. More detailed discussion on the Cl incorporation to the perovskite is given in Section S1 of the ESI. For the hole transport layer, for the first time, it has been demonstrated that the proper amount doping of Cu in the NiO_x film is of great benefit in boosting the hole transport to the electrode with a low resistance and trap states, and thus a high transport mobility. This effect results from the reduced distance of the valence band positions between optimized Cu-NiO_x film and the Cl-based perovskite material, which is energetically favorable to reduce the energy loss of the hole transport at the interface. The reduced energy loss is relevant to the 0.05V improvement in V_{oc} in comparison with the device without Cu doping. The enhancement in hole mobility and reduction in hole trap states are responsible for the improvement in FF and J_{sc}, and are similar to the discussion on electron transport. In addition, the thermally stable FTO substrate with a high conductivity and transmission is also of importance in realizing a high PCE. The high conductivity of the substrate is of significance for reducing the recombination of the charge carriers, and is particularly beneficial for improving FF.

3 Conclusions

In summary, a systematic approach was used to optimize the performance of PSCs, using a Cl-doped $CH_3NH_3PbI_3$ active layer, modifying the electron capture using a ZrAcac buffer layer, boosting the hole transport by doping the NiO_x film with Cu and adopting the use of a FTO substrate with a high thermal stability and conductivity. Consequently, a PCE about 20.5% has been realized and such a value is almost the highest one for the $CH_3NH_3PbI_{3-x}Cl_x$-based photovoltaic devices. Insights on how to manage the extraction of charge carriers efficiently are illustrated using multiple techniques and guidelines on how to construct highly efficient perovskite solar cells are given in this paper.

4 Experimental

4.1 Materials and solution preparation

Methylammonium iodide (MAI; >99.5%) and PbI_2 (>99.99%) were purchased from Xi'an Polymer Light Technology Corp. $PC_{61}BM$ (>99.9%) and anhydrous *N*, *N*-dimethylformamide (DMF, 99.8%) were purchased from Sigma-Aldrich. Nickel acetate tetrahydrate ($Ni(ac)_2$; $Ni(CH_3COO)_2 \cdot 4H_2O$) and cupric acetate monohydrate ($Cu(ac)_2$; $Cu(CH_3COO)_2 \cdot H_2O$) were obtained from Sinopharm Chemical Reagent Co., Ltd. Zirconium (IV) acetylacetonate (ZrAcac; $ZrC_{20}H_{28}O_8$) was provided by Alfa Aesar. All the solvents were purchased from the Beijing Chemical Works. Unless otherwise stated, all the chemicals were used as received.

MAI, PbI_2 and $PbCl_2$ were mixed in the ratio of (1 : 0.9 : 0.1) in anhydrous DMF. The concentration of the perovskite precursor was 1.2M. The mixture was then stirred overnight at 60℃ under a nitrogen atmosphere. The final solution of the perovskite precursor was stored in a glove box, and filtered using 0.45μm PTFE filters before use. $Ni(ac)_2$ and $Cu(ac)_2$ were dissolved in 0.1M anhydrous ethanol. Cu doped NiO_x films were prepared by varying the volume ratio of the solution (Ni : Cu, 100 : 0, 97.5 : 2.5, 95 : 5, 90 : 10, 85 : 15, 80 : 20). ZrAcac was dissolved in anhydrous ethanol at a series of concentrations from 0 to 4mg/mL.

4.2 Device fabrication

Devices were fabricated on prepatterned ITO (FTO) substrates cleaned using ultrasonication in deionized water, acetone, and isopropanol for 20min each and then dried under a nitrogen flow. The mixed NiO_x solution was spin-coated onto the ITO (FTO) surface at 3000rpm for 40s and then annealed at 340℃ for 60min in an ambient atmosphere. To deposit perovskite films, these substrates were transferred into an inert glove box under a nitrogen atmosphere (H_2O and O_2 <1ppm). The perovskite solution was dropped onto the center of the substrates. Then substrates were spun at 6000rpm for 30s, and after 5s anhydrous toluene was quickly dropped onto the center of substrate. The instant color change of the films from yellow to brown was observed upon dropping toluene solvent. The resulting dark brown films were dried at 100℃ for 10min. Next, $PC_{61}BM$ (20mg/mL) in chlorobenzene solution was coated onto the perovskite layer at 1000rpm for 60s. The next step was spin-coating the ZrAcac and the solution was coated onto the $PC_{61}BM$ layer at 2000rpm for 60s. Finally, 100nm of Al was deposited using thermal evaporation under a high vacuum.

4.3 Characterization

The absorbance spectrum was measured on a TU-1950 ultravioletvisible spectrophotometer (Persee, Beijing) in transmission mode. UPS data were obtained using an Axis Ultra DLD (Kratos Analytical). XPS data were obtained using an ESCALAB 250Xi (Thermo Fisher Scientific). The cross-sectional SEM image of the device structure was measured using a JSM-7401F (Jeol). The current density-voltage ($J-V$) characteristics were obtained using a 2450 source measure unit (Keithley) under AM1.5 illumination with an intensity of 100mW/cm. IPCE measurements were performed using a QEPVSI-b Measurement System (Newport Corporation). Impedance spectroscopy measurement was performed using a CHI660E electrochemical workstation (CH Instruments) under dark conditions, with an oscillating voltage of 10mV and frequency scanning range of 100Hz to 1M Hz. The transient PL measurements were performed using a F900 time-correlated single photon counting system (Edinburgh Instruments) and the samples were excited using a pulsed laser with the wavelength of 485nm. The XRD spectra were measured using a V2500 (Rigaku) X-ray diffractometer.

5 Conflicts of interest

There are no conflicts to declare.

Acknowledgements

This work was mostly supported by the National Basic Research Program of China (Grant No. 2014CB643503), the National Key Research and Development Program of China (Grant No. 2017YFA0206600), the Key Research Program of Frontier Science, Chinese Academy of Sciences (Grant No. QYZDB-SSW-SLH006), the National Natural Science Foundation of China (Contract No. 61674141, 61504134 and 21503209), the Beijing Natural Science Foundation (2162042), the European Research Council (ThreeDsurface, 240144 and HiNaPc, 737616), the German Federal Ministry of Education and Research (BMBF, ZIK-3DNanoDevice, 03Z1MN11), and the German Research

Foundation(DFG, LE 2249_4-1). ZW appreciates the support from Hundred Talents Program(Chinese Academy of Sciences).

References

[1] Q. Dong, Y. Fang, Y. Shao, P. Mulligan, J. Qiu, L. Cao and J. Huang, Science, 2015, 347, 967-970.

[2] J. P. Correa-Baena, A. Abate, M. Saliba, W. Tress, T. J. Jacobsson, M. Gratzel and A. Hagfeldt, Energy Environ. Sci., 2017, 10, 710-727.

[3] M. Saliba, T. Matsui, J. Y. Seo, K. Domanski, J. P. Correa-Baena, M. K. Nazeeruddin, S. M. Zakeeruddin, W. Tress, A. Abate, A. Hagfeldt and M. Gratzel, Energy Environ. Sci., 2016, 9, 1989-1997.

[4] H. Tan, A. Jain, O. Voznyy, X. Lan, F. P. Garcia de Arquer, J. Z. Fan, R. Quintero-Bermudez, M. Yuan, B. Zhang, Y. Zhao, F. Fan, P. Li, L. N. Quan, Y. Zhao, Z. H. Lu, Z. Yang, S. Hoogland and E. H. Sargent, Science, 2017, 355, 722-726.

[5] W. S. Yang, J. H. Noh, N. J. Jeon, Y. C. Kim, S. Ryu, J. Seo and S. I. Seok, Science, 2015, 348, 1234-1237.

[6] W. Nie, H. Tsai, R. Asadpour, J. C. Blancon, A. J. Neukirch, G. Gupta, J. J. Crochet, M. Chhowalla, S. Tretiak, M. A. Alam, H. L. Wang and A. D. Mohite, Science, 2015, 347, 522-525.

[7] H. L. Kim, K. G. Lee and T. W. Lee, Energy Environ. Sci., 2016, 9, 12-30.

[8] Y. Shao, Y. Yuan and J. Huang, Nat. Energy, 2016, 1, 15001.

[9] Y. Li, S. Ye, W. Sun, W. Yan, Y. Li, Z. Bian, Z. Liu, S. Wang and C. Huang, J. Mater. Chem. A, 2015, 3, 18389-18394.

[10] D. Liu and T. L. Kelly, Nat. Photonics, 2014, 8, 133-138.

[11] M. M. Lee, J. Teuscher, T. Miyasaka, T. N. Murakami and H. J. Snaith, Science, 2012, 338, 643-647.

[12] P. W. Liang, C. Y. Liao, C. C. Chueh, F. Zuo, S. T. Williams, X. K. Xin, J. Lin and A. K. Y. Jen, Adv. Mater., 2014, 26, 3748-3754.

[13] Z. He, C. Zhong, S. Su, M. Xu, H. Wu and Y. Cao, Nat. Photonics, 2012, 6, 591-595.

[14] J. Y. Jeng, Y. F. Chiang, M. H. Lee, S. R. Peng, T. -F. Guo, P. Chen and T. C. Wen, Adv. Mater., 2013, 25, 3727-3732.

[15] N. J. Jeon, J. H. Noh, Y. C. Kim, W. S. Yang, S. Ryu and S. Il Seol, Nat. Mater., 2014, 13, 897-903.

[16] Q. Jiang, L. Zhang, H. Wang, X. Yang, J. Meng, H. Liu, Z. Yin, J. Wu, X. Zhang and J. You, Nat. Energy, 2016, 1, 16177.

[17] F. Jiang, Y. Rong, H. Liu, T. Liu, L. Mao, W. Meng, F. Qin, Y. Jiang, B. Luo, S. Xiong, J. Tong, Y. Liu, Z. Li, H. Han and Y. Zhou, Adv. Funct. Mater., 2016, 26, 8119-8127.

[18] Y. C. Kim, N. J. Jeon, J. H. Noh, W. S. Yang, J. Seo, J. S. Yun, A. Ho-Baillie, S. Huang, M. A. Green, J. Seidel, T. K. Ahn and S. I. Seok, Adv. Energy Mater., 2016, 6, 1502104.

[19] W. B. Yan, S. Y. Ye, Y. L. Li, W. H. Sun, H. X. Rao, Z. W. Liu, Z. Q. Bian and C. Huang, Adv. Energy Mater., 2016, 6, 1600474.

[20] J. W. Li, Q. S. Dong, N. Li and L. D. Wang, Adv. Energy Mater., 2017, 7, 1602922.

[21] S. Lu, K. Liu, D. Chi, S. Yue, Y. Li, Y. Kou, X. Lin, Z. Wang, S. Qu and Z. Wang, J. Power Sources, 2015, 300, 238-244.

[22] S. Guo, B. Cao, W. Wang, J. F. Moulin and P. Muller-Buschbaum, ACS Appl. Mater. Interfaces, 2015, 7, 4641-4649.

[23] Q. Dong, Y. Fang, Y. Shao, P. Mulligan, J. Qiu, L. Cao and J. Huang, Science, 2015, 347, 967-970.

[24] D. Yang, R. Yang, X. Ren, X. Zhu, Z. Yang, C. Li and S. F. Liu, Adv. Mater., 2016, 28, 5206-5213.

[25] J. H. Kim, P. W. Liang, S. T. Williams, N. Cho, C. C. Chueh, M. S. Glaz, D. S. Ginger and A. K. Jen, Adv. Mater., 2015, 27, 695-701.

[26] J. W. Jung, C. C. Chueh and A. K. Jen, Adv. Mater., 2015, 27, 7874-7880.

[27] H. Yan, D. Zhang, J. Xu, Y. Lu, Y. Liu, K. Qiu, Y. Zhang and Y. Luo, Nanoscale Res. Lett., 2014, 9, 424.

[28] M. A. Abbasi, Z. H. Ibupoto, M. Hussain, Y. Khan, A. Khan, O. Nur and M. Willander, Sensors, 2012, 12, 15424-15437.

[29] C. Guan, Y. Wang, Y. Hu, J. Liu, K. H. Ho, W. Zhao, Z. Fan, Z. Shen, H. Zhang and J. Wang, J. Mater. Chem. A, 2015, 3, 23283-23288.

[30] P. Wang, Y. H. Ng and R. Amal, Nanoscale, 2013, 5, 2952-2958.

[31] A. A. Dubale, W. N. Su, A. G. Tamirat, C. J. Pan, B. A. Aragaw, H. M. Chen, C. H. Chen and B. J. Hwang, J. Mater. Chem. A, 2014, 2, 18383-18397.

[32] Z. Zhang and P. Wang, J. Mater. Chem., 2012, 22, 2456-2464.

[33] D. Yang, X. Zhou, R. X. Yang, Z. Yang, W. Yu, X. L. Wang, C. Li, S. Z. Liu and R. P. H. Chang, Energy Environ. Sci., 2016, 9, 3071-3078.

[34] M. T. Greiner, M. G. Helander, W. M. Tang, Z. B. Wang, J. Qiu and Z. H. Lu, Nat. Mater., 2011, 11, 76-81.

[35] S. Yue, S. Lu, K. Ren, K. Liu, M. Azam, D. Cao, Z. Wang, Y. Lei, S. Qu and Z. Wang, Small, 2017, 13, 1700007.

[36] S. Y. Chang, Y. C. Lin, P. Sun, Y. T. Hsieh, L. Meng, S. H. Bae, Y. W. Su, W. Huang, C. Zhu, G. Li, K. H. Wei and Y. Yang, Solar RRL, 2017, 1, 1700139.

[37] W. Chen, L. Xu, X. Feng, J. Jie and Z. He, Adv. Mater., 2017, 29, 1603923.

[38] D. Bartesaghi, C. Perez Idel, J. Kniepert, S. Roland, M. Turbiez, D. Neher and L. J. Koster, Nat. Commun., 2015, 6, 7083.

[39] A. Guerrero, L. F. Marchesi, P. P. Boix, S. Ruiz-Raga, T. Ripolles-Sanchis, G. Garcia-Belmonte and J. Bisquert, ACS Nano, 2012, 6, 3453-3460.

[40] Q. Xue, Y. Bai, M. Liu, R. Xia, Z. Hu, Z. Chen, X. F. Jiang, F. Huang, S. Yang, Y. Matsuo, H. L. Yip and Y. Cao, Adv. Energy Mater., 2017, 7, 1602333.

Room temperature continuous wave quantum dot cascade laser emitting at 7.2μm

Ning Zhuo,[1,3] Jin-Chuan Zhang,[1,3] Feng-Jiao Wang,[1,2,3] Ying-Hui Liu,[1,2,3] Shen-Qiang Zhai,[1,3] Yue Zhao,[1,2,3] Dong-Bo Wang,[1,2,3] Zhi-wei Jia,[1,2,3] Yu-Hong Zhou,[1,2,3] Li-Jun Wang,[1,2,3] Jun-Qi Liu,[1,2,3] Shu-man Liu,[1,2,3] Feng-Qi Liu,[1,2,3] Zhan-Guo Wang,[1,3] Jacob B. Khurgin,[4] And Greg Sun[5]

([1]Key Laboratory of Semiconductor Materials Science, Institute of Semiconductors, Chinese Academy of Sciences, Beijing 100083, China)

([2]College of Materials Science and Opto-Electronic Technology, University of Chinese Academy of Sciences, Beijing 10083, China)

([3]Beijing Key Laboratory of Low Dimensional Semiconductor Materials and Devices, Beijing 100083, China)

([4]Department of Electrical and Computer Engineering, Johns Hopkins University, Baltimore, MD 21218, USA)

([5]Department of Engineering, University of Massachusetts at Boston, Boston, MA 02125, USA)

Abstract We demonstrate a quantum cascade laser with active regions consisting of InAs quantum dots deposited on GaAs buffer layers that are embedded in InGaAs wells confined by InAlAs barriers. Continuous wave room temperature lasing at the wavelength of 7.2μm has been demonstrated with the threshold current density as low as 1.89kA/cm^2, while in pulsed operational mode lasing at temperatures as high as 110℃ had been observed. A phenomenological theory explaining the improved performance due to weak localization of states had been formulated.

1 Introduction

Over the past two decades, quantum cascade lasers (QCLs) have been constantly improved in their performance and at this point have matured into the preferred choice of coherent sources in the mid-infrared (mid-IR) spectral region for a wide range of applications[1-10]. In recent years, it appears the rate of progress in their continuous improvement has stalled and their

原载于: Optics Express, 2017, 25(12): 13807-13815.

performance indicators have plateaued that left foremost the even lower threshold currents and higher wall-plug efficiencies to be desired[11-13]. The main attribution to this lack of continued progress has long been identified as the extremely short non-radiative lifetimes commonly associated with the intersubband transitions in the quantum wells(QWs) that form the basis for QCLs. While optimization of subband design and QCL fabrication has offered some remedy to the situation, effectiveness in slowing down the fast intersubband transition rate is intrinsically limited by the strong phonon scattering between subbands with parabolic in-plane dispersions. Proposals have been put forward to drastically reduce the optical phonon scattering by removing the subband in-plane dispersion with the use of quantum dots(QDs) in place of QWs[14-16]. The fundamental idea behind this is the three-dimensional (3D) confinement of QDs that gives arise to truly discrete energy states without in-plane dispersion and therefore energy conservation required for such scattering to take place cannot be satisfied for as long as these discrete levels are not separated with energy that is equal to that of an optical phonon. The proposed schemes of QD-based QCLs all required nicely patterned QDs of equal size repeated over a large number of periods which has proven to be technically impossible with even the most advanced material growth methods. Nevertheless, efforts have been made by many groups to study its feasibility[17-22]. Results published so far are mostly based on self-assembled InAs QDs and present a challenge in bringing these devices to lase and most demonstrated electroluminescence(EL). First of all, there are significant technical difficulties in growing nicely shaped InAs QD layers in InGaAs/InAlAs-the material system of choice for mid-IR QCLs, and repeating them consistently over many periods to construct the needed cascade structure. This is because the lower level of strain induced by the smaller lattice-mismatch in this material combination tends to produce quantum dashes other than dots when grown in the Stranski-Krastanov mode[21-23]. Second, QDs and their barriers are designed so that at least one of the laser states is confined inside the QDs to take advantage of their nondispersive discrete levels. The challenge presented here is the difficulty in either efficiently injecting electrons into the upper laser state or extracting them from the lower state because the 3D confined QD states have little overlap with the extended wavefunctions of the chirped superlattice transport region.

2 Laser design, fabrication and measurements

Recently, we have demonstrated a successful QD cascade laser (QDCL) with the active regions consisting of InAs QDs embedded in InGaAs QWs with GaAs and InAlAs barriers[24]. Lasing from such a device is credited with a combination of quality QDs embedded in active

QWs and localization of lower laser and extraction subbands. This device with a cavity length of 3mm has produced pulsed lasing at the wavelength of 6.15μm up to 250K with threshold current density of 5.36kA/cm^2 and a characteristic temperature of about 400K. Since then we have re-engineered the QDCL structure and improved the assembly of QDs and the result is a much improved device that delivered room temperature continuous-wave (CW) operation at the wavelength of 7.2μm. The threshold current density has been reduced to 1.34kA/cm^2 under pulsed operation at room temperature-four times of magnitude reduction, even with a shorter cavity length of 2mm now adopted that increased mirror loss. The key to this improvement is the insertion of a thin GaAs buffer layer on top of the InAlAs barrier upon which more uniform InAs QDs are self-assembled with precise control and tunability. In this work we present the results of this latest QDCL that we have fabricated and characterized and explain the mechanism with which the QDs contribute to its improved performance.

InAs QDs grown on InP-based InGaAs/InAlAs material system tend to be quantum dashes due to the lower strain induced by the underlying buffer material. While InAs QDs can be readily grown on GaAs/AlGaAs because of their larger lattice mismatch, injection and extraction of electrons in and out of QDs prove to be problematic in this material system. In our first QDCL[24], we used a thin GaAs capping layer on top of the InAs QDs grown on tensile strained $In_{0.44}Al_{0.56}As$ to increase the lattice mismatch between InAs and its top and bottom layers, and this technique indeed produced QDs instead of dashes. While lasing was achieved, its performance was not on par with those of mid-IR QCLs based on QWs[25], primarily due to the extreme broadening of electroluminescence spectrum as a result of the size inhomogeneity of QDs compared with that of QW-based QCLs. In our new design one important modification is the insertion of a thin GaAs buffer layer on top of $In_{0.44}Al_{0.56}As$ right before the deposition of InAs to take advantage of the large lattice mismatch between InAs and GaAs and alloying effect originating from different buffer layer composition[23] compared with the previous design for the formation of self-assembled InAs QDs, and the result is the dramatically improved size uniformity of QDs. The three QD layers in each period are obtained by depositing between 2.5 to 3.5 monolayers (MLs) of InAs on the thin GaAs buffer layers (0.5–0.65nm), all are above the critical thickness. The thicknesses of remaining QW and barriers are such that the QDCL structure is strain compensated and its injection and extraction are properly lined up with the respective upper and lower laser states. Fig. 1(a) illustrates the active region consisting of three layers of coupled InAs QDs in the QDCL with progressively varying QD layer thicknesses allowing for their proper energy alignment under electrical bias.

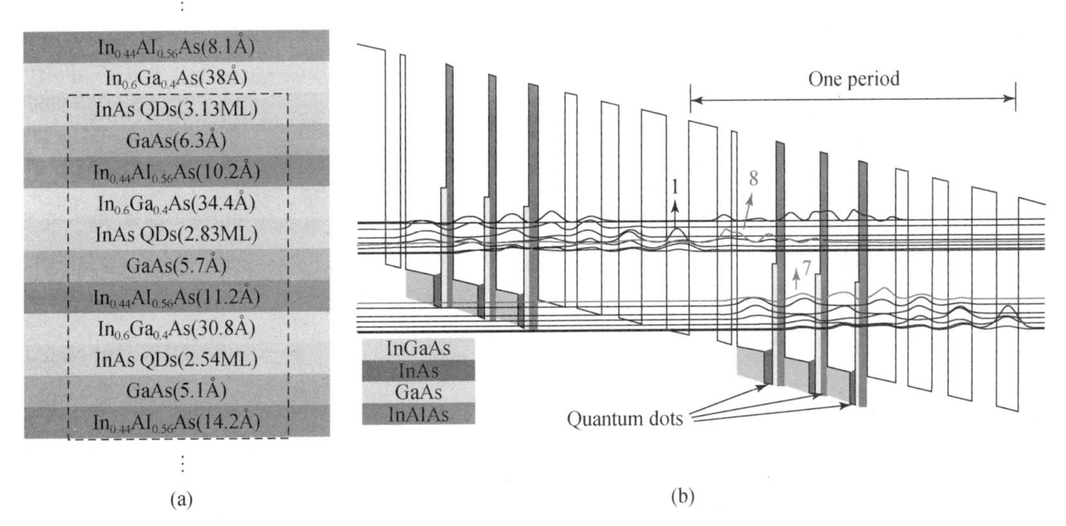

Fig. 1 (a) QDCL active region layer structure. (b) Calculated conduction band diagram and subband wavefunctions using 1D model under electric bias

The QDCL structure was grown by molecular beam epitaxy (MBE) combined with metalorganic chemical vapor deposition (MOCVD). The epitaxial layer sequence starting from the n-doped InP substrate was as follows: 4μm InP cladding layer (Si, 2.2e16cm^{-3}), 0.3μm-thick-$In_{0.53}Ga_{0.47}As$ layer (Si, 4e16cm^{-3}), 35 QDCL stages, 0.3μm thick n-$In_{0.53}Ga_{0.47}As$ layer (Si, 4e16cm^{-3}), 3μm upper cladding (Si, 2.6e16cm^{-3}), and 0.75μm cap cladding (Si, 1e19cm^{-3}). The QDCL operating mechanism is based on a bound-to-continuum design as shown in Fig. 1 (b) in which the lasing transition takes place from the upper laser state (subband 8 in red) to the lower state (subband 7 in green). The short injectors and closely coupled injector levels, characteristics of this new design, have led to the increase of the dynamic range of the alignment voltage that can be applied to the QDCL relative to that of the former bound-to-continuum design reported in[24]. The layer sequence, with four material compositions, starting from the injection barrier is as follows (in angstroms, and InAs in ML): **40.6**/18.3/**8.1**/*38/3.13ML(InAs)*/**6.3**/**10.2**/34.4/*2.83ML(InAs)*/**5.7**/**11.2**/30.8/*2.54ML (InAs)*/**5.1**/**14.2**/36.5/**17.3**/33.5/**24.4**/31.5/34.5/29.4, with $In_{0.44}Al_{0.56}As$ in bold, $In_{0.6}Ga_{0.4}As$ in regular, GaAs in bold and italic, and InAs QD layer in italic, and underlined layers are doped with Si at 1.5e17cm^{-3}. Only InP was grown by MOCVD, and the growth parameters for QDs is given in [24].

The wafer was processed into double-channel ridge waveguides using conventional photolithography and wet chemical etching. Fabrication details are identical to [26]. The average core width is 13μm, and the waveguides were cleaved into 2mm long bars. The laser spectral measurements were carried out using a standard Fourier transform infrared

(FTIR) spectrometer (Bruker Equinox55, Bruker Corporation, Billerica, MA, USA). The emitted optical power from laser was measured with a calibrated thermopile detector placed directly in front of the laser facet. For comparison, we have also grown and subsequently fabricated a regular QCL without the InAs QDs and the underlying GaAs buffer layers but with an increased number of cascading periods of 50.

3 Experimental results and analysis

The transmission electron microscopy (TEM) image in Fig. 2 shows the QDs embedded in the QDCL structure. Dark regions correspond to sites of atoms with higher atomic number (indium in this case). The image clearly reveals that there are three QD layers in a period that are nicely stacked along the growth direction, making them perpendicularly coupled and that these QDs are uncoupled horizontally with average in-plane separation of about 35nm. The atomic force microscopy (AFM) image of the last layer of InAs QDs in the QDCL (indicated by the dashed rectangle in Fig. 1 (a)) is shown as the inset in Fig. 2 with the dimension of 2.5μm × 2.5μm. These QDs are of elliptical shape with a short axis of ~45nm, long axis of ~90nm and height of ~3nm, and slightly elongated along the $[1\bar{1}0]$ direction. The accumulation of strain around the QD sites has resulted in some interface undulation in this QDCL sample. This can potentially be improved in future with the finer control of QD sizes as well as the thicknesses and compositions of the InGaAs cladding layers as we have achieved in a single QD layer sample but not presented here.

Fig. 2 TEM image of a portion of the cleaved cross section of a QDCL active region (Inset: AFM image of the last InAs QD layer in the QDCL)

Using the Bruker Equinox 55 FTIR spectrometer, we measured at room temperature the spontaneous and stimulated emission spectra from both QDCL and QCL samples as shown in Fig. 3 inset. While the small blue shift of the QDCL emission relative to that of the QCL can be accounted for by the presence of QDs, the spontaneous FWHM of QDCL is just a little bit bigger than that of QCL, indicating a high level of QD uniformity. Fig. 3 shows the *P–I–V* curves of QDCL and QCL operating in CW mode around room temperature. At 20℃, the CW threshold current is 490mA for a 2mm long laser, corresponding to 1.89kA/cm^2, with a slope efficiency of about 1.1W/A, from which we can deduce its internal quantum efficiency of about 55%. All these indicators are comparable with the QCL sample. Meanwhile, in pulsed mode operation at 25℃ the slope efficiency is 1.34W/A and 1.76W/A for QDCL and conventional QCL respectively, and considering more periods of active region in the later(50 vs 35) this result indicates that the QDCL delivers higher efficiency owing to its better electron extraction from the lower states because of the hybridization of the QD states with the extended QW states described later.

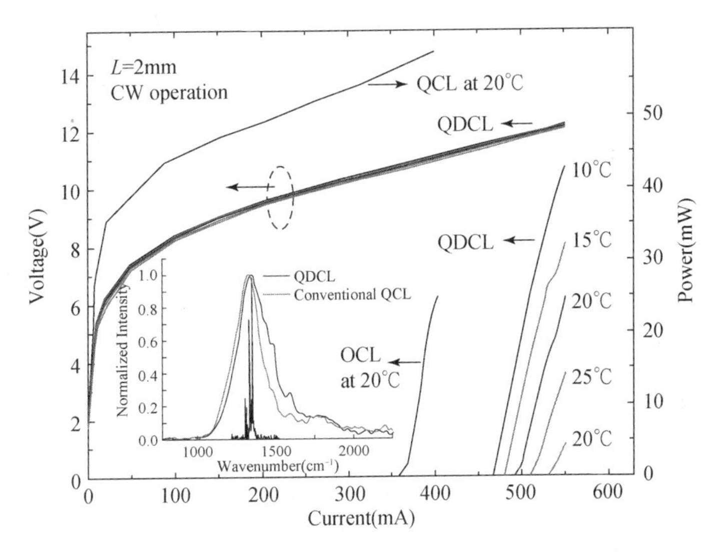

Fig. 3 *P–I–V* curves of QDCL and QCL operating in CW mode around room temperature. Inset: Spontaneous and stimulated emission spectra from both QDCL and QCL samples at room temperature

Fig. 4 plots the temperature dependence of the threshold current for both QDCL and QCL when driven in pulsed mode with a pulsed current source (PCX-7410). The QDCL lased slightly above 110℃, while the QCL stopped working at 90℃. At 283K QDCL and QCL have the threshold density of 1.31kA/cm^2 and 0.96kA/cm^2 respectively, and considering roughly 30% reduction in the optical confinement for QDCL due to the smaller number of periods in the QDCL compared with QCL, the threshold current density of QDCL is on par with that of QCL. Within the temperature range of 283-363K (Fig. 4(a)), T_0 of QDCL reaches 265K, higher than 181K of the QCL. In the lower temperature range of 80–190K (Fig. 4(b)), T_0 of QDCL reaches a higher value of 437K, while the inset shows the thermal-tuning behavior of lasers at threshold with $\Delta\nu/\Delta T$ = –0.123cm^{-1}/K. This value is about half of that of normal FP QCLs and reveals the temperature insensitivity of spectra of QDCL, which can be attributed to relatively stronger quantum confinement effect originating from hybridization between the QD and QW states which compensates the band gap shrinking due to temperature.

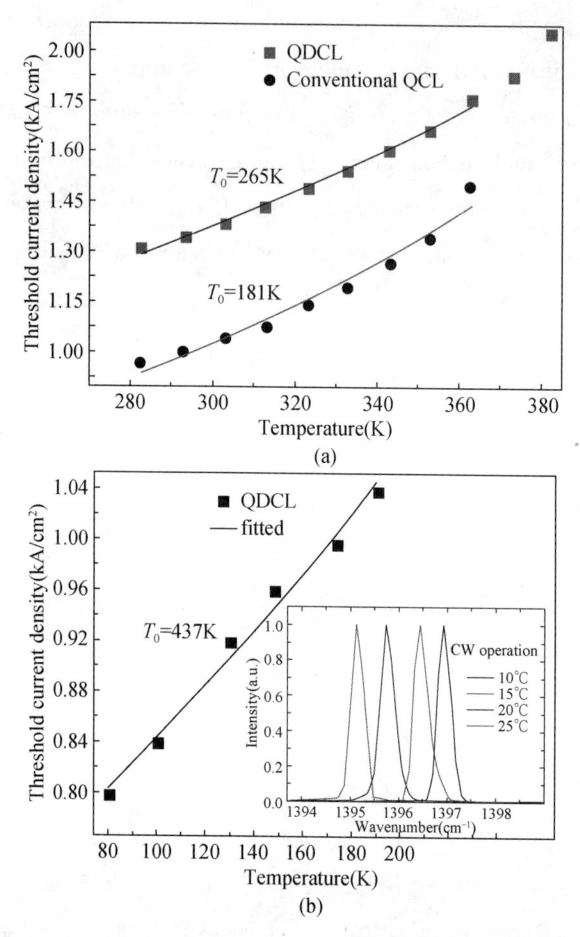

Fig. 4　Temperature dependence of the threshold current when driven in pulsed mode for (a) both QDCL and QCL within the temperature range of 283-363K and (b) QDCL within the temperature range of 80-190K with inset of variable temperature spectra

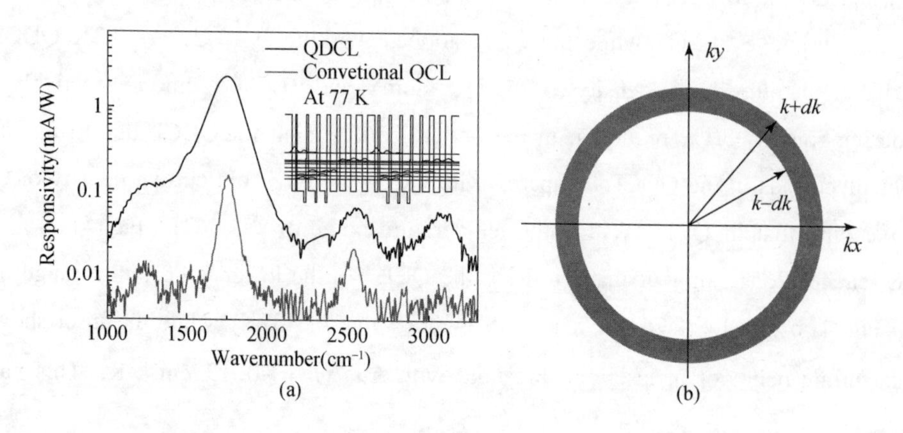

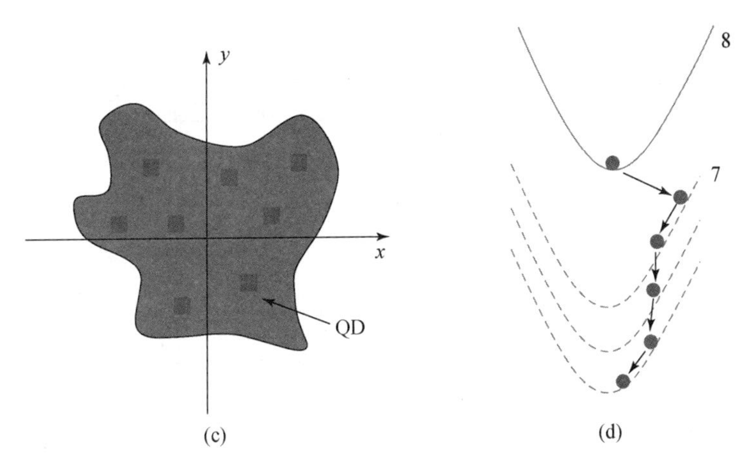

Fig. 5 (a) Responsivity of the QDCL and QCL operated as QCD at 77K under normal incident light. Illustration of (b) states in k-space within the ring of radius $\bar{k}$ and thickness $2\delta k$ that are mixed by QD scattering, (c) localized "amoeba-like" wavefunction that extends over multiple QD sites, and (d) dispersions of upper subband 8 not influenced by the QDs and lower subbands (dashed curves) that are hybridized due to QD scattering of the electrons inside QW, and electron relaxation from subband 7 into lower subbands

The overall improvement of the lasing characteristics can be attributed to placing QDs inside the active region of the QDCL. Because the sizes and band offsets of the QDs are not sufficient to provide the strong in-plane confinement, the QD states hybridize with extended QW states and subbands are formed. The hybridization has been verified with the photocurrent measurement on the same QDCL operating as a quantum cascade detector (QCD) with zero bias under the normal incidence illumination with a mid-IR source. It is well known that for optical transitions to occur between subbands in QWs, the electric field must be polarized in the direction that is perpendicular to the QW plane, eliminating the possibility that a QCD based only on QWs would be responsive to a normal incident light whose optical field is always polarized in parallel to the QW plane. This is indeed what we have observed in photocurrent measurements of the QCL sample operated as a QCD at 77K as shown in Fig. 5 (a). In contrast, the QDCL sample has produced a rather appreciable photo-response that clearly indicates the hybridized nature of the states in the active region with contributions from both QW and QD characteristics. This hybridization can be thought of as being the result of QD scattering of the electrons inside a QW and the degree of hybridization reduces with the increase of electron energy simply because electrons of higher energies will be scattered less by the QDs. It is therefore not difficult to see those higher subbands consisting of injection and upper laser states should preserve mostly the characteristics of the completely delocalized QW subbands, namely the plane-wave description of the in-plane wavefunctions. In contrast to it, the lower subbands consisting of extraction and lower laser states, do get hybridized by acquiring some QD characteristics with some degree of in-plane localization. The effect of hybridization between the QD and QW states in the lower subbands is illustrated in Fig. 5 (b). The role played by QDs is conceptually similar to the role played by any strong non-periodic perturbation as described in [27]. Depending on the dimension and density of the QDs the plane wave states within a certain energy range $\overline{E}\pm\delta E$, located in the k-space within the ring of radius $\overline{k}=(2m_c\overline{E}/\hbar^2)^{1/2}$ and thickness $2\delta k$, where $\delta k/\overline{k}=\delta E/2\overline{E}$ get mixed by QD scattering and combine into new hybrid states. Each hybrid state contains wavevectors in this range of $2\delta k$ and has a localized wavefunction that extends over multiple QD sites on the scale of $\pi/\delta k$ with the probability density having an "amoeba-like" shape as depicted in Fig. 5(c). The average momenta $\overline{k}$ of different states run from 0 all the way to the maximum wavevector of the Brillouin zone, hence their kinetic energies $\overline{E}$ still follow the parabolic dispersion (dashed curves) as shown in Fig. 5(d). Obviously, transport between these states occurs mostly by hopping. Let us now consider an electron gets scattered by, say, an optical phonon, from the upper laser state in subband 8 to some high kinetic energy state in the lower subband 7. In the absence of QD scattering, this electron will relax towards the bottom of subband 7 via the ultrafast intrasubband process to populate the lower laser state. With QDs, in order for the electron to relax within the same subband, it must jump through a series of amoeba-like states. This process is hindered by the smaller overlap between different amoeba-like states in comparison to the large overlap of extended plane-wave functions in the pure QW. In addition, this slowed-down intrasubband process has to compete with electron relaxation into states of other lower subbands by jumping through their amoeba-like states as illustrated in Fig. 5(d). A positive consequence of this is the reduction of the probability of electron rolling down subband 7 and reaching its bottom relative to the probability of this electron falling down to the lower subbands and eventually into the injector. Therefore, the lower laser state 7 ends up less populated even at elevated temperatures which can be a plausible explanation for the increase in T_0 of the QDCL in comparison to that of QCL.

4 Summary

In conclusion, we have demonstrated a QDCL made of 35 cascading periods each with an active region of three layers of InAs QDs grown on GaAs buffer layers that are embedded in InGaAs QWs and InAlAs barriers. This QDCL has been operated at room temperature in CW mode at 7.2μm. In comparison with a QCL sample without QDs but with otherwise same layer structure, the QDCL has shown better temperature performance, which we attribute to the effect of hybridization between the QD and QW states and then the reduction of lower-laser-state population because of the QD scattering of electrons.

Funding

National Basic Research Program of China (Grant Nos. 2013CB632800); National Key Research and Development Program (Grant Nos. 2016YFB0402303); National Natural Science Foundation of China (NSFC) (Grant Nos. 61404131, 61435014, 61627822, 61574136, 61674144); Key projects of Chinese Academy of Sciences (Grant Nos. ZDRWXH-2016-4, QYZDJ-SSW-JSC027); Beijing Natural Science Foundation (Grant Nos. 4162060, 4172060).

Acknowledgments

We thank Ping Liang and Ying Hu for their help in device processing.

References

[1] J. Faist, F. Capasso, D. L. Sivco, C. Sirtori, A. L. Hutchinson, and A. Y. Cho, "Quantum cascade laser," Science 264(5158), 553-556(1994).

[2] M. Beck, D. Hofstetter, T. Aellen, J. Faist, U. Oesterle, M. Ilegems, E. Gini, and H. Melchior, "Continuous Wave Operation of a Mid-Infrared Semiconductor Laser at Room Temperature," Science 295(5553), 301-305(2002).

[3] J. C. Shin, M. D'Souza, Z. Liu, J. Kirch, L. J. Mawst, D. Botez, I. Vurgaftman, and J. R. Meyer, "Highly temperature insensitive, deep-well 4.8μm emitting quantum cascade semiconductor lasers," Appl. Phys. Lett. 94(20), 201103(2009).

[4] P. Q. Liu, A. J. Hoffman, M. D. Escarra, K. J. Franz, J. B. Khurgin, Y. Dikmelik, X. Wang, J. Fan, and C. F. Gmachl, "Highly power-efficient quantum cascade lasers," Nat. Photonics 4(2), 95-98(2010).

[5] Y. Bai, S. Slivken, S. Kuboya, S. R. Darvish, and M. Razeghi, "Quantum cascade lasers that emit more light than heat," Nat. Photonics 4(2), 99-102(2010).

[6] Y. Bai, N. Bandyopadhyay, S. Tsao, S. Slivken, and M. Razeghi, "Room temperature quantum cascade lasers with 27% wall plug efficiency," Appl. Phys. Lett. 98(18), 181102(2011).

[7] K. Fujita, S. Furuta, T. Dougakiuchi, A. Sugiyama, T. Edamura, and M. Yamanishi, "Broad-gain ($\Delta\lambda/\lambda_0 \sim 0.4$), temperature-insensitive($T_0 \sim 510$K) quantum cascade lasers," Opt. Express 19(3), 2694-2701(2011).

[8] A. Lyakh, R. Maulini, A. Tsekoun, R. Go, and C. K. N. Patel, "Tapered 4.7μm quantum cascade

lasers with highly strained active region composition delivering over 4.5 watts of continuous wave optical power," Opt. Express 20(4), 4382-4388(2012).

[9] M. Bahriz, G. Lollia, A. N. Baranov, and R. Teissier, "High temperature operation of far infrared ($\lambda \approx 20\mu m$) InAs/AlSb quantum cascade lasers with dielectric waveguide," Opt. Express 23(2), 1523-1528(2015).

[10] A. Lyakh, M. Suttinger, R. Go, P. Figueiredo, and A. Todi, "5.6μm quantum cascade lasers based on a two-material active region composition with a room temperature wall-plug efficiency exceeding 28%," Appl. Phys. Lett. 109(12), 121109(2016).

[11] Y. Yao, A. J. Hoffman, and C. F. Gmachl, "Mid-infrared quantum cascade lasers," Nat. Photonics 6(7), 432-439(2012).

[12] M. S. Vitiello, G. Scalari, B. Williams, and P. De Natale, "Quantum cascade lasers: 20 years of challenges," Opt. Express 23(4), 5167-5182(2015).

[13] M. Razeghi, Q. Y. Lu, N. Bandyopadhyay, W. Zhou, D. Heydari, Y. Bai, and S. Slivken, "Quantum cascade lasers: from tool to product," Opt. Express 23(7), 8462-8475(2015).

[14] N. S. Wingreen and C. A. Stafford, "Quantum-dot cascade laser: proposal for an ultralow-threshold semiconductor laser," IEEE J. Quantum Electron. 33(7), 1170-1173(1997).

[15] C. Hsu, J. O, and P. Zory, "Intersubband quantum-box semiconductor lasers," IEEE J. Sel. Top. Quantum Electron. 6(3), 491-503(2000).

[16] I. A. Dmitriev and R. A. Suris, "Quantum cascade lasers based on quantum dot superlattice," Phys. Status Solidi 202(6), 987-991(2005).

[17] D. Wasserman and S. A. Lyon, "Midinfrared luminescence from InAs quantum dots in unipolar devices," Appl. Phys. Lett. 81(15), 2848-2850(2002).

[18] S. Anders, L. Rebohle, F. F. Schrey, W. Schrenk, K. Unterrainer, and G. Strasser, "Electroluminescence of a quantum dot cascade structure," Appl. Phys. Lett. 82(22), 3862-3864(2003).

[19] N. Ulbrich, J. Bauer, G. Scarpa, R. Boy, D. Schuh, G. Abstreiter, S. Schmult, and W. Wegscheider, "Midinfrared intraband electroluminescence from AlInAs quantum dots," Appl. Phys. Lett. 83(8), 1530-1532(2003).

[20] D. Wasserman, T. Ribaudo, S. A. Lyon, S. K. Lyo, and E. A. Shaner, "Room temperature midinfrared electroluminescence from InAs quantum dots," Appl. Phys. Lett. 94(6), 061101(2009).

[21] V. Liverini, A. Bismuto, L. Nevou, M. Beck, and J. Faist, "Midinfrared electroluminescence from InAs/InP quantum dashes," Appl. Phys. Lett. 97(22), 221109(2010).

[22] V. Liverini, L. Nevou, F. Castellano, A. Bismuto, M. Beck, F. Gramm, and J. Faist, "Room-temperature transverse-electric polarized intersubband electroluminescence from InAs/AlInAs quantum dashes," Appl. Phys. Lett. 101(26), 261113(2012).

[23] J. Brault, M. Gendry, G. Grenet, G. Hollinger, Y. Desieres, and T. Benyattou, "Role of buffer surface morphology and alloying effects on the properties of InAs nanostructures grown on InP(001)," Appl. Phys. Lett. 73(20), 2932-2934(1998).

[24] N. Zhuo, F. Q. Liu, J. C. Zhang, L. J. Wang, J. Q. Liu, S. Q. Zhai, and Z. G. Wang, "Quantum dot cascade laser," Nanoscale Res. Lett. 9(1), 144(2014).

[25] N. Bandyopadhyay, Y. Bai, S. Slivken, and M. Razeghi, "High power operation of $\lambda \sim 5.2-11\mu m$ strain balanced quantum cascade lasers based on the same material composition," Appl. Phys. Lett. 105(7), 071106(2014).

[26] J. C. Zhang, F. Q. Liu, S. Tan, D. Y. Yao, L. J. Wang, L. Li, J. Q. Liu, and Z. G. Wang, "High-performance uncooled distributed-feedback quantum cascade laser without lateral regrowth," Appl. Phys. Lett. 100(11), 112105(2012).

[27] J. B. Khurgin, "Inhomogeneous origin of the interface roughness broadening of intersubband transitions," Appl. Phys. Lett. 93(9), 091104(2008).

第三篇

学 术 贡 献

忍受辐照伤痛，换来我国空间用硅太阳电池的定型投产

科研工作犹如深海行船，表面看似很平静、很安全，其实却荆棘丛生，到处都会遇上险滩暗礁，时刻都要迎接狂风暴雨。

1962年，王占国大学毕业后就到了中国科学院半导体研究所，开始了他的科学探索之路。1967年，正当“文化大革命”在全国展开的时候，王占国受命负责651任务中的关键部件——人造卫星用硅太阳电池辐照效应研究。

不少人对辐照实验存有恐惧心理。因为在做辐照实验时，除来自高能电子加速器的电子束会给人带来致命的辐射损伤外，高能电子束与其他物质相互作用产生的X射线也会给人的健康造成严重损害。当时，由于受实验条件限制，实验用的测量设备只得安装在加速器机房外的走廊里。在停机时，人工将试样放置在加速器的靶心位置上，人离开机房后，开机进行电子辐照。试样经过一定剂量的电子辐照后，停机取出进行太阳电池转换效率的测量，然后再放入靶室辐照，再取出测量，如此反复，每天要进行几十次这样的操作。这就是说，若一个人做实验，就要进出机房几十次！

为了保证实验任务的按期完成，在其他人不愿进入加速器机房取放样品的情况下，王占国挺身而出，一人承担了辐照样品的放取和测试任务。实验的头两天，一切进展都很顺利，但到了第3天，他感到右手有隐隐烧伤的痛感，并本能地想到了是否与反复进出辐照机房而受到电子辐照有关？为此，王占国请来了该实验室主任，一位曾在苏联留过学的副研究员。在听了王占国的汇报后，他不加任何思索地断然否决了可能与电子辐照有关的看法。这位副研究员说，他在苏联就干这一行，并在这个实验室工作了多年，做过很多次实验，从未发生过这样的事情。权威的话打消了王占国的怀疑，而错误地把已隐隐作痛的右手归结为是由于受实验用的强红外灯辐射所致。

实验继续进行着，可王占国右手的烧伤感非但没有丝毫减轻，而且出现了红肿。为了确保任务的按时完成，他把在嘉定做实验的尹永龙同志请来帮忙。王占国回忆说：“是我‘害’了他，两天后，他的右手也出现了与我类似的现象。”而尹永龙却未曾受到过红外灯的辐射，这才使王占国又怀疑到了电子辐照损伤的可能性。这时，王占国受伤的手已由红肿变成了很大的水疱。

王占国被立即送往上海中山医院诊治，同时也惊动了上海市防疫站的领导，并派技术人员到加速器机房进行模拟实验测试。测试结果发现，机房的自锁机关虽然工作正常，但被高达百万伏特电压加速的电子通过真空放电需要一个相当长的过程；从关断电压到允许进入机房时，加速器上仍然带有数万伏的高压，未完全冷却的灯丝，仍有高能电子束流射向靶区。在正常的情况下，射向靶区的电子被放置在灯丝和靶心之

间的继电器金属挡板所阻断，不会造成对人身的损害；不巧的是隔离电子束的继电器常常失灵，无法阻挡剩余电子束向靶区的发射，这是造成电子辐射受伤的直接原因。所幸运的是剩余X射线的辐射已构不成对人的严重损害。反复不断地用毫无保护的手去放取样品时，就会受到剩余电子束的辐射，随着受辐射手上的电子束剂量不断地积累，导致了辐照损伤事故的发生。看来真正的“凶手”就是电子辐射。入住中山医院后，院方十分重视，针对病情邀请了上海华山医院有名的皮肤科专家会诊。但由于种种原因，感染的伤口在一天天地恶化；如此下去，有可能带来截去手指的危险！然而，王占国从来没有被困难吓倒过。住院期间，在给一位北京的朋友回信时，他试着用左手写道：“即使失去了一只手，但能为发展我国的航天事业贡献一分力量，也在所不惜！”

为了能更及时有效地控制病情发展，王占国在领导的安排下于7月中旬由上海乘机回到北京，并直接入住有治疗辐射损伤经验的中国人民解放军307医院。经过40多天的治疗，伤口终于愈合。出院时，医生在诊断书上写了“右手Ⅲ度电子辐射损伤，左手Ⅱ度，全身受超剂量辐射；建议半年内不要接触放射性，全休三天”的证明。

出院后，王占国抓紧机会，一头钻进实验室，开始整理已付出惨痛代价而得到的宝贵实验资料。实验结果使他惊奇地发现，*NP*结硅太阳电池抗电子辐照的能力要比*PN*结硅电池大几十倍！这立刻使他回想起了在1964–1965年做硅材料的中子和γ射线辐照实验时，得到的*P*型硅比*N*型硅更耐辐照的结论，使他对*NP*电池（与*PN*电池比）抗辐照的机理有了进一步的认识，也更加坚定了对所得结果可靠性的信心。

在1967年年底，由651任务电源总体组召开的人造卫星用硅太阳电池定型会上，王占国报告了实验结果，并提出应将*PN*结改为*NP*结硅电池定型投产的建议。实验结果引起了电源总体组负责同志的重视，但也有人借当时*PN*结电池工艺比较成熟、太阳能转换效率比*NP*结电池高等理由加以反对。会上，经过激烈的辩论后，651设计院电源总体组采纳了王占国的建议，做出了将原定的硅*PN*结电池改为*NP*结定型投产的决定。这不仅避免了弯路，节约了经费，而更重要的是保证了我国人造卫星长期安全的运行，具有重大的社会效益。

只有英雄的战船归港时才会受到热情的欢迎，也只有风雨过后的彩虹才是最美丽的。虽然在完成这个任务时，在王占国的右手上留下了永久的伤痕，但他却为这个工作能为我国人造卫星用硅太阳电池的定型投产做出关键性的贡献，感到欣慰和自豪。

后来，王占国在任国防科委第14研究院辐照实验组业务组长期间，对该院研制的电子材料、元件、器件和组件进行了系统的电子、质子、中子和γ射线辐照效应研究，将其实验结果，与同事一起汇编成册，这一开拓性工作，为我国航天事业、电子对抗以及核突围等国防工程做出了积极贡献。

挑战国际权威，澄清 GaAs 和硅中深能级物理本质

创新是发展的动力，是时代进步的不竭源泉。在科学的道路上，正是由于创新的精神和敢于向权威挑战的勇气，才能结出累累硕果。

1978 年以后，王占国紧跟世界学术发展的潮流，开展了半导体深能级物理的实验，取得了一定的成绩。研究所所长黄昆，很赏识王占国的才华，推荐他到自己的同行、瑞典国际深能级研究中心负责人哥尔马斯手下深造。1980 年 10 月，王占国来到了瑞典南部的大学城——隆德，开始了他挑战国外权威的征程。

王占国到隆德的第二天，就风尘仆仆地来到学校的实验大楼，应约前去会见自己的导师，听候工作安排。初来乍到的王占国，连导师的面都没机会见，就被安排在了一个无人问津、条件很差的实验室里。这一切，王占国看在眼里，却没放在心上。占据他那心田的，是祖国的委托和所内同志们美好的祝愿。他简单收拾了一下房间，便埋头去阅读有关的实验资料。

在这里，给王占国印象最深的是人际关系淡漠。已经好多天了，没人来跟他攀谈和互道学识，只有技术员受命来讲仪器操作，那也只是粗略演示了一下就匆匆离去了。又过了几天，工程师送来了样品，这就是在告诉王占国：你可以正式开始工作了。王占国深深地预感到，在这样的环境中从事研究工作，一切都得靠自己去摸索、去探求。

半导体材料中的深能级研究，是半导体材料中的杂质、缺陷行为研究的一个重要方面，是当时世界上固体物理和材料物理学家广泛关注的前沿课题。研究结果不但对固体物理本身的发展具有重要意义，而且对进一步提高半导体材料的质量、制备完美晶体以及提高半导体器件的稳定性和成品率等，都起着至关重要的作用。王占国接手的是研究液相外延砷化镓材料中 A、B 两个能级的性质。这个题目，一些研究生也曾涉猎过，但都是半途而废；一些有名望的学者对此也甚感头疼，因而被搁置了多年。这不仅因为 A、B 系统自身关系甚为复杂，很难理清眉目；同时，在这里还不具备 *P* 型样品的情况下，要利用 *N* 型样品来研究少数载流子空穴的行为规律，更是一件困难的事。因为要在多数载流子的系统中，研究少数载流子的行为，这本身就是个尚待解决的课题。

王占国在仔细查阅了有关文献资料、总结前人失败的教训之后，制订了切实可行的实验方案。他不贪图捷径，从最基本的实验做起，绘制有关的曲线，再进行理论分析。他连续做了好多次，由于设备条件太差，实验误差很大，描绘出的曲线，一次一个样，很不理想。为了保证测量的精确可靠，中途不能停机，他就带上干粮，连续工作 20 多个小时，直到这组数据测完为止。他在摸熟仪器的性能之后，大胆地改进实验

条件，创造性地运用实验技巧，终于有了理想的结果。经过近两个月的摸索、探求，王占国打开了研究工作的局面。他运用*PN*结耗尽层宽度随偏压而改变的基本原理，总结出一套克服来自结区边缘的自由载流子尾俘获而导致的慢瞬态过程的实验新方法，解决了在多子系统中研究少子陷阱性质的难题。这一实验方法上的创新，既节省了做测量的时间，也提高了实验的精度和可靠性，还简化了对实验样品的制备要求。这个难题的解决，引起了物理系同行的注目。王占国那个被冷落的实验室，陡然热闹起来。好几位研究生，有的来向王占国祝贺，有的来求教实验中的种种疑难。

圣诞节前夕，四处奔波忙碌的导师哥尔马斯回到了隆德，在听说了王占国的出色表现后，次日上班就把王占国叫到自己的办公室，要他汇报工作。王占国把自己近两个多月来的实验设想、实验过程、实验结果，所做的理论分析等，一一地做了回答。王占国讲的是英语，虽不那么娴熟，但叙述得清清楚楚。哥尔马斯默默地听着，不轻易打断王占国的话。渐渐地，他的眉宇舒展了，对王占国变得异乎寻常地亲热起来。他沉思片刻，突然站起身来，对王占国说："走，去看看你的实验室。"进了王占国的实验室，哥尔马斯显得有些吃惊，说："怎么能在这样的实验条件下做出第一流的工作呢？给你另建一个，要与我的实验室不相上下，甚至还可以好一点的实验室。跟我工作，所有的东西都要是第一流的。"

实验条件的好坏，对一个科学研究工作者说来，虽不是第一位的，但也是至关重要的因素。有了新的实验室，王占国如鱼得水，如虎添翼，他满怀胜利的喜悦，开始了新的征程。

对A、B中心是否相关这一课题的研究，不仅对识别缺陷的化学组成，而且对提出新的理论模型都有着十分重要的意义。王占国深知这一点。他夜以继日，甚至节假日，都不肯离开自己的实验室。他用了好几个月的时间，做了数十次的实验，不断取得新的结果，研究工作获得了突破性进展。然而，这些新的结果与导师等权威人士的结论大相径庭；为了慎重起见，王占国未将其马上公开，而是写信向半导体研究所所长黄昆院士和副所长林兰英院士汇报。不久，回信收到了："物理学上没有不变的规律，不要迷信权威，要相信自己。科学，从来就是后人对前人结论的不断修正才得以发展起来的。"王占国受到鼓舞，他再次用他提出的测量深中心上电子占有率随时间的变化新方法取代这里采用的测量整个电容瞬态的传统方法，对液相外延GaAs中的A、B能级和硅中金施主与金受主能级的行为进行了严格地测量，结合理论分析和计算机模拟验证，进一步肯定了液相外延GaAs中A、B能级是同一缺陷的两个不同能态的双受主中心。其化学组成，很可能是镓占砷位的反结构缺陷。

硅中金施主及金受主能级也同样是同一缺陷的两个不同能态。两者都不是两个不同化学起源的相互独立的能级中心。这与导师哥尔马斯等在很有名望的《物理评论》上发表的很有分量的文章中所得的结论相反，这无疑是对权威学者的挑战，是对自己导师所下结论的否定。

当王占国把这一系列实验结果和得出的结论交给哥尔马斯审阅时，看得出，他的心情既惊讶、激动，又蕴含着忐忑不安。作为知名学者的哥尔马斯，在学术见解上也同样是个大度的人，他在听了王占国的汇报并详细阅读了王占国的实验报告后，起身热情地握住王占国的手说："密斯特王，了不起，祝贺你，祝贺你！"经过近半年的深入、系统地实验研究，其结果从另一个角度进一步证实了王占国原来的结论是正确的。哥尔马斯不得不再次承认"实验结果是无可挑剔的"。

瑞典首都有个奥林匹克运动场，是创造体育运动世界纪录最多的地方。在隆德大学固体物理系，王占国是出高水平论文最多的学者。3 年的进修时间，王占国与合作者一起前后共撰写了 20 多篇科学论文。多数已在美国《物理评论》《应用物理快报》《应用物理》、英国的《物理杂志》《固体通讯》等国际和我国的《半导体学报》《中国物理》等有影响的刊物上发表。这些研究成果，引起了同行的好评和重视。王占国出色的工作成绩，为祖国争得了荣誉，为民族增添了光彩。

1983 年 11 月，王占国载誉归来，立即开始了报效祖国的新征途。从 1984 年开始致力于深能级物理实验室建设和材料物理研究。脚踏实地，不迷信权威；实事求是，不轻易下结论。这应是每个科研工作者要具备的品格。

“863”十年，掌舵我国新型半导体材料与器件发展

1　建立 MBE 基地，引领我国低维半导体材料和器件发展

1991 年 5 月，王占国在缺席未参加答辩的情况下，被推荐、选举和聘任为国家“863”高技术新材料领域第二届专家委员会委员，后又连任第三届专家委员会委员（1994. 5 - 1996. 5）和第四届专家委员会委员、常委，功能专家组组长（1996. 5 - 2001. 10），这前后共经历了 10 年的时间。在此期间，他为推动我国高技术新材料的发展做出了重要的贡献，尤其在分子束外延（MBE）超薄层微结构材料基地（下简称 MBE 基地）建设中发挥了核心作用。

王占国正式开始“863”工作是从 1991 年 8 月 13 日在国谊宾馆召开的第 2 届专家委员会第三次会议开始的。在这次会议后，经过近 1 个月对“863”计划的方针、政策和新材料领域的详细了解，针对半导体光电子专题下设的课题多（20 个），支持强度低（每个课题平均每年仅 8 万元）、低水平重复问题严重和材料制备的设备落后等问题，王占国提出了引进先进的分子束外延和金属有机化学气相沉积（MOCVD）等关键设备，建立光电子信息材料研发基地的设想。在 1991 年 10 月 13 日召开的专家委员会和专家组的联席会上，根据国家科学技术委员会（以下简称国家科委）关于建立“863”研发基地或中心的要求精神，把基地建设列入了每次专家委员会的议事日程。在随后的多次会议上，就基地建设章程、筛选条件和申请、批准程序等进行了议论。1992 年 3 月 25-31 日的杭州会议还就中心、基地的关键设备引进暂行条例做了规定。

1992 年上半年，王占国等 3 人受专家委员会派遣，考察了苏联解体后独联体的俄罗斯，与先进技术公司（ATC）就 MBE 技术生产大功率半导体量子阱激光器签订了合作意向书。1992 年年底王占国等代表中方以“863”光电子功能材料中心（OEFMC）名义与俄罗斯先进技术公司半导体部（ATC-SD）达成在北京高技术开发区建立联合风险公司（JVC）的双方协议。按协议规定，第一步，双方合作首先在北京研究和开发大功率半导体量子阱材料和量子阱激光器，俄方将利用中方引进的 MBE 设备和 307 主题所属的国家光电子工艺中心的器件工艺条件，第一年负责完成连续输出光功率 1W、寿命为 3000h 的 808nm 量子阱激光器，第二年负责完成准连续光功率输出 15W、寿命 2000h 的量子阱激光器阵列的研制任务。为此，中方向俄方支付 10 万美元的技术转让费。第二步，在第一步取得实质性的进展后，双方组建 JVC。这个协议分别于 1993 年 2 月和 4 月由双方的上级领导部门正式批准生效。

对俄合作意向书的签订，加快了我国 MBE 设备的引进和 MBE 基地组建的步伐。

1992 年 8 月，经国家科委批准，同意在中国科学院半导体研究所引进 MBE 设备，筹建 MBE 基地，并于同年 10 月通过了专家组的论证。1993 年 5 月引进法国 Riber 的 MBE 设备的安装、调试和验收完毕。MBE 基地的 3 个辅助实验室（光学性质、电学性质和结构性能检测实验室）陆续到 1995 年年底全部建成。MBE 基地，包括对俄合作项目（131.5 万元）和引进 MBE 设备（300 万元）的费用一起，共要投入人民币 430 多万元。当时要下这个决心，非常不易！

MBE 设备的成功引进和顺利运转，为执行对俄合作研发项目创造了条件。从 1993 年 10 月到 11 月和 1994 年元月到 2 月，俄专家两批 4 人到 MBE 基地会同我方研究人员组成联合实验小组，开展大功率半导体量子阱材料的生长，工作进展顺利，如期完成了阶段目标。其间，俄专家熟练地驾驭 MBE 生长技术的能力和认真、负责的工作态度给中方人员留下了深刻的印象。

然而在执行后续协议时，俄方提出多种理由不派专家来华进行后续的量子阱激光器制造。与俄方多次交涉未果后，专家委员会派王占国等于 1994 年 9 月 29 日到 10 月 13 日再次赴俄与 ATC-SD 谈判。经过 4 次谈判双方达成协议，然而针对协议内容条款执行的时间表的备忘录，俄方始终未有签字，猜测是俄方并不愿意将大功率激光器的关键制备技术转让我方。面对这样的情形，王占国向专家委员会和国家科委建议依靠国内单位来完成项目，并得到了批准。

1994 年年底开始，经过一年半的刻苦攻关，MBE 基地、上海光学精密机械研究所光电子器件实验室和国家光电子工艺中心分工协作完成了“八五”期间对俄合作项目中所规定的大部分任务。“九五”期间，MBE 基地和光电子工艺中心联合完成了国家“863”重大项目“人工晶体和全固态激光器”中的两个课题任务，研制成功可实用化的 808nm 大功率量子阱激光器和线列阵器件。利用我们的 808nm 大功率量子阱激光器，又研制成功室温连续输出功率超过 25W 的大功率半导体激光光纤耦合模块，替代进口同类产品，打破了国外的禁运，并成功地用于全固态激光器泵浦光源。808nm 大功率激光器泵源的研制成功，为全固态激光器在国防、制造、科研、显示和医疗等方面的应用奠定了基础。

除了完成了多项国家“863”任务外，他还先后承担了国家自然科学基金重大、重点项目，国家重点科技攻关项目和中国科学院基础研究重大、重点项目等近 20 个，在低维半导体结构生长和量子器件研制方面取得了令人瞩目的成绩。他领导的 MBE 组研制成功低温电子迁移率超百万的 GaAs/AlGaAs 二维电子气（2DEG）材料，使我国的 MBE 技术一举跨入国际先进行列；在应变自组织量子点材料生长和大功率量子点激光器研制方面获得了突破性进展；应变补偿量子级联激光器研究处于国际前列。此外，在 808nm 大功率量子阱激光材料、器件和线列阵以及大功率激光光纤耦合模块（>25W）等方面也取得了重要进展，打破了国外对我国的封锁，并成功地用于全固态激光器泵浦光源。值得指出的是，808nm 大功率激光器泵源的研制成功，极大地增强了新材料领域独立自主发展我国的全固态激光事业的信心。结合我国人工晶体，特别是非线性光学晶体的优势，新材料领域在此基础上，将“人工晶体和全固态激光器”作为领域的重大项目予以重点支持，取得了丰硕成果，并为全固态激光器在国防、制造、科研、显示和医疗等方面的应用打下了基础。

多年来，以 MBE 基地为基础，王占国及其领导的实验室在 GaAs/AlGaAs 二维电子气材料、大功率半导体量子阱激光材料与器件、量子点材料和器件、量子级联材料和器件等几个国际前沿研究领域取得了国际先进水平的成果，为提高我国材料科学研究水平和创新能力以及高级研究人才的培养等做出了重要贡献。

MBE 基地的这些成绩充分证明了王占国在新材料领域发展战略上的前瞻性眼光，为我国半导体低维结构材料和新型量子器件的发展起到了很好的保驾护航作用。

2 力推氮化物研究，为我国半导体白光照明奠定基础和谋划未来

在 20 世纪 90 年代初中期，国际上处在热衷研制Ⅱ-Ⅵ蓝光激光器的大环境下，王占国和蒋民华受新材料领域专家委员会派遣赴美考察光电子材料和器件的研究动态。他们用了两周时间，分别参观了明尼苏达大学、哈佛大学、佛罗里达大学和得克萨斯大学等单位，发现Ⅱ-Ⅵ蓝光激光器虽已实现激射，但寿命极短，其原因主要与Ⅱ-Ⅵ材料原子结合键能弱有关，很难得到实际应用。相反，Ⅲ族氮化物当时并不被人们看好，特别是 GaN 的纯度和 *P* 型掺杂进展缓慢，使多数人望而却步，材料领域的专家也不看好这个材料体系。根据考察结果和分析，认为宽带隙的 GaN 基材料体系是未来蓝绿光、紫光和高频大功率电子器件与电路的基础材料，如果不给以重视，国际上一旦获得突破，我国就将在此新兴领域失去机会。王占国和蒋民华于 1994 年年底，率先在新材料领域设立了两个 GaN 基光电子材料与器件的“863”课题，每个课题支持强度为 20 万元/年，分别由半导体研究所和北京大学的科研组承担。

经过近 1 年的攻关，半导体研究所和北京大学在自制的生长设备上，都做出了好成绩，研制出 MIS 结构的蓝光发光管，南昌大学江风益教授也取得了不错的成绩。1995 年年底在制定“九五”计划时，又把该课题上升为重点项目，经费强度增加到近 200 万元。经费虽然增加了，但自制的 MOCVD 设备落后，稳定性差，要想制备出高发光效率的材料，难度很大。针对严重制约材料发光性质的 MOCVD 设备问题，王占国等建议引进国外先进 MOCVD 设备，以实现在该领域的跨越发展。然而遗憾的是，这个有利于我国大陆在该领域跨越发展的建议由于种种原因没有被采纳。这使我国大陆白白地失去了宝贵的 5 年时间，本来 GaN 基材料与器件研发水平落后的韩国和台湾地区反而超越了我国大陆，损失之大，难以估算。直到 2001 年，有关公司介入后，才购置了进口的 MOCVD 设备。

尽管购买进口设备受阻，王占国等继续推动 GaN 基发光材料与器件作为领域重大项目给予更大支持（经费总强度约 600 万），目标是实现蓝绿光 LED 的小批量生产，并试制蓝光加黄光荧光粉的白光照明光源。在新材料领域“863”计划 15 周年成果展览会上，GaN 基蓝绿紫 LED 和基于 GaN 基蓝光加黄光荧光粉的半导体白光照明光源成为一个亮点，这也为 2003 年科学技术部紧急启动的半导体白光照明工程打下了技术基础。

从 2009 开始，国家发展和改革委员会、财政部、中国人民银行、税务总局、住建部、交通运输部等国家机关，通过发布有关文件、建立示范工程等措施大力推进 LED 产业化发展，各省（自治区、直辖市）地方政府投入巨资大力扶植相关产业，各地兴起了 LED 产业园建设热潮。当时预计，4–5 年，仅国内新增的生产型 MOCVD 设备就高达 1000 多台（费用 20 亿–30 亿美元），我国将成为世界 LED 第一生产大国。

2010 年，王占国又敏锐地指出：中国 LED 产业发展步入关键期，有过热趋势，背后隐藏的风险值得关注。

王占国认为，一个新兴产业需要政府的支持和引导，而现在这种情况很像前几年一哄而上的光伏热，令人担心，会不会又来一次产能过剩？LED 照明产业是个系统工程，需要配套企业、配套设备，需要培养专业人员。王占国认为，企业有做大、做强的目标是对的，但是我国企业所需的 MOCVD 等关键设备依赖进口，外延片、芯片和封装等相关核心工艺技术基本上是跟踪国外，缺乏自主知识产权，产品的国际竞争力较差。这是我国企业做大、做强 LED 产业首先需要解决的大问题。

目前 LED 照明产业上、中和下游分布不太平衡，作为 LED 照明基础的外延和芯片所占比例偏低（在近 4000 企业中仅有 65 家）。更重要的是研发，特别是创新能力不足，应当给以足够重视和加大支持力度。另外衬底材料、MO 源和高纯氨等电子气体的规模生产也应布点安排。当然，建立 LED 产业联盟是件好事，但要真正做到集成创新、成果共享和互通有无，尚需体制创新和企业认可的规章制度来保证。从我国氮化物材料和 LED 照明产业的历史来看，经验教训到今天还会使人警醒，希望能得到相关部门的重视。

王占国认为，在“十二五"期间要实现 LED 产业健康发展，第一，要发挥 LED 产业联盟的作用，通过联盟的机制创新，做到相互取长补短，避免企业间无序竞争和内耗，实现知识产权共享和集成创新，这是提升我国 LED 产业创新能力，做大、做强的一个关键措施。第二，从国家层面来讲，加强对基础研究长期、稳定的支持至关重要，比如应该积极开展 GaN、AIN 单晶衬底的研发等。第三，应该研制我国的生产型 GaN MOCVD 设备。采用国产设备，成本就会下降，对提高产品竞争力有利。另外，材料质量提高和新材料探索与设备功能改进与创新息息相关，两者相辅相成。第四，企业切忌急功近利，要采取有力措施，加大引进科技人才力度，建立研发团队，并给以稳定支持，为企业的创新发展打下基础。

任“S-863”专家组长，开展新材料领域战略研究

1996 年 7 月至 1999 年 6 月期间，王占国被国家科委（科学技术部）聘为新材料技术领域发展战略研究（即“S-863”计划软课题研究）专家组组长，与专家组其他 20 位成员一起，历时 3 年，进行了大量的调查和分析研究，向 22 个部（委）以及 250 个单位或著名科学家发了调查函，其中 14 个部（委）通过不同形式反映了许多很好的意见，200 多名著名学者也对研究报告提出了修改意见。同时委派专家组成员亲赴沈阳、西安和上海等地征求专家对研究报告 1.0 版本的修改意见；召开了 20 余次产业部门、两院院士座谈会，分析了国内需求和国外（如美国、英国、德国、日本、韩国等）对高技术新材料领域的研究现状和发展趋势；专家组 8 次全体会议讨论，以王占国为组长的执笔专家组经 10 余次反复讨论和修改，形成了（并通过了专家论证）本领域研究报告 4.0 版本，其中包括对新材料技术领域 5 个重大项目建议书。本研究报告为国家“973”项目的顺利启动和第二期“863”项目的实施做出了贡献。

任咨询组组长，为“973”材料领域发展做出重要贡献

王占国于2007年12月至2013年7月被科学技术部聘任为国家“973”计划材料科学领域第三届专家咨询组组长。在这期间，他以饱满的工作热情和科学决断力，带领和团结来自不同单位的11位各学科专家勤奋工作，很好地完成了科学技术部给咨询组确定的各项任务。

咨询组的主要职责是对本领域实施中的项目进行过程管理，跟踪了解项目执行情况，定期向科学技术部提出咨询工作报告，对项目实施过程中存在的问题向科学技术部提出咨询意见和建议等。当时，“973”计划材料科学领域正在实施的项目共有44项，涉及材料学科的方方面面，工作量很大，同时各项目所研究的内容均处于该研究领域的最前沿，需要学习的知识很多；更为重要的是，国家要求以基础研究为主的“973”计划研究项目也要有应用目标和实用价值。因此在项目实施过程中，把握好基础研究与实际应用之间的关系颇费心思。作为咨询组组长的王占国充分发挥他的领导魄力和睿智，更难能可贵的是，他襟怀坦诚，敢于直言，受到咨询组所有专家的尊敬。在此期间，专家咨询组向科学技术部提供了4份内容充实的咨询报告，每年都获得科学技术部的好评。

作为专家咨询组组长，王占国院士在任期内还参与了“973”计划材料科学领域每年的研究方向指南讨论和制订。指南的制订需要把握材料科学领域重要研究方向的国际前沿和发展趋势，了解我国的在这些重要研究方向的优势和不足，尤其是学科带头人的基本情况，此外还要考虑材料领域相关前沿学科的布局和平衡。王占国认为此前几年中光电信息功能材料的研究安排较少，应引起重视。在他的建议下，先后安排了“高性能近红外InGaAs探测材料”“全组分可调Ⅲ族氮化物半导体光电功能材料及其器件”“硅芯片光互连用发光材料及器件”“人工微结构红外光电耦合材料”“红外量子级联激光材料和探测材料”等一批前瞻性光电信息功能材料研究项目。目前这些项目都已按时通过了科学技术部组织的结题验收，在相关科学问题的突破等基础研究和器件验证等方面都取得了重要进展。

王占国在“973”计划的发展进程中的另一项重要贡献是，作为“973”计划材料科学领域“十二五”战略研究组成员，参与了“十二五”战略规划的研究并任“信息功能材料”研究组组长。他负责编写的1.5万字的“信息功能材料”战略研究报告指出：下一阶段应以新型沟道材料、高K材料和超低K材料等微电子材料与器件；电荷捕获存储材料和新型金属氧化物存储介质材料等新型超高密度存储材料与器件；微结构光子材料、硅基光电子发光材料和硅基光电子集成器件及集成技术的下一代信息系

统光互连的光电子材料与器件；人工晶体材料和超小型高功率全固态激光器；铁基和二硼化镁等前瞻性超导材料物理及超导材料应用；OLED 和激光显示材料与器件；全固态半导体照明智能通信的材料与器件；生化敏感微结构材料和智能敏感材料等敏感材料与器件 8 个方面的基础研究作为重点研究方向。现在，无论从国际上信息功能材料的发展趋势或是后来落实的项目研究进展来看，王占国领导的研究组提出的战略发展方向是完全正确的。

王占国担任“973”计划材料科学领域专家咨询组组长虽然只有 5 年多的时间，但他学识渊博，品德高尚，平易近人，待人诚恳，在项目管理中宽容失败，并鼓励青年学者提出不同意见，给我们专家组的同事们留下了深刻印象，他是我们工作中的学习榜样，也是我们的良师益友。

（本文由时任国家“973”计划材料科学领域专家咨询组副组长欧阳世翕撰稿）

开拓创新，解决“信息功能材料相关基础问题”

进入21世纪后，数字化和网络化信息技术迅猛发展，超大容量信息传输、超快实时信息处理和超高密度信息存储已成为信息技术追求的目标。然而我国无论是光纤通信还是移动通信，甚至国防建设所需的关键器件和电路芯片（材料）都靠进口，严重地制约了我国信息产业的发展，威胁着国家安全。面对这样的严峻局面，1999年由王占国执笔、联合邹世昌和王圩等向科学技术部提出了开展“信息功能材料相关基础问题”的国家重点基础研究发展规划（即“973”计划）项目建议书，获科学技术部立项。2000年王占国组织中国科学院半导体研究所、中国科学院微电子中心、中国科学院上海微系统与信息技术研究所、北京工业大学和浙江大学等6家单位共同申请该项目。期间，王占国召集了相关研究人员进行多次项目预备会，确定了项目的关键科学问题、5年研究目标、主要研究内容和预期成果等，亲自准备了项目申请书和答辩报告，项目最终获科学技术部批准，预算为3000万元。王占国作为项目的首席科学家负责整个项目的组织管理。

项目2000年10月启动，2005年9月结题，在王占国的领导下，各参加单位紧密围绕“结构、电子行态与信息功能的内在关联和规律”以及“异质结构材料生长动力学规律”这两个关键科学问题开展研究，最终在光通信用的关键微结构材料和器件、中红外量子级联激光器材料与器件、THz物理与器件、应变自组织量子点材料和器件、大失配异质结材料衬底技术、Ⅲ族氮化物微结构材料和器件、ZnO的p型掺杂等多个方面取得了突破性的进展，超额完成了项目的5年预期目标，有关研究成果获得国家自然科学奖二等奖2项，获得省部级自然科学奖和科学技术进步奖4项。

该项目实施效果显著。可调谐DBR激光器和EA调制器芯片与HBT激光驱动电路构成一个完整的10Gb/s光通信发射模块，该模块以及研制的宽带偏振不灵敏半导体光放大器可应用于我国新一代光通信系统。新型大功率半导体激光材料和器件的研究提升了我国光电子材料与器件的原始创新能力，有关器件可直接应用于军工、通信和材料加工等方面。应变自组装半导体纳米结构的工作推动了我国纳米光电子器件的研究，使得我国在国际纳米光电子器件领域占据了一席之地，其中研制的大功率量子点激光器和超辐射发光管等可用于未来光通信系统和光纤陀螺等。高性能中远红外量子级联激光器的研制成功，则拓宽并提升了我国红外技术应用水平，对国家安全和国民经济建设意义重大。自支撑GaN衬底和柔性衬底技术不仅可直接用于支撑我国半导体白光照明工程和蓝光DVD技术的发展，而且还将引领未来大失配半导体异质外延生长技术的研发方向。

另外，项目开展过程中，引进了多台生长和测试设备，研发、积累了大量的材料生长、器件工艺和测试表征新技术，建立了各种形式的数据库，形成了多个面向半导体信息功能材料应用的研发平台，这些数据库和平台为研发新型半导体光电功能材料奠定了坚实的基础。

推动材料基础研究，实施光电信息功能材料重大研究计划

“十五”开始，国家自然科学基金委员会（以下简称基金委）根据国家重大战略需求和重大科学前沿需要，希望通过顶层设计、凝练科学目标、凝聚优势力量，形成具有相对统一目标或方向的项目集群，以提升我国基础研究的原始创新能力，为国民经济、社会发展和国家安全提供科学支撑。为此设立了重大研究计划——光电信息功能材料。

这是基金委设立的第一个重大研究计划项目，由沈家骢和王占国主持，闵乃本、周其凤、顾炳林和朱鹤逊等为组员（后调整许宁生和张国义进入专家组替代闵和顾）。王占国和沈家骢及有关专家一起，负责组织了多次专家研讨会，确定了研究目标、主要研究内容和预期成果等，项目立项得到基金委初步认可后，由王占国执笔和项目组成员一起精心撰写了答辩提纲，并亲自向基金委领导和评审专家作了报告，获得通过。项目经过 7 年多实施，顺利完成了预定任务。

该研究计划于 2001 年启动，涉及光电信息功能材料的设计、制备、结构与性能及加工过程中的基础和应用基础领域，共设立研究项目 124 项，主持和参加的科研单位和高等学校共 32 个，资助经费 5757 万元。研究计划于 2008 年 5 月之前完成。在项目实施过程中，王占国亲力亲为，除了参与重大研究计划中项目的评审、现场考察、学术研讨会等事务活动，作为研究计划协调组的组长专门负责研究计划的宏观管理、协调和中期评估，组织相关领域项目的集成，以及协助组织重大计划的专题研讨会等，有力地促进了学科交叉和项目交流。

重大研究计划经过 7 年的实施，在可控量子点材料制备及光电性质研究、基于楔形微结构和 InP/GaAs 异质集成工艺的新型光电子器件、光信息处理用 InP 基光电子集成材料研究、ZnO 基材料生长和 *P* 型掺杂及其光电应用研究、抗光损伤近化学计量比酸锂晶体与全光微结构研究、微纳结构加工与制备技术、分子器件与单分子磁体、分子信息存储材料、有机微纳结构材料的制备、双层膜有机场效应晶体管、有机发光软屏与全印刷的高分子彩屏技术、新型高性能电致发光材料等诸多前沿探索和技术领域取得了一批具有创新性的重要成果，解决了光电信息功能材料应用于国民经济发展及国防建设具有重大意义所涉及的关键材料、关键技术等问题，取得了一系列国内外发明专利，为我国信息技术产业发展奠定了基础。

第四篇

回 忆

半导体材料科学实验室的筹建与初期发展历程回顾

王占国，钱家骏

20世纪80年代中期，国际上以硅基超大规模集成电路（VLSI）为基础的微电子技术发展十分迅速，那时日本和美国已先后研制出兆位（10^6比特）动态随机存取存储（DRAM）芯片，并且预计在10年之后，即到90年代中期，一片集成电路的计算能力将相当于80年代中期十余部价值400万美元超级计算机的能力。面临这种挑战，半导体研究所瞄准硅基VLSI和化合物半导体材料与光电子器件，以80年代初攻克的4K和16K DRAM为基础，把新目标定位在64K和256K DRAM芯片的研制上。当时任副所长的林兰英院士，及时组织原一室的有关人员，分析国际、国内微电子技术发展态势，并依据中国科学院和半导体研究所的发展方向，认为硅单晶材料（也包括砷化镓、磷化铟等）的内在质量是制约VLSI发展的重要因素，特别是随VLSI集成度不断提高以后，材料中的杂质与微缺陷的影响成为关键，因此，提出了筹建“半导体材料物理重点实验室”的设想，目标是针对材料中的杂质与缺陷行为的研究。后来，由于中国科学院整体结构的调整，研究所内与大规模集成电路相关的研究室和工艺线与109厂合并，成立了中国科学院微电子中心，半导体研究所的研究方向集中到光电子方面。在这种形势下，面对大容量、超高速光通信技术发展的需要，就材料方面而言，单纯的材料物理研究已不适应，在林兰英院士的领导下，在原有实验室建议书的基础上，重新起草了名为“半导体材料科学重点实验室”的建议书，向中国科学院提出组建申请。新的建议书将实验室的研究方向定位为以应用基础及应用研究为主，材料体系主要是Ⅲ-Ⅴ族化合物的异质结构材料、薄层和超薄层材料，以及空间微重力条件下GaAs单晶的生长和性质研究等。依据这个研究方向，新组建的实验室由原半导体材料研究室的材料物理组、MOCVD组、离子束外延组、空间材料组、SOI硅外延组以及原三室的分子束外延、化学束外延和GeSi小组为基础组成。

1990年6月25日，在半导体研究所召开了“半导体材料科学开放研究实验室”开放评议会。会议由中国科学院技术科学局严陆光局长主持，邀请柯俊、万群、秦国刚、薛明伦、陈难先等有关方面的12位专家组成评议组，国家自然科学基金委员会、中国科学院等有关部委负责同志也参加了会议。实验室筹备组负责人王占国代表实验室向专家评议组作了工作汇报。经过专家们的热烈讨论，评议组认为从我国“四化”建设的长远需要出发，建立有关半导体材料科学实验室是完全必要的，具有重要意义；实验室的研究方向和研究内容符合我国国情和学科发展方向，具有先进性；并认为在著名半导体材料专家林兰英先生指导下，实验室培养了一支以中青年为主体的有较高理论水平和研究经验的科技队伍；在薄层和超薄层材料，以及空间材料的发展和性质研究等方面，取得了一批具有自己特色的研究成果，正式建立半导体材料科学实验室的时机已经成熟。

1990 年 9 月 15 日，院长办公会议对经各专业局组织同行专家讨论、评议的实验室开放申请进行了审议，决定批准“半导体材料科学实验室”正式对国内外开放，王占国任实验室主任，钱家骏为副主任。1991 年 9 月 6 日，按《中国科学院开放实验室管理办法》的规定，根据开放实验室学术委员会民主选举结果，经院长审议批准，聘任林兰英为实验室首届学术委员会主任，屠海令、孔梅影为副主任。至此，经过几年的不懈努力，实验室正式对外开放，踏上了新的历程。

这里还要指出，筹建实验室的申请是在“六五”末期上报的，由于种种原因，直到“七五”末期才被批准。这时半导体研究所已先后建立了包括表面物理、传感器实验室、超晶格实验室和光电集成（半导体所区）实验室的 4 个国家重点实验室，因此，材料科学实验室在争取基本建设投资和运转费用上遇到很多困难。在自费开放的情况下，实验室采取多层面、多渠道筹集经费，首先从国家科学技术委员会争取到 48 万元的“开门”费，使实验室有能力启动第一批十余项开放研究课题，接着从国家发展计划委员会和中国科学院争取到一笔用于建立高分辨率傅里叶红外光谱测量实验室的费用约 150 万人民币（35 万美元）。1992 年 10 月 15 日，在中国科学院技术科学部主任师昌绪院士和国家科学技术委员会工业司汪宗荣副司长主持下，通过了国家“863”新材料领域 MBE 基地的论证，争取到 360 余万元的经费支持，购置了法国 Riber 公司的 MBE 系统。1995 年 10 月又从中国科学院获得离子束外延设备技术改造费 250 万元，以及获得航天部“921”项目设备更新费 200 多万元等。在这些经费支持下，大大改善了实验室仪器设备条件。

1996 年实验室学术委员会换届，选举林兰英院士为第二届学术委员会名誉主任，蒋民华担任主任，王圩、屠海令任副主任。实验室主任仍由王占国担任，副主任由钱家骏和陈伟担任。同年，实验室参加了中国科学院开放实验室评估，名列前茅，并获运转费支持。

1999 年实验室进入研究所创新工程，调整了结构，精简了人员，实验室平均年龄为 38 岁，实验室的管理也做了相应调整，增加了陈诺夫、陈涌海和王玲 3 名年轻的副主任。

王占国 2002 年 7 月任实验室学术委员会主任，2009 年任实验室学术委员会名誉主任，在宏观上把握实验室发展方向，支持和指导青年学术骨干开展科研工作。近年来，实验室在半导体材料和器件诸多研究领域取得了一系列创新性成果，为我国半导体科学与技术的发展做出了重要贡献。

深情厚谊，历久弥坚

林耀望

王占国与我，在20世纪60年代初，前后来所工作，历经50多年，我们结下了深厚的情谊。他善于学习，刻苦钻研，为半导体材料科学发展打下了深厚的物理理论基础；他高瞻远瞩，具有宽阔的国际视野；他才思敏捷，知人善任，具有很强的组织管理能力。他带领半导体材料科研团队，推动了半导体材料从体单晶材料→外延材料→超薄层量子结构纳米材料发展，开拓了一片崭新的研究领域，为祖国半导体材料科学发展建立了丰功伟绩，成为一名享誉国内外半导体材料科学著名的科学家。他为人正直朴实，待人诚恳，平易近人，处事多谋善断，勇于担当，在我从事科学研究漫长的人生道路上，留下了许多值得回味而又美好的记忆。

（一）

在提高GaAs材料质量研究过程中，当时国内外同行共同认识到，从理论上GaAs比Si有许多优越性，但在材料制备上有许多技术难关需要突破。我作为气相外延材料生长负责人，非常需要材料测量为我们提供准确可靠的电学参数，以作为我们工艺不断改进的依据。王占国负责电学测量，他创造性地开展电学测量工作，在电极制作和测量方法上精益求精，为我们提供了大量准确的、可靠的数据。此外，他查阅了大量国外文献资料，对GaAs材料物理问题做了深入而透彻的分析，为破解GaAs制备中存在的难题提出了自己独特的见解，推动了我国GaAs材料研究工作从国内领先走向国际领先。1981年，我作为中国科学界的代表，在出席美国圣地亚哥召开的国际学术会议上宣读论文，正式向同行宣告，我国GaAs材料电学参数的各项指标超过了美国与日本水平，居国际领先。王占国在瑞典异国他乡给我送来了深深的祝福。我让人转达，这项研究成果有王占国的心血与智慧。可是，我心中十分遗憾，论文署名与得奖名单申报中“王占国”3个字没有显现。他对我的困难处境深表理解，从不使我为难。

（二）

1996–2001年国家光电子工艺中心MBE组，潘中博士从日本学成回国，负责开展GaInNAs/GaAs材料与光电子器件及集成研究工作，取得重大突破，在国内外重量级刊物上发表了20多篇论文。美国同行高度关注我们的研究进展，多次派人来洽谈合作事宜。合作洽谈，只是幌子，其真实目的是挖掘中国的科技精英、青年才俊到美国工作。潘中和我的合作研究十分默契，可以说达到亲密无间程度。有关领导多次提示我，要留住他。美国优越的科研环境，丰厚的薪资报酬，还是把他吸引走了。虽然他在多次公开场合说：“没有林老师地点头同意，我是不会放弃现有工作去美国的。”但我心中明白，我区区个人力量，哪能阻止年轻人对美好前途的向往。2001 年 10 月 7-11 日，由美国海军实验室资助的国际窄带氮化物学术研讨会（The International Narrow Gap Nitride Workshop）在新加坡举行。我国出席会议的代表，除王占国和我外，还有江德生、陆大成、徐仲英和南京大学的郑有炓。我的报告题目是“1. 3μm GaInNAs/GaAs quantum well lasers and photodetectors”。王占国院士以中国科学家代表参加会议的筹备工作。会议决定安排两位代表在大会上作特邀报告，一位是美方代表，一位是中方代表。潘中虽然为我办理了相关申请手续并为我做了充分准备，但是美方并不认可，在筹备会上仍然提出希望让潘中作报告。王占国对美方干涉中国事务十分不满，他仍耐心地向美方介绍我研究的工作背景与学术成就。最终，美方接受了王占国的意见，会议主席 Dr. Y. S. Park 向我发出邀请，让我在大会上作特邀报告。王占国在学术上对我的赏识和信赖令我感动。王占国进一步要求会议的一切费用、参加会议代表的住宿费都由美方承担。在美国人强势面前，王占国以自己渊博的学识、敏捷的思维、锐利的洞察力，为中国学者参加会议争取到最大的利益，展现了中国人的骨气与志气，给我留下了深刻而美好的印象。

在大会上，我用英文成功地作完报告，并准确无误地回答了代表们的质疑与提问，得到了与会代表的认可与赞赏。会后，郑有炓院士（我大学时代半导体专业的老师）对我说：“你和王院士配合得真好！为祖国争光了，我为你们高兴。”最后，在大会评选唯一一篇优秀论文的过程中，王占国竭力为我争取。然而，美国人多势众，我落选了，这是意料之中的事儿。王占国为我惋惜，但我却欣然接受。2002 年，王占国邀请我到中国科学院半导体材料重点实验室工作。

王占国院士科研事迹回顾

何春藩

王占国来到半导体研究所后，分在材料研究室从事材料特性的测量分析工作。他独自研制了高阻静电计及其测试系统，解决了低温高阻测量问题，在国内首先实现了20–400K硅的变温霍尔系数的测试。他在GaAs及其他Ⅲ–Ⅴ族化合物半导体材料的电学和光学性质的研究中，特别在GaAs体材料的强场特性、热稳定性的研究方面所取得的成果，以及与林兰英一起提出的“杂质控制”观点，为我国高纯GaAs材料的研制由体材料转移到外延材料上来，并为20世纪70年代末我国GaAs外延材料达到国际先进水平起了重要的作用。

1967年5月，正当“文化大革命”在全国开展的时候，王占国受651设计院的委托，负责人造卫星用硅太阳电池电子和质子辐照效应研究的任务。他亲手制定了实验方案和实施计划。5月底，他与同事一起，赴上海中国科学院有机化学研究所和中国科学院原子核研究所，分别做硅太阳电池的电子辐照和质子辐照效应实验。实验结果发现：硅*NP*结要比*PN*结更耐电子辐照。在这年年底召开的硅太阳能电池定型会上，651设计院电源总体组采纳了王占国的建议，将原定的*PN*结电池改为*NP*结电池定型投产。王占国的这一研究成果，不仅节约了经费，还保证了人造卫星的长期安全运行，具有重大的经济效益和社会效益。

1970年7月，王占国受命起草代号为2100（空爆试验）的实施方案，并被任命为空爆核试验小组副组长、实验项目技术负责人。7月底，他完成了实施方案的起草和14研究院参试实验小组的组建。8月初，实验组乘专列出发，一周后到达罗布泊参试现场。在现场的近两个多月里，王占国不但要为实验小组设计瞬态的实验方案和安排样品的布点，还要负责对实验样品和装置的拍照，工作非常忙碌和辛苦，经常加班加点到深夜。

1970年10月中旬的一日，戈壁滩上炸雷般的一声巨响，氢弹点爆成功了。实验小组的同志们，获得了核爆释放出的脉冲中子流、光辐射和冲击波以及强电磁波辐射对半导体材料、半导体器件和半导体集成电路的影响的第一手宝贵数据，顺利完成了核裂变辐照试验。

1980年10月，王占国来到了瑞典南部的隆德城，作为访问学者在这里从事半导体深能级物理和光谱物理的研究。王占国接手的是研究液相外延GaAs材料中A、B两个能级的性质。王占国夜以继日忘我地工作，研究工作获得了突破性进展。得出了液相外延GaAs中的A、B能级，是同一缺陷的两个不同能态的双受主中心；其化学组成很可能是 Ga 占据了 As 位的反结构缺陷。硅中金施主及金受主能级也同样是同一缺陷的两个不同能态；两者都不是两个不同化学起源的相互独立的能级中心。这一结论，与导师哥尔马斯和郎格等学者联名在很有名望的《物理评论》上发表的文章中所得的结论相反，这无疑是对权威学者的挑战，是对自己导师所下结论的否定。

王占国用心血写成的《硅中金施主和金受主光电性质的系统研究》这篇有分量的学术文章，哥尔马斯似乎不大愿意让其发表。过了 3 年之后，王占国独自署名，在我国的《半导体学报》上发表了。文章发表后，国际上深能级研究的另一大权威，美籍华人萨支唐教授在美国《应用物理》杂志上感慨地写道：“王用哥尔马斯的实验室测得的实验结果，否定了哥氏的结论。”

在这 3 年的进修日程里，王占国做出了有国际影响的工作。他与合作者提出了识别两个深能级共存系统两者是否同一缺陷不同能态的新方法，解决了国际上对 GaAs 中 A、B 能级和硅中金受主及金施主能级本质的长期争论；提出了混晶半导体中深能级展宽和光谱谱线分裂的物理新模型，解释了它们的物理实质；澄清和识别了一些长期被错误指派的 GaAs 中与铜等相关的发光中心等问题。这些研究成果，引起了同行们的好评和重视。据 *SCI* 检索，到 1995 年 5 月，这些论文先后被引用了 200 余次。王占国的出色工作成绩，为祖国争得了荣誉，为民族增添了光彩。

1983 年 11 月，王占国载誉归来，立即开始了报效祖国的征途，先后承担了国家科技重点攻关的“863”项目和国家自然科学基金等多个项目的研究任务。在祖国这块希望的田野上，王占国奋力拼搏，为祖国默默地耕耘了 30 多个年头，在半导体材料及材料物理研究中，做出了一系列重要贡献。自 1984 年始，王占国致力于深能级物理实验室建设和材料物理的研究，获得了中国科学院科学技术进步奖二等奖（1989）、三等奖（1990）以及“七五”国家科技重点攻关奖（个人）（1990）。

1987 年，在林兰英院士的倡导与组织下，在航天部和国家科学技术委员会等部门的大力支持下，利用我国发射的返回式地球卫星，做了拉制 GaAs 单晶的试验，首次在空间从熔体中生长出了 GaAs 单晶，开拓了我国微重力材料科学研究新领域，其成果荣获中国科学院科学技术进步奖一等奖（1989）和国家科学技术进步奖三等奖（1990）。王占国除了协助林兰英院士制定和签署与航天部 501 所合作协议、组织项目实验方案论证和实施外，还主持了空间生长 GaAs 材料的光、电性质的研究，取得了国际先进水平的研究成果。如在微重力条件下，碲在 GaAs 中的平衡分凝系数有明显减小趋势；实现了由表面张力梯度驱动的马拉高尼（Marangoni）对流在晶体中产生杂质条纹理论预言的实验验证等。这些研究成果，对微重力条件下的晶体生长理论研究有着重要的意义；根据在太空生长 GaAs 的深能级种类和密度减小的实验结果，提出了“太空中由于重力驱动的溶质对流消失，可使化合物材料的化学配比得以能精确控制”的新观点，这对化合物半导体中的本征缺陷、杂质及其相互作用问题的研究有着指导性的作用。

王占国还同时承担了其他一些研究任务。例如：在直拉硅中新施主的电子组态和微观结构研究中，摒弃了新施主与氧直接相关的传统观点，与合作者一起提出了“新施主是由热处理过程中氧的沉淀导致硅自间隙发射、聚集、滑移而形成的60°刃型位错环”的结构新模型，不仅解释了新施主的电学、热学性质，而且还预示了它的新行为，具有重要的学术意义。通过对半绝缘 GaAs 中深能级行为的实验研究和理论分析，提出了“半绝缘 GaAs 五能级补偿新模型和电学补偿新判据”，为提高半绝缘 GaAs 的质量提供了科学依据。与同事一起，在 GaAlAs/GaAs 二维电子气材料的性质、高电子迁移率晶体管的结构材料制备和大功率半导体量子阱激光材料以及 GaAs 材料质量与器件性能关系等研究中，也取得了优异成绩，获中国科学院科学技术进步奖二等奖 2 项（1995，1997），国家“八五”重点科技攻关奖和先进个人（1996）等荣誉。

早在 20 世纪 90 年代初，王占国在任国家“863”计划新材料领域专家委员会委员、常委和功能材料专家组组长期间，他与时任专家委员会首席科学家的蒋民华院士一起，率先倡议在新材料领域的“863”计划中，设立了“氮化镓基材料和蓝绿光器件研究”课题，继而还以重点项目和重大专项任务给予了有力支持，为我国半导体固态照明技术的发展奠定了基础；与此同时，他还坚持把全固态激光器，作为材料领域的优先研发课题，继而以重大专项任务给以持续、稳定的支持，使合作者们取得了一批具有国际先进水平的成果，培养了一支高素质的研发人才队伍，为我国在国际该领域占有一席之地做出了贡献。为此，在纪念“863”计划 15 周年之际，王占国被科学技术部授予先进个人称号。

由于王占国在半导体材料科学领域的卓越贡献，1986 年他被破格晋升为研究员，1995 年当选为中国科学院院士。这位曾在国外遭受过冷遇的学者，饱含崇高的爱国热情，在中华大地崛起，成为我国第二代著名的半导体材料及材料物理学家。

近 20 年来，王占国的研究工作重点又集中在半导体低维结构和量子器件这一国际前沿研究方面。

一段往事

钱家骏

2017 年 11 月份，半导体研究所老年科协分会举办了一场“空间用硅太阳能电池研制与回顾”的科技沙龙活动。与会的有 20 世纪 60 年代研究所参与硅太阳能电池研制的科技人员。大家回忆当年那些往事，依然历历在目，津津乐道。

太阳能电池，主要是用来做太空人造卫星的长寿命电源的。其关键特性是，*PN* 结的光电转换效率及其使用寿命。20 世纪 60 年代，研究所研制的硅太阳能电池，采用在 *N* 型硅衬底上扩散 *P* 型层，制成 *PN* 结型光电池。1964 年的电池转换效率突破了 10%，1965 年达到了 15%，与国外发表的实验室转换效率持平。然而，当时国外曾有文献报道，在 *P* 型硅衬底上扩散 *N* 型层，制成 *NP* 型光电池，尽管初始的光电转换效率，略低于 *PN* 结型光电池的转换效率，但长期处于太空的环境中，受高能电子和质子等粒子辐照的影响，*NP* 型太阳能电池的退化速率，比 *PN* 型电池要慢许多。因此成为太空中太阳能电池应用的选项之一。

问题出现了，对于我国的人造卫星，到底应使用何种类型硅太阳能电池？特别是今后定型批量生产的硅太阳能电池板，应该采用哪种结构更合适？为了解决此问题，1967 年 5 月，651 设计院，委托王占国担任硅太阳能电池受高能电子等辐照影响的课题组业务组长。他带领研究组制定方案、实施计划，并赴外地做实验。为了加快进度，实验小组日夜加班，很快得出了结果。实验数据表明，*NP* 结硅太阳能电池抗电子辐照的能力比 *PN* 结太阳能电池提高了几十倍。这就是说，*NP* 型太阳能电池在恶劣的太空环境下，其使用寿命会更长。于是在 1967 年年底召开的硅太阳能电池定型会上，651 设计院电源总体组采纳了王占国的提议。将我国原先使用的 *PN* 结型硅太阳能电池，改型为 *NP* 结硅太阳能电池，定型投入批量生产。

这项研究成果，不仅保证了人造卫星长期安全可靠运行，还节约了研究经费，为我所在 20 世纪 80 年代，把硅太阳能电池生产转让给生产厂家打下基础。具有可喜的经济效益和社会效益。

在科技沙龙会上，当年参加 *NP* 型硅太阳能电池研制的陈宗圭，回忆了 50 年前的一段往事。当年为了尽快得出 *NP* 硅太阳能电池受高能电子辐照影响的可靠结果，在实验测试中，王占国直接用右手的拇指、食指和中指拿镊子，装取受高能电子辐照的样品。然而未曾想到，他的手臂和手指却暴露在泄漏出的高能电子辐照中，受到严重灼伤。几经住院治疗也未能彻底痊愈。

如今，50 年过去了，王占国手指上仍然存有清晰的伤疤烙印。这个烙印记录了他当年为我国航天事业艰苦奋斗的历程，为曾经在一起工作过的人们所记忆。

王占国院士支持南昌大学 GaN 研究记事

江风益

1996 年

“863”新材料技术领域“GaN 基半导体材料生长与蓝光器件研制”重点项目立项，课题负责人：陆大成、范广涵等，项目责任专家：王占国院士。

1998 年

南昌大学该课题负责人更换为江风益。

2000 年 1 月 17 日

王占国先生发邮件告知江风益，南昌大学蓝光课题经费追加 20 万元，表示对该课题取得的技术进展给予鼓励。

2000 年 3 月 10 日

江风益与王占国先生通电话，汇报南昌大学蓝光阶段性成果将与企业合作，实施蓝光 LED 材料产业化。王占国先生表示同意，并给予鼓励。

2000 年 6 月 14 日

北京会议中心天安门宾馆，“863”计划 715-001 专题交流会报告，王占国先生主持会议。江风益报告课题研制的 382nm 紫光 LED 结果（0.5mW，半宽 10nm），引起了包括王占国先生在内的与会者浓厚兴趣。用此紫 LED 激发红色和黄色荧光粉，看到红光和黄光，照射 100 元人民币可见水印。

2000 年 6 月 28 日

王占国先生给江风益发邮件，认可江风益起草的“863”蓝光课题补充和修改合同书。

2000 年 12 月 4 日

北京西郊宾馆，“863”计划新材料技术领域项目验收会。江风益第一个汇报，王占国先生问江风益课题所突破的关键技术，哪个最主要？江风益回答是 *P* 型 GaN 掺杂和激活技术。

2004 年 1 月 18 日

王占国先生给江风益电子邮件，让江风益对国家中长期科技发展规划中关键信息功能材料和器件专项提出建议，2 月 1 日江风益提交。

2004 年 8 月 16 日

王占国院士、陈皓明教授、丁文江教授（现为院士）等专家应教育部科技司邀请，参加依托南昌大学建立的教育部发光材料与器件工程研究中心验收，专家结论是“业绩一流，圆满完成了建设任务”，验收组长丁文江、副组长王占国分别在专家验收意见书上签字。

1997–2017 年

20 年来，王占国先生一直高度关心、热情鼓励和大力支持南昌大学 GaN 材料与器件的研究、开发和产业化。南昌大学在完成了该校承担的第一个“863”课题（蓝宝石衬底蓝光 LED）之后，创造性突破了硅衬底蓝光、绿光和黄光 LED 材料生长与芯片制造技术，并实现了量产，其中 GaN/Si 蓝光项目成果获 2015 年度国家技术发明奖一等奖，GaN/Si 黄光项目成果获 2017 年国际半导体照明联盟年度新闻奖。王占国先生对江风益教授团队每取得的技术新突破，均表示高兴和祝贺，曾称赞江风益“原创技术、为国争光”。

2017 年 1 月

王占国先生等专家推荐江风益教授参选院士，先后经中国光学学会会评、中国科协会评和信息技术科学部通信评审，江风益成为 2017 年度中国科学院院士初步候选人。

我生命中的贵人

李瑞钢

我是王占国老师带的第一个博士研究生，他是我生命中的贵人。

我是1986年大学毕业的，毕业之后就分配到了电子工业部第13研究所，从事GaAs超高速集成电路的测试及设计工作。1989年萌发了考研究生的念头，1990年在山东石岛召开“八五”攻关论证会有幸认识了王占国老师，在烟台机场聊起了报考他的博士研究生的可能性。我是从本科破格考他的博士研究生的。为了证明我有硕士研究生同等学力，从没入学就得到了王老师的巨大帮助。他拿着我的一些科研项目获奖证书带我找到了电子科技大学的陈星弼院士和清华大学研究生院院长介绍我的情况，并联系了中国科学院教育局有关主管。所以我顺利得到了中国科学院博士研究生报考资格。我很幸运，各种考试都通过了，终于成为王老师的博士研究生。

在学期间，我得到王老师无微不至的关怀和指导，无论是在半导体材料物理、半导体材料表征方面还是在科学研究的思想方法方面都取得了长足的进步。因为我的英文不好，我的第一篇国际会议论文在发出之前经历了很多曲折，王老师给我一点一点地梳理实验和理论内容，归纳逻辑，修改英文，最后被选为大会宣讲论文。王老师虽然在工作上非常严肃认真，但在生活方面非常平易近人，我读研究生的时候他家就住在半导体研究所后院，我们学生经常往他家跑，他就给我们讲家里猫和狗如何斗的故事，也讲家里松鼠如何藏花生的故事。1993年因为我参加半导体研究所组织的卡拉OK比赛，所以很多人都认识了我。有一次王老师去财务报销，听别人说我唱歌的事，他骄傲地说，“这是我的学生”；后来大家又说某某唱的不行，他又赶紧说，“某某不是我的学生”。

1995年我博士毕业了，王老师也搬了新家，我和王老师都住进了科学园南里，有时在院子里散步还可以碰到。后来王老师就搬到了现在中国科学院黄庄院士大楼。2008年北京奥运会期间我回国探亲，和王老师约到他家吃晚饭。那天中午和朋友去了奥体公园，因为鸟巢里还在比赛，我们被困在国家奥体中心怎么也出不来，晚上12点才到王老师的家，王老师一直等着我们，还在家里给我们准备了夜宵。2011年回国，得知王老师做了脑部手术，去看望他之前李树深院士叮嘱我不要和他聊太长时间，他身体还比较虚弱。但我到了他家以后一聊就是3个多小时，聊新型半导体材料，聊国家科研投资的合理使用，聊科学技术与产业的资源配置，也聊他在全世界各地的学生……2017年8月份我们组织了一个中国三代半导体项目的评审会，想邀请王老师给指导意见，他很快给我回了信：“瑞钢，我因身体欠佳，一直在家休息，无法参会，还请见谅。王占国。”收到以后我内心感到非常难过，会一结束我马上去了他家。王老师这几年一直被病痛折磨，确实很虚弱，但我们谈起过去的时候，他很快就精神起来了。我这才突然意识到他已经把接力棒交到了我们手中。他很难再回到 DLTS 旁边测量 GaAs 的 *EL*2 的寿命和发射率了，很难再回到 FTIR 旁去研究杂质吸收了，很难……但我们，我们的学生和我们学生的学生会继续探索凝聚态物理的奥秘，继续发现新的功能材料，继续制造新型的半导体器件，继续让材料科学为人类社会发挥更大的作用。

盼望王老师早日康复，继续为固体材料科学做出更多的贡献。

贺王占国老师80寿辰

傅建明

1985年夏天，正值改革开放初期，我在玉泉路中国科学院研究生院学习一年基础课后，来到半导体研究所，师从王占国老师攻读硕士学位。王老师那时刚从瑞典回国，带来了国际上最新的科研方法和器材，建立了国内最先进的半导体材料和器件实验室。我最初的主要工作是熟悉使用深能级瞬态谱仪（DLTS），承担日常测量，建立光电容测量装置。我当时21岁，初出大学校门，没有认识到自己的不足，还自视甚高，但在一个简单的热电偶安装和使用中就出了差错，被王老师批评了一回；不过王老师批评归批评，他观察到我的不足以后细心指导我的动手实验能力，从最基本的画图纸开始，我跑所属加工厂加工，组装真空、光学及电子测量设备，调试等，终于建立起国内第一台光电容测试装置。接着王老师又指导我利用这台新设备，结合其他如DLTS和在物理所做的光荧光等，测试了钯掺杂的硅材料，而且他根据推测分析，大胆提出我们可能发现了一个新能级；我们发现实验结果完全与他的这个推断吻合，很快我们的论文发表在美国的《应用物理杂志》上。一个初出茅庐的学生在顶级专业杂志上作为第一作者发表论文，这对我是莫大的鼓舞，也体现了王老师对后辈的提携和鼓励，王老师的教诲让我一生受益。以后的博士论文和实际工作及事业中，我对自己的动手能力信心十足，解决了许多难题，也有多达80项专利的发明，这都是源自于王老师对我的训练和创新性思维的启发。

对我们这些年轻学生，王老师也从生活上关心，逢年过节，他和师母把我和当时还是女朋友的太太叫到他们家吃饭，嘘寒问暖。有次冬天去广州开半导体年会，我们在火车上两天，王老师也和我们有说有笑，还聊起当年在上海做实验由于条件有限缺乏探测器导致手上被粒子束辐射受伤的经历，让我紧张的心情放松不少，也感受到前辈为国赶超先进科技水平的艰辛付出。记得1987还是1988年有日全食，王老师也和我们讨论安全观赏方法，真的是团结紧张，严肃活泼。

我在研究所里的短短3年时间，王老师就承担了国家多项科研课题如微重力晶体生长、Ⅲ-Ⅴ族材料生长和外延等，还招收了好几个硕士和博士生。一晃30多年过去了，王老师现在是桃李满天下，我作为其中的一员也小有成就。我在取得博士学位后加入了世界第一大半导体设备公司——美国应用材料公司，研究生学习时从王老师那里学习的仪器设计及装配能力让我如鱼得水，领导开发了多项关键产品，曾任职物理气相淀积（PVD）产品部门总经理。2007年我和半导体研究所研究生同学徐征博士共同创立了太阳能公司Silevo，成功开发了利用非晶硅/晶硅异质结的太阳能电池，高性能低成本的铜栅线为世界首创；2011 年又在国内建立了首条高效率太阳能电池及模组生产线并成功量产，23% 的转换效率为国内首创，也处于世界领先地位，更重要的是展示了新技术的量产化，产品成功销售全球，这个过程中王老师也给予我们不少的指导和帮助。2014 年以马斯克为董事长的 SolarCity 公司认为 Silevo 公司技术先进、产品一流，于 2014 年 9 月以 3. 5 亿美元并购 Silevo 公司；SolarCity 随后于 2016 年 11 月为了和特斯拉汽车公司的储能结合并入了特斯拉汽车公司。在王老师熏陶下的创新能力和设备自力更生是我们成功的法宝；王老师对我们国家的半导体事业贡献巨大，也是我们这些弟子们的最重要的人生导师。王老师为国强盛，辛勤耕耘；奋斗 60 载，霜雪双鬓。桃李天下，常感师恩。衷心祝愿王老师身体健康，幸福吉祥。

往事点滴

董建荣

时间飞逝，往事如烟，转眼间毕业已20多年了，但跟随先生学习的几年间的点点滴滴如昨日之事，历历在目，难以忘却。

1992年4月中旬，我去北京参加博士入学考试，这也是我第一次去北京，心情确实有点激动，毕竟是众人向往的首都北京，又是报考中国最高研究机构中国科学院王先生的博士生。从西安乘火车到达北京后已是黄昏，乘地铁转大巴，晕头转向地摸到了半导体研究所的招待所。第二天一早惴惴不安地打通了先生家的电话，约好了见面时间。按约定时间在半导体研究所西大门口见到了先生。先生身着浅灰色的西装外套，感觉先生非常朴实随和，想象中的那种和中国科学院研究员之间的距离感顿时减少了许多，也让我紧张的心情得以放松几分。在去王先生当时行政楼办公室的道上和他聊了几句，先生话语中透露着淡定与沉着，说话声音虽不是很洪亮但中气十足。在办公室简单交流了一下报考博士的有关事项及我在学校的学习情况后，先生给了我一些资料，是有关MOCVD和MBE的中文介绍材料，也是我第一次接触到这些外延技术，未曾想到这些技术成为我后来职业发展中所依赖的重要工具。

先生平时工作非常忙，我没有事也尽量避免打扰先生，只有讨论论文研究计划、工作进展汇报、论文修改和发表论文需导师签字等事才去找先生，先生有时也找我了解论文工作进展。先生对学生要求很严格，尤其是学术问题，对技术细节非常较真。记得我发表第一篇国外期刊论文时，将写好的论文初稿拿去请先生过目，先生看完后给我指出论文中论据的不足，点出英文表述上的欠缺，后来我花了很长时间增补内容和修改语言，几次请先生指点，经过半年多的修改论文稿终于投了出去并被接受。虽然反复修改论文的过程有点痛苦，但让我基本明白了如何写科技论文，尤其在英文写作能力方面得到了很大的提高，这为以后的英文写作奠定了扎实的基础，尽管我当时还是入门的水准。我后来投递的英文论文稿件，不管审稿人对论文稿的技术水平评价如何，没有一篇指出语言存在问题的。

还有一次是1994年12月底欧中材料学会在友谊宾馆召开的一次学术会议上，我当时作了一个口头报告，下来后先生告知我“temperature”这一词的发音不对，那么多年我竟浑然不知自己的发音错误，足见先生的细致认真。

1995年在放暑假前的一次研究生工作进展简要汇报会上，我提到了用一种合金材料中深能级的微观构型解释当时正在做的镓铟磷材料深能级实验的结果，没有表达清楚，先生点出问题所在并指出那些构型在很多年前已阐述得很清楚了。会后我才得知那些深能级微观构型最早是先生提出的，让我深感自己做研究工作的肤浅和不求甚解。

先生平时对我们学生并没有讲太多的大道理，而是在传授专业知识的同时通过一件件小事教我们做人的道理与为人的原则。虽然在读博士的三年半期间学到的专业知识非常有限，但通过点点滴滴的事让我明白做科研需要坚持、耐心和认真细致的态度，这一点对我确实印象深刻，铭记在心，受益匪浅。博士毕业十多年后，自己也回到科学院的研究所工作，也开始指导研究生。在先生的言传身教之下，深感作为研究生导师的职责和义务，也更能体会到当时先生指导我们时的辛勤付出。受先生的影响，我一直要求学生做事耐心认真、注重细节。

工作之后，也偶尔有机会拜访先生并讨教一些技术问题，比如后来我做半导体多结太阳电池方面的工作，虽然先生当时没有做这方面的研究项目，但先生对国际上多结和聚光太阳电池的研究水平及各种技术路线与技术细节了如指掌，点评技术高屋建瓴。一次次的交流让我更加敬佩先生渊博的学识、活跃的思维和极强的记忆力，这也是为什么如果不是别人问起，我都不敢主动说出导师的尊名，生怕作为学生言行不慎有损先生的声誉。

每当回想起师从王先生期间的点点滴滴，倍感温馨，从内心感激先生的指导和教诲。在王先生的生日到来之际，祝先生生日快乐，健康平安！

在王占国导师身边的日子

杨 斌

我是1992年王占国院士和林兰英院士合招的博士研究生。

1992年金秋，北京花团锦簇，秋高气爽，到处洋溢着欢快的国庆节日气氛和丰收的喜悦。26岁的我，揣着博士研究生录取通知书，带着对未来无比美好的憧憬，载着家人和乡亲厚厚的期盼，怀着激动的心情来到中国科学院半导体研究所报到。对于我来说，从偏远的大西北兰州到全国人民向往的首都北京，能在号称国家队的中国科学院从师世界驰名的两位科学家门下做博士研究生是多么令人羡慕和向往的机会。我暗自下决心一定要尽自己最大的努力，争取做出最出色的成绩，不辜负家人和乡亲的期望。

报到后我去见王老师，紧张的心情不言而喻。然而，王老师直率的性格、谈话时不时流露出的笑容、谦虚和蔼的态度和不厌其烦的讲解缓解了我紧张的心情，使我很快地适应了在北京的学习和研究工作。同时，王老师对未来科技发展方向精准的指点和高瞻远瞩的预测给予了我极大的激励，指导我尽快地选定了研究方向和课题。

王占国老师是世界著名的半导体深能级缺陷研究专家。1980–1983年，作为访问学者，他曾赴国际著名的深能级研究中心瑞典隆德大学固体物理系，从事半导体深能级物理和光谱物理研究。与国际该领域的权威H. G. Grimmeiss教授合作，做出了多项有国际影响的工作，如提出了识别两个深能级共存系统两者是否同一缺陷不同能态新方法，解决了国际上对GaAs中A、B能级和硅中金受主及金施主能级本质的长期争论；提出了混晶半导体中深能级展宽和光谱谱线分裂的物理新模型，解释了它们的物理实质；澄清和识别了一些长期被错误指派的GaAs中与铜等相关的发光中心等，他的论文被引用200余次。他在半导体材料、器件辐照效应和光学、电学性质研究，尤其是硅太阳电池电子、质子辐照效应研究领域成果累累，曾为我国人造卫星用硅太阳电池定型（由*PN*改为*NP*）投产起到了关键作用。比如，受国防科委第14研究院的委托，他负责制定了我国电子材料、元件、器件和组件辐照效应研究方案和实施计划，他在电子材料、器件和集成电路的电子、质子、中子和γ射线的静态、动态和核爆瞬态辐照等方向取得了丰硕的成果，为中国航天事业、核加固、核突围和电子对抗等国防工程做出了突出贡献。

记得王老师给我在半导体研究所上的第一课就是学习和深刻理解深能级瞬态谱仪、光荧光、电化学C–V、霍尔效应测试的物理原理，使用方法以及在研究GaAs外延材料性能和缺陷方面的应用，为我在博士论文研究中碰到的DX中心、GaAs/AlGaAs界面平整度分析打下了坚实的理论和实验基础。

20 世纪 80 年代末，王老师将研究重点集中到了半导体材料生长、性质研究和新型电学和光学器件的研究领域。他先后负责承担多项国家自然科学基金、国家重点科技攻关和国家“863”研究课题。比如他提出了直拉硅中新施主微观结构新模型，摒弃了新施主微观结构直接与氧相关的传统观点，成功地解释了现有的实验事实，预示了它的新行为。另外他还在国内率先开展了超长波长锑化物材料生长和性质研究，并首先在国内研制成功 InGaAsSb、AlGaAsSb 材料及红外探测器和激光器原型器件。他还协助林兰英先生，开拓了我国微重力半导体材料科学研究新领域，首次在太空从熔体中生长出 GaAs 单晶，并对其光、电性质做了系统研究，受到国内外同行的高度评价。

20 世纪 90 年代初，王老师研究工作的重点集中在半导体低维结构和量子器件这一国际前沿研究方向，先后主持和参与负责十多个国家“863”项目、国家重点科技攻关项目，国家自然科学基金重大、重点项目，以及中国科学院重点、重大研究项目。

1992 年，我有幸在王老师安排下加入了由梁基本研究员率领的超薄层分子束外延低维材料研究小组，小组研究人员有廖奇为、徐波、朱战平、我和另外一位博士生李伟。当时，我国在Ⅲ－Ⅴ异质结半导体二维电子气材料研究领域仍然比较落后。这种高质量Ⅲ－Ⅴ异质结半导体二维电子气材料是研制现代高科技电子器件的核心和基础，有着十分广阔的军事和民用前景，可制成广泛应用于通信、相控雷达、电子对抗等领域的高频、微波、毫米波、低噪声和大功率器件。为了大幅度提升我国在Ⅲ－Ⅴ异质结半导体高质量二维电子气材料领域的研究水平，王老师亲自主持挂帅，设定了 1 年内达到国际最好水平的科研攻关项目和卓越目标，指导我负责从理论上模拟计算如何优化 GaAs/AlGaAs 二维电子气异质结结构参数来提升电子迁移率。经过 3 个月紧张的分子束外延设备调试和另外 3 个月日日夜夜 24 小时不停机的奋战，终于在 1993 年夏天，王老师领导的研究小组提前 3 个月取得了低温 4. 8K 电子迁移率超越 100 万的优异成果。这个结果远远超越了当时国内水平，跻身国际寥寥可数的几个曾经取得 GaAs/AlGaAs 二维电子气超越 100 万电子迁移率的研究机构之列。此项研究成果获得了 1995 年中国科学院科学技术进步奖二等奖。王老师指导我优化 GaAs/AlGaAs 二维电子气异质结结构的论文也于 1994 年获得了北京市青年优秀科技论文奖，我在 1996 年荣获中国科学院院长奖学金。在王老师的精心指导下，我们的论文先后发表在美国《应用物理快报》（1995，2 篇）杂志，并在 1994 美国秋季 MRS 国际会议上作报告。

近年来，王老师又发展了应变自组装 In(Ga)As/GaAs，InAlAs/AlGaAs/GaAs，InAs/InAlAs/InP 和 InAs/InGaAs/InP 等量子点、量子线和量子点（线）超晶格材料生长技术，并初步在纳米尺度上实现了对量子点（线）尺寸、形状和密度的可控生长；首次发现 InP 基 InAs 量子线空间斜对准的新现象，被国外评述文章大段引用；成功地制备了从可见光到近红外的量子点（线）材料，并研制成功室温连续工作输出光功率达 4W（双面之和）的大功率量子点激光器，为目前国际上报道的最好结果之一；红光量子点激光器的研究水平也处在国际的前列；2000年，他作为国家重点基础研究发展计划（“973”计划）项目“信息功能材料相关基础问题”的首席科学家，又提出了柔性衬底的概念，为大失配异质结构材料体系研制开辟了一个可能的新方向。

1995年我博士毕业后，经王老师和江德生教授推荐，有幸赴德国柏林Paul Drude Institute做博士后研究，在著名教授Ploog指导下从事氮化镓异质结外延材料研究。在27个月博士后研究期间共发表国际级杂志和会议论文16篇，被Ploog教授誉为他学生中出成果最快的研究人员之一。

1998–2010年，我先后在美国Agere Systems（吉尔系统公司）任高级研究员，Advanced Micro Devices（超微）任高级工程师。自2010年冬，我入职世界移动通信龙头高科技公司Qualcomm（高通），目前任美国高通总部高技术研发部总经理，主管14nm/10nm/7nm超高精尖Advanced CMOS研发和产业化，以及CMOS建模（CMOS Spice Model）工作。

在美国工业界20年研发和产业化的经历中，我总共荣获十多项公司大奖，获得美国已授权专利63项，受邀3次在国际大会作报告；是3个重大国际会议组委会委员，7个国际杂志的评委；发表论文62篇（含博士、博士后阶段发表论文）。所有这些成绩都与早期王老师注重培养学生综合科研能力密不可分。我记忆最深刻的是，王老师要求我们学习要勤动脑筋，做到概念清楚，基础扎实，见解独到。从博士研究阶段培养的这些良好的科研素质，使我在以后的工作和研究中受益匪浅，才有了我今天取得的这点儿成绩。

在王老师80大寿之际，我想对王老师说一句：谢谢您对我的辛勤培养，祝您健康长寿，桃李满天下！

我们的大导师王占国院士

周树云，陈广超，滕　枫

2004 年，以学科为基础、跨单位的大导师制度开始试行，我们有幸成为中国科学院院士王占国先生的学生，作为“北京市科技新星”这一群体的一员，可以说我们是幸运儿里的幸运儿，感谢北京市科学技术委员会为我们安排了一位好导师！

科研工作最为重要的是选题，即确立研究方向。如果研究方向有问题，即便付出再多的努力，也不会有多大的收获，更不要说对社会的贡献。作为青年科研工作者，虽然在热情、精力和体力方面有着巨大的优势，但是，对社会需求、领域动向、研究历史这些对研究工作有着巨大影响的因素还是很难把握。如果没有前辈智者的指点，青年科研工作者的优势可能会被消耗殆尽，而工作成果却可能是微乎其微。先生绝对是北京市科学技术委员会为我们解决这一问题而安排的最好的前辈！在和先生进行第一次交流时，先生就像洞悉了我们的困惑，从信息材料研究的历史切入，比对国内外的社会条件，纵论各个研究方向，剖析各个研究现象的物理本质，在先生细心的讲解中，我们每个人都好像得到了一张专属于自己的研究线路图，通过这张图，我们清晰地看到了自己领域的研究范畴、自己的研究在整个研究体系中的位置、所从事研究的前生和来世。在先生的讲解中，我们都有一种豁然开朗的感觉。著名社会心理学家马斯洛指出：每个人都有归属的需求，这是人能够构成社会的最基本的心理基础。科研也一样，像似暗夜行舟对灯塔的渴求！不能不说，先生是北京市科学技术委员会送给我们的一座灯塔！两个多小时的时间，在不知不觉中流过，从先生办公室出来，在每个人兴奋的脸上，都分明多了一份自信和坚定！

实验室对于科研人员来说是至关重要的，所有的科学设想、理论预言首先都要经得起实验室的检验和印证。任何一个国际一流的学者无一例外地都依靠着一个国际一流的实验室。先生是深谙此道的，在实验室建设方面给了我们大力支持。

古人云：“流水不腐，户枢不蠹”，交流成为当前科研工作中不可或缺的部分。但是，作为青年科研人员，往往因为知名度不高，位卑言轻，很难和业内实力派人士及相关部门领导进行有效沟通和交流。对于这一点，先生给了我们充分的理解，他一方面鼓励我们踏踏实实地多做工作、做好工作；另一方面利用各种学术会议，把我们介绍给相关的“VIP”们，不遗余力地为我们构建国内外交流渠道。有先生作引荐，这无疑是最好的桥梁和最好的通行证！我们得以结识许多在研究领域中做出巨大贡献的优秀的科研工作者，这里有许多我们早想拜见却无缘结识的科研前辈、有早有耳闻却没机会交流的同辈！记得有一次，先生在广西北海主持全国半导体材料的系列年会，我们本打算像其他会议一样在宣读完自己的论文后就去“欣赏”北海美丽的海景和“体味”洁净鲜美的海鲜。可是，大会一开始，我们就感到我们的“欣赏-体味”计划要泡汤！经先生精心的设计，大会的会序紧凑而丰富，一批领域的中坚人物到会，他们的精彩报告被错落有致地分布于会序之中。像早知道我们的心理一样，当我们急急赶到有精彩报告的会议室时，先生早已安稳端坐在那里，报告人做完报告，都会过来向先生致意，这时先生总会非常自然地把我们介绍给报告人。一周的会议，就这样紧张地开完了，然而我们却感到前所未有的充实。虽然没有领略北海海水的温润和海鲜的美味，可我们却见识了先生在驾驭大场面时的气魄和精致，更重要的是我们得到了先生的关怀和信任，并被这种关怀和信任所鼓舞和感动！

先生是无私的、宽厚的、仁爱的，他像大海托举着我们这些敢于为科学而冒险的小船，他又像大树为我们这些奋飞的鸟在疲惫时伸出可依靠的虬枝。在我们检测遇到困难时，先生毫无保留地为我们开放了他的研究室。

对于院士，一般人理解必然是放眼全球来考量事情的人，院士是不会关注于某一地域性质的问题的。但是，和先生的交流，让我们觉得这种理解是多么的片面。先生对北京市信息产业相关的研究团体和产业实体非常了解，这些单位所面临的问题和它们对首都的贡献都洞若观火。先生对北京市的关注和关心在言谈间不时流露，可以看出，先生虽然纵览国内外，但是对首都的感情还是非常深厚和真挚的。

和先生交流越多，越觉得先生人格魅力的非凡。我们 3 个人都有不同角度和程度的感受。每次聚会，先生都是我们的主要话题。虽然我们的体会各有不同，但是，先生治学和治研的特点，我们都认为是：精确和前沿！先生对自己的领域了若指掌，对其他领域也是明察秋毫。在一次交流中，先生指出目前能做的最大的 CVD 金刚石单晶的尺寸精确到了小数点后，这让我们震惊！每次交流，先生谈及的都是各个领域尖端的前沿，对某些时下的热点先生却不以为然，表现出的魄力和睿智让人瞠目！在日常生活中，作为年事渐高的前辈，先生却有着不同寻常的能力，先生竟然会用短信和我们沟通!！这是大大出乎我们意料的，大家都知道，手机短信是年轻人喜欢的沟通方式，上年纪的人一般不习惯使用，但是先生却运用自如。有一次，我们想给先生汇报事情，电话打过去，没有接线，正当我们疑惑时，短信过来“正在开会!”我们在惊诧之余，不禁对先生肃然起敬。先生事务繁忙，场合严肃而重大，在不方便应线时，为了不使联系人尴尬，先生已高龄竟学会了短信。由此先生体贴关怀的仁爱性格可见一斑！

和先生的交流已 14 年的时间了，在先生的指导下，我们在各自的工作中都取得了很大的进步。先生在半导体材料领域的丰功伟绩以及汗牛充栋的著述是我辈的榜样，无时无刻地激励着我们。我们由衷地感谢先生为我们付出的时间和精力，我们真诚地祝愿先生能够健康长寿！

我眼中的王占国院士

李　伟

1992 年我开始在中国科学院半导体研究所读博士。第一次见到王占国院士，他就给我就留下了深刻的印象，说话直爽、严谨，知识渊博又平易近人。

平时，王院士对我们的学习、实验课题抓得很紧，要求非常严格。我们除了白天上课和实验室工作，晚上也会去图书馆看资料。每个星期组里都有例行会议，我们需要准备汇报文献综述、实验计划方案以及研究进展。每次王院士都给我们提出意见和指导，对我们的研究工作帮助非常大。过去的经历仍历历在目，心中充满感激之情。

记得每次写论文时，王院士都仔细阅读，然后提出修改意见。第一篇论文修改了好几次，我几乎失去信心。后来，论文终于在国际知名期刊上发表了。经过王院士的耐心指导，我不断学习、积累经验，信心也得到提升，两年多的时间里在国际著名期刊发表了多篇论文。

王院士还鼓励我们博士生参加国际会议，拓展视野，开展国际交流与合作，了解国际最新研究动态。这些都对我博士期间研究工作起到了极大的推动作用。

除了学业，王院士也很关心我们的生活。我们也经常去王院士家聚会。王院士以身作则，教导我们要做一个正直、有爱心的人。记得有一次，我们用自己的博士研究经费领了一点学习用具，王院士知道后，暂停了经费。

在王院士悉心指导帮助下，经过 3 年的努力，我顺利完成了博士生的学习和博士论文工作。博士论文通过答辩，并获得了优秀。毕业时，我获得了中国科学院亿利达奖学金一等奖。随后，经王老师热情推荐，去瑞典做博士后。

永远不会忘记王院士的教育和指导。衷心感谢和祝福王院士。

德高望重，仰之弥高

王元立

心中的千言万语还是从2003年7月说起，我从北京科技大学材料物理与化学系在柳得橹教授的指导下博士毕业后，通过了中国科学院半导体研究所组织的博士后差额答辩，很有幸来到王老师领导的材料科学重点实验室，在王老师的指导下进行博士后的研究工作。虽然博士后的研究工作时间很短，只有两年时间，但是，王老师从我博士后研究工作的选题、课题研究、日常生活和出站后的工作选择方面都给予了大量的指导和帮助，为我在半导体材料领域的职业生涯打下了良好的基础。

至今，一晃出站都12个春秋了，但是博士后工作期间的往事如在眼前。王老师因材施教，提倡在学术面前人人平等的思想让我记忆犹新。记得第一次到王老师办公室报告自己以往工作的情况，王老师很随和地招呼我坐到座位上，听我介绍了博士研究生期间主要工作的情况后，因为我在博士研究生期间做了大量的微合金钢中和低碳钢中的位错和纳米级析出相的透射电子显微镜的分析工作，王老师建议我今后的研究侧重于Ⅲ-Ⅴ族半导体外延自组装纳米结构的微观分析工作，这样既能发挥我在博士研究生期间显微分析方面的优势，又容易在自组装纳米结构的微观分析研究方面获得创新性结果。王老师告诉我吴巨老师在半导体外延纳米组织的微观分析研究方面已经开展了大量的研究工作，通过透射电子显微镜对自组装纳米结构和外延超晶格结构进行了很多的研究，所以把我安排在吴老师办公室，建议我多向吴老师学习半导体外延纳米结构的微观结构分析工作。在吴巨老师的帮助和指导下，我逐步熟悉了半导体外延材料的微观分析工作，吴老师对待研究工作严谨认真，一丝不苟的风格给我留下了深刻的印象。随后，陈涌海老师给了我3个分子束外延的磷化铟（100）衬底上不同InAs沉积厚度的自组装量子线超晶格外延样品，在吴巨老师的帮助和指导下，通过对这几个样品进行透射电镜样品的准备、通过衍衬像进行样品平面和截面微观结构分析以及截面高分辨电子显微分析，获得了对半导体外延纳米结构微观分析的经验，高分辨透射电子显微镜的研究工作也发表在荷兰《晶体生长杂志》上。

入站几个月后，通过对实验室已有研究工作的学习和了解以及文献调研，发现在磷化铟（100）衬底邻位面上，也就是从（100）偏向｛111｝A富铟表面或从（100）偏向｛111｝B富磷表面不同角度对其上外延的自组装纳米组织的影响的研究不多，并且未见到系统研究的报道，但是自己也不知道对于这种邻位面衬底上外延的研究是否有实际意义。在博士后定选题前和王老师报告讨论过程中，王老师非常鼓励学生自己多提想法，对于在磷化铟邻位面衬底上进行纳米组织的外延工作非常支持，要求对磷化铟（100）偏转｛111｝A 富铟表面 和｛111｝B 富磷表面进行系列化的细致研究。按照王老师的建议，购买了磷化铟（100）偏转｛111｝A 富铟表面 2°、4°、6°与 8°衬底和（100）偏转｛111｝B 富磷表面 2°、4°、6°与 8°衬底，安排金鹏师兄帮我在 Riber32 分子束外延设备上进行自组装纳米结构的生长工作。后续通过对样品进行透射电子显微镜的分析工作，以及和雷文、黄秀颀等同学合作，对外延样品进行了光学性质和同步辐射 X 射线衍射的研究工作，获得了创新的实验结果，在美国《应用物理快报》和荷兰《晶体生长杂志》等期刊上发表了系列化的研究结果。还有很多上面没有提及的实验室的老师和同学，在我博士后工作期间也给了许多帮助和支持，整个实验室和睦友好的协作氛围给我留下了深刻而美好的记忆。

桃李不言，下自成蹊！在王老师 80 寿诞来临之际，附诗一首，衷心祝愿王老师生日快乐，身体安康，阖家幸福！

深耕Ⅲ－Ⅴ六十载，桃李芬芳遍神州；
八十寿辰喜相聚，弟子友朋乐心头；
德高望重授道业，诲人不倦解惑由；
寿比南山松不老，福如东海水长流！

超宽禁带半导体材料研究组发展历程

杨少延，刘祥林，汪连山，韩培德

超宽禁带半导体材料研究组（原MOCVD组）成立于1990年，是半导体材料开放实验室成立时最早的4个研究组之一，也是国内最早开展GaN材料研究的团队之一，第一任组长是陆大成研究员。研究组成立初期，昝育德副研究员首先建立了基本的MOCVD框架结构，后经陆大成等改进，研制出国产的MOCVD设备。

时任国家“863”高技术新材料领域专家委员会委员的王占国院士高瞻远瞩，在当时大部分人关注于Ⅱ-Ⅵ族蓝光材料的时候，他敏锐地预见到GaN基材料体系是未来蓝绿光、紫光和高频大功率电子器件的基础材料。在王占国院士建议下，1994年年底率先在新材料领域设立了两个GaN基光电子材料与器件的“863”课题，每个课题经费为20万元/年，其中一个由本组承担。在此经费支持下，研究组采用自建的MOCVD设备，利用掺Zn的GaN作为绝缘层、*N*型GaN为半导体层，1995年在国内率先研制成功室温工作的MIS结构蓝光二极管。在此基础上，1996年获得了“863”重点项目“氮化镓基蓝光材料生长及LED研制”支持（陆大成为负责人、王占国院士为责任专家）。经过3年多的攻关，先后实现蓝宝石衬底GaN单晶薄膜高质量异质外延生长、GaN有效可控*P*型掺杂及InGaN/GaN多量子阱材料制备，并率先在国内实现绿光GaN基LED。当时正在攻读林兰英先生在职博士的刘祥林助理研究员，经过上百次实验和无数次通宵工作，攻克了GaN材料有效可控*P*型掺杂这一关键工艺难题，为LED器件的研制奠定了基础。当时的副研究员韩培德针对InGaN有源层In组分的问题，将乙基二甲基铟换成三甲基铟，获得了In组分从0.1到0.5的可控调节，发光波长从410nm拓展到了580nm，实现了绿光发光。

上述结果形成了“GaN蓝绿光发光二极管外延片生产技术”成果，该成果2000年被科学技术部作为国家“十五”期间“新材料高技术产业化专项”示范项目，经财政部评估折价1010万元。2010年5月1日，王占国院士代表中国科学院半导体研究所在福州招商大会上亲自签字将该成果转让给福日集团进行产业化推广。

1997年，王占国院士的博士生汪连山与昝育德副研究员合作，在Si衬底上采用γ-Al_2O_3单晶薄膜为过渡层，实现了国内最早的Si基GaN外延生长。

研究组第二任组长为刘祥林研究员，他预见到MOCVD设备将会成为制约我国GaN基LED产业发展的最主要技术瓶颈。在王占国院士大力支持下，“十五”期间先后承担了“863”预研课题“GaN专用MOCVD设备实用化的可行性研究”，“863”课题“用于GaN的生产型MOCVD设备”、北京新材料基地重点项目子课题“MOCVD

设备研制”等有关科研任务。2004 年年底，自主研制出适合 GaN 基 LED 外延片生长的 3 片 2 英寸 MOCVD 设备。2005 年“863”课题顺利通过验收，不仅获得了“863”专家和科学技术部领导好评，并奖励追加了 100 万元经费。中央电视台、北京电视台、《科学时报》等国内多家媒体都给予了报道，《中国科技成果》杂志将其列入“2004 年中国重大科技成果”。世界最著名的半导体装备制造商美国应用材料公司曾先后两次到访研究组洽谈合作开发事宜。2006 年，该项成果被作为半导体照明工程重大装备项目估价 4000 万元，转让给秦皇岛市鹏远集团。

2003 年，由中国科学院半导体研究所牵头，在以王占国院士为代表的国内多名半导体学家联名倡议下，科学技术部紧急启动了“国家半导体照明工程”，推动了以 GaN 基 LED 为基础的中国半导体照明产业蓬勃发展。2017 年半导体照明已发展成为中国年产值超 5000 亿元的半导体产业集群。

王占国院士预见到，超宽禁带半导体材料以其优越的物理性质必将会成为未来重要的发展方向，于是在 2008 年将研究组人员和设备重新整合，成立了现在的“超宽禁带半导体材料”研究组，王占国院士担任新组建的研究组组长，增设“大失配异质外延衬底”和“AlN 单晶厚膜材料生长”两个研究方向。

2013 年王占国院士的博士生杨少延晋升为研究员并担任研究组组长，同年将 1998 年毕业的博士研究生汪连山研究员引进研究组。2016 年汪连山将研究组 GaN 基 LED 器件的发光波长拓展到橙黄光，2017 年获得南昌大学国家硅基 LED 工程技术研究中心 100 万元开放课题资助进行相关的产业化技术合作开发。

除了材料生长和器件研制方面的研究外，在王占国院士和朱勤生研究员指导下，研究组还注重宽禁带半导体材料物理研究，在有关二维电子气输运性质和异质结带阶研究方面做出了出色的成果，共发表 *SCI* 论文 60 余篇。

自 1990 年以来，研究组在王占国院士的支持和领导下，已累计承担各类项目 50 多项，经费累计 5000 多万元，发表学术论文 280 多篇，获得授权国家发明专利 46 项，培养研究生 50 多名，其中包括王占国院士指导的 15 名博士研究生和 3 名硕士研究生。

高山仰止　心念恩师

陆　沉

王占国老师是中外半导体材料物理学界的泰斗，我有幸追随先生多年，受到的教益颇多，虽然现在身处海外，一件件难忘的往事却常常萦绕心间。

1999年年初，我清华大学毕业的时候，看到中国科学院半导体研究所硕士招生简章，当时最吸引我的就是王占国老师实验室的量子点生长课题，于是我打算报考他的研究生。第一次见到王老师是招生面试的时候，当时他和陈涌海老师问了我一个半导体物理最基本的概念问题："什么是载流子的迁移率？"我一时想不起来，就自作聪明地回答："迁移率就是电子从价带跃迁到导带的概率。"结果两位老师哈哈大笑。然后，王老师语重心长地说："基础物理概念一定要牢记，不能不懂装懂，迁移率是描述载流子在电场作用下移动快慢程度的物理量。"这件事给我的印象很深，从此我在学习和科研中，迈开的每一步都注重从基础物理概念的明晰开始发力。

在半导体研究所读博期间，我在王老师所在的大实验室下面的分室从事硅基氮化镓生长的研究。虽然王老师不是我直接的论文指导老师，并且他的工作和院士身份承担的事务十分繁忙，但他还是经常告诉我们可以找他讨论课题进展。有一次，我的材料生长有较大的进展，就高兴地向他请教，没料到，他马上决定说："过两天你来做场报告。正好我邀请几个材料生长方面的专家来所里，可以一起帮助你讨论课题。"我大喜过望，非常认真地作了准备。报告那天让我出乎意料的是竟然来了好几位院士参加讨论。由于硅和氮化镓存在非常大的晶格常数差异，异质外延生长一直是一个难题，所以王老师提出了一个"柔性衬底"的概念，希望能解决这个问题，因此他对我的报告非常重视。在报告中，我提出了一个新型自组织横向外延的模型来解释实验现象。王老师说："这很好，但是必须要有严谨的实验证明，不能凭空想象。"后来，他建议通过观察透射电镜来了解样品的结构。而当时我还没有接触过透射电镜。为了帮助我完成从制样到观察透射电镜的研究，王老师专门返聘了已退休的范缇文老师。在整个论文期间，我得到了王老师不遗余力的支持。他治学严谨的态度、洞察敏锐的眼光令我终生难忘。

博士论文答辩结束，当我准备找工作之际，是王老师的举荐改变了我的人生轨迹。当时，考虑过去公司，也考虑过去清华大学范守善老师组工作。这时，曾经和王老师实验室有过合作关系的法国雷恩实验室发来消息，希望推荐合适博士后人选，于是王老师和陈涌海老师帮我做了推荐。在经过几个月的耐心等待之后，我终于得到了去法国做博士后的机会。到法国后，我的研究课题与我的博士课题虽然有所不同，但是得益于博士期间对材料生长、测试的扎实训练，很快适应了新的工作环境。在此之后，我逐步进入了一个新的研究领域——自旋电子学，在第一个博士后结束之后我来到另外一个在巴黎的 Thales/法国科学研究中心联合物理实验室工作。这个实验室是法国最好的自旋电子学实验室，2007 年物理诺贝尔奖得主 Albert Fert 就工作在此实验室。

2007 年 5 月，我接到王老师的邮件说他要到德国和英国开会，于是我高兴地建议他顺路到我所在的法国实验室参观访问，王老师非常爽快地答应了。因为 Thales 公司与军工有关系，所以参观必须提前申请并审查护照。6 月 6 日一大早，我接到王老师一起到实验室访问参观，由我的博士后导师 Jean- Marie George 接待，他为王老师进行了细致的介绍，王老师对于其中采用 AFM 探针来制作纳米尺度的隧穿结的技术非常感兴趣，进行了详细的讨论。中午，我们与 Albert Fert 一起共进午餐。Albert 说他对中国文化非常感兴趣，希望以后能够到中国访问交流。王老师非常热情地向他邀请，希望他能访问半导体研究所。午餐后，Albert 与王老师一起在实验大楼前留影纪念。参观结束后，我和王老师游览了巴黎的几个景区。当时王老师快 70 岁了，仍是健步如飞，我都赶不上。他嘱咐我说，要想做好学术，没有一个好的身体可不行。让我印象非常深刻的是，虽然王老师是院士，但是在生活上非常简朴，平易近人。我希望请王老师去餐馆吃饭，王老师却让我早点回去，照顾家庭，他竟独自一人在楼下中餐馆吃了一碗面条。

王老师的这次访法让我深切感受到中法交流合作是一个非常重要的领域，而这正是我的优势所在。2007 年注定是不平常的一年，在王老师访法、Albert 获诺贝尔奖之后，我得到了一个法国科学研究中心的研究工程师的固定位置。这个位置在法国的另外一个城市南锡。虽然是研究工程师，但是主要时间是维护仪器，进行技术更新。为了能够有更多的自主科研时间，我申请了一年到美国马里兰大学做访问学者的研究机会，主要学习硅基自旋输运。从美国回到法国之后，我开始准备申请中法合作项目。这个项目主要研究将自旋电子注入Ⅲ-Ⅴ族半导体 LED 中，通过发光的圆偏振度来检测注入电子的自旋极化率。考虑到半导体研究所的优势在于Ⅲ-Ⅴ族材料生长，而我们的优势在于自旋检测，这个项目将可以中法双方强强联合。当我把这个想法和王老师商量之后，他非常支持这个项目的申请，同时答应担任中方项目负责人。在王老师的大力支持之下，我们顺利申请成功了这个项目。如今，这个项目取得了丰硕的成果，发表多篇高质量一流的论文。在王老师的倡议下，2013 年我们还组织了中法在半导体自旋领域的研讨会。每次回半导体研究所访问，王老师都会在百忙之中挤出时间，在他的办公室会见我，与我亲切交谈最近的工作进展并告诉我他想到的一些新的课题方向。

王老师虽然年届高龄，学术思想仍然十分活跃，他常常谆谆教诲我做科研一定要创新，要多交流，不能固步自封。时间如梭，从刚刚步入科研，到如今有自己的实验组和课题项目，我的每一步成长都离不开王老师的大力扶持和提携。记得有一位名人说过，评价一位老师是否成功，最重要的不是他自己的成就，更要看他的学生有多少能够成为教授。如今王老师的学生桃李满天下，这就是最好的证明。涓流汇沧海，一篑成山丘。王老师的一言一行无愧大家风范、育人楷模，他是我们景行一生的榜样。

师道山高

王智杰

何谓师道？师者，所以传道授业解惑也。从小学到大学，再到研究生及出国留学生涯，对我帮助最大、影响最深的老师当属王占国院士。王老师是半导体材料领域的资深院士，为中国半导体材料与器件的发展做出了卓越的贡献。能做王老师的弟子，是我一生的荣耀。

王老师的教导是全方位的。老师所传之道，既包括有言之道，也有无言之道。老师是一个知识渊博、专业精深的著名材料学家，能把深奥的理论讲得深入浅出，使初学者能立即领悟其中的要领。每次与老师交谈，他都会一语中的点明相关学科的核心理论、重点问题及发展方向，使我们这些学生能在极短时间内领会该学科的核心，拓展学术视野；同时，老师又会针对我们的表述，指出我们在科研工作中的问题，并提出相应的解决办法，避免我们误入歧途。

老子曰："是以圣人处无为之事，行不言之教，万物作焉而不辞，生而不有，为而不恃，功成而弗居。夫惟弗居，是以不去。"我们材料重点实验室的发展壮大，与老师的不言之教是分不开的。老师是一个坚持真理、不畏艰险、敢于挑战权威的勇士。在我刚入学的时候就听说老师这方面的很多事迹。老师早年在做卫星用太阳电池辐照实验的时候被电子束辐射伤手；他力排众议，坚持正确的太阳能电池设计方案，促进了卫星用太阳能电池的定型投产；他访学瑞典，推翻了物理学权威的观点，澄清了硅中金杂质能级的本质。老师是一位勤奋的科学家，只要不在外面开会，我们总会看到老师伏案工作的场景。老师是一个无私的领导者，经常把自己应得的奖励分给实验室成员，有时候自掏腰包激励实验室的先进分子，促进大家进步。老师是一位谦逊和蔼的人，丝毫没有大科学家的傲气，待人非常随和，和我们这些学生的通信中都称呼"您"，这让我们这些晚辈汗颜。更多时候，老师在我们眼中是一位慈祥的长者，主动帮我们解决生活中遇到的困难，不辞劳苦。

老师所授之业，不仅深且广，是我们的立身之业。其中，既有如何做学术做科研的方法论知识，也有如何操作实验设备的具体技术，还有如何看透一个学科本质和核心的独到视角。这些足以让我们在科研工作中得心应手。我记得老师曾经说过：一个合格的博士生要做到经过半年的学习和调研能在任何一个陌生领域成为专家。"业精于勤，荒于嬉；行成于思，毁于随。"老师经常以身作则教导我们，事业兴旺的根基在于勤奋。即使老师年过七旬，我们总是能看到他在电脑前阅读大量文献的场景，老师的记忆力和思维力从来不输给我们这些年轻人。

先生所解之惑，既有学业之惑，也有人生之惑。学海无涯，我们难免碰到力所不能解之惑，经过老师的一番讲解，大有拨云见日恍然大悟之快感。人生苦短，大千世界迷惑繁多，老师教导我们要化繁为简、生命不息、学业不止。

师道如山，师道永存，祝老师健康长寿。

人生楷模　学习的榜样

皮　彪

喜闻王院士今年80大寿，不觉思绪万千。十几年前的往事就像昨天发生的那样历历在目。记得2003年，“非典”刚过，我来到了还在筹建中的南开大学泰达学院MBE实验室。这一切对我来说都是那么新奇和富有挑战，我们即将接手的是世界一流的MBE设备，我也将与国内做MBE一流的科学家一同工作和学习，心里既高兴又忐忑，可没高兴几天就惴惴不安了，MBE太难了，看了几天资料我还是云里雾里，不知MBE是啥？能干啥？我开始怀疑自己是否有能力掌握这套复杂的科技含量很高的设备，带着这些疑问我敲开了林耀望老师办公室的门。那年林老师已66岁，但从外表和精神上看一点也不显老，说话做事与年轻人一样有精神。他知道我的来意之后，很热情地跟我说起了他搞科研遇到的种种困难，特别跟我讲起了改革开放之初，中国在半导体这一领域还很落后，是因为有一批像王院士这样一流的科学家带领的团队通过多年不懈努力，克服种种困难，做出世界一流的科研成绩，得到外国同行的一致认可。林老师还讲，他之所以帮助南开大学建MBE实验室，是受王院士之托想帮助母校南开大学培养半导体研究人才，同时探索一条大学向研究型发展之路。听了林老师的一席话使我悠然升起了对王院士、林老师这样的老科学家的敬佩之情；同时，点燃了我内心战胜困难，迎接挑战的激情。

我们这个团队，在王院士和许京军老师的领导下，在开发区领导和泰达学院领导的大力支持下，在半导体研究所林老师、徐波老师和曲胜春老师等的大力帮助下，初战取得了非常好的成绩，MBE生长的二维电子气（2DEG）AlGaAs/GaAs异质结构材料，电学参数突破国内最高水平；MBE生长的InP/InP外延材料，在同类系统中，电学参数突破国际最高水平。

但是，这套法国MBE设备在除磷技术方面当时还很不成熟，存在很大的隐患，所以，在法国工程师离开3个月后，MBE生长室出现了磷污染，为此王院士跟我们一样焦急，多次打电话来询问情况，并不顾自己年事已高，路途劳累亲自来泰达学院帮助查找原因，解决P炉使用中存在的问题，并委派林老师、徐波老师协助舒永春老师与法国Riber公司进行协商，最后获得圆满解决，法国公司免费赠送给我们一套价值一二十万元的除磷冷阱。

接下来我们这个组的发展又遇到了困难，我们不知该做什么？不知该向那个方向发展？就在我们组为下一步如何发展不知所措时，是王院士高瞻远瞩从国家战略需求和我们的MBE设备的实际情况出发，经与舒永春教授、李伟博士、林老师等反复讨论，最后确定我们 MBE 组下一步的发展方向是开展 GaAs 基垂直外腔面发射激光器、超快激光器、无铝大功率激光器等的研究。

2006 年舒永春老师拿下了一个国家“863”项目“光泵浦 1064nm 半导体垂直外腔表面发射激光器芯片材料的制备”，为了保证项目研究顺利进行，王院士又亲自来泰达学院听取项目研究进展情况和研究中存在的问题，并帮助我们提出解决方案。总之，泰达学院 MBE 实验室的每一次进步都离不开王院士的指导和像林老师、徐波等许多老师的帮助，通过与这些老师的合作，我从他们身上学到了好多做科研的宝贵经验。

王院士在指导我做博士论文研究中，多次指出一个好的科研人员要把主要精力用在对实验细节的把握上，多注意观察，遇到问题时多思考，多从物理的角度考虑问题。

王院士不仅是我学业的导师，也是我人生的楷模，王院士爱党爱国，敬业献身的精神值得我们一生去学习！

谆谆教诲　润物无声

黎大兵

今年适逢恩师王占国院士80大寿。想起十几年前，有幸在先生门下求学，并在先生的引领下走进半导体材料这个充满挑战和神奇的领域。先生言传身教、整躬率物，不仅是我在半导体学科中的启蒙者，也是我人生道路上的引路人。先生渊博的学识，敏锐的思维、严谨的治学风格和科学的工作方法，深深地影响着我并让我难以忘怀，从先生的身上，我不仅学会了如何做学问，同时也学会了如何做人。一直以来先生对我的教诲言犹在耳，从治学到修身都是我人生的绳尺。

如约翰·杜威所说，“教师总是真正上帝的代言者，真正天国的引路人”。能够得到恩师的指引，是我一生的幸事。我还记得与先生初识的日子，那是2000年年底，我硕士刚毕业想进一步深造，通过调研我决定改专业报考半导体方向的博士研究生，而先生正是国内半导体材料领域的著名科学家，于是我十分冒昧的拨通了先生办公室的电话。还记得我当时心怀忐忑，一来先生当时已经是中科院的院士，著作等身，桃李天下；二来我在硕士期间的研究工作与先生的研究领域尚有出入。电话接通后，我简单介绍了一下自己，说明了自己想报考先生的博士研究生，并小心地提出能否去北京拜访先生，没想到先生毫不犹豫地答应了。和先生见面，给我的感觉是先生十分平易近人，先生详细地介绍了他的研究领域并告诉我该如何去准备博士研究生入学考试，同时先生也谈到博士研究生是十分艰辛的，要靠灌注心血才能够有所产出，让我要有思想准备。初识先生给我留下了深刻的印象，也消除了很多顾虑，先生的鼓励和鞭策，也增加了我报考的决定和信心。回到学校，我认真准备，后来也如愿地考取了先生的博士研究生。

在先生的指导下，我在半导体材料学科的研究步入了正轨，开始了我在氮化物领域上的科研生涯。攻读博士期间，先生在学术上的高屋建瓴给了我很多启发和灵感，先生在科研上的严谨缜密也让我获益良多。当我写完小论文和博士论文的初稿后，先生逐字逐句地帮助我修改文章，并且把我叫到了办公室，逐一告诉我每一个修改的地方改动的原因和目的。在关于博士论文中创新点的归纳总结时，先生对我说过一句话：“在研究工作中，要踏踏实实，不要好高骛远；随着现代科技的发展，要做出完全原创性的工作很难，但是只要能做出一点点创新，就是对科学发展做出了贡献。”这句话对我后来的研究工作有了极大的启发，现在每做一项工作，我都要认真地思考我做的这个工作是否有那么一点点是创新的。因为哪怕是一点点的创新，也会对氮化物领域的研究发展起到推动作用，当所有的科研工作一起推动这个领域的发展，就会对领域产生革命性的进步。现在，我博士毕业多年之后，也成了博士生的导师。我也用先生的这句话去教育我的学生，把先生的桃李之教传扬开来。

先生对学生们的科研上要求很严格，而在生活中，先生像慈父一样关爱着每一个学生的发展，十几年来，先生一直关心、支持着我，一如求学之时。从我远赴日本三重大学留学，到回国后对我研究工作的指导；从帮助我修改项目申请书，到学科前沿研究方向提出建议，我的科研生涯一直都离不开先生的指引和鞭策，我在科研上取得的一点成绩也离不开先生的教导和熏陶。当得知我获批“国家杰出青年基金”资助，先生在祝贺的同时也提醒我：“作为杰青，你的活动会多起来，要注意不要影响研究工作。”每次去先生家里探望时，先生谈得最多的还是科研，大到国家层面的科技发展、小到我的科研，先生都格外关切用心，每次总是叮嘱我，不要把研究丢了。先生不但对学生有着殷切的期望，更对国家科研的发展投以极大的热忱。然而，先生对学生的帮助，并不是无条件地护着自己的学生，而是站在为国家培养人才、发现人才的高度，如果先生自己学生的水平不够，先生也绝不会护短。我记得刚回国的时候，我自己牵头申请一项国家级项目，同时也参与申请了一项国家级重点项目，先生是项目会评专家组组长。而且那次会评是在我工作单位所在地——长春召开，我申请和参与的项目都通过了初评，进入了会评，因此我侥幸地认为我的项目肯定能中，但结果是我的项目都落选了。后来，我才得知在我的项目评审时，先生都主动提出回避，尽管当时从申请书中看不出来先生是我的导师（当然现在的项目申请书中都明确要求列出各阶段的指导老师）。再后来，先生仔细地帮我分析了申请书中存在的问题，让我好好修改来年再报，我也顺利地拿到了研究生涯中的第一桶金。

先生是河南南阳人，正是武侯诸葛亮的躬耕之地。先生对于国家的科研事业也正如诸葛武侯一般，做到了“鞠躬尽瘁”。学生得入师门，受到恩师的时雨春风，是学生之幸。国家有先生这样的科学大家，为祖国的科研事业殚精竭虑，朝乾夕惕，何尝不是国家之幸！《礼记 · 王制》云，“五十杖于家，六十杖于乡，七十杖于国，八十杖于朝……”先生多年来执杖于家，言传身教，为我们这些学生指明了科研的道路和发展的方向，早已润物无声，桃李满门。如今先生掌朝之年，仍兢兢业业工作于科研一线。我衷心祝愿先生健康长寿，松鹤延年，为我国半导体事业的发展创造新的辉煌，续写新的篇章！

桃李遍天下

丁　飞

前几天智杰联系我，说大家张罗着给王占国院士准备文集。这个是好事，我能够尽自己的一分力量当然非常激动、高兴！今天不需要授课，汉诺威阳光也难得地不错。泡了一杯咖啡坐在书桌前，本以为会文思泉涌。但到真正提笔之时，却发觉颇难。这倒不是因为我旅居德国十年，忘了如何书写中文，原因在于，王占国院士可以书写的地方实在是太多了。

也许我可以作为半导体物理研究的后学，来评价王先生在半导体材料生长、光学性质、器件等领域的重要贡献；也许我可以作为一个旅人，来转述我从几位欧洲教授那里听到关于王先生在瑞典隆德大学的工作轶事；又也许，我只是作为一个年轻人，记下王先生满怀热情并严谨的治学态度。最终，想起了在国内外工作的几位师兄弟，那我权且记录下王先生培养学生的几件事吧。

“桃李满天下”，这个词用在王先生身上应是毫无争议的。前几天得知智杰当选为半导体研究所材料重点实验室主任，衷心为他高兴！而他只是王先生在半导体研究所近30年来培养的无数优秀学生中的一位。我和智杰是同年进入半导体所的，当然研究生入学，智杰考了个第二名，我却是第一名，同年级的还有汤晨光、赵超、郝亚非等。当时，高年级的师兄有不少已经在欧美的一流实验室深造了。

我虽说考试很厉害，但其实是个不折不扣的书呆子，对半导体研究是个十足的门外汉。这么多年过去了，我在半导体量子点研究领域的几个小方向上也取得了一些成绩，并靠着这些获得了德国汉诺威大学建校186年历史上第一个中国籍讲席正教授位置。毋庸置疑，我是幸运的。幸运之处却在于进入王先生领导的材料重点实验室之后获得了良好的基础训练！

研究生考试成绩出来后，我带着激动且不安的心情敲开了王先生在二楼办公室的门。那时，我是个20出头的毛头小伙，也不知道哪来的勇气去打扰一位中国科学院院士、知名半导体材料及材料物理学家。很诧异的是，王先生很热情地打开门接待了我。显然，他没有介意这个“毛头小伙”的鲁莽。坐下后，王先生问明来意，知道了我想咨询一下导师的问题。他很热情地介绍了一下实验室的研究方向和各个老师的情况。先生了解了我的各方面情况之后，表示非常满意。王先生当时就给实验室里的学术骨干之一陈涌海研究员（中国科学院百人计划入选者、国家杰出青年基金获得者、“973”项目首席，也就是我后来的直接导师）拨通了电话，请他要认真负责我研究生的培养工作。这个第一次会面让我对半导体纳米材料领域有了一个重要的初步了解，并深刻地影响了我的学术生涯。

陈涌海老师其实也是王先生的学生之一，私下我们几个都叫他“海哥”。虽然在这个过度娱乐化的社会里，“海哥”是以一首“将进酒”闻名江湖，以“摇滚科学家”著称。但其实，“海哥”的科研功底是毋庸置疑的，他在偏振反射差分光谱等领域的研究在国际上都是独树一帜的。同时也毋庸置疑的是，他也是王先生培养出来的最优秀的学生之一。私下，我也听说了不少王先生严格（甚至是严厉）培养陈老师的趣事，待之后仔细查证再说吧。不能靠猜想（和传言），要相信数据和实验，这也是我从王先生实验室学到的最重要东西之一。

我在半导体研究所的日子是充满兴奋以及担心的。兴奋的是，研究的半导体量子点当时是固体物理领域最火热的课题之一（和当下的拓扑绝缘体、二维材料相似的热）；担心的是，每个月的组会该如何交差。我在所里学习之时，王先生身体矍铄，他每月的组会必定是要参加的。他会对学生的工作报告提出独到的见解，有时候也会有严厉的批评（年轻的学生总是不懂导师的焦虑啊）。那时候我对自组织定位生长量子点产生了浓厚的兴趣，有一次沾沾自喜地在组会上讲解了该领域最新进展，讨论了德国马普学会 Oliver G. Schmidt 小组的几篇工作。没想到，王先生对这些工作非常熟悉，他提出了一些可能的改进思路。肯定了我的工作进展之后，他提出不能只是复制别人的工作，要想法超过、要创新。这些建议也影响了我后来的学术之路。

也说说王智杰和赵超。他俩和我是一届的，与我相比，他俩在王老师的指导下学习时间更久。虽然博士期间没有在国外实验室工作，他俩也取得了傲人的成绩，这也让一直在德国打拼的我非常汗颜！这从另外一个方面也说明了材料重点实验室的学生培养是成功的。他俩在博士毕业后也在国外优秀的小组深造，做得非常成功！我也为他们感到高兴。这么些年来，王先生领导的材料重点实验室培养出了非常多优秀的科研人员，包括英国伦敦大学学院的刘会赟教授、法国国家科学研究中心的陆沅博士、中国科学院苏州纳米技术与纳米研究所的张子旸研究员、中国科学院宁波工业技术研究院的廖梅勇研究员、中国科学院化学所的王吉政研究员……不胜枚举。王先生真的不愧于“桃李满天下”这个称谓！

“你来自 Z. G. Wang 的实验室，那也是我们非常高兴接收你的一个重要原因！”这是 Oliver G. Schmidt 教授见我第一面说的。感谢中国科学院与德国马普学会的联合培养项目，我得以到德国的斯图加特的马普固体物理所 Klaus von Klitzing 教授（由于发现量子霍尔效应而获得诺贝尔物理学奖）的研究组进行博士生联合培养。Oliver 是当时 MBE 小组负责人，很显然，他对这个来自中国科学院王先生小组的年轻人高看一眼。这，当然归功于王先生多年来在这个领域的深耕，使得我们小组获得了不小的国际知名度！

我博士课题的一部分是在荷兰代尔夫特工业大学（江湖传言，欧洲麻省理工）进行的。当时联合培养的导师 Val Zwiller 教授和我见面之后问“你真的是跟从

Z. G. Wang 教授学习的吗？我当年在瑞典隆德大学学习的时候，实验室里的墙上还挂着他的照片呢！”无疑，王先生是隆德这个欧洲半导体物理重镇培养的知名学者之一！我当时心里惴惴，“这要是没有学好，岂不是坏了王先生和陈老师的名声？！”

过去十年，我走过了许多地方。其实心里还是希望能有机会回到让我感到亲切的材料室大家庭工作的。先生也一直很热情地与我保持联系，每次发 E-mail 过去，先生总是能及时回复。先生也会经常嘱咐陈涌海和刘峰奇老师一定要仔细给我讲解各项支持政策。有时候回复晚了，先生也没有长辈的架子，竟然与我说抱歉！我心里感动的时候，又更加感到不安，觉得没有能回国报效自己的一分力量。材料重点实验室，这是先生挥洒了一生辛勤汗水的学术沃土。我希望今后有机会与实验室里的年轻人们一起合作，使得实验室的学术水平以及国际知名度更上一层楼。

先生这几年身体不好，但他还是关注着领域的发展。前两年我回国作黄昆论坛报告，虽然我已经做 PI 多年，但面对台下的王先生竟然还是略有紧张，好似又回到了学生之时。尽管先生年事已高且身体不如之前，但听我报告之时，他还是非常非常认真。他也点评了量子光源发展的一些瓶颈、未来的前景等。毕竟先生在半导体领域工作了一生，眼光非常独到，他的建议让我收获匪浅。

时光荏苒，虽然我在德国学术界已经可以独当一面，但是在王先生面前我依然只是那个鲁莽敲门的“毛头小子”。此生我很庆幸的是，先生为我打开了半导体纳米材料研究的这扇大门。感谢先生，也祝先生身体健康！

记与王老师交往的二三事

赵　超

今年欣逢王老师 80 华诞，作为长期受王老师教导的学生，向王老师致以最热烈的祝贺！恭祝王老师身体健康，笑口常开。

我和王老师的交往是从 15 年前第一次给王老师发邮件开始的。2003 年的时候我还在天津大学读大三，到了快毕业考研的时候，自己坚持认为研究生阶段要换一个不同的环境。所以放弃了保送本校研究生的名额，开始准备考研同时找合适的学校。由于一直对光电子材料及器件感兴趣，所以开始关注本领域最强的中国科学院半导体研究所，前辈师兄向我推荐了王老师领导的材料开放重点实验室。由于本科阶段没有相关的基础知识，于是我怀着惴惴不安的心情第一次给王老师发邮件，询问报考王老师博士生的可能性。没想到王老师非常鼓励我报考，让我好好准备，使我深受鼓舞。在备考专业课期间遇到了很多问题，给王老师写邮件，王老师也都非常耐心地一一解答，让我非常感动。这是我从事半导体研究的起点，王老师是把我带进这个领域的人。

2004 年我如愿进入了半导体研究所研究生招生考试的复试，在复试中见到了王老师。当时每人都要准备一段英文的自我介绍，当我讲完我的老家湖北有很多历史名人比如屈原、伍子胥等，王老师马上接了一句说还有诸葛亮，然后参加复试的老师们都笑了起来。几年后有一次组里组织石林峡活动的时候，王老师又谈起此事，说诸葛亮躬耕地之争已经在王老师家乡河南南阳和湖北襄阳间持续了很久。当时对此知之甚少，王老师又给我讲解了这个争议的由来，一位德高望重的学者，仍然那么关注文化动态，十分令人敬佩。

2005 年我在中国科学研究生院完成了基础课程的学习，开始进入半导体所进行科研工作。虽然王老师平时工作很忙，但对每一个学生的情况都非常关心。为了帮助记忆新学生的名字，那一年的王老师生日合影之后，老师特意嘱咐我把每个同学的名字标在照片上面。我的办公室和王老师在同一层楼，每次王老师见到我都要询问最近的工作进展，有时候还会走进我们办公室与学生挨个交谈。

在王老师的指导下，我当时有幸接触到了液滴外延这一新颖的材料生长方式。经过几年的努力，在这一领域发表了一些文章。2008 年，我第一次去国外参加会议作口头报告。回来以后王老师详细询问了参会的情况，别人问了什么问题，是怎么样回答的。我都一一向王老师做了汇报，老师也非常满意。

除了科研工作，每次集体活动，王老师都和大家打成一片。我们常组织的爬山活动，王老师都走在前面带领大家，边走边聊，我们这些年轻人追都追不上，实在令人惭愧。不仅如此，在河北怀来的时候，看到古城保护不力，王老师也非常关心地询问当地人情况，希望可以提供一些帮助。

王老师在教学科研岗位耕耘几十年，特别关心青年学生的成长，给我国半导体事业培养了大批优秀科技人才。毕业后我长期在国外，都定期给王老师发邮件汇报工作的一些情况，每次到北京的时候都去看望王老师。有时候工作中需要推荐信了，王老师不管有多忙总是第一时间准备好不同版本发给我。一位长者那么快就做出回应，甚至是利用宝贵的休息时间，就是为了学生的工作，实在令我非常感动。王老师对学生的爱护，由此可见。最近几次去拜访王老师，老师不仅在学术上进行指导，给我指点本领域发展的大方向，还关心个人的家庭发展，叮嘱我工作生活都要顾好，让我感受到了家庭般的温暖。

王老师以自身的行动，言传身教，我们这些学生都觉得能受到王老师长期指导，非常幸运。祝恩师健康长寿，生活快乐如意，幸福美满！

第五篇

附　　录

个 人 简 历

王占国，男，汉族，1938 年 12 月 29 日生于河南省镇平县，1962 年毕业于南开大学物理系。中国科学院半导体研究所研究员，半导体材料及材料物理学家，曾任中国科学院半导体研究所副所长，国家“863”高技术新材料领域专家委员会委员、常委、功能材料专家组组长，中国材料研究学会副理事长，国家自然科学基金重大研究计划“光电信息功能材料”专家组副组长，“973”计划材料领域咨询专家组组长。现任中国科学院半导体研究所研究员、半导体材料科学重点实验室学委会名誉主任、中国电子学会半导体和集成技术分会名誉主任、中国电子材料行业协会半导体材料分会第四届理事会名誉理事长、北京市人民政府第八届专家顾问团顾问、天津市人民政府特聘专家和多个国际会议顾问委员会委员以及多所高校特聘和兼职教授。1995 年当选为中国科学院院士，曾先后任信息技术科学部常委、信息技术科学部副主任，任第七届中国科学院学部主席团成员。

长期从事半导体材料和材料物理研究。其中，人造卫星用硅太阳电池辐照效应和电子材料、元件、器件和组件的静态、动态和核瞬态辐照实验结果，为我国的两弹一星事业发展做出了贡献。从 1980 年起，主要从事半导体深能级物理和光谱物理研究，提出了识别两个深能级共存系统两者是否同一缺陷不同能态的新方法，解决了国际上对 GaAs 中 A、B 能级和硅中金受主及金施主能级本质的长期争论；提出了混晶半导体中深能级展宽和光谱谱线分裂的物理模型，解释了它们的物理起因；提出了 GaAs 电学补偿五能级模型和电学补偿新判据；协助林兰英先生，首次在太空从熔体中生长出 GaAs 单晶，并对其光、电性质做了系统研究；领导的课题组在应变自组装 In（Ga）As/GaAs，In（Ga）As/InAlAs/InP 等量子点（线）与量子点（线）超晶格材料和量子级联激光器和探测器材料生长、大功率量子点激光器、量子点超辐射发光管、量子级联激光器和探测器以及太赫兹激光器研制方面获得突破；提出了柔性衬底的概念，开拓了大失配材料体系研制的新方向。提出了开展超宽禁带半导体材料研究的建议，得到有关部门的重视。上述成果先后获国家自然科学奖二等奖（2001）和国家科学技术进步奖三等奖（1990），中国科学院自然科学奖一等奖（2000）和中国科学院科学技术进步奖一（1989）、二（1989，1995，1997）和三等奖（1990），何梁何利科学与技术进步奖（2001），国家重点科技攻关奖（个人）和优秀研究生导师奖等。与合作者一起在国际著名学术刊物发表论文 700 余篇。先后培养硕士、博士和博士后 170 余名。

大 事 记

开始时间	结束时间	事　件
1938. 12. 29		出生于河南省镇平县贾宋区马庄乡东黄楝扒村
1946. 9	1948. 9	河南省邓县红庙小学读书
1948. 9	1951. 7	河南省镇平县贾宋区马家小学和马庄小学读书
1951. 9	1954. 7	河南省镇平县候集中学读书
1954. 9	1957. 7	河南省南阳二中读书
1957. 9	1962. 7	南开大学物理系读书
1962. 7		南开大学物理系固体物理专业毕业，同年分配到中国科学院半导体研究所半导体材料室工作
1962. 9	1963. 9	与佘觉觉合作研制半导体少子寿命测试仪
1963. 9	1964. 7	研制成功基于真空静电管高阻静电计的半导体低温(20–300K)电学测量系统
1964. 8	1965. 8	半导体硅材料电学性质和中子、γ 射线辐照效应研究。1964 年年底与徐寿定一起，受委托对我国第一颗原子弹(1964 年 10 月 16 日)辐照的硅材料电学性质进行了秘密测量
1965. 8	1966. 12	和尹永龙、李永常和彭万华一起组成半导体材料电学和光学测试仪器研制小组，后因“文化大革命”而停止
1967. 5	1968. 5	受组织委托负责人造卫星(651 任务)用半导体硅太阳能电池电子和质子辐照效应研究。电子辐照实验研究结果为我国卫星用电池的定型生产起了关键作用。右手在实验中受到电子辐照烧伤，先后多次住中国人民解放军 307 医院治疗
1968. 6	1970. 6	受国防科委第 14 研究院委托，负责制定了 1469 电子材料、元件、器件和组件辐照实验方案，并作为小组业务组长，在上海、北京、兰州，对国产电子材料、元件、器件和组件实施了电子、质子、中子和 γ 射线静态和动态辐照实验。实验结果汇编成我国第一本《电子材料、元件、器件和组件抗辐照性能手册》
1970. 8. 5	1970. 11	作为国防科委第 14 研究院辐照实验组业务组长，负责组织、实施了氢弹空爆瞬态核辐射对电子材料、元件、器件和组件性能影响的研究工作(21 基地)

续表

开始时间	结束时间	事 件
1971.9	1978.6	负责设计、建成了低温电学测量和光致发光实验系统，并对 GaAs 和 InP 等化合物半导体材料的光电性质做了实验研究，与林兰英先生一起提出“杂质控制”的论点。半导体材料室的材料组由体材料转向 LPE 和 VPE GaAs 外延生长，为材料质量，特别是材料的纯度取得国际先进水平的成果打下了基础。与此同时，对体 GaAs 开展了强场和热学性质研究，实验结果否定了此种材料作为 LSA 器件的基础材料的看法，不久所里便终止了此项研究
1978.7	1978.9	因电子辐照损伤的右手碰破，伤口久不愈合，再次住进中国人民解放军 307 医院诊治，成功地进行了植皮手术
1978.8.1	1978.8.15	参加在江西庐山召开的中国物理学会年会。这是“文化大革命”后首次物理学界的盛会，也是众多物理学家及其领导人物的亮相大会。半导体研究所的主要业务领导人黄昆所长，王守武、林兰英副所长等以及王占国等几位青年科技工作者（还包括郑厚植、沈光地）也有幸获准参加了会议
1980.10.14	1983.11	赴瑞典隆德大学固体物理系（国际半导体深能级研究中心）进修，主要从事半导体深能级物理和光谱物理研究
1982.5.1		在瑞典隆德大学由中国驻瑞大使馆委派党小组主持下加入中国共产党
1984.1	1985.10	回国后，在林兰英先生的支持下，组建了半导体深能级瞬态谱仪（DLTS）和深能级光电容实验室；提出了半绝缘 GaAs 五能级补偿模型，对发展我国的高频大功率 GaAs 器件和电路研制起到一定作用，为此曾两次被授予国家重点科技攻关奖（个人）
1986.5		破格晋升为研究员，并任中国科学院半导体材料研究室主任
1986.5	1991.12	从事 CZ-Si 中新施主，等电子杂质在 GaAs 中行为和空间 GaAs 生长及光电性质研究，先后负责承担多项国家自然科学基金和国家重点科技攻关任务。1987 年，协助林兰英先生（利用返地卫星）开辟了我国半导体微重力材料科学研究新领域，并在国际上首次在太空从熔体中生长出 GaAs 单晶，并对其光、电学性质做了研究
1988.1		与陈廷杰等一起，开始筹建半导体材料物理国家（重点）实验室。遗憾的是，由于种种原因至今未得到国家批准；1991 年中国科学院半导体材料科学重点实验室对国内外开放
1989.11		“微重力条件下从熔体生长 GaAs 单晶及其性质研究”获中国科学院科学技术进步奖一等奖（排名第四位）
1989.10		“LPE InGaAsSb/InP 材料和性质研究”获中国科学院科学技术进步奖二等奖（排名第二位）。其中，研制成功国内第一支四元锑化物 2μm 激光器

续表

开始时间	结束时间	事　件
1990. 9		荣获 1990 年年度中国科学院优秀研究生导师奖
1990. 10		"等电子杂质 In 在 GaAs 中行为的研究"获中国科学院科学技术进步奖三等奖(排名第一位)和国家"七五"重点科技攻关奖(个人)
1990. 12		"微重力条件下从熔体生长 GaAs 单晶及其性质研究"获国家科学技术进步奖三等奖(排名第四位)
1990. 5	1994. 10	任中国科学院半导体研究所副所长
1990		被国务院学位办批准为博士研究生导师
1991. 5	2001. 10	先后任第二、三和四届国家"863"高技术新材料领域专家委员会委员、常委、功能材料专家组组长。其间对我国全固态激光材料与器件、红绿蓝发光材料与器件,特别对 GaN 基材料与器件研究的及时立项,半导体白光照明技术发展做出了贡献
1991. 3	2002. 7	中国科学院半导体材料科学重点实验室正式被中国科学院批准向国内外开放,任首届、第二届实验室主任
1993. 1		获国务院特殊津贴
1995. 10		"高质量 GaAs/AlGaAs 二维电子气材料研制及其器件应用"获中国科学院科学技术进步奖二等奖(排名第七位)
1995. 10		当选为中国科学院院士(技术科学部)
1996. 12		获电子行业国家"八五"重点科技攻关奖(HBT 和 HEMT 基础技术研究)
1996. 7		他领导的 MBE 小组研制成功我国首支量子点激光器。有源区采用 InAs/GaAs 量子点,室温脉冲工作,波长 ~1μm,阈值电流密度为 590A/cm^2
1996. 7	1999. 6	被国家科学技术委员会(科学技术部)聘为新材料技术领域发展战略研究("S-863"计划软课题研究)专家组组长。研究报告为国家"973"基础研究规划项目的顺利启动和第二期"863"的实施做出了贡献
1996	2002	先后任国家自然科学基金委员会信息学部半导体学科评审专家组组长,国家科技奖励评审专家、天津经济技术开发区、河南郑州经济技术开发区等顾问
1997. 12		"砷化镓材料、器件与电路研究"获中国科学院科学技术进步奖二等奖(排名第二位)

续表

开始时间	结束时间	事 件
1998.	2002	他指导的博士生孙中哲于1998年在国际上首先提出以应变自组装量子点为有源区来研制超辐射发光管的思想,并进行了初步的探索。随后,徐波和博士生张子旸于2002年年底在国际上率先研制成功以量子点超辐射发光管。器件室温连续工作,中心波长~1.0 μm,光谱宽度~60 nm,最大输出功率可达200 mW
1999.7-		当选国际半导体缺陷识别、成像与物理(DRIP)国际会议顾问委员会委员
2000.5.1	2000.5.3	代表半导体所在福州招商大会上与福日集团草签了全国第一个以GaN外延材料和LED器件为主要产品的公司,并得到国家发展和改革委员会产业化示范项目的支持
2000.6		国务院学位委员会授予2000年百篇优秀博士论文(李含轩)导师奖
2000.12		"自组织生长量子点激光材料和器件研究"获中科学院自然科学奖一等奖(排名第一位)
2000.12-		当选北京市人民政府第八届专家顾问团顾问
2001.2		"863"计划15周年时,被科学技术部授予先进个人称号
2001.3		"自组织生长量子点激光材料和器件研究"获国家自然科学奖二等奖(排名第一位)
2001.9		荣获何梁何利科学与技术进步奖,宝洁优秀研究生导师奖
2001.10	2005.10	任国家"973"项目"信息功能材料相关基础研究"(2001-2005年)首席科学家
2002.7		受聘中国科学院微电子技术战略指导委员会专家
2003.3	2004.12	任国家中长期科技发展战略研究新材料专家组组长
2003.9.1	2004.7	作为脱产集中规划建议研究人员,参加国家中长期科学和技术发展规划第十三专题(战略高技术和高技术产业化)和第十四专题(基础研究)研究,并获荣誉证书
2004	2006	应国家自然科学基金委员会工程和材料学部邀请,任"十一五"无机非金属材料学科发展战略研究报告《无机非金属材料科学》的编写组副组长并兼任"信息功能材料"组组长。该书于2006年10月出版
2004.2	2005.9	负责组织并完成了《纳米半导体技术》一书的编写工作,该书于2006年4月正式出版发行
2004.3		任《中国材料工程大典》副主编,以及《中国材料工程大典·信息功能材料工程》(第11卷、第12卷和第13卷)主编(王占国、陈立泉和屠海令),共撰写300余万字,该书于2006年正式出版

续表

开始时间	结束时间	事　件
2004. 7	2005. 5	受国家发展和改革委员会委托,中国工程院和中国科学院组织开展了对高技术产业8个重点领域“十一五”专项规划发展重点咨询研究工作,作为新材料领域专家组组长,确定了新材料领域17个“十一五”高技术产业发展重点,并执笔撰写了领域综合报告
2006. 6. 4	2006. 6. 9	院士大会,当选信息技术科学部常委
2007. 5. 22	2007. 5. 23	任中国科学院信息技术科学部和技术科学部第25次技术科学论坛主席,并作“半导体太阳能电池研发现状与趋势”大会邀请报告
2008. 5. 10		被授予中国科学院研究生院杰出贡献教师荣誉称号
2008		被评为中国科学院半导体研究所2006-2008年度优秀党员
2011. 4. 15		任科学技术部、教育部和中组部千人计划评审会工程与材料组组长
2012. 6. 14		受邀参加白春礼院长召开的2010-2020年科技发展战略研讨会,并提出两条建议:①近中远红外激光材料和器件;②高效太阳能光伏电池基础研究
2012. 6. 14		当选第七届中国科学院学部主席团成员
2013. 5. 24		参加第七届中国科学院学部主席团第5次会议
2015. 9. 7		参加第七届中国科学院学部主席团第15次会议
2016. 9. 22	2016. 9. 23	参加“高功率高光束质量半导体激光技术发展战略”的第570次香山科学会议。作为执行主席之一主持22日下午的报告讨论会
2017. 10. 13		受中国科学技术协会视频采访,题目为“无处不在的半导体”

学术交流目录

开始时间	结束时间	事 件
1978. 8. 1	1978. 8. 15	参加在江西庐山召开的中国物理学会年会。这是“文化大革命”后首次物理学界的盛会,也是众多物理学家及其领导人物的亮相大会。半导体研究所的主要业务领导人黄昆所长,王守武、林兰英副所长等以及王占国等几位青年科技工作者(还包括郑厚植、沈光地)也有幸获准参加了会议
1980. 10. 14	1983. 11	赴瑞典隆德大学固体物理系(国际半导体深能级研究中心)进修,主要从事半导体深能级物理和光谱物理研究
1982. 9	1982. 10	参加在法国蒙皮里埃召开的国际半导体物理会议和在意大利 Cagliari 召开的三元-四元化合物半导体材料国际会议,会后顺访了德国马克斯-普朗起研究所
1987. 9		中日第二次晶体生长双边会议在日本筑波召开,半导体研究所王占国、许振嘉、王圩和南京大学蒋树声、山东大学蒋民华参加
1988. 5	1988. 7	瑞典隆德大学固体物理系短期学术访问,之前应邀顺访匈牙利布达佩斯半导体物理研究所
1988. 9	1988. 10	受 Pankove 教授邀请,与林兰英先生一起访问美国科罗拉多州立大学,并作“空间微重力 GaAs 材料光电性质研究”报告,会后在芝加哥参加美国 NASA 主持的空间材料会议
1989. 9		受日本学习院的小川智哉教授邀请,作为分会主席,参加在日本东京召开的第3届半导体缺陷识别、成像和物理(DRIP-3)国际会议,并作“空间材料光电性质研究”大会报告
1990. 5. 7	1990. 5. 22	参加在加拿大多伦多召开的第6届半导体和绝缘体材料(SIMC-6)国际会议,并当选该会议国际顾问委员会委员。会后受刘焕明等教授邀请到西安大略大学表面物理中心访问
1990. 6. 18	1990. 6. 22	参加在北京劳动大厦召开的中国材料研究学会(C-MRS)会议。会间参与同澳大利亚材料代表团(威廉姆和加德葛士等教授)会谈合作事宜
1990. 9. 9	1990. 9. 22	隆德大学固体物理系 Par Omling 博士来访,在访问半导体研究所后,由我陪同他参观了中国科学院长春物理研究所、天津电子部46所等
1991. 2. 15	1991. 2. 26	代表新材料领域与郭景坤等一起到印度孟买参加国际金属材料大会,会后访问安娜大学。随后访问新德里大学

续表

开始时间	结束时间	事　件
1991. 5. 4	1991. 5. 15	参加在匈牙利布达佩斯举行的欧洲第三届晶体生长会议,会后访问匈牙利首都布达佩斯、保加利亚首都索菲亚等,后赴瑞典隆德大学固体物理系做短期访问
1991. 5. 15	1991. 7	到瑞典隆德大学固体物理系短期学术访问。7 月底 8 月初顺访苏联圣彼得堡约飞技术物理研究所
1991. 11. 26	1991. 11. 30	受小川智哉教授邀请,赴美国夏威夷参加由日本科学家主持召开的硅材料会议
1992. 4. 初	1992. 4. 中	受“863”新材料领域专家委员会派遣,与熊家炯、金亿鑫共同考察了俄罗斯。与先进技术公司的半导体部(ATC-SD)的 V. E. Myachin 和 V. Chaly 博士签下了研制大功率量子阱激光器合作意向书
1992. 4. 19	1992. 4. 27	参加在墨西哥海滨小城镇 Ixtapa(20-24 日)召开的 SIMC-7 国际会议。会后应 D. C. Look 邀请,经美国国防部批准,到俄亥俄空军基地访问
1993. 10. 12	1993. 10. 16	第 7 次中日晶体生长双边会议在日本召开,期间(14-16 日)和蒋民华一起受 Sasaki 教授邀请,访问了大板工业大学
1994. 2. 5	1994. 2. 20	受“863”新材料领域专家委员会委托,与蒋民华一起就蓝、绿光 LED 材料与器件发展趋势问题访问美国。这次访问为在新材料领域“863”计划的Ⅲ族氮化物首先立项打下了基础
1994. 6. 6	1994. 6. 13	作为该国际会议顾问委员会委员,参加在波兰首都华沙召开的 SIMC-8 国际会议。会后受哈弗雷迪教授邀请赴冰岛参加北欧光电材料会议
1994. 6. 13	1994. 7. 12	作为 1991-1995 年中国科学院和英国皇家科学院合作研究项目中方主持人,受邀到英国萨里大学电气工程系短期学术访问
1994. 7	1994. 7	访问萨里大学期间,又与英国诺丁汉大学物理系(C. A. Bates 教授)签订了新的合作研究协议
1994. 9. 29	1994. 10. 13	受“863”新材料领域专家委员会派遣,与蒋民华、熊家炯赴俄罗斯与 ATC-SD 公司谈判
1995. 7. 1	1995. 7. 1	在日本仙台召开的中日晶体生长会议
1995. 8. 1	1995. 8. 2	与萨里大学合作协议结题,再次访问英国
1995. 12		与姚林生、夏冠群和中国科学院的盛海涛、国家自然科学基金委员会的孟太生女士一行 5 人赴美考察,并应邀在怀特州立大学 D. C. Look's 实验室作“空间 GaAs 晶体性质研究”报告
1996. 4. 29	1996. 5. 3	参加在法国图鲁兹召开的 SIMC-9 国际会议。会后参观西班牙巴塞罗那微电子研究所

续表

开始时间	结束时间	事　件
1996.8.2	1996.8.1.15	与陈伟博士一起参加在美国丹佛召开的国际晶体生长会议，会后参观位于匹兹堡奥克兰区的卡内基·梅隆大学(CMU)
1996.9		受会议主席邀请，在河北承德召开的第9届全国化合物半导体、微波器件、光电器件学术会议作大会邀请报告，报道了他领导的MBE组研制出国内首支量子点激光器的结果
1997.11.1	1997.11.2	参加在英国格拉斯哥召开的Ⅱ-Ⅵ族化合物半导体国际会议，并访问诺丁汉大学物理系(17-19日)
1997.12.1	1997.12.2	受"863"新材料领域专家委员会派遣，与清华大学陈浩明、南京大学刘治国教授和工业信息部郑敏政副司长一起访问澳大利亚墨尔本大学、悉尼大学等，主要考察太阳能光伏电池等新能源的研发现状
1998.5		主持在中国科学院半导体研究所召开的第一次中瑞纳米科学和技术双边会议
1998.7.5	1998.7.20	金属学会组团参加在美国夏威夷召开的环太平洋电子材料大会，并受D. C. Look 教授邀请，到怀特州立大学访问
1999.4.21	1999.4.30	受韩国科技部门邀请与石力开教授访问韩国首都汉城(现名首尔)、釜山等高校和研究院(所)
1999.10.7	1999.10.9	受邀参加深圳第一届中国国际高新技术成果交易会会，并作"半导体材料研究进展"报告
2000.5.23	2000.6.6	C-MRS组团访问法国科学院。随后参加2000EMRS会议，应邀作"中国电子材料研究进展"大会应缴报告
2000.7.3	2000.7.7	作为SIMC国际会议顾问委员会委员，和吴巨一起参加在澳大利亚堪培拉召开的SIMC-11国际会议。会后去悉尼大学访问，并在校作了学术报告
2000.8.29	2000.9.6	参加在日本仙台召开的中日双边晶体生长会议。会后受高军思和邹德春博士邀请，访问九州大学
2000.11.27	2000.12.3	中国电子学会第一届全国纳米技术与应用学术会议在厦门大学召开。应邀作"半导体量子点激光器研究进展"的大会特邀报告
2001.1.7	2001.1.10	中国专家网组织国内数十位院士、校长访问香港中文大学、城市大学和香港科技大学等
2001.3.24	2001.3.28	同闵乃本、蒋民华等一起参加日本学术振兴会第161委员会第24次年会，并作大会特邀报告；26日在东京为小川智哉教授祝寿
2001.7.1	2001.7.8	与"863"新材料领域原专家委员李成功教授等一起参加在新加坡召开的新加坡MRS会议，并作分会邀请报告

续表

开始时间	结束时间	事　件
2001.9.16	2001.9.21	受中国工程院邀请,参加在西安召开的化工、冶金与材料工程学第三届学术年会议和第四届中国功能材料及其应用学术会议(任后者大会副主席)
2001.9.25	2001.10.5	参加在意大利里米尼召开的DRIP-9国际会议
2002.6.24	2002.6.28	参加在瑞典马尔默召开的第7届国际纳米科学与技术会议和第21届欧州表面科学会议(NANO-7/ECOSS-21)
2002.6.30	2002.7.5	受F. Dubesky教授邀请,参加在斯洛伐克一个古城堡召开的SIMC-12国际会议
2002.7.8	2002.7.15	参加在西安市召开的国际材联电子材料会议,任会议副主席兼程序委员会主席,并受邀作大会报告
2002.12.6	2002.12.15	参加在巴西福塔莱萨召开的低维半导体材料与器件国际会议。会后参观巴西首都巴西利亚
2003.9.27	2003.10.8	参加在法国Batz-sur-Mer召开的DRIP-10会议;应邀参观瑞典隆德大学和丹麦技术大学,并作报告
2003.10.中	2003.10末	受中国工程院邀请,参加在湖南长沙召开的化工、冶金与材料工程学部第四届学术年会,作“光电信息功能材料的研究进展”大会报告
2003.11.13	2003.11.16	应中国运载火箭技术研究院邀请,赴西昌卫星发射中心,并在发射场附近观看卫星发射
2004.3.7	2004.3.14	受韩国KIST李精一博士邀请访问韩国,就量子点激光器材料与器件、量子级联激光器材料与器件讲学;后参加第12届首尔半导体物理和应用国际会议(作邀请报告,并任分会主席)
2004.4.2	2004.4.17	应阿肯色州立大学萨尔茂教授邀请到该校学术交流,作“低维结构材料和量子器件”的学术报告。在美国旧金山代表C-MRS参加IUMRS会议
2004.9.20	2004.9.25	参加在北京友谊宾馆召开的SIMC-13国际会议,任会议主席,约100位国外代表参加
2005.4.27	2005.4.30	第九届全国金属有机化学气相沉积(MOCVD)学术会议在杭州召开,任大会联合主席
2005.9.8	2005.9.13	受总装政治部邀请,参加纪念两弹结合实验成功40周年两院院士西部行
2005.9.13	2005.9.19	参加在北京香山饭店召开的DRIP-11国际会议,任会议主席,有100多位国外代表参会
2005.10.12		参观酒泉卫星发射中心及观看“神舟六号”发射

续表

开始时间	结束时间	事　件
2005. 11		参加中国宁波新材料及产业化国际论坛,任论坛学术委员会主任,并应邀作“半导体白光照明:材料与器件”报告
2005. 9		任第三届亚洲晶体生长和技术会议(CGCT-3)的联合主席之一和宽带隙半导体材料分会主席(北京)
2005. 12. 8	2005. 12. 14	受中国工程院化工、冶金和材料工程学部邀请(刘建文同往),参加在海南博鳌召开的该学部第五届年会,并作题为“半导体太阳能光伏电池的发展现状与趋势”报告
2005. 12. 20	2005. 12. 24	受邀参加第二届全国 ZnO 材料学术研讨会任大会主席之一,应邀作“ZnO基半导体材料与器件的研究现状、发展趋势和对策建议”大会特邀报告
2006. 6. 14	2006. 6. 16	受西安硅产业高技术论坛邀请,作“硅基微电子技术极限对策探讨”报告
2006. 6. 25	2006. 6. 30	任 2006 北京国际材料周论坛副主席和 C-MRS 年会光电子材料分会主席,并在 2006 年中国材料研讨会上作 “半导体光电信息功能材料研究进展”大会特邀报告
2006. 7. 12	2006. 7. 14	任 2006 中国(深圳)国际半导体照明论坛大会主席之一,并主持 12 日上午部分会议
2006. 7. 18	2006. 7. 21	作为山东大学微电子学院特聘教授,受邀到青岛理工大学和青岛大学作“硅基微电子技术极限对策探讨”报告
2006. 8. 23	2006. 8. 25	参加国家自然科学基金委员会信息学部组织的“以宽禁带半导体材料与器件”为主题的“双清论坛”,并作题目为“ZnO 基半导体材料的研究现状与发展趋势”的大会邀请报告
2006. 9. 16	2006. 9. 18	受湖北大学物理学院王浩教授邀请,作“后摩尔时代微电技术发展探讨”报告
2006. 9. 18	2006. 9. 23	受总装政治部邀请,参加纪念两弹结合实验成功 40 周年两院院士西部行,也是王占国院士 36 年后重返核试验基地(21 基地)之行
2006. 10. 18	2006. 10. 20	任中国科学院信息技术科学部常委。作“后摩尔时代微电子技术探讨”报告,后应钱鹤研究员邀请参观苏州和杭州三星公司研究院
2006. 11. 4	2006. 11. 11	参加第 14 届全国化合物半导体材料、微波器件和光电子器件学术会议(北海),任大会会议主席,并作“后摩尔时代微电子技术探讨”大会特邀报告
2006. 11. 20	2006. 11. 27	应邀访问台湾材料科学研究会,在两岸华人前瞻技术与应用论坛上应邀作“半导体纳米结构制备和量子器件应用”邀请报告;会后到台湾各地参观访问

续表

开始时间	结束时间	事　件
2007.4.26	2007.4.28	受中部六省郑州招商大会南阳光电产业发展高层论坛邀请,作"光电功能材料产业发展热点"大会特邀报告
2007.5.22	2007.5.23	任中国科学院信息技术科学部和技术科学部第25次技术科学论坛(连云港)主席,并作"半导体太阳能电池研发现状与趋势"大会邀请报告
2007.6.1	2007.6.15	出访瑞典隆德大学固体物理系、德国弗朗霍夫学会太阳能研究所、法国国家科学研究中心、英国射菲尔德大学、剑桥大学和苏格兰格拉斯哥大学和达拉谟大学等
2007.7.8	2007.7.9	受江苏新材料产业发展高层论坛邀请,作"'十一五'新材料高技术产业发展重点建议"大会报告
2007.8.23	2007.8.25	中国科学院信息技术科学部和技术科学部第26次技术科学论坛(威海),受邀作"半导体光电信息功能材料研究新进展"报告
2007.9.12	2007.9.19	受信息产业部中国电子材料行业协会半导体材料分会年度学术会议(兰州)邀请,作"光电信息功能材料与器件研究进展"大会报告
2007.11.1	2007.11.3	受第15届全国半导体集成电路、硅材料学术会议(重庆)邀请,作"硅微电子技术物理极限的可能对策探讨"大会报告
2007.11.15	2007.11.19	受第六届全国功能材料及其应用学术会议暨2007国际功能材料专题论坛邀请,作"Semiconductor nanostructures and quantum device applications"大会报告
2007.12.20	2007.12.23	受中澳纳米双边会议(厦门大学)邀请,作"Semiconductor nanostructures and quantum device applications"大会报告
2008.4.18	2008.4.22	受邀参加2008年中瑞纳米科学与技术论坛(北京),并作量子级联激光器有关的报告
2008.6.9	2008.6.12	受邀参加2008中美材料国际研讨会暨MRS国际材料研究大会,并作题为"Semiconductor nanostructures and quantum devices"大会特邀报告
2008.8.13	2008.8.15	受邀参加第二届微纳电子技术行业交流会,作"半导体纳米结构生长及器件应用"大会特邀报告
2008.9.21	2008.9.24	作为中方代表团团长参加US-China Workshop on Nano-structured Materials for Global Energy and Environmental Challenge (Northwestern University, Evanston, Chicago)会议,并作"RDS characterization of wetting layer around InAs/GaAs quantum dots by SK growth mode"报告
2008.9.24	2008.9.26	受王中林邀请,出访美国亚特兰大佐治亚理工学院
2008.9.26	2008.10.2	出访美国University of Texas Dallas at Arlington, 作"Semiconductor Nanostructures Growth and Quantum Device Fabrication"报告

续表

开始时间	结束时间	事　件
2009.3.30	2009.4.3	参加在京西宾馆召开的中国科学院和中国科学技术协会2010年院士初选推荐会。为清华大学研究生作“半导体激光器”的讲座
2009.4.7	2009.4.8	应河北大学电子学院邀请，作有关太阳能光伏电池报告
2009.8.26	2009.8.29	受黄伟其教授邀请赴贵州大学光电学院（任光电学院名誉主任和贵大兼职教授），期间作“硅基发光问题”报告
2009.9.10		受河北省保定英利多晶硅有限责任公司邀请，作“太阳能光伏电池”报告
2009.9.22	2009.9.24	参加河南省科协洛阳年会，作“光电信息材料与器件研究现状与趋势”报告
2009.10.15	2009.10.17	有关省光伏材料重点实验室事宜赴河南大学物理学院，并作“第三代太阳能光伏电池”报告
2009.10.20	2009.10.22	参加第10届中俄双边新材料新工艺研讨会，并作“Semiconductor nanostructure growth and device application”报告
2009.10.27	2009.10.30	参加第16届全国半导体集成电路、硅材料学术会议，作为大会主席并作“新概念太阳能光伏电池研究进展”报告，后去浙江师范大学，并作报告
2010.4.20		参加中国人民解放军总装备部科学技术委员会主持召开的我国中长期装备发展材料需求咨询会，作“超宽禁带半导体材料与器件”报告
2010.8.8	2010.8.13	参加在北京国际会议中心举行的第16届国际晶体生长会议暨第14届国际气相生长与外延会议，任国际气相外延大会主席
2010.10.25	2010.10.29	参加第16届全国化合物半导体材料、微波器件和光电子器件学术会议（西安），作“第三代半导体太阳能光伏电池研究进展”报告
2011.3.18		作为共同主席参加在北京航空航天大学举办的2011年全国光电子与量子电子学技术会议，作“第三代太阳能光伏电池状研究现状”大会报告
2011.4.24	2011.4.26	受邀去东北大学电气自动化研究所参观，作“半导体光电信息材料和器件研究进展” 报告
2011.10.23	2011.10.31	参加中国科学院7人代表团访美国、加拿大科学院。26日上午与美国科学院、工程院和科学艺术院座谈；28日上午与加拿大皇家学会座谈
2012.2.18		在中国科学技术馆作“半导体的过去、现在和将来”科普报告
2012.4.19	2012.4.23	应邀参加工业信息部电子材料行业分会（金坛）年度会员大会，作“超宽带隙半导体材料及应用”大会邀请报告

续表

开始时间	结束时间	事　件
2012. 6. 14		受邀参加白春礼院长主持召开的 2010-2020 年科技发展战略研讨会，并提出两条建议；①近中远红外激光材料和器件；②高效太阳能光伏电池基础研究
2012. 11. 6	2012. 11. 11	参加第 17 届全国化合物半导体、微波器件和光电子器件学术会议（开封），作"半导体量子级联激光器材料及其器件应用"大会邀请报告
2012. 12. 4	2012. 12. 13	作为会议执行主席之一，参加香山科学会议第 448 次国际学术讨论会：压电电子学和纳米发电机发展前沿
2012. 12. 9		河南卫视《根在中原》节目播放"半导体材料和材料物理学家王占国"节目
2014. 12. 29		在半导体材料科学重点实验室为王占国院士举办的 76 岁生日会上作"第三代半导体材料及应用"报告
2015. 4. 23		应南开大学今日物理小组邀请，在南开大学作"第三代半导体材料及应用简介"报告
2016. 5. 3		英国伦敦大学学院刘会赟教授（王占国院士 2001 年毕业的博士）来访，作"在 Si 衬底上生长 InAs/GaAs 量子点激光器"报告
2016. 12. 29		在半导体材料科学重点实验室为王占国院士举办的 78 岁生日会上作"二维半导体材料研究现状"报告
2017. 10. 13		中国科学技术协会视频采访，作"半导体材料物理——杂质和缺陷在半导体材料和器件中行为研究"的科普报告

获奖目录

获奖时间	奖项
1989. 10	“LPE InGaAsSb/InP 材料和性质研究”获中国科学院科学技术进步奖二等奖(排名第二位)
1989. 11	“微重力条件下从熔体生长 GaAs 单晶及其性质研究”获中国科学院科学技术进步奖一等奖(排名第四位)
1990. 9	荣获 1990 年度中国科学院优秀研究生导师奖
1990. 10	“等电子杂质 In 在 GaAs 中行为的研究”获中国科学院科学技术进步奖三等奖(排名第一位)和国家“七五”重点科技攻关奖(个人)
1990. 12	“微重力条件下从熔体生长 GaAs 单晶及其性质研究”获国家科学技术进步奖三等奖(排名第四位)
1993. 1	获国务院特殊津贴
1995. 10	“高质量 GaAs/AlGaAs 二维电子气材料研制及其器件应用”获中国科学院科学技术进步奖二等奖(排名第七位)
1996. 12	获电子行业国家“八五”重点科技攻关奖(HBT 和 HEMT 基础技术研究)
1997. 12	“砷化镓材料、器件与电路研究”获中国科学院科学技术进步奖二等奖(排名第二位)
2000. 6	荣获国务院学位委员会授予的 2000 年百篇优秀博士论文(李含轩)导师奖
2000. 12	“自组织生长量子点激光材料和器件研究”获中国科学院自然科学奖一等奖(排名第一位)
2001. 2	荣获国家“863”计划 15 周年先进个人称号
2001. 3	“自组织生长量子点激光材料和器件研究”获国家自然科学奖二等奖(排名第一位)
2001. 9	荣获何梁何利科学与技术进步奖,宝洁优秀研究生导师奖
2008. 5	荣获中国科学院研究生院杰出贡献教师荣誉称号
2008	荣获中国科学院半导体研究所 2006-2008 年度优秀党员称号
2016. 6	拍摄的“神六发射”照片获中国科学院庆祝建党 95 周年书法、绘画、摄影和微电影比赛特等荣誉奖
2017. 12	荣获国家知识产权局第 19 届中国专利奖评选工作最佳推荐奖

论 著 目 录

截至本书出版,据不完全统计共公开发表科技论文700余篇、出版学术专著5部(章),撰写会议报告60篇、工作报告5篇、讲义1篇。

部分科技论文

1. 周洁,王占国,刘志刚,王万年和尤兴凯. 硅单晶低温电学性质. 物理学报,1966,22:404;科学通报(外文版),1966,17(5):206.
2. 王占国. GaAs体效应器件击穿现象的观察和分析. 半导体通讯,1975,6:14.
3. 王占国,林兰英. *N*-GaAs单晶热学稳定性的研究. 半导体通讯,1977,(3):32.
4. 林兰英,王占国. 杂质和缺陷在GaAs中的行为. 稀有金属,1978,(1):3.
5. L. Ledebo and Z. G. Wang. Evidence that the gold donor and acceptor in silicon are two levels of the same defect. Appl. Phys. Lett.,1983,42(8):680-682.
6. L. Jansson, Z. G. Wang, L. Å. Ledebo, and H. G. Grimmeiss. Composition dependence and random alloy effects for Cu and Fe in the semiconductor alloy AlGaAs. IL Nuovo Cimento D,1983,2(6):1718-1722.
7. E. Meijer, L. Å. Ledebo, and Z. G. Wang. Influence from free-carrier tails in Deep Level Transient Spectroscopy(DLTS). Solid State Commun.,1983,46(3):255-258.
8. Z. G. Wang, L. Ledebo, and H. G. Grimmeiss. Electronic properties of native deep level defects in liquid-phase epitaxial GaAs. J. of Phys. C: Solid State Phys.,1984,17(2):259-272.
9. L. Samuelson, S. Nilsson, Z. G. Wang, and H. G. Grimmeiss. Direct evidence for random-alloy splitting of Cu level in $GaAs_{1-x}P_x$. Phys. Rev. Lett.,1984,53(15):1501-1503.
10. Z. G. Wang, L. Ledebo, and H. G. Grimmeiss. Optical properties of iron doped $Al_xGa_{1-x}As$ alloys. J. Appl. Phys.,1984,56(10):2762-2767.
11. A. R. Peaker, U. Kaufmann, Z. G. Wang, R. Worner, B. Hamilton, and H. G. Grimmeiss. Electrical and optical-properties of the neutral nickel acceptor in gallium-phosphide. J. of Phys. C: Solid State Phys.,1984,17(34):6161-6167.
12. Z. G. Wang, H. P. Gislason, and B. Monemar. Acceptor associates and bound excitons in GaAs:Cu. J. Appl. Phys.,1985,58(1):230-239.
13. B. Monemar, H. P. Gislason, and Z. G. Wang. Localization of excitons to Cu-related defects in GaAs. Phys. Rev. B,1985,31(12):7919-7924.

14. H. P. Gislason, B. Monemar, Z. G. Wang, Ch. Uihlein, and P. L. Liu. Direct evidence for the acceptor-like character of the Cu-related C and F bound-exciton centers in GaAs. Phys. Rev. B, 1985, 32(6): 3723-3729.

15. Z. G. Wang, Z. W. Shi, S. D. Xu, and L. Y. Lin. Studies on optical and electrical properties of GaAs single crystal grown in space. Chinese J. of Semicond., 1988, 9: 317.

16. L. Zhong, Z. G. Wang, S. Wan, and L. Y. Lin. Absorption peaks at 2663 and 2692 cm^{-1} observed in neutron-transmutation-doped silicon. J. Appl. Phys., 1989, 66(9): 4275-4278.

17. L. Zhong, Z. G. Wang, S. K. Wan, and L. Y. Lin. On the correlation between high-order bands and some photoluminescence lines in neutron-irradiated FZ silicon. J. Appl. Phys., 1989, 66(8): 3787-3791.

18. J. J. Qian, Z. G. Wang, S. K. Wan, and L. Y. Lin. A novel model of new donors in Czochralski-grown Silicon. J. Appl. Phys., 1990, 68(3): 954-957.

19. Z. G. Wang, C. J. Li, S. K. Wan, and L. Y. Lin. Spatial distribution of impurities and defects in Te- and Si-doped GaAs grown in a reduced gravity environment. J. Cryst. Growth, 1990, 103(1-4): 38-45.

20. Z. G. Wang, C. J. Li, F. N. Cao, Z. W. Shi, B. J. Zhou, X. R. Zhong, S. K. Wan, S. D. Xu, and L. Y. Lin. Electrical characteristics of GaAs grown from the melt in a reduced-gravity environment. J. Appl. Phys., 1990, 67(3): 1521-1524.

21. L. Zhong, Z. G. Wang, S. Wan, J. B. Zhu, and F. Shimura. Hydrogen-related donor in silicon-crystals grown in a hydrogen atmosphere. Appl. Phys. A-Materials Science & Processing, 1992, 55(4): 313-316.

22. Q. H. Du, Z. G. Wang, and J. M. Mao. Simulation of lateral confinement in very narrow channels. Phys. Rev. B, 1994, 49(24): 17452-17455.

23. G. B. Ren, Z. G. Wang, B. Xu, and B. Zhou. Theoretical investigation of the dynamic process of the illumination of GaAs. Phys. Rev. B, 1994, 50(8): 5189-5195.

24. V. L. Dostov and Z. G. Wang. Effect of image forces on electrons confined in low dimensional structures under a magnefic field. Semicond. Sci. Technol., 1994, 9(10): 1781-1786.

25. B. Yang, Y. H. Chen, Z. G. Wang, J. B. Liang, Q. W. Liao, L. Y. Lin, Z. P. Zhu, B. Xu, and W. Li. Interface roughness scattering in GaAs-AlGaAs modulation-doped heterostractures. Appl. Phys. Lett., 1994, 65(26): 3329-3331.

26. W. Li, Z. G. Wang, J. B. Liang et al. Investigation of epitaxial growth of $In_xGa_{1-x}As$ on GaAs (001) and extension of two dimensional-three dimensional growth mode

transition. Appl. Phys. Lett. ,1995,66(9): 1080-1082.

27. B. Yang,Y. H. Cheng,Z. G. Wang et al. Influence of DX centers in the $Al_xGa_{1-x}As$ barrier on the low-temperature density and mobility of the two-domensional electron gas in GaAs/AlGaAs modulation-doped heterostructure. Appl. Phys. Lett. ,1995,66(11): 1406-1408.
28. W. Li,Z. G. Wang,J. B. Liang,B. Xu,Z. P. Zhu,Z. L. Yuan,and J. A. Li. Photoluminescence studies of single submonolayer InAs structures grown on GaAs (001) matrix. Appl. Phys. Lett. ,1995,67(13): 1874-1876.
29. J. R. Dong,Z. G. Wang,X. L. Liu,D. C. Lu,D. Wang,and X. H. Wang. Photoluminescence of ordered $Ga_{0.5}In_{0.5}P$ grown by metalorganic vapor-phase epitaxy. Appl. Phys. Lett. ,1995,67(11): 1573-1575.
30. R. G. Li,Z. G. Wang,J. B. Liang,G. B. Ren,T. W. Fan,and L. Y. Lin. Backgating and light sensitivity in GaAs metal-semiconductor field-effect transistors. J. Cryst. Growth,1995,150(1-4): 1270-1274.
31. W. Chen,Z. G. Wang et al. New observation on the formation of PbS clusters in zeolite-Y. Appl. Phys. Lett. ,1996,68(14): 1990-1992.
32. W. Chen,Z. J. Lin,Z. G. Wang,and L. Y. Lin. Some new observation on the formation and optical properties of CdS clusters in zeolite-Y. Solid State Commun. , 1996,100(2): 101-104.
33. J. R. Dong,Z. G. Wang et al. Ordering along <111> and <100> directions in GaInP demonstrated by photoluminescence under hydrostatic pressure. Appl. Phys. Lett. , 1996,68(12): 1711-1713.
34. J. R. Dong,G. H. Li,Z. G. Wang,D. C. Lu,X. L. Liu,X. B. Li,D. Z. Sun,M. Y. Kong,and Z. J. Wang. Photoluminescence of GaInP under high pressure. J. Appl. Phys. ,1996,79(9): 7177-7182.
35. J. Wu,Z. G. Wang,and L. Y. Lin. Influence of the semi-insulating GaAs schottky pad on the schottky barrier in the active layer. Appl. Phys. Lett. , 1996, 68 (18): 2550-2552.
36. N. F. Chen,Y. T. Wang,H. J. He,Z. G. Wang,L. Y. Lin,and O. Oda. Nondestructive measurements of stoichiometry in undoped semi-insulating gallium arsenide by X-ray bond method. Appl. Phys. Lett. ,1996,69(25): 3890-3892.
37. Z. H. Chen,P. L. Liu,W. Lu,Z. H. Chen,X. H. Shi,G. L. Shi,S. C. Shen,B. Yang,Z. G. Wang,and L. Y. Lin. Magnetospectroscopy of bound phonons in high purity GaAs. Phys. Rev. Lett. ,1997,79(6): 1078-1081.
38. W. Chen,Z. G. Wang,Z. J. Lin,and L. Y. Lin. Thermoluminescene of ZnS

Nanoparticles. Appl. Phys. Lett. ,1997,70(9-12): 1465-1467.

39. D. H. Zhu, Z. G. Wang et al. 808 nm high-power laser grown by MBE through the control of Be diffusion and use of superlattice. J. Cryst. Growth, 1997, 175-176(Part 2): 1004-1008.

40. C. Jiang, B. Xu, H. X. Li, F. Q. Liu, Q. Gong, W. Zhou, D. H. Zhu, J. B. Liang, and Z. G. Wang. Molecular beam epitaxy growth of InAlAs/InGaAs/InAlAs/InP P-HEMTs with enhancement conductivity using an intentional nonlattice-matched buffer layer. J. Vac. Sci, Technol. B, 1997, 15(16): 2021-2025.

41. W. Chen, Z. G. Wang, and L. Y. Lin. Thermoluminescence of CdS clusters in zeolite-Y. J. Lumin. ,1997,71(2): 151-156.

42. H. X. Li, Z. G. Wang, J. B. Liang, B. Xu, J. Wu, Q. Gong, C. Jiang, F. Q. Liu, and W. Zhou. The third subband population in modulation-doped InGaAs/InAlAs heterostructures. J. Appl. Phys. ,1997,82(12): 6107-6109.

43. Y. H. Chen, Z. Yang, R. G. Li, Y. Q. Wang, and Z. G. Wang. Reflectance-difference spectroscopy study of the Fermi-level position of low-temperature grown GaAs. Phys. Rev. B, 1997, 55(12): 7379-7382.

44. Y. H. Chen, Z. Yang, Z. G. Wang, B. Xu, J. B. Liang, and J. J. Qian. Optical anisotropy of InAs submonolayer quantum wells in a (311) GaAs matrix. Phys. Rev. B, 1997, 56(11): 6770-6773.

45. W. Chen, Z. G. Wang, Z. J. Lin, and L. Y. Lin. Absorption and luminescence of the surface states in ZnS nanoparticles. J. Appl. Phys. ,1997,82(6): 3111-3115.

46. Z. J. Lin, Z. G. Wang, W. Chen, L. Y. Lin, G. H. Li, Z. X. Liu, H. X. Han, and Z. P. Wang. Influence of the ring-ring interaction in Se-8-ring clusters on their absorption. J. Mater. Sci. Lett. ,1997,16(9): 732-734.

47. X. H. Zhang, B. X. Jiang, Y. F. Yang, and Z. G. Wang. Spectroscopic properties of anisotropic absorption in a neodymium-doped $YAlO_3$ laser crystal. J. Appl. Phys. , 1997, 81(10): 6939-6942.

48. W. Chen, Z. G. Wang et al. Photostimulated Luminescence of AgI cluster in zeolite-Y. J. Appl. Phys. ,1998,83(7): 3811-3815.

49. W. Chen, Z. G. Wang, L. Y. Lin, and M. Z. Su. New Color Centers and Photostimulated Luminescence of BaFCl: Eu^{2+}. J. Phys. Chem. Solids. ,1998,59(1): 49-53.

50. F. Q. Liu, Z. G. Wang, G. H. Li, and G. H. Wang. Photoluminescence from Ge clusters embedded in porous silicon. J. Appl. Phys. ,1998,83(6): 3435-3437.

51. H. X. Li, J. Wu, B. Xu, C. Jiang, Q. Gong, F. Q. Liu, W. Zhou, J. B. Liang, and

Z. G. Wang. Growth and characterization of InGaAs/InAlAs high electron mobility transistors towards high channel conductivity. J. Cryst. Growth, 1998, 186 (3): 309-314.

52. H. X. Li, J. B. Liang, B. Xu, J. Wu, and Z. G. Wang. Ordered InAs quantum dots in InAlAs matrix grown by molecular beam epitaxy. Appl. Phys. Lett., 1998, 72 (17): 2-2125.

53. H. X. Li, Z. G. Wang, J. B. Liang, B. Xu, J. Wu, Q. Gong, C. Jiang, F. Q. Liu, and W. Zhou. InAs quantum dots in InAlAs matrix grown by molecular beam epitaxy. J. Cryst. Growth, 1998, 187(3-4): 564-568.

54. Q. Gong, J. B. Liang, B. Xu, D. Ding, H. Li, C. Jiang, W. Chen, F. Q. Liu, and Z. G. Wang. Analysis of atomic force microscopic results of InAs islands formed by MBE. J. Cryst. Growth, 1998, 192(3): 376-380.

55. J. Wu, H. X. Li, T. W. Fan, and Z. G. Wang. Alignment of misfit dislocations in the $In_{0.52}Al_{0.48}As/In_xGa_{1-x}As/In_{0.52}Al_{0.48}As/InP$ heterostructure. Appl. Phys. Lett., 1998, 72(3), 311-313.

56. L. S. Wang, X. L. Liu, Y. D. Zan, D. Wang, D. C. Lu, and Z. G. Wang. The growth and characterization of GaN grown on an Al_2O_3 coated (001) Si substrate by MOVCD. J. Cryst. Growth, 1998, 193(4): 484-490.

57. W. Chen, Y. Xu, Z. J. Lin, Z. G. Wang, and L. Y. Lin. Formation, structure and fluorescence of CdS clusters in a mesoporous zeolite. Solid State Commun., 1998, 105 (2): 129-134.

58. H. X. Li, Z. G. Wang, J. B. Liang, B. Xu, C. Jiang, Q. Gong, F. Q. Liu, and W. Zhou. Photoluminescence study of $In_xGa_{1-x}As/In_yAl_{1-y}As$ one-side-modulation-doped asymmetric step quantum wells. Solid State Commun., 1998, 106(12): 811-814.

59. L. W. Lu, J. Wang, Y. Wang, W. K. Ge, G. W. Yang, and Z. G. Wang. Conduction-band offset in a pseudomorphic $GaAs/In_{0.2}Ga_{0.8}As$ quantum well determined by capacitance-voltage profiling and deep-level transient spectroscopy techniques. J. Appl. Phys., 1998, 83(4): 2093-2097.

60. L. S. Wang, X. L. Liu, Y. D. Zan, D. Wang, J. Wang, D. C. Lu, and Z. G. Wang. Wurtzite GaN epitaxial growth on a Si (001) substrate using γ-Al_2O_3 as an intermediate layer. Appl. Phys. Lett., 1998, 72(1): 109-111.

61. H. X. Li, J. Wu, Z. G. Wang et al. High-density InAs nanowires realized in situ on (100) InP. Appl. Phys. Lett., 1999, 75(8): 1173-1175.

62. H. Q. Guo, S. M. Liu, L. Zhu, Z. H. Zhang, W. Chen, and Z. G. Wang. Synthesis and luminescence of CdS/ZnS core/shell nanocrystals. Molecular Crystals and Liquid

Crystals Science and Technology Section a-Molecular Crystals and Liquid Crystals, 1999,337: 197-200.

63. F. Q. Liu, Z. G. Wang, J. Wu, B. Xu, Q. Gong, and J. B. Liang. Structure and photoluminescence of InGaAs self-assembled quantum dots grown on InP(001). J. Appl. Phys., 1999, 85(1): 619-621.

64. Z. Z. Sun, D. Ding, Q. Gong, W. Zhou, B. Xu, and Z. G. Wang. Quantum-dot superluminescent diode: A proposal for an ultra-wide output spectrum. Opt. Quantum Electron., 1999, 31(12): 1235-1246.

65. H. X. Li, T. Daniels-Race, and Z. G. Wang. Growth mode and strain relaxation of InAs on InP (111) grown by MBE. Appl. Phys. Lett., 1999, 74(10): 1388-1390.

66. X. L. Sun, H. Yang, L. X. Zheng, D. P. Xu, J. B. Li, Y. T. Wang, G. H. Li, and Z. G. Wang. Stability investigation of cubic GaN films grown by metalorganic chemical vapor deposition on GaAs (100). Appl. Phys. Letter., 1999, 74 (19): 2827-2829.

67. F. Q. Liu, Y. Z. Zhang, Q. S. Zhang, D. Ding, B. Xu, Z. G. Wang, D. S. Jiang, and B. Q. Sun. High-performance strain-compensated InGaAs/InAlAs quantum cascade lasers. Semicond. Sci. Technol., 2000, 15(12): 44-46.

68. H. X. Li, Q. D. Zhuang, Z. G. Wang, and T. Daniels-Race. Influence of indium composition on the surface morphology of self-organized $In_xGa_{1-x}As$ quantum dots on GaAs substrates. J. Appl. Phys., 2000, 87(1): 188-191.

69. F. Q. Liu, Y. Z. Zhang, Q. S. Zhang, D. Ding, B. Xu, Z. G. Wang, D. S. Jiang, and B. Q. Sun. Room temperature (34℃) operation of strain-compensated quantum cascade lasers. Electronics Lett., 2000, 36(20): 1704-1706.

70. S. M. Liu, F. Q. Liu, H. Q. Guo, Z. H. Zhang, and Z. G. Wang. Surface states induced photoluminescence from Mn^{2+} doped CdS nanoparticles. Solid State Commun., 2000, 115(11): 615-618.

71. F. Q. Liu, D. Ding, B. Xu, Y. Z. Zhang, Q. S. Zhang, Z. G. Wang, D. S. Jiang, and B. Q. Sun. Strain-compensated quantum cascade lasers operating at room temperature, J. Cryst. Growth, 2000, 220(4): 439-443.

72. Y. F. Li, X. L. Ye, F. Q. Liu, B. Xu, D. Ding, W. H. Jiang, Z. Z. Sun, H. Y. Liu, Y. C. Zhang, and Z. G. Wang. Structural and optical characterization of InAs nanostructures grown on (001) and high index InP substrates. Appl. Surf. Sci., 2000, 167(3-4): 191-196.

73. Y. F. Li, F. Q. Liu, B. Xu, X. L. Ye, D. Ding, Z. Z. Sun, W. H. Jiang, H. Y. Liu, Y. C Zhang, and Z. G. Wang. Two dimensional ordering of self-assembled InAs

quantum dots grown on (311) B InP substrates, J. Cryst. Growth, 2000, 219(1-2): 17-21.

74. Y. F. Li, X. L. Ye, B. Xu, F. Q. Liu, D. Ding, W. H. Jiang, Z. Z. Sun, Y. C. Zhang, H. Y. Liu, and Z. G. Wang. Room temperature 1.55 μm emission from InAs quantum dots grown on (001) InP substrate by molecular beam epitaxy. J. Cryst. Growth, 2000, 218(2): 451-454.

75. H. Y. Liu, X. D. Wang, B. Xu, D. Ding, W. H. Jiang, J. Wu, and Z. G. Wang. Effect of In-mole-fraction in InGaAs overgrowth layer on self-assembled InAs/GaAs quantum dots. J. Cryst. Growth, 2000, 213(1): 193-197.

76. H. Y. Liu, B. Xu, Y. H. Chen, D. Ding, and Z. G. Wang. Effects of seed layer on the realization of larger self-assembled cohenrent InAs/GaAs quantum dots. J. Appl. Phys., 2000, 88(9): 5433-5436.

77. H. Y. Liu, X. D. Wang, J. Wu, B. Xu, Y. Q. Wei, W. H. Jiang, D. Ding, X. L. Ye, F. Lin, J. F. Zhang, J. B. Liang, and Z. G. Wang. Structural and optical properties of self-assembled InAs/GaAs quantum dots covered by $In_xGa_{1-x}As$ ($0<x<0.3$). J. Appl. Phys., 2000, 88(6): 3392-3395.

78. H. Y. Liu, W. Zhou, D. Ding, W. H. Jiang, B. Xu, J. B. Liang, and Z. G. Wang. Self-organized type-II InGaAs/AlGaAs quantum dots realized on GaAs(311)A. Appl. Phys. Lett., 2000, 76(25): 3741-3743.

79. Z. Z. Sun, J. Wu, F. Q. Liu, H. Z. Xu, Y. H. Chen, X. L. Ye, W. H. Jiang, B. Xu, and Z. G. Wang. Structural and photoluminescence properties of $In_{0.9}(Ga/Al)_{0.1}As$ self-assembled quantum dots on InP substrate. J. Appl. Phys., 2000, 88(1): 533-536.

80. W. H. Jiang, X. L. Ye, B. Xu, H. Z. Xu, D. Ding, J. B. Liang, and Z. G. Wang. Anomalous temperature dependence of photoluminescence from InAs quantum dots. J. Appl. Phys., 2000, 88(5): 2529-2532.

81. W. Zhou, B. Xu, H. Z. Xu, W. H. Jiang, F. Q. Liu, Q. Gong, D. Ding, J. B. Liang, and Z. G. Wang. Photoluminescence study of InAlAs quantum dots grown on differently oriented surface. J. Vac. Sci. Technol. B, 2000, 18(1): 21-24.

82. H. Z. Xu, Z. G. Wang, I. Harrison, A. Bell, B. J. Ansell, A. J. Winser, T. S. Cheng, C. T. Foxon, and M. Kawabe. Photoluminescence and optical quenching of photoconductivity studies on undoped GaN grown by molecular beam epitaxy. J. Cryst. Growth, 2000, 217(3): 228-232.

83. Z. G. Wang, F. Q. Liu, J. B. Liang, and B. Xu. Self-assembled InAs/GaAs quantum dots and quantum dot laser. Science in China (series A), 2000, 43(8): 861-870.

84. Z. G. Wang. Research and development of electronic and optoelectronic materials in

China. Chin. J. of Semiconductors,2000,21(11): 1041-1049.

85. Y. F. Li,J. Z. Wang,X. L. Ye,B. Xu,F. Q. Liu,D. Ding,J. F. Zhang,and Z. G. Wang. InAs self-assembled quantum dots grown on an InP(311)B substrate by molecular beam epitaxy. J. Appl. Phys. ,2001,89(7): 4186-4188.

86. H. Y. Liu,B. Xu,J. J. Qian,X. L. Ye,Q. Han,D. Ding,J. B. Liang,X. R. Zhong,and Z. G. Wang. Effect of growth temperature on luminescence and structure of self-assembled InAlAs/AlGaAs quantum dots. J. Appl. Phys., 2001, 90 (4): 2048-2050.

87. F. Lin,J. Wu,W. H. Jiang,H. Cui,and Z. G. Wang. Structural anisotropy and optical properties of $In_xGa_{1-x}As$ quantum dots on GaAs (001). J. Cryst. Growth,2001, 223(1-2): 55-60.

88. Z. G. Wang,Y. H. Chen,F. Q. Liu and,B. Xu. Self-assembled quantum dots,wires and quantum-dot lasers. J. Cryst. Growth,2001,227-228: 1132-1139.

89. M. Y. Liao,F. G. Qin,J. H. Zhang,Z. K. Liu,S. Y. Yang,Z. G. Wang,and S. T. Lee. Ion bombardment as the initial stage of diamond film growth. J. Appl. Phys., 2001,89(3): 1983-1985.

90. F. Q. Liu,Q. S. Zhang,Y. Z. Zhang,D. Ding,B. Xu,and Z. G. Wang. Growth and characterization of InGaAs/InAlAs quantum cascade lasers. Solid-State Electron., 2001,45(10): 1831-1835.

91. S. M. Liu,F. Q. Liu,and Z. G. Wang. Relaxation of carriers in terbium-doped ZnO nanoparticles. Chemical Physics Letters,2001,343(5): 489-492.

92. H. Y. Liu,B. Xu,Y. Q. Wei,D. Ding,J. J. Qian,Q. Han,J. B. Liang,and Z. G. Wang. High-power and long-lifetime InAs/GaAs quantum-dot laser at 1080 nm. Appl. Phys. Lett. ,2001,79(18): 2868-2870.

93. P. D. Han,Z. G. Wang,X. F. Duan,and Z. Zhang. Polarity dependence of hexagonal GaN films on two opposite c faces of Al_2O_3 substrate. Appl. Phys. Lett., 2001,78(25): 3974-3976.

94. W. H. Jiang,H. Z. Xu,B. Xu,W. Zhou,Q. Gong,D. Ding,J. B. Liang,and Z. G. Wang. Substrate dependence of InGaAs quantum dots grown by molecular beam epitaxy. J. Vac. Sci. Technol. ,B,2001,19(1): 197-201.

95. M. Y. Liao,C. L. Chai,Z. Y. Yao,S. Y. Yang,Z. K. Liu,and Z. G. Wang. Carbonization process of Si (100) by ion-beam bombardment. J. Cryst. Growth,2001, 233(3): 446-450.

96. H. Y. Liu,B. Xu,Z. G. Wang et al. Room-temperature,ground-state lasing for red-emitting vertically aligned InAlAs/AlGaAs quantum dots grown on a GaAs(100)

substrate. Appl. Phys. Lett. ,2002,80(20): 3769-3771.

97. S. C. Qu,W. H. Zhou,F. Q. Liu,N. F. Chen,and Z. G. Wang. Photoluminescence properties of Eu^{3+}-doped ZnS nanocrystals prepared in a water/methanol solution. Appl. Phys. Lett. ,2002,80(19): 3605-3607.

98. Z. Y. Zhang,B. Xu,P. Jin,X. Q. Meng,C. M. Li,X. L. Ye,and Z. G. Wang. Photoluminescence study of self-assembled InAs/GaAs quantum dots covered by InAlAs and InGaAs combination layer. J. Appl. Phys. ,2002,92(1): 511-514.

99. X. Q. Meng,B. Xu,P. Jin,X. L. Ye,Z. Y. Zhang,and Z. G. Wang. Dependence of Optical Properties on the Structure of Multi-Layer Self-Organized InAs Quantum Dots Emitting Near 1.3μm. J. Cryst. Growth,2002,243(3-4): 432-438.

100. B. Gu,Y. Xu,F. W. Qin,S. S. Wang,Y. Sui,and Z. G. Wang. ECR plasma in growth of cubic GaN by low pressure MOCVD. Plasma Chem. Plasma Process. , 2002,22(1): 159-174.

101. X. L. Ye,Y. H. Chen,B. Xu,and Z. G. Wang. Detection of indium segregation effects in InGaAs/GaAs quantum wells using reflectance-difference spectrometry. Mater. Sci. Eng. ,B,2002,91: 62-65.

102. C. Zhen,D. C. Lu,H. R. Yuan,P. D. Han,X. L. Liu,Y. F. Li,X. H. Wang,Y. Lu,and Z. G. Wang. A new method to fabricate InGaN quantum dots by metalorganic chemical vapor deposition. J. Cryst. Growth,2002,235(1): 188-194.

103. Z. Chen,D. C. Lu,P. Han,X. L. Liu,X. H. Wang,Y. F. Li,H. R. Yuan,Y. Lu,L. D. Bing,Q. S. Zhu,Z. G. Wang,X. F. Wang,and L. Yan. The structure and current-voltage characteristics of multi-sheet InGaN quantum dots grown by a new multi-step method. J. Cryst. Growth,2002,243(1): 19-24.

104. Z. Chen,D. C. Lu,X. H. Wang,X. L. Liu,H. R. Yuan,P. D. Han,D. Wang,Z. G. Wang,and G. H. Li. The mechanism of blueshift in excitation-intensity-dependent photoluminescence spectrum of nitride multiple quantum wells. J. Lumin. ,2002,99(1): 35-38.

105. Z. Chen,H. R. Yuan,D. C. Lu,X. H. Sun,S. K. Wan,X. L. Liu,P. D. Han,X. H. Wang,Q. S. Zhu,and Z. G. Wang. Nitrogen vacancy scattering in GaN grown by metal-organic vapor phase epitaxy. Solid-State Electron. ,2002,46(12): 2069-2074.

106. M. Y. Liao,Z. H. Feng,C. L. Chai,S. Y. Yang,Z. K. Liu,and Z. G. Wang. Violet/blue emission from hydrogenated amorphous carbon films deposited from energetic CH^{3+} ions and ion bombardment. J. Appl. Phys. ,2002,91(4): 1891-1893.

107. M. Y. Liao,Z. H. Feng,S. Y. Yang,C. L. Chai,Z. K. Liu,J. L. Yang,and Z. G. Wang. Anomalous temperature dependence of photoluminescence from a-C : H

film deposited by energetic hydrocarbon ion beam. Solid State Commun. ,2002,121 (5): 287-290.

108. L. W. Lu,H. Yan,C. L. Yang,M. H. Xie,Z. G. Wang,J. Wang,and W. K. Ge. Study of GaN thin films grown on vicinal SiC (0001) substrates by molecular beam epitaxy. Semicond. Sci. Technol. ,2002,17(9): 957-960.

109. M. Y. Liao,C. L. Chai,S. Y. Yang,Z. K. Liu,F. G. Qin,and Z. G. Wang. Surface morphology of ion-beam deposited carbon films under high temperature. J. Vac. Sci. Technol. ,A,2002,20(6): 2072-2074.

110. Y. Lu,X. L. Liu,D. C. Lu,H. R. Yuan,Z. Chen,T. W. Fan,Y. F. Li,P. D. Han,X. H. Wang,D. Wang,and Z. G. Wang. Investigation of GaN layer grown on Si(111) substrate using an ultrathin AlN wetting layer. J. Cryst. Growth,2002,236 (1-3): 77-84.

111. Z. Z. Sun, S. F. Yoon, J. Wu, and Z. G. Wang. Origin of the vertical-anticorrelation arrays of InAs/InAlAs nanowires with a fixed layer-ordering orientation. J. Appl. Phys. ,2002,91(9): 6021-6026.

112. Z. C. Zhang,S. Y. Yang,F. Q. Zhang,D. B. Li,Y. H. Chen,and Z. G. Wang. Strain relaxation of InP film directly grown on GaAs patterned compliant substrate. J. Cryst. Growth,2002,243(1): 71-76.

113. J. Y. Kang,Y. W. Shen,and Z. G. Wang. Effects of residual C and O impurities on photoluminescence in undoped GaN epilayers. Mater. Sci. Eng. , B-Solid State Materials for Advanced Technology,2002,91: 303-307.

114. C. M. Li, F. Q. Liu, Z. Y. Zhang, X. Q. Meng, P. Jin, and Z. G. Wang. Correlation of structural and optical investigation of InGaAs/InAlAs quantum cascade laser. J. Cryst. Growth,2003,253(1-4): 198-202.

115. J. He, Y. C. Zhang, B. Xu, and Z. G. Wang. Effects of seed dot layer and thin GaAs spacer layer on the structure and optical properties of upper In(Ga)As quantum dots. J. Appl. Phys. ,2003,93(11): 8898-8902.

116. B. Z. Qu,Z. Chen,D. C. Lu,P. Han,X. G. Liu,X. H. Wang,D. Wang,Q. S. Zhu,and Z. G. Wang. Structure characteristics of InGaN quantum dots fabricated by passivation and low temperature method. J. Cryst. Growth,2003,252(1-3): 19-25.

117. C. M. Li, F. Q. Liu, P. Lin, and Z. G. Wang. Realization of low threshold of InGaAs/InAlAs quantum cascade laser. Chin. Phys. Lett. ,2003,20(9): 1478-1481.

118. P. Jin,X. Q. Meng,Z. Y. Zhang,C. M. Li,B. Xu,F. Q. Liu,Z. G. Wang,Y. G. Li,C. Z. Zhang, and S. H. Pan. Effect of InAs quantum dots on the Fermi level pinning of undoped-n^+ type GaAs surface studied by contactless electroreflectance. J.

Appl. Phys. ,2003,93(7):4169-4172.

119. C. M. Li,F. Q. Liu,P. Jin,and Z. G. Wang. Realization of quantum cascade laser operating at room temperature. J. Cryst. Growth,2003,250(3-4):285-289.

120. Z. C. Zhang,B. Y. Ren,Y. H. Chen,S. Y. Yang,and Z. G. Wang. Effects and numerical analysis of argon gas flow on the oxygen concentration in Czochralski silicon single crystal growth. Microelectron. Eng. ,2003,66(1-4):504-509.

121. D. B. Li,X. Dong,J. S. Huang,X. L. Liu,Z. Y. Xu,X. H. Wang,Z. Zhang,and Z. G. Wang. Structural and optical properties of InAlGaN films grown directly on low-temperature buffer layer with (0001) sapphire substrate. J. Cryst. Growth,2003, 249(1-2):72-77.

122. Y. Lu,X. L. Liu,D. C. Lu,H. R. Yuan,G. Q. Hu,X. H. Wang,and Z . G,X. F. Duan. The growth morphologies of GaN layer on Si(111) substrate. J. Cryst. Growth,2003,247(1-2):91-98.

123. Z. Chen,D. C. Lu,X. L. Liu,X. H. Wang,P. D. Han,D. Wang,H. R. Yuan,Z. G. Wang,G. H. Li,and Z. L. Fang. Luminescence study of $(11\bar{2})$ GaN film grown by metalorganic chemical-vapor deposition. J. Appl. Phys. ,2003,93(1):316-319.

124. Z. G. Wang and J. Wu. Semiconductor nanometer structures and devices. J. of the Korean Physical Society,2004,45:S877-S880.

125. Y. Lu,G. W. Cong,X. L. Liu,D. C. Lu,Z. G. Wang,and M. F. Wu. Depth distribution of the strain in the GaN layer with low-temperature AlN interlayer on Si (111) substrate studied by Rutherford backscattering/channeling. Appl. Phys. Lett. , 2004,85(23):5562-5564.

126. Z. Z. Sun,S. F. Yoon,J. Wu,and Z. G. Wang. Size self-scaling effect in stacked InAs/InAlAs nanowire multilayers. Appl. Phys. Lett. ,2004,85(21):5061-5063.

127. J. Sun,P. Jin,and Z. G. Wang. Extremely low density InAs quantum dots realized in situ on (100) GaAs. Nanotechnology,2004,15(12):1763-1766.

128. X. X. Han,D. B. Li,H. R. Yuan,X. H. Sun,X. L. Liu,X. H. Wang,Q. S. Zhu,and Z. G. Wang. Dislocation scattering in a two-dimensional electron gas of an $Al_xGa_{1-x}N$/GaN heterostructure. Phys. Status Solidi B,2004,241(13):3000-3008.

129. F. A. Zhao,Y. H. Chen,X. L. Ye,P. Jin,B. Xu,Z. G. Wang,and C. L. Zhang. Growth of nano-structures on composition-modulated InAlAs surfaces. J. Phys.: Condens. Matter,2004,16(43):7603-7610.

130. X. X. Han,Z. Chen,D. B. Li,J. J. Wu,J. M. Li,X. H. Sun,X. L. Liu,P. D. Han,X. H. Wang,Q. S. Zhu,and Z. G. Wang. Structural and optical properties of 3D growth multilayer InGaN/GaN quantum dots by metalorganic chemical vapor

deposition. J. Cryst. Growth,2004,266(4): 423-428.

131. J. J. Wu,D. B. Li,Y. Lu,X. X. Han,J. M. Li,H. Y. Wei,T. T. Kang,X. H. Wang,X. L. Liu,Q. S. Zhu,and Z. G. Wang. Crack-free InAlGaN quaternary alloy films grown on Si (111) substrate by metalorganic chemical vapor deposition. J. Cryst. Growth,2004,273(1-2): 79-85.

132. Y. Lu,G. W. Cong,X. L. Liu,D. C. Lu,Q. S. Zhu,X. H. Wang,J. Wu,and Z. G. Wang. Growth of crack-free GaN films on Si (111) substrate by using Al-rich AlN buffer layer. J. Appl. Phys. ,2004,96(9): 4982-4988.

133. P. Jin,C. M. Li,Z. Y. Zhang,F. Q. Liu,Y. H. Chen,X. L. Ye,B. Xu,and Z. G. Wang. Quantum-confined Stark effect and built-in dipole moment in self-assembled InAs/GaAs quantum dots. Appl. Phys. Lett. ,2004,85(14): 2791-2793.

134. X. Q. Huang, F. Q. Liu, X. L. Che, J. Q. Liu, W. Lei, and Z. G. Wang. Characterization of InAs quantum dots on lattice-matched InAlGaAs/InP superlattice structures. J. Cryst. Growth,2004,270(3-4): 364-369.

135. G. X. Shi,P. Jin,B. Xu,C. M. Li,C. X. Cui,Y. L. Wang,X. L. Ye,J. Wu,and Z. G. Wang. Thermal annealing effect on InAs/InGaAs quantum dots grown by atomic layer molecular beam epitaxy. J. Cryst. Growth,2004,269(2-4): 181-186.

136. X. L. Ye,Y. H. Chen,B. Xu,Y. P. Zeng,and Z. G. Wang. Investigation of GaAs/AlGaAs interfaces by reflectance-difference spectroscopy. Eur. Phys. J. Appl. Phys. ,2004,27(1-3): 297-300.

137. C. H. Wang,Y. H. Chen,G. Yu,and Z. G. Wang. Formation of ferromagnetic clusters in GaAs matrix and GaAs/AlGaAs superlattice through Mn ion implantation at two different temperatures. J. Cryst. Growth,2004,268(1-2): 12-17.

138. J. He,B. Xu,and Z. G. Wang. Effect of $In_{0.2}Ga_{0.8}As$ and $In_{0.2}Al_{0.8}As$ combination layer on band offsets of InAs quantum dots. Appl. Phys. Lett. , 2004, 84 (25): 5237-5239.

139. C. L. Zhang,Z. G. Wang,F. A. Zhao,B. Xu,and P. Jin. Ordering growth of InAs quantum dots on ultra-thin InGaAs strained layer. J. Cryst. Growth,2004,265(1-2): 60-64(2004).

140. Y. Lu,X. L. Liu,X. H. Wang,D. C. Lu,D. B. Li,X. X. Han,G. W. Cong,and Z. G. Wang. Influence of the growth temperature of the high-temperature AlN buffer on the properties of GaN grown on Si (111) substrate. J. Cryst. Growth,2004,263(1-4): 4-11.

141. Z. Y. Zhang, Z. G. Wang, B. Xu, P. Jin, Z. Z. Sun, and F. Q. Liu. High-performance quantum-dot superluminescent diodes. IEEE Photon. Technol. Lett. ,

2004,16(1): 27-29.

142. B. Xu, Z. G. Wang, Y. H. Chen, P. Jin, X. L. Ye, and F. Q. Liu. Controlled growth of Ⅲ-Ⅴ compound semiconductor nanostructures and their application in quantum devices. Mater. Sci. Forum, 2005, 475-479: 1783-1786.

143. G. W. Cong, Y. Lu, W. Q. Peng, X. L. Liu, X. H. Wang, and Z. G. Wang. Design of the low-temperature AlN interlayer for GaN grown on Si (111) substrate. J. Cryst. Growth, 2005, 276(3-4): 381-388.

144. X. X. Han, J. M. Li, J. J. Wu, X. H. Wang, D. B. Li, X. L. Liu, P. D. Han, Q. S. Zhu, and Z. G. Wang. Effects of different modified underlayer surfaces on growth and optical properties of InGaN quantum dots. Vacuum, 2005, 77(3): 307-314.

145. H. L. Xiao, X. L. Wang, J. X. Wang, N. H. Zhang, H. X. Liu, Y. P. Zeng, J. M. Li, and Z. G. Wang. Growth and characterization of InN on sapphire substrate by RF-MBE. J. Cryst. Growth, 2005, 276(3-4): 401-406.

146. X. X. Han, J. M. Li, J. J. Wu, G. W. Cong, X. L. Liu, Q. S. Zhu, and Z. G. Wang. Intersubband optical absorption in quantum dots-in-a-well heterostructures. J. Appl. Phys., 2005, 98(5): 053703.

147. J. J. Wu, X. X. Han, J. M. Li, D. B. Li, Y. Lu, H. Y. Wei, G. W. Cong, X. L. Liu, Q. S. Zhu, and Z. G. Wang. Crack-free GaN/Si (111) epitaxial layers grown with InAlGaN alloy as compliant interlayer by metalorganic chemical vapor deposition. J. Cryst. Growth, 2005, 279(3-4): 335-340.

148. C. L. Zhang, Z. G. Wang, Y. H. Chen, et al. Site controlling of InAs quantum wires on cleaved edge of AlGaAs/GaAs superlattice. Nanotechnology, 2005, 16(8): 1379.

149. W. Lei, Y. H. Chen, B. Xu, X. L. Ye, Y. P. Zeng, and Z. G. Wang. Raman study on self-assembled InAs/InAlAs/InP (001) quantum wires. Nanotechnology, 2005, 16: 974-1977.

150. C. X. Cui, Y. H. Chen, C. Zhao, P. Jin, G. X. Shi, Y. L. Wang, B. Xu, and Z. G. Wang. Cleaved-edge overgrowth of aligned InAs islands on GaAs (110). Nanotechnology, 2005, 16(11): 2661.

151. W. Lei, Y. H. Chen, B. Xu, P. Jin, C. Zhao, L. K. Yu, and Z. G. Wang. Interband and intraband photocurrent of self-assembled InAs/InAlAs/InP nanostructures. Nanotechnology, 2005, 16: 2785-2789.

152. C. Zhao, Y. H. Chen, C. X. Cui, B. Xu, J. Sun, W. Lei, L. K. Yu, and Z. G. Wang. Quantum dot growth simulation on periodic stress of substrate. J. Chem. Phys., 2005, 123: 094708.

153. W. Lei, Y. H. Chen, Y. L. Wang, B. Xu, X. L. Ye, Y. P. Zeng, and Z. G.

Wang. Influence of rapid thermal annealing on InAs/InAlAs/InP quantum wires with different InAs deposited thickness. J. Crystal Growth, 2005, 284: 20-27.

154. R. Y. Li, Z. G. Wang, B. Xu, P. Jin, X. Guo, and M. Chen. Time dependence of wet oxidized AlGaAs/GaAs distributed Bragg reflectors. J. Vac. Sci. Technol., B, 2005, 23(5): 2137-2140.

155. B. Xu, Z. G. Wang, Y. H. Chen, P. Jin, X. L. Ye, and F. Q. Liu. Controlled growth of Ⅲ-Ⅴ compound semiconductor nano-structures and their application in quantum-devices. Mater. Sci. Forum, 2005, 475-479: 1783-1786.

156. Y. L. Wang, Y. H. Chen, J. Wu, Z. G. Wang, and Y. P. Zeng. Self-organized superlattices along the [001] growth direction in $In_{0.52}Al_{0.48}As$ layers grown on nominally(001) InP substrates by molecular beam epitaxy. Superlattices Microstruct. 2005, 38(3): 151-160.

157. P. Jin, X. L. Ye, and Z. G. Wang. Growth of low-density InAs/GaAs quantum dots on a substrate with an intentional temperature gradient by molecular beam epitaxy. Nanotechnology, 2005, 16(12): 2775-2778.

158. N. Liu, P. Jin, and Z. G. Wang. InAs/GaAs quantum-dot superluminescent diodes with 110 nm bandwidth. Electron. Lett, 2005, 41(25): 1400-1401.

159. B. C. Cheng, and Z. G. Wang. Synthesis and optical properties of Europium-doped ZnS long-lasting phosphorescence from aligned nanowires. Adv. Funct. Mater., 2005, 15: 1883-1890.

160. W. Q. Peng, G. W. Cong, S. C. Qu, and Z. G. Wang. Synthesis of shuttle-like ZnO nanostructures from precursor ZnS nanoparticles. Nanotechnology, 2005, 16: 1469.

161. X. Z. Lu, F. Q. Liu, J. Q. Liu, P. Jin, and Z. G. Wang. Room temperature operation of strain-compensated 5.5 μm quantum cascade lasers. Chin. J. Semiconductors, 2005, 26(12): 267-2270.

162. Y. Guo, F. Q. Liu, J. Q. Liu, C. M. Li, and Z. G. Wang. 8μm strain-compensated quantum cascade laser operating at room temperature. Semicond. Sci. Technol., 2005, 2: 844-846.

163. J. Q. Liu, F. Q. Liu, X. Z. Lu, Y. Guo, and Z. G. Wang. Realization of GaAs/AlGaAs quantum-cascade lasers with high average optical power. Solid-State Electron., 2005, 49: 1961-1964.

164. F. Q. Liu, Y. Shao, X. Q. Huang, and Z. G. Wang. Artificial nanograting woven by self-assembled nanowires. Nanotechnology, 2005, 16: 2077-2081.

165. Y. Yao, M. Y. Liao, Z. G. Wang, Y. Lifshitz, and S. T. Lee. Nucleation of

diamond by pure carbon ion bombardment—a transmission electron microscopy study. Appl. Phys. Lett. ,2005,87:063103.

166. W. Q Peng, S. C Qu, G. W. Cong, and Z. G Wang. Synthesis and temperature-dependent near-band-edge emission of chain-like Mg-doped ZnO nanoparticles. Appl. Phys. Lett. ,2006,88(10): 101902.

167. B. C. Cheng, S. C. Qu, H. Y. Zhou, and Z. G. Wang. Al_2O_3: Cr^{3+} nanotubes synthesized via homogenization precipitation followed by heat treatment. J. Phys. Chem. B,2006,110(32): 15749-15754.

168. Y. H. Chen, X. L. Ye, B Xu, Z. G. Wang, and Z. Yang. Large g factors of higher-lying excitons detected with reflectance difference spectroscopy in GaAs-based quantum wells. Appl. Phys. Lett. ,2006,89(5): 051903.

169. Y. H. Chen, P. Jin, L. Y. Liang, X. L. Ye, and Z. G. Wang. A. I. Martinez. Evolution of the amount of InAs in wetting layers in a InAs/GaAs quantum dot system studied by reflectance difference spectroscopy. Nanotechnology,2006,17:2207-2211.

170. Y. H. Chen, X. L. Ye, B. Xu, and Z. G. Wang. Strong in-plane anisotropy of asymmetric (001) quantum wells. J. Appl. Phys. ,2006,99: 096102.

171. B. C. Cheng, S. C. Qu, H. Y. Zhou, and Z. G. Wang. Porous $ZnAl_2O_4$ spinel nanorods doped with Eu^{3+}: synthesis and photoluminescence. Nanotechnology,2006, 17(12): 2982-2987.

172. J. Sun, P. Jin, C. Zhao, L. K. Yu, X. L. Ye, B. Xu, Y. H. Chen, and Z. G. Wang. Electron resonant tunneling through InAs/GaAs quantum dots embedded in a Schottky diode with an AlAs insertion layer. J. Electrochem. Soc. ,2006,153(7): G703-G706.

173. J. Sun, R. Y. Li, C. Zhao, L. K. Yu, X. L. Ye, B. Xu, Y. H. Chen, and Z. G. Wang. Room-temperature observation of electron resonant tunneling through InAs/AlAs quantum dots. Electrochem. Solid-State Lett. ,2006,9(5): G167-G170.

174. C. M. Wang, X. L. Wang, G. X. Hu, J. X. Wang, J. P. Li, and Z. G. Wang. Influence of AlN interfacial layer on electrical properties of high-Al-content $Al_{0.45}Ga_{0.55}N$/GaN HEMT structure. Appl. Surf. Sci. ,2006,253(2): 762-765.

175. J. J. Wu, X. X. Han, J. M. Li, H. Y. Wei, G. W. Cong, X. L. Liu, Q. S. Zhu, Z. G. Wang, Q. J. Jia, L. P. Guo, T. D. Hu, and H. H. Wang. Crack control in GaN grown on silicon (111) using In doped low-temperature AlGaN interlayer by metalorganic chemical vapor deposition. Opt. Mater. ,2006,28(10): 1227-1231.

176. J. J. Wu, J. M. Li, G. W. Cong, H. Y. Wei, P. F. Zhang, W. G. Hu, X. L. Liu, Q. S. Zhu, Z. G. Wang, Q. J. Jia, and L. P. Gu. Temperature dependence of the

formation of nano-scale indium clusters in InAlGaN alloys on Si (111) substrates. Nanotechnology,2006,17(5): 1251-1254.

177. C. L. Zhang, L. Tang, Y. L. Wang, Z. G. Wang, and B. Xu. Influence of dislocation stress field on distribution of quantum dots. Physica E-Low-Dimensional Systems & Nanostructures,2006,33(1): 130-133.

178. X. F. Zhu and Z. G. Wang. Nanoinstabilities as revealed by shrinkage of nanocavities in silicon during irradiation. Int. J. Nanotech. ,2006,3(4): 492-516.

179. X. L. Che, L. Li, F. Q. Liu, X. Q. Huang, and Z. G. Wang. Porous InP array-directed assembly of InAs nanostructure. Appl. Phys. Lett. ,2006,88(26): 263107.

180. Y. L. Wang, P. Jin, X. L. Ye, C. L. Zhang, G. X. Shi, R. Y. Li, Y. H. Chen, and Z. G. Wang. High uniformity of self-organized InAs quantum wires on InAlAs buffers grown on misoriented InP (001). Appl. Phys. Lett. ,2006,88(12): 123104.

181. W. Lei, Y. H. Chen, P. Jin, X. L. Ye, Y. L. Wang, B. Xu, and Z. G. Wang. Shape and spatial correlation control of InAs-InAlAs-InP (001) nanostructure superlattices. Appl. Phys. Lett. ,2006,88(6): 063114.

182. J. Sun, D. W. Patrik, I. Maximov, Z. G. Wang, and H. Q. Xu. Frequency mixing and phase detection functionalities of three-terminal ballistic junctions. Nanotechnology,2007,18: 195205.

183. R. B. Liu, Y. C. Shu, G. J. Zhang, J. M. Sun, X. D. Xing, B. Pi, J. H. Yao, Z. G. Wang, and J. J. Xu. Study of nonlinear absorption in GaAs/AlGaAs multiple quantum wells using the reflection Z-scan. Opt. Quantum Electron. ,2007,39(14): 1207-1214.

184. J. X. Ran, X. L. Wang, G. X. Hu, J. P. Li, B. Z. Wang, H. L. Xiao, J. X. Wang, Y. P. Zeng, J. M. Li, and Z. G. Wang. Effects of doping on the crystalline quality and composition distribution in InGaN/GaN structure grown by MOCVD. J. Cryst. Growth,2007,298: 235-238.

185. X. L. Wang, G. X. Hu, Z. Y. Ma, J. X. Ran, C. M. Wang, H. L. Mao, H. Tang, H. P. Li, J. X. Wang, Y. P. Zeng, J. M. Li, and Z. G. Wang. AlGaN/AlN/GaN/SiC HEMT structure with high mobility GaN thin layer as channel grown by MOCVD. J. Cryst. Growth,2007,298: 835-839.

186. B. Pi, Y. C. Shu, Y. W. Lin, J. M. Sun, S. C. Qu, J. H. Yao, X. D. Xing, B. Xu, Q. Shu, Z. G. Wang, and J. J. Xu. Morphological and electrical properties of InP grown by solid source molecular beam epitaxy. J. Cryst. Growth, 2007, 299(2): 243-247.

187. J. J. Wu, G. Y. Zhang, X. L. Liu, Q. S. Zhu, Z. G. Wang, Q. J. Jia, and L. P.

Guo. Effect of an indium-doped barrier on enhanced near-ultraviolet emission from InGaN/AlGaN: In multiple quantum wells grown on Si (111). Nanotechnology, 2006,18(1): 015402.

188. X. L. Wang, G. X. Hu, Z. Y. Ma, J. X. Ran, C. M. Wang, H. L. Mao, H. Tang, H. P. Li, J. X. Wang, Y. P. Zeng, J. M. Li, and Z. G. Wang. AlGaN/AlN/GaN/SiC HEMT structure with high mobility GaN thin layer as channel grown by MOCVD. J. Cryst. Growth, 2007, 298: 835-839.

189. R. Q. Zhang, P. F. Zhang, T. T. Kang, H. B. Fan, X. L. Liu, S. Y. Yang, H. Y. Wei, Q. S. Zhu, and Z. G. Wang. Determination of the valence band offset of wurtzite InN/ZnO heterojunction by X-ray photoelectron spectroscopy. Appl. Phys. Lett., 2007, 91(16): 162104.

190. W. Lei, Y. H. Chen, and Z. G. Wang. Ordering of self-assembled quantumwires on InP (001) surfaces. One-Dimensional Nanostructures, Chapter 2007, 12: 291, Springer.

191. J. Z. Zhao, Z. J. Lin, T. D. Corrigan, Z. Wang, Z. D. You, and Z. G. Wang. Electron mobility related to scattering caused by the strain variation of AlGaN barrier layer in strained AlGaN/GaN heterostructures. Appl. Phys. Lett., 2007, 91 (17): 173507.

192. X. Q. Meng, Z. Q. Chen, P. Jin, C. G. Wang, and L. Wei. Defects around self-organized InAs quantum dots measured by slow positron beam. Appl. Phys. Lett., 2007, 91: 093510.

193. 王占国. 微电子技术极限对策探讨. 中国科学院院刊, 2007, 22(6): 480.

194. H. G. Fan, G. S. Sun, S. Y. Yang, P. F. Zhang, R. Q. Zhang, H. Y. Wei, C. M. Jiao, X. L. Liu, Y. H. Chen, Q. S. Zhu, and Z. G. Wang. Valence band offset of ZnO/4H-SiC heterojunction measured by X-ray photoelectron spectroscopy. Appl. Phys. Lett., 2007, 92: 192107.

195. B. C. Cheng, X. M. Yu, H. J. Liu, and Z. G. Wang. Zn_2SiO_4/ZnO core/shell coaxial heterostructure nanobelts formed by an epitaxial growth. J. Phys. Chem. C, 2008, 112(42): 16312-16317.

196. Z. J. Lin, J. Z. Zhao, T. D. Corrigan, Z. Wang, Z. D. You, Z. G. Wang, and W. Lu. The influence of Schottky contact metals on the strain of AlGaN barrier layers. J. Appl. Phys., 2008, 103(4): 044503.

197. G. H. Liu, Y. H. Chen, C. H. Jia, and Z. G. Wang. Spin precession induced by an effective magnetic field in a two-dimensional electron gas. Appl. Phys. Lett., 2008, 93 (23): 233108.

198. G. H. Liu, Y. H. Chen, Y. Liu, C. H. Jia, and Z. G. Wang. Spin-dependent transport and spin polarization in coupled quantum wells. J. Appl. Phys., 2008, 104 (6): 064321.

199. Z. J. Wang, S. C. Qu, X. B. Zeng, C. S. Zhang, M. J. Shi, F. R. Tan, Z. G. Wang, J. P. Liu, Y. B. Hou, F. Teng, and Z. H. Feng. Synthesis of MDMO-PPV capped PbS quantum dots and their application to solar cells. Polymer, 2008, 49(21): 4647-4651.

200. J. J. Wu, L. B. Zhao, G. Y. Zhang, X. L. Liu, Q. S. Zhu, Z. G. Wang, Q. J. Jia, L. P. Guo, and T. D. Hu. Effect of indium-doped interlayer on the strain relief in GaN films grown on Si (111). Phys. Status Solidi A, 2008, 205(2): 294-299.

201. X. Q. Xu, X. L. Liu, X. X. Han, H. R. Yuan, J. Wang, Y. Guo, H. P. Song, G. L. Zheng, H. Y. Wei, S. Y. Yang, Q. S. Zhu, and Z. G. Wang. Dislocation scattering in $Al_xGa_{1-x}N$/GaN heterostructures. Appl. Phys. Lett., 2008, 93 (18): 182111.

202. X. B. Zhang, X. L. Wang, H. L. Xiao, C. B. Yang, J. X. Ran, C. M. Wang, Q. F. Hou, J. M. Li, and Z. G. Wang. Theoretical design and performance of $In_xGa_{1-x}N$ two-junction solar cells. J. Phys. D: Appl. Phys., 2008, 41(24): 245104.

203. R. Q. Zhang, Y. Guo, H. P. Song, X. L. Liu, S. Y. Yang, H. Y. Wei, Q. S. Zhu, and Z. G. Wang. Band alignment of InN/GaAs heterojunction determined by X-ray photoelectron spectroscopy. Appl. Phys. Lett., 2008, 93(12): 122111.

204. X. Q. Lv, N. Liu, P. Jin, and Z. G. Wang, Broadband emitting superluminescent diodes with InAs quantum dots in AlGaAs matrix. IEEE Photon. Technol. Lett., 2008, 20: 1742.

205. P. F. Zhang, X. L. Liu, R. Q. Zhang, H. B. Fan, H. B. Song, H. Y. Wei, C. M. Jiao, S. Y. Yang, Q. S. Zhu, and Z. G. Wang. Valence band offset of MgO/InN heterojunction measured by X-ray photoelectron spectroscopy. Appl. Phys. Lett., 2008, 92: 042906.

206. P. F. Zhang, X. L. Liu, R. Q. Zhang, H. B. Fan, A. L. Yang, H. Y. Wei, P. Jin, S. Y. Yang, Q. S. Zhu, and Z. G. Wang. Valence band offset of ZnO/GaAs heterojunction measured by X-ray photoelectron spectroscopy. Appl. Phys. Lett., 2008, 92: 012104.

207. Z. Y. Zhang, R. A. Hong, P. Jin, T. L. Choi, B. Xu, and Z. G. Wang. High-power quantum dot superluminescent LED with broadband drive current insensitive emission spectra using a tapered active region. IEEE photon. Technol. Lett., 2008, 20 (10): 782.

208. Z. Y. Zhang, R. A. Hong, B. Xu, P. Jin, and Z. G. Wang. Realization of extremely broadband quantum-dot superluminescent light-emitting diodes by rapid thermal-annealing process. Opt. Lett., 2008, 33(11): 1210.

209. J. Wu, Y. H. Chen, Z. G. Wang. Epitaxial semiconductor quantum wires. J. Nanosci. Nanotechnol., 2008, 7: 3300-3314.

210. X. W. He, B. Shen, Y. H. Chen, and Z. G. Wang. Anomalous photogalvanic effect of circularly polarized light incident on the two-dimensional electron gas in $Al_xGa_{1-x}N$/GaN heterostructures at room temperature. Phys. Rev. Lett., 2008, 101: 147402.

211. H. P. Song, A. L. Yang, H. Y. Wei, Y. Guo, B. Zhang, G. L. Zheng, S. Y. Yang, X. L. Liu, Q. S. Zhu, Z. G. Wang, T. Y. Yang, and H. H. Wang. Determination of wurtzite InN/cubic In_2O_3 heterojunction band offset by X-ray photoelectron spectroscopy. Appl. Phys. Lett., 2009, 94(22): 222114.

212. B. C. Cheng, B. X. Tian, W. Sun, Y. H. Xiao, S. J. Lei, and Z. G. Wang. Ordered zinc antimonate nanoisland attachment and morphology control of ZnO nanobelts by Sb doping. J. Phys. Chem. C, 2009, 113(22): 9638-9643.

213. F. Q. Liu, L. Li, L. J. Wang, J. Q. Liu, W. Zhang, Q. D. Zhang, W. F. Liu, Q. Y. Lu, and Z. G. Wang. Solid source MBE growth of quantum cascade lasers. Appl. Phys. A-Materials Science & Processing, 2009, 97(3): 527-532.

214. J. Q. Liu, F. Q. Liu, L. Li, L. J. Wang, and Z. G. Wang. A mini-staged multi-stacked quantum cascade laser for improved optical and thermal performance. Semicond. Sci. Technol., 2009, 24(7): 075023.

215. Z. J. Wang, S. C. Qu, X. B. Zeng, J. P. Liu, C. S. Zhang, M. J. Shi, F. R. Tan, and Z. G. Wang. The synthesis of MDMO-PPV capped PbS nanorods and their application in solar cells. Current Appl. Phys., 2009, 9: 1175-1179.

216. M. J. Shi, F. R. Tan, and Z. G. Wang. The synthesis of MDMO-PPV capped PbS nanorods and their application in solar cells. Current Appl. Phys., 2009, 9(5): 1175-1179.

217. Z. J. Wang, S. C. Qu, X. B. Zeng, J. P. Liu, C. S. Zhang, F. R. Tan, L. Jin, and Z. G. Wang. The application of SnS nanoparticles to bulk heterojunction solar cells. J. Alloys Compd., 2009, 482(1-2): 203-207.

218. Q. S. Zhu, Z. G. Wang, A. L. Yang, H. P. Song, H. Y. Wei, X. L. Liu, J. Wang, X. Q. Lv, P. Jin, and S. Y. Yang. Measurement of polar C-plane and nonpolar A-plane InN/ZnO heterojunctions band offsets by X-ray photoelectron spectroscopy. Appl. Phys. Lett., 2009, 94(6): 163301.

219. X. Q. Xu, X. L. Liu, S. Y. Yang, J. M. Liu, H. Y. Wei, Q. S. Zhu, and Z. G.

Wang. Dislocation core effect scattering in a quasitriangle potential well. Appl. Phys. Lett. ,2009,94(11): 112102.

220. A. L. Yang, H. P. Song, X. L. Liu, H. Y. Wei, Y. Guo, G. L. Zheng, C. M. Jiao, S. Y. Yang, Q. S. Zhu, and Z. G. Wang. Determination of MgO/AlN heterojunction band offsets by X-ray photoelectron spectroscopy. Appl. Phys. Lett. , 2009,94(5): 052101.

221. Q. Lu, W. Zhang, L. Wang, F. Q. Liu, and Z. G. Wang. Photonic crystal distributed feedback quantum cascade laser fabricated with holographic technique. Electron. Lett. ,2009,45(1): 53.

222. Y. F. Hao, Y. H. Chen, Y. Liu, Z. G. Wang. Spin splitting of conduction subbands in $Al_{0.3}Ga_{0.7}As/GaAs/Al_xGa_{1-x}As/Al_{0.3}Ga_{0.7}As$ step quantum wells. Eur. Phys. J. Appl. Phys. B,2009,85(3): 37003.

223. G. H. Liu, Y. H. Chen, C. H. Jia, and Z. G. Wang. Spin precession and electron spin polarization wave in [001]-grown quantum wells. Eur. Phys. J. Appl. Phys. B, 2009,70(3): 397-401.

224. C. G. Tang, Y. H. Chen, B. Xu, X. L. Ye, and Z. G. Wang. Well- width dependence of in- plane optical anisotropy in (001) GaAs/AlGaAs quantum wells induced by in- plane uniaxial strain and interface asymmetry. J. Appl. Phys. ,2009, 105(10): 103108.

225. Q. Y. Lu, W. Zhang, L. J. Wang, Y. Gao, W. Yin, Q. D. Zhang, W. F. Liu, F. Q. Liu, and Z. G. Wang. Design of low-loss surface-plasmon quantum cascade lasers. Jpn. J. Appl. Phys. ,2009,48(12): 122101.

226. J. Wang, S. S. Li, Y. W. Lue, X. L. Liu, S. Y. Yang, Q. S. Zhu, and Z. G. Wang. Binding energy and spin-orbit splitting of a hydrogenic donor impurity in AlGaN/GaN triangle-shaped potential quantum well. Nanosc. Res. Lett. , 2009, 4 (11): 1315-1318.

227. Y. S. Peng, B. Xu, X. L. Ye, J. B. Niu, R. Jia, and Z. G. Wang. Fabrication of high quality two-dimensional photonic crystal mask layer patterns. Opt. Quantum Electron. ,2009,41(3): 151-158.

228. Q. Y. Lu, W. Zhang, L. J. Wang, J. Q. Liu, L. Li, F. Q. Liu, and Z. G. Wang. Holographic fabricated photonic-crystal distributed-feedback quantum cascade laser with near-diffraction-limited beam quality. Opt. Express,2009,17(21): 18900-18905.

229. X. Q. Xu, X. L. Liu, Y. Guo, J. Wang, H. P. Song, S. Y. Yang, H. Y. Wei, Q. S. Zhu, and Z. G. Wang. Influence of band bending and polarization on the valence band offset measured by X-ray photoelectron spectroscopy. J. Appl. Phys. ,2010,107

(10): 104510.

230. A. L. Yang, H. P. Song, D. C. Liang, H. Y. Wei, X. L. Liu, P. Jin, X. B. Qin, S. Y. Yang, Q. S. Zhu, and Z. G. Wang. Photoluminescence spectroscopy and positron annihilation spectroscopy probe of alloying and annealing effects in nonpolar m-plane ZnMgO thin films. Appl. Phys. Lett., 2010, 96(15): 151904.

231. J. M. Liu, X. L. Liu, X. Q. Xu. J. Wang, C. M. Li, H. Y. Wei, S. Y. Yang, Q. S. Zhu, Y. M. Fan, X. W. Zhang, and Z. G. Wang. Measurement of w-InN/h-BN heterojunction band offsets by X-ray photoemission spectroscopy. Nanoscal. Res. Lett., 2010, 5(8), 1340-1343.

232. X. Q. Lv, P. Jin, W. Y. Wang, and Z. G. Wang. Broadband external cavity tunable quantum dot lasers with low injection current density. Opt. Express, 2010, 18(9): 8916-8922.

233. S. M. Liu, W. Chen, and Z. G. Wang. Luminescence nanocrystals for solar cell enhancement. Journal of Nanoscience and Nanotechnology, 2010, 10(3): 1418-1429.

234. P. F. Xu, T. Yang, H. M. Ji, Y. L. Cao, Y. X. Gu, Y. Liu, W. Q. Ma, and Z. G. Wang. Temperature-dependent modulation characteristics for 1.3μm InAs/GaAs quantum dot lasers. J. Appl. Phys., 2010, 107(1): 013102.

235. Q. Y. Lu, W. H. Guo, W. Zhang, L. J. Wang, J. Q. Liu, L. Li, F. Q. Liu, and Z. G. Wang. Room temperature operation of photonic-crystal distributed-feedback quantum cascade lasers with single longitudinal and lateral mode performance. Appl. Phys. Lett., 2010, 96(5): 051112.

236. Y. S. Peng, B. Xu, X. L. Ye, J. B. Niu, R. Jia, and Z. G. Wang. Characterization and analysis of two-dimensional GaAs-based photonic crystal nanocavities at room temperature. Microelectron. Eng., 2010, 87(10): 1834-1837.

237. Y. Liu, Y. H. Chen, and Z. G. Wang. Photoexcited charge current for the presence of pure spin current. Appl. Phys. Letts., 2010, 96: 262108.

238. Q. D. Zhang, F. Q. Liu, W. Zhang, Q. Y. Lu, L. J. Wang, L. Li, and Z. G. Wang. Thermal induced facet destructive feature of quantum cascade lasers. Appl. Phys. Lett., 2010, 96: 141117.

239. J. Q. Liu, N. Kong, L. Li, F. Q. Liu, L. J. Wang, J. Y. Chen, and Z. G. Wang. High resistance AlGaAs/GaAs quantum cascade detectors grown by solid source molecular beam epitaxy operating above liquid nitrogen temperature. Semicond. Sci. Technol., 2010, 25: 075011.

240. W. Zhang, L. Wang, L. Li, J. Liu, F. Q. Liu, and Z. Wang. Small-divergence single-mode emitting tapered distributed-feedback quantum cascade lasers. Electron.

Lett. ,2010,46: 528-529.

241. J. Q. Liu,N. Kong,L. Li,F. Q. Liu,L. J. Wang,J. Y. Chen,and Z. G. Wang. High resistance AlGaAs/GaAs quantum cascade detectors grown by solid source molecular beam epitaxy operating above liquid nitrogen temperature. Semicond. Sci. Technol. ,2010,25(7): 075011.

242. Y. Liu,Y. H. Chen,and Z. G. Wang. Kondo effect in a triangular triple quantum dots ring with three terminals. Solid State Commun. ,2010,150(25-26): 1136-1140.

243. Y. Liu,Y. H. Chen,and Z. G. Wang. Photoexcited charge current for the presence of pure spin current. Appl. Phys. Lett. ,2010,96(26): 262108.

244. Y. F. Song,Y. W. Lu,B. Zhang,X. Q. Xu,J. Wang,Y. Guo,K. Shi,Z. W. Li, X. L. Liu, S. Y. Yang, Q. S. Zhu, and Z. G. Wang. Intersubband absorption energy shifts in 3-level system for asymmetric quantum well terahertz emitters. J. Appl. Phys. ,2010,108(8): 083112.

245. F. R. Tan,S. C. Qu,X. B. Zeng,C. S. Zhang,M. J. Shi,Z. J. Wang,L. Jin,Y. Bi,J. Cao,Z. G. Wang,Y. B. Hou,F. Teng,and Z. H. Feng. Photovoltaic effect of tin disulfide with nanocrystalline/amorphous blended phases. Solid State Commun. , 2010,150(1-2): 58-61.

246. Z. J. Wang,L. Jin,Y. Bi,J. Cao,Z. G. Wang,Y. B. Hou,F. Teng,and Z. H. Feng. Photovoltaic effect of tin disulfide with nanocrystalline/amorphous blended phases. Solid State Commun. ,2010,150(1-2): 58-61.

247. A. H. Tang,S. C. Qu,Y. B. Hou,F. Teng,H. R. Tan,J. Liu,X. W. Zhang,Y. S. Wang,and Z. G. Wang. Electrical bistability and negative differential resistance in diodes based on silver nanoparticle-poly (N-vinylcarbazole) composites. J. Appl. Phys. ,2010,108(9): 094320.

248. A. H. Tang,S. C. Qu,K. Li,Y. B. Hou,F. Teng,J. Cao,Y. S. Wang,and Z. G. Wang. One-pot synthesis and self-assembly of colloidal copper (I) sulfide nanocrystals. Nanotechnology,2010,21(28): 285602.

249. A. H. Tang,F. Teng,Y. B. Hou,Y. S. Wang,F. R. Tan,S. C. Qu,and Z. G. Wang. Optical properties and electrical bistability of CdS nanoparticles synthesized in dodecanethiol. Appl. Phys. Lett. ,2010,96(16): 163112.

250. Z. J. Wang,S. C. Qu,X. B. Zeng,J. P. Liu,F. R. Tan,Y. Bi,and Z. G. Wang. Organic/inorganic hybrid solar cells based on SnS/SnO nanocrystals and MDMO-PPV. Acta Mater. ,2010,58(15): 4950-4955.

251. Z. J. Wang,S. C. Qu,X. B. Zeng,J. P. Liu,F. R. Tan,L. Jin,and Z. G. Wang. Influence of interface modification on the performance of polymer/Bi_2S_3 nanorods bulk

heterojunction solar cells. Appl. Surf. Sci. ,2010,257(2): 423-428.

252. Q. D. Zhang, F. Q. Liu, W. Zhang, Q. Y. Lu, L. J. Wang, L. Li, and Z. G. Wang. Thermal induced facet destructive feature of quantum cascade lasers. Appl. Phys. Lett. ,2010,96(14): 14117.

253. Y. Guo, H. Y. Wei, C. M. Jiao, S. Y. Yang, Q. S. Zhu, and Z. G. Wang. Valence band offset of MgO/TiO_2 (rutile) heterojunction measured by X-ray photoelectron spectroscopy. Appl. Surf. Sci. ,2010,256(23): 7327-7330.

254. G. L. Zheng, A. L. Yang, H. Y. Wei, X. L. Liu, H. P. Song, Y. Guo, C. H. Jia, C. M. Jiao, S. Y. Yang, Q. S. Zhu, and Z. G. Wang. Effects of annealing treatment on the formation of CO_2 in ZnO thin films grown by metal-organic chemical vapor deposition. Appl. Surf. Sci. ,2010,256(8): 2606-2610.

255. Z. W. Li, X. Q. Xu, J. Wang, J. M. Liu, X. L. Liu, S. Y. Yang, Q. S. Zhu, and Z. G. Wang. Cluster scattering in two-dimensional electron gas investigated by born approximation and partial-wave methods. Physica E. ,2010,43(1): 543-546.

256. Q. An, J. D. Lin, P. Jin, and Z. G. Wang. Theoretical investigation of a surface-emitting superluminescent diode with a circular grating. Semicond. Sci. Technol. , 2011,26(8): 085036.

257. Q. W. Deng, X. L. Wang, H. L. Xiao, C. M. Wang, H. B. Yan, H. Chen, Q. F. Hou, D. F. Lin, J. M. Li, Z. G. Wang, and X. Hou. An investigation on $In_xGa_{1-x}N$/GaN multiple quantum well solar cells. J. Phys. D: Appl. Phys. ,2011,44(26): 265103.

258. Q. W. Deng, X. L. Wang, H. L. Xiao, C. M. Wang, H. B. Yan, H. Chen, D. F. Lin, L. J. Jiang, C. Chun, J. M. Li, Z. G. Wang, and X. Hou. Comparison of as-grown and annealed GaN/InGaN: Mg samples. J. Phys. D: Appl. Phys. ,2011,44(34): 345101.

259. Q. W. Deng, X. L. Wang, H. L. Xiao, C. M. Wang, H. B. Yan, H. Chen, D. F. Lin, J. M. Li, Z. G. Wang, and X. Hou. Behavioural investigation of InN nanodots by surface topographies and phase images. J. Phys. D: Appl. Phys. , 2011, 44(44): 445306.

260. Q. W. Deng, X. L. Wang, C. B. Yang, H. L. Xiao, C. M. Wang, H. B. Yan, Q. F. Hou, J. M. Li, Z. G. Wang, and X. Hou. Theoretical study on $In_xGa_{1-x}N$/GaN quantum dots solar cell. Physica B-Condensed Matter. ,2011,406(1): 73-76.

261. Y. Liu, Y. H. Chen, and Z. G. Wang. Chiral splitting and polarization-dependent optical absorption of exicton for Rashba spin-orbit interaction. Phys. Lett. A, 2011, 375: 3025-3031.

262. Y. X. Gu, T. Yang, H. M. Ji, P. F. Xu, and Z. G. Wang. Redshift and discrete enengy level separation of self-assembled quantum dots induced by strain-reduction layer. J. Appl. Phys., 2011, 109(6): 064320-1.

263. W. H. Guo, J. Q. Liu, J. Y. Chen, L. Li, L. J. Wang, F. Q. Liu, and Z. G. Wang. Single-mode surface-emitting distributed feedback quantum-cascade lasers based on hybrid waveguide structure. Chin. Opt. Let., 2011, 9(6): 061404.

264. J. Wang, X. L. Liu, A. L. Yang, G. L. Gao, S. Y. Yang, H. Y. Wei, Q. S. Zhu, and Z. G. Wang. Measurement of wurtzite ZnO/rutile TiO_2 heterojunction band offsets by X-ray photoelectron spectroscopy. Appl. Phys. A- Materials Science & Processing, 2011, 103(4): 1099-1103.

265. Y. B. Guo, Y. H. Chen, Y. Xiang, S. C. Qu, and Z. G. Wang. Photorefractive effects in ZnO nanorod doped liquid crystal cell. Appl. Opt., 2011, 50(8): 1101-1104.

266. C. H. Jia, Y. H. Chen, Y. Guo, X. L. Liu, S. Y. Yang, W. F. Zhang, and Z. G. Wang. Valence band offset of InN/$BaTiO_3$ heterojunction measured by X-ray photoelectron spectroscopy. Nanosc. Res. Lett., 2011, 6: 316.

267. L. Jin, H. Y. Zhou, S. C. Qu, and Z. G. Wang. Ordered InAs nanodots formed on the patterned GaAs substrate by molecular beam epitaxy. Mater. Sci. Semicond. Process., 2011, 14(2): 108-113.

268. M. Wei, X. L. Wang, X. Pan, H. L. Xiao, C. M. Wang, Q. F. Hou, and Z. G. Wang. Effect of AlN buffer thickness on GaN epilayer grown on Si (111). Mater. Sci. Semicond. Process., 2011, 14(2): 97-100.

269. M. Wei, X. L. Wang, X. Pan, H. L. Xiao, C. M. Wang, C. B. Yang, and Z. G. Wang. Effect of high temperature AlGaN buffer thickness on GaN epilayer grown on Si (111) substrates. J. Mater. Sci. Mater. Med., 2011, 22(8): 1028-1032.

270. P. F. Xu, H. M. Ji, T. Yang, B. Xu, W. Q. Ma, and Z. G. Wang. The research progress of quantum dot lasers and photodetectors in China. J. Nanosci. Nanotechno., 2011, 11(11): 9345-9356.

271. T. F. Li, Y. H. Chen, W. Lei, X. L. Zhou, S. Luo, Y. Z. Hu, L. J. Wang, T. Yang, and Z. G. Wang. Effect of growth temperature on the morphology and phonon properties of InAs nanowires on Si substrates. Nanosc. Res. Lett., 2011, 6: 463.

272. B. Zhang, H. P. Song, X. Q. Xu, J. M. Liu, J. Wang, X. L. Liu, S. Y. Yang, Q. S. Zhu, and Z. G. Wang. Well-aligned Zn-doped tilted InN nanorods grown on rplane sapphire by MOCVD. Nanotechnology, 2011, 22(23): 235603.

273. J. M. Zhang, S. C. Qu, L. S. Zhang, A. H. Tang, and Z. G. Wang. Quantitative

surface enhanced Raman scattering detection based on the " sandwich" structure substrate. Spectrochimica Acta Part a-Molecular and Biomolecular Spectroscopy, 2011, 79(3): 625-630.

274. B. Zhang, H. P. Song, J. Wang, C. H. Jia, J. M. Liu, X. Q. Xu, X. L. Liu, S. Y. Yang, Q. S. Zhu, and Z. G. Wang. MOCVD growth of a-plane InN films on r-Al_2O_3 with different buffer layers. J. Cryst. Growth, 2011, 319(1): 114-117.

275. T. F. Li, Y. H. Chen, W. Lei, X. L. Zhou, and Z. G. Wang. Optical properties of InAsSb nanostructures embedded in InGaAsSb strain reducing layer. Physica E., 2011, 43(4): 869-873.

276. X. K. Li, P. Jin, Q. An, Z. C. Wang, X. Q. Lv, H. Wei, J. Wu, J. Wu, and Z. G. Wang. A high- performance quantum dot superluminescent diode with a two-section structure. Nanosc. Res. Lett., 2011, 6: 1-5.

277. S. L. Huang, Y. Wu, X. F. Zhu, L. X. Li, Z. G. Wang, L. Z. Wang, and G. Q. Lu. VLS growth of SiO_x nanowires with a stepwise nonuniformity in diameter. J. Appl. Phys., 2011, 109(8): 084328.

278. G. H. Liu, Y. H. Chen, C. H. Jia, G. D. Hao, and Z. G. Wang. Spin splitting modulated by uniaxial stress in InAs nanowires. J. Phys. Condensed Matter, 2010, 23(1): 015801.

279. G. P. Liu, J. Wu, Y. W. Lu, Z. W. Li, Y. F. Song, C. M. Li, S. Y. Yang, X. L. Liu, Q. S. Zhu, and Z. G. Wang. Scattering due to spacer layer thickness fluctuation on two dimensional electron gas in AlGaAs/GaAs modulation- doped heterostructures. J. Appl. Phys., 2011, 110(2): 023705.

280. G. P. Liu, J. Wu, Y. W. Lu, B. Zhang, C. M. Li, L. Sang, Y. F. Song, K. Shi, X. L. Liu, S. Y. Yang, Q. S. Zhu, and Z. G. Wang. A theoretical calculation of the impact of GaN cap and $Al_xGa_{1-x}N$ barrier thickness fluctuations on two- dimensional electron gas in a GaN/$Al_xGa_{1-x}N$/GaN heterostructure. IEEE Trans. Electron. Devices, 2011, 58(12): 4272-4275.

281. S. Y. Liu, X. B. Zeng, W. B. Peng, H. B. Xiao, W. J. Yao, X. B. Xie, C. Wang, and Z. G. Wang. Improvement of amorphous silicon n-i-p solar cells by incorporating double-layer hydrogenated nanocrystalline silicon structure. J. Non- Cryst. Solids, 2011, 357(1): 121-125.

282. J. M. Liu, X. L. Liu, C. M. Li, H. Y. Wei, Y. Guo, C. M. Jiao, Z. W. Li, X. Q. Xu, H. P. Song, S. Y. Yang, Q. S. Zhu, Z. G. Wang, A. L. Yang, T. Y. Yang, and H. H. Wang. Investigation of cracks in GaN films grown by combined hydride and metal organic vapor-phase epitaxial method. Nanosc. Res. Lett., 2011, 6(1): 69.

283. Y. J. Lv, Z. J. Lin, L. G. Meng, Y. X. Yu, C. B. Luan, Z. F. Cao, H. Chen, B. Q. Sun, and Z. G. Wang. Evaluating AlGaN/AlN/GaN heterostructure Schottky barrier heights with flat-band voltage from forward current-voltage characteristics. Appl. Phys. Lett., 2011, 99(12): 123504.

284. F. R. Tan, S. C. Qu, J. Wu, K. Liu, S. Y. Zhou, and Z. G. Wang. Preparation of SnS_2 colloidal quantum dots and their application in organic/inorganic hybrid solar cells. Nanosc. Res. Lett., 2011, 6: 1-8.

285. X. Pan, X. L. Wang, H. L. Xiao, C. M. Wang, C. B. Yang, W. Li, W. Y. Wang, P. Jin, and Z. G. Wang. Characteristics of high Al content AlGaN grown by pulsed atomic layer epitaxy. Appl. Surf. Sci., 2011, 257(20): 8718-8721.

286. F. R. Tan, S. C. Qu, J. Wu, Z. J. Wang, L. Jin, Y. Bi, J. Cao, K. Liu, J. M. Zhang, and Z. G. Wang. Electrodepostied polyaniline films decorated with nano-islands: Characterization and application as anode buffer layers in solar cells. Sol. Energy Mater. Sol. Cells, 2011, 95(2): 440-445.

287. A. H. Tang, S. C. Qu, F. Teng, Y. B. Hou, Y. S. Wang, and Z. G. Wang. Recent developments of hybrid nanocrystal/polymer bulk heterojunction solar cells. J. Nanosci. Nanotechno., 2011, 11(11): 9384-9394.

288. A. H. Tang, S. C. Qu, Y. B. Hou, F. Teng, Y. S. Wang, and Z. G. Wang. One-pot synthesis, optical property and self-assembly of monodisperse silver nanospheres. Journal of Solid State Chemistry, 2011, 184(8): 1956-1962.

289. Y. J. Lv, Z. J. Lin, Yu Zhang, L. G. Meng, C. B. Luan, Z. F. Cao, H. Chen, and Z. G. Wang. Polarization Coulomb field scattering in AlGaN/AlN/GaN heterostructure field-effect transistors. Appl. Phys. Lett., 2011, 98(12): 123512.

290. Q. F. Hou, X. L. Wang, H. L. Xiao, C. M. Wang, C. B. Yang, H. B. Yan, Q. W. Deng, J. M. Li, Z. G. Wang, and X. Hou. Influence of electric field on persistent photoconductivity in unintentionally doped *n*-type GaN. Appl. Phys. Lett., 2011, 98 (10): 102104.

291. H. Y. Zhou, S. C. Qu, P. Jin, B. Xu, X. L. Ye, J. P. Liu, and Z. G. Wang. Study of molecular-beam epitaxy growth on patterned GaAs (100) substrates by masked indium ion implantation. J. Cryst. Growth, 2011, 318(1): 572-575.

292. Q. S. Zhu, J. Wu, C. M. Li, and Z. G. Wang. Conduction and valence band discontinuities in some new semiconductor heterojunctions. J. Nanoscience Nanotechol. 2011, 11(11): 9368-9383.

293. X. F. Zhu, L. X. Li, S. L. Huang, Z. G. Wang, G. Q. Lu, C. H. Sun, and L. Z. Wang. Nanostructural instability of single-walled carbon nanotubes during electron

beam induced shrinkage. Carbon,2011,49(9):3120-3124.

294. G. Y. Zhou, Y. H. Chen, C. G. Tang, L. Liang, P. Jin, and Z. G. Wang. The transition from two-stage to three-stage evolution of wetting layer of InAs/GaAs quantum dots caused by postgrowth annealing. Appl. Phy. Lett. ,2011,98:071914.

295. X. K. Li, P. Jin, Q. Z. Wang, X. Q. Lv, H. Wei, J. Wu, J. Wu, and Z. G. Wang. A high performance quantum dot superluminescent diode with two-section structure. Nanosc. Res. Lett. ,2011,6:625.

296. X. Q. Xu, Y Guo, X. L. Liu, J. M. Liu, H. P. Song, B. Zhang, J. Wang, S. Y. Yang, H. Y. Wei, Q. S. Zhu, and Z. G. Wang. GaN growth with InGaN as a weakly bonded layer. Cryst. Eng. Comm. ,2011,13:1580.

297. J. C. Zhang, L. J. Wang, S. Tan, W. F. Liu, L. H. Zhao, F. Q. Liu, J. Q. Liu, L. Li ,Z. G. Wang. Low threshold CW operation of DFB quantum cascade laser at $\lambda \sim 4.6\mu m$. IEEE Photon. Technol. Lett. ,2011,23:1334-1339.

298. J. Y. Chen, J. Q. Liu, W. H. Guo, T. Wang, J. C. Zhang, L. Li, L. J. Wang, F. Q. Liu, and Z. G. Wang. High-power surface-emitting surface-plasmon-enhanced distributed feedback quantum cascade lasers. IEEE Photon. Technol. Lett. ,2012,24(11):972-974.

299. Q. An, P. Jin, J. Wu, and Z. G. Wang. Optical loss in bent-waveguide superluminescent diodes. Semicond. Sci. Technol. ,2012,27(5):055003.

300. L. Dong, G. S. Sun, L. Zheng, X. F. Liu, F. Zhang, G. G. Yan, W. S. Zhao, L. Wang, X. G. Li, and Z. G. Wang. Infrared reflectance study of 3C-SiC epilayers grown on silicon substrates. J. Phys. D: Appl. Phys. ,2012,45(24):245102.

301. L. L. Gong, S. M. Liu, S. Luo, T. Yang, L. J. Wang, F. Q. Liu, X. L. Ye, B. Xu, and Z. G. Wang. Metalorganic chemical vapor deposition growth of InAs/GaSb type II superlattices with controllable As_xSb_{1-x} interfaces. Nanosc. Res. Lett. ,2012,7:1-7.

302. D. Ji, B. Liu, Y. W. Lu, G. P. Liu, Q. S. Zhu, and Z. G. Wang. Polarization-induced remote interfacial charge scattering in Al_2O_3/AlGaN/GaN double heterojunction high electron mobility transistors. Appl. Phys. Lett. , 2012, 100(13):132105.

303. L. L. Gong, S. M. Liu, S. Luo, T. Yang, L. J. Wang, J. Q. Liu, F. Q. Liu, X. L. Ye, B. Xu, and Z. G. Wang. Effect of growth temperature on surface morphology and structure of InAs/GaSb superlattices grown by metalorganic chemical vapor deposition. J. Cryst. Growth,2012,359:55-59.

304. S. Q. Zhai, J. Q. Liu, F. Q. Liu, and Z. G. Wang. A normal incident quantum

cascade detector enhanced by surface plasmons. Appl. Phys. Lett., 2012, 100(18): 181104.

305. X. K. Li, P. Jin, Q. An, Z. C. Wang, X. Q. Lv, H. Wei, J. Wu, and Z. G. Wang. Experimental investigation of wavelength-selective optical feedback for a high-power quantum dot superluminescent device with two-section structure. Opt. Express, 2012, 20(11): 11936-11943.

306. Z. W. Li, H. Y. Wei, X. Q. Xu, G. J. Zhao, X. L. Liu, S. Y. Yang, Q. S. Zhu, and Z. G. Wang. Growth of a-plane GaN on r-plane sapphire by self-patterned nanoscale epitaxial lateral overgrowth. J. Cryst. Growth, 2012, 348(1): 10-14.

307. B. Liu, Y. W. Lu, Y. Huang, G. P. Liu, Q. S. Zhu, and Z. G. Wang. Interfacial misfit dislocation scattering effect in two-dimensional electron gas channel of GaN heterojunction. Phys. Lett. A, 2012, 376(10-11): 1067-1071.

308. G. P. Liu, J. Wu, Y. W. Lu, G. J. Zhao, C. Y. Gu, C. B. Liu, L. Sang, S. Y. Yang, X. L. Liu, Q. S. Zhu, and Z. G. Wang. Two-dimensional electron gas mobility limited by barrier and quantum well thickness fluctuations scattering in $Al_xGa_{1-x}N$/GaN multi-quantum wells. Appl. Phys. Lett., 2012, 100(16): 162102.

309. G. P. Liu, J. Wu, G. J. Zhao, S. M. Liu, W. Mao, Y. Hao, C. B. Liu, S. Y. Yang, X. L. Liu, Q. S. Zhu, and Z. G. Wang. Impact of the misfit dislocations on two-dimensional electron gas mobility in semi-polar AlGaN/GaN heterostructures. Appl. Phys. Lett., 2012, 100(8): 082101.

310. Y. Z. Hu, L. J. Wang, J. C. Zhang, L. Li, J. Q. Liu, F. Q. Liu, and Z. G. Wang. Facet temperature distribution of a room temperature continuous-wave operating quantum cascade laser. J. Phys. D: Appl. Phys., 2012, 45(32): 325103.

311. L. L. Gong, S. M. Liu, S. Luo, T. Yang, L. J. Wang, F. Q. Liu, X. L. Ye, B. Xu, and Z. G. Wang. Formation of As_xSb_{1-x} mixing interfaces in InAs/GaSb superlattices grown by metalorganic chemical vapor deposition. EPL (Europhysics Letters), 2012, 97(3): 36001.

312. D. Ji, Y. W. Lu, B. Liu, G. R. Jin, G. P. Liu, Q. S. Zhu, and Z. G. Wang. Electric field-induced scatterings in rough quantum wells of AlGaN/GaN high-mobility electronic transistors. J. Appl. Phys., 2012, 112(2): 024515.

313. G. J. Li, X. L. Liu, J. X. Wang, D. D. Jin, H. Zhang, S. Y. Yang, S. M. Liu, W. Mao, Y. Hao, Q. S. Zhu, and Z. G. Wang. Calculation of discrepancies in measured valence band offsets of heterojunctions with different crystal polarities. J. Appl. Phys., 2012, 112(11): 113712.

314. X. K. Li, P. Jin, Q. An, Z. C. Wang, X. Q. Lv, H. Wei, J. Wu, J. Wu, and Z. G.

Wang. Improved continuous-wave performance of two-section quantum-dot superluminescent diodes by using epi-down mounting process. IEEE Photon. Technol. Lett. ,2012,24(14): 1188-1190.

315. C. B. Luan,Z. J. Lin,Z. H. Feng,L. G. Meng,Y. J. Lv,Z. F. Cao,Y. X. Yu, and Z. G. Wang. Polarization Coulomb field scattering in $In_{0.18}Al_{0.82}N$/AlN/GaN heterostructure field-effect transistors. J. Appl. Phys. ,2012,112(5): 054513.

316. C. B. Luan,Z. J. Lin,Y. J. Lv,L. G. Meng,Y. X. Yu,Z. F. Cao,H. Chen,and Z. G. Wang. Influence of the side-Ohmic contact processing on the polarization Coulomb field scattering in AlGaN/AlN/GaN heterostructure field-effect transistors. Appl. Phys. Lett. ,2012,101(11): 113501.

317. Y. J. Lv,Z. J. Lin,L. G. Meng,C. B. Luan,Z. F. Cao,Y. X. Yu,Z. H. Feng, and Z. G. Wang. Influence of the ratio of gate length to drain-to-source distance on the electron mobility in AlGaN/AlN/GaN heterostructure field-effect transistors. Nanosc. Res. Lett. ,2012,7(1): 434.

318. K. F. Wang,Y. X. Gu,X. G. Yang,T. Yang,and Z. G. Wang. Si delta doping inside InAs/GaAs quantum dots with different doping densities. J. Vac. Sci. Technol. ,B,2012,30(4): 04D108.

319. H. Y. Zhang,Y. H. Chen,G. Y. Zhou,C. G. Tang,and Z. G. Wang. Wetting layer evolution and its temperature dependence during self-assembly of InAs/GaAs quantum dots. Nanosc. Res. Lett. ,2012,7(1): 600.

320. J. C. Zhang,L. J. Wang,S. Tan,J. Y. Chen,S. Q. Zhai,J. Q. Liu,F. Q. Liu, and Z. G. Wang. Room temperature continuous-wave operation of top metal grating distributed feedback quantum cascade laser at $\lambda \sim 7.6\mu m$. IEEE Photon. Technol. Lett. ,2012,24(13): 1100-1102.

321. L. H. Zhao,F. Q. Liu,J. C. Zhang,L. J. Wang,J. Q. Liu,L. Li,and Z. G. Wang. Improved performance of quantum cascade laser with porous waveguide structure. J. Appl. Phys. ,2012,112(1): 013111.

322. J. Y. Chen,J. Q. Liu,T. Wang,F. Q. Liu,and Z. G. Wang. Monolithically integrated terahertz quantum cascade array laser. Electron. Lett. ,2013,49(25): 1632-1633.

323. L. Jing,H. L. Xiao,X. L. Wang, C. M. Wang,Q. W. Deng,Z. D. Li,J. Q. Ding,Z. G. Wang,and X. Hou. Enhanced performance of InGaN/GaN multiple quantum well solar cells with patterned sapphire substrate. J. Semicond. ,2013,34(12): 124004.

324. D. Y. Yao,J. C. Zhang,F. Q. Liu,N. Zhou,F. L. Yan,J. Wang,J. Q. Liu,and

Z. G. Wang. Surface emitting quantum cascade laser operating in CW mode above 70℃ at λ ~4.6μm. Appl. Phys. Lett. ,2013,103(4): 041121.

325. D. Chi,C. Liu,S. C. Qu,Z. G. Zhang,Y. J. Li,Y. L. Li,J. Z. Wang,and Z. G. Wang. Photovoltaic performance optimization of methyl 4-[6,6]-C61-benzoate based polymer solar cells with thermal annealing approach. Synth. Met., 2013, 181: 117-122.

326. L. Gong,Y. C. Shu,J. J. Xu,Q. S. Zhu,and Z. G. Wang. Numerical analysis on quantum dots-in-a-well structures by finite difference method. Superlattices Microstruct. ,2013,60: 311-319.

327. L. Gong, Y. C. Shu, J. J. Xu, and Z. G. Wang. Numerical computation of pyramidal quantum dots with band non-parabolicity. Superlattices Microstruct. ,2013, 61: 81-90.

328. L. Dong,G. S. Sun,L. Zheng,X. F. Liu,F. Zhang,G. G. Yan,L. X. Tian,X. G. Li, and Z. G. Wang. Analysis and modeling of localized faceting on 4H-SiC epilayer surfaces. Phys. Status Solidi A,2013,210(11): 2503-2509.

329. Y. Z. Hu,L. J. Wang,F. Q. Liu,J. C. Zhang,J. Q. Liu,and Z. G. Wang. Micro-Raman study on chirped InGaAs-InAlAs superlattices. Phys. Status Solidi A,2013,210(11): 2364-2368.

330. L. Sang,S. Y. Yang,G. P. Liu,G. J. Zhao,C. B. Liu,C. Y. Gu,H. Y. Wei,X. L. Liu, Q. S. Zhu, and Z. G. Wang. Dislocation scattering in ZnMgO/ZnO heterostructures. IEEE Trans. Electron. Devices,2013,60(6): 2077-2079.

331. G. J. Li,G. P. Liu,H. Y. Wei,C. M. Jiao,J. X. Wang,H. Zhang,D. D. Jin,Y. X. Feng,S. Y. Yang,L. S. Wang,Q. S. Zhu,and Z. G. Wang. Scattering due to Schottky barrier height spatial fluctuation on two dimensional electron gas in AlGaN/GaN high electron mobility transistors. Appl. Phys. Lett. ,2013,103(23): 232109.

332. S. Q. Zhai,J. Q. Liu,X. J. Wang,N. Zhuo,F. Q. Liu,Z. G. Wang,X. H. Liu, N. Li, and W. Lu. 19μm quantum cascade infrared photodetectors. Appl. Phys. Lett. ,2013,102(19): 191120.

333. T. Wang,J. Q. Liu,Y. F. Li,J. Y. Chen,F. Q. Liu,L. J. Wang,J. C. Zhang, and Z. G. Wang. High-power distributed feedback terahertz quantum cascade lasers. IEEE Electron Device Letters,2013,34(11): 1412-1414.

334. D. D. Jin,S. Y. Yang,L. W. Zhang,H. J. Li,H. Zhang,J. X. Wang,T. Yang, X. L. Liu, Q. S. Zhu, and Z. G. Wang. Electron scattering in GaAs/InGaAs quantum wells subjected to an in-plane magnetic field. J. Appl. Phys., 2013, 113 (21): 213711.

335. D. Ji, Y. W. Lu, B. Liu, G. P. Liu, Q. S. Zhu, and Z. G. Wang. Dielectric and barrier thickness fluctuation scattering in Al_2O_3/AlGaN/GaN double heterojunction high-electron mobility transistors. Thin Solid Films, 2013, 534: 655-658.

336. D. Ji, Y. W. Lu, B. Liu, G. P. Liu, Q. S. Zhu, and Z. G. Wang. Theoretical calculation of the interfacial charge-modulated two-dimensional electron gas mobility in Al_2O_3/AlGaN/GaN double heterojunction high-electron mobility transistors. Solid State Commun., 2013, 153(1): 53-57.

337. C. H. Jia, Y. H. Chen, X. L. Liu, S. Y. Yang, W. F. Zhang, and Z. G. Wang. Control of epitaxial relationships of ZnO/$SrTiO_3$ heterointerfaces by etching the substrate surface. Nanosc. Res. Lett., 2013, 8: 23.

338. T. F. Li, L. Z. Gao, W. Lei, L. J. Guo, T. Yang, Y. H. Chen, and Z. G. Wang. Raman study on zinc-blende single InAs nanowire grown on Si (111) substrate. Nanosc. Res. Lett., 2013, 8: 27.

339. Y. C. Li, S. Yu, X. Q. Meng, Y. H Liu, Y. H Zhao, F. Q. Liu, and Z. G. Wang. The effect of magnetic ordering on light emitting intensity of Eu-doped GaN. J. Phys. D: Appl. Phys., 2013, 46(21): 215101.

340. C. B. Liu, S. Y. Yang, K. Shi, G. P. Liu, H. Zhang, D. D. Jin, C. Y. Gu, G. J. Zhao, L. Sang, X. L. Liu, Q. S. Zhu, and Z. G. Wang. Two dimensional electron gas mobility limited by scattering of quantum dots with indium composition transition region in quantum wells. Physica E-Low-Dimensional Systems & Nanostructures, 2013, 52: 150-154.

341. C. B. Liu, G. J. Zhao, G. P. Liu, Y. F. Song, H. Zhang, D. D. Jin, Z. W. Li, X. L. Liu, S. Y. Yang, Q. S. Zhu, and Z. G. Wang. Scattering due to large cluster embedded in quantum wells. Appl. Phys. Lett., 2013, 102(5): 052105.

342. J. Q. Liu, J. Y. Chen, T. Wang, Y. F. Li, F. Q. Liu, L. Li, L. J. Wang, and Z. G. Wang. High efficiency and high power continuous-wave semiconductor terahertz lasers at similar to 3.1 THz. Solid State Electron., 2013, 81: 68-71.

343. K. Liu, S. C. Qu, F. R. Tan, Y. Bi, S. D. Lu, and Z. G. Wang. Ordered silicon nanowires prepared by template-assisted morphological design and metal-assisted chemical etching. Mater. Lett., 2013, 101: 96-98.

344. K. Liu, S. C. Qu, X. H. Zhang, F. R. Tan, and Z. G. Wang. Improved photovoltaic performance of silicon nanowire/organic hybrid solar cells by incorporating silver nanoparticles. Nanosc. Res. Lett., 2013, 8: 88.

345. K. Liu, S. C. Qu, X. H. Zhang, and Z. G. Wang. Anisotropic characteristics and morphological control of silicon nanowires fabricated by metal-assisted chemical

etching. J. Mater. Sci. Lett. ,2013,48(4): 1755-1762.

346. E. C. Peng, X. L. Wang, H. L. Xiao, C. M. Wang, H. B. Yan, H. Chen, C. Chun, L. J. Jiang, X. Hou, and Z. G. Wang. Growth and characterization of AlGaN/AlN/GaN/AlGaN double heterojunction structures with AlGaN as buffer layers. J. Cryst. Growth, 2013, 383: 25-29.
347. E. C. Peng, X. L. Wang, H. L. Xiao, C. M. Wang, H. B. Yan, H. Chen, C. Chun, L. J. Jiang, X. Hou, and Z. G. Wang. Bipolar characteristics of AlGaN/AlN/GaN/AlGaN double heterojunction structure with AlGaN as buffer layer. J. Alloys Compd. ,2013,576: 48-53.
348. E. C. Peng, X. L. Wang, H. L. Xiao, C. M. Wang, H. B. Yan, H. Chen, C. Chun, L. J. Jiang, S. Q. Qu, H. Kang, X. Hou, and Z. G. Wang. Tunable density of two-dimensional electron gas in GaN-based heterostructures: The effects of buffer acceptor and channel width. J. Appl. Phys. ,2013,114(15): 154507.
349. F. R. Tan, S. C. Qu, W. F. Zhang, X. W. Zhang, and Z. G. Wang. Conjugated molecule doped polyaniline films as buffer layers in organic solar cells. Synth. Met. , 2013,178: 18-21.
350. F. R. Tan, S. C. Qu, X. W. Zhang, K. Liu, and Z. G. Wang. Synthesis of silver quantum dots decorated TiO_2 nanotubes and their incorporation in organic hybrid solar cells. J. Nanopart. Res. ,2013,15(8): 1844.
351. K. F. Wang, S. C. Qu, D. W. Liu, K. Liu, J. Wang, L. Zhao, H. L Zhu, and Z. G. Wang. Large enhancement of sub-band-gap light absorption of sulfur hyperdoped silicon by surface dome structures. Mater. Lett. ,2013,107: 50-52.
352. D. D. Jin, C. Chao, G. D. Li, L. W. Zhang, T. Yang, X. L. Liu, S. Y. Yang, Q. S. Zhu, and Z. G. Wang. Scattering due to anisotropy of ellipsoid quantum dots in GaAs/InGaAs single quantum well. J. Appl. Phys. ,2013,113(3): 033701.
353. X. Q. Xu, Yang Li, J. M. Liu, H. Y. Wei, X. L. Liu, S. Y. Yang, Z. G. Wang, and H. H. Wang. X-ray probe of GaN thin films grown on InGaN compliant substrates. Appl. Phys. Lett. ,2013,102(13): 132104.
354. X. G. Yang, K. F. Wang, Y. X. Gu, H. Q. Ni, X. D. Wang, T. Yang, and Z. G. Wang. Improved efficiency of InAs/GaAs quantum dots solar cells by Si-doping. Sol. Energy Mater. Sol. Cells, 2013, 113: 144-147.
355. J. P. Zeng, W. Li, J. C. Yan, J. X. Wang, P. P. Cong, J. M. Li, W. Y. Wang, P. Jin, and Z. G. Wang. Temperature-dependent emission shift and carrier dynamics in deep ultraviolet AlGaN/AlGaN quantum wells. Phys. Status Solidi-Rapid Research Letters, 2013, 7(4): 297-300.

356. H. Y. Zhang, Y. H. Chen, X. L. Zhou, Yanan Jia, X. L. Ye, B. Xu, and Z. G. Wang. Observation of photo darkening in self assembled InGaAs/GaAs quantum dots. J. Appl. Phys. ,2013,113(17): 173508.

357. J. C. Zhang, F. Q. Liu, L. J. Wang, J. Y. Chen, S. Q. Zhai, J. Q. Liu, and Z. G. Wang. High performance surface grating distributed feedback quantum cascade laser. IEEE Photon. Technol. Lett. ,2013,25(7): 686-689.

358. S. Z. Zhang, W. F. Zhou, X. L. Ye, B. Xu, and Z. G. Wang. Cavity-mode calculation of L3 photonic crystal slab using the effective index perturbation method. Opt. Rev. ,2013,20(5): 420-425.

359. N. Zhuo, J. C. Zhang, F. Q. Liu, L. J. Wang, S. Tan, F. L. Yan, J. Q. Liu, and Z. G. Wang. Tunable distributed feedback quantum cascade lasers by a sampled bragg grating. IEEE Photon. Technol. Lett. ,2013,25(11): 1039-1042.

360. D. Chi, S. C. Qu, and Z. G. Wang, et al. , High efficiency P3HT: PCBM solar cells with an inserted PCBM layer. J. Mater. Chem. C,2014,2(22): 4383-4387.

361. D. Chi, B. Y. Qi, J. Z. Wang, S. C. Qu and Z. G. Wang. High-performance hybrid organic-inorganic solar cell based on planar *n*-type silicon. Appl. Phys. Lett. ,2014, 104(19): 193903.

362. Y. X. Feng, G. P. Liu, S. Y. Yang, H. Y. Wei, X. L. Liu, Q. S. Zhu, and Z. G. Wang. Interface roughness scattering considering the electrical field fluctuation in undoped $Al_xGa_{1-x}N$/GaN heterostructures. Semicond. Sci. Technol. , 2014, 29 (4): 045015.

363. D. D. Jin, L. S. Wang, S. Y. Yang, L. W. Zhang, H. J. Li, H. Zhang, J. X. Wang, R. F. Xiang, H. Y. Wei, C. M. Jiao, X. L. Liu, Q. S. Zhu, and Z. G. Wang. Anisotropic scattering effect of the inclined misfit dislocation on the two-dimensional electron gas in Al(In)GaN/GaN heterostructures. J. Appl. Phys. ,2014, 115(4): 043702.

364. H. Li, C. Liu, G. Liu, H. Wei, C. Jiao, J. Wang, H. Zhang, D. D. Jin, Y. Feng, S. Yang, L. Wang, Q. Zhu, and Z. G. Wang. Single-crystalline GaN nanotube arrays grown on c-Al_2O_3 substrates using InN nanorods as templates. J. Cryst. Growth, 2014, 389: 1-4.

365. X. J. Wang, S. Q. Zhai, N. Zhuo, J. Q. Liu, F. Q. Liu, S. M. Liu, and Z. G. Wang. Quantum dot quantum cascade infrared photodetector. Appl. Phys. Lett. , 2014, 104(17): 171108.

366. J. X. Wang, L. S. Wang, S. Y. Yang, H. J. Li, G. J. Zhao, H. Zhang, H. Y. Wei, C. M. Jiao, Q. S. Zhu, and Z. G. Wang. Effects of V/III ratio on a-plane GaN

epilayers with an InGaN interlayer. Chin. Phys. B,2014,23(2): 026801.

367. W. N. Du,X. G. Yang,X. Y. Wang,H. Y. Pan,H. M. Ji,S. Luo,T. Yang,and Z. G. Wang. The self-seeded growth of InAsSb nanowires on silicon by metal-organic vapor phase epitaxy. J. Cryst. Growth,2014,396: 33-37.

368. G. J. Li,X. L. Liu,L. Sang,J. X. Wang,D. D. Jin,H. Zhang,S. Y. Yang,S. M. Liu,W. Mao,Y. Hao,Q. S. Zhu,and Z. G. Wang. Determination of polar C-plane and nonpolar A-plane AlN/GaN heterojunction band offsets by X-ray photoelectron spectroscopy. Phys. Status Solidi B-Basic Solid State Physics,2014,251(4): 788-791.

369. K. Liu,S. C. Qu,X. H. Zhang,F. R. Tan,Y. Bi,S. D. Lu,and Z. G. Wang. Sulfur-doped black silicon formed by metal-assist chemical etching and ion implanting. Appl. Phys. A-Materials Science & Processing,2014,114(3): 765-768.

370. Y. Liu,Y. H. Chen,C. Y. Wang,and Z. G. Wang. Effective period potential in a hybrid mesoscopic ring with Rashba spin-orbit interaction. Phys. Lett. A,2014,378(5-6): 584-589.

371. W. Li,X. L. Wang,S. Q. Qu,Q. Wang,H. L. Xiao,C. M. Wang,E. C. Peng, X. Hou, and Z. G. Wang. Numerical simulation of two-dimensional electron gas characteristics of a novel ($In_x Al_{1-x}N$/AlN) MQWs/GaN high electron mobility transistor. J. Alloys Compd. ,2014,605: 113-117.

372. F. R. Tan,S. C. Qu,L. Wang,Q. W. Jiang,W. F. Zhang,and Z. G. Wang. Core/shell-shaped CdSe/PbS nanotetrapods for efficient organic-inorganic hybrid solar cells. J. Mater. Chem. A,2014,2(35): 14502-14510.

373. J. C. Zhang,Y. H. Liu,Z. W. Jia,D. Y. Yao,F. L. Yan,F. Q. Liu,L. J. Wang, J. Q. Liu, and Z. G. Wang. Complex-coupled edge-emitting photonic crystal distributed feedback quantum cascade lasers at $\lambda \sim 7.6\mu m$. Solid State Electron. , 2014,94: 20-22.

374. X. F. Zhu,J. B. Su,Y. Wu,L. Z. Wang,and Z. G. Wang. Intriguing surface-extruded plastic flow of SiO_x amorphous nanowire as athermally induced by electron beam irradiation. Nanoscale,2014,6(3): 1499-1507.

375. K. Liu,Y. Bi,S. C. Qu,F. R. Tan,D. Chi,S. D. Lu,Y. P. Li,Y. L. Kou,and Z. G. Wang. Efficient hybrid plasmonic polymer solar cells with Ag nanoparticle decorated TiO_2 nanorods embedded in the active layer. Nanoscale,2014,6(11): 6180-6186.

376. C. B. Luan,Z. J. Lin,Y. J. Lv,J. T. Zhao,Y. T. Wang,H. Chen,and Z. G. Wang. Theoretical model of the polarization Coulomb field scattering in strained

AlGaN/AlN/GaN heterostructure field- effect transistors. J. Appl. Phys. ,2014,116 (4): 044507.

377. P. Sun,Y. C. Li,X. Q. Meng,S. Yu,Y. H. Liu,F. Q. Liu,and Z. G. Wang. The magnetic field effect on optical properties of Sm-doped GaN thin films. J. Mater. Sci. Mater. Med. ,2014,25(7): 2974-2978.
378. F. R. Tan,S. C. Qu,Q. W. Jiang,J. P. Liu,Z. J. Wang,F. M. Li,G. T. Yue,S. J. Li,C. Chen,W. F. Zhang,and Z. G. Wang. Interpenetrated inorganic hybrids for efficiency enhancement of PbS quantum dot solar cells. Adv. Energy Mater. ,2014,4 (17): 1400512.
379. S. Tan,J. C. Zhang,L. J. Wang,F. Q. Liu,N. Zhuo,F. L. Yan,J. Q. Liu,and Z. G. Wang. Index-coupled multi-wavelength distributed feedback quantum cascade lasers based on sampled gratings. Opt. Quantum Electron. , 2014, 46 (12): 1539-1546.
380. J. D. Yan,X. L. Wang,Q. Wang,S. Q. Qu,H. L. Xiao,E. C. Peng,H. Kang,C. M. Wang,C. Chun,H. B. Yan,L. J. Jiang,B. Q. Li,Z. G. Wang,and X. Hou. Two-dimensional electron and hole gases in $In_xGa_{1-x}N/Al_yGa_{1-y}N/GaN$ heterostructure for enhancement mode operation. J. Appl. Phys. ,2014,116(5): 054502.
381. D. Y. Yao,J. C. Zhang,F. Q. Liu,Z. W. Jia,F. L. Yan,L. J. Wang,J. Q. Liu, and Z. G. Wang. 1. 8W room temperature pulsed operation of substrate-emitting quantum cascade lasers. IEEE Photon. Technol. Lett. ,2014,26(4): 323-325.
382. S. Q. Zhai,J. Q. Liu,X. J. Wang,S. Tan,F. Q. Liu,and Z. G. Wang. Study on the thermal imaging application of quantum cascade detectors. Infrared Phys. Technol. ,2014,63: 17-21.
383. J. C. Zhang,D. Y. Yao,N. Zhuo,F. L. Yan,F. Q. Liu,L. J. Wang,J. Q. Liu, and Z. G. Wang. Directional collimation of substrate emitting quantum cascade laser by nanopores arrays. Appl. Phys. Lett. ,2014,104(5): 052109.
384. F. R. Tan,S. C. Qu,P. Yu,F. M Li,C. Chen,W. F. Zhang,and Z. G. Wang. Hybrid bulk-heterojunction solar cells based on all inorganic nanoparticles. Sol. Energy Mater. Sol. Cells,2014,120: 231-237.
385. D. Y. Yao,J. C. Zhang,Y. H. Liu,N. Zhou,Z. W. Jia ,F. Q. Liu,and Z. G. Wang. Small divergence substrate emitting quantum cascade laser by subwavelength metallic grating. Opt. Express,2015,23: 11462-11469.
386. D. Chi,S. D. Lu,R. Xu,K. Liu,D. W. Cao,L. Y. Wen,Y. Mi,Z. J. Wang,Y. Lei,S. C. Qu,and Z. G. Wang. Fully understanding the positive roles of plasmonic nanoparticles in ameliorating the efficiency of organic solar cells. Nanoscale,2015,7

(37): 15251-15257.

387. Y. H. Liu, J. C. Zhang, F. L. Yan, F. Q. Liu, N. Zhuo, L. J. Wang, J. Q. Liu, and Z. G. Wang. Coupled ridge waveguide distributed feedback quantum cascade laser arrays. Appl. Phys. Lett., 2015, 106(14): 142104.

388. Y. Z. Hu, F. Q. Liu, L. J. Wang, J. C. Zhang, L. H. Zhao, and Z. G. Wang. Broad area single mode operation of quantum cascade lasers by integrating porous waveguide and distributed feedback grating. Opt. Quantum Electron., 2015, 47(3): 515-521.

389. Y. F. Song, Q. S. Zhu, X. L. Liu, S. Y. Yang, and Z. G. Wang. Plasmon mode coupling and depolarization shifts in AlGaAs/GaAs asymmetric step quantum wells with and without electric field. Int. J. Mod Phys B, 2015, 29(29): 1550212.

390. H. Kang, Q. Wang, H. L. Xiao, C. M. Wang, L. J. Jiang, C. Chun, H. Chen, H. B. Yan, S. Q. Qu, E. C. Peng, J. M. Gong, X. L. Wang, B. Q. Li, Z. G. Wang, and X. Hou. Effects of a GaN cap layer on the reliability of AlGaN/GaN Schottky diodes. Phys. Status Solidi A, 2015, 212(5): 1158-1161.

391. G. J. Li, G. P. Liu, G. J. Zhao, H. Y. Wei, L. S. Wang, S. Y. Yang, Z. Chen, and Z. G. Wang. Theoretical study of the anisotropic electron scattering by steps in vicinal AlGaN/GaN heterostructures. Physica E: Low-Dimensional Systems & Nanostructures, 2015, 66: 116-119.

392. Y. H. Liu, J. C. Zhang, Z. W. Jia, D. Y. Yao, F. Q. Liu, L. J. Wang, J. Q. Liu, and Z. G. Wang. Development of low power consumption DFB quantum cascade lasers. IEEE Photon. Technol. Lett., 2015, 27(22): 2335-2338.

393. Y. H. Liu, J. C. Zhang, Z. W. Jia, D. Y. Yao, F. Q. Liu, S. Q. Zhai, N. Zhuo, and Z. G. Wang. Top grating surface-emitting DFB quantum cascade lasers in continuous-wave operation. IEEE Photon. Technol. Lett., 2015, 27(17): 1829-1832.

394. Y. H. Liu, J. C. Zhang, Z. W. Jia, D. Y. Yao, F. L. Yan, F. Q. Liu, L. J. Wang, J. Q. Liu, and Z. G. Wang. Stable single mode operation of quantum cascade lasers by complex-coupled second-order distributed feedback grating. Solid-State Electron., 2015, 103: 79-82.

395. S. D. Lu, K. Liu, D. Chi, S. Z. Yue, Y. P. Li, Y. L. Kou, X. C. Lin, Z. J. Wang, S. C. Qu, and Z. G. Wang. Constructing bulk heterojunction with componential gradient for enhancing the efficiency of polymer solar cells. J. Power Sources, 2015, 300: 238-244.

396. Z. D. Ning, S. M. Liu, F. Ren, F. J. Wang, L. J. Wang, F. Q. Liu, Z. G. Wang, and L. C. Zhao. Metal organic chemical vapor deposition growth and characterization

of InAs/GaSb type-II superlattices on GaAs (001) substrates. Mater. Lett., 2015, 143: 223-225.

397. S. L. Peng, D. S. Wang, F. H. Yang, Z. G. Wang, and F. Ma. Grown low-temperature microcrystalline silicon thin film by VHF PECVD for thin films solar cell. J. Nanomater., 2015, 2015: 327596.

398. K. F. Wang, P. G. Liu, S. C. Qu, Y. X. Wang, and Z. G. Wang. Optical and electrical properties of textured sulfur-hyperdoped silicon: a thermal annealing study. J. Mater. Sci. Lett., 2015, 50(9): 3391-3398.

399. K. F. Wang, H. Z. Shao, K. Liu, S. C. Qu, Y. X. Wang, and Z. G. Wang. Possible atomic structures responsible for the sub-bandgap absorption of chalcogen-hyperdoped silicon. Appl. Phys. Lett., 2015, 107(11): 112106.

400. T. Wang, J. Q. Liu, F. Q. Liu, L. J. Wang, J. C. Zhang, and Z. G. Wang. High-power single-mode tapered terahertz quantum cascade lasers. IEEE Photon. Technol. Lett., 2015, 27(14): 1492-1494.

401. D. Y. Yao, J. C. Zhang, O. Cathabard, S. Q. Zhai, Y. H. Liu, Z. W. Jia, F. Q. Liu, and Z. G. Wang. 10W pulsed operation of substrate emitting photonic-crystal quantum cascade laser with very small divergence, Nanosc. Res. Lett., 2015, 10: 1-6.

402. D. Y. Yao, J. C. Zhang, Z. W. Jia, F. L. Yan, L. J. Wang, J. Q. Liu, and Z. G. Wang. 1.8W room temperature pulsed operation of substrate emitting quantum cascade lasers. IEEE Photon. Technol. Lett., 2015, 26: 323.

403. Z. W. Jia, L. J. Wang, J. C. Zhang, F. Q. Liu, N. Zhuo, S. Q. Zhai, J. Q. Liu, and Z. G. Wang. Stable single-mode distributed feedback quantum cascade lasers at $\lambda \sim 4.25\mu m$ with low power consumption. Solid-State Electron., 2016, 124: 42-45.

404. J. M. Gong, Q. Wang, J. D. Yan, F. Q. Liu, C. Feng, X. L. Wang, and Z. G. Wang. Comparison of GaN/AlGaN/AlN/GaN HEMTs grown on sapphire with Fe-modulation-doped and unintentionally doped GaN buffer: Material Growth and Device Fabrication. Chin. Phys. Lett., 2016, 33(11): 117303.

405. W. Li, Q. Wang, C. M. Wang, H. B. Yan, Junda Yan, J. M. Gong, B. Q. Li, X. L. Wang, and Z. G. Wang. Self-consistent simulation of two-dimensional electron gas characteristics of a novel ($In_xAl_{1-x}N$/AlN) MQWs/InN/GaN heterostructure. Phys. Status Solidi A, 2016, 213(5): 1263-1268.

406. Y. Y. Li, J. Q. Liu, F. Q. Liu, J. C. Zhang, S. Q. Zhai, N. Zhuo, L. J. Wang, S. M. Liu, and Z. G. Wang. High power-efficiency terahertz quantum cascade laser. Chin. Phys. B, 2016, 25(8): 084206.

407. Y. L. Liu, P. Jin, G. P. Liu, W. Y. Wang, Z. Q. Qi, C. Q. Chen, and Z. G.

Wang. Exciton-phonon interaction in $Al_{0.4}Ga_{0.6}N/Al_{0.53}Ga_{0.47}N$ multiple quantum wells. Chin. Phys. B,2016,25(8): 087801.

408. W. Li,Q. Wang,X. M. Zhan,J. D. Yan,L. J. Jiang,H. B. Yan,J. M. Gong,X. L. Wang,F. Q. Liu,B. Q. Li,and Z. G. Wang. Impact of dual field plates on drain current degradation in InAlN/AlN/GaN HEMTs. Semicond. Sci. Technol. ,2016,31(12): 125003.

409. Y. Y. Li,J. Q. Liu,F. Q. Liu,J. C. Zhang,and Z. G. Wang. High-performance operation of distributed feedback terahertz quantum cascade lasers. Electron. Lett. , 2016,52(11): 945-946.

410. Z. D. Ning,S. M. Liu,S. Luo,F. Ren,F. Wang,T. Yang,F. Q. Liu,Z. G. Wang,and L. C. Zhao. Growth and characterization of InAs/InAsSb superlattices by metal organic chemical vapor deposition for mid-wavelength infrared photodetectors. Mater. Lett. ,2016,164: 213-216.

411. Y. Y. Li,T. Wang,S. Q. Zhai,J. Q. Liu,F. Q. Liu,and Z. G. Wang. High-power epitaxial-side down mounted terahertz quantum cascade lasers. Electron. Lett. , 2016,52(16): 1401-1402.

412. F. Ren,F. J. Wang,S. M. Liu,Z. D. Ning,N. Zhuo,X. L. Ye,J. Q. Liu,L. J. Wang, F. Q. Liu, and Z. G. Wang. Dual-wavelength intersubband electroluminescence from double-well active layers in InGaAs/InAlAs quantum cascade structures. Appl. Phys. Express,2016,9(5): 052104.

413. F. J. Wang,N. Zhuo,S. M. Liu,F. Ren,Z. D. Ning,X. L. Ye,J. Q. Liu,S. Q. Zhai,F. Q. Liu,and Z. G. Wang. Temperature independent infrared responsivity of a quantum dot quantum cascade photodetector. Appl. Phys. Lett. , 2016, 108(25): 251103.

414. K. Liu,S. D. Lu,S. Z. Yue,K. K. Ren,M. Azam,F. R. Tan,Z. J. Wang,S. C. Qu, and Z. G. Wang. Wrinkled substrate and indium tin oxide-free transparent electrode making organic solar cells thinner in active layer. J. Power Sources,2016, 331: 43-49.

415. Z. D. Ning,S. M. Liu,S. Luo,F. Ren,F. J. Wang,T. Yang,F. Q. Liu,Z. G. Wang,and L. C. Zhao. Structural and optical properties of InAs/InAsSb superlattices grown by metal organic chemical vapor deposition for mid-wavelength infrared photodetectors. Appl. Surf. Sci. ,2016,368: 110-113.

416. F. R. Tan,Z. J. Wang,S. C. Qu,D. W. Cao,K. Liu,Q. W. Jiang,Y. Yang,S. Pang,W. F. Zhang,Y. Lei,and Z. G. Wang. A CdSe thin film: a versatile buffer layer for improving the performance of TiO_2 nanorod array: PbS quantum dot solar

cells. Nanoscale,2016,8(19):10198-10204.

417. Z. J. Wang,D. W. Cao,R. Xu,S. C. Qu,Z. G. Wang,and Y. Lei. Realizing ordered arrays of nanostructures:a versatile platform for converting and storing energy efficiently. Nano Energy,2016,19:328-362.

418. Z. B. Zhao,L. J. Wang,Z. W. Jia,J. C. Zhang,F. Q. Liu,N. Zhuo,S. Q. Zhai,J. Q. Liu,and Z. G. Wang. Low-threshold external-cavity quantum cascade laser around 7.2μm. Opt. Eng.,2016,55(4):046116.

419. Y. H. Zhou,S. Q. Zhai,F. J. Wang,J. Q. Liu,F. Q. Liu,S. M. Liu,J. C. Zhang,N. Zhuo,L. J. Wang,and Z. G. Wang. High-speed,room-temperature quantum cascade detectors at 4.3μm. AIP Advances,2016,6(3):035305.

420. F. L. Yan,J. C. Zhang,Z. W. Jia,N. Zhuo,S. Q. Zhai,S. M. Liu,F. Q. Liu,and Z. G. Wang. High-power phase-locked quantum cascade laser array emitting at $\lambda \sim 4.6$μm. AIP Advances,2016,6(3):035022.

421. Y. H. Zhou,S. Q. Zhai,J. Q. Liu,F. Q. Liu,J. C. Zhang,N. Zhuo,L. J. Wang,and Z. G. Wang. High-speed quantum cascade laser at room temperature. Electron. Lett.,2016,52(7):548-549.

422. Z. S. Ji,L. S. Wang,G. J. Zhao,Y. L. Meng,F. Z. Li,G. J. Li,S. Y. Yang,and Z. G. Wang. Growth and characterization of AlN epilayers using pulsed metal organic chemical vapor deposition. Chin. Phys. B,2017,26(7):078102.

423. F. M. Cheng,J. C. Zhang,Z. W. Jia,Y. Zhao,D. B. Wang,N. Zhuo,S. Q. Zhai,L. J. Wang,J. Q. Liu,S. M. Liu,F. Q. Liu,and Z. G. Wang. High power substrate-emitting quantum cascade laser with a symmetric mode. IEEE Photon. Technol. Lett.,2017,29(22):1994-1997.

424. W. Li,P. Jin,W. Y. Wang,D. F. Mao,X. Pan,X. L. Wang,and Z. G. Wang. Enhancing redshift phenomenon in time-resolved photoluminescence spectra of AlGaN epilayer. Chin. Phys. B,2017,26(7):077802.

425. C. W. Liu,J. C. Zhang,Z. W. Jia,Y. Zhao,N. Zhuo,S. Q. Zhai,L. J. Wang,J. Q. Liu,S. M. Liu,F. Q. Liu,and Z. G. Wang. Coupled ridge waveguide substrate-emitting DFB quantum cascade laser arrays. IEEE Photon. Technol. Lett.,2017,29(2):213-216.

426. C. W. Liu,J. C. Zhang,F. L. Yan,Z. W. Jia,Z. B. Zhao,N. Zhuo,F. Q. Liu,and Z. G. Wang. External cavity tuning of coherent quantum cascade laser array emitting at ~7.6μm. Chin. Phys. Lett.,2017,34(3):034209.

427. X. F. Jia,L. J. Wang,N. Zhuo,Z. W. Jia,J. C. Zhang,F. Q. Liu,J. Q. Liu,S. Q. Zhai,and Z. G. Wang. Single-mode quantum cascade laser at 5.1μm with slotted

refractive index modulation. IEEE Photon. Technol. Lett., 2017, 29 (22): 1959-1962.

428. Z. W. Jia, L. J. Wang, S. Tan, J. C. Zhang, F. Q. Liu, N. Zhuo, S. Q. Zhai, J. Q. Liu, and Z. G. Wang. Improvement of buried grating DFB quantum cascade lasers by small-angle tapered structure. IEEE Photon. Technol. Lett., 2017, 29 (10): 783-785.

429. G. Q. Yang, S. Z. Zhang, B. Xu, Y. H. Chen, and Z. G. Wang. Anomalous temperature dependence of photoluminescence spectra from InAs/GaAs quantum dots grown by formation-dissolution-regrowth method. Chin. Phys. B, 2017, 26(6): 071.

430. Y. Zhao, J. C. Zhang, Y. H. Zhou, Z. W. Jia, N. Zhuo, S. Q. Zhai, L. J. Wang, J. Q. Liu, S. M. Liu, F. Q. Liu, and Z. G. Wang. External-cavity beam combining of 4-channel quantum cascade lasers. Infrared Phys. Technol., 2017, 85: 52-55.

431. X. M. Zhan, Q. Wang, K. Wang, W. Li, H. L. Xiao, C. Feng, L. J. Jiang, C. M. Wang, X. L. Wang, and Z. G. Wang. Fast electrical detection of carcinoembryonic antigen based on AlGaN/GaN high electron mobility transistor aptasensor. Chin. Phys. Lett., 2017, 34(9): 097302.

432. C. W. Liu, J. C. Zhang, F. M. Cheng, Y. Zhao, N. Zhuo, S. Q. Zhai, L. J. Wang, J. Q. Liu, S. M. Liu, F. Q. Liu, and Z. G. Wang. High efficiency quantum cascade lasers based on excited-states injection. IEEE Photon. Technol. Lett., 2018, 30(4): 299-302.

433. X. M. Zhan, M. L. Hao, Q. Wang, W. Li, H. L. Xiao, C. Feng, L. J. Jiang, C. M. Wang, X. L. Wang, and Z. G. Wang. Highly sensitive detection of deoxyribonucleic acid hybridization using Au-gated AlInN/GaN high electron mobility transistor-based sensors. Chin. Phys. Lett., 2017, 34(4): 047301.

434. Z. W. Jia, L. Wang, J. C. Zhang, Y. Zhao, C. W. Liu, S. Q. Zhai, N. Zhuo, J. Q. Liu, L. J. Wang, ShuMan Liu, F. Q. Liu, and Z. G. Wang. Phase-locked array of quantum cascade lasers with an intracavity spatial filter. Appl. Phys. Lett., 2017, 111 (6): 061108.

435. F. Z. Li, L. S. Wang, G. J. Zhao, Y. L. Meng, G. J. Li, S. Y. Yang, and Z. G. Wang. Performance enhancement of AlGaN-based ultraviolet light-emitting diodes by inserting the last quantum well into electron blocking layer. Superlattices Microstruct., 2017, 110: 324-329.

436. S. M. Liu, Q. Wang, H. L. Xiao, K. Wang, C. M. Wang, X. L. Wang, W. K. Ge, and Z. G. Wang. Optimization of growth and fabrication techniques to enhance the InGaN/GaN multiple quantum well solar cells performance. Superlattices

Microstruct. ,2017,109：194-200.

437. S. D. Lu,Jie Lin,K. Liu,S. Z. Yue,K. K. Ren,F. R. Tan,Z. J. Wang,P. Jin,S. C. Qu,and Z. G. Wang. Large area flexible polymer solar cells with high efficiency enabled by imprinted Ag grid and modified buffer layer. Acta Mater. , 2017, 130：208-214.

438. K. K. Ren,L. Huang,S. Z. Yue,S. D. Lu,K. Liu,M. Azam,Z. J. Wang,Z. M. Wei, S. C. Qu, and Z. G. Wang. Turning a disadvantage into an advantage：synthesizing high-quality organometallic halide perovskite nanosheet arrays for humidity sensors. J. Mater. Chem. C,2017,5(10)：2504-2508.

439. W. Z. Xu,F. R. Tan,Q. Liu,X. S. Liu,Q. W. Jiang,L. Wei,W. F. Zhang,Z. J. Wang,S. C. Qu,and Z. G. Wang. Efficient PbS QD solar cell with an inverted structure. Sol. Energy Mater. Sol. Cells,2017,159：503-509.

440. S. Z. Yue,S. D. Lu,K. K. Ren,K. Liu,M. Azam,D. W. Cao,Z. J. Wang,Y. Lei,S. C. Qu, and Z. G. Wang. Insights into the influence of work functions of cathodes on efficiencies of perovskite solar cells. Small,2017,13(19)：1700007.

441. J. Q. Liu, F. J. Wang, S. Q. Zhai, J. C. Zhang, S. M. Liu, J. Q. Liu, L. J. Wang, F. Q. Liu, and Z. G. Wang. Normal-incidence quantum cascade detector coupled by nanopore structure. Appl. Phys. Express,2018,11(4)：042001.

442. G. J. Zhao,G. J. Li,L. S. Wang,Y. L. Meng,F. Z. Li,H. Y. Wei,S. Y. Yang, and Z. G. Wang. Measurement of semi-polar (11-22) plane AlN/GaN heterojunction band offsets by X-ray photoelectron spectroscopy. Appl. Phys. A-Materials Science & Processing,2018,124(2)：130.

443. G. J. Zhao,L. S. Wang,G. J. Li,Y. L. Meng,F. Z. Li,S. Y. Yang,and Z. G. Wang. Structural and optical properties of semi- polar (11- 22) InGaN/GaN green light- emitting diode structure. Appl. Phys. Lett. ,2018,112(5)：052105.

学术专著

444. 王占国,郑有炓. 半导体材料研究进展（第一卷）. 北京：高等教育出版社,2012.

445. 王占国. 半导体材料与器件技术新进展//高技术发展报告(中国科学院),北京：科学出版社,2008.

446. 王占国,陈立泉和屠海令. 中国材料工程大典 · 信息功能材料工程（第 11 卷、第 12 卷和第 13 卷）. 北京：化学工业出版社,2006.

447. 王占国,陈涌海和叶小玲. 纳米半导体技术. 北京：化学工业出版社,2006.

448. 王占国. 材料科学与工程手册（第十篇）. 北京：化学工业出版社,2014.

会议报告

449. 王占国. 硅单晶高温电学性质研究. 北京:第一届全国物理学会年会,1965.
450. 王占国,尹永龙和何良. 中子和 γ 辐照硅的电学性质. 北京:第二届全国半导体物理会议,1965.
451. 王占国,林兰英. GaAs 材料质量的初步探讨. 上海:1972 年 GaAs 学术会议,1972.
452. 王占国. 一个研究 GaAs 器件(材料)强场性质的方法. 上海:1972 年 GaAs 学术会议,1972.
453. 王占国,林兰英. 砷化镓材料热学稳定性的初步研究——n^+外延工艺热循环引起的退化. 柳州:1977 年 GaAs 及其他Ⅲ–Ⅴ族化合物半导体会议,1977,4,5-12.
454. 王占国. *N*-GaAs 补偿度的计算和散射机制的分析. 庐山:第二届全国物理学会会议,1978.
455. 王占国,林兰英. 杂质和缺陷在 GaAs 中的行为. 柳州:1977 年 GaAs 及其他Ⅲ–Ⅴ族化合物半导体会议,1977,4,5–12.
456. Z. G. Wang, Y. J. Dai, C. J. Li, H. P. Lv, Z. Y. Deng, S. K. Wan, S. X. Xu, M. F. Sun and L. Y. Lin. Effect of micro-uniformity on electron mobility of LEC undoped SI-GaAs crystals. Istanbul: 7th Conference on Semi-insulating Materials, 1992,4,21-24.
457. Z. G. Wang, C. H. Wang, Y. L. Liu, S. K. Wan, G. H. Li and L. Y. Lin. Evidence for the antisite defect B_{As} in Si and B co-implanted and B implanted undoped SI LEC-GaAs crystal. Istanbul: 7th Conference on Semi-insulating Materials, 1992, 4, 21-24.
458. 王占国. 半导体材料性能检测技术的新进展. 昆明:1993 年全国砷化镓及有关化合物学术会议(大会特邀报告),1993,10,26-11, 1.
459. Z. G. Wang, V. L. Dostov, R. G. Li, J. B. Liang, T. W. Fan, L. Y. Lin. Experimental and theoretical treatment of nucleation of As-related defects in low-temperature grown GaAs. Warsaw: The 8th Conference on Semi-insulating Ⅲ–Ⅴ Materials, 1994,6,6-10.
460. 王占国. 垂直耦合应变自组装 In(Ga)As/GaAs 量子点激光材料生长及器件应用探索. 无锡: 第四届全国 MBE 会议,1997,9.
461. Z. G. Wang, Q. Gong, W. Zhou, R. Y. Xin and H. X. Li. Self-organized quantum dots: material growth and device application. Honolulu: The 3rd Pacific Rim International Conference on Advanced Materials & Processing(分会邀请报告),1998, 7,12-16.
462. Z. G. Wang, F. Q. Liu, B. Xu et al. . Self-assembled quantum dots and high power

quantum dot laser. Beijing: The 5th IUMRS International Conference on Advanced Materials (ICAM'99)(分会邀请报告),1999,6,13-18.

463. Z. G. Wang. High power continuous-ware operation of self-organized In(Ga)As/GaAs quantum dot lasers. Hong kong: 1999 IEEE Hong Kong Electron Devices Meeting(特邀报告),1999,6,26.

464. Zhan-guo Wang. Optical properties of self-assembled InAlAs/AlGaAs quantum dots on GaAs. Lund: The Second Sweden-China Meeting on Nanometer-scale Science and Technology,1999,10,18-21.

465. Z. G. Wang. Research and development of electronic materials in China. 2000 EMRS/IUMRS International Conference on Electronic Materials (ICEM2000)(大会特邀报告),Strasboung: 2000,5,30-6,2.

466. Z. G. Wang. Tailoring of self-assembled quantum dots and quantum wires. Sendai: The 1st Asian Conference on Crystal Growth and Crystal Technology(分会邀请报告),2000,8,29-9,1.

467. Z. G. Wang,Y. H. Chen,F. Q. Liu,J. Wu. Tailoring of self-assembled quantum dots and quantum wires. Beijng: 11th International Conference on Molecular Beam Epitaxy(分会邀请报告),2000,9,10-15.

468. 王占国. 半导体量子点激光器研究进展. 厦门: 第一届全国纳米技术与应用学术会议(大会特邀报告),2000,11,27-12,1.

469. Z. G. Wang,Y. H. Chen,F. Q. Liu,J. Wu. Self-assembled quantum dots,wires and quantum-dot lasers. Beijing: China-European Union Workshop on Materials Strategies for the New Millennium,2000,12,8-9.

470. Z. G. Wang. Research and development of electronic and optoelectronic materials in China. Tokyo: 日本学术振兴会第161委员会第24次年会(大会特邀报告),2001,3,26.

471. Zhanguo Wang. Self-assembled quantum dot materials and quantum dot lasers. Singapore: International Conference on Materials for Advanced Technologies(分会邀请报告),2001,7,1-6.

472. 王占国. 提高我国信息功能材料创新能力的探讨. 西安: 新材料发展现状及21世纪发展趋势研讨会暨中国工程院化工、冶金与材料工程学部第三届学术会议(大会特邀报告),2001,9,16-20.

473. 王占国. 关于提高我国信息功能材料创新能力的探讨. 厦门:第四届中国功能材料及其应用学术会议(大会特邀报告), 2001,10.

474. 王占国. 纳米半导体结构及量子器件. 西安: 西部地区纳米技术与应用研讨会(特邀报告),2002,4.

475. 王占国. 半导体纳米结构的可控生长. 北京：中国物理学会2002年秋季学术会议暨中国物理学会成立70周年庆祝会(分会邀请报告),2002,8,29-31.

476. Wang Zhanguo. Self-assembled growth of semiconductor nano-structure and device application. Hong Kong：Workshop on Selective, Patterned and Self-assembled Growth of Nano-Structures(大会特邀报告),2003,1,6-8.

477. 王占国. 光电信息功能材料的研究进展. 长沙：中国工程院化工、冶金与材料工程学部第四届学术会议(大会特邀报告),2003,10,20-22.

478. Zhanguo Wang, Ju Wu, Fengai Zhao. Boston：Self-assembled nanostructures and quantum devices in InGaAs/GaAs and InAs/InGaAlAs/InP. 2003 MRS Fall Meeting, 2003,12,1-5.

479. Z. G. Wang, J. Wu. Semiconductor nanometer structures and devices. Qingzhou：The 12th Seoul International Symposium on the Physics of Semiconductors and Applications-2004,2004,3,14-16.

480. 王占国. 光电信息功能材料研究新进展. 大连：第13届全国化合物半导体材料、微波器件和光电器件学术会议暨第9届全国固体薄膜学术会议(大会特邀报告),2004,8,2-6.

481. Wang Zhanguo. Fabrication of semiconductor nanostructures and quantum devices in InGaAs/GaAs and InAs/InGaAlAs/InP. Hong Kong：Inter-Pacific Workshop on Nanoscience and Nanotechnology(邀请报告),2004,11,22-24.

482. 王占国. "十一五"我国新材料高技术产业可持续发展战略. 北京：第八届中国北京国际科技产业博览会"中国可持续发展战略院士论坛"(特邀报告),2005,5,22.

483. 王占国. "十一五"我国新材料领域高技术产业发展重点建议. 沈阳：2005年东北亚高新技术博览会"东北老工业基地振兴高层论坛"暨第二届沈阳科学学术年会(大会特邀报告),2005,9,21-24.

484. 王占国. "十一五"我国新材料领域高技术产业发展重点建议. 乌鲁木齐：中国电子材料行业协会半导体材料分会年会(特邀报告),2005,10,16-19.

485. 王占国. 半导体白光照明：材料与器件. 宁波：中国宁波新材料及产业化国际论坛(特邀报告),2005,11.

486. 王占国. 半导体太阳能光伏电池的发展现状与趋势. 博鳌：中国工程院化工、冶金与材料工程学部第五届学术会议(特邀报告),2005,12,8-14.

487. 王占国. ZnO基半导体材料和器件的研究现状、发展趋势和对策建议. 南京：第二届全国ZnO材料学术研讨会(大会特邀报告),2005,12,20-23.

488. 王占国. 纳米半导体研究与国际合作. 北京：Europe-Chinese Workshop on China Frontier：China's Realities from a Frontier Research Perspective,2006,6,22-23.

489. 王占国. 半导体光电信息功能材料研究进展. 北京：2006北京国际材料周

(2006BIMW)暨2006中国材料研讨会(大会特邀报告),2006,6,25-30.

490. 王占国. 后摩尔时代的微电子技术探讨. 无锡：中国科学院信息技术科学部和技术科学部第24次技术科学论坛(特邀报告),2006,10.

491. 王占国. 后摩尔时代的微电子技术探讨. 北海：第14届全国化合物半导体材料、微波器件和光电器件学术会议(大会特邀报告),2006,11,4-11.

492. 王占国. 半导体纳米结构制备和量子器件应用. 台湾：两岸华人前瞻材料技术与应用论坛,2006,11,24-25.

493. 王占国. 光电功能材料产业发展热点. 南阳：中部六省郑州招商大会南阳光电产业发展高层论坛(大会特邀报告),2007,4,27.

494. 王占国. 半导体太阳能电池研发现状与趋势. 连云港：中国科学院信息技术科学部和技术科学部第25次技术科学论坛(特邀报告),2007,5,23.

495. 王占国. "十一五"新材料高技术产业发展重点建议. 南京：江苏新材料产业发展高层论坛,2007,7,9.

496. 王占国. 半导体光电信息功能材料研究新进展. 威海：中国科学院信息技术科学部和技术科学部第26次技术科学论坛(特邀报告),2007,8,23-25.

497. 王占国. 光电信息功能材料与器件研究进展. 兰州：信息产业部中国电子材料行业协会半导体材料分会年度学术会议(大会特邀报告),2007,9,13-15.

498. 王占国. 硅微电子技术物理极限的可能对策探讨. 重庆：第15届全国半导体集成电路、硅材料学术会议(大会特邀报告),2007,11,1-3.

499. Wang Zhanguo. Semiconductor nanostructures and quantum device applications. 武汉：第六届全国功能材料及其应用学术会议暨2007国际功能材料专题论坛(大会特邀报告),2007,11,16-19.

500. Zhanguo Wang. GaAs/AlGaAs and strain-compensated $In_x Ga_{1-x} As/In_y Al_{1-y} As$ quantum cascade lasers. Beijing：The China-Sweden Workshop on Nanoscience and Technology 2008,2008,4,18-22.

501. Wang Zhanguo. Semiconductor nanostructure growth and device application. 重庆：2008中美材料国际研讨会暨MRS International Materials Research Conference(IMRC 2008)(大会特邀报告),2008,6,9-12.

502. 王占国. 半导体纳米结构生长及器件应用. 黄山：第二届中国微纳电子技术行业交流会(大会特邀报告),2008,8,13-15.

503. Wang Zhanguo. RDS characterization of wetting layer around InAs/GaAs quantum dots by SK growth mode. Chicago：US-China Workshop on Nano-structured Materials for Global Energy and Environmental Challenge,2008,9,21-24.

504. 王占国. 激光显示发展现状与趋势. 海口：中国电子行业协会半导体材料分会年会,2008,11,12-18.

505. 王占国. 激光显示技术的发展现状与趋势. 广州：第十五届全国化合物半导体材料、微波器件和光电器件学术会议(大会特邀报告),2008,11,30-12,5.

506. Wang Z. G., Liu F. Q., Chen Y. H., Xu B., Jin P., Wu J. Semiconductor nanostructure growth and device application. 嘉兴:第10届中俄双边新材料新工艺研讨会(大会特邀报告), 2009,10,20-24.

507. 王占国. 新概念太阳能光伏电池研究进展. 杭州：第16届全国半导体集成电路、硅材料学术会议(大会特邀报告),2009,10,27-29.

508. 王占国. 第三代太阳能光伏电池研究进展. 西安：第16届全国化合物半导体材料、微波器件和光电器件学术会议(大会特邀报告),2010,10,25-29.

509. 王占国. 半导体量子级联激光器材料及其器件应用. 开封：第17届全国化合物半导体材料、微波器件和光电器件学术会议(大会特邀报告),2012,11,7-10.

工作报告及内部资料

510. 王占国. CZ和FZ Si少子寿命测量和研究. 工作报告,1963.

511. 王占国. 重掺硅的电学性质. 工作报告,1965.

512. 王占国等. 电子辐照对*PN*和*NP*硅太阳电池的影响(内部资料). 1967.

513. 王占国等. 低、高能质子对硅太阳电池辐照效应研究(内部资料). 1969.

514. 王占国,14研究院辐照试验小组人员. 电子材料、元件、器件及组件的电子、质子、中子和γ射线辐照效应研究(内部资料). 1969.

515. 王占国,14研究院辐照试验小组人员. 半导体材料、元件、器件及组件瞬态核辐射效应的研究(内部资料). 1970.

516. 王占国. 恒流源的设计、制作和理论分析. 工作报告,1978.

517. 王占国. *N*-GaAs电子迁移率的一个半经验公式. 工作报告,1979.

518. 王占国,万寿科. 掺杂和纯度GaAs、InP光致发光研究. 工作报告,1979.

专 利 目 录

申请国家发明专利 185 项、实用新型专利 1 项。截至本书出版，获得授权发明专利 101 项、实用新型专利 1 项，获得授权专利如下：

1. 发明名称：Ⅲ族氮化物单/多层异质应变薄膜的制作方法. 发明人：陈振，陆大成，刘祥林，王晓晖，袁海荣，王占国. 专利号：ZL 01100454.1. 授权公告日：2004 年 06 月 16 日.
2. 发明名称：一种制作白光发光二极管的方法. 发明人：陈振，韩培德，陆大成，刘祥林，王晓晖，朱勤生，王占国. 专利号：ZL 02128518.7. 授权公告日：2005 年 09 月 14 日.
3. 发明名称：制备低温超薄异质外延用柔性衬底的方法. 发明人：张志成，杨少延，黎大兵，陈涌海，朱勤生，王占国. 专利号：ZL 02142452.7. 授权公告日：2005 年 08 月 24 日.
4. 发明名称：一只自组织量子点为有源区的超辐射发光管，发明人：张子旸，王占国，徐波，金鹏，刘峰奇，专利号：ZL 02147587.3，授权公告日：2005 年 10 月 26 日.
5. 发明名称：一种氢致解耦合的异质外延用柔性衬底. 发明人：张志成，杨少延，黎大兵，刘祥林，陈涌海，朱勤生，王占国. 专利号：ZL 03155388.5. 授权公告日：2007 年 10 月 24 日.
6. 发明名称：电泵浦边发射半导体微腔激光器的制作方法. 发明人：陆秀真，常秀兰，李成明，刘峰奇，王占国. 专利号：ZL 200310123450.1. 授权公告日：2007 年 03 月 14 日.
7. 发明名称：在硅衬底上生长无裂纹三族氮化物薄膜的方法. 发明人：陆沅，刘祥林，陆大成，王晓晖，王占国. 专利号：ZL 200410004028.9. 授权公告日：2007 年 05 月 02 日.
8. 发明名称：低能氧离子束辅助脉冲激光沉积氧化物薄膜的方法. 发明人：杨少延，刘志凯，柴春林，陈诺夫，王占国. 专利号：ZL 200410044605.7. 授权公告日：2007 年 08 月 15 日.
9. 发明名称：自适应柔性层制备无裂纹硅基Ⅲ族氮化物薄膜的方法. 发明人：吴洁君，黎大兵，陆沅，韩修训，李杰民，王晓辉，刘祥林，王占国. 专利号：ZL 200410048229.9. 授权公告日：2008 年 02 月 13 日.
10. 发明名称：电化学腐蚀制备多孔磷化铟半导体材料的方法. 发明人：车晓玲，刘峰奇，黄秀颀，雷文，刘俊岐，王占国. 专利号：ZL 200410049950.X. 授权公告日：2008 年 01 月 23 日.
11. 发明名称：真空镀膜机加热装置. 发明人：路秀真，常秀兰，刘峰奇，王占国. 专利号：ZL 200420077769.5. 授权公告日：2005 年 07 月 06 日.
12. 发明名称：在解理面上制作半导体纳米结构的方法. 发明人：陈涌海，张春玲，崔草香,徐波,金鹏,刘峰奇,王占国. 专利号:ZL 200410069295.4. 授权公告日:2008 年 03 月 05 日.

13. 发明名称:利用离子束外延生长设备制备氮化锆薄膜材料的方法. 发明人:杨少延,柴春林,刘志凯,陈涌海,陈诺夫,王占国. 专利号:ZL 200410009815.2. 授权公告日:2008 年 10 月 08 日.
14. 发明名称:利用离子束外延生长设备制备氮化铪薄膜材料的方法. 发明人:杨少延,柴春林,刘志凯,陈涌海,陈诺夫,王占国. 专利号:ZL 200410009816.7. 授权公告日:2008 年 10 月 08 日.
15. 发明名称:用于电调制光致发光光谱测量的样品架. 发明人:丛光伟,彭文琴,吴洁君,魏宏源,刘祥林,王占国. 专利号:ZL 200410098998.X. 授权公告日:2007 年 12 月 05 日.
16. 发明名称:一种制备金属铪薄膜材料的方法. 发明人:杨少延,柴春林,刘志凯,陈涌海,陈诺夫,王占国. 专利号:ZL 200410098996.0. 授权公告日:2008 年 07 月 09 日.
17. 发明名称:一种制备金属锆薄膜材料的方法. 发明人:杨少延,柴春林,刘志凯,陈涌海,陈诺夫,王占国. 专利号:ZL 200410101885.0. 授权公告日:2008 年 08 月 13 日.
18. 发明名称:一种制备二元稀土化合物薄膜材料的方法. 发明人:杨少延,柴春林,刘志凯,周建平,陈诺夫,王占国. 专利号:ZL 200410101884.6. 授权公告日:2008 年 10 月 08 日.
19. 发明名称:利用砷化铟-铟铝砷叠层点制备砷化铟纳米环的生长方法. 发明人:李凯,叶小玲,王占国. 专利号:ZL 200510054469.4. 授权公告日:2007 年 10 月 24 日.
20. 发明名称:氧化锌一维纳米材料的制备方法. 发明人:彭文琴,曲胜春,王占国. 专利号:ZL 200510076326.3. 授权公告日:2008 年 03 月 05 日.
21. 发明名称:以原位垂直超晶格为模板定位生长量子线或点的方法. 发明人:王元立,吴巨,金鹏,叶小玲,张春玲,黄秀颀,陈涌海,王占国. 专利号:ZL 200510012105.X. 授权公告日:2008 年 07 月 02 日.
22. 发明名称:一种半导体晶片亚表面损伤层的测量方法. 发明人:陈涌海,王占国. 专利号:ZL 200510012173.6. 授权公告日:2009 年 02 月 11 日.
23. 发明名称:砷化镓衬底上制备纳米尺寸坑的方法. 发明人:李凯,叶小玲,王占国. 专利号:ZL 200510084357.3. 授权公告日:2008 年 01 月 23 日.
24. 发明名称:一种亚分子单层量子点激光器材料的外延生长方法. 发明人:于理科,徐波,王占国,金鹏,赵昶,张秀兰. 专利号:ZL 200510086313.4. 授权公告日:2008 年 02 月 13 日.
25. 发明名称:1.02-1.08 微米波段 InGaAs/GaAs 量子点外延结构及其制造方法. 发明人:于理科,徐波,王占国,金鹏,赵昶,张秀兰. 专利号:ZL 200510086314.9. 授权公告日:2009 年 07 月 08 日.

26. 发明名称:1.02-1.08 微米波段 InGaAs/GaAs 量子点外延结构及其制造方法. 发明人:于理科,徐波,王占国,金鹏,赵昶,张秀兰. 专利号:ZL 200910119057.2. 授权公告日:2010 年 12 月 08 日.
27. 发明名称:单模 *F-P* 腔量子级联激光器的器件结构. 发明人:郭瑜,刘峰奇,刘俊岐,王占国. 专利号:ZL 200510126478.X. 授权公告日:2008 年 05 月 28 日.
28. 发明名称:宽光谱砷化铟/砷化铟镓/砷化镓量子点材料生长方法. 发明人:刘宁,金鹏,王占国. 专利号:ZL 200610002667.0. 授权公告日:2009 年 05 月 13 日.
29. 发明名称:利用可协变衬底制备生长氧化锌薄膜材料的方法. 发明人:杨少延,陈涌海,李成明,范海波,王占国. 专利号:ZL 200610003072.7. 授权公告日:2008 年 10 月 08 日.
30. 发明名称:宽光谱砷化铟/砷化铝镓量子点材料生长方法. 发明人:刘宁,金鹏,王占国. 专利号:ZL 200610064883.8. 授权公告日:2009 年 02 月 04 日.
31. 发明名称:砷化镓/砷化铝分布布拉格反射镜的湿法腐蚀方法. 发明人:李若园,徐波,王占国. 专利号:ZL 200610011792.8. 授权公告日:2009 年 03 月 04 日.
32. 发明名称:长波长砷化铟/砷化镓量子点材料. 发明人:刘宁,金鹏,王占国. 专利号:ZL 200610088947.8. 授权公告日:2010 年 11 月 10 日.
33. 发明名称:一种用于氧化锌外延薄膜生长的硅基可协变衬底材料. 发明人:杨少延,范海波,李成明,陈涌海,王占国. 专利号:ZL 200610169750.7. 授权公告日:2009 年 09 月 30 日.
34. 发明名称:一种生长氧化锌薄膜的装置及方法. 发明人:杨少延,刘祥林,赵凤瑗,焦春美,董向芸,张晓沛,范海波,魏宏源,张攀峰,王占国. 专利号:ZL 200610169751.1. 授权公告日:2010 年 02 月 17 日.
35. 发明名称:砷化铟和砷化镓的纳米结构及其制作方法. 发明人:赵超,徐波,陈涌海,金鹏,王占国. 专利号:ZL 200610171666.9. 授权公告日:2009 年 03 月 11 日.
36. 发明名称:在半导体衬底上制备有序砷化铟量子点的方法. 发明人:周慧英,曲胜春,金鹏,徐波,王赤云,刘俊朋,王智杰,王占国. 专利号:ZL 200710063705.8. 授权公告日:2009 年 08 月 05 日.
37. 发明名称:在半导体衬底上制备量子环结构的方法. 发明人:周慧英,曲胜春,金鹏,徐波,王赤云,刘俊朋,王占国. 专利号:ZL 200710063706.2. 授权公告日:2011 年 10 月 05 日.
38. 发明名称:δ 掺杂制备 *P* 型氧化锌薄膜的方法. 发明人:魏鸿源,刘祥林,张攀峰,焦春美,王占国. 专利号:ZL 200710065182.0. 授权公告日:2009 年 07 月 08 日.
39. 发明名称:以二氧化硅为掩模定位生长量子点的方法. 发明人:任芸芸,徐波,周惠英;刘明,李志刚,王占国. 专利号:ZL 200710099863.9. 授权公告日:2009 年 12 月 02 日.

40. 发明名称:量子点光调制器有源区结构. 发明人:梁志梅,金鹏,王占国. 专利号:ZL 200710175972.4. 授权公告日:2010 年 02 月 10 日.
41. 发明名称:MDMO-PPV 包裹 PbS 量子点和纳米棒材料及电池的制备方法. 发明人:王智杰,曲胜春,刘俊朋,王占国. 专利号:ZL 200810057181.6. 授权公告日:2012 年 03 月 28 日.
42. 发明名称:控制自组织铟镓砷量子点成核的生长方法. 发明人:梁凌燕,叶小玲,金鹏,陈涌海,徐波,王占国. 专利号:ZL 200810102198.9. 授权公告日:2010 年 09 月 01 日.
43. 发明名称:一种生长可控量子点和量子环的方法. 发明人:赵暕,陈涌海,王占国,徐波. 专利号:ZL 200810104761.6. 授权公告日:2011 年 02 月 02 日.
44. 发明名称:集成肋片式红外半导体激光器结构. 发明人:刘俊岐,刘峰奇,李路,王利军,王占国. 专利号:ZL 200810116832.4. 授权公告日:2011 年 01 月 26 日.
45. 发明名称:边发射二维光子晶体分布反馈量子级联激光器及制备方法. 发明人:陆全勇,张伟,王利军,高瑜,尹雯,张全德,刘万峰,刘峰奇,王占国. 专利号:ZL 200810119796.7. 授权公告日:2011 年 04 月 13 日.
46. 发明名称:用于电子束蒸发薄膜的样品夹具. 发明人:刘万峰,王利军,刘峰奇,王占国. 专利号:ZL 200810240352.9. 授权公告日:2011 年 08 月 31 日.
47. 发明名称:有机染料分子敏化非晶硅/微晶硅太阳电池的制备方法. 发明人:张长沙,王占国,石明吉,彭文博,刘石勇,肖海波,廖显伯,孔光临,曾湘波. 专利号:ZL 200910076560.4. 授权公告日:2011 年 09 月 28 日.
48. 发明名称:硅基薄膜叠层太阳能电池隧道结的制作方法. 发明人:石明吉,曾湘波,王占国,刘石勇,彭文博,肖海波,张长沙. 专利号:ZL 200910078560.8. 授权公告日:2011 年 08 月 17 日.
49. 发明名称:一种测量光子晶体孔洞侧壁垂直度的方法. 发明人:彭银生,徐波,叶小玲,王占国. 专利号:ZL 200910081986.9. 授权公告日:2012 年 04 月 25 日.
50. 发明名称:用于半导体光放大器的宽增益谱量子点材料结构. 发明人:刘王来,叶小玲,徐波,王占国. 专利号:ZL 200910081987.3. 授权公告日:2012 年 07 月 04 日.
51. 发明名称:一种调节 GaAs 基二维光子晶体微腔共振模式的方法. 发明人:彭银生,徐波,叶小玲,王占国. 专利号:ZL 200910083495.8. 授权公告日:2012 年 04 月 25 日.
52. 发明名称:斜腔面二维光子晶体分布反馈量子级联激光器及制备方法. 发明人:陆全勇,张伟,王利军,刘俊岐,李路,刘峰奇,王占国. 专利号:ZL 200910092876.2. 授权公告日:2012 年 03 月 28 日.
53. 发明名称:锥形光子晶体量子级联激光器及其制作方法. 发明人:张伟,王利军,刘俊岐,李路,张全德,陆全勇,高瑜,刘峰奇,王占国. 专利号:ZL 200910237094.3. 授权公告日:2013 年 09 月 04 日.

54. 发明名称:光伏型InAs量子点红外探测器结构. 发明人:唐光华,徐波,叶小玲,金鹏,刘峰奇,陈涌海,王占国,姜立稳,孔金霞,孔宁. 专利号:ZL 200910242347.6. 授权公告日:2011年09月28日.
55. 发明名称:一种改进非晶硅太阳电池性能的方法. 发明人:刘石勇,曾湘波,彭文博,肖海波,姚文杰,谢小兵,王超,王占国. 专利号:ZL 201010117751.3. 授权公告日:2012年05月23日.
56. 发明名称:利用非极性ZnO缓冲层生长非极性InN薄膜的方法. 发明人:郑高林,杨安丽,宋华平,郭严,魏鸿源,刘祥林,朱勤生,杨少延,王占国. 专利号:ZL 201010157517.3. 授权公告日:2011年10月19日.
57. 发明名称:一种生长高质量富In组分InGaN薄膜材料的方法. 发明人:郭严,宋华平,郑高林,魏鸿源,刘祥林,朱勤生,杨少延,王占国. 专利号:ZL 201010157637.3. 授权公告日:2011年10月05日.
58. 发明名称:垂直发射量子级联激光器结构. 发明人:郭万红,刘俊岐,陆全勇,张伟,江宇超,李路,王利军,刘峰奇,王占国. 专利号:ZL 201010171511.1. 授权公告日:2012年03月28日.
59. 发明名称:带有准光子晶体波导阵列的量子级联激光器及其制作方法. 发明人:尹雯,陆全勇,张伟,刘峰奇,张全德,刘万峰,江宇超,李路,刘俊岐,王利军,王占国. 专利号:ZL 201010175432.8. 授权公告日:2012年01月25日.
60. 发明名称:量子级联探测器结构. 发明人:孔宁,刘俊岐,李路,刘峰奇,王利军,王占国. 专利号:ZL 201010191861.4,授权公告日:2011年12月07日.
61. 发明名称:磷化镓铝应力补偿的砷化铟量子点太阳电池制作方法. 发明人:王科范,杨晓光,杨涛,王占国. 专利号:ZL 201010217374.0. 授权公告日:2011年12月07日.
62. 发明名称:短波长光栅面发射量子级联激光器结构及制备方法. 发明人:江宇超,刘俊岐,陆全勇,张伟,郭万红,李路,刘峰奇,王利军,王占国. 专利号:ZL 201010231191.4. 授权公告日:2011年12月07日.
63. 发明名称:不同粒径银纳米颗粒修饰二氧化钛纳米管的制备方法. 发明人:谭付瑞,曲胜春,王占国. 专利号:ZL 201010256808.8. 授权公告日:2012年06月27日.
64. 发明名称:InP基长波长2-3μm准量子点激光器结构. 发明人:孔金霞,徐波,王占国. 专利号:ZL 201010591575.7. 授权公告日:2012年03月28日.
65. 发明名称:定量检测痕量罗丹明6G的方法. 发明人:张君梦,曲胜春,张利胜,王占国. 专利号:ZL 201010591599.2. 授权公告日:2012年07月04日.
66. 发明名称:深紫外激光光致发光光谱仪. 发明人:金鹏,王占国. 专利号:ZL 201110087894.9. 授权公告日:2012年08月15日.
67. 发明名称:利用温度周期调制生长氧化锌材料的方法. 发明人:时凯,刘祥林,魏鸿源,焦春美,王俊,李志伟,宋亚峰,杨少延,朱勤生,王占国. 专利号:ZL 201110113282.2. 授权公告日:2012年07月04日.

68. 发明名称:在非极性蓝宝石衬底上生长水平排列氧化锌纳米线的方法. 发明人:桑玲,王俊,魏鸿源,焦春美,刘祥林,朱勤生,杨少延,王占国. 专利号:ZL 201110119981.8. 授权公告日:2012年12月21日.

69. 发明名称:条纹相机反射式离轴光学耦合装置. 发明人:金鹏,王占国. 专利号:ZL 201110133353.5. 授权公告日:2012年10月03日.

70. 发明名称:同步实现阻止GaAs盖层氧化和提高氧化层热稳定性的方法. 发明人:周文飞,徐波,叶小玲,张世著,王占国. 专利号:ZL 201210083214.0. 授权公告日:2014年03月12日.

71. 发明名称:一种有多色响应的量子点红外探测器. 发明人:贾亚楠,徐波,孔金霞,王占国. 专利号:ZL 201210086429.8. 授权公告日:2014年09月03日.

72. 发明名称:量子点级联激光器. 发明人:刘峰奇,卓宁,李路,邵烨,刘俊岐,张锦川,王利军,王占国. 专利号:ZL 201210105753.X. 授权公告日:2014年01月08日.

73. 发明名称:一种在蓝宝石衬底上生长自剥离氮化镓薄膜的方法. 发明人:王建霞,李志伟,赵桂娟,桑玲,刘长波,魏鸿源,焦春美,杨少延,刘祥林,朱勤生,王占国. 专利号:ZL 201210325765.3. 授权公告日:2015年09月09日.

74. 发明名称:制备超饱和硫系元素掺杂硅的方法. 发明人:王科范,刘孔,曲胜春,王占国. 专利号:ZL 201210484770.9. 授权公告日:2015年05月06日.

75. 发明名称:双全息曝光制备量子级联激光器掩埋双周期光栅方法. 发明人:姚丹阳,张锦川,闫方亮,刘俊岐,王利军,刘峰奇,王占国. 专利号:ZL 201310028764. 7,授权公告日:2014年12月21日.

76. 发明名称:用于超薄半导体芯片接触式曝光的方法. 发明人:梁平,胡颖,刘俊岐,刘峰奇,王利军,张锦川,王涛,姚丹阳,王占国. 专利号:ZL 201310057280.5. 授权公告日:2015年05月06日.

77. 发明名称:滤波式波长可调谐外腔激光器. 发明人:魏恒,金鹏,王占国. 专利号:ZL 201310071764.5. 授权公告日:2015年04月15日.

78. 发明名称:一种超饱和掺杂半导体薄膜的制备方法. 发明人:王科范,张华荣,彭成晓,曲胜春,王占国. 专利号:ZL 201310157821.1. 授权公告日:2015年10月21日.

79. 发明名称:复合构型可调谐光栅外腔双模激光器. 发明人:金鹏,魏恒,吴艳华,陈红梅,王占国,王占国. 专利号:ZL 201310193627.9. 授权公告日:2015年12月02日.

80. 发明名称:光栅分布反馈量子级联激光器. 发明人:张锦川,刘峰奇,卓宁,王利军,刘俊岐,王占国. 专利号:ZL 201310308727.1. 授权公告日:2016年01月20日.

81. 发明名称:可调谐分布反馈量子级联激光器阵列器件及其制备方法. 发明人:闫方亮,张锦川,姚丹阳,谭松,王利军,刘峰奇,王占国. 专利号:ZL 201310447427.1. 授权公告日:2016 年 01 月 20 日.

82. 发明名称:一种低发散角分布反馈量子级联激光器结构及制作方法. 发明人:张锦川,刘峰奇,闫方亮,姚丹阳,王利军,刘俊岐,王占国. 专利号:ZL 201310504071.0. 授权公告日:2016 年 04 月 20 日.
83. 发明名称:一种低发散角的面发射量子级联激光器结构. 发明人:张锦川,姚丹阳,闫方亮,刘峰奇,王利军,刘俊岐,王占国. 专利号:ZL 201310503782.6. 授权公告日:2015 年 12 月 30 日.
84. 发明名称:提高 AlN 外延薄膜荧光强度的方法. 发明人:王维颖,毛德丰,李维,金鹏,王占国. 专利号:ZL 201310585868.8. 授权公告日:2016 年 08 月 24 日.
85. 发明名称:利用蓝宝石衬底制备垂直结构氮化镓基发光二极管的方法. 发明人:杨少延,张恒,魏鸿源,焦春美,赵桂娟,汪连山,刘祥林,王占国. 专利号:ZL 201310651999.1. 授权公告日:2016 年 03 月 02 日.
86. 发明名称:高功率低发散角的半导体太赫兹垂直面发射激光器. 发明人:王涛,刘俊岐,刘峰奇,张锦川,王利军,王占国. 专利号:ZL 201310652143.6. 授权公告日:2016 年 06 月 01 日.
87. 发明名称:利用硅衬底制备垂直结构氮化镓基发光二极管器件的方法. 发明人:杨少延,冯玉霞,魏鸿源,焦春美,赵桂娟,汪连山,刘祥林,王占国. 专利号:ZL 201310652125.8. 授权公告日:2016 年 03 月 30 日.
88. 发明名称:一种可调谐衬底发射量子级联激光器阵列器件. 发明人:闫方亮,张锦川,姚丹阳,卓宁,王利军,刘峰奇,王占国. 专利号:ZL 201310705313.2. 授权公告日:2016 年 06 月 01 日.
89. 发明名称:硫族元素超饱和掺杂硅红外探测器及其制作方法. 发明人:王科范,彭成晓,刘孔,谷城,曲胜春,王占国. 专利号:ZL 201410035355.4. 授权公告日:2016 年 03 月 02 日.
90. 发明名称:输出功率和光谱形状独立可调的发光二极管的制作方法. 发明人:陈红梅,金鹏,王占国. 专利号:ZL 201410083290.0. 授权公告日:2016 年 04 月 20 日.
91. 发明名称:半极性面氮化镓基发光二极管及其制备方法. 发明人:项若飞,汪连山,赵桂娟,金东东,王建霞,李辉杰,张恒,冯玉霞,焦春美,魏鸿源,杨少延,王占国. 专利号:ZL 201410147977.6. 授权公告日:2016 年 08 月 17 日.
92. 发明名称:一种改善面发射半导体激光器慢轴远场的金属天线结构. 发明人:姚丹阳,张锦川,周予虹,贾志伟,闫方亮,王利军,刘俊岐,刘峰奇,王占国. 专利号:ZL 201410520059.3. 授权公告日:2017 年 08 月 17 日.
93. 发明名称:碳化硅中间带太阳电池及其制备方法. 发明人:王科范,张光彪,丁丽,刘孔,曲胜春,王占国. 专利号:ZL 201410612819.3. 授权公告日:2017 年 09 月 08 日.
94. 发明名称:一种高反射率的垂直结构发光二极管芯片及其制备方法. 发明人:赵桂

娟,汪连山,杨少延,刘贵鹏,魏鸿源,焦春美,刘祥林,朱勤生,王占国. 专利号:ZL 201410641928.8. 授权公告日:2018 年 03 月 23 日.

95. 发明名称:一种双面散热量子级联激光器器件结构. 发明人:闫方亮,张锦川,刘峰奇,卓宁,刘俊岐,王利军,王占国. 专利号:ZL 201410687356.7. 授权公告日:2017 年 07 月 14 日.

96. 发明名称:一种氮化铝一维纳米结构材料的制备方法. 发明人:孔苏苏,李辉杰,冯玉霞,赵桂娟,魏鸿源,杨少延,王占国. 专利号:ZL 201410764375.5. 授权公告日:2016 年 08 月 24 日.

97. 发明名称:硅基高晶体质量 InAsSb 平面纳米线的生长方法. 发明人:杨涛,杜文娜,杨晓光,王小耶,季祥海,王占国. 专利号:ZL 201410785433.2. 授权公告日:2017 年 04 月 19 日.

98. 发明名称:采用表面等离激元效应的倒置太阳电池结构及制备方法. 发明人:卢树弟,曲胜春,刘孔,池丹,李彦沛,寇艳蕾,岳世忠,王占国. 专利号:ZL 201510270129.9. 授权公告日:2017 年 10 月 24 日.

99. 发明名称:大尺寸非极性 A 面 GaN 自支撑衬底的制备方法. 发明人:李辉杰,杨少延,魏鸿源,赵桂娟,汪连山,李成明,刘祥林,王占国. 专利号:ZL 201510270779.3. 授权公告日:2017 年 04 月 12 日.

100. 发明名称:红外光电探测器及其制造方法. 发明人:任飞,刘舒曼,王风娇,翟慎强,梁平,刘峰奇,王占国. 专利号:ZL 201510300412.1. 授权公告日:2017 年 06 月 20 日.

101. 发明名称:制备高选择比量子级联激光器脊波导结构的湿法腐蚀方法. 发明人:梁平,刘峰奇,张锦川,闫方亮,胡颖,王利军,刘俊岐,王占国. 专利号:ZL 201610948341.0. 授权公告日:2018 年 03 月 30 日.

102. 发明名称:电子输运通道为斜跃迁-微带型的量子级联红外探测器. 发明人:王风娇,任飞,刘舒曼,翟慎强,刘俊岐,梁平,刘峰奇,王占国. 专利号:ZL 201510329218.6. 授权公告日:2018 年 02 月 13 日.

培养学生简况

王占国院士从1985年开始招收研究生，在任半导体研究所研究员以及多所高等院校的特聘教授或岗位教授期间，据不完全统计，共计培养(包括毕业和在读)博士后9名、博士研究生152名、硕士研究生35人、北京市科技新星4人，这些学生在各自的科研、教学、生产等岗位上做出了巨大成绩，为我国半导体事业的发展做出了很大贡献。

1　培养的学生目录

1.1　硕士研究生

姓　名	在学时间	学位论文题目	现工作地/职务
付建明 (林兰英、王占国)	1984.9–1987.7	硅中与钯相关的 E_{TA}、E_{TB} 能级的研究	–
张　芊 (林兰英、王占国)	1985.9–1988.7	LEC-GaAs 中几个重要缺陷中心的研究	–
何　定 (林兰英、王占国)	1986.9–1989.7	–	–
刘　清 (代培)	1986.9–1990.7	–	–
戴元筠 (林兰英、王占国)	1989.9–1992.7	–	–
徐　波 (林兰英、王占国)	1989.9–1992.7	Si-GaAs 中 *EL2* 的光淬灭效应及有关深能级行为的研究	中国科学院半导体研究所/研究员
邓航军	1990.9–1993.7	低温 GaAs 的生长及性质研究	中国科学院半导体研究所/–
童玉珍 (代培)	1992.9–1995.7	–	–
李胜英	1993.9–1996.7	InAs/GaAs 低维半导体材料的制备及其特性研究	–
莫庆伟	1995.9–1998.7	–	大连德豪光电科技有限公司/–
阎　华	1997.9–2000.7	GaN 基材料的光学及电学性质研究	–

续表

姓　名	在学时间	学位论文题目	现工作地/职务
魏永强	1998.9-2001.7	GaAs基In(Ga)As量子点材料生长及器件应用	-
朱天伟(与兰州大学联合培养)	2000.9-2003.7	-	-
董　逊	2000.9-2003.7	AlInGaN四元合金的生长及性质研究	-
曲宝壮(王占国、朱勤生)	2000.9-2003.7	Ⅲ族氮化物材料的性质研究	-
王春华(王占国、陈涌海)	2001.9-2004.7	离子注入法生长GaMnAs材料性质研究	-
李　凯(王占国、叶小玲)	2002.9-2005.7	GaAs基In(Ga)As量子环的材料生长和性质研究	-
金　灿(王占国、金鹏)	2004.9-2007.7	InAs/GaAs量子点光学性质及激光器研究	-
周振宇(王占国、陈涌海)	2005.9-2008.7	半导体材料平面光学各向异性研究	厦门银行/外汇交易员
刘王来(王占国、徐波)	2006.9-2009.7	量子点半导体光放大器研究	-
唐光华(王占国、徐波)	2007.9-2010.7	光伏型InAs量子点红外探测器研究	中国电子科技集团公司第五十五研究所/-
孔　宁(王占国、刘峰奇)	2007.9-2010.7	量子级联探测器的研究	华为技术有限公司/-
姜立稳(王占国、叶小玲)	2007.9-2010.7	InAs/GaAs量子点材料生长及双区结构激光器研究	-
王佐才(王占国、金鹏)	2007.9-2010.7	双注入区量子点超辐射发光管的研究	阿里巴巴网络技术有限公司/高级产品专家
胡发杰(王占国、金鹏)	2012.9-2015.7	外腔扫频量子点激光器的研究	普联技术有限公司/工程师
龚　猛(王占国、金鹏)	2014.9-2017.12	微波等离子体化学气相沉积法生长金刚石和金刚石离子注入的研究	京东方科技集团股份有限公司/-
唐天义	2017.9-	-	在读硕士生
吴　静(王占国、陈涌海、金鹏)	2017.9-	-	在读硕士生

1.2　博士研究生

姓　名	在学时间	学位论文题目	现工作地/职务
杨保华（林兰英、王占国）	1987.9-1990.7	等电子杂质在Ⅲ-Ⅴ族化合物半导体中的行为研究	-
库意沃（代培，中山大学莫党）	1988.9-1993.7	LEC半绝缘GaAs中EL2缺陷的亚稳性和光恢复效应与深能级缺陷的性质	-
李瑞钢	1991.9-1995.1	低温分子束外延GaAs材料性质及相关器件研究	AZ Power Inc., USA/CEO
杨　斌（王占国、林兰英）	1992.9-1995.7	Ⅲ-Ⅴ半导体低维结构材料分子束外延生长及其性质的研究	美国高通公司/高技术研发部总经理
李　伟	1992.9-1995.7	应变异质结构(InGaAs/(Al)GaAs)材料的分子束外延生长性质及其相关器件的研究	芬兰-中国智慧健康股份公司/-
邝国魁（与德国大众联合培养）	1993.9-1996.7	-	-
邹吕凡	1993.9-1996.7	$Si_{1-x}Ge_x$/Si应变异质结构材料的气态分子束外延生长及性质研究	-
董建荣	1993.9-1996.7	Ⅲ-Ⅴ族可见光半导体材料生长及其性质研究	中国科学院苏州纳米技术与纳米仿生研究所/研究员
龚　谦	1993.9-1998.7	In(Ga)As量子点材料的分子束外延生长性质及激光器的研究	中国科学院上海微系统与信息技术研究所/研究员
林兆军	1994.9-1997.7	沸石分子筛中半导体团簇的组装及性质研究	山东大学/教授
张兴宏（杨玉芬、王占国）	1994.9-1997.7	HEMT材料界面特性及其器件的二维数值模拟	-
朱东海	1994.9-1997.7	AlGaAs/GaAs，InGaAs/GaAs体系光电子材料的MBE生长及相关性质研究	-
陈涌海	1995.9-1998.7	半导体及其低维结构的平面光学各向异性研究	中国科学院半导体研究所/研究员
江　潮	1995.9-1998.7	应变异质外延材料生长和低维结构制备研究	国家纳米科学中心/研究员
李含轩	1995.9-1998.7	应变异质结构In(GaAs/InAlAs/InP)材料的分子束外延生长、性质及其相关器件的研究	Spectral Physics/-

续表

姓　名	在学时间	学位论文题目	现工作地/职务
汪连山	1995.9–1998.7	氮化镓材料的外延生长及性质研究	中国科学院半导体研究所/研究员
周　伟	1996.9–1999.7	自组织 InAlAs/AlGaAs 红光量子点材料生长及其性质研究	电子科技大学/教授
梁建军	1997.9–2000.7	掺铒硅基材料光学性质和微结构研究	–
姜卫红	1997.9–2000.7	GaAs 基应变自组装 $In_xGa_{1-x}As$ 量子点的生长、结构和光学性质的研究	加拿大国家研究委员会(NRC)/–
孙中哲	1997.9–2000.7	InP 基半导体低维结构材料分子束外延生长、表征及相关器件研究	新加坡南洋理工大学/–
刘舒曼	1997.9–2000.7	Ⅱ–Ⅵ族半导体纳米团簇的制备及性质研究	中国科学院半导体研究所/研究员
王吉政	1997.9–2000.7	一些低维Ⅲ–Ⅴ半导体材料光学性质的研究	中国科学院化学研究所/研究员
孙小玲	1997.9–2000.7	GaN 基材料的生长、结构性能及光电性质研究	–
李月法	1998.9–2001.7	InP 基半导体低维结构材料的分子束外延生长和性质的研究	–
张元常	1998.9–2001.7	GaAs 基 In(Ga)As 量子点及其复合结构的材料生长和性质研究	–
刘会赟	1998.9–2001.7	GaAs 基应变自组织 In(Al)As 量子点材料生长、性质及器件应用	英国伦敦大学学院(UCL)/–
叶小玲	1998.9–2001.7	Ⅲ–Ⅴ族半导体低维结构材料的平面光学各向异性	中国科学院半导体研究所/副研究员
陈　振	1999.9–2002.7	Ⅲ族氮化物材料生长和特性研究	
李昱峰	1999.9–2002.7	InGaN 量子点的 MOCVD 生长及特性研究	
廖梅勇	1999.9–2002.7	类金刚石系列薄膜的离子束生长、微观结构与性能	日本国立材料研究所(NIMS)/终身研究员
陆　沅	1999.9–2004.7	Si(111)衬底 GaN 生长及特性研究	法国洛林大学让·拉莫尔研究所/–
孟宪权	2000.9–2003.7	应变自组织 InAs 量子点材料及垂直腔面发射激光器研究	武汉大学/教授

续表

姓 名	在学时间	学位论文题目	现工作地/职务
张子旸	2000.9-2003.7	GaAs 基 In(Ga, Al)As 复合量子点材料的生长及量子点超辐射发光管的研制	中国科学院苏州纳米技术与纳米仿生研究所/研究员
何 军	2000.9-2003.7	1.3μm GaAs 基 InAs 量子点材料生长及器件应用	国家纳米科学中心/研究员
李成明	2001.9-2004.7	量子级联激光器材料生长及器件制作	中国科学院半导体研究所/助理研究员
赵凤瑷	2001.9-2004.7	InP 基半导体量子点、量子线的分子束外延生长、表征及生长机理研究	-
黎大兵	2001.9-2004.7	(In,Al)GaN 材料的 MOCVD 生长及其物性研究	中国科学院长春光学精密机械与物理研究所/研究员
张志成（王占国、陈涌海）	2001.9-2003.12	大失配异质外延柔性衬底的研究	-
张春玲（王占国、刘峰奇）	2002.9-2005.7	GaAs 基 InAs 量子点和量子线的控位生长研究	南开大学/副教授
崔草香（王占国、陈涌海）	2002.9-2005.7	GaAs 基 InAs 量子点的可控定位生长研究	京东方科技集团股份有限公司/-
史桂霞（王占国、徐波）	2002.9-2005.7	1.3μm 长波长 InAs/GaAs 自组织量子点材料及激光器研究	-
路秀真（王占国、刘峰奇）	2002.9-2005.7	应变补偿量子级联激光器的材料与器件研究	上海大学/-
郭 瑜（王占国、刘峰奇）	2002.9-2005.7	应变补偿分布反馈量子级联激光器材料生长及器件性能研究	北京汇德信科技有限公司/部门经理
韩修训（王占国、朱勤生）	2002.9-2005.7	Ⅲ族氮化物低维结构的生长及物性研究	-
杨少延（王占国、陈涌海）	2002.9-2005.7	大失配异质外延超薄中间层柔性衬底研究	中国科学院半导体研究所/研究员
舒永春（与南开大学联合培养）	2002.9-2005.7	GaAs 基和 InP 基化合物半导体材料 MBE 生长与特性的研究	郑州大学/教授
李杰民（王占国、朱勤生）	2003.9-2006.7	量子阱子带间跃迁及相关物理特性研究	-
于理科（王占国、徐波）	2003.9-2006.7	不同发光波长的 GaAs 基 In(Ga)As 量子点的材料生长和性质研究	-
李若园（王占国、徐波）	2003.9-2006.7	GaAs 基量子点垂直腔面发射激光器的研究	中芯国际集成电路制造有限公司/-

续表

姓　名	在学时间	学位论文题目	现工作地/职务
孙　捷 （与隆德大学联合培养）	2003.9-2007.7	低维Ⅲ-Ⅴ族半导体中的电子输运	北京工业大学/教授
彭文琴 （王占国、曲胜春）	2003.9-2006.7	硫化锌和氧化锌低维结构的制备及物性研究	日本国家先进工业科学和技术研究所/-
丛光伟 （王占国、朱勤生）	2003.9-2006.7	ZnO纳米棒阵列和薄膜的生长控制及物性研究	日本国家先进工业科学和技术研究所/-
吴洁君 （王占国、刘祥林）	2003.4-2006.3	Si基GaN及InAlGaN四元合金生长和特性研究	北京大学
赵　暕 （王占国、陈涌海）	2005.9-2008.7	图形衬底结合液滴外延生长InAs/GaAs量子点、量子环的研究	-
范海波 （王占国、陈涌海）	2003.9-2008.7	Si基ZnO材料MOCVD生长及物性研究	-
梁凌燕 （王占国、叶小玲）	2003.9-2008.7	长波长自组装InAs/GaAs量子点材料和激光器结构生长及性质研究	中国科学院宁波材料技术与工程研究所/副研究员
赵　昶 （王占国、陈涌海）	2003.9-2006.7	半导体量子点生长的动力学蒙特卡罗模拟研究	-
雷　文 （王占国、陈涌海）	2003.9-2006.7	InP基InAs纳米结构材料的生长、光电性质与其器件应用研究	-
张冠杰 （与南开大学联合培养）	2003.9-2006.7	GaAs基多量子阱和自组织量子点材料的MBE优化生长及其特性的研究	中芯国际集成电路制造有限公司/-
胡良均 （王占国、陈涌海）	2004.9-2007.7	注MnInAs/GaAs量子点光电磁性质的研究	-
杨新荣 （王占国、徐波）	2004.9-2007.7	InP基InAs纳米结构形貌控制及激光器研究	邯郸学院/副教授
任芸芸 （王占国、徐波）	2004.9-2007.7	GaAs基InAs量子点的定位有序生长研究	国家知识产权局/-
周慧英 （王占国、曲胜春）	2004.9-2007.7	GaAs基图形衬底上GaAs和InAs纳米结构的控位生长	中南林业科技大学/-
刘俊朋 （王占国、曲胜春）	2004.9-2007.7	有机/无机复合半导体材料与光电转换器件研究	英国诺丁汉大学/-
刘　宁 （王占国、金鹏）	2004.9-2007.7	量子点超辐射发光管材料与器件研究	国家知识产权局专利局专利审查协作北京中心/副研究员
焦玉恒 （王占国、吴巨）	2004.9-2007.7	分子束外延生长InAs/InGaAs/GaAs长波长量子点	-

续表

姓　名	在学时间	学位论文题目	现工作地/职务
魏鸿源（王占国、刘祥林）	2004.9–2007.7	氧化锌低维结构制备及物性研究	中国科学院半导体研究所/副研究员
皮　彪（与南开大学联合培养）	2004.9–2007.7	含磷Ⅲ–Ⅴ族化合物半导体材料MBE生长与特性研究	南开大学/高级工程师
贾国治（与南开大学联合培养）	2004.9–2007.7	低维半导体结构的设计、可控性制备及性质研究	天津城建大学/–
舒　强（与南开大学联合培养）	2004.9–2007.7	调制掺杂GaAs/AlGaAs 2DEG和InAs量子点材料的制备及特性研究	中芯国际集成电路制造有限公司/–
汤晨光（王占国、陈涌海）	2004.9–2009.7	Ⅲ–Ⅴ族低维半导体材料偏振相关的光学性质	国家知识产权局专利局专利审查协作北京中心/–
赵　超（王占国、陈涌海）	2004.9–2009.7	半导体纳米结构的液滴外延生长及性质研究	阿卜杜拉国王科技大学/–
王智杰（王占国、曲胜春）	2004.9–2009.7	有机/无机复合半导体材料与体异质结太阳能电池的研究	中国科学院半导体研究所/研究员
石礼伟（王占国、陈涌海）	2005.9–2008.7	In(Ga)As量子点材料生长及激光器稳态和瞬态特性研究	中国矿业大学/–
王志成（王占国、陈涌海）	2005.9–2008.7	InAs量子点红外探测器的研究	工业和信息化部电子科学技术情报研究所/–
梁志梅（王占国、金鹏）	2005.9–2008.7	自组装InAs/GaAs量子点光学性质中几个重要问题研究	英特尔半导体(大连)有限公司/–
刘如彬（与南开大学联合培养）	2005.9–2008.7	GaAs基低维应变异质结和多周期结构的MBE制备及特性研究	中国电子科技集团公司第十八研究所/–
刘建庆（王占国、陈涌海）	2005.9–2010.7	GaAs基有序量子点和量子线生长研究	三安光电股份有限公司/–
宋华平（王占国、朱勤生）	2005.9–2010.7	InN掺Zn纳米棒的生长及相关物性研究	–
高　瑜（王占国、刘峰奇）	2006.9–2009.7	短腔长与锥形量子级联激光器研究	国家知识产权局/–
郝亚非（王占国、陈涌海）	2006.9–2009.7	半导体低维结构的能带自旋分裂	浙江师范大学/–
石明吉（王占国、曾湘波）	2006.9–2009.7	纳米硅P^+层在非晶/微晶叠层电池隧道结中的应用	–
张长沙（王占国、曾湘波）	2006.9–2009.7	有机染料敏化非晶硅材料及其太阳电池的研究	–
阮　军（王占国、李晋闽）	2006.9–2009.7	氮化物LED材料生长及垂直结构器件制备技术研究	–

续表

姓　名	在学时间	学位论文题目	现工作地/职务
刘石勇（王占国、曾湘波）	2006.9-2011.7	硅薄膜太阳电池 i/p 界面的研究	-
刘万峰（王占国、刘峰奇）	2006.9-2011.7	量子级联激光器输出特性研究及器件应用	英特尔半导体（大连）有限公司/-
张全德（王占国、刘峰奇）	2007.9-2010.7	量子级联激光器室温连续工作及失效分析	恩智浦半导体有限公司/-
彭银生（王占国、徐波）	2007.9-2010.7	二维光泵浦/电注入 GaAs 基光子晶体微腔的研究	浙江工业大学/-
吕雪芹（王占国、金鹏）	2007.9-2010.7	宽增益谱量子点材料与器件研究	厦门大学/副教授
安　琪（王占国、金鹏）	2007.9-2012.7	量子点超辐射发光管理论模拟	中国电子技术标准化研究院/高级工程师
贾彩虹（王占国、陈涌海）	2007.9-2012.7	$SrTiO_3$衬底上 MOCVD 生长 ZnO 和 InN 薄膜	河南大学/副教授
周文飞（王占国、徐波）	2007.9-2012.7	GaAs 基光子晶体平板微腔及其激光器的研究	-
梁德春（王占国、金鹏）	2008.9-2011.7	短波长量子点超辐射发光管材料与器件研究	北京航天控制仪器研究所/高级工程师
龚　亮（与南开大学联合培养）	2008.9-2013.7	量子点结构与微波烧结粉体 ZnO 的数值模拟研究	河北联合大学/-
张世著（王占国、徐波）	2008.9-2013.7	低密度的长波长量子点和量子点分子制备及性质研究	南京绿旗喷墨技术有限公司/-
陈剑燕（王占国、刘峰奇）	2008.9-2013.7	中红外面发射及太赫兹量子级联激光器研究	国家知识产权局专利局专利审查协作广东中心/-
卓　宁（王占国、刘峰奇）	2008.9-2013.7	量子点级联激光器的研究	中国科学院半导体研究所/助理研究员
孔金霞（王占国、徐波）	2009.9-2012.7	量子点红外激光器和探测器研究	中国科学院半导体研究所/助理研究员
李天峰（王占国、陈涌海）	2009.9-2012.7	Ⅲ-Ⅴ族半导体纳米结构的 MOCVD 生长和光学性质研究	河南大学/副教授
李立功（王占国、刘舒曼）	2009.9-2012.7	InAs/GaSb Ⅱ型超晶格红外探测材料的 MOCVD 生长特性研究	-
李新坤（王占国、金鹏）	2009.9-2012.7	量子点超辐射发光管器件及工艺研究	北京航天控制仪器研究所/高级工程师
翟慎强（王占国、刘峰奇）	2009.9-2014.7	量子级联红外探测器及其应用研究	中国科学院半导体研究所/助理研究员

续表

姓　名	在学时间	学位论文题目	现工作地/职务
李　维 (王占国、金鹏)	2009.9-2014.7	Ⅲ族氮化物载流子局域特性研究	中国航空工业集团公司北京长城计量测试技术研究所/高级工程师
杨　萍 (王占国、曾湘波)	2009.9-2014.7	硅纳米线阵列 SiNWs/Si:H 异质结太阳电池的研究	中国工程物理研究院微系统与太赫兹中心/-
宁振动 (与哈尔滨工业大学联合培养)	2009.9-2014.7	锑化物红外探测材料的 MOCVD 生长及光电性能研究	三安光电股份有限公司天津分公司/-
魏　恒 (王占国、金鹏)	2010.9-2013.7	可调谐外腔量子点激光器的研究	中国电子科技集团公司第五十四研究所/高级工程师
池　丹 (王占国、曲胜春)	2010.9-2015.7	聚合物和硅基有机/无机杂化太阳电池研究	浙江师范大学/讲师
姚丹阳 (王占国、刘峰奇)	2010.9-2015.7	面发射量子级联激光器及其光束整形研究	华为技术有限公司/-
毛德丰 (王占国、金鹏)	2011.9-2014.7	AlGaN/GaN 高电子迁移率晶体管结构的光学性质研究	京东方科技集团股份有限公司/-
任　飞 (王占国、刘峰奇)	2011.9-2016.7	长波红外 InGaAs/InAlAs 量子级联发光及探测器件的研究	中芯国际集成电路新技术研发(上海)有限公司/-
闫方亮 (王占国、刘峰奇)	2011.9-2016.7	大功率及宽调谐量子级联激光器阵列的研究	北京聚睿众邦科技有限公司/董事长
杜文娜 (王占国、杨涛)	2011.9-2016.7	Si 基 InAsSb 纳米线 MOCVD 生长及机理研究	国家纳米科学中心/-
吴艳华 (王占国、金鹏)	2012.9-2015.7	被动锁模量子点激光器的研究	北京北方华创微电子装备有限公司/工程师
周予虹 (王占国、刘峰奇)	2012.9-2017.7	高速量子级联激光和探测器件研究	华进半导体封装先导技术研发中心有限公司/-
王风娇 (王占国、刘峰奇)	2012.9-2017.7	InP 基量子点量子级联探测器研究	北京航天控制仪器研究所/-
李媛媛 (王占国、刘峰奇)	2013.9-2018.7	高性能太赫兹量子级联激光器的研究	中国科学院微电子研究所/-
王东博 (王占国、刘峰奇)	2014.9-		在读博士生
杨　诚	2014.9-		在读博士生
郝美兰 (王占国、王晓亮)	2015.9-2018.7	AlGaN/GaN HEMT 功率开关器件漏电与陷阱效应研究	
程凤敏 (王占国、刘峰奇)	2015.9-		在读博士生
黄艳宾	2015.9-		在读博士生

1.3 博士后

姓　名	在站时间	出站报告题目	现工作地/职务
V. L. Dostov	1992.8-1994.4		莫斯科约飞技术物理研究所/-
刘峰奇	1996.7-1998.6	InP 基 InGaAs/InAlAs 微结构材料生长及相关物理问题研究	中国科学院半导体研究所/研究员
林　峰	1998.12-2000.10	在 GaAs(InP)衬底上生长 InAs 量子点和量子线的结构和光学性质研究	-
曲胜春	1999.9-2001.8	半导体纳米晶的化学制备和结构、物性研究	中国科学院半导体研究所/研究员
金　鹏	2001.9-2003.9	InAs/GaAs 自组装量子点结构和光学性质研究	中国科学院半导体研究所/研究员
王元立	2003.9-2005.7	InP 基 InAs/InAlAs 自组织量子线的生长、结构与性质以及 InAlAs 自发组分调制的研究	北京通美晶体技术有限公司/高级工程师
程抱昌	2004.11-2006.6	准一维掺杂半导体纳米材料的制备与研究	南昌大学/教授
张利胜 (王占国、曲胜春)	2008-2010	纳米阵列上表面增强拉曼光谱及镧锶镍氧薄膜激光感生热电电压的研究	-
唐爱伟 (王占国、曲胜春)	2009.9-2011.4	无机纳米材料的制备、自组装及光电性能研究	-
王　超 (王占国、曾湘波)	2009.10-2011.11	硅基太阳能电池前电极材料的制备及相关性能研究	-
王科范 (王占国、杨涛、曲胜春)	2009.10-2012.6	量子点中间带和黑硅太阳电池研究	河南大学/-
项若飞 (王占国、汪连山)	2012.12-2015.1	大尺寸硅基氮化镓电力电子器件的 MOCVD 外延生长及半极性黄绿光氮化镓基 LED 的外延研究	苏州慧通汇创科技有限公司/-
赵立华 (王占国、刘峰奇)	2010.7-2012.5	量子级联激光器输出光特性及散热性能优化	-
彭尚龙	2012.2-2014.7	金属诱导晶化多晶硅薄膜及作为薄膜晶体管的特性研究	-
刘建奇 (王占国、刘峰奇)	2017.6-2018.4	基于纳米孔结构量子(点)级联探测器研究	-

1.4　北京市科技新星

姓　名	起止时间	现工作地/职务
陈广超	2003.8-2006.7	中国科学院大学/材料学院常务副院长、研究员
滕　枫	2003.8-2006.7	北京交通大学/理学院院长、教授
周树云	2003.8-2006.7	中国科学院理化技术研究所/室主任,研究员
曲胜春	2004.8-2007.7	中国科学院半导体研究所/研究员

2　部分代表性的学生简历

李瑞钢,博士,中组部千人计划国家特聘专家,IEEE 教育协会副主席,AZ Power Inc. USA. CEO。

教育背景

1982-1986 年,西安电子科技大学,学士;

1991-1995 年,中国科学院科学院半导体研究所,理学博士;

1995-1996 年,中国科学院科学院微电子中心,博士后;

1996-1997 年,香港科技大学,博士后;

1998-2001 年,美国加利福尼亚大学 洛杉矶分校,博士后。

工作经历

1986-1991 年,电子工业部十三所,助理工程师;

1997-1998 年,中国科学院微电子中心,副研究员;

2001-2002 年,Motorola 公司,高级工程师;

2002-2004 年,美国加利福尼亚大学,高级顾问;

2002-2006 年,JAZZ 半导体公司,项目经理;

2006-2007 年,AMD 公司,技术委员会委;

2007-2012 年,EXAR 半导体公司,高级总监;

2012-现在年,AZ Power Inc 半导体公司,总执行长。

设计的"208 门砷化镓门阵列"获中国"七五"十大科技成果。

"中国第一代砷化镓门电路"(主要研究人员)获电子工业部科学技术进步奖一等奖。

"新结构异质结化合物场效应晶体管"获中国科学院科学技术进步奖二等奖。

在国际上首次系统研究低温分子束外延砷化镓材料及相关器件光敏和背栅效应,建立器件模型。

在国际上首次系统研究自组织量子点多值逻辑存储行为和材料在超低温强磁场下的电子能谱。

在 Motorola 公司工作期间，负责 90nm SOI CMOS 工艺集成。4Mbit SOI SRAM 成品率达 79.8%（> 5 million USD）。建立 90nm 技术逻辑电路库的（KARMA）自动效验系统。

在 AMD 公司和 IBM 联合研发中心工作期间，定义 65nm、45nm 以及 32nm 设计规则，校验 DRC、LVS、DFM 等电路设计 QA。应用于双核和四核服务器 CPU 批量生产。

在 JAZZ 半导体公司锗硅晶圆代工厂研发部工作期间，负责研制开发的 0.18um SiGe BiCMOS 工艺平台，2.5V 和 2.2V 异质结 *NPN* 分别达到 120GHz 和 200GHz 截止频率，超过 IBM 和 TSMC 技术。

成功转让 0.35nm SiGe BiCMOS 工艺平台到上海先进半导体制造股份有限公司，利用该技术平台制造出世界第一块 100mW、802.11g 无线局域网集成电路。该平台也是中国第一条 8 寸 SiGe BiCMOS 生产线。带领团队利用 PFA/EFA 等失效分析和工艺优化使成品率从零改进为 75%（批次）-85%（晶片）。

作为技术负责人，转让 8 寸 0.25μm CMOS 数字、模拟、射频和 CMOS 影像，转让 0.18μm CMOS 数字、模拟、射频工艺平台到上海华虹 NEC（100 million USD）。

在 Exar 集成电路设计公司工作期间，负责产品生产，涉及亚洲和美国 8 个晶圆代工厂和 4 个封装厂。通过产能和产率的改进，降低 Exar 及有关公司成本近千万美元。在 18 个月内，使 Exar 在杭州 Silan、Chartered、TSMC、Episil 和 Jazz 生产的近百种产品（平均 7000 片/月）成品率从 30%-50% 改进到 90%。

创建 ICDS 半导体公司，研发出 *i*-Flow 现代生产线的信息化管理平台：国际标准化 MES，应用于生产效率管理、质量控制、信息安全以及生产和市场开发标准化。研发出 i-Data 动态数据库及数据分析工具，应用于半导体产品及工艺研发、建模、失效分析及生产监控。通过推广 SPC 及 PCM 等质量控制方法和半导体芯片生产模型及生产力资源优化服务，致力于半导体工艺和设计研发向规模生产的转换。

2013 年筹建 AZ Power Inc. 任 CEO，建立 SiC 二极管及三极管设计平台、研发平台、测试平台和可靠性检测平台。研制的 JBS、SBD 性能超过国际 4 代产品水平，器件性能价格比超过国际先进水平为碳化硅电力电子器件的量产化和产业链的形成打下基础。

发表学术论文 40 多篇；多次受到美国及中国大学、公司和学术机构邀请进行演讲和学术交流。

杨斌，博士，美国高通总部高技术研发部总经理。

1992-1995 年在王占国和林兰英共同指导下，进行 GaAs/AlGaAs 二维电子气的分子束外延生长研究。1993 年夏天，王老师领导的研究小组取得了低温 4.8K 电子迁移率超越 100 万的优异成果。这个结果远远超越了当时国内水平，跻身国际寥寥可数的几个曾经取得 GaAs/AlGaAs 二维电子气超越 100 万电子迁移率的研究机构之列。此项研究成果获得了 1995 年国家科学技术进步奖二等奖。在王老师指导下，“优化 GaAs/AlGaAs 二

维电子气异质结结构”的论文于1994年获得了北京市青年优秀科技论文奖，并在1996年荣获中国科学院院长奖学金。

1995年博士毕业后，经王老师和江德生教授推荐，赴德国柏林Paul Drude Institute做博士后研究，在著名教授Ploog指导下从事氮化镓异质结外延材料研究，被Ploog教授誉为“他学生中出成果最快的研究人员之一”。

1998-2010年，先后在美国吉尔系统(Agere Systems)公司任高级研究员，超微(Advanced Micro Devices)任高级工程师。自2010年冬，入职世界移动通信龙头高科技公司高通(Qualcomm)，目前任美国高通总部高技术研发部总经理，主管14nm/10nm/7nm超高精尖Advanced CMOS研发和产业化，以及CMOS建模(CMOS Spice Model)工作。

在美国工业界20年研发和产业化的经历中，共荣获十多项公司大奖，获得美国已授权专利63项，受缴3次在国际大会作报告；为3个重大国际会议组委会委员，7个国际杂志的评委，发表论文62篇(含博士、博士后阶段发表论文)。

陈涌海，博士，研究员，博士生导师，国家杰出青年科学基金获得者、中国科学院百人计划入选者、新世纪百千万人才工程国家级人选、国务院政府特殊津贴获得者。

1990年于北京大学物理系获学士学位；1993年于北京科技大学物理系获硕士学位(导师陈难先院士)；1993年至今在中国科学院半导体研究所从事半导体材料物理研究工作。1998年在中国科学院半导体研究所获材料物理与化学专业博士学位(导师王占国院士)；曾多次前往香港科技大学物理系进行合作研究；曾任中国科学院半导体材料重点实验室主任。主要从事半导体光学性质研究，具体研究方向包括低维结构材料和纳米材料的光学性质、半导体偏振调制光谱技术及应用以及半导体自旋光电流谱等研究。先后主持了国家“973”项目和国家自然科学基金重大项目等十余项，在*Physical Review Letters*、*Nano Letters*、*Applied Physics Letters*等刊物发表论文200多篇，获得国家发明专利10余项。

刘峰奇，博士，研究员，博士生导师，新世纪百千万人才工程国家级人选，国家杰出青年科学基金获得者。

1985年6月河南师范大学物理系本科毕业；1990年7月中国科学技术大学物理系硕士毕业；1996年6月南京大学物理系(固体微结构物理国家重点实验室)博士毕业；1996

年6月进入中国科学院半导体研究所博士后流动站,师从王占国院士开展量子级联激光器研究。1997年6月被聘为半导体研究所副研究员;1998年6月从博士后流动站出站后入编半导体材料科学重点实验室,一直从事量子级联激光器研究;2000年研制出国际上第一个室温激射的波长3.54μm的应变补偿InGaAs/InAlAs量子级联激光器;2001年9月被聘为研究员;2002年年初被聘为博士生导师;2004年将量子级联激光器的材料体系从InGaAs/InAlAs拓展到GaAs/AlGaAs体系,并开展量子级联探测器研究;2005年获得国家杰出青年科学基金资助;2006年研制出THz量子级联激光器材料;2007年入选新世纪百千万人才工程国家级人选。近年来先后研制出波长4.3-11μm的一系列大功率、低阈值、高光束质量的中远红外量子级联激光器,频率2.95-3.25THz的一系列THz量子级联激光器,以及中心响应波长4.0-19μm的一系列低噪声、高探测率的中远红外量子级联探测器;研制出国际上第一个量子点级联激光器;研制出国际上第一个可室温工作的正入射响应的量子点级联探测器。2012年10月至2017年9月任半导体材料科学重点实验室主任。

王智杰,博士,研究员,博士生导师,中国科学院百人计划入选者,德国物理学会会员。

教育经历

2000年9月-2004年7月,浙江大学,本科;

2004年9月-2009年7月,中国科学院半导体研究所,博士。

工作经历

2009年8月-2011年3月,美国怀俄明大学,博士后研究员;

2011年4月-2011年5月,美国迈阿密大学,访问学者;

2011年6月-2013年6月,美国密歇根大学,博士后研究员;

2013年9月-2015年2月,德国伊尔梅诺理工大学,研究组长;

2015年3月至今,中国科学院半导体研究所,研究员。

从2005年至今,致力于新能源器件研究。积极创新,注重实验研究、理论探索与实际应用相结合。在先进纳米结构、光解水与太阳能电池、纳米激光器等领域取得了诸多前沿性研究成果。发表*SCI*论文50余篇,其中JCR TOP期刊论文28篇。担任*Polymer*、*The Journal Physical of Chemistry*、*Journal of the Electrochemical Society*、*Nanoscale*等杂志审稿人。多次参与主办国内、国际会议,出任组委秘书、会议主席等职务,并在国际学术会议上作邀请报告。承担科研项目多项,涵盖美国自然科学基金项目、中国科学院引进人才项目、国家自然科学基金项目、国家重点研发计划等。

董建荣,博士,研究员,中国科学院百人计划入选者。

1989 年和 1992 年在西安电子科技大学分别获得学士和硕士学位;1996 在中国科学院半导体研究所获半导体物理与器件专业博士学位。1996 年 3 月-1997 年 8 月在中国科学院半导体研究所任助理研究员;1997 年 9 月-2008 年 1 月在新加坡科技研究局材料研究院(IMRE)先后任 Research Associate、Research Fellow 和 Research Scientist;2008 年 1 月进入中国科学院苏州纳米技术与纳米仿生研究所工作,任研究所第一届学位委员会主席。2009 年入选中国科学院百人计划,2010 年获得苏州工业园区金鸡湖双百人才计划科教领军人才奖。多年来一直从事 InP 和 GaAs 基光电子材料、材料表征及器件研究,包括半导体多结光伏器件和半导体激光器方面的研究工作。发表论文 90 多篇,拥有 40 多项发明专利。

黎大兵,博士,中国科学院特聘研究员,博士生导师,国家杰出青年基金获得者,享受国务院政府特殊津贴。

2001 年 9 月-2004 年 7 月,师从王占国院士在中国科学院半导体研究所攻读博士学位。现工作于中国科学院长春光学精密机械与物理研究所任研究员。主要研究领域为 GaN 半导体材料与器件。作为项目负责人承担国家重点研发计划课题、国家自然科学基金、中国科学院科研项目 10 余项。曾获中国科学院优秀导师、吉林省青年领军人才、吉林省第五批拔尖创新人才、中国科学院青年创新促进会优秀会员、中国科学院卢嘉锡青年人才奖、吉林省自然科学学术成果奖一等奖(排名第一位)、吉林省优秀海外归国人才优秀奖等奖项和荣誉。担任 *Scientific Reports*、《中国科学:物理学、力学、天文学》《半导体学报》《发光学报》编委,中国金属学会宽禁带委员会委员,OSA 高级会员。在 *Physical Review Letters*, *Advanced Materials*, *Light: Science & Applications*, *Applied Physics Letters* 等国际期刊上发表 *SCI* 论文 50 余篇,申请发明专利 10 余项。

王元立,博士后,教授级高级工程师,美国晶体技术集团(AXT Inc.)全资子公司北京通美晶体技术有限公司技术研发副总监。

1994 年 9 月–1998 年 7 月,在北京科技大学物理系获得理学学士学位;1998 年 9 月–2003 年 6 月,在北京科技大学材料物理与化学系柳得橹教授指导下硕博连读获得工学博士学位;2003 年 6 月–2005 年 7 月,在中国科学院半导体研究所材料科学重点实验室王占国院士指导下进行博士后研究工作。在博士后研究工作期间,通过分子束外延技术在化合物半导体衬底表面生长自组装量子点和量子线等纳米材料,研究了分子束外延生长条件对纳米组织的结构和光电性能的影响,对纳米组织的形成机理、形核长大动力学、空间有序性和微观结构等进行了深入的研究,获得了原创性的研究成果。2005 年 7 月至今,在北京通美晶体技术有限公司工作,担任技术副总监。在砷化镓、磷化铟和锗单晶衬底生产的新工艺研发和工艺改进方面开展了大量的研究工作。2017 年入选首批通州区高层次人才发展支持计划领军人才。至今在中国、美国、日本等国家获得 6 项发明专利授权,在国内外杂志上发表了 37 篇研究论文。

陈广超,博士,教授,中国科学院大学材料科学与光电技术学院执行院长。

从 1994 年起,一直从事金刚石的制备及其光、电性能研究,在国际上率先发明了动态单晶 CVD 金刚石的等离子体喷射生长技术,形成了由学术论文、学术编著、发明专利和实用新型专利共同构成的完整独立的自主知识产权体系,打破了毫米级单晶金刚石只有微波法生长的技术格局。2002 年回国以来,先后主持军工“863”项目 1 项、国家自然科学基金 3 项、北京市自然科学基金 1 项、中国科学院科研装备研制项目 1 项,以及教育部留学回国人员、北京市科技新星、中国科学院百人计划等人才项目各 1 项。

发表学术论文 60 多篇,授权发明专利 5 项、实用新型 6 项,编著 3 部。组建运行了碳量子材料及器件实验室,拥有超净间、多型气相沉积设备(总功率超过 100kW)、多种表征设备(如场发射扫描电镜、X 射线仪等)以及微纳加工设备。

廖梅勇,博士,日本国立材料研究所主干研究员(终身制)。

1996 年和 1999 年,在兰州大学分别获得学士和硕士学位;2002 年,在中国科学院半导体研究所(香港城市大学联合培养)取得博士学位;2008 年,获日本国立材料研究所主任研究员(终身制)。十几年来一直从事宽禁带半导体材料的制备、物性、半导体器件,MEMS 及其物理和应用研究,在这些领域做出了一系列原创和具有特色的工作。

在半导体光电器件与物理领域主要原创性学术贡献有:①提出了宽禁带半导体光电导物理机制;提出并设计了多种新型光电传感器件概念。为发展半导体新型光电器件提供了物理原理和技术依据。②提出了一种全新的低能耗场效应晶体管,某种程度上克服了宽禁带半导体双极型晶体管的制备难点。丰富了半导体电子器件家族。③研发了目前为止的唯一单晶金刚石 MEMS/NEMS 开关器件;实现了单晶金刚石 MEMS/NEMS 器件和电子器件的集成;解决了金刚石 MEMS/NEMS 领域加工难的挑战课题,为发展下一代高可靠性高性能 MEMS/NEMS 奠定了基础。

学术研究成果受到国际同行广泛关注:多次受邀在金刚石领域最著名的国际会议如 International Conference on Diamond and Carbon Materials 等作大会特邀报告,受邀为美国材料学会 *MRS Bulletin* 和 *Japanese Journal of Applied* 撰稿,并多次被邀为日本 *New Diamond* 撰写亮点文章。还受到日本 NHK 电视台、日刊工业新闻、日经产业新闻等 20 多家新闻媒体关注。研究工作在国际著名刊物发表 *SCI* 论文 160 余篇,论文被引用 7500 余次,H 因子 43。并在 *Chemical Society Reviews* 等著名杂志上发表相关领域的综述论文 5 篇,出版专著 1 部。已获美国专利 4 项,日本专利 13 项。主持和参与日本文部省自然科学基金 5 项,日本新能源组织项目 1 项,日本科技厅专项 2 项,欧盟项目 2 项,日本科技厅中日交流项目 1 项,科学技术部中日政府间科技项目 1 项,国家自然科学基金委员海外合作基金 1 项。获省部级科学技术进步奖一等奖 1 项。

后　记

2018 年 12 月 29 日，值此中国科学院院士王占国先生 80 华诞之际，中国科学院半导体研究所编辑出版《**占“新”为民　兴“材”报国**　王占国院士文集》一书，系统梳理王占国院士近 60 年的科研历程，谨以本书表达对王占国院士的敬仰之情。

本书从各个方面、不同角度反映了王占国院士近 60 年来在半导体科学技术领域里辛勤耕耘、不断开拓，取得的一系列开创性成果的奋斗历程。本书的编写是一项复杂的工程，凝聚了许多人的共同心血。中国科学院副院长、中国科学院大学党委书记、校长李树深院士在百忙之中给予了大力支持，并提出了宝贵意见和建议；王占国院士亲自撰写了部分文集内容，并提供了大量图文资料；半导体材料科学重点实验室金鹏研究员编写了框架目录并统稿全书；半导体材料科学重点实验室王智杰研究员、陈涌海研究员、徐波研究员、刘峰奇研究员、杨少延研究员、曲胜春研究员、刘舒曼研究员、叶小玲副研究员、翟慎强助理研究员，综合办公室主任慕东、宣传主管高艳等相关人员都为本书的出版做了大量的工作；陈亚男等 9 位同学对文集进行了认真负责的校对工作；原中国建筑材料科学研究院院长、时任国家“973”计划材料科学领域专家咨询组副组长的欧阳世翕教授撰文记述了王占国院士任职国家“973”计划材料科学领域专家咨询组组长期间的重要贡献；半导体研究所退休职工林耀旺、钱家骏和何春藩研究员，南昌大学江风益教授，AZ Power Inc. 的 CEO 李瑞刚博士，北京通美晶体技术有限公司研发副总监王元立博士等近 20 人撰写了感情真挚的回忆文章。在此，一并向他们表示衷心的感谢！

本书中的照片、论文、获奖情况、学生名单均反映了王占国院士几十年来的成绩。每一幅历史图片，都展示了王占国院士的人生轨迹；每一篇学术论文，都凝结着王占国院士的心血与汗水；每一次获奖，都体现了王占国院士勇攀科技高峰的精神；每一位毕业学生，都寄托着王占国院士真切的教诲和殷切的期望。

由于编辑时间仓促，资料搜集还不够全面，书中难免存在疏漏及错误之处，加之编者水平有限，编写会有不尽如人意之处，敬请读者见谅。

中国科学院半导体研究所综合办公室
2018 年 10 月